DIGITAL COMMUNICATIONS

Dr. SANJAY SHARMA
B.E., M.Tech., Ph.D., MIETE
Professor and Head
Department of Electronics & Communication Engg,
Krishna Institute of Engg. & Technology
Ghaziabad, India

Edited by
DEEKSHA SHARMA
&
V. SHARMA

S.K. KATARIA & SONS®
Publisher of Engineering Books
4885/109, Prakash Mahal, Dr. Subhash Bhargav Lane,
Opposite Delhi Medical Association, Daryaganj, New Delhi–110002 (INDIA)
Phone: +91-11-43551243; *Mobile:* +91-9871775858
e-mail: katariabook@gmail.com, katariabooks@yahoo.com
Website: www.skkatariaandsons.com
(e-books, subscriptions and bulk buying also available on our website)

Published by:

S.K. Kataria & Sons®

4885/109, Prakash Mahal, Dr. Subhash Bhargav Lane,
Opposite Delhi Medical Association, Daryaganj, New Delhi-110002 (INDIA)
Phone: +91-11-43551243, *Mobile:* +91-9871775858
e-mail: katariabook@gmail.com, katariabooks@yahoo.com
Website: www.skkatariaandsons.com

First Edition : March, 2003; Eighth Edition: January, 2012; Reprint: 2024

Price: ₹ 595/-

Laser Type Setting:
PLUS COMPUTERS, Shahdara, Delhi

Printed at:
New A.S. Printing Press, Delhi

PREFACE TO THE EIGHTH EDITION

I feel happy in presenting the Eighth revised and enlarged edition of this book to the students. In fact, the favourable and warm reception which the first, second, third, fourth, fifth, sixth and seventh editions and subsequent reprints have earned has been a matter of great satifaction for me. The author wishes to express his deep gratitude to the large number of readers who have used it and a particular to those of them, who sent valuable suggestions for the improvements of the book.

In this addition, the author has made sincere efforts to make the book up-to-date.

The present edition has been thoroughly revised and a lot of useful material has been added. To make it more useful, several solved examples and problems have been added from the point of view of Examination. The mistakes and misprints, which had crept in last edition, have been eliminated in this edition. The author hopes that all these features will make the book even more useful and attractive to the readers

Any errors, misprints and suggestions for the improvements of the book, brought to my notice would be gratefully and thankfully acknowledged and incorporated in the next edition.

Dr. Sanjay Sharma

KIET, Ghaziabad

January, 2012

PREFACE TO THE FIRST EDITION

Digital Communications is a core subject for undergraduate students in Electronics and Communication Engineering, Electrical Engineering and Electronics and Instrumentation Engineering. This book is primarily intended to serve as a textbook in accordance with the syllabus of Digital communications offered by various universities in India as well as abroad.

In this book, a significant effort has been made to find simple ways to develop theoretical aspects of digital communications. A strong emphasis has been given on numerical examples. Neat and Clear diagrams have been used for explanation.

Every effort has been made to make the treatments simple and comprehensive. The, whole text has been divided in ten chapters :

Chapter 1 introduces the concepts of digital communications with their classification and basic properties.

Chapter 2 introduces the Sampling and their types.

Chapter 3 explains the PCM modulation and its different modified form.

Chapter 4 explains digital multiplexers.

Chapter 5 introduces the base band transmission.

Chapter 6 discusses the digital modulation techniques.

Chapter 7 deals with Information theory.

Chapter 8 discuss Error Control Coding.

Chapter 9 explains detection of digital signals

Chapter 10 gives introduction to spread spectrum techniques

Chapter 11 provides an introduction to Cellular and Mobile Communication

The author hope that the book will fulfill the need of readers and would welcome any suggestions towards the improvement of the book.

Dr. Sanjay Sharma
KIET, Ghaziabad
March, 2003

ACKNOWLEDGEMENT

I take this opportunity to express my gratitude and thanks to the respected Director General of KIET, **REAR ADMIRAL Dr. AJAY SHARMA** for his valuable technical suggestions and constant encouragement, without which this book would not have come into existence.

I am specially grateful to **Dr. S. NASEEM AHMAD**, Professor and Head, Department of Electronics and Communication Engg., Jamia Millia Islamia, New Delhi for his time-to-time, much needed valuable guidance.

I would like to express my sincere thanks to **Mr. Vibhav Sachan** (Associate Professor, E&C, KIET), and **Mr. Manish Gupta** (Lecturer, E&C, MIT, Modinagar) for constant inspiration and encouragement.

Special thanks to **Dr. B.P. Singh** (Former Professor, Electronics & Comm. Engg., MMMEC, Gorakhpur), **Dr. R.K. Chauhan** (Professor, Electronics & Comm. Engg., MMMEC, Gorakhpur), **Dr. G.S. Tripathi** (Electronics & Comm. Engg., MMMEC, Gorakhpur), **Dr. Arun Kumar** (Former Professor, Electronics & Comm. Engg., MMMEC, Gorakhpur) and **Dr. C.B. Tripathi** (Professor Electronics & Comm. Engg., MMMEC Gorakhpur) and **Dr. Vipin Kumar** (H.O.D. Physics, KIET), and **Mr. Manish Gupta** (Lecturer, MIT, Modinagar) for their guidance and valuable suggestions.

I am grateful to my younger brothers **Rajeev** and **Rakesh,** daughter **Ritika**, for inspiring me for this project.

I deeply express my heartful thanks to the publishers **S.K. Kataria & Sons** for publishing this book in such a beautiful get-up and well in time.

Lastly, I am also grateful to **Mr. Manoj Tyagi** and his staff from **Plus Computer's** for their hard work.

Dr. Sanjay Sharma

DEDICATED

TO

MY LOVING SON

MRIDUL

CONTENTS

UNIT-IV 327–369

UNIT-V 371–519

UNIT– I

Digital Data transmission, Line coding review, Pulse shaping, Scrambling, Digital receivers, Eye diagram, Digital carrier system, Method of generation and detection of coherent & non-coherent binary ASK, FSK & PSK, Differential phase shift keying, quadrature modulation techniques (QPSK and MSK), M-ary Digital carrier Modulation.

Above Syllabus has been covered in following chapters:

CHAPTER 1

INTRODUCTION

SYLLABUS

Introduction: Elements of Communication System, modulation, Analog & Digital Communication, EM Spectrum, Fundamental limitations of Communication Systems.

1.1 INTRODUCTION

Communication is the process of establishing connection or link between two points for information exchange.

OR

Communication is simply the process of conveying message at a distance or communication is the basic process of exchanging information.

The electronic equipments which are used for communication purpose, are called communication equipments. Different communication equipments when assembled together form a communication system.

Typical examples of communication system are line telephony and line telegraphy, radio telephony and radio telegraphy, radio broadcasting, point-to-point communication and mobile communication, computer communication, radar communication, television broadcasting, radio telemetry, radio aids to navigation, radio aids to aircraft landing etc.

The earliest communication system namely line-telegraphy originated in eighteen fourties (1840s). In addition to this, line telephony came a few decades later whereas radio-communication could become possible in the beginning of twentieth century on invention of triode valve. Radio communication was further greatly improved during World War II. It became more widely used through the invention of transistor, integrated circuits (ICs) and other semiconductor devices in the subsequent years. Also, in

Chapter Outline

- Introduction
- The Communication Process: Elements of a Communication System
- Concept of Bandwidth
- Sources of Information
- Communication Channels
- Baseband and Passband Signals
- Representations of Signals
- Classification of Communication
- The Modulation Process
- Primary Communication Resources
- Survey of Communication Applications
- Analog and Digital Signals
- Conversion of Analog Signals to Digital Signals
- Analog and Digital Communication
- Advantages and Disadvantages of Digital Communication
- The Electromagnetic (EM) Spectrum
- Fundamental Limitations of Communication System
- Historical Perspective

recent years, communication has become more widespread with the use of satellites and fiber optics. Today, there has been an increasing emphasis on the use of computers in communication.

1.2 THE COMMUNICATION PROCESS: ELEMENTS OF A COMMUNICATION SYSTEM

In the most fundamental sense, communication involves the transmission of information from one point to another through a succession of process as listed ahead:

(i) The generation of a thought pattern or image in the mind of an originator.
(ii) The description of that image, with a certain measure of precision, by a set of oral visual symbols.
(iii) The encoding of these symbols in a form that is suitable for transmission over a physical medium of interest.
(iv) The transmission of the encoded symbols to the desired destination.
(v) The decoding and reproduction of the original symbols.
(vi) The recreation of the original thought pattern or image, with a definable degradation in quality, in the mind of a recipient.

> **DO YOU KNOW?**
>
> Electric power is provided to homes from a group of interconnected regional plants. This means that the 50 Hz frequency of all the generators must be synchronized. Your power may arrive at your home via a loop hundreds of miles long rather than by the shortest distance.

Hence, as discussed above, the purpose of a communication system is to transmit an information-bearing signal, from a source, located at one point, to a user or destination, located at another point some distance away. Figure 1.1 shows the block diagram of a general communication system* in which the different functional elements are represented by blocks. The essential components of a communication system are information source, input transducer, transmitter, communication channel, receiver and destination.

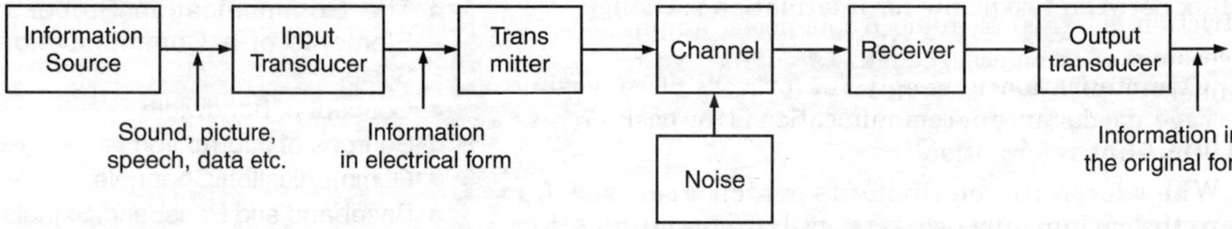

Fig. 1.1 Block diagram of a communication system

Now we shall discuss the functioning of these blocks.

1. Information Source

We know that a communication system serves to communicate a message or information. This message or information originates in the information source. In general, there can be various messages in the form of words, groups of words, code, symbols, sound signal etc. However, out of these messages, only the desired message is selected and conveyed or communicated.

In short, we can say that the function of information source is to produce required message which has to be transmitted.

2. Input Transducer

A Transducer is a device which converts one form of energy into another form. The message from the information source may or may not be electrical in nature. In a case when the message produced by the information source is not electrical in nature, an input transducer is used to convert it into a time-varying electrical signal. For example, in case of radio-broadcasting, a microphone converts the information or message which is in the form of sound waves into corresponding electrical signal.

* It is also known as elements of a communication system.

Pioneer in Communications : Heinrich Rudolph Hertz (1857–1894)

Born in Hamburg, Germany, and educated at the University of Berlin, Heinrich Rudolph Hertz was a physicist who verified experimentally British physicist James Clark Maxwell's Electromagnetic theory by demonstrating the nature of electromagnetic (EM) wave propagation. Hertz's research led to the discovery of radio waves (also called as Electromagnetic waves), which are a form of Electromagnetic radiation. Hertz also demonstrated that these waves travel at the speed of light and exhibit several other properties of light, including reflection, refraction, and polarization. Hertz's discovery led directly to radio, television and radar. The unit of frequency, the hertz Hz, which is a measure in cycles per second, has been named in honour of him.

Interestingly, Maxwell and Hertz (as did many late 19th-and early 20th-century scientists) believed that Electromagnetic waves were like sound waves and hence, propagated through some previously unknown medium (as sound waves propagate through air).

3. Transmitter

The function of the transmitter is to process the electrical signal from different aspects. For example in radio broadcasting the electrical signal obtained from sound signal, is processed to restrict its range of audio frequencies (upto 5 KHz in amplitude modulation radio broadcast) and is often amplified. In wire telephony, no real processing is needed. However, in long-distance radio communication or broadcast, signal amplification is necessary before modulation. Modulation is the main function of the transmitter. In modulation, the message signal is superimposed upon the high-frequency carrier signal. In short, we can say that inside the transmitter, signal processings such as restriction of range of audio frequencies, amplification and modulation are achieved. All these processings of the message signal are done just to ease the transmission of the signal through the channel*.

4. The Channel and the Noise

With reference to the block diagram of a communication system in figure 1.1, the term channel means the medium through which the message travels from the transmitter to the receiver. In other words, we can say that the function of the channel is to provide a physical connection between the transmitter and the receiver.

There are two types of channels, namely point-to-point channels and broadcast channels. Examples of point-to-point channels are wire lines, microwave links and optical fibers. Wire-lines operate by guided electromagnetic waves and they are used for local telephone transmission. In case of microwave links, the transmitted signal is radiated as an electromagnetic wave in free space. Microwave links are used in long-distance telephone transmission. An optical fiber is a low-loss, well-controlled, guided optical medium. Optical fibers are used in optical communications. Although these three channels operate differently, they all provide a physical medium for the transmission of signals from one point to another point. Therefore, for these channels, the term point-to-point is used.

On the other hand, the broadcast channels provide a capability where several receiving stations can be reached simultaneously from a single transmitter. An example of a broadcast channel is a satellite in geostationary orbit, which covers about one third of the earth's surface. During the process of transmission and reception, the signal gets distorted due to noise introduced in the

* The three major fields of Electronics are computers, communications, and control. The computer segment is the largest, communications is the second largest.

system. Noise is an unwanted signal which tend to interfere with the required signal. Noise signal is always random in character. Noise may interfere with signal at any point in a communication system. However, the noise has its greatest effect on the signal in the channel.

5. Receiver

The main function of the receiver is to reproduce the message signal in electrical form from the distorted received signal. This reproduction of the original signal is accomplished by a process known as the demodulation or detection. Demodulation is the reverse process of modulation carried out in transmitter.

> **REMEMBER**
>
> ISDN (Integrated services digital network) is a technology for transmitting more information over ordinary telephone lines.

6. Destination

Destination is the final stage which is used to convert an electrical message signal into its original form. For example in radio broadcasting, the destination is a loudspeaker which works as a transducer i.e. it converts the electrical signal in the form of original sound signal.

1.3 CONCEPT OF BANDWIDTH

Bandwidth may be defined as the portion of the electromagnetic spectrum occupied by a signal. We may also define the bandwidth as the frequency range over which an information signal is transmitted. Bandwidth is the difference between the upper and lower frequency limits of the signal. We already know different types of passband signals such as voice signal, music signal, TV signal etc. Each of these signals will have its own frequency range. This frequency range of a signal is known as its bandwidth. As an example, the range of music signal is 20 Hz to 15 kHz. Therefore, as shown in figure 1.2, the bandwidth is $(f_2 - f_1)$

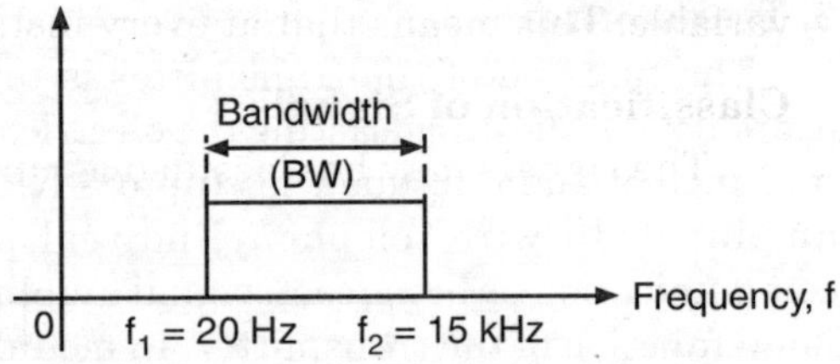

Fig. 1.2 Illustration of bandwidth of music signal

Thus, we write $BW = f_2 - f_1$

$$= 15000 - 20 = 14980 \text{ Hz}$$

The bandwidth of different signals have been listed in table 1.1.

TABLE 1.1.

S.No.	Type of the signal	Range of frequency in Hz	Bandwidth in Hz
1.	Voice signal (speech) for telephony	300 – 3400	3,100
2.	Music signal	20 – 15000	14, 980
3.	TV signals (picture)	0 – 5 MHz	5 MHz
4.	Digital data	300 – 3400 (it is using the telephone line for its transmission)	3,100

Important Point : *In fact, the required bandwidth in the data transmission depends upon the rate at which the data is being transmitted. The BW increases with increase in the rate of data transmission.*

1.3.1. Concept of Frequency Spectrum

Frequency spectrum may be defined as the representation of a signal in the frequency domain.

It can be obtained by using either Fourier series or Fourier transform. It consists of the amplitude and phase spectrums of the signal. The frequency spectrum indicates the amplitude and phase of various frequency components present in the given signal. The frequency spectrum enables us to analyze and synthesize a signal.

DO YOU KNOW?

Digital Signal Processing (DSP) can be used to compress video signals and to identify finger-prints.

1.4 SOURCES OF INFORMATION

Some of the important sources of information in the communication environment may be listed as under :

1. Speech
2. Music
3. Pictures
4. Computer data

A source of information is basically a signal which carries the information.*

1.4.1. Signal

A signal may be defined as the single valued function of time. Time plays the role of an independent variable. This means that at every instant of time, the signal has a unique value.

Classification of Signals

The signals may be classified as under :

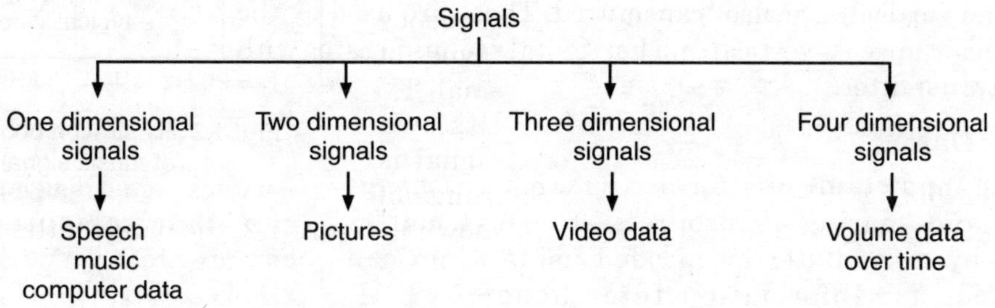

Fig. 1.3 Classification of signals

Let us discuss these sources one by one.

1. Speech

Speech involves transfer of information from the speaker to the listener. Such a transfer of information takes place in following three stages:

(i) Production
(ii) Propagation and
(iii) Perception

The band of frequencies considered to be essential for the speech communication is 300 Hz to 3100 Hz. This band is utilized for the commercial telephonic communication**.

2. Music

Music signal is originated from the instruments such as the piano, violin, flute etc. The musical

---

* Electronic communication plays a vital role in all our lives and is essential to the success of our information society.

** Communication is the process of exchanging information. Most human communication is oral, but, a great deal of it is also in written or printed form.

note may list for a short time or for a longer time depending upon the instrument being used for its origination.

Music signal has following two possible structures:

(i) Melodic structure

(ii) Harmonic structure

The melodic structure consists of a time sequence of sounds, whereas the harmonic structure consists of a set of simultaneous sounds. Music signal is a bipolar signal and requires a bandwidth of about 15 kHz. Hence, the channel bandwidth required for the transmission of music signal is much larger than that required for the transmission of speech signal.

3. Picture

The picture can be either static or dynamic. Example of static picture is the picture sent by fax machine and that of a dynamic picture is the one produced on T.V. The T.V. camera produces a dynamic picture and picture tube reproduces it at the receiving end. The electrical equivalent of dynamic picture is produced at the output of the camera and it is called as the typical video signal.

Figure 1.4 shows a typical video signal. The bandwidth required for this video signal is 0 to 5 MHz. The video signal consists of a luminance signal (Y) and chrominance signal. The luminance signal conveys the brightness information while chrominance signal conveys the colour information. Alongwith them the synchronizing signals (horizontal and vertical) are also transmitted. These are used by the receiver to synchronize its vertical and horizontal scanning rates with those at the transmitter.

Fig. 1.4 Illustration of a typical Video signal

4. Computer Data*

Personal computers are used for electronic mail, exchange of software, and sharing of resources. The text matter transmitted by a computer is encoded using American Standard Code for Information Interchange (ASCII) Each character in ASCII is represented by seven data bits. Hence, total $2^7 = 128$ characters can be represented by using ASCII.

Table 1.2 gives complete listing of ASCII characters. The seven data bits are ordered starting with the MSB b_7 and LSB b_1. An extra eighth bit b_8 is used as a parity bit for the purpose of error detection. When ASCII data are transmitted over these ports, a start bit (which is set to 0) and a stop bit (which is set to 1) are added as shown in figure 1.5.

DO YOU KNOW?

An unexpected development is the role that computers play in communications. E-mail now allows individuals with PCs to communicate with one another within and between messages transmitted over networks with office buildings, nation wide or worldwide. And interestingly, e-mail has not replaced the telephone or the fax. It has simply added a new way for people to interact with one another.

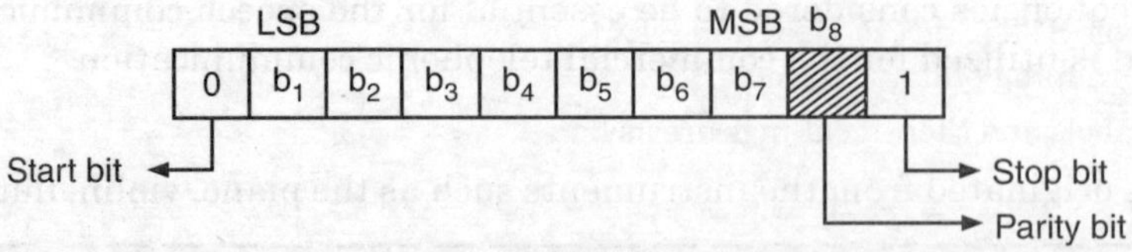

Fig. 1.5 Bit format for sending asynchronous serial data on RS-232 port

* Data communication is the transmission of computer and other digital data via the telephone systems, microwave links, or satellite.

TABLE 1.2 ASCII code set

Bit numbers	7	0	0	0	0	1	1	1	1
	6	0	0	1	1	0	0	1	1
	5	0	1	0	1	0	1	0	1
4321		0	1	2	3	4	5	6	7
0000	0	NUL	DLE	SPACE	0	@	P		p
0001	1	SOH	DC1	!	1	A	Q	a	q
0010	2	STX	CD2	“	2	B	R	b	r
0011	3	ETX	DC3	#	3	C	S	c	s
0100	4	EOT	DC4	$	4	D	T	d	t
0101	5	ENQ	NAK	%	5	E	U	c	u
0110	6	ACK	SYN	&	6	F	V	f	v
0111	7	BEL	ETB	,	7	G	W	g	w
1000	8	BS	CAN	(	8	H	X	h	x
1001	9	HT	EM	)	9	I	Y	i	y
1010	A	LF	SUB	*	:	J	Z	j	z
1011	B	VT	ESC	+	;	K	[	k	{
1100	C	FF	FS	,	<	L	\	*l*	|
1101	D	CR	GS	-	=	M	]	m	}
1110	E	SO	RS		>	N	^	n	~
1111	F	SI	US	/	?	O	—	o	DEL

EXAMPLE 1.1. Using the table 1.2, obtain the ASCII code word for the character "K".

Solution: Looking at the table 1.2, we come to know that the co-ordinates of character "K" are (4, B) therefore its code is 100 10 11 i.e., 4 B.

Other symbols and their meanings

(i) The control symbols are codes reversed for special functions. The control symbols are as listed in table 1.3 CR (Carriage Return) and LF (Line Feed) are the symbols used for basic operations of printers or displays.

(ii) The symbols ACK (Acknowledgement) or NAK (Negative Acknowledgement) are used for the error control.

(iii) And the symbols STX (Start to Text) and ETX (End to Text) are used for grouping of data character.

(iv) The other symbols and their meanings are as listed in table 1.3.

REMEMBER

Well-known forms of electronic communications, such as the telephone, radio, and television, have increased our ability to share information. Today, they are a major part of our lives.

TABLE 1.3 Control symbols used in ASCII code set

ACK	Acknowledgement	FF	Form Feed
BEL	Bell	FS	File separator
CAN	Cancel	HT	Horizontal tabulation
CR	Carriage Return	LF	Line Feed
DC1	Device control 1	NAN	Negative Acknowledgement
DC2	Device control 2	ACK	Acknowledgement
DC3	Device control 3	RS	Record separation
DC4	Device control 4	SI	Shift-in
DEL	Delete	SO	Shift-Out
DLE	Data Line Escape	SOH	Start of Heading
EM	End of Medium	STX	Start to Text
ENQ	Enquiry	SUB	Substitute Character
EOT	End of Transmission	SYN	Synchronous Idle
ESC	Escape	US	Unit separator
ETB	End of transmission block	VT	Vertical tabulation
ETX	End of Text		

Important Point: *Computer-generated data and television signals are both wide-band signals in that their power content occupiès a wide range of frequencies. Another important characteristics of data communication between personal computers is burstiness, which means that information is usually transmitted from one terminal to another in bursts with silent periods between bursts.*

1.5 COMMUNICATION CHANNELS

The medium over which the information is passed from the transmitter to the receiver is called as a communication channel. Depending on the mode of transmission, we may classify the communication channels into following two categories*.

(i) Channels based on guided propagation

(ii) Channels based on free propagation.

The classification of channels has been shown in figure 1.6.

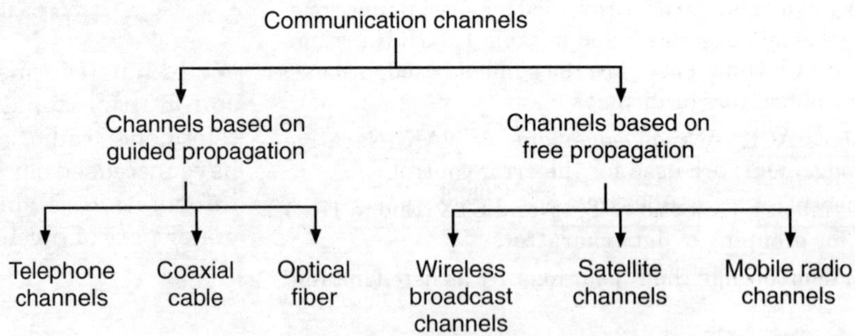

Fig. 1.6. Classification of communication channels

* The three primary communications media are wires, free space, and fiber-optic cable.

Some of the important characteristics of a channel are as under:

(i) Power required to achieve the desired S/N ratio.

(ii) Bandwidth of the channel.

(iii) Amplitude and phase response of channel.

(iv) Type of channel (Linear or non-linear)

(v) Effects of external interference on the channel.

DO YOU KNOW?

Cellular phones transmitting above 3 W, such as some portable and mobile cellular phones, should be held at least 12 inch away from a pacemaker, or they could interfere with its function. Even some high-power stereo speakers could be dangerous to pacemaker users, because they contain large magnets.

1.5.1. Telephone Channels

It is designed for providing service to voice signals such as telephones. The telephone channels are also used for the worldwide Internet connection. Therefore telephone channel is the best possible option for the data communication over long distances.*

Salient Features

Features of telephone channel are as follows :

(i) Bandpass characteristics over 300 to 3400 Hz as shown in figure 1.7.

(ii) High signal to noise ratio of about 30 dB.

(iii) Approximately linear response.

The amplitude response is flat over the entire passband as shown in figure 1.7.

However, no particular attention is given to the phase response because human ears are not very sensitive to the phase delay variations. But the data and images (i.e., pictures) are strongly affected by the phase delay variations. Therefore, for digital transmission over the telephone channels, it is essential to use an equalizer as shown figure 1.8 (a).

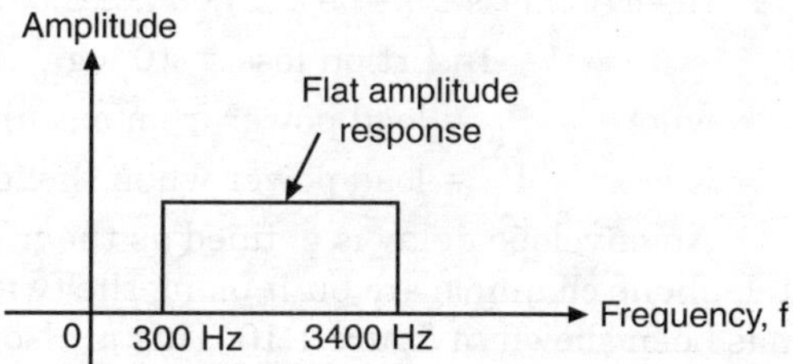

Fig. 1.7 Characteristics of telephone channel

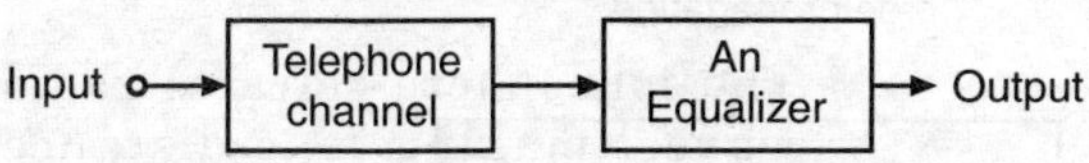

(a) Use of equalizer

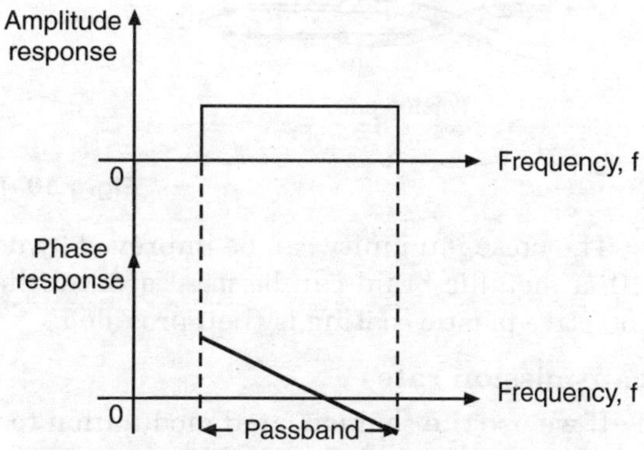

(b) Amplitude and phase response of the telephone channel with equalizer

Fig. 1.8.

* Devices called modems permit digital data to be transmitted over the analog telephone networks.

As depicted in figure 1.8 (b), the equalizer maintains a flat amplitude response and linear phase response over the passband. Also the telephone channel is a bandwidth limited channel. Figure 1.9 (a) depicts the variation of insertion loss with frequency and figure 1.9 (b) illustrate the variation of envelope delay with frequency, for a telephone channel.*

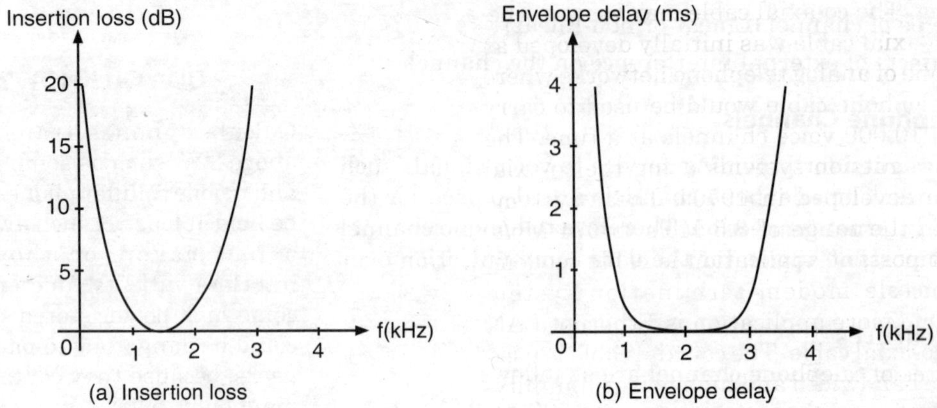

Fig. 1.9 Characteristics of a typical telephone channel

Insertion loss may be defined as under:

$$\text{Insertion loss} = 10 \log_{10} (P_o | P_L)$$

where P_L = load power from a source to load via channel

P_o = load power when the load is connected directly to source

An envelope delay is defined as the negative of the phase response with respect to $\omega = 2\pi f$. The telephone channels are built using the twisted pair of wires. The construction of twisted pair cables has been shown in figure 1.10. This is also commonly used medium and it is quite cheaper than the co-axial cable. A twisted pair consists of two insulated conductor twisted together in the spiral form as shown in figure 1.10. It can be shielded or unshielded. The unshielded twisted pair cables are very cheap and easy to install. However, they are badly affected by the noise interference.

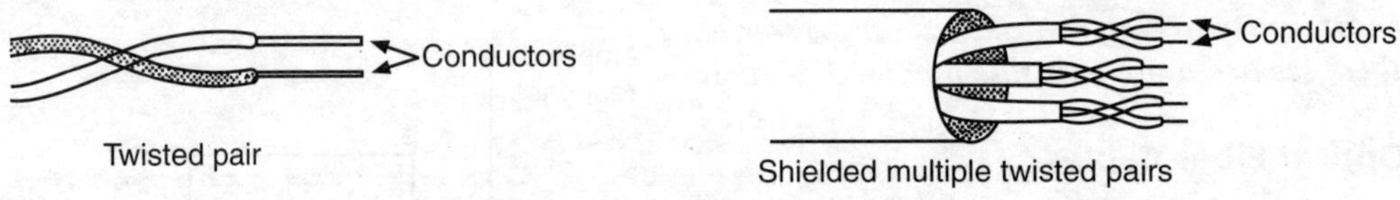

Fig. 1.10 Twisted pair cables

The noise immunity can be improved by using a shielded twisted pair cable. As shown in figure 1.10, a metallic braid can be used around the twisted pair. Protective plastic coating is then provided.

DO YOU KNOW?

The ionosphere refracts radio waves, making long-distance communications possible for some frequencies at certain times.

Transmission rates

If we use the sophisticated modulation techniques along with equalizer, then it is possible to attain the transmission rates upto 16.8 kilobits/sec (kb/s).

* The telegraph (1844) and telephone (1876) were the first two long-distance communication systems.

1.5.2. Co-axial Cables

Figure 1.11 shows the construction of co-axial cable. It consists of two concentric conductors separated by a dielectric material. The external conductor is metallic braid and used for the purpose of shielding. The co-axial cable may contain one or more co-axial pairs.

The co-axial cable was initially developed as the backbone of analog telephone networks where a single telephone cable would be used to carry more than 10,000 voice channels at a time. The digital transmission systems using the co-axial cable were developed in 1970s. These systems operated in the range of 8.5 Mb/s to 274 Mb/s. Another important application is cable modem, with the cable modem termination system (CMTS). One more application is Ethernet LAN using the co-axial cable. The co-axial cable is used for its large bandwidth and high noise immunity.

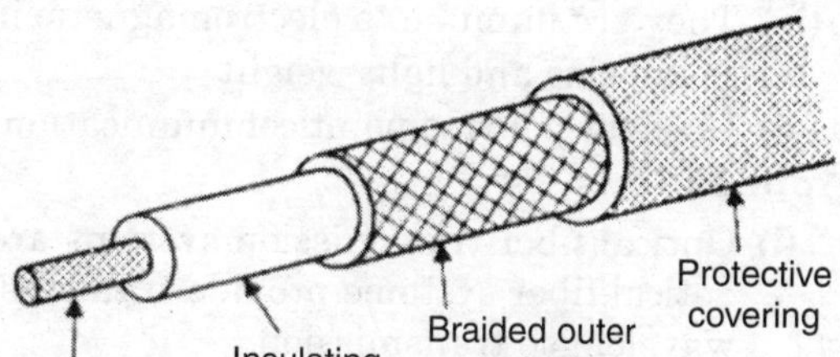

Fig. 1.11 Construction of a co-axial cable

Salient Features

The important characteristics of a co-axial cable may be listed as under :

(i) Two types of cables having 75 Ω and 50 Ω impedance are available.
(ii) Because of the shield provided, this cable has excellent noise immunity.
(iii) It has a large bandwidth and low losses.
(iv) This cable is suitable for point to point or point to multipoint applications. Actually, this is the most widely used medium for local area networks (LANs).
(v) These cables are costlier than twisted pair cables, however, they are cheaper than the optical fiber cables.
(vi) It is essential to use closely spaced (after every 1 km) repeaters to achieve the data rates of 8.5 Mb/s to 274 Mb/s.

Important Point : *As compared to the twisted pair of wires, the coaxial cable provides a higher immunity to electromagnetic interference (EMI). Because of their higher bandwidths, the coaxial cables are more suitable for the digital signal transmission.*

1.5.3. Optical Fiber Cables

1. Definition and Construction

The construction of an optical fiber cable has been shown in figure 1.12. It consists of an inner glass core surrounded by a glass cladding which has a lower refractive index. Digital signals are transmitted in the form of intensity-modulated light signal which is trapped in the glass core.

Light is launched into the fiber using a light source such as a Light Emitting Diode (LED) or Laser. It is detected on the other side using a photo detector such as a phototransistor. The optical fiber cables are costlier than the other two types but they have several advantages over the other two types.*

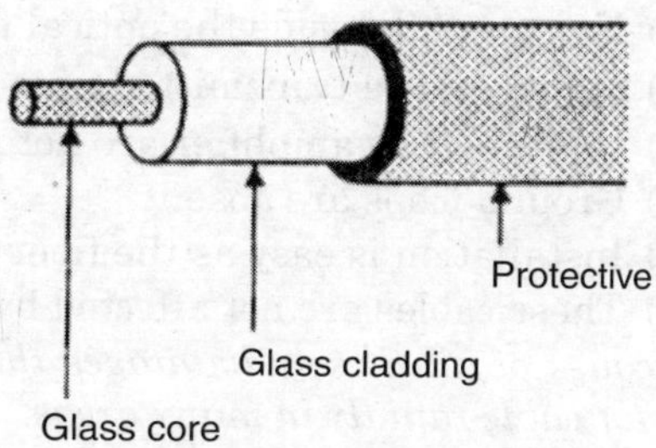

Fig. 1.12 Construction of optical fiber cable

* The transmission medium greatly attenuates and degrades the transmitted signal.

2. Special Characteristics

The characteristics of optical fiber cables may be listed as under :

(i) Higher bandwidth therefore, it can operate at higher data rates
(ii) Reduced losses as the signal attenuation is low
(iii) Distortion is reduced hence better quality is assured.
(iv) They are immune to electromagnetic interferences.
(v) Small size and light weight
(vi) Used for point to point communication

3. Applications

(i) Optical fiber transmission systems are widely used in the backbone of networks. Current optical fiber systems provide transmission rates from 45 Mb/s to 9.6 Gb/s using the single wavelength transmission.
(ii) The installation cost of optical fibers is higher than that for the co-axial or twisted wire cables.
(iii) Optical fibers are now used in the telephone systems.
(iv) In the local area networks (LANs).

4. Advantages of Optical Fibers

Some of the advantages of fiber optic communication over the conventional means of communication may be listed as under :

(i) Small Size and Light Weight : The size (diameter) of the optical fibers is very small (it is comparable to the diameter of human hair). Therefore, a large number of optical fibers can fit into a cable of small diameter.

> **REMEMBER**
>
> The SHF and EHF bands are used primarily for satellite communications and radar.

(ii) Easy availability and low cost : The material used for the manufacturing of optical fibers is silica glass. This material is easily available. Hence, the optical fibers cost lower than the cables with metallic conductors.

(iii) No electrical or Electromagnetic interference : Since the transmission takes place in the form of light rays the signal is not affected due to any electrical or electromagnetic interference.

(iv) Large bandwidth : As the light rays have a very high frequency in the GHz range, the bandwidth of the optical fiber extremely large. This allows transmission of more number of channels. Therefore, the information carrying capacity of an optical fiber is much higher than that of a co-axial cable.

(v) Other advantages : In addition to the advantages discussed earlier, the optical fiber communication has the following other advantages:

(i) No crosstalk inside the optical fiber cable
(ii) Signal can be transmitted upto 100 times faster
(iii) Intermediate amplifier are not required as the transmission losses in the fiber are low
(iv) Ground loops are absent
(v) Installation is easy as the fiber optic cables are flexible
(vi) These cables are not affected by the drastic environmental conditions

Because of all these advantages the optical fiber cable is replacing the conventional metallic conductor cable rapidly in many areas.

5. Drawbacks of Optical Fiber

Some of the drawbacks of optical communication system may be listed as under:

(i) Sophisticated plants are required for manufacturing optical fibers
(ii) The initial cost incurred is high
(iii) Joining the optical fibers is a difficult job.

1.5.4. Wireless Broadcast Channels

These channels are used for the transmission of radio and TV signals. The information signal or message signal which represents the speech, music etc. modulates a carrier frequency. The carrier frequency is different for every transmitting station. A transmitting antenna radiates the modulated signal in the form of electromagnetic radiator into the free space. These waves are radiated in all the directions or in some specific directions. The transmitting antenna is mounted on a tower or a hall in order to reach the farther receiver. The ground wave, sky wave and space wave are the three types of propagation techniques used for the propagation of EM waves. At the receiving end, the receiving antenna is used for picking up the transmitted signal. The receivers are superheterodyne type.

1.5.5. RF Link (Microwave Link)

Long form of RF link is radio frequency link.* This is actually a type of point to point wireless communication. The radio frequencies used for RF links are in microwave range, therefore, RF links are also called as microwave links. This has been shown in figure 1.13.

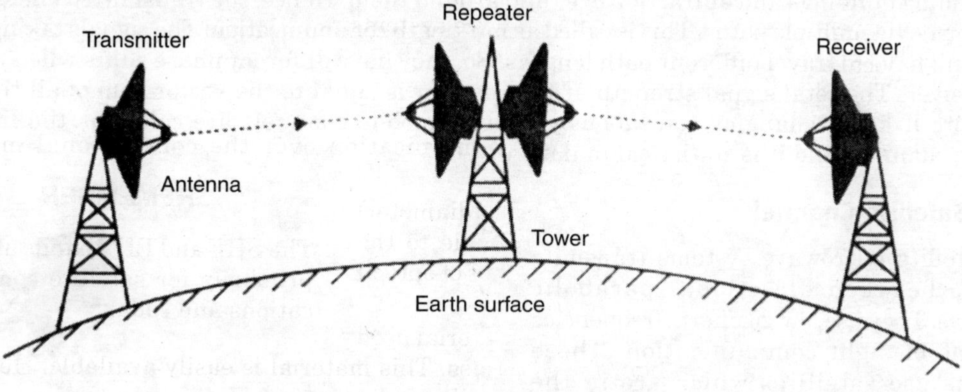

Fig. 1.13 Illustration of microwave link

Although many wire communication systems use copper wires or optical fiber, some just send the signal into the air. This happens when infrared, lasers, microwaves and radio are used for the transmission of data, as they do not need any physical medium. For long distance communication, microwave radio transmission is widely used as an alternative to coaxial cable. The signal transmission takes place in the form of electromagnetic waves which have wavelengths of few centimeters. Parabolic antennas can be mounted on the towers to send a beam of waves to another antenna, tens of kilometers away. The transmitting and receiving antennas are highly directional to enable a point to point communication. This system is widely used for both telephone and television transmission. The higher this tower which holds the antenna, the greater is the range. With a 100 meter high tower, the distances of 100 km can be easily covered.

Salient Features of microwave link

Some of the important features may be listed as under:

(i) Installation of towers and associated equipments is cheaper than laying down a cable of 100 km length.

(ii) Less maintenance as compared to cables.

(iii) Repeaters can be used. Therefore, effect of noise is reduced.

(iv) No adverse effects such as cable breakage etc.

---

* Frequencies above 1 GHz are called microwaves. The SHF and EHF bands are used primarily for satellite communications and radar.

(v) Because of the use of highly directional antenna, these links do not make any interference with other communication systems.

(vi) Size of transmitter and receiver reduces because of the use of high frequency.

Drawbacks

(i) Signal strength at the receiving antenna reduces due to multipath reception.

(ii) The transmission will be affected by the thunderstorms, and other atmospheric phenomenons.

Range of Frequencies

Generally, the microwave transmission takes place at frequencies between 2 and 40 GHz. This will correspond to a wavelength of 15 cm to 0.75 cm.

1.5.6. A Mobile Radio Channel

In mobile communication, the sender and the receiver both are allowed to move with respect to each other. The radio propagation takes place due to scattering of EM waves from the surfaces of the surrounding buildings and diffraction over and around them. Hence, the transmitted energy reaches the receiver via multiple paths. This is called as multipath communication. The signals taking different paths will have to travel different path lengths. So, they have different phase shifts when they reach the receiver. The total signal strength at the receiver is equal to the vector sum of all the signals. Therefore it keeps changing continuously. Hence, mobile channels are called as the linear time varying channels and it is statistical in nature.

1.5.7. Satellite Channel

Satellite microwave systems transmits signals between directional parabolic antennas. They use low gigahertz frequencies and line of sight communication. These systems use satellites which are in the geostationary orbit (36000 km above the earth). The satellites act as repeaters with receiving antenna, transponder and transmitting antenna. Satellite microwave systems can reach the most remote places on earth and communicate with mobile devices. This systems works in the following way: signal is sent through cable media to an antenna which beams the signal to the satellite. The satellite then transmits the signal back to another location on earth as shown in figure 1.14.

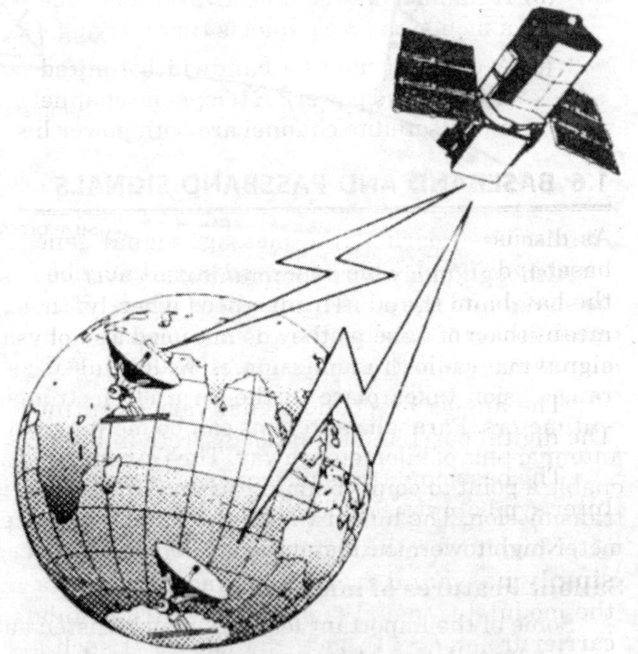

Fig. 1.14 Illustration of satellite system

Satellite microwave systems experience delays between the transmission of a signal and its reception back to the earth (540 ms).

Characteristics

Satellite microwave systems have the following characteristics:

1. It uses frequency range between 11 and 14 GHz.
2. Attenuation depends on frequency, power, antenna size and atmospheric condition.
3. The signals are affected by EMI effect, jamming and eavesdropping.
4. The installation of satellites is extremely difficult and the alignment of earth station antennas must be perfectly aligned.

5. The cost of building and launching is very high.
6. The satellites can provide point to point or broadcast services.
7. The message signal transmitted by the earth station to the satellite is called as an uplink signal
8. It is amplified and down converted in frequency by the transponder (electronic circuitry) and then retransmitted back to various earth stations.
9. The signal from satellite to earth station is called as the downlink signal.
10. The uplink signal frequency is 6 GHz and downlink signal frequency is 4 GHz.
11. In the 6/4 GHz band, a typical satellite is assigned a 500 MHz bandwidth which is divided among 12 transponders on the satellite.
12. Each transponder uses approximately 36 MHz bandwidth. A transponder can carry at least one TV channel, 1200 voice channels or a digital data at the rate of 50 Mb/s.

Important Point: *To summarize, we can conclude that a communication channel is central to the operation of a communication system. Its properties determine both that information-carrying capacity of the system and the quality of service offered by the system.*

As a matter of fact, communication channels may be classified in different ways as under:

(a) A channel may be linear or non-linear. A telephone channel is linear, whereas a satellite channel is usually (but not always) non-linear.

(b) A channel may be time invariant or time varying. An optical fiber is time invariant whereas a mobile radio channel is time varying.

(c) A channel may be bandwidth limited or power limited (i.e., limited in the available transmitted power). A telephone channel is bandwidth limited whereas an optical fiber link and a satellite channel are both power limited.

1.6 BASEBAND AND PASSBAND SIGNALS

As discussed earlier, the message signal generated from the information source is known as baseband signal. This baseband signal may be a combination of two or more message signals. If the baseband signal is transmitted directly, then it is known as **baseband transmission**. This means that the baseband transmission does not used modulators and demodulators. The baseband signal may be both analog as well as digital*.

The analog baseband signal varies continuously with time and has continuous amplitude. The digital baseband signal is discrete in both time and amplitude.

The baseband transmission is preferred at low frequencies and for short distances. However, Inter symbol Interference (ISI) is the major problem associated with the baseband transmission.

Now, if the modulated signal is transmitted over the channel, it is known as **bandpass or simply passband transmission**. Whenever a modulating signal is impressed upon a carrier, the modulated signal is produced. This modulated signal has fixed band of frequencies around carrier frequency. In fact, the nature of such a signal is bandpass type. Therefore, modulated signals are known as passband or bandpass signals. As an example, if a 1 kHz modulating signal is impressed upon a 100 kHz carrier signal by using amplitude modulation technique then two sidebands will be generated. Lower sideband will be at (100-1) i.e., 99 kHz and upper sideband will be at (100 + 1) i.e., 101 kHz.

---

* The baseband signal cannot usually be transmitted through space by radio because the antennas required are too long and multiple baseband signals transmitting simultanously would interfere with one another.

Thus, the amplitude modulated signal will have frequencies from 99 kHz to 101 kHz. These frequencies are bandpass type. This means that the modulated signals are also bandpass i.e., passband type. It may be noted that the bandpass transmission is generally used at high frequencies and for long distances. All television and satellite transmission are bandpass type. The bandpass transmission may be both analog and digital. If the digital signal modulates the carrier, then it is known as digital passband transmission.

1.7 REPRESENTATIONS OF SIGNALS

In order to develop knowledge of the operation of a communication system, we require some mathematical tools for the representations of signals and systems. As a matter of fact, the most useful method of signal representation depends upon the particular type of signal under consideration.

Therefore, depending upon the feature of interest, we may distinguish following three different classes of signals which arise in the study of communication systems:

(i) Periodic and non periodic signals
(ii) Deterministic and random signals
(iii) Energy and power signals.

1.7.1. Periodic and Non periodic Signals

A periodic signal g(t) is a function of time which satisfies the following condition:

$$g(t) = g(t + T_0) \text{ for all values of } t$$

where, t denotes time, and

T_0 is a constant ...(1.1)

Now, the smallest value of T_0 for which this condition is satisfied is called the period of g(t). Accordingly, the period T_0 defines the duration of one complete cycle of g(t). Also, any signal for which equation (1.1) does not hold for any T_0 is called a non-periodic or simply aperiodic signal.

1.7.2. Deterministic and random signals

A deterministic signal is a signal about which there is no uncertainty with respect to its value at any time, past, present or future.

Accordingly, deterministic signals may be modelled as completely specified functions of time.

On the other hand, a random signal is a signal about which there is some degree of uncertainty before it actually occurs.

1.7.3. Energy and Power Signals

We already know that in electrical systems, a signal may represent a voltage or a current. Now, let us consider a voltage v(t) developed across a resistor R, producing a current i(t).

Now, the instantaneous power dissipated in this resistor may be defined as under:

$$p(t) = \frac{|v(t)|^2}{R} \quad ...(1.2)$$

above expression may also be modified as

$$p(t) = R|i(t)|^2 \quad ...(1.3)$$

It may be noted that in both the above cases, the instantaneous power p(t) is proportional to the squared amplitude of the signal. Further, for a resistor R equal to 1 ohm, equations (1.2) and (1.3) both take on same mathematical form.

Accordingly, in signal analysis, it is customary to normalize the calculations by working with a 1-ohm resistor, so that, regardless of whether a given signal g(t) represents a voltage or a current, we may express the instantaneous power associated with the signal simply as under:

$$p(t) = |g(t)|^2 \qquad ...(1.4)$$

Now, based upon above convention, the total energy of a signal g(t) may be defined as

$$E = \lim_{T\to\infty} \int_{-T}^{T} |g(t)|^2\, dt = \int_{-\infty}^{\infty} |g(t)|^2\, dt \qquad ...(1.5)$$

Correspondingly, the average power of a signal g(t) is defined as

$$P = \lim_{T\to\infty} \frac{1}{2T} \int_{-T}^{T} |g(t)|^2\, dt \qquad ...(1.6)$$

Now, we can say that the signal g(t) is an energy signal if and only if the total energy of the signal satisfies the condition

$$0 < E < \infty$$

Also, it can be concluded that the signal g(t) is a power signal if and only if the average power of the signal satisfies the condition

$$0 < P < \infty$$

Important Point: *It may be noted that the energy and power classifications of signals are mutually exclusive. In particular, an energy signal has zero average power, whereas a power signal has infinite energy. Also, it will be interesting to note that, usually, periodic signals and random signals are power signals and signal that are both deterministic and non-periodic are energy signals.*

Theoretically, there are several possible methods for the representation of signals. However, practically, Fourier analysis, involving the resolution of signals into sinusoidal components, is most important among all other methods. Basically, this is a consequence of the well-known fact that the response of a system to a sine-wave input is another sine wave of the same frequency under following two conditions:

(i) The system is linear in that it obeys the principle of superposition.

This means that if $y_1(t)$ and $y_2(t)$ denote the responses of a system to the inputs $x_1(t)$ and $x_2(t)$, respectively, the system is linear if the response to the composite input $a_1x_1(t) + a_2\,x_2(t)$ is equal to $a_1y_1(t) + a_2y_2(t)$, where a_1 and a_2 are arbitrary constants.

(ii) The system is time invariant

This means that if y(t) is the response of a system to the input x(t), the system is time invariant if the response to the time-shifted input $x(t - t_0)$ is equal to $y(t - t_0)$ where t_o is constant.

Important Point : *There are several methods of Fourier analysis available for the representation of signals. The particular method that is used, in practice, basically depends upon the type of signal under consideration. As an example, if the signal is periodic, then the logical choice is to use the Fourier series to represent the signal in terms of a set of harmonically related sine waves.*

On the other hand, if the signal is an energy signal, then it is customary to use the Fourier Transform to represent the signal. By using the Fourier series or the Fourier Transform, we obtain the frequency domain description or spectrum of the signal.

1.8 CLASSIFICATION OF COMMUNICATION

Regarding the mode of propagation, communication may be divided in the following two forms* :

(i) Line Communication

(ii) Wireless or Radio Communication

* Major electrical discoveries in the mid and the late nineteenth century made possible the development of electronic communications over long distances.

1.8.1. Line Communication

In line communication, the medium of transmission is a pair of conductors called transmission line. This is also called as line channel. This means that in line communication, the transmitter and the receiver are connected through a wire or line. However, the installation and maintenance of a transmission line is not only costly and complex but also overcrowds the open space. Apart from this, its message transmission capability is also limited.

1.8.2. Wireless or Radio Communication*

In wireless or radio communication, a message is transmitted through open space by electromagnetic waves called as radio waves. Radio waves are radiated from the transmitter in open space through a device called antenna. A receiving antenna intercepts the radio waves at the receiver. All the radio, T.V. and satellite broadcasting are wireless or radio communication. The advantages of wireless communication are cost effectiveness, possible long distance communication and simplicity.

1.9 THE MODULATION PROCESS (UPTU Semester Exam., 2004-05) (5 marks)

As discussed earlier, the purpose of a communication system is to deliver a message signal from an information source in recognizable form to a user destination, with the source and the user being physically separated from each other.

To do this, the transmitter modifies the message signal into a form which is suitable for transmission over channel. This modification is achieved by means of a process known as modulation, which involves varying some parameter of a carrier wave in accordance with the message signal. Also the receiver recreates the original message signal from a degraded version of the transmitted signal after propagation through the channel. This recreation is achieved by using a process called as **demodulation**. In fact, the demodulation is the reverse of the modulation process used in the transmitter. We can define modulation process in a more systematic form as under:

Modulation may be defined as the process by which some characteristic of a signal called carrier is varied in accordance with the instantaneous value of another signal called modulating signal. Signals containing information or intelligence are referred as modulating signals. This information bearing signal is also called baseband signal. The carrier frequency is greater than the modulating frequency. The signal resulting from the process of modulation is called modulated signal.

1.9.1. Types of Modulation

Modulation is basically of two types:

(i) Continuous Wave Modulation: When the carrier wave is continuous in nature, the modulation process is known as continuous wave (CW) modulation or analog modulation. Examples of continuous wave modulation are Amplitude Modulation and Angle Modulation. When the amplitude of the carrier is varied in accordance with the message signal, it is known as **amplitude modulation (AM)**. Also, when the angle of the carrier is varied according to the instantaneous value of the modulating signal, it is called **angle modulation**. Angle modulation may be further subdivided into Frequency Modulation (FM) and Phase Modulation (PM), in which the instantaneous frequency and phase of the carrier, respectively, are varied in accordance with the message signal.

(ii) Pulse Modulation : When the carrier wave is a pulse-type waveform, the modulation process is known as pulse modulation. In pulse modulation, the carrier consists of a periodic sequence of rectangular pulses. Pulse modulation can be of an analog or digital type.

---

* Radio was discovered in 1887, and wireless telegraphy was demonstrated in 1895.

In analog pulse modulation, the amplitude, duration or position of a pulse is varied in accordance with sample values of the message signal. The analog pulse modulation may be of following three types:

(i) pulse-amplitude modulation (PAM)
(ii) pulse-duration modulation (PDM).
(iii) pulse-position modulation (PPM).

On the other hand, the digital form of pulse modulation is known as **pulse-code modulation (PCM).**

1.9.2. Need for Modulation or Benefits of Modulation

(UPTU, Semester Exam. 2003-04; 2004-05)

As discussed earlier, the message signal or baseband signal is used to modulate a high frequency carrier signal inside the transmitter. After modulation, the resulting modulated signal is transmitted with the help of an antenna which is connected at the output side of the transmitter. This modulated signal then travels down the channel to reach at the input of the receiver*.

Now, one question can arise why we use modulation in communication system or what will happen if we transmit message signal or audio signal without modulation. The answer is that the modulation serves several purpose in communication system as discussed below :

(i) Practicality of Antenna : We know that in case when free space is used as a transmitting medium (i.e. channel), messages are transmitted and received with the help of antennas. For efficient radiation and reception, the transmitting and receiving antennas must have lengths comparable to a quarter-wavelength of the frequency used. For example, In AM broadcast systems, the maximum audio frequency transmitted from a radio station is of the order of 5 KHz. If this message audio signal were to be transmitted without modulation, then the height of the antenna required for an effective radiation and reception will be 1/4th of the wavelength given as

$$l = \frac{\lambda}{4} = \frac{c}{4f} = \frac{3\times10^8}{4\times5\times10^3} = 5 \text{ km.}$$

Obviously, it will be totally impracticable to construct and install an antenna of such a height. However, this height of the antenna may be reduced by modulation technique and yet effective radiation and reception is achieved. In modulation process, low frequency or audio signal at radio stations are translated to higher frequency spectrum i.e., radio frequency range. These higher radio frequencies with the small wavelength act as carrier for the audio frequencies (i.e. modulating signal). Thus the height of the antenna required is much reduced and becomes practical.

As an example, if an audio frequency is translated to a radio frequency carrier of frequency 3 MHz, the antenna height required would be

$$l = \frac{\lambda}{4} = \frac{c}{4f} = \frac{3\times10^8}{4\times3\times10^6}$$

$$l = \frac{1}{4}\times10^2 = 0.25 \times 100 = 25 \text{ metres}$$

This antenna height may be achieved practically.

(ii) To remove interference : Another reason for not radiating modulating signal itself is that the frequency range of audio signal is from 20 Hz to 20 kHz. In radio-broadcasting, there are several radio stations. In case, there is no modulation, all these stations transmit audio or sound signals in the range of 20 Hz to 20 kHz. Due to this transmission over same range, the programmes of different stations will get mixed up.

---

* In most communications systems, the baseband signal is used to modulate a higher-frequency carrier signal then it is transmitted by radio.

Hence, in order to keep the various signals separate, it is necessary to translate or shift them to different portions of the electromagnetic spectrum. Thus, each station is allocated a band of frequency. This also overcomes the drawback of poor radiation efficiency at low frequency.

As an example, in Amplitude Modulation radio-broadcast, the maximum modulating signal frequency permitted is 5 kHz. Amplitude Modulation requires a bandwidth of 10 kHz for each station or channel. Therefore, broadcast channels can be placed adjacent to each other, each channel occupying 10 kHz bandwidth. Hence, different stations may be allotted bandwidths say from 790 to 800 kHz, 800 to 810 kHz and so on. In radio receiver, a tuned circuit at the input selects the desired station and rejects all other stations.

(iii) Reduction of noise : Noise is the major limitation of any communication. Although noise can not be eliminated completely, but with the help of several modulation schemes, the effect of noise can be minimized.

1.10 PRIMARY COMMUNICATION RESOURCES

As a matter of fact, in a communication system, there are following two primary communication resources to be employed:

(i) transmitted power, and

(ii) channel bandwidth*

The transmitted power refers to the average power of the transmitted signal.

The channel bandwidth is defined as the band of frequencies allocated for the transmission of the message signal.

Now, the most important system design objective is to use these two resources as efficiently as possible. In most communication channels, one resource may be considered more important than the other. Because of this, we may classify communication channels as power limited or band limited.

For example, the telephone circuit is a typical band-limited channel, whereas a space communication link or a satellite channel is typically power-limited channel.

Let us consider a particular case as under:

When the spectrum of a message signal extends down to zero or low frequencies, we define the bandwidth of the signal as that upper frequency above which the spectral content of the signal is negligible and therefore unnecessary for transmitting information. For example, the average voice spectrum extends well beyond 10 kHz, though most of the average power is concentrated in the range of 100 to 600 Hz, and a band from 300 to 3100 Hz gives good articulation.

Accordingly, we find that telephone circuits that respond well to this latter range of frequencies give quite satisfactory commercial telephone service.

Another aspect of utmost importance is the unavoidable presence of noise in a communication system. A quantitative way of accounting for the effect of noise is to introduce signal-to-noise ratio (SNR) as a system parameter. For example, we may define the SNR at the receiver input as the ratio of the average signal power to the average noise power, both being measured at the same point.

A fundamental question that arises in the study of modulation schemes is the following:

With channel bandwidth and signal-to-noise ratio being the two principal parameters that are available to the designer of a communication system, which particular modulation scheme provides for their most efficient use? or in other words we can say for a specified channel bandwidth, which modulation scheme requires the smallest signal to nose ratio for a prescribed level of system performance.

The answer to this important question lies with information capacity theorem by Shannon.

---

* Bandwidth is the spectrum space occupied by a signal, the frequency range of a transmitted signal, or the range of frequencies accepted by a receiver. It is the difference between the upper and lower frequencies of the range in question.

Information Capacity Theorem

The information capacity theorem, given by Shannon, deals with a continuous channel. According to this theorem, channel bandwidth and signal-to-noise ratio are exchangeable in the sense that we may trade off one for the other for a prescribed system performance. The choice of one modulation scheme over another for the transmission of a message signal in the communication system is often dictated by the nature of this trade off. Infact, the interplay between channel bandwidth and signal-to-noise ratio, and the limitation that they impose on communication, is highlighted by the information capacity theorem.

Let B denote the channel bandwidth, and SNR denote the received signal-to-noise ratio. The information capacity theorem states that ideally these two parameters are related by

$$C = B \log_2 (1 + SNR) \text{ b/s}$$

where C is the information capacity.

The information capacity is defined as the maximum rate at which information may be transmitted without error through the channel. It is measured in bits per second (b/s). Above equation clearly shows that for a prescribed information capacity, we may reduce the required SNR by increasing the channel bandwidth, hence the advantage of using a broad bandwidth to transmit messages. Moreover, above equation provides an idealized framework for comparing the noise performance of one modulation scheme against another.

1.11 SURVEY OF COMMUNICATION APPLICATIONS

In this article, let us discuss various communication applications. Most of the applications are very familiar to us. The most common applications are telephone, radio and television (TV). We may also use other forms of communications such as cellular telephones, CB radios, pagers or remote controls. There are many more applications. Figure 1.15 summarizes various major applications of electronic communications. These applications are divided basically into two types namely simplex applications and duplex applications. Simplex applications correspond to one way applications whereas duplex applications correspond to two way applications.

Communication applications

Simplex (One way)

- AM and FM broadcasting
- TV broadcasting
- Cable TV
- Facsimile (Fax)
- Wireless remote control
- Pagers
- Navigation and direction finding
- Telemetry
- Radio astronomy
- Surveillance
- Teletext and view data
- Music services

Duplex (Two way)

- Telephones
- Two-way radio
- Radar
- Sonar
- Amateur radio
- Citizens radio
- Data communications
- Local area networks

Fig. 1.15 Survey of communication applications

1.12 ANALOG AND DIGITAL SIGNALS

Before we discuss Analog and digital communications, let us have a quick view of analog and digital signals.

The analog signal is that type of signal which varies smoothly and continuously with time*. This means that analog signals are defined for every value of time and they take on continuous values in a given time interval. Thus, we can say that analog messages are characterized by data whose values vary over a continuous range. The signal depicted in figure 1.16 is an analog signal. In fact, for analog signal, the name derives from the fact that such a signal is analogous to the physical signal that it represents. The vast majority of signals in the world around us are analog. For example, the temperature or the atmospheric pressure of a certain location may vary over a continuous range and may assume an infinite number of possible values. Similarly, a speech waveform is an analog signal since it has amplitudes that vary over a continuous range.

REMEMBER

Devices called modems permit digital data to be transmitted over the analog telephone networks.

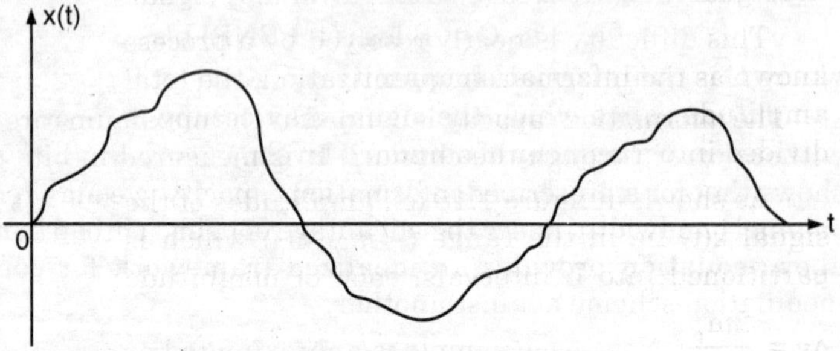

Fig. 1.16 An analog signal

An alternative form of signal representation is that of a sequence of numbers, each number representing the signal magnitude at any instant of time. The resulting signal is called a digital signal. Digital messages are constructed with a finite number of symbols. For example, the printed language consists of 26 letters, 10 numbers, a space and several punctuation marks. Therefore, any text is a digital message constructed from about 50 symbols.

Now, since a digital signal is represented only by digits, therefore, we can use any number system to represent a digital signal. However, in practice, we generally use binary number system to represent a digital signal. In a binary system, each digit in the number takes on one of only two possible values, denoted 0 and 1. Correspondingly, the digital signals in binary systems need have only two voltage levels which may be labelled low and high.

Figure 1.17 shows a digital signal. Here, it may observed that the waveform is a pulse train with 0 V representing a '0' signal or logic '0' and + 5 V representing logic '1'.

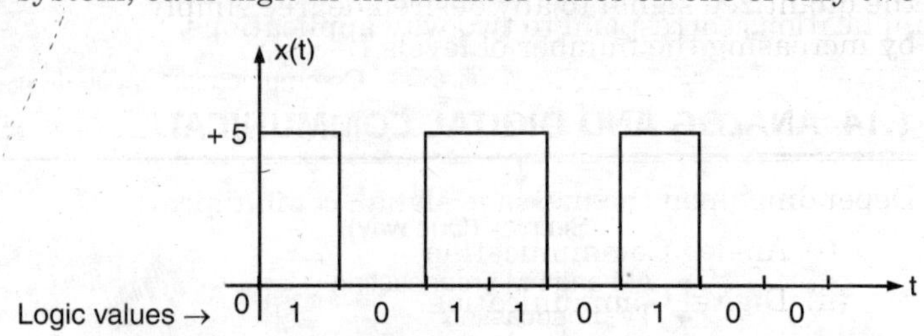

Fig. 1.17 Variation of a binary digital signal with time.

1.13 CONVERSION OF ANALOG SIGNALS TO DIGITAL SIGNALS (Important)

In communication systems, sometimes it happens that we are available with analog signal, however, we have to transmit a digital signal for a particular application. In such cases, we have to convert an analog signal into digital signal. This means that we have to convert a continuous time signal in the form of digits. To see how a signal can be converted from analog to digital form, let us consider an analog signal as shown in figure 1.18(a). First of all, we get sample of this signal according to sampling theorem. For this purpose, we mark the time-instants t_0, t_1, t_2 and so on, at equal time-intervals along the time axis. At each of these time-instants, the magnitude of the signal is measured and

* Information signals may be either analog or digital. Analog signals are smooth, continuous voltage variations such as voice or video. Digital signals are binary pulses or codes.

thus samples of the signal are taken. Figure 1.18(b) shows a representation of the signal of figure 1.18 (a) in terms of its samples.

Now, we can say that the signal in figure 1.18(b) is defined only at the sampling instants. This means that it no longer is a continuous function of time, but rather, it is a discrete-time signal. However, since the magnitude of each sample can take any value in a continuous range, the signal in figure 1.18 (b) is still an analog signal.

This difficulty is neatly resolved by a process known as quantization. In quantization, the total amplitude range which the signal may occupy is divided into a number of standard levels.

As shown in figure 1.18 (c), amplitudes of the signal x(t) lie in the range ($-m_p$, m_p) which is partitioned into L intervals, each of magnitude $\Delta v = \frac{2m_p}{L}$. Now, each sample is approximated or rounded off to the nearest quantized level as shown in figure 1.18(c). Since each sample is now approximated to one of the L numbers, therefore the information is digitized.

The quantized signal is an approximation of the original one. We can improve the accuracy of the quantized signal to any desired degree simply by increasing the number of levels L.

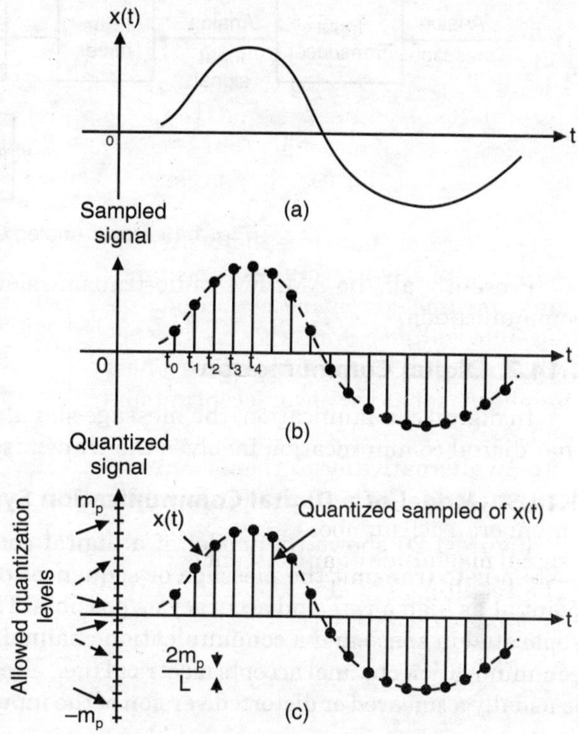

Fig. 1.18 (a) An analog signal, (b) Samples of analog signal, (c) quantization

1.14 ANALOG AND DIGITAL COMMUNICATION

Depending upon the message signal, communication may also be classified as under:

(i) Analog Communication,

(ii) Digital Communication.

1.14.1. Analog Communication

Analog communication is that type of communication in which the message or information signal to be transmitted is analog in nature. This means that in analog communication the modulating signal (i.e., baseband signal) is an analog signal. This analog message signal may be obtained from sources such as speech, video shooting etc.

In Analog communication, the analog message signal modulates some high carrier frequency inside the transmitter to produce modulated signal. This modulated signal is then transmitted with the help of a transmitting antenna to travel through the transmission channel. At the receiver, this modulated signal is received and processed to recover the original message signal. Figure 1.19 shows the block diagram of an analog communication system.

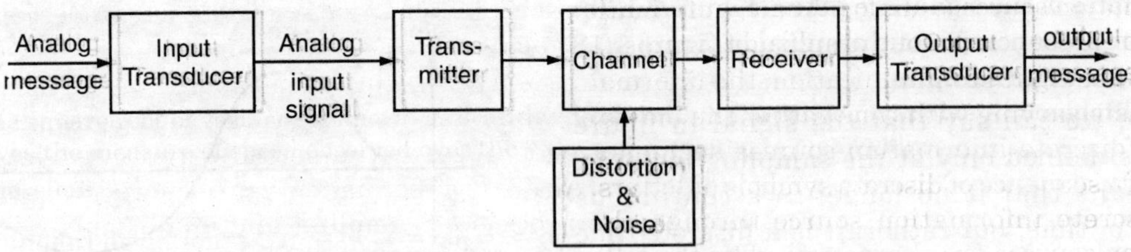

Fig. 1.19 Basic Analog Communication System.

Presently all the AM, FM radio transmission and T.V. transmission are examples of analog communication.

1.14.2. Digital Communicaton

In digital communication, the message signal to be transmitted is digital in nature. This means that digital communication involves the transmission of information in digital form.

1.14.3. Model of a Digital Communication System (U.P.Tech., Semester Exam; 2005-06)

Figure 1.20 shows the model of a digital communication system. The overall purpose of the system is to transmit the message or sequences of symbols coming out of a source to a destination point at as high a rate and accuracy as possible. The source and the destination point are physically separated in space and a communication channel connects the source to the destination point. The communication channel accepts electrical (i.e., electromagnetic) signals and the output of the channel is usually a smeared or distorted version of the input due to the non-ideal nature of the communication

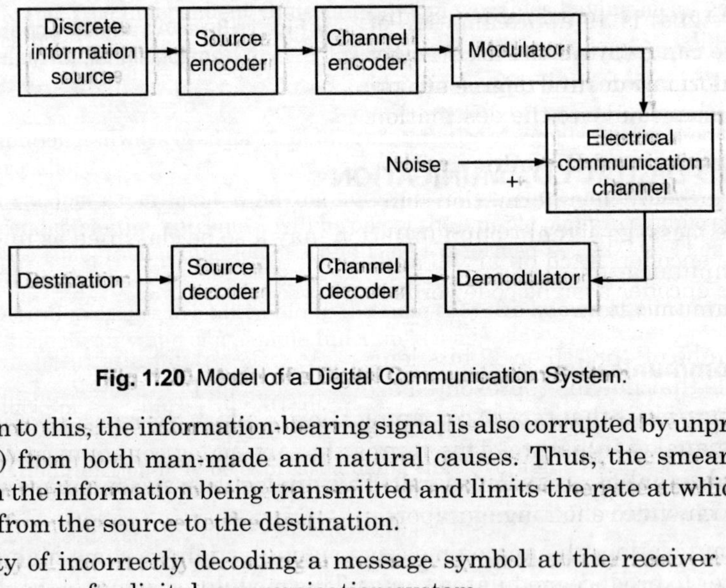

Fig. 1.20 Model of a Digital Communication System.

channel. In addition to this, the information-bearing signal is also corrupted by unpredictable electrical signals (i.e., noise) from both man-made and natural causes. Thus, the smearing and the noise introduce errors in the information being transmitted and limits the rate at which information can be communicated from the source to the destination.

The probability of incorrectly decoding a message symbol at the receiver is often used as a measure of performance of a digital communication system.

Now let us have a detailed look at each of the functional blocks in a digital communication system.

1.14.3.1. Discrete Information Source

Information source may be classified into two categories based upon the nature of their output i.e., analog information sources and discrete information sources. In case of analog communication, the information source is analog. Analog information sources, such as microphone actuated by speech, emit one or more continuous amplitude signals.

In case of digital communication, the information source produces a message signal which is not continuously varying with time rather the message signal is intermittent with respect to time. The output of discrete information sources such as a teletype or the numerical output of a computer consists of a sequence of discrete symbols or letters. An analog information source may be transformed into a discrete information source through the process of sampling and quantizing. Discrete information sources are characterized by the following parameters:

(i) **Source alphabet :** These are the letters, digits or special characters available from the information source.

(ii) **Symbol rate :** It is the rate at which the information source generates source alphabets. It is generally represented in symbols/sec unit.

(iii) **Source alphabet probabilities :** Each source alphabet from the source has independent occurrence rate in the sequence. As an example, letters A, E, I etc., occur frequently in the sequence. Hence, probability of the occurrence of each source alphabet can become one of the important property which is useful in digital communication.

(iv) **Probabilistic dependence of symbols in a sequence :** The information carrying capacity of each source alphabet is different in a particular sequence. This parameter defines average information content of the symbols. The entropy of a source describes the average information content per symbol in long messages. Entropy may be defined in terms of bits per symbol. Bit is the abbreviation for a binary digit.

This means that the source information rate is the product of symbol rate and source entropy i.e.,

$$\underset{\text{(Bits/sec.)}}{\text{Information rate}} = \underset{\text{(Symbols/sec.)}}{\text{Symbol rate}} \times \underset{\text{(Bits/Symbol)}}{\text{Source entropy}}$$

Thus, the information rate represents minimum average data rate required to transmit information from source to the destination.

1.14.3.2. Source Encoder and Decoder

The symbols produced by the information source are given to the source encoder. These symbols cannot be transmitted directly. They are first converted into digital form (i.e., binary sequence of 1's and 0's) by the source encoder. Each binary '1' and '0' is known as a bit. The group of bits is called a codeword. The source encoder assigns codewords to the symbols. For each distinct symbol, there is an unique codeword.

The codeword can be of 4, 8, 16 or 32 bits length. As the number of bits are increased in each codeword, the symbols that may be represented are also increased.

As an example, 8 bits would have 2^8 i.e., 256 distinct codewords. This means that 8-bits may be used to represent 256 symbols and similarly 16 bits may represent $2^{16} = 65536$ symbols and so on. Some typically source encoders are pulse code modulators, delta modulators, vector quantizers etc. Source encoders must have following important parameters :

(i) Block size

Block size describes the maximum number of distinct codewords which can be represented by a source encoder. This depends on the number of bits in the codeword. As an example, the block size of 8 bits source encoder will be 2^8 i.e., 256 codewords.

(ii) Codeword length

Codeword length is the number of bits used to represent each codeword. As an example, if 8 bits are assigned to each codeword, then the codeword length will be 8 bits.

(iii) Average data rate

Average data rate is the output bits per second from the source encoder. In fact, the source encoder assigns multiple number of bits to each input symbol. Hence the data rate is generally higher than the symbol rate. As an example, if we consider that the symbols are given to the source encoder and the length of codeword is 8 bits, then the output data rate from the source encoder would be given as

$$\text{Data rate} = \text{Symbol rate} \times \text{Codeword length}$$

$$\text{Data rate} = 10 \times 8$$

$$\text{Data rate} = 80 \text{ bits/seconds}$$

Also, since the information rate is the minimum number of bits per second needed to convey information from source to destination, therefore the optimum data rate is equal to the information rate. However, due to practical limitations, designing such type of source encoder is quite difficult. Thus, the average data rate is higher than the information rate and hence symbol rate also.

(iv) Efficiency of the Encoder

The efficiency of the encoder is the ratio of minimum source information rate to the actual output data rate of the source encoder.

In last, it may be noted that at the receiver end, some sort of decoder is used to perform the reverse operation to that of source encoder. It converts the binary output of the channel decoder into a symbol sequence. Some decoders also use memory to store codewords. The decoders and the encoders can be synchronous or asynchronous.

1.14.3.3. Channel Encoder and Decoder

After converting the message or information signal in the form of binary sequence by the source encoder, the signal is transmitted through the channel. The communication channel adds noise and interference to the signal being transmitted*. Hence errors are introduced in the binary sequence received at the receiver end.

Therefore, the errors are also introduced in the symbols generated from these binary codewords. Thus channel coding is done to avoid these types of errors. In fact, the channel encoder adds some redundant binary bits to the input sequence. Also, these redundant bits are always added with some properly defined logic. As an example, let us consider that the codeword from the source encoder to make it 4-bits long. This fourth bit is added (i.e., 1 or 0) in such a manner that the number of 1's in the encoded word remain even (also known as even parity). Table 1.4 gives the output of a source encoder, the fourth bit depending on the parity and the output of channel encoder.

It may be observed from the table that each codeword at the output of channel encoder contains even number of 1's. Now, at the receiver end, if odd number of 1's are detected, then the receiver comes to know that there is an error in the received signal. The channel decoder at the receiver is thus able to reconstruct error free accurate bit sequence and reduce the effects of channel noise and distortion.

---

* Noise is any interference that disturbs the legible transmission of a signal. Noise is produced by the atmosphere, heavenly bodies, manufactured electrical equipment, and thermal agitation in electronic components.

TABLE 1.4.

Output of source encoder			Bit to be added by the channel encoder for an even parity	Output of a channel encoder			
b_3	b_2	b_1	b_0	b_3	b_2	b_1	b_0
1	0	0	0	1	1	0	0
0	1	0	1	0	1	0	1
0	0	0	0	0	0	0	0
1	1	1	1	1	1	1	1
	:		:		:		
	:		:		:		

This means that the channel encoder and decoder serve to increase the reliability of a received signal. However the extra bits which are added by the channel encoders carry no information, rather, they are used by the channel decoder to detect and correct errors if any.

The coding and decoding operation at the encoder and decoder needs the memory and processing of binary data. However, in the modern time, due to use of microcontrollers and computers, the complexity of the encoders and decoders is much reduced.

A channel encoder must have the following important parameters:

(i) The coding rate that depends upon the redundant bits added by the channel encoder.
(ii) The coding method used.
(iii) Coding efficiency which is the ratio of data rate at the input to the data rate at the output of the encoder.
(iv) Error control capabilities.
(v) Feasibility of the encoder and decoder.

1.14.3.4. Digital Modulators and Demodulators

Earlier in this chapter, we discussed, why modulaton is needed in communication system. Now, if the modulating signal is digital (i.e. binary codewords), then digital modulation techniques are used. The carrier signal used by digital modulators is always continuous sinusoidal wave of high frequency. In fact, the digital modulators map the input binary sequence of 1's and 0's to the analog signal waveforms. For example, if one bit at a time is transmitted, then digital modulator signal is $s_1(t)$ to transmit binary '0' and $s_2(t)$ to transmit binary '1' as shown in figure 1.21.

Here the signal $s_1(t)$ has low frequency compared to signal $s_2(t)$. Hence, here, even though the modulated signal seems to be continuous, the modulation is discrete (i.e., in steps). This means that a signal carrier is converted into two waveforms $s_1(t)$ and $s_2(t)$ because of digital modulation.

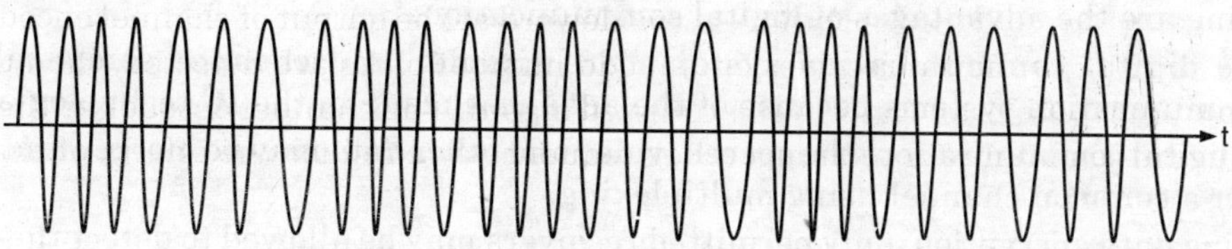

Fig. 1.21 The output of a digital modulator.

Now, if the codeword consists of two bits and they are to be transmitted at a time, then there would be 2^2 i.e., 4 distinct symbols i.e., codewords. Thus, these codewords will require four distinct waveforms for transmission purpose. Such types of modulators are known as M-ary modulators.

Amplitude shift keying (ASK), phase shift keying (PSK), frequency shift keying (FSK), differential phase shift keying (DPSK) and minimum shift keying (MSK) are the examples of various digital modulators.

However, since these modulators use a continuous carrier wave, therefore they are also known as digital CW modulators.

At the receiver end, the digital demodulator converts the input modulated signal into the sequence of binary bits.

A digital modulation method must have following important parameters:

(i) Bandwidth needed to transmit the signal,

(ii) Probability of symbol or bit error,

(iii) Synchronous or asynchronous method of detection,

(iv) Complexity of implementation.

1.14.3.5. Communications Channel

As discussed earlier, the connection between transmitter and receiver is established through a communication channel. The communication can take place through wirelines, wireless or fiber optic channels. The other media such as optical disks, magnetic tapes and disks etc. may also be called as a communication channel since they can also carry data through them. However, it may be noted that each and every communication channel has some inherent problems. These are :

(i) Signal Attenuation : The signal attenuation in channel occurs due to the internal resistance of the channel and fading of the signal.

(ii) Amplitude and Phase distortion : The transmitted signal is distorted in amplitude and phase due to the non-linear characteristics of the communication channel.

(iii) Additive noise interference : Additive noise interference is produced due to internal solid state devices and resistors etc., used to implement a communication system.

(iv) Multipath distortion : The multipath distortion occurs mostly in wireless communication channels.

In fact, the signals coming from different paths tend to interfere with each other.

1.15 ADVANTAGES AND DISADVANTAGES OF DIGITAL COMMUNICATION (Important)

In this section, we shall discuss the advantages and disadvantages of digital communication briefly.

Advantages:

Following are the advantages of digital communication:

(i) The digital communication systems are simpler and cheaper compared to analog communication systems because of the advances made in the IC technologies.

(ii) In digital communication, the speech, video and other data may be merged and transmitted over a common channel using multiplexing.

(iii) Using data encryption, only permitted receivers may be allowed to detect the transmitted data. This property is of its most importance in military applications.

(iv) Since the transmission is digital and the channel encoding is used, therefore the noise does not accumulate from repeater to repeater in long distance communications.

(v) Since the transmitted signal is digital in nature, therefore a large amount of noise interference may be tolerated.

(vi) Since in digital communication, channel coding is used, therefore the errors may be detected and corrected in the receivers.

(vii) Digital communication is adaptive to other advanced branches of data processing such as digital signal processing, image processing and data compression etc.

Disadvantages

Although digital communication offers, so many advantages as discussed above, it has some drawbacks also. However, the advantages of digital communication outweight disadvantage.

The disadvantages may be listed as under:

(i) Due to analog to digital conversion, the data rate becomes high. Therefore more transmission bandwidth is required for digital communication.

(ii) Digital communication needs synchronization in case of synchronous modulation.

1.15.1. Performance Comparison of Analog and Digital Modulation

Table 1.5 presents the comparison of analog and digital modulation

TABLE 1.5.

S.No.	Analog modulation	Digital modulation
1.	Transmitted modulated signal is analog in nature.	Transmitted signal is digital, *i.e.*, train of digital pulses.
2.	Amplitude, frequency or phase variations in the transmitted signal represent the information or message.	Amplitude, width or position of the transmitted pulses is constant. The message is transmitted in the form of code words.
3.	Noise immunity is poor for AM, but improved for FM and PM.	Noise immunity is excellent.
4.	It is not possible to separate out noise and signal. Therefore, repeaters cannot be used.	It is possible to separate signal from noise. Therefore, repeaters can be used.
5.	Coding is not possible.	Coding techniques can be used to detect and correct the errors.
6.	Bandwidth required is lower than that for the digital modulation methods.	Due to higher bit rates, higher channel bandwidth is required.
7.	FDM is used for multiplexing.	TDM is used for multiplexing.
8.	Not suitable for transmission of secret information in military applications.	Due to coding techniques, it is suitable for military applications.
9.	Analog modulation systems are AM, FM, PM, PAM, PWM etc.	Digital modulation systems are PCM, DM, ADM, DPCM etc.

1.16 THE ELECTROMAGNETIC (EM) SPECTRUM

As a matter of fact, the information signal should be first converted into an electromagnetic signal before transmission because the wireless transmission takes place using the electromagnetic waves. The electromagnetic waves consist of both electric and magnetic fields. The electromagnetic waves can travel a long distance through space. The electromagnetic signals are also called as **radio frequency (RF) waves.** The EM waves oscillate, they are sinusoidal and their frequency is measured in Hz. The frequency of EM signal can be very low or it can be

DO YOU KNOW?

The electromagnetic spectrum is that range of frequencies from approximately 30 Hz to visible light over which electronic communications take place.

extremely high. This entire range of frequencies of EM waves is called as **Electromagnetic spectrum.***

The electromagnetic spectrum consists of signals such as 50 Hz line frequency and voice signals at the lower end. The radio frequencies which are used for the two way communication reside at the center of the EM spectrum. These frequencies are used for the applications such as radio or TV broadcasting as well. The infrared and visible light are at the upper end of the EM spectrum. Figure 1.22 shows the entire electromagnetic spectrum**.

DO YOU KNOW?

The high-frequency range (3 to 30 MHz), or shortwave, is used for worldwide two-way communications and broadcasting.

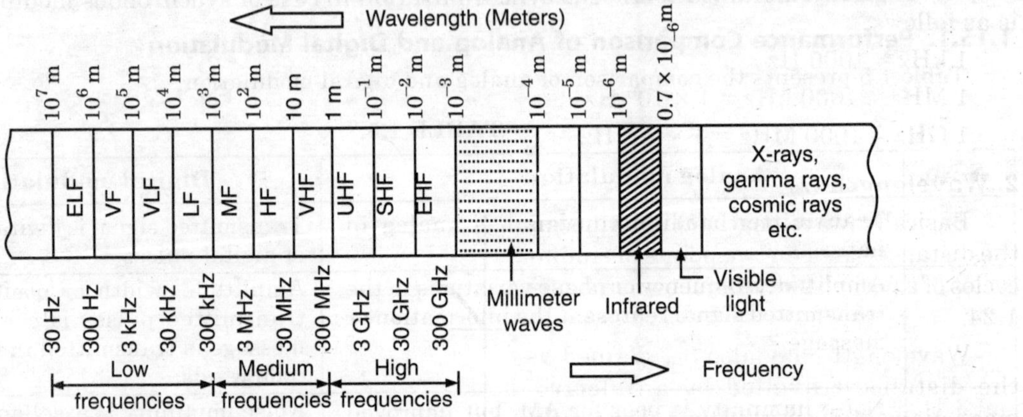

Fig. 1.22. Illustration of complete electromagnetic (EM) spectrum

Here, it may be noted that the short forms used in the EM spectrum of figure 1.22 have the detailed meanings as depicted in table 1.6.

TABLE 1.6. Segments of the electromagnetic spectrum.

S.No.	Name	Frequency	Wavelength
1.	Extremely low frequency (ELF)	30-300 Hz	10^7 to 10^6 m
2.	Voice frequencies (VF)	300-3000 Hz	10^6 to 10^5 m
3.	Very low frequencies (VLF)	3-30 kHz	10^5 to 10^4 m
4.	Low frequencies (LF)	30-300 kHz	10^4 to 10^3 m
5.	Medium frequencies (MF)	300 kHz-3 MHz	10^3 to 10^2 m
6.	High frequencies (HF)	3-30 MHz	10^2 to 10 m
7.	Very high frequencies (VHF)	30-300 MHz	10 to 1 m
8.	Ultra high frequencies (UHF)	300 MHz-3GHz	1 to 10^{-1} m
9.	Super high frequencies (SHF)	3-30 GHz	10^{-1} to 10^{-2} m
10.	Extremely high frequencies (EHF)	30-300 GHz	10^{-2} to 10^{-3} m
11.	Infrared	—	0.7 to 10 μm
12.	Visible light	—	0.4 μm to 0..8 μm

* The electromagnetic spectrum is that range of frequencies from approximately 30 Hz to visible light over which electronic communications take place.

** The greatest portion of the spectrum covers radio waves, which are oscillating electric and magnetic fields that radiate for long distances.

1.16.1. Frequency and Wavelength

In the EM spectrum, we have used frequency as well as wavelength in order to define various segments. Now, therefore, let us define these terms and the relation between them.

1. Frequency

Basically, frequency is defined as the number of cycles of a waveform per second. It is expressed in hertz (Hz). We can use kilohertz (kHz), Megahertz (MHz) and Gigahertz (GHz) to measure higher frequencies. Their relation with the basic unit Hz is as follows:

1 kHz = 1000 Hz

1 MHz = 1000 kHz = 1×10^6 Hz

1 GHz = 1000 MHz = 1×10^9 Hz

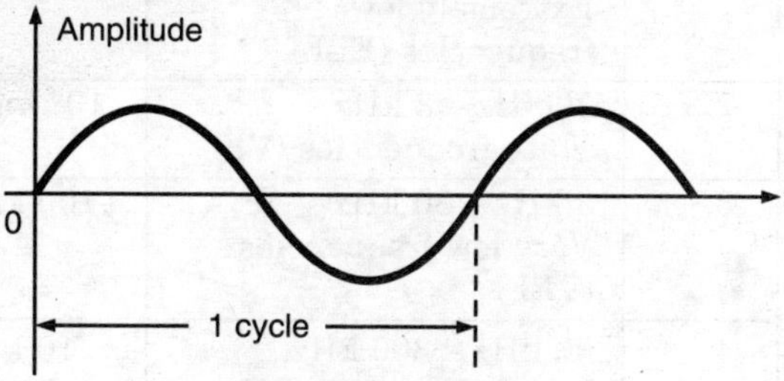

Fig. 1.23 One cycle

2. Wavelength (λ)

Basically, wavelength (λ) is defined as the distance between two points of similar cycles of a periodic wave as shown in figure 1.24.

Wavelength may also be defined as the distance travelled by an electromagnetic wave during the time of one cycle. Since, EM waves travel at the speed of light in the free space or vacuum, their wavelength is given by,

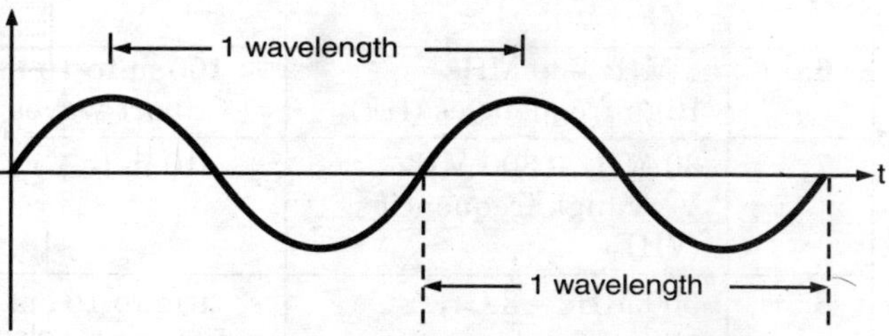

Fig. 1.24 Definition of wavelength

$$\lambda = \frac{\text{Speed of light}}{\text{Frequency}} = \frac{3 \times 10^8 \text{ m/s}}{f} \qquad ...(1.7)$$

Hence, wavelength decreases with increase in frequency.

1.16.2. EM Spectrum and Communication Applications

In the radio communication system, the frequencies ranging from a few kilohertz to many gigahertz all are being used for various purposes. Let us discuss the applications of various frequency bands. The frequencies most commonly used in early days were from about 300 kHz to 3 MHz and were called as **medium frequencies (MF).** The frequencies in the range 30 kHz to 300 kHz are known as the **low frequencies (LF)**. The frequencies in the range 3 kHz to 30 KHz are called as **very low frequencies (VLF)**. On the higher frequency side, high frequencies (HF) will cover the frequency range from 3 MHz to 30 MHz. Then **very high frequency (VHF)** from 30 MHz to 300 MHz and so on*.

DO YOU KNOW?

Electromagnetic radiation does not propagate underwater the way it does in air. Long-range communication underwater has mostly been based on sound waves.

DO YOU KNOW?

Electromagnetic signals produced primarily by heat sources are called infrared. They cover the 0.7-to-100 μm range.

Table 1.7 presents the details of entire usable frequency spectrum and its applications.

* The range of human hearing is approximately 20 to 20,000 Hz. The voice frequency range is 300 to 3000 Hz. The high-frequency range (3 to 30 MHz), or short wave, is used for worldwide two-way communications and broadcasting.

TABLE 1.7. The radio frequency (RF) spectrum

S.No.	Frequency band	Wavelength	Applications
1.	30 Hz – 300 Hz. Extremely low frequencies (ELF).	10^4 km to 10^3 km	Power transmission
2.	300 Hz – 3 kHz. Voice frequencies (VF)	10^3 km to 100 km	Audio applications
3.	3 kHz – 30 kHz Very low frequencies (VLF)	100 km to 10 km	Submarine communications. Navy, Military communications
4.	30 kHz – 300 kHz. Low frequencies (LF)	10 km to 1 km Long waves.	Aeronautical and marine. navigation, these frequencies act as sub carriers.
5.	300 kHz – 30 MHz Medium frequencies (MF)	1 km to 100 m. Medium waves.	AM radio broadcast, Marine and aeronautical communi-cations.
6.	3 MHz –30 MHz High frequencies (HF)	100 m to 10 m. Short waves	Shortwave transmission, Amateur and CB communication.
7.	30 MHz – 300 MHz Very high frequencies (VHF)	10 m to 1 m	TV broadcasting FM broadcasting
8.	300 MHz – 3 GHz Ultra high frequencies (UHF)	1 m to 10 cm. Microwaves.	UHF TV channels, Cellular phones, Military applications.
9.	3 GHz – 30 GHz (SHF)	10^{-1} m to 10^{-2} m	Satellite communications and Radar
10.	30 – 300 GHz (EHF)	10^{-2} m to 10^{-3} m	Satellites and specialized radars.

1.16.3. Infrared Signals*

The EM signals having frequencies above 300 GHz are not referred as radio waves. The signal occupying the range between 0.1 mm and 700 nanometers (nm) are called **infrared signals**. These are used in various special kinds of communications as listed below.

(i) In astronomy to detect stars and other heavenly bodies.

(ii) For guidance in weapon systems.

(iii) TV remote control.

REMEMBER

In radio applications, keep resistor leads short to minimize inductive and capacitive effects.

1.16.4. Visible Light

Basically, light is a special type of electromagnetic radiation. It has wavelength in the range of 0.4 to 0.8 μm. Light is used for various kinds of communications. Light waves can be modulated and transmitted through the glass fibers in the optical fiber communication system. Light signals can also be transmitted through free space. Laser is a type of light, which can be easily modulated with voice, video and data information. Beyond the visible light, there are X-rays, gamma rays and cosmic rays. But, they are not useful for the communication applications.

* Electromagnetic signals produced primarily by heat sources are called infrared. They cover the 0.7-to 100-μm range.

1.17 FUNDAMENTAL LIMITATIONS OF COMMUNICATION SYSTEM

While designing a communication system, an engineer generally faces several constraints or limitations. These are noise limitation, bandwidth limitation and equipment limitation.

1.17.1. Noise Limitation

The noise may be defined as an unwanted form of energy which tend to interfere with the transmission and reception of the desired signals in a communication system. The noise cannot be eliminated completely. However, the effect of noise on desired signals can be minimised with the help of several techniques. Noise can be classified into two broad categories depending upon its source: external noise and internal noise. External noise is that type of noise whose sources are external to a communication system. The examples of external noise are atmospheric noise, galactic noise and industrial noise. On the other hand, internal noise is that type of noise whose sources are internal to a communication system. The examples of internal noise are thermal noise and shot noise. This type of noise is inevitable in every communication system and forms a basic limitation on transmission and reception of signals.

Infact, the noise limits our ability to identify the intended or desired message correctly and thus, limits information transmission. Typical noise variations are measured in microvolts. If the signal variations are quite large in comparison, then the effect of noise is very small and may be ignored. Infact, in many practical communication systems, the effect of noise is generally unnoticeable. However, in long-distance communication systems operating with a limited amount of signal power, the signal may be as small as noise or even smaller than the noise. Thus, in such cases, the presence of noise severally limits the capabilities of a communication system.

1.17.2. Bandwidth Limitation*

In addition to noise, the bandwidth limitation is another major constraint in a communication system. The frequency range or the band of frequency needed for a particular given transmission is known as **bandwidth**. This band of frequencies required for a particular transmission is also called channel. This band of frequencies or bandwidth for a particular transmission is always allocated by some International regulatory agencies. This type of regulation is essential to avoid interference among the signals having same frequency. But, for a given transmission, this allocated bandwidth may not be sufficient to convey the entire information.

Infact, the information theory states that the greater is the transmission bandwidth of a communication system, the more is the information that can be transmitted. Let us consider a simple example. Suppose one is listening to a music in an AM radio. The complete amount of information available to the human ear is contained in a frequency range upto 15 KHz, i.e., musical information extends upto a frequency of 15 KHz. However, in AM radio the maximum modulating frequency is restricted upto 5 KHz and hence, the maximum bandwidth of AM transmission is 10 KHz.

Therefore, an AM radio receiver cannot reproduce all the information contained in the music because this will require a bandwidth of 30 KHz. On the other hand, the bandwidth allocated to a FM transmission is about 200 KHz. Thus, on FM receiver can easily reproduce the transmitted information without any distortion. This means that a FM system has a better fidelity than an AM system. Also, it may be observed in common life that one prefers to listen to a FM radio than an AM radio. Thus, we can conclude that bandwidth is a major fundamental limitation of a communication system.

---

* Spectrum space is a precious natural resource. In the United States, the FCC regulates the use of the spectrum and most forms of Electronic communications according to the communications Act of 1934.

1.17.3. Equipment Limitation

The noise and bandwidth limitations dictate theoretically what can or cannot be achieved in terms of performance in a communication system. However, this theoretical limit may not be realised in a practical system due to equipment limitations. For example, the theory might require a band-pass filter with a quality factor of 100 at a centre frequency of 1 KHz. Such a filter cannot be realised in practice. Even if a filter with nearly identical characteristics is built, the cost may exceed what the user of the communication system wants to pay. Thus, equipment limitation is another major problem in a communication system.

1.18 HISTORICAL PERSPECTIVE

The Evolution of Modern Electronic Communication

The electronic communication has undergone through a chain of dramatic developments even since the first electronic communication invented by Morse in 1838. However, it may be noted that this development became possible due to the parallel development in the field of electronic devices, circuits etc. To provide historical perspective, the important developments in the field of electronic communication are listed ahead :

TABLE 1.8. A Chronology of Electronic Communication

Year	Development
1800-1837	Preliminary developments
1838	The invention of telegraphy
1845	The Kirchoff's circuit laws enunciated
1864	Maxwell's equations predict electromagnetic radiation
1876	The invention of telephony
1887	The invention of wireless telegraphy
1904	Communication Electronics
1920	Transmission theory
1923	Invention of television
1927	Federal communications commission established
1931	Teletypewriter service initiated
1934	H.S. Black develops the negative-feedback amplifier
1936	Armstrong's paper states the case for FM radio
1937	Pulse code modulation
1938	Radar and microwave communication
1948	Information theory and coding
1950	Time-division multiplexing applied to telephony
1953	Colour T.V. Standards established in United States
1956	First transoceanic states telephone cable (36 voice channels)
1958	Long-distance data transmission system developed for military purposes.
1960	Maiman demonstrated the first Laser
1961	Integrated circuits go into commercial production.
1962	Satellite communication begins with Telestar I
1962-1966	High-speed digital communication.
1963	Solid-state microwave oscillators
1964	Full electronic telephone switching system goes into service.
1965	Mariner IV transmits pictures from Mars to earth.
1966-1975	Wideband communication systems
1979	Intercity optical links
1988	Under sea fiber optical link

SUMMARY

- Communication is the process of establishing connection or link between two points for information exchange.
- The electronic equipments which are used for communication purpose, are called communication equipments. Different communication equipments when assembled together form a communication system.
- Typical examples of communication system are line telephony and line telegraphy, radio telephony and radio broadcasting, point-to-point communication and mobile communication, computer communication, radar communication, television broadcasting, radio telemetry, radio aids to navigation, radio aids to aircraft landing etc.
- In the most fundamental sense, communication involves the transmission of information from one point to another through a succession of process.
- The function of information source is to produce required message which has to be transmitted.
- A transducer is a device which converts one form of energy into another form. The message from the information source may or may not be electrical in nature. In a case when the message produced by the information source is not electrical in nature, an input transducer is used to convert it into a time-varying electrical signal.
- The function of the transmitter is to process the electrical signal from different aspects. For example, in radio broadcasting the electrical signal obtained from sound signal, is processed to restrict its range of audio frequencies (upto 5 KHz in amplitude modulation radio broadcast) and is often amplified.
- The function of the channel is to provide a physical connection between the transmitter and the receiver.
- There are two types of channels, namely point-to-point channels and broadcast channels. Examples of point-to-point channels are wirelines, microwave links and optical fibers. Wirelines operate by guided electromagnetic waves and they are used for local telephone transmission. In case of microwave links, the transmitted signal is radiated as an electromagnetic wave in free space.
- Noise is an unwanted signal which tend to interfere with the required signal. Noise signal is always random in character. Noise may interfere with signal at any point in a communication system. However, the noise has its greatest effect on the signal in the channel.
- The main function of the receiver is to reproduce the message signal in electrical form from the distorted received signal. This reproduction of the original signal is accomplished by a process known as the demodulation or detection. Demodulation is the reverse process of modulation carried out in transmitter.
- Destination is the final stage which is used to convert an electrical message signal into its original form. For example in radio broadcasting, the destination is a loudspeaker which works as a transducer i.e. it converts the electrical signal in the form of original sound signal.
- The telecommunication environment is dominated by the following four important sources of information:
  (i) speech
  (ii) television
  (iii) facsimile, and
  (iv) personal computers
- Several sources of information which are responsible for the generation of message signals. These message signals need to be transmitted to the destination point. Now, let us discuss various communication channel that carryout the transmission.
- Basically, we shall consider following channels:
  (i) telephone channels
  (ii) optical fibers
  (iii) mobile radio channels
  (iv) satellite channels.
- A telephone network makes use of a switching mechanism. This switching mechanism is known as circuit switching and it is used to establish an end-to-end communication link on a temporary basis.
- The second communication channel, an optical fiber, is a dielectric waveguide which transports light signals from one place to another just as a metallic wire pair or a co-coxial cable, transports electrical signals.

- An optical fiber consists of a central core within which the propagating electromagnetic field is confined and which is surrounded by a cladding layer, which is itself surrounded by a thin protective jacket.
- The third communication channel, mobile radio channel, extends the capability of the public telecommunications network by introducing mobility into the network by virtue of its ability to broadcast.
- A satellite channel provides broad-area coverage in a continental as well as intercontinental sense. Moreover, access to remove areas not covered by conventional cable or fiber communications is also a distinct feature of satellites.
- Communications satellites in geostationary orbit offer the following unique system capabilities:
  (a) broad-area coverage
  (b) reliable transmission links
  (c) wide transmission bandwidth.
- The message signal generated from the information source is known as baseband signal. This baseband signal may be a combination of two or more message signals. If the baseband signal is transmitted directly, then it is known as baseband transmision. This means that the baseband transmission does not used modulators and demodulators. The baseband signal may be both analog as well as digital.
- The analog baseband signal varies continuously with time and has continuous amplitude. The digital baseband signal is discrete in both time and amplitude. The baseband transmission is preferred at low frequencies and for short distances. However, Inter Symbol Interference (ISI) is the major problem associated with the baseband transmission.
- Regarding the mode of propagation, communication may be divided in the following two forms :
  (i) Line Communication
  (ii) Wireless or Radio Communication
- In line communication, the medium of transmission is a pair of conductors called transmission line. This is also called as line channel.
- In wireless or radio communication, a message is transmitted through open space by electromagnetic waves called as radio waves. Radio waves are radiated from the transmitter in open space through a device called antenna.
- Modulation may be defined as the process by which some characteristic of a signal called carrier is varied in accordance with the instantaneous value of another signal called modulating signal. Signals containing information or intelligence are referred as modulating signals. This information bearing signal is also called baseband signal. The carrier frequency is greater than the modulating frequency. The signal resulting from the process of modulation is called modulated signal.
- When the carrier wave is continuous in nature, the modulation process is known as continuous wave (CW) modulation or analog modulation. Examples of continuous wave modulation are Amplitude Modulation and Angle Modulation.
- When the carrier wave is a pulse-type waveform, the modulation process is known as pulse modulation.
- In a communication system, there are following two primary communication resources to be employed:
  (i) transmitted power,
  (ii) channel bandwidth
- The transmitted power refers to the average power of the transmitted signal.
  The channel bandwidth is defined as the band of frequencies allocated for the transmission of the message signal.
- In communication systems, sometimes it happens that we are available with analog signal, however, we have to transmit a digital signal for a particular application. In such cases, we have to convert an analog signal into digital signal.
- Depending upon the message signal, communication may also be classified as under:
  (i) Analog Communication,
  (ii) Digital Communication.
- Analog communication is that type of communication in which the message or information signal to be transmitted is analog in nature. This means that in analog communication the modulating signal (i.e., baseband signal) is an analog signal. This analog message signal may be obtained from sources such as speech, video shooting etc.
- In digital communication, the message signal to be transmitted is digital in nature. This means that digital communication involves the transmission of information in digital form.
- Following are the advantages of digital communication:

(i) The digital communication systems are simpler and cheaper compared to analog communication systems because of the advances made in the IC technologies.

(ii) In digital communication, the speech, video and other data may be merged and transmitted over a common channel using multiplexing.

(iii) Using data encryption, only permitted receivers may be allowed to detect the transmitted data. This property is of its most importance in military applications.

(iv) Since the transmission is digital and the channel encoding is used, therefore the noise does not accumulate from repeater to repeater in long distance communications.

(v) Since the transmitted signal is digital in nature, therefore a large amount of noise interference may be tolerated.

(vi) Since in digital communication, channel coding is used, therefore the errors may be detected and corrected in the receivers.

(vii) Digital communication is adaptive to other advanced branches of data processing such as digital signal processing, image processing and data compression etc.

GLOSSARY

1. **Baseband:** The band of frequencies occupied by an information signal before it modulates the carrier.
2. **Carrier:** A signal that can be modulated by an information signal.
3. **Channel:** A path for the transmission of signals.
4. **Distortion:** Any undesirable change in an information signal.
5. **Modulation:** The process by which some characteristics of a carrier is varied by an information signal.
6. **Frequency translation:** Movement of a signal from one frequency to another using a mixer-oscillator combination.
7. **Demodulator:** Circuit to recover the baseband signal from a modulated signal.
8. **Detector:** See demodulator.

REVIEW QUESTIONS

1. What are the basic constituents of a communication system ?
2. Draw the block diagram of a communication system and explain the function of each block.
3. What is meant by the term 'Channel' as applied to a communication system.
4. Why is a high-frequency carrier needed in a communication system ?
5. Explain the need for modulation in a communication system ?
6. Explain the radio frequency spectrum used in communication system.
7. Explain the fundamental limitations of a communication system.
8. Explain the difference between Analog and Digital communications.
9. How will you convert an analog signal into a digital signal ?
10. Write the advantages and disadvantages of a digital communication system.
11. Explain the difference between baseband transmission and bandpass transmission.

OBJECTIVE TYPE QUESTIONS

1. The highest modulating frequency used in AM broadcast system is
   (a) 10 kHz (b) 15 kHz
   (c) 5 kHz (d) 2 MHz
2. The high frequency (HF) range extends from
   (a) 300-3000 kHz (b) 3-30 MHz
   (c) 30-300 MHz (d) 300-3000 MHz.
3. The very high frequency (VHF) range extends from
   (a) 3-30 MHz (b) 30-300 MHz
   (c) 300-3000 MHz (d) 3000-30,000 MHz
4. The ultra high frequency (UHF) range extends from
   (a) 3-30 MHz (b) 30-300 MHz
   (c) 300-3000 MHz (d) 3000-30,000 MHz
5. The function of the input transducer in a communication system is
   (a) to transmit the message signal
   (b) to modulate the message signal
   (c) to convert message sound signal into electrical signal
   (d) none of the above

6. Communication is the process of
   (a) keeping in touch
   (b) broadcasting
   (c) exchanging information
   (d) entertainment by electronics
7. Two key barriers to human communication are
   (a) distance (b) cost
   (c) ignorance (d) language
8. Electronics communication was discovered in which century ?
   (a) sixteenth (b) eighteenth
   (c) nineteenth (d) twentieth
9. Which of the following is not a major communications medium ?
   (a) free space (b) water
   (c) wires
   (d) fiber-optic cable
10. The communications medium causes the signal to be
    (a) amplified (b) modulated
    (c) attenuated (d) interferred
11. The process of transmitting two or more information signals simultaneously over the same channel is called
    (a) multiplexing (b) telemetry
    (c) detection (d) modulation
12. Recovering information from a carrier is known as
    (a) demultiplexing (b) modulation
    (c) detection (d) carrier recovery
13. Radio signals are made up of
    (a) voltages and currents
    (b) electric and magnetic fields
    (c) electrons and protons
    (d) noise and data

Answers

1. (c)	2. (b)	3. (b)	4. (c)	5. (c)	6. (c)	7. (a & b)
8. (c)	9. (c)	10. (c)	11. (a)	12. (c).	13. (b)	

□□□

CHAPTER 2

DIGITAL DATA TRANSMISSION

SYLLABUS

Digital Data transmission, Line coding review, Pulse shaping, Scrambling, Digital receivers, Eye diagram.

2.1 INTRODUCTION

Although a significant portion of communication, these days, is in analog form, it is now being replaced rapidly by digital communication. This chapter deals with the problems of transmitting digital data over a channel. Hence, the starting messages are assumed to be digital. Infact, to begin with, we shall consider the binary case where the data consists of only two symbols i.e. 1 and 0. We assign a distinct waveform (i.e. pulse) to each of these two symbols. The resulting sequence of these pulses is transmitted over a channel. At the receiver end, these pulses are detected and are converted back to binary data (i.e., 1 s and 0 s).

As a matter of fact, the pulse code Modulation (PCM), Delta modulation (DM), Adaptive delta modulation (ADM) etc. are used to convert an analog signal to digital data. This digital data is the sequence of binary symbols. This digital data needs to be transmitted over the channel. In digital data stream, the occurrence of 1's and 0's is not exactly equal. Therefore, the signal contains frequencies right from very low frequencies to high frequencies. This means that the baseband transmission of digital data requires a low-pass channel. In other words, this channel should transmit very low frequencies as well as highest frequency that can be represented by the signal. Thus, the low-pass channel used to transmit digital data can be approximated by the low-pass filter (LPF). When the digital data is transmitted over a channel, each received pulse is affected somewhat by adjacent pulses. This means that a

Inside this Chapter

signal received at a particular time is due to the transmitted pulse and some interference due to the adjacent pulses. This interference due to adjacent pulses is known as **Intersymbol Interference (ISI)**. Because of this intersymbol interference (ISI), the received signal is distorted from the expected value. Therefore, errors are introduced in the detection of the symbols. To control the effect of ISI, care is taken so that the interference between the transmitted symbols is minimum. This needs pulse shaping. Further, we know that any transmission channel adds noise to the signal. This noise is known as the channel noise. The channel noise is normally additive white noise. Thus, a noisy signal is received at the receiver. In the presence of such noise, the receiver should decide which of the possible waveform was transmitted. When this decision is made, then the symbol (1 or 0) is recovered exactly without any noise. Therefore, the channel noise can introduce errors in the signal and decision for a particular symbol can be wrong. This means that the ISI and channel noise simultaneously affect the received signal and create errors in the signal. Special cares must be taken to reduce the noise and increase the signal to noise ratio. In this chapter, first, we shall discuss the digital representation of signals. Then, we shall discuss ISI and method to minimize it.

Inside this Chapter

Also, it may be noted that the optimum detection of such a pulse signal involves the use of a linear time invariant (LTI) filter known as a **matched filter.**

2.2 A BASEBAND DIGITAL COMMUNICATION SYSTEM

A baseband digital communication system is made up of several elements as described under:

1. Source

The input to a digital system is in the form of sequence of digits. The input can be the output from such sources as a data set, a computer, a digitized voice signal (PCM or DM), a digital facsimile or tele-vision or telemetry equipment. Most of the discussion in this chapter is restricted to the binary case (i.e. communication schemes using only two symbols).

2. Multiplexer

Generally speaking, the capacity of a practical channel of transmitting data is much larger than the data rate of individual sources. To utilize this capacity effectively, we combine several sources through a digital multiplexer using the process of interleaving. Thus, a channel is time-shared by several messages simultaneously.

3. Line Coder

The output of a multiplexer is coded into electrical pulses or waveforms for the purpose of transmission over the channel. This process is known as **line coding or transmission coding**. There are several possible ways of assigning waveforms (i.e. pulses) to the digital data. In the binary case (two symbols), for example, conceptually the simplest line code is **on-off**, where a 1 is transmitted by a pulse p(t) and a 0 is transmitted by no pulse (i.e.

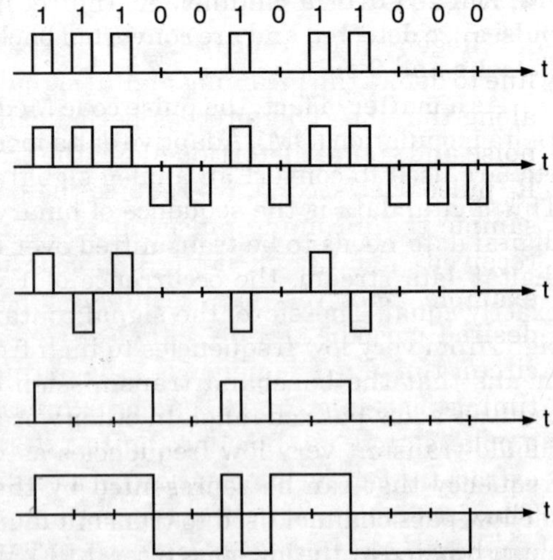

Fig. 2.1. Some line codes (a) On-off (RZ). (b) Polar (RZ), (c) Bipolar (RZ), (d) On-off NRZ), (e) Polar (NRZ)

zero signal), as shown in figure 2.1(a). Another commonly used code is **polar**, where 1 is transmitted by a pulse p(t) and 0 is transmitted by a pulse –p(t) as shown in figure 2.1(b). The polar scheme is the most power efficient code, since for a given noise immunity (i.e. error probability) this code needs the least power. Another popular code in PCM is **bipolar**, also known as **pseudoternary** or **alternate mark inversion (AMI)**, where 0 is encoded by no pulse and 1 is encoded by a pulse p(t) or –p(t), depending on whether the previous 1 is encoded by –p(t) or p(t). In short, pulses representing consecutive 1's alternate in sign, as shown in figure 2.1(c). This code has the advantage that if an error is made in the detection of pulses, the received pulse sequence will violate the bipolar rule and the error is immediately detected (although not corrected).

DO YOU KNOW?

Binary 1's and 0's, such as in PCM signaling, may be represented in various serial-bit signaling formats called line codes.

Another line code that in the past appeared promising is the duobinary (and modified duobinary) proposed by Lender. Although this code is better than the bipolar in terms of bandwidth efficiency, it has lost its appeal due to some practical problems and will not be discussed here.

In our discussion so far, we have used half-width pulses just for the sake of illustration. We can select other widths. However, full-width pulses are often used in some applications. Whenever full-width pulses are used, the pulse amplitude is held to a constant value throughout the pulse interval (it does not have a chance to go to zero before the next pulse begins). For this reason these schemes are known as **non-return-to-zero** (NRZ) schemes in contrast to **return-to-zero** (RZ) schemes as shown in figure 2.1(a), (b), (c). Figure 2.1(d) shows an on-off NRZ signal, whereas figure 2.1(e) illustrates a polar NRZ signal.

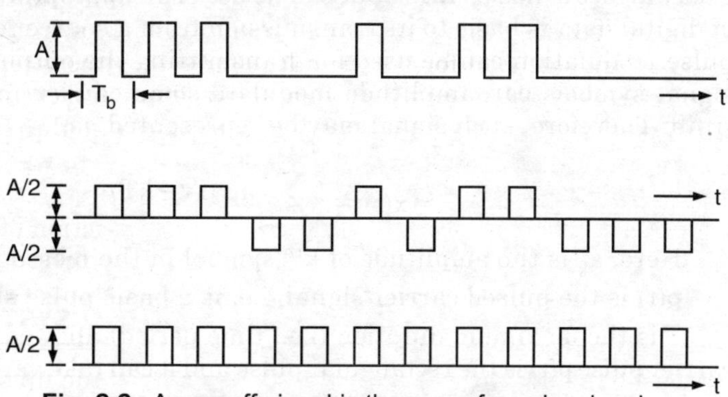

Fig. 2.2. An on-off signal is the sum of a polar signal and a clock frequency periodic signal

4. Regenerative Repeater

Regenerative repeaters are used at regularly spaced intervals along a digital transmission line to detect the incoming digital signal and regenerate new clean pulses for further transmission along the line. This process periodically eliminates, and thereby combats, the accumulation of noise and signal distortion along the transmission path. If the pulses are transmitted at a rate of R_b pulses per second, we require the periodic timing information—the clock signal at R_b Hz—to sample the incoming pulses at a repeater. This timing information can be extracted from the received signal itself if the line code is chosen properly. The polar signal in figure 2.1(b), for example, when rectified, results in a periodic signal of clock frequency R_b Hz which contains the desired periodic timing signal of frequency R_b Hz. When this signal is applied to a resonant circuit tuned to frequency R_b, the output, which is a sinusoid of frequency R_b Hz, can be used for timing. The on-off signal can be expressed as the sum of a periodic signal (of clock frequency) and a polar signal, as shown in figure 2.2. Because of the presence of the periodic component, we can extract the timing information from this signal using a resonant circuit tuned to the clock frequency. Thus, a bipolar signal, when rectified, becomes an on-off signal. Hence, its timing information can be extracted the same way as that for an on-off signal.

The timing signal (i.e., the resonant circuit output) is sensitive to the incoming bit pattern. In the on-off or bipolar case, a '0' is transmitted by "no pulse". Hence, if there are too many 0's in a sequence (no pulses), there is no signal at the input of the resonant circuit and the sinusoidal output of the resonant circuit starts decaying, thus, causing error in the timing information. We shall discuss later the ways of overcoming this problem. A line code in which the bit pattern does not affect the accuracy of the timing information is said to be a **transparent** line code. The polar scheme (where each bit is transmitted by some pulse) is transparent, whereas, the on-off and bipolar schemes are non-transparent.

2.3 INTRODUCTION TO DISCRETE PAM SIGNALS (DIGITAL DATA FORMATS)*

There are various techniques used to convert the analog signal to digital signal. But, the other way of obtaining digital data is from the source such as computers. The information from such a source is inherently discrete in nature. If such a discrete signal is transmitted over a bandlimited channel, then the signal gets dispersed. This causes the pulses to overlap and cause distortion. Such a distortion is called as intersymbol interference (ISI). In order to avoid this, we should not transmit the discrete signal as it is on the transmission medium. This data is first converted into a PAM suitable format and then transmitted over a communication channel. The various formats used are also called as **line codes**. The use of an appropriate waveform for baseband representation of digital data is basic to its transmission from a source to a destination. This means that digital pulse modulation can be used for transmitting the output of a digital source. We know that the signal symbols can amplitude modulate some carrier to generate amplitude modulated pulse train. Therefore, such signal may be represented as,

$$x(t) = \sum_{k=-\infty}^{\infty} a_k \, p(t-kT) \qquad \text{...(2.1)}$$

Here, a_k is the amplitude of kth symbol in the message sequence.

$p(t)$ is the pulsed carrier signal, i.e. is a basic pulse shape. Its pulses are modulated by a_k.

T is the maximum duration (i.e., time period) allowed for the carrier pulse. The unmodulated carrier pulse $p(t)$ is the rectangular pulse and it can take variable duty cycle. It can be represented as,

$$p(t) = \begin{cases} 1 & \text{for} \quad t = 0 \\ 0 & \text{for} \quad t = \pm T, \pm 2T, \ldots\ldots \end{cases} \qquad \text{...(2.2)}$$

$x(t)$ is the pulse waveform. To recover the original digital signal from $x(t)$ we have to sample $x(t)$ at some fixed intervals and check the signal in these intervals. This checking is the detection of the transmitted symbols. From equation (2.1), we observe that if $p(t)$ is zero, then $x(t)$ is also zero. Therefore, it is preferable to sample $x(t)$ when $p(t)$ is zero. This means that at this time $[p(t) = 0]$ no digital information is present/transmitted in the pulse waveform. Therefore, $x(t)$ can be sampled periodically at $t = kT$ where $k = 0, \pm 1, \pm 2, \ldots$etc.

Also, $p(t)$ is the rectangular pulse and may be written as,

$$p(t) = \text{rect}\left(\frac{t}{\tau}\right) \qquad \text{...(2.3)}$$

Because, the pulse to pulse interval is 'T', therefore the width of the pulse τ should be less than or equal to T.

i.e., $\tau \leq T$

The signalling rate is given as,

$$r = \frac{1}{T} \qquad \text{...(2.4)}$$

If 'T' represents the duration of one bit, then $T = T_b$ and signalling rate will be,

$$r = \frac{1}{T_b} \qquad \text{...(2.5)}$$

* The important parameters involved in the design of a PAM system are data rate, error rate, transmitted power, noise power spectral density, and system complexity,

2.4 DEFINITION OF LINE CODING

It is the process of converting binary data, a sequence of bits to a digital signal. The data, text, numbers, graphical images, audio and video which are stored in computer memory are in the form of sequences of bits. Line coding converts these sequences into digital signals as shown in figure 2.3.

Fig. 2.3.

2.4.1 Important Characteristics of Line Coding

Some of the important characteristics of line coding are:

(i) Signal level and data level
(ii) Pulse rate and bit rate
(iii) DC component
(iv) Self synchronization

2.4.2 Signal Level and Data Level

A digital signal has a limited number of values. However, all of them are not used to represent the data. Only some of them can be used to represent the data. The remaining values are used for some other purpose.

(i) Signal levels

The number of values allowed in a particular signal is defined as the number of signal levels.

(ii) Data levels

The number of values to represent data is called as the number of data levels. Figure 2.4 gives a clear idea about signal levels and data levels. Figure 2.4 (a) has two signal levels (0 and A) and two data levels (0 and 1) whereas figure 2.4 (b) has three signal levels and two data levels.

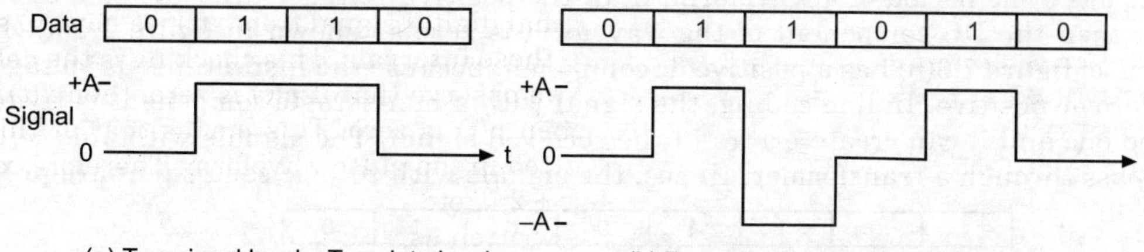

(a) Two signal levels, Two data levels (b) Two data levels, Three signal levels

Fig. 2.4.

Figure 2.4 (b) has three signal levels (+ A, 0, – A) and two data levels (0 and 1).

2.2.3 Pulse Rate and Bit Rate

Pulse Rate is defined as the number of pulses per second and a pulse is defined as the minimum amount of time required to transmit a symbol. Bit rate is defined as the number of bits per second. If one pulse corresponds to one bit then the pulse rate is equal to the bit rate. But if a pulse carries more than 1 bit then the pulse rate is lower than bit rate. The relation between bit rate and pulse rate is as follows:

$$\text{Bit rate} = \text{Pulse rate} \times \log_2 L \qquad \text{...(2.6)}$$

EXAMPLE 2.1. For the signals shown in figures 2.5(a) and (b), determine the bit rate and pulse rate.

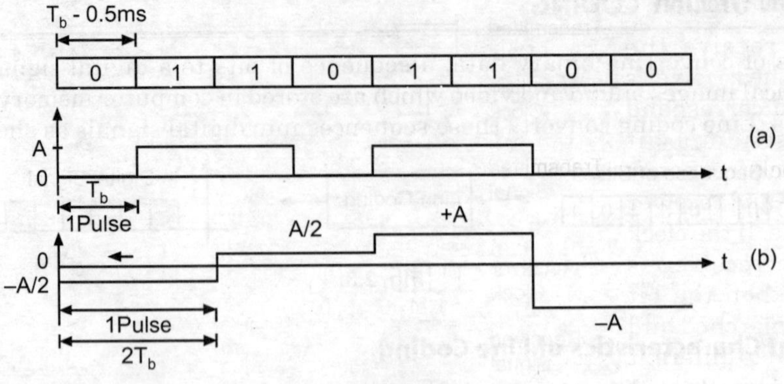

Fig. 2.5.

Solution: The data has two levels and the bit duration T_b = 0.5 ms.

We consider figure 2.5(a) which depicts that there are two signal levels (0 or A).

Therefore, $\text{Pulse Rate} = \dfrac{1}{0.5\times10^{-3}} = 2000$ pulses/sec.

and $\text{Bit rate} = 2000 \times \log_2(2) = 2000$ bps

We consider figure2.5 (b). Now, one pulse corresponds to a duration of 2 T_b.

Hence, pulse duration = 2 T_b = 1 ms

Therefore, $\text{Pulse rate} = \dfrac{1}{1\times10^{-3}} = 1000$ pulses/sec.

and $\text{Bit rate} = \text{Pulse rate} \times \log_2 L$

Here $L = 4$ levels

$\text{Bit rate} = 1000 \times \log_2 4 = 2000$ bps.

2.4.4 DC Component

Over one cycle period of a waveform, if all the positive voltages are cancelled by negative voltages then the DC component of the waveform is zero as shown in figure 2.6(a). But, the waveform of figure 2.6(b) has a positive dc component because the instantaneous voltage can be either zero or positive. In line coding, the signal with a non-zero dc component is treated as a distorted one and it can create errors in the received signal. The signals with a dc component cannot pass through a transformer. Hence, the signals with zero dc component are preferred.

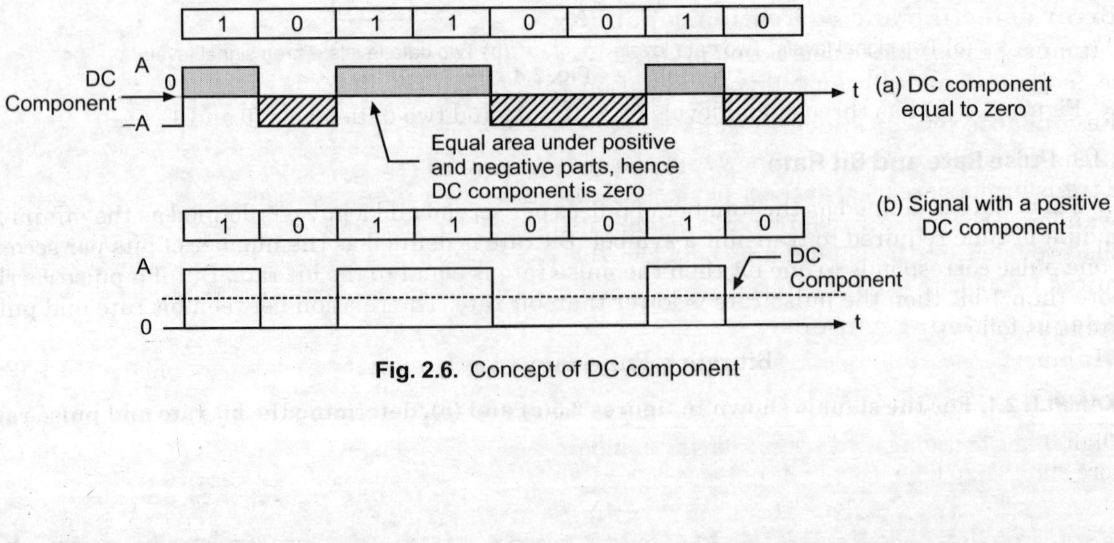

Fig. 2.6. Concept of DC component

2.4.5 Self Synchronization

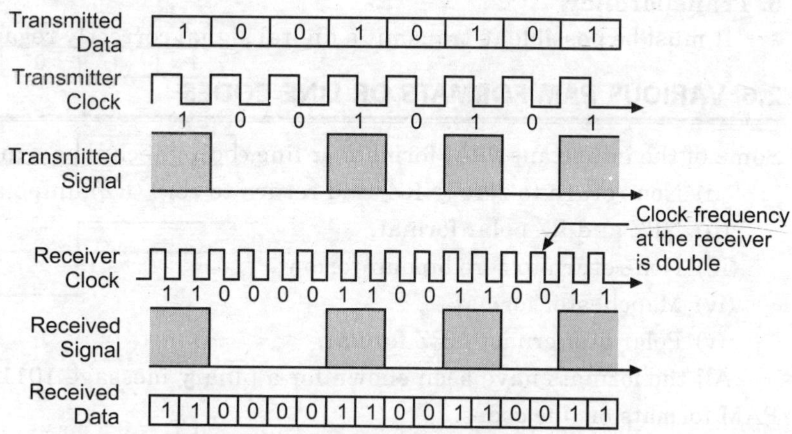

Fig. 2.7. Concept of synchronization

In order to receive the signal correctly, the receiver's bit intervals must correspond exactly to the sender's bit intervals. The clock frequency of the transmitter and receiver should be the same. If the clock frequency at the receiver is slower or faster then the bit intervals are not matched and the received signal is different that the transmitted one. Figure 2.7 shows the effect of change in clock frequency. The receiver clock frequency is twice that of the transmitter frequency. Hence, received data is totally different than the transmitted one. Such thing would not happen if the receiver clock is synchronized with the transmitter clock. To achieve this, the transmitted digital signal includes the timing information. This will force the self synchronization. For achieving synchronization, the transmitted signal should cross the zero frequently. So if the transmitted signal consists of long trains of 0's or 1's, then the synchronization is affected.

2.5 LINE CODING REVIEW : LINE CODING AND ITS PROPERTIES

The digital data can be transmitted by various transmission or line codes such as on-off, polar, bipolar and so on. This is called **line-coding***. Each type of line-code has its advantages and disadvantages.

Thus, among other desirable properties, a line code must have the following properties:

1. Transmission bandwidth

For a line-code, the transmission bandwidth must be as small as possible.

2. Power efficiency

For a given bandwidth and a specified detection error probability, the transmitted power for a line code should be as small as possible.

3. Error detection and correction capability

It must be possible to detect and preferably correct detection errors. For example, in a bipolar case, a signal error will cause bipolar violation and thus can easily be detected.

4. Favourable power spectral density

It is desirable to have zero power spectral density (PSD) at $\omega = 0$ (i.e., dc) since ac coupling and transformers are used at the repeaters. Significant power in low-frequency components causes dc wander in the pulse stream when ac coupling is used. The a.c. coupling is required since the dc paths provided by the cable pairs between the repeater sites are used to transmit the power required to operate the repeaters.

5. Adequate timing content

It must be possible to extract timing or clock information from the signal.

*Digital line encoding involves converting standard logic levels (TTL, CMOS, and the like) to a form more suitable to telephone line transmission.

6. Transparency

It must be possible to transmit a digital signal correctly regardless the pattern of 1's and 0's.

2.6 VARIOUS PAM FORMATS OR LINE CODES

Some of the important PAM formats or line coding techniques are as under:

(i) Non-return to zero (NRZ) and return to zero (RZ) unipolar format.

(ii) NRZ and RZ polar format.

(iii) Non-return to zero bipolar format.

(iv) Manchester format.

(v) Polar quaternary NRZ format.

All the formats have been shown for a binary message 10110100. Figure 2.8 shows various PAM formats or line codes.

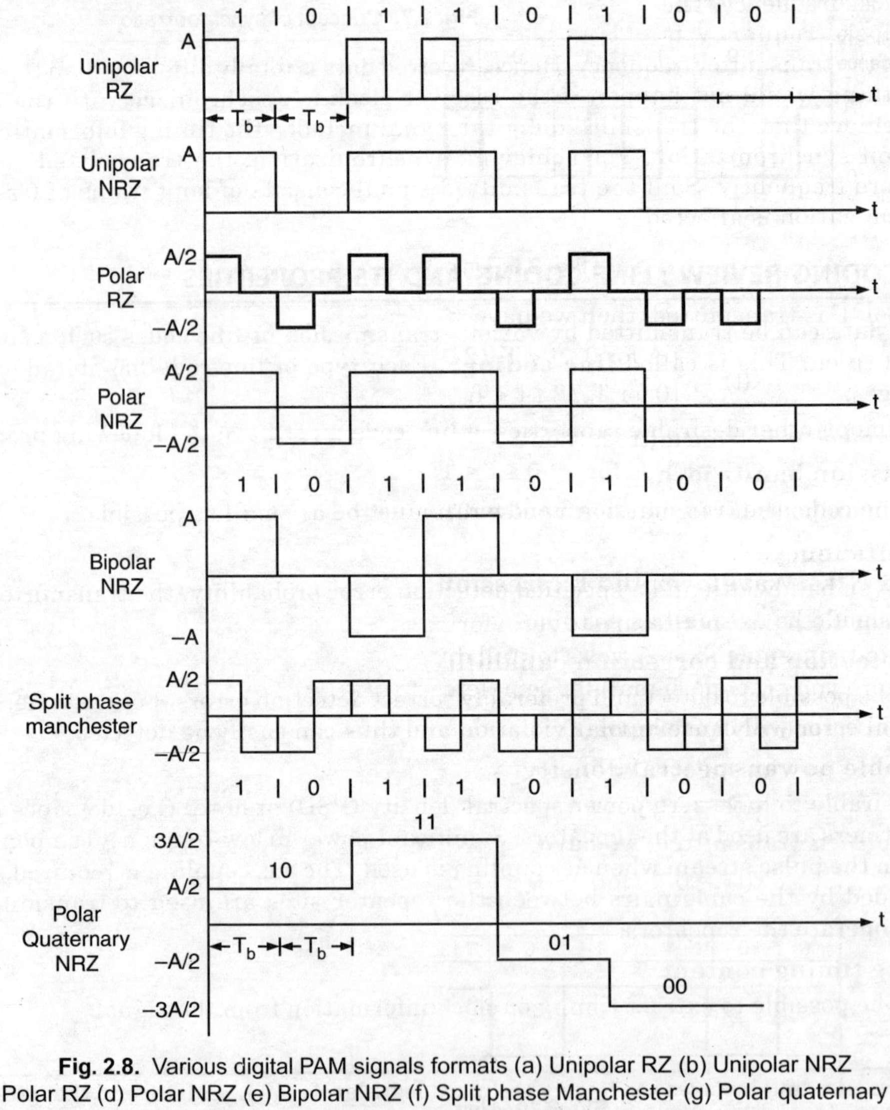

Fig. 2.8. Various digital PAM signals formats (a) Unipolar RZ (b) Unipolar NRZ (c) Polar RZ (d) Polar NRZ (e) Bipolar NRZ (f) Split phase Manchester (g) Polar quaternary NRZ

2.7 UNIPOLAR RZ AND NRZ

DO YOU KNOW?

Binary 1's are represented by alternately positive or negative values. The binary 0 is represented by a zero level. The term pseudoternary refers to the use of three encoded signal levels to represent two-level (binary) data. This is also called alternate mark inversion (AMI) signaling.

1. Definition

In unipolar format, the waveform has a single polarity. The waveform can have +5 or +12 volts when high. The waveform is simple on-off.

2. Unipolar RZ : Waveform and Expression

In the unipolar RZ form, the waveform has zero value when symbol '0' is transmitted and waveform has 'A' volts when '1' is transmitted. In RZ form, the 'A' volts is present for $T_b/2$ period if symbol '1' is transmitted and for remaining $T_b/2$, waveform returns to zero value, i.e., for unipolar RZ form, we have

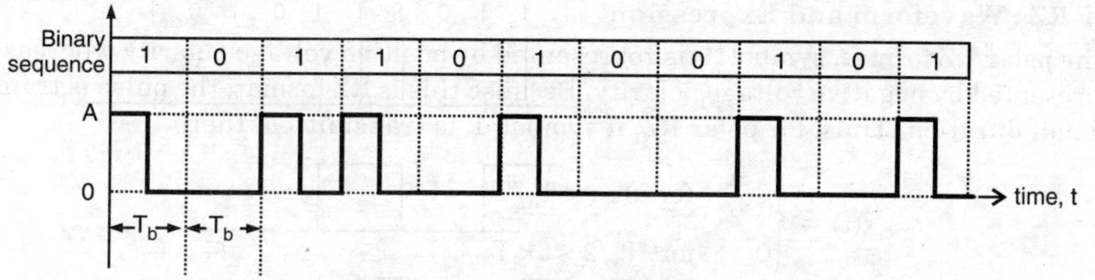

Fig. 2.9. Unipolar RZ format

If symbol '1' is transmitted, then we have

$$x(t) = \begin{cases} A \text{ for } 0 \le t < T_b/2 & \text{(Half interval)} \\ 0 \text{ for } T_b/2 \le t < T_b & \text{(Half interval)} \end{cases} \quad ....(2.7)$$

and if symbol '0' is transmitted, then

$$x(t) = 0 \quad \text{for} \quad 0 \le t < T_b \quad \text{(complete interval)} \quad ...(2.8)$$

Hence, in unipolar RZ format, each pulse returns to a zero value. Figure 2.9 shows this signal format.

3. Unipolar NRZ: Waveform and Expression

A unipolar NRZ (i.e., not return to zero) format is shown in figure 2.10. When symbol '1' is to be transmitted, the signal has 'A' volts for full duration. When symbol '0' is to be transmitted, the signal has zero volts (i.e. no signal) for complete symbol duration.

Thus, for unipolar NRZ format,

If symbol '1' is transmitted, we have

$$x(t) = A \quad \text{for} \quad 0 \le t < T_b \quad \text{(complete interval)} \quad ...(2.9)$$

If symbol '0' is transmitted, we have

$$x(t) = 0 \quad \text{for} \quad 0 \le t < T_b \quad \text{(complete interval)} \quad ...(2.10)$$

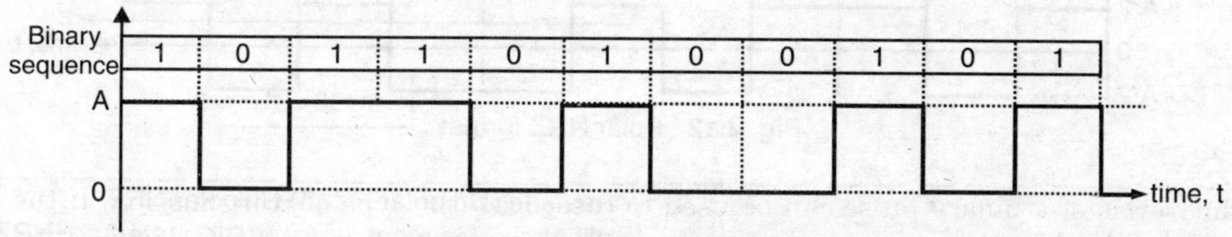

Fig. 2.10. Unipolar NRZ format

4. Important Points

(i) For NRZ format, it may be observed that the pulse does not return to zero on its own. If symbol '0' is to be transmitted, then pulse becomes zero.

(ii) Internal computer waveforms are usually of unipolar NRZ type.

(iii) Because, there is no separation between the pulses, therefore, the receiver needs synchronization to detect unipolar NRZ pulse.

(iv) As compared to RZ format, NRZ pulse width (pulse to pulse interval is same) is more. Thus, energy of the pulse is more.

(v) However, unipolar format has some average DC value. This DC value does not carry any information.

2.8 POLAR RZ AND NRZ*

1. Polar RZ : Waveform and Expression

In the polar RZ format, symbol '1' is represented by positive voltage polarity whereas symbol '0' is represented by negative voltage polarity. Because this is RZ format, the pulse is transmitted only for half duration. Thus, for polar RZ, if symbol '1' is transmitted, then

$$x(t) = \begin{cases} +\dfrac{A}{2} & \text{for } 0 \le t < T_b/2 \\ 0 & \text{for } T_b/2 \le t < T_b \end{cases} \quad ...(2.11)$$

and if symbol '0' is transmitted, then

$$x(t) = \begin{cases} -\dfrac{A}{2} & \text{for } 0 \le t < T_b/2 \\ 0 & \text{for } T_b/2 \le t < T_b \end{cases} \quad ...(2.12)$$

Polar RZ waveform has been shown in figure 2.11.

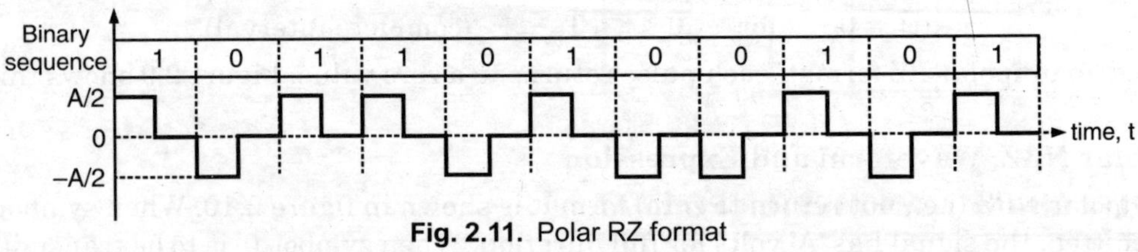

Fig. 2.11. Polar RZ format

2. Polar NRZ : Waveform and Expression

The polar NRZ is shown in figure 2.12. In polar NRZ format, symbol '1' is represented by positive polarity whereas symbol '0' is represented by negative polarity. These polarities are maintained over the complete pulse duration i.e., for polar NRZ, we have

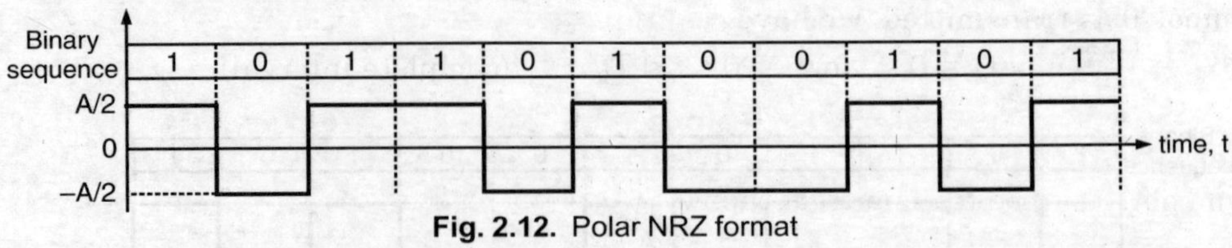

Fig. 2.12. Polar NRZ format

* The duty cycle of a binary pulse can be used to categorize the type of transmission. If the binary pulse is maintained for the entire bit time, this is called non return to zero (NRZ). If the active time of the binary pulse is less than 100%. of the bit time, this is called return to zero (RZ).

If symbol '1' is transmitted, then

$$x(t) = +\frac{A}{2} \quad \text{for} \quad 0 \le t < T_b \qquad ...(2.13)$$

and if symbol '0' is transmitted, then

$$x(t) = -\frac{A}{2} \quad \text{for} \quad 0 \le t < T_b \qquad ...(2.14)$$

3. Important Points

(i) Since polar RZ and NRZ formats are bipolar, therefore, the average DC value is minimum in these waveforms.

(ii) If probabilities of occurrence of symbols '1' and '0' are same, then average DC components of the waveform would be zero.

2.9 BIPOLAR NRZ [ALTERNATE MARK INVERSION (AMI)]

(UPTU, Sem. Exam., 2005-06) (06 marks)

1. Definition

In this format, the successive '1's are represented by pulses with alternate polarity and '0's are represented by no pulses.

2. Waveform

Figure 2.13 illustrates the Bipolar NRZ or AMI waveform. If there are even number of 1's, the DC component of the waveform would be zero. The advantage of this format is that the ambiguities due to transmission sign inversion are eliminated.

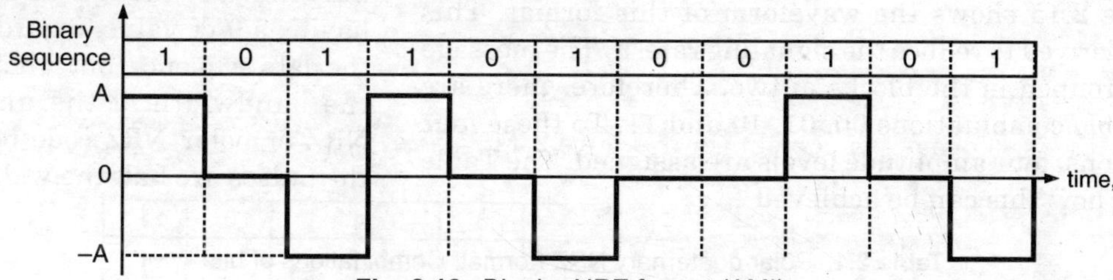

Fig. 2.13. Bipolar NRZ format (AMI)

2.10 SPLIT PHASE MANCHESTER FORMAT

1. Definition and Waveform

This type of waveform is shown in figure 2.14. In this case, if symbol '1' is to be transmitted, then a positive half interval pulse is followed by a negative half interval pulse. If symbol '0' is to be transmitted, then a negative half interval pulse is followed by a positive half interval pulse. Hence, for any symbol the pulse takes positive as well as negative value.

DO YOU KNOW?

The various line codes are also known by other names. For example, polar NRZ is also called NRZ-L, where L denotes the normal logical level assignment. Bipolar RZ is also called RZ-AMI, where AMI denotes alternate mark (binary 1) inversion.

2. Mathematical Expressions

If symbol '1' is to be transmitted, then

$$x(t) = \begin{cases} \dfrac{A}{2} & \text{for } 0 \le t < \dfrac{T_b}{2} \\ -\dfrac{A}{2} & \text{for } \dfrac{T_b}{2} \le t < T_b \end{cases} \qquad ...(2.15)$$

and if symbol '0' is to be transmitted, then

$$x(t) = \begin{cases} -\dfrac{A}{2} & \text{for } 0 \le t < \dfrac{T_b}{2} \\ \dfrac{A}{2} & \text{for } \dfrac{T_b}{2} \le t < T_b \end{cases} \quad ...(2.16)$$

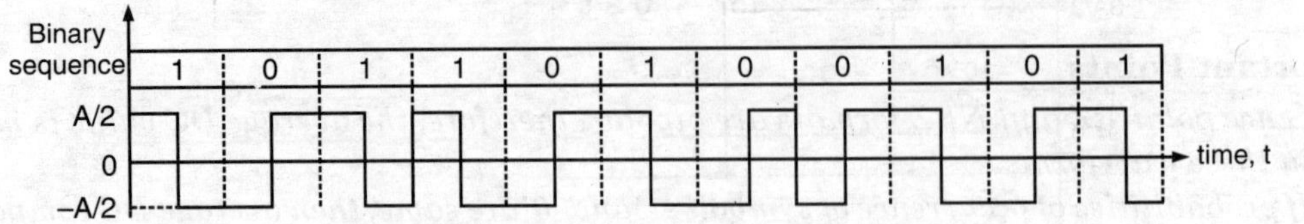

Fig. 2.14. Split phase manchester format

3. Important Points

(i) The primary advantage of this format is that irrespective of the probability of occurrence of symbol '1' and '0' the waveform has zero average value. Therefore by this mode, the power saving is quite more.

(ii) However, the drawback of this format is that it requires absolute sense of polarity at the receiver end.*

2.11 POLAR QUATERNARY NRZ FORMAT

1. Definition and Waveform

Figure 2.15 shows the waveform of this format. This format is derived to reduce the signaling rate 'r'. The message bits are grouped in the blocks of two. Therefore, there are four possible combinations 00, 01, 10 and 11. To these four combinations, four amplitude levels are assigned. The Table 2.1 shows how this can be achieved

DO YOU KNOW?

The Manchester NRZ line code has the advantage of always having a 0-dc value, regardless of the data sequence, but it has twice the bandwidth of the unipolar NRZ or polar NRZ code because the pulses are half the width.

Table 2.1. Polar quaternary NRZ Format: Combinations of bits

Message combination	$x(t) = a_n$
00	$-\dfrac{3A}{2}$
01	$-\dfrac{A}{2}$
10	$\dfrac{A}{2}$
11	$\dfrac{3A}{2}$

In the waveform of figure 2.15, the first combination of two bits is 10. Thus, from Table 2.1, we may observe that the level should be $\dfrac{A}{2}$. The second combination in figure 2.15 is 11, hence

*Unipolar and bipolar transmission voltages can be combined with either RZ or NRZ in several ways to achieve a particular line-encoding scheme.

from Table 2.1, the level taken is $\frac{3A}{2}$. Similarly other levels are selected. Hence, for two message bits only one pulse is transmitted with duration $2T_b$, i.e.,

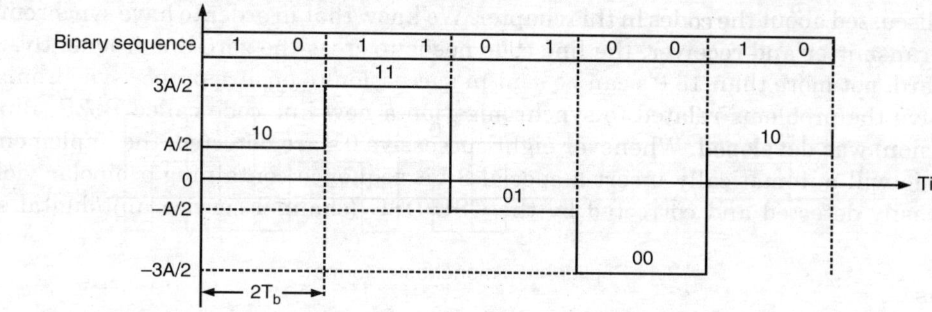

Fig. 2.15. Polar Quaternary NRZ format

$$T_s = 2T_b$$

and signaling rate is given as,

$$r = \frac{r_b}{2} = \frac{1}{2T_b} \quad ...(2.17)$$

2.12 HIGH DENSITY BIPOLAR (HDB) SIGNALLING

(U.P.Tech, Sem. Exam., 2006-07)(08 marks)

1. Definition

In case of bipolar NRZ or AMI signal, the transmitted signal is equal to zero when a binary 0 is to be transmitted. This is true even for the unipolar RZ and unipolar NRZ signals. The absence of transmitted signal can cause problems in synchronization at the receiver, if long sequence of binary 0's are being transmitted. This problem can be solved by adding (transmitting) pulses when long strings of 0's exceeding a number n are being transmitted. This type of coding is called as **High Density Bipolar coding**. It is denoted by **HDBN**. Here, N = 1, 2, 3, ... The most widely used HDB format is with N = 3 i.e., HDB3.

2. Waveforms

In the string of message bits, when (N + 1) or minimum number of zeros occur, they are replaced by special binary sequences of (N + 1) length. As shown in figure 2.16, these sequences contain some binary 1's which are necessary for synchronization at the receiver end. The (N + 1) long special sequences for the HDB3 coding are 000V and B00V where B and V both are considered to be binary 1's. When the number of consecutive zeros exceed (N + 1) i.e., 4 in case of HDB3, the above mentioned special sequences are inserted as shown in figure 2.16.

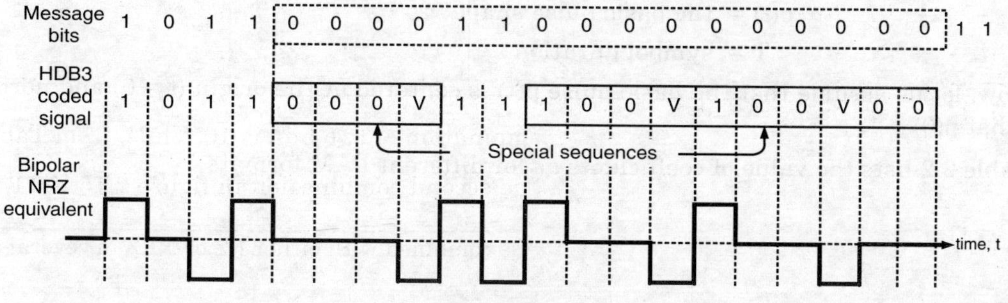

Fig. 2.16.

2.13 B8ZS LINE CODE

1. Definition and Working Principle

We have discussed about the codes in this chapter. We know that in order to have synchronization between the transmitter and receiver, the line code needs to cross the zero line frequently. As per U.S. T_1 standard, not more than 15 0's can be sent in succession to ensure proper synchronization. In order to solve the problems related to synchronization a new line code called B8ZS (Binary 8-zeros suppression) was developed. Whenever eight successive 0's are detected, the implementation of this line code will automatically insert a special 8 bit sequence containing a bipolar violation. This can be easily detected and corrected by the CSU/DSU (channel service unit/digital service unit).

2. Waveforms

Figure 2.17 gives a clear idea about the B8ZS line code. The violations (BPV in figure 2.17), will distinguish a byte substituted for all 0's from a normal byte which contains 1's. The B8ZS does not allow more than 8, consecutive 0's and the bipolar violation pattern uniquely identifies the eight 0's. It may that the voltage levels in the violating byte, has a zero average (dc) value.

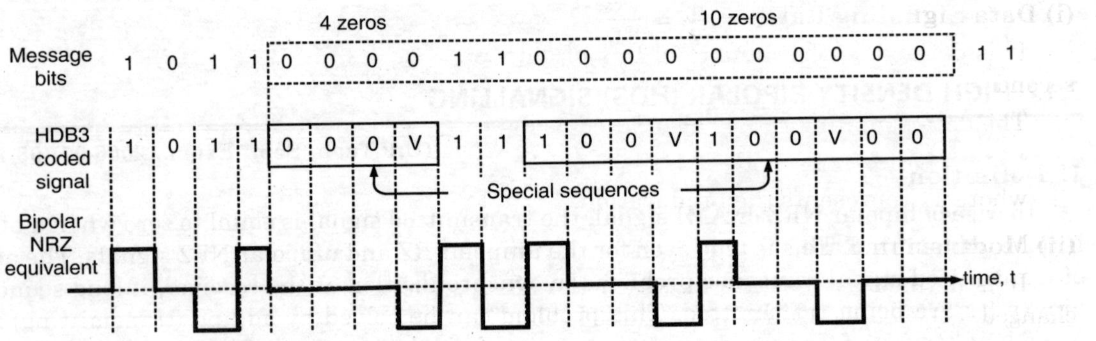

Fig. 2.17.

2.14 POWER SPECTRA OF DISCRETE PAM SIGNALS (VARIOUS LINE CODES)

As discussed earlier, we can represent all the discrete PAM signals with the help of a single expression as under:

$$x(t) = \sum_{k=-\infty}^{\infty} a_k p(t - kT) \qquad ...(2.18)$$

where

a_k = coefficient

$p(t)$ = the basic pulse shape

T = symbol duration

Now, let us assume that the basic pulse p(t) is centered at the origin (t = 0) and normalized such that p(0) = 1.

Table 2.2 lists the value of coefficients a_k for different PAM formats.

Table 2.2. Values of a_k for various data formats.

S.No.	NRZ formats	Coefficient a_k		Basic pulse p(t)
		for symbol 0	for symbol 1	
1.	Unipolar NRZ	$a_k = 0$	$a_k = A$	Basic pulse p(t) is a rectangular pulse of unit amplitude and duration T_b.
2.	Polar NRZ	$a_k = -A$	$a_k = A$	Basic pulse p(t) is a rectangular pulse of unit amplitude and duration T_b.
3.	Bipolar NRZ	$a_k = 0$	$a_k = A$ or $-A$	Alternately for 1s.
4.	Manchester	$a_k = -A$	$a_k = +A$	p(t) consists of double pulses of amplitude ± 1 and duration T_b.
5.	Polar Quaternary	$a_k = -3A/2$ for 00 and $+3A/2$ for 11	$-A/2$ for 01 and $+A/2$ for 10	p(t) is a rectangular pulse of unit amplitude and duration 2 T_b.

Now, let us discuss few important terms as under:

(i) Data Signaling Rate

It is defined as the number of bits of data transmitted per second. It is measured in bits/second.

The data signaling rate is also called as data rate and it is defined as follows:

$$R_b = 1/T_b \qquad \text{...(2.19)}$$

where T_b represents the bit duration.

(ii) Modulation Rate

It is defined as the rate at which the signal level is changed. The modulation rate is measured in bauds or symbols per second.

DO YOU KNOW?

The particular type of waveform selected for digital signaling depends on the application. The advantages and disadvantages of each signal format are also encountered.

(iii) Power Spectra

We can represent any PAM format as follows:

$$x(t) = \sum_{k=-\infty}^{\infty} a_k p(t - kT)$$

Each format x(t) may be considered as a random process and each coefficient a_k as a random variable. Let these coefficients be generated by a discrete stationary random source which is characterized by the following relation between autocorrelation and ensemble average i.e.,

$$R_\tau(n) = E[a_k \, a_{k-n}] \qquad \text{...(2.20)}$$

where $R_\tau(n)$ = Autocorrelation function

and E = Expectation operator

Also, power spectral density (psd) of PAM signal x(t) is given by

$$S(f) = \frac{1}{T}|P(f)|^2 \sum_{n=-\infty}^{\infty} R_\tau(n)\, e^{-j2\pi nfT} \qquad \text{...(2.21)}$$

where P(f) = Fourier transform of the basic pulse p(t).

It may be noted that the values of P(f) and $R_\tau(n)$ depends upon the type of PAM format.

Anger is a brief madness. – HORACE, Epistles.

2.15 POWER SPECTRAL DENSITY (PSD) OF NRZ UNIPOLAR FORMAT

For simplicity, let us assume that the 0s and 1s are equally likely to happen or have equal probability.

$$P(a_k = 0) = P(a_k = A) = 1/2$$

We know that a discrete stationary random source is characterized by the following autocorrelation:

$$R_\tau(n) = [a_k\, a_{k-n}]$$

Hence, for n = 0, above expression becomes

$$R_\tau(n) = E[a_k\, a_k] = E[a_k{}^2] \quad ...(2.22)$$

or,
$$E[a_k]^2 = 0^2 P(a_k = 0) + (A)^2\, P(a_k = A)$$

or
$$E[a_k]^2 = A^2 \times \frac{1}{2} = \frac{A^2}{2} \quad ...(2.23)$$

Now, let us obtain the product $a_k\, a_{k-n}$ for $n \neq 0$. The four possible values of this product are 0, 0, 0 and A^2. With the assumption that the successive symbols in a binary sequence are equiprobable (with a probability equal to 1/2), the probability of the four values 0, 0, 0 and a^2 will be equal to 1/4 each.

Therefore, for $n \neq 0$, we have

$$E[a_k\, a_{k-n}] = (3 \times 0 \times 1/4) + (A^2 \times 1/4)$$

or
$$E[a_k\, a_{k-n}] = \frac{A^2}{4} \quad \text{for } n \neq 0 \quad ...(2.24)$$

Thus, the autocorrelation function $R_\tau(n)$ can be expressed as under:

$$R_\tau(n) = \begin{cases} A^2/2 \text{ for } n = 0 & ...(2.25) \\ A^2/4 \text{ for } n \neq 0 & ...(2.26) \end{cases}$$

The basic pulse p(t) is a rectangular pulse of unity amplitude and duration T_b. Hence, its Fourier transform is given by

$$P(f) = T_b \text{ sinc } (fT_b) \quad ...(2.27)$$

where the sinc function is expressed as,

$$\text{sinc } x = \frac{\sin (\pi x)}{\pi x}$$

Now, substituting the expression for P(f) and $R_\tau(n)$ in equation (2.21) and substituting $T = T_b$ to get power spectral density (psd) as under:

$$S(f) = \frac{1}{T_b} \times [T_b \text{ sinc } (fT_b)]^2 \left[\sum_{\substack{n=-\infty \\ n \neq 0}}^{\infty} \frac{A^2}{4} e^{-j2\pi nfT_b} + \sum_{n=0} \frac{A^2}{2} e^{-j2\pi nfT_b} \right] \quad ...(2.28)$$

or
$$S(f) = T_b \text{ sinc}^2 (fT_b) \left[\frac{A^2}{2} e^0 + \sum_{\substack{n=-\infty \\ n \neq 0}}^{\infty} \frac{A^2}{4} e^{-j2\pi nfT_b} \right]$$

or
$$S(f) = \frac{A^2 T_b}{2} \text{ sinc}^2 (fT_b) + \frac{A^2 T_b}{4} \text{ sinc}^2 (fT_b) \sum_{\substack{n=-\infty \\ n \neq 0}}^{\infty} e^{-j2\pi nfT_b} \quad ...(2.29)$$

According to Poisson's expression, we can write,

$$\sum_{n=-\infty}^{\infty} e^{-j2\pi nfT_b} = \frac{1}{T_b}\sum_{m=-\infty}^{\infty}\delta\left(f-\frac{m}{T_b}\right) \quad ...(2.30)$$

where δ(f) is the unit impulse at f = 0 in the frequency domain.

Substituting equation (2.30) into equation (2.29) and noting that the sinc function passes through zero at $f = \pm 1/T_b, \pm 2/T_b,...$, we can write equation (2.29) in the simplified form as under:

$$S(f) = \frac{A^2T_b}{4}\,\text{sinc}^2(fT_b) + \frac{A^2}{4}\delta(f) \quad ...(2.31)$$

This expression may be plotted as shown in figure 2.18 (curve a).

2.16 POWER SPECTRAL DENSITY (PSD) OF NRZ POLAR FORMAT

(UPTU, Sem. Exam., 2004-05) (08 marks)

Here, let us follow a similar procedure as followed in the previous section to obtain the psd of NRZ polar format.

For the NRZ polar format, we have

$$R_\tau(n) = \begin{cases} a^2 & \text{for } n = 0 \\ 0 & \text{for } n \neq 0 \end{cases} \quad ...(2.32)$$

We know the basic pulse p(t) is same for all the PAM formats, therefore, its Fourier transform will also remain same.

Hence, $P(f) = T_b\,\text{sinc}(fT_b)$

Thus, the power spectral density (psd) is given by,

$$S(f) = A^2T_b\,\text{sinc}^2(fT_b) \quad ...(2.33)$$

The normalized form of this expression has been plotted in figure 2.18 as curve b.*

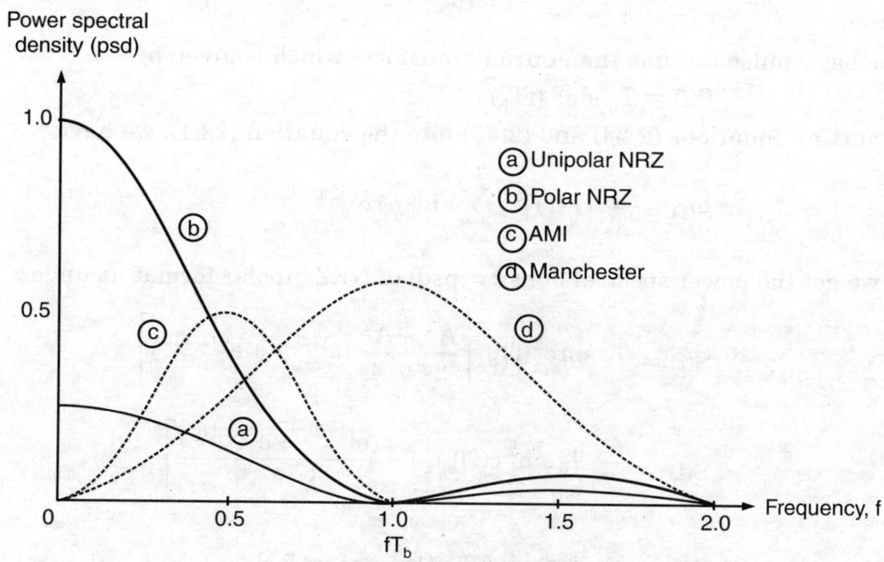

Fig. 2.18. Power spectra of different line codes

* Curve *b* of figure 2.18 shows that most of the power of polar NRZ format lies inside the main lobe of the sinc shaped curve extending upto the bit rate $1/T_b$.

Similarly, it is possible to obtain the expressions for the power spectral densities (psds) of other PAM formats.

2.17 POWER SPECTRAL DENSITY (PSD) OF NRZ BIPOLAR FORMAT

We know that the bipolar NRZ format has three levels A, 0 and – A. If we assume that the 1s and 0s in the binary input data occur with equal probability then the probabilities of occurrence of these levels will be given by

$$P(a_k = A) = 1/4$$
$$P(a_k = 0) = 1/2$$
$$P(a_k = -A) = 1/4$$

Therefore, for n = 0, we can have

$$E\left[a_k^2\right] = A^2P(a_k = A) + (0)^2P(a_k = 0) + (-A)^2\,P(a_k = -A) = A^2/2$$

Similarly, for n = 1, the input sequence can assume four possible formats as (0, 0), (0, 1), (1, 0) and (1, 1). The corresponding values of the product $a_k\,a_{k-1}$ are 0, 0, 0 and 1 respectively.

Now, assuming that the 0s and 1s in the binary sequence have equal probability, we can write

$$E[a_k\,a_{k-1}] = 3(0)\,(1/4) + (-A^2)\,(1/4) = -A^2/4$$

Further, for n > 1 we find that

$$E[a_k\,a_{k-1}] = 0$$

DO YOU KNOW?

The spectral efficiencies for all the line codes can be easily evaluated from their PSDs.

Therefore, for the bipolar NRZ format, we have

$$R_\tau(n) = \begin{cases} A^2/2 & \text{for } n = 0 \\ -A^2/4 & \text{for } n = \pm 1 \\ 0 & \text{for } n > 1 \end{cases} \qquad ...(2.34)$$

Also, the basic pulse p(t) has the Fourier transform which is given by

$$P(f) = T_b\,\text{sinc}\,(fT_b) \qquad ...(2.35)$$

Hence, putting equations (2.34) and (2.35) into the equation (2.21), we have

$$S(f) = \frac{1}{T_b}|P(f)|^2 \sum_{n=-\infty}^{\infty} R_\tau(n)\,e^{-j2\pi nfT_b}$$

Hence, we get the power spectral density (psd) of NRZ bipolar format as under:

$$S(f) = T_b\,\text{sin}\,c^2(fT_b)\left[\frac{A^2}{2} - \frac{A^2}{4}(e^{j2\pi fT_b} + e^{-j2\pi fT_b})\right]$$

or $$S(f) = \frac{A^2}{2}T_b\,\text{sin}^2\,c(fT_b)\left[1 - \frac{(e^{j2\pi fT_b} + e^{-j2\pi fT_b})}{2}\right]$$

or $$S(f) = \frac{A^2T_b}{2}\,\text{sin}^2\,c(fT_b)\,[1 - \cos\,(2\pi fT_b)] \qquad ...(2.36)$$

or $$S(f) = A^2T_b\,\text{sin}^2\,c\,(fT_b)\,\sin^2\,(\pi fT_b)$$

The normalized form of this equation has been shown plotted in 'c' in figure 2.18.

2.18 POWER SPECTRAL DENSITY (PSD) OF THE MANCHESTER FORMAT

Here, let us assume that the input binary data consists of 0s and 1s of equal probability (i.e., $\frac{1}{2}$). Then the autocorrelation function of this format is same as that for the NRZ polar format.

First, let us express the manchester signal x(t) mathematically in time domain.

The basic pulse p(t) of the Manchester signal is as shown in figure 2.19. We can express the Manchester signal in time domain as under:

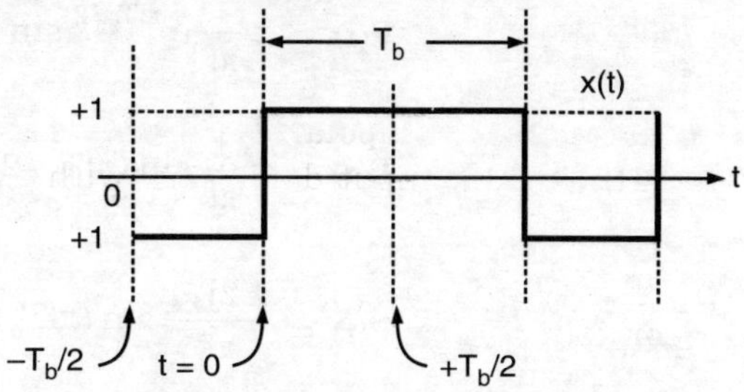

Fig. 2.19. Manchester signal

$$p(t) = \begin{cases} \pm 1 & \text{for } -T_b/2 \le t \le 0 \\ \mp 1 & \text{for } 0 \le t \le +T_b/2 \end{cases}$$

The Fourier transform of p(t) is given by

$$P(f) = \int_{-T_b/2}^{0} \pm 1\, e^{-j2\pi ft}\, dt + \int_{0}^{T_b/2} \mp 1\, e^{-j2\pi ft}\, dt \qquad ...(2.37)$$

or
$$P(f) = \frac{\pm 1}{-j2\pi f}\left[e^{-j2\pi ft}\right]_{-T_b/2}^{0} \mp \frac{1}{-j2\pi f}\left[e^{-j2\pi ft}\right]_{0}^{T_b/2}$$

or
$$P(f) = \frac{\pm 1}{-j2\pi f}\left[e^{0} - e^{-j\pi fT_b}\right] \mp \frac{1}{-j2\pi f}\left[e^{-j\pi fT_b} - e^{0}\right]$$

or
$$P(f) = \underbrace{\frac{\pm 1}{-j2\pi f}\left[1 - e^{j\pi fT_b}\right]}_{\text{Term 1}} \mp \underbrace{\frac{1}{-j2\pi f}\left[e^{-j\pi fT_b} - 1\right]}_{\text{Term 2}}$$

Now, let us separately solve Term 1 and Term 2 as under:

Term 1:

$$\frac{\pm 1}{-j2\pi f}\left[e^{j\pi fT_b/2} \times e^{-j\pi fT_b/2} - e^{j\pi fT_b/2} \times e^{j\pi fT_b/2}\right]$$

$$= \frac{\pm 1}{-j2\pi f}\left[e^{j\pi fT_b/2} \times (e^{-j\pi fT_b/2} - e^{j\pi fT_b/2})\right]$$

$$= \frac{\pm 1}{-j2\pi f} e^{j\pi fT_b/2}\left(\frac{e^{j\pi fT_b/2} - e^{-j\pi fT_b/2}}{2j}\right) = \pm \frac{1}{\pi f} e^{j\pi fT_b/2} \sin \pi fT_b/2$$

Term 2:

$$\mp \frac{1}{-j2\pi f}\left[e^{-j\pi fT_b} - 1\right]$$

$$= \mp \frac{1}{-j2\pi f}\left[e^{-j\pi fT_b/2} \cdot e^{-j\pi fT_b/2} - e^{j\pi fT_b/2} \times e^{-j\pi fT_b/2}\right]$$

$$= \mp \frac{1}{-j2\pi f}\left[e^{-j\pi fT_b/2} (e^{-j\pi fT_b/2} - e^{j2\pi fT_b/2})\right]$$

$$= \mp \frac{1}{\pi f} e^{-j\pi T_b/2}\left[\frac{e^{j\pi fT_b/2} - e^{-j\pi fT_b/2}}{2j}\right] = \mp \frac{1}{\pi f} e^{-j\pi T_b/2} \sin(\pi fT_b/2) \qquad ...(2.38)$$

Now, adding both the terms, we obtain

$$P(f) = \pm\frac{1}{\pi f}e^{j\pi f T_b/2}\sin(\pi f T_b/2) \mp \frac{1}{\pi f}e^{-j\pi f T_b/2}\sin(\pi f T_b/2)$$

or
$$P(f) = \pm\frac{1}{\pi f}\sin(\pi f T_b/2)\left[e^{j\pi f T_b/2} - e^{-j\pi f T_b/2}\right]$$

or
$$P(f) = \frac{\pm 2j.1}{\pi f}\sin(\pi f T_b/2)\left[\frac{e^{j\pi f T_b/2} - e^{-j\pi f T_b/2}}{2j}\right]$$

or
$$P(f) = \frac{\pm 2j.1}{\pi f}\sin(\pi f T_b/2)\left[\frac{e^{j\pi f T_b/2} - e^{-j\pi f T_b/2}}{2j}\right] = \pm\frac{2j1}{\pi f}\sin^2(\pi f T_b/2)$$

Further, multiplying and dividing by $T_b/2$, we obtain

$$P(f) = \pm\frac{2j(T_b/2)}{\pi f\, T_b/2}\sin^2(\pi f T_b/2) = \pm j(T_b)\left[\frac{\sin^2(\pi f T_b/2)}{\pi f T_b/2}\right]$$

Rearranging terms, we obtain

$$P(f) = \pm j(T_b)\frac{\sin(\pi f T_b/2)}{(\pi f T_b/2)}\sin(\pi f T_b/2) = \pm(T_b)\,\mathrm{sinc}\frac{(fT_b)}{2}\sin\frac{(\pi f T_b)}{2}$$

This is the required Fourier transform.
Now, let us obtain the expression for S(f).
We know that the power spectral density (psd) S(f) is given by

$$S(f) = \frac{1}{T_b}|P(f)|^2\sum_{n=-\infty}^{\infty} R_\tau(n)\,.\,e^{-j2\pi n f T_b}$$

Substituting the expressions for P(f), and $R_\tau(n)$, we obtain

$$S(f) = T_b\,\mathrm{sinc}^2\frac{(fT_b)}{2}\sin^2\frac{(\pi f T_b)}{2}\,.\,A^2$$

or
$$S(f) = A^2 T_b\,\mathrm{sinc}^2\frac{(fT_b)}{2}\sin^2\frac{(\pi f T_b)}{2} \quad ...(2.39)$$

Important Point: *This is the required expression for the power spectral density (psd). The normalized form of this power spectral density (psd) is plotted as curve d in figure 2.18. Most of the power lies inside the bandwidth equal to $2/T_b$ which is twice as large as that of unipolar, polar and bipolar formats of NRZ type.*

2.18.1. Salient Features of the psd of a Manchester Format

(i) The main lobes extend from 0 to $2f_b$ rather than 0 to f_b (as NRZ).
(ii) At f = 0, S(f) = 0 as shown in figure 2.20.

One friend in a life is much, two are many, three are hardly possible. – **Henry Brooks Adams.**

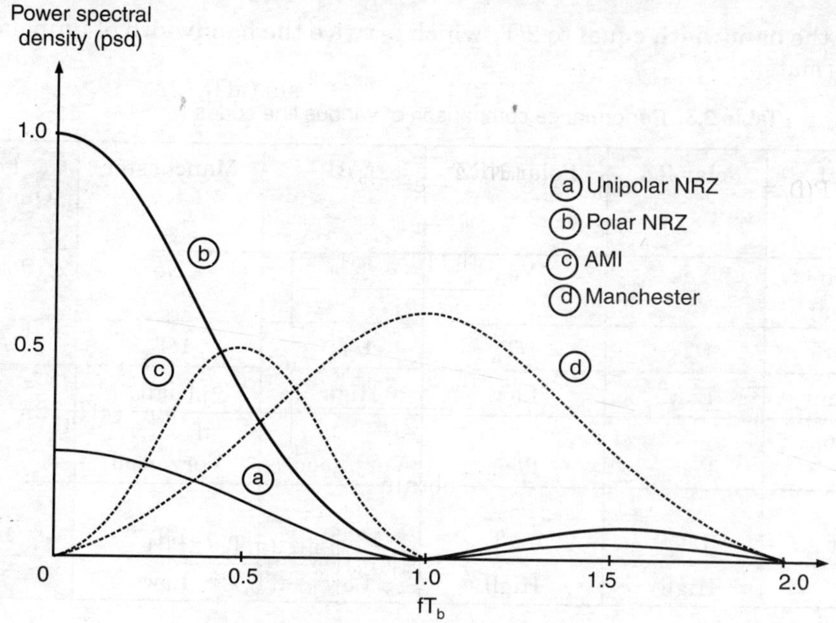

Fig. 2.20. Power spectra of different line codes

(iii) If the Manchester signal is transmitted through an ideal low pass filter with a cut-off frequency of $2f_b$ then about 95% power will be transmitted but if the cut-off frequency is reduced to f_b then only about 70% power is passed.

(iv) The bandwidth required is $2f_b$ which is double the bandwidth of the NRZ signal.

2.19 PERFORMANCE COMPARISON OF VARIOUS DISCRETE PAM FORMATS ON THE BASIS OF POWER SPECTRA

As a matter of fact, power spectra is a graph power spectral density (psd) plotted against frequency. It gives the distribution of power of different frequency components. The power spectra for four important line codes have been shown in figure 2.20.

(i) Power Spectra of NRZ Unipolar

Curve a in figure 2.20 shows a normalized graph. Both the power spectral density (psd) and frequency are normalized. Most of the power of the NRZ unipolar format is concentrated between dc and the bit rate of input data.

DO YOU KNOW?

Bipolar signals also have single-error detection capabilities built in, since a single error will cause a violation of the bipolar line code rule. Any violations can easily be detected by receiver logic.

(ii) Power Spectra of NRZ Polar Format

Curve b in figure 2.20 shows a normalized power spectra of NRZ polar format. Here, also like unipolar NRZ format, most of the power lies in main lobe of the sinc shaped curve. This main lobe extends upto a bit rate of $1/T_b$.

(iii) NRZ Bipolar (AMI) Format

Curve c in figure 2.20 shows the normalized power spectra for the AMI format. It can be seen that almost all the power lies inside the bandwidth equal to $1/T_b$. Similar to the two NRZ formats discussed earlier, however the spectral content of the NRZ bipolar format is relatively small around the zero frequency.

(iv) Manchester Format

Curve d in figure 2.20 shows the normalized power spectra of Manchester format. Here most of power lies inside the bandwidth equal to $2/T_b$ which is twice the bandwidth of unipolar, polar and bipolar NRZ formats.

Table 2.3. Performance comparison of various line codes

S. No.	Parameter of Comparison	Polar RZ	Polar NRZ	AMI	Manchester	Polar Quaternary NRZ
1.	Transmission of DC component	Yes	Yes	No	No	Possible
2.	Signaling rate	$1/T_b$	$1/T_b$	$1/T_b$	$1/T_b$	$1/2\ T_b$
3.	Noise immunity	Low	Low	High	High	High
4.	Synchronizing capability	Poor	Poor	Very good	Very good	Poor
5.	Bandwidth requirement	$1/T_b$	$1/2\ T_b$	$1/2\ T_b$	$1/T_b$	$1/2T_b$
6.	Crosstalk	High	High	Low	Low	Low

EXAMPLE 2.1. Encode the following binary data stream into return to zero (RZ), Non-return to zero (NRZ), AMI and Manchester codes.

Data stream: 1 1 0 0 0 0 1 0

Solution: The line coding formats for the given data stream will be as shown in figure 2.21.

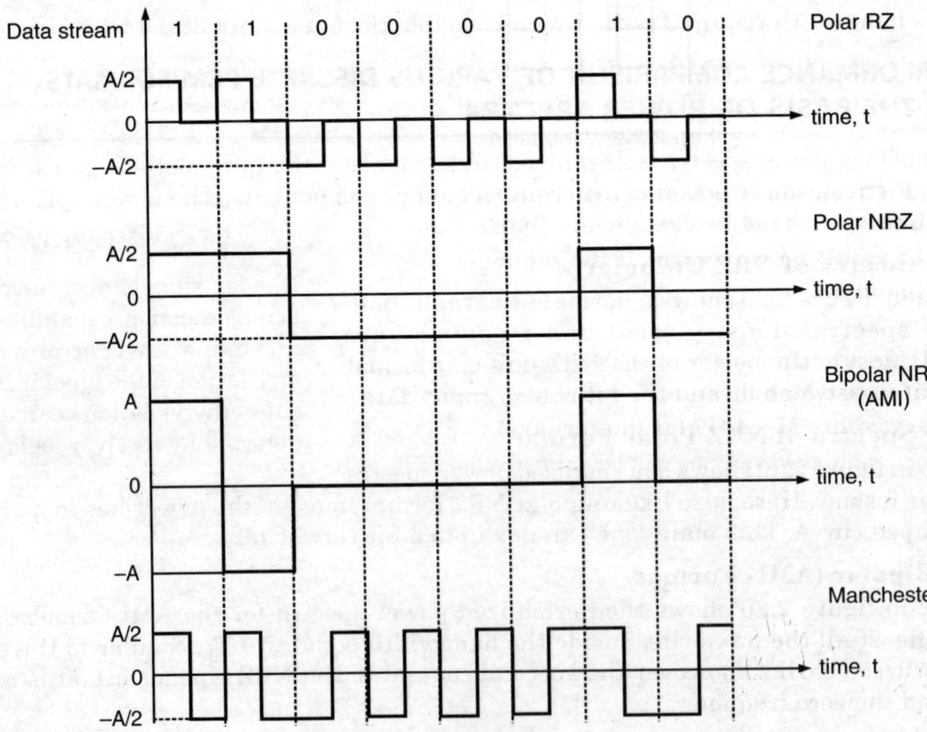

Fig. 2.21.

EXAMPLE 2.2. **Draw the following data formats for the bit stream 1 1 0 0 1 1 0:**

(i) Polar NRZ

(ii) Unipolar RZ

(iii) AMI

(iv) Manchester

Solution: The data formats will as shown in figure 2.22.

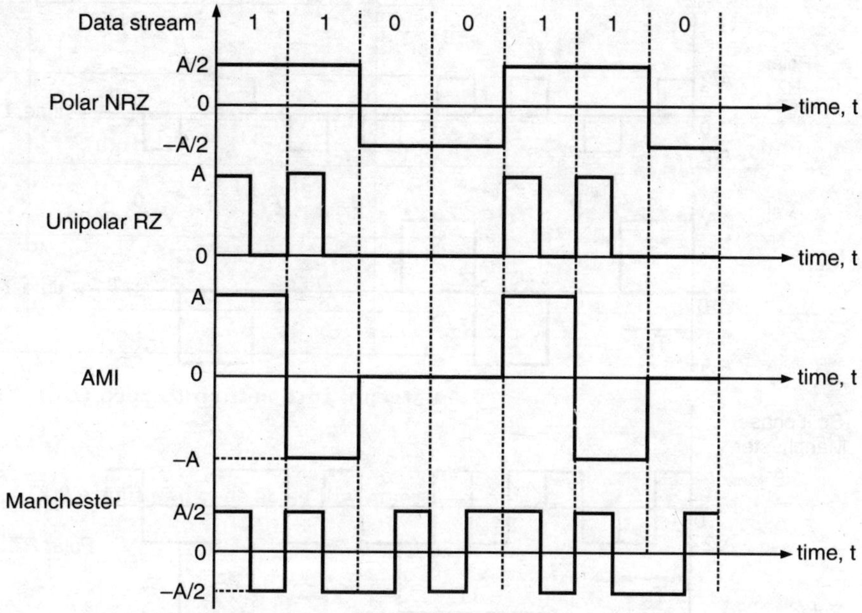

Fig. 2.22. Various line codes for example 2.2

EXAMPLE 2.3. **Given that the bit sequence given below is to be transmitted**

Bit sequence = 10110010

Draw the resulting waveform, if the sequence is transmitted using:

(i) Unipolar RZ

(ii) Polar RZ

(iii) AMI

(iv) Split Phase Manchester

(v) M-ary where M = 4 (Polar quaternary) **(UPTU, Sem. Exam., 2003-04)**

Solution: The required waveforms will be as shown in figure 2.23.

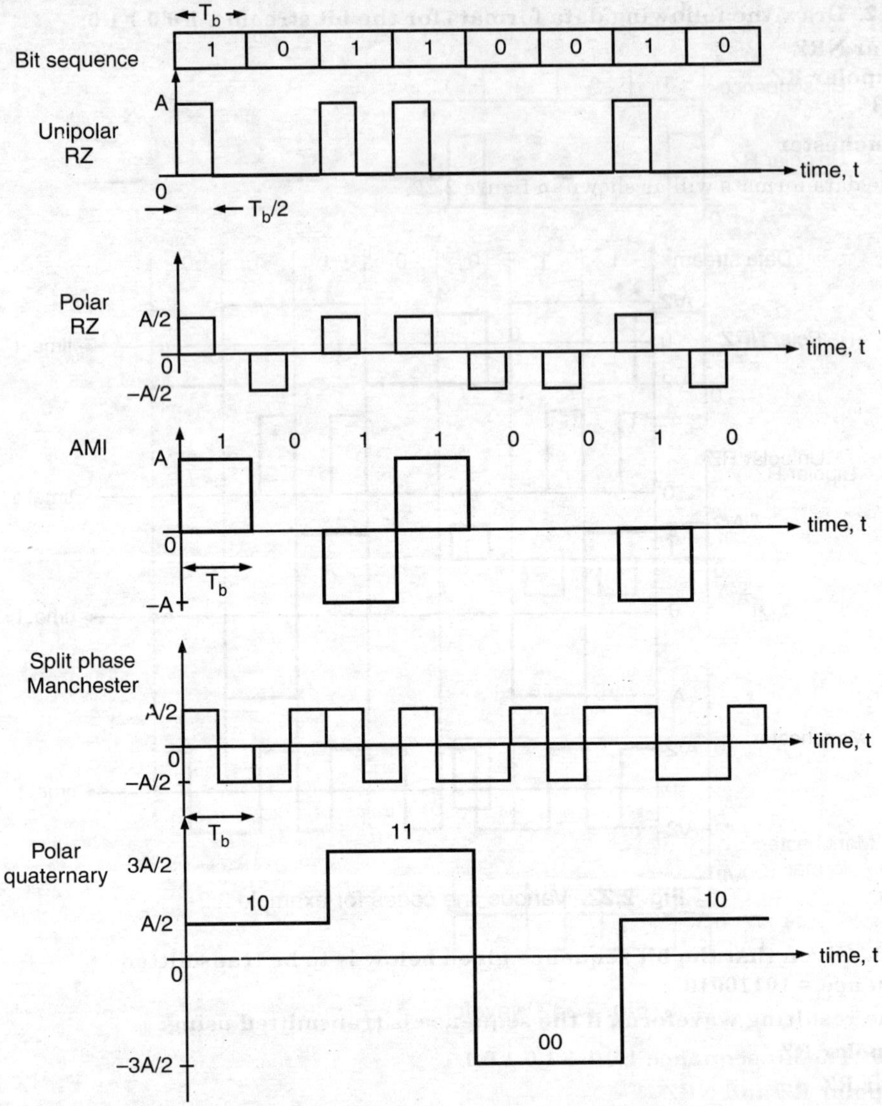

Fig. 2.23. Waveforms for example 2.3

EXAMPLE 2.4. Which are the desirable properties of digital waveform ? To transmit a bit sequence 10011011, draw the resulting waveforms using:

(i) Unipolar RZ
(ii) Unipolar NRZ
(iii) Bipolar RZ
(iv) AMI RZ
(v) Manchester.

Solution: For the properties of the line codes please study the related text.

The required waveforms have been drawn in figure 2.24.

Every generation revolts against its fathers and makes friends with its grand fathers.

– Mumford Lewis

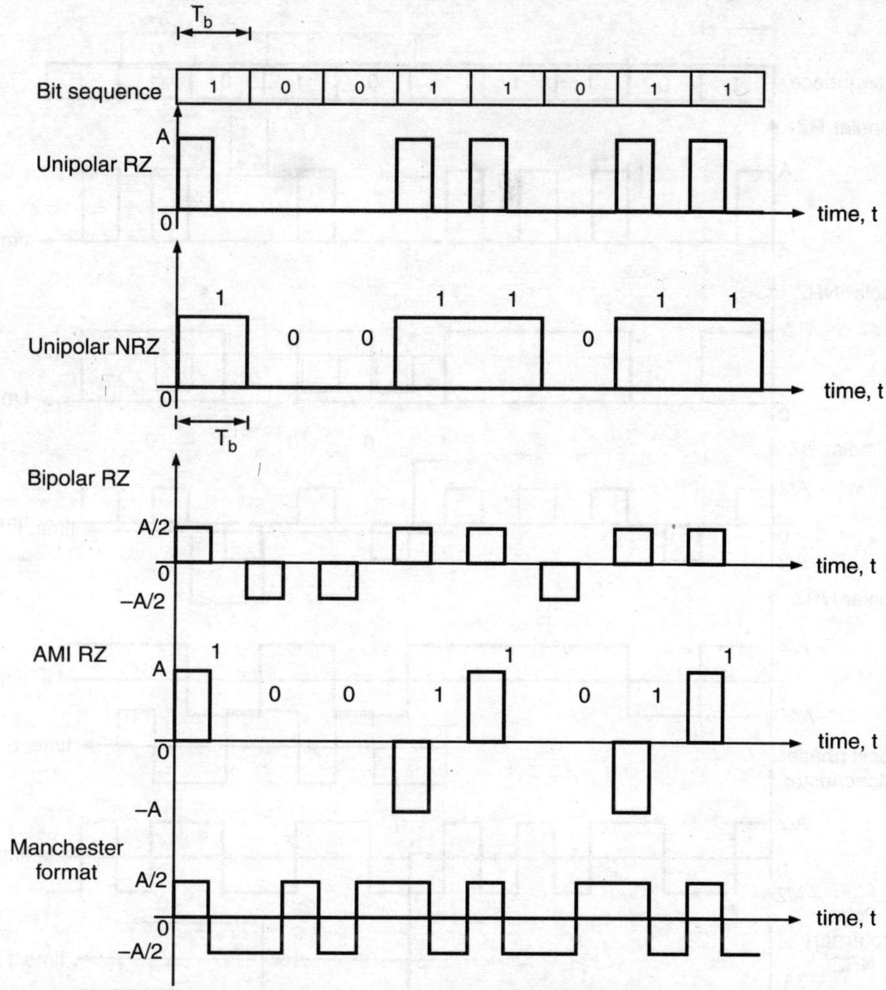

Fig. 2.24. Waveforms for example 2.4.

EXAMPLE 2.5. The bit sequence 1 0 1 1 1 0 1 0 1 1 is to be transmitted using following formats:

(i) Unipolar RZ and NRZ

(ii) Bipolar RZ and NRZ

(iii) Split-phase Manchester

(iv) Polar quaternary NRZ.

Draw all the waveforms.

Solution: The required waveforms will be as shown in figure 2.25.

The important thing in life is not the victory but the contest; the essential thing is not to have won but to have run the race.
– Coubertin

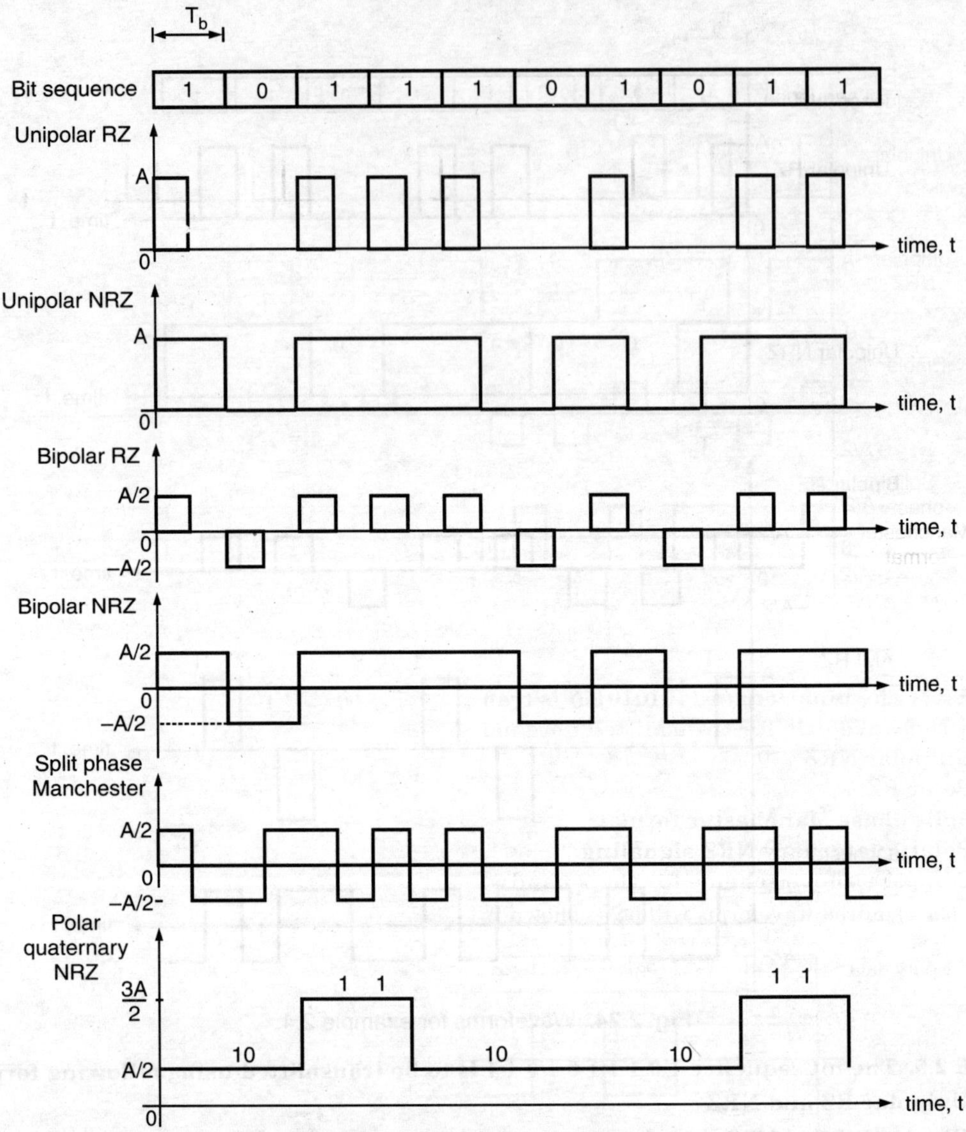

Fig. 2.25. Waveforms for example 2.5.

EXAMPLE 2.6. The binary data 101100110101 is transmitted over a baseband channel. Draw the waveform for the transmitted data using following formats:

(i) Unipolar RZ

(ii) Unipolar NRZ

(iii) Bipolar RZ

(iv) Split-phase Manchester

Compare above schemes for their BW requirements.

Solution: The required waveforms will be as shown in figure 2.26.

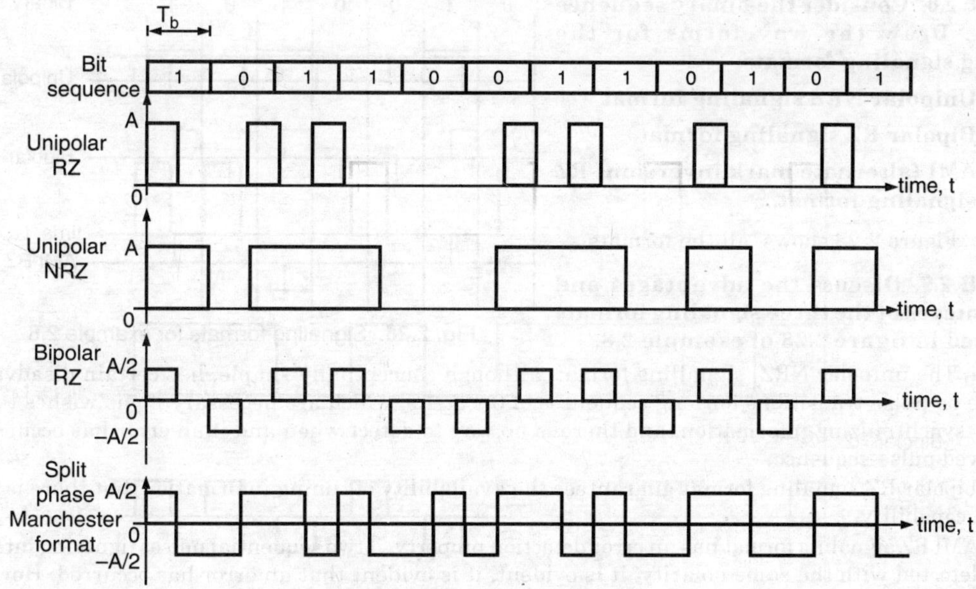

Fig. 2.26. Waveforms for example 2.6

EXAMPLE 2.7. The binary data 110101010110 is transmitted over a baseband channel. Draw the waveform for transmitted data using following format:

(i) Unipolar NRZ

(ii) Polar RZ

(iii) Split phase Manchester format

(iv) Polar quaternary NRZ signaling.

Compare above scheme for their bandwidth requirement.

Solution: The required waveforms will be as shown in figure 2.27.

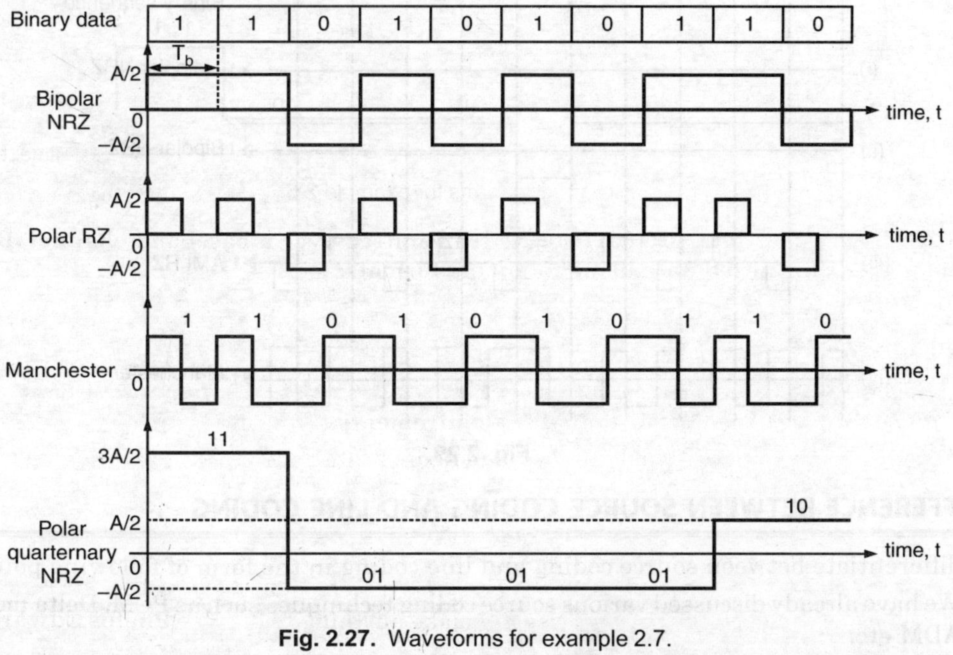

Fig. 2.27. Waveforms for example 2.7.

EXAMPLE 2.8. Consider the binary sequence 0100101. Draw the waveforms for the following signaling formats:

(i) Unipolar NRZ signaling format

(ii) Bipolar RZ signaling format

(iii) AMI (alternate mark inversion) RZ signaling format.

Solution: Figure 2.28 shows all the formats.

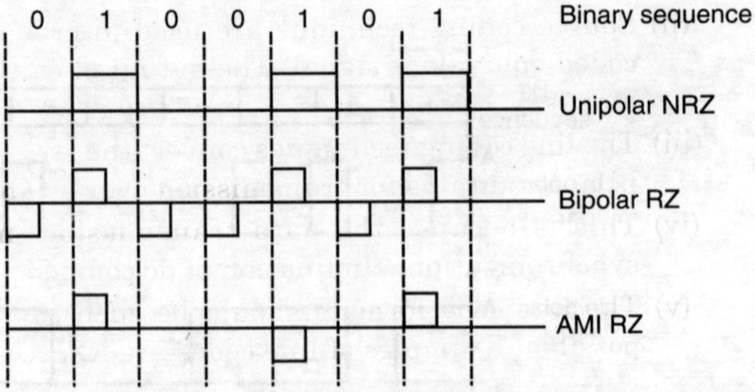

Fig. 2.28. Signaling formats for example 2.8

EXAMPLE 2.9. Discuss the advantages and disadvantages of the three signaling formats illustrated in figure 2.28 of example 2.8.

Solution: The unipolar NRZ signalling format, although conceptually simple, has certain disadvantages. There are no pulse transitions for long sequences of 0 s or 1 s, which are necessary if one wishes to extract timing or synchronizing information, and there is no way to detect when and if an error has occurred from the received pulse sequence.

The bipolar RZ signaling format guarantees the availability of timing information, but there is no error detection capability.

The AMI RZ signaling format has an error detection property, if two sequential pulses (ignoring intervening 0 s) are detected with the same polarity, it is evident, it is evident that an error has occurred. However, to guarantee the availability of timing information, it is necessary to restrict the allowable number of consecutive 0 s.

EXAMPLE 2.10. Consider a binary sequence with a long sequence of 1 s followed by a single 0 and then a long sequence of 1 s. Draw the waveforms for this sequence, using the following signaling formats:

(i) Unipolar NRZ signaling

(ii) Bipolar NRZ signaling

(iii) AMI RZ signaling

(iv) Split-Phase (Manchester) signaling

Solution: Figure 2.29 shows all the required formats.

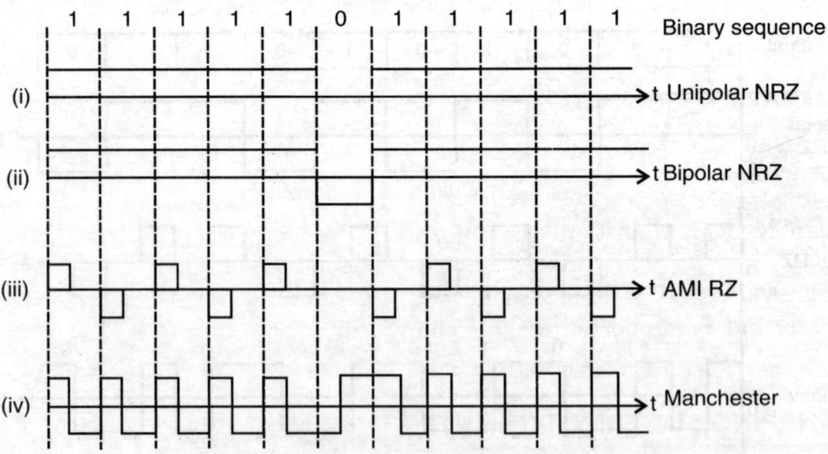

Fig. 2.29.

2.20 DIFFERENCE BETWEEN SOURCE CODING AND LINE CODING

We can differentiate between source coding and line coding in the form of following points :

(i) We have already discussed various source coding techniques such as PCM, Delta modulation ADM etc.

(ii) Source coding technique are used in order to convert an analog signal to its digitally coded equivalent signal. The output of a source coder is a train of binary digits, i.e., 0s and 1s.

(iii) The line coding techniques convert the stream of binary digits into a format or code which is more suitable for transmission over a cable or any other medium.

(iv) This is because the data transmission must satisfy certain requirements such as synchronization, elimination of dc component, low cross talk etc.

(v) The line code formats should be such that the bandwidth requirements be as low as possible. Examples of line codes are RZ, NRZ, AMI, Manchester and other formats.

2.21. PULSE TRANSMISSION

In all the digital systems, the pulses are transmitted from the transmitter to receiver. But the channels carrying these pulses never have enough bandwidth required to ensure the preservation of the shape of the pulse. Such channels are called as the band limited channels. Figure 2.30 depicts the output of a bandlimited channel when a narrow pulse is applied at its input.

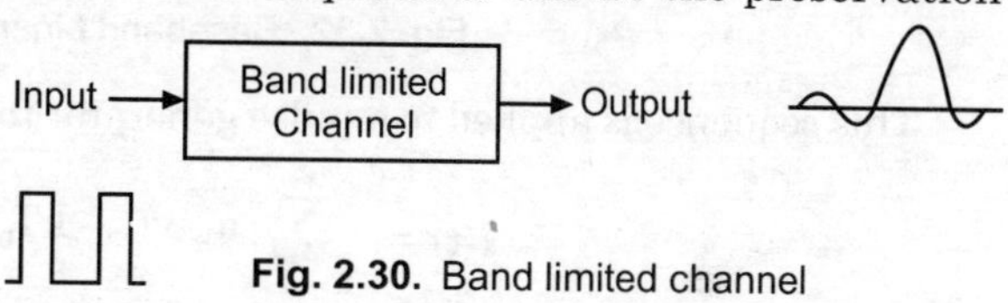

Fig. 2.30. Band limited channel

Figure 2.31 shows that the bandlimited channel tends to spread the narrow pulses. This produces the secondary lobes as shown in figure 2.31. The secondary lobes are called as ringing tails. Figure 2.31 shows the energy distribution in the output of a bandlimited channel. It shows that almost 90% energy is contained in the main lobe.

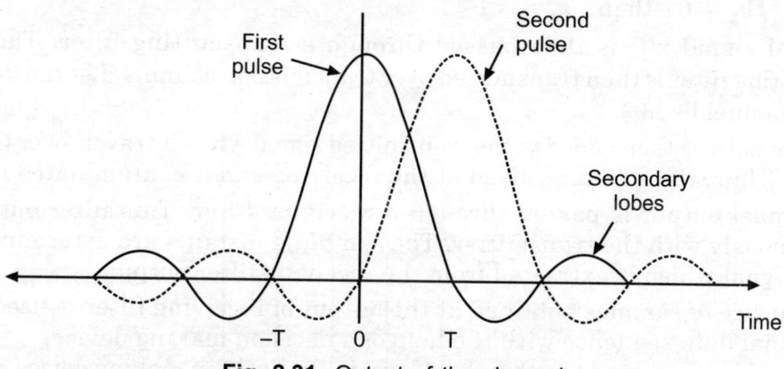

Fig. 2.31. Output of the channel

Thus, we can recover almost all the information even if we allow the main lobe to pass through. This means that even if the bandwidth is equal to $\frac{1}{T}$, i.e., $B = \frac{1}{T}$, the corresponding maximum bit rate that is supported by this channel is,

$$R = 2B$$

2.22. PULSE SHAPING : INTERSYMBOL INTERFERENCE (ISI)

In a communication system, when the data is being transmitted in the form of pulses (i.e., bits), the output produced at the receiver due to other bits or symbols interferes with the output produced by the desired bit. This is known as **intersymbol interference (ISI)**. The intersymbol interference will introduce errors in the detected signal. Let us consider figure 2.32 which shows the elements

binary PAM system. The input signal consists of a binary data sequence $\{b_k\}$ with a bit duration of T_b seconds.

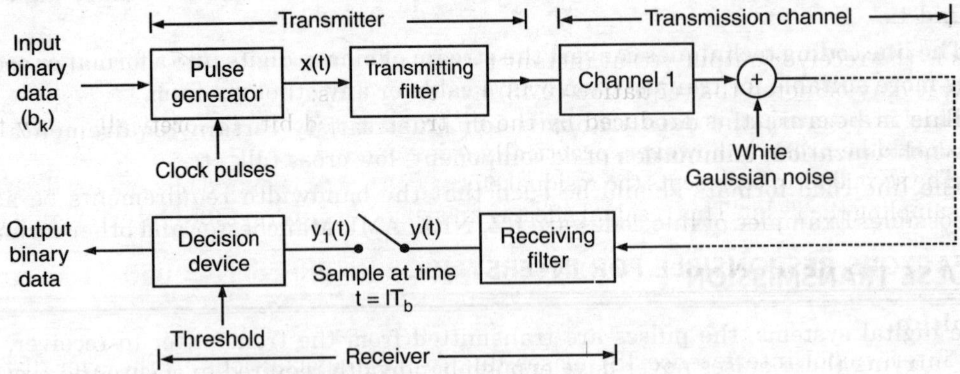

Fig. 2.32. Baseband binary data transmission system

This sequence is applied to a pulse generator to produce a discrete PAM signal which is given by

$$x(t) = \sum_{k=-\infty}^{\infty} a_k\, v(t - kT_b) \qquad ...(2.40)$$

where v(t) denotes the basic pulse, normalized such that v(0) = 1. The first block of the system i.e., pulse amplitude modulator converts this input sequence into polar form as under:

if $\quad b_k = 1$ then $a_k = 1$

and $\quad b_k = 0$ then $a_k = -1$

(i) The PAM signal x(t) is then passed through a transmitting filter. The output of the transmitting filter is then transmitted over transmission channel. Let the impulse response of this channel be h(t).

(ii) A random noise is then added to the transmitted signal when it travels over the transmission channel. Thus, the signal received at the receiving end is contaminated with noise.

(iii) The channel output is passed through a receiving filter. This filter output is sampled synchronously with the transmitter. The sampling instants are determined by a clock or timing signal which is extracted from the receiving filter output.

(iv) The sequence of samples obtained at the output of receiving filter is used to reconstruct the original data sequence with the help of a decision making device.

(v) Each sample is compared to a threshold level in the decision making device. If the amplitude of the sample is higher than the threshold level then it is decided that a symbol 1 is received. On the other hand, if the signal has an amplitude lower than the threshold, then the decision is that a 0 is received.

The receiving filter output can be written as under:

$$y(t) = \mu \sum_{k=-\infty}^{\infty} a_k\, p(t - kT_b) + n(t) \qquad ...(2.41)$$

where μ is a scaling factor and the noise n(t) is the noise at the output of the receiving filter due to the added noise. The term $p(t - kT_b)$ represents the combined impulse response of the receiving filer.

The receiving filter output y(t) is sampled at time $t_i = iT_b$ with i taking on integer values to provide following expression:

$$y(t_i) = \mu \sum_{k=-\infty}^{\infty} a_k\, p(iT_b - kT_b) + n(t_i) \qquad ...(2.42)$$

or

$$y(t_i) = \mu a_i + \mu \sum_{\substack{k=-\infty \\ k \neq i}}^{\infty} a_k\, p(iT_b - kT_b) + n(t_i) \qquad ...(2.43)$$

This is the receiver output y(t) at instant $t = t_i$.

Now, it may be noted that equation (2.43) has two terms:

(i) The first term μa_i is produced by the ith transmitted bit. Theoretically, only this term should be present, however, practically it is not so.

(ii) The second term represents the residual effect of all the transmitted bits, obtained at the time of sampling the ith bit. This residual effect is known as the **intersymbol interference (ISI).**

2.23. FACTORS RESPONSIBLE FOR INTERSYMBOL INTERFERENCE (ISI)

1. Definition

The intersymbol interference ISI arises due to the imperfections in the overall frequency response of the system. When a short pulse of duration T_b seconds is transmitted through a bandlimited system, then the frequency components contained in the input pulse are **differentially attenuated** and more importantly **differentially delayed** by the system. Due to this, the pulse appearing at the output of the system will be dispersed over an interval which is longer than T_b seconds. Due to this dispersion, the symbols each of duration T_b will interfere with each other when transmitted over the communication channel. This will result in the intersymbol interference (ISI).

DO YOU KNOW?

The raised cosine filter is also called a Nyquist filter. It is only one of a more general class of filters that satisfy Nquist first criterion.

The transmitted pulse of duration T_b seconds and the dispersed pulse of duration more than T_b seconds are shown in figure 2.33.

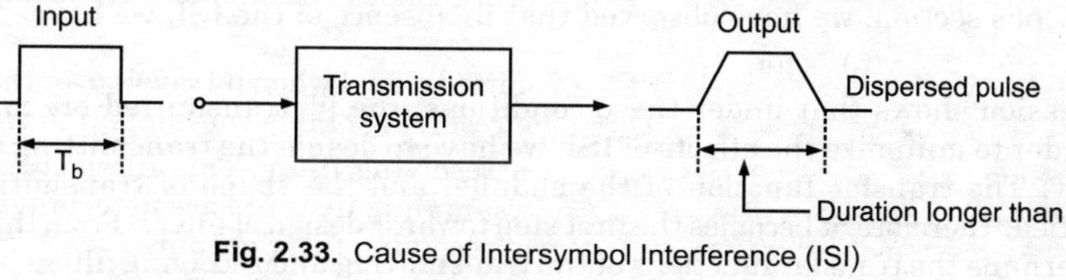

Fig. 2.33. Cause of Intersymbol Interference (ISI)

2. Effect of Intersymbol Interference (ISI)

Following are the effects of ISI:

(i) In the absence of ISI and noise, the transmitted bit can be decoded correctly at the receiver.

(ii) The presence of ISI will introduce errors in the decision at the receiver output.

(iii) Hence, the receiver can make an error in deciding whether it has received a logic 1 or a logic 0.

3. Remedy to Reduce ISI

(i) It has been proved that the function which produces a zero intersymbol interference is a sinc function. Hence, instead of a rectangular pulse if we transmit a sinc pulse then the ISI can be reduced to zero.

(ii) This is known as Nyquist Pulse Shaping. The sinc pulse transmitted to have a zero ISI has been shown in figure 2.34 (a).

(iii) Further, we know that Fourier transform of a sinc pulse is a rectangular function. Hence, to preserve all the frequency components, the frequency response of the filter must be exactly flat in the pass band and zero in the attenuation band as shown in figure 2.34 (b).

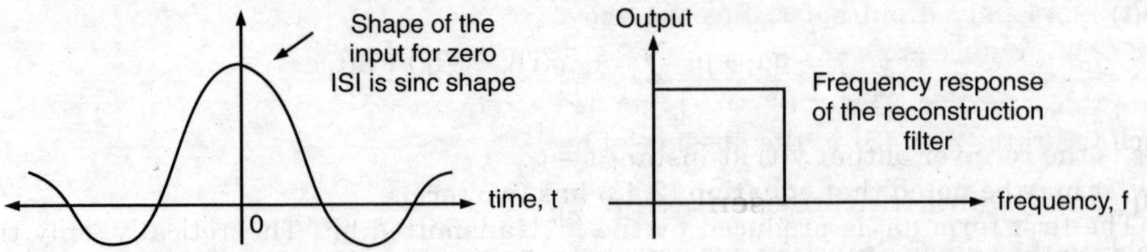

(a) Ideal pulse shape for zero ISI (b) Frequency response of the filter

Fig. 2.34.

✍ **Important Point:** *However, type of filter is practically not present or not possible. Hence, in practice, the frequency response of the filter is modified as shown in figure 2.35 with different roll off factors α to obtain the achievable filter response curves.*

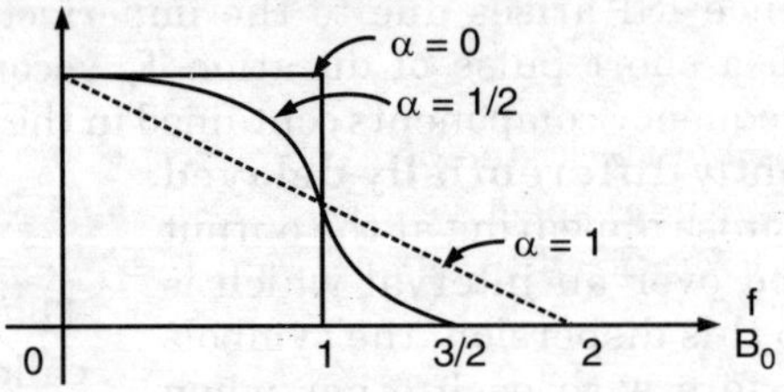

Fig. 2.35. Illustration of practical filter characteristics

2.24 NYQUIST'S CRITERION FOR DISTORTIONLESS BASEBAND BINARY TRANSMISSION

1. Basic concept

In the previous section, we have observed that in absence of the ISI, we have

$$y(t_i) = \mu a_i \quad \text{...(2.44)}$$

This expression shows that under these conditions, the ith transmitted bit can be decoded correctly. In order to minimize the effects of ISI, we have to design the transmitting and receiving filters properly. The transfer function of the channel and the shape of transmitted pulse are generally specified. Therefore, it becomes the first step towards design of filters. From this information we have to determine the transfer functions of the transmitting and receiving filters, to reconstruct the transmitted data sequence $\{b_k\}$. This is achieved by first extracting and then decoding the corresponding sequence of weights from the output y(t).

As expressed by following equation

$$y(t) = \mu \sum_{k=-\infty}^{\infty} a_k\, p(t - kT_b) \quad \text{...(2.45)}$$

This shows that output y(t) is dependent on a_k, the received pulse p(t) and the scaling factor μ.

2. Extraction

Extraction is basically the process of sampling. The signal y(t) is sampled at $t = iT_b$.

3. Decoding

The decoding should be such that the contribution of the weighted pulse i.e., $a_k\, p(iT_b - kT_b)$ for i = k be free from ISI. This can be stated mathematically as under:

$$p(iT_b - kT_b) = \begin{cases} 1 & \text{for } i = k \\ 0 & \text{for } i \neq k \end{cases} \quad \text{...(2.46)}$$

where p(0) = 1 due to normalizing.

If p(t) i.e., received pulse satisfies the above expression, then the receiver output given by equation (2.46) reduces to

$$y(t_i) = \mu a_i \qquad ...(2.47)$$

which indicates zero ISI in the absence of noise.

4. Frequency Domain Representation

To obtain the transfer function, we have to transform the condition stated in equation (2.46) into frequency domain. Let us consider that the sequence of samples is represented by $p(nT_b)$ where n = 0, ± 1, ± 2... after the process of extraction. The Fourier transform of such a sequence of samples can be obtained, as we did for the sampling process of low pass signal. Therefore, the frequency domain representation of $p(nT_b)$ will be as under:

$$P_\delta(f) = F[p(nT_b)] = R_b \sum_{n=-\infty}^{\infty} P(f - nR_b) \qquad ...(2.48)$$

where $R_b = 1/T_b$ i.e., bit rate.

But, $P_\delta(f)$ also represents the Fourier transform of an infinite periodic sequence of unit impulses whose strengths are weighted by the respective sample values of p(t). Hence, $P_\delta(f)$ can be written as under:

$$P_\delta(f) = \int_{-\infty}^{\infty} \sum_{m=-\infty}^{\infty} [p(mT_b)\, \delta(t - mT_b)]\, e^{-j2\pi ft} dt \qquad ...(2.49)$$

where $\sum_{m=-\infty}^{\infty} p(mT_b)\, \delta(t - mT_b)$ represents the sequence of unit impulses weighted by the respective sample values.

Now, let m = i – k, therefore, if i = k then m = 0 and if i ≠ k then m ≠ 0.

Hence, let us apply condition of equation (2.46) to equation (2.49) to get, for m = 0.

$$P_\delta(f) = \int_{-\infty}^{\infty} p(0)\, \delta(t)\, e^{-j2\pi ft}\, dt$$

Using the shifting property of delta function, we have

$$P_\delta(f) = p(0)$$

But, p(0) = 1 due to normalization, hence $P_\delta(f) = 1$.

Substituting this into equation (2.48), we obtain

$$1 = R_b \sum_{n=-\infty}^{\infty} P(f - nR_b)$$

Therefore, we have

$$\sum_{n=-\infty}^{\infty} P(f - nR_b) = 1/R_b = T_b \qquad ...(2.50)$$

This expression is called as the Nyquist criterion for distortionless baseband transmission in the absence of noise.

2.25 IDEAL SOLUTION

1. Basic concept

It may be noted that the L.H.S. of equation (2.50) represents a series of shifted spectrums. For n = 0, the LHS corresponds to P(f) and it represents a frequency function with the narrowest band which satisfies equation (2.50). The range of frequencies for P(f) will extend from $-B_0$ to B_0 where B_0 corresponds to half the bit rate.

Hence, $$B_0 = R_b/2 \qquad ...(2.51)$$

Therefore, P(f) can be specified in the following form:

$$P(f) = \frac{1}{2B_0} \text{rect}\left(\frac{f}{2B_0}\right) \qquad ...(2.52)$$

and it has been shown graphically in figure 2.36(a). This is the spectrum of a signal which produces zero ISI. Hence, the signal that produces zero ISI can be obtained by taking the IFT of P(f).

This means that we have

$$p(t) = F^{-1}[P(f)]$$

$$p(t) = F^{-1}\left[\frac{1}{2B_0} \text{rect}\left(\frac{f}{2B_0}\right)\right]$$

$$p(t) = \text{sinc}(2B_0 t) \qquad ...(2.53)$$

Figure 2.36 (b) shows the plot of this function. This function p(t) can be regarded as the impulse response of an ideal low pass filter (LPF) with bandwidth B_0.

Thus, the shape of a pulse should be a sinc pulse rather than being a rectangular one to eliminate the ISI.

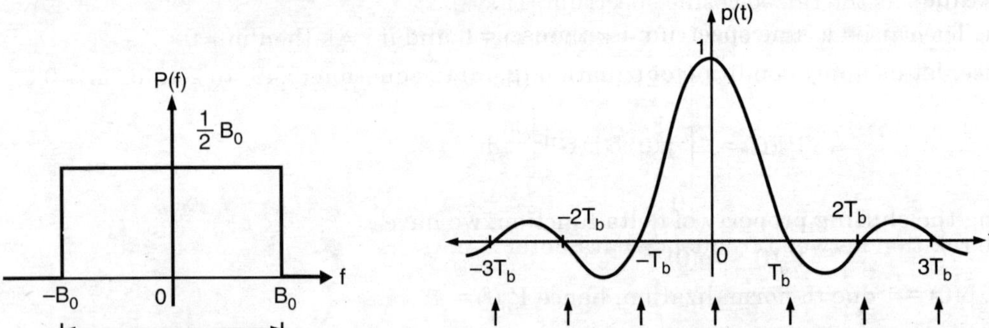

(a) Graphical representation of P(f) (b) Time-domain representation

Fig. 2.36.

2. Advantages of Using the Sinc Pulse

(i) Bandwidth requirement (of the channel) is reduced.

(ii) ISI is reduced to zero.

3. Possible Difficulties

(i) It is necessary that the amplitude characteristics of P(f) should be flat from $-B_0$ to B_0 and zero outside this band. But, abrupt transition at $\pm B_0$ is not physically realizable.

(ii) Due to discontinuity of P(f) at $\pm B_0$, there is practically no margin of error in sampling times at the receiver end.

2.26. RAISED COSINE SPECTRUM

1. Basic concept and Mathematical Expression

The two difficulties experienced by the ideal Nyquist channel can be overcome by increasing the bandwidth from its minimum value $B_0 = R_b/2$ to an adjustable value between B_0 and $2\,B_0$. A condition is put on the overall frequency response P(f) to satisfy the given condition.

As per equation (2.50), we have

$$\sum_{n=-\infty}^{\infty} P(f - nR_b) = T_b = \frac{1}{R_b}$$

Expanding the summation sign, we get

$$\ldots P(f + R_b) + P(f) + P(f - R_b) + P(f - 2R_b) \ldots = T_b$$

But $\qquad B_0 = \dfrac{R_b}{2}, \quad$ therefore, $\quad R_b = 2B_0$

Thus, we write

$$\ldots P(f + 2B_0) + P(f) + P(f - 2B_0) + P(f - 4B_0) + \ldots = \frac{1}{2B_0}$$

We retain only the three terms on LHS which correspond to n = – 1, n = 0 and n = 1 and restrict the frequency band of interest to $(-B_0, B_0)$ to get,

$$P(f + 2B_0) + P(f) + P(f - 2B_0) = \frac{1}{2B_0} \quad \text{and given } -B_0 \le f \le B_0$$

It is possible to devise several bandlimited functions which will satisfy above equation, one of them is called as the raised cosine spectrum. This spectrum consists of a flat portion and a roll off portion. The raised cosine spectrum is expressed mathematically as under:

$$P(f) = \begin{cases} 1/2B_0 & \text{(flat portion)} \quad \ldots\, 0 \le |f| < f_1 \\ \dfrac{1}{4B_0}\left\{1 - \sin\left[\dfrac{\pi(|f| - B_0)}{2B_0 - 2f_1}\right]\right\} & \ldots\, f_1 \le |f| < 2B_0 - f_1 \\ 0 & \ldots\, |f| \ge 2B_0 - f_1 \end{cases}$$

The relation between frequency parameter f_1 and the bandwidth B_0 are related as under :

$$\alpha = 1 - \frac{f_1}{f_m}$$

where α is called as the roll off factor. It indicates the excess bandwidth over the ideal solution B_0. The transmission bandwidth B_T is defined as under :

$$B_T = 2B_0 - f_1 = B_0(1 + \alpha)$$

The normalized frequency response of raised cosine function is obtained by multiplying P(f) by $2B_0$ and it is plotted in figure 2.37(a), for different values of α. The corresponding time response p(t) is shown in figure 2.37(b).

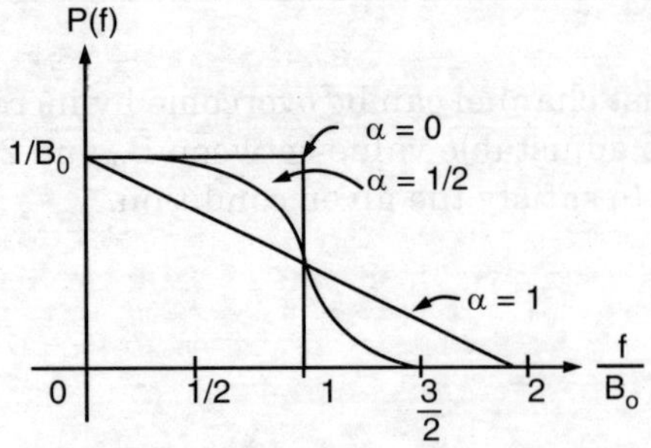

(a) Frequency response

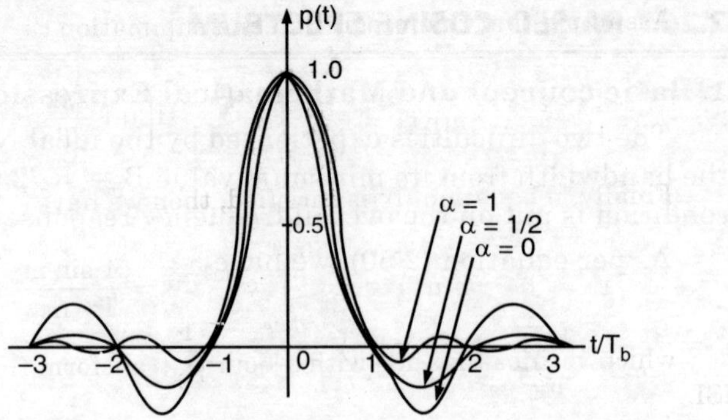

(b) Time response

Fig. 2.37 Responses for different roll-off factors, α

2. Observations

(i) For $\alpha = 0.5$ and 1, the characteristics of P(f) changes gradually with respect to frequency. Hence, it is easier to realize this characteristics practically.

(ii) The time response has a sinc shape and all the sinc functions pass through zero at $t = \pm T_b, \pm 2T_b, \ldots\ldots$

(iii) The amplitude of side lobes increases with reduction in the value of α.

(iv) With $\alpha = 0$, the bandwidth requirement is maximum equal to $2B_0$.

EXAMPLE 2.11. Show that (a pulse-shape, function) h(t), with Fourier transform given by H(ω), that satisfies the criterion

$$\sum_{k=-\infty}^{\infty} H\left(\omega + \frac{2\pi k}{T}\right) = 1 \text{ for } |\omega| \le \frac{\pi}{T} \quad \text{...(i)}$$

has h(nT) which is given by

$$h(nT) = \begin{cases} \frac{1}{T} & \text{for } n = 0 \\ 0 & \text{for } n \ne 0 \end{cases} \quad \text{...(ii)} \quad \text{[GATE Examination 1999]}$$

Solution: The criterion given in equation (i) is known as Nyquist's pulse-shaping criterion.

Taking the inverse Fourier transform of H(ω), we have

$$h(t) = \frac{1}{2\pi}\int_{-\infty}^{\infty} H(\omega)e^{j\omega t}d\omega$$

The range of integration in the above equation can be divided into segments of length 2π/T as under:

$$h(t) = \frac{1}{2\pi}\sum_{k=-\infty}^{\infty}\int_{(2k-1)\pi/T}^{(2k+1)\pi/T} H(\omega)e^{j\omega t}d\omega$$

and we can write h(nT) as

$$h(nT) = \frac{1}{2\pi}\sum_{k=-\infty}^{\infty}\int_{(2k-1)\pi/T}^{(2k+1)\pi/T} H(\omega)e^{j\omega nT}d\omega \quad \text{...(iii)}$$

Changing variable $u = \omega - 2\pi(k/T)$, equation (iii) becomes

$$h(nT) = \frac{1}{2\pi}\sum_{k=-\infty}^{\infty}\int_{-\pi/T}^{\pi/T} H\left(u + \frac{2\pi k}{T}\right)e^{j(u + 2\pi k/T)nT}du$$

Assuming that the integration and summation can be interchanged, we have

$$h(nT) = \frac{1}{2\pi}\int_{-\pi/T}^{\pi/T}\sum_{k=-\infty}^{\infty} H\left(u+\frac{2\pi k}{T}\right)e^{junT}du$$

Finally, if equation (i) is satisfied, then we have

$$h(nT) = \frac{1}{2\pi}\int_{-\pi/T}^{\pi/T} e^{junT}du = \frac{1}{T}\frac{\sin n\pi}{n\pi} = \begin{cases}\frac{1}{T} & \text{for } n = 0\\ 0 & \text{for } n \neq 0\end{cases}$$

which verifies that h(t) with a Fourier transform H(ω) satisfying criterion in equation (i) produces zero ISI.

Ans.

EXAMPLE 2.12. A communication channel of bandwidth 75 kHz is required to transmit binary data at a rate of 0.1 Mb/s using raised-cosine pulses. Determine the roll-off factor α.

Solution: We have

$$T_b = \frac{1}{0.1(10^6)} = 10^{-5}\text{ s}$$

or $f_B = 75\text{ kHz} = 75\ (10^3)\text{ Hz}$

We know that $1 + \alpha = 2\, f_B T_b = 2(75)(10^3)(10^{-5}) = 1.5$

Hence, we obtain $\alpha = 0.5$ **Ans.**

EXAMPLE 2.13. In a certain telemetry system, eight message signals having 2 kHz bandwidth each are time-division multiplexed using a binary PCM technique. The error in sampling amplitude cannot be greater than 1 percent of the peak amplitude. Determine the minimum transmission bandwidth required if raised-cosine pulses with roll-off factor α = 0.2 are used. The sampling rate must be at least 25 percent above the Nyquist rate.

Solution: We know that the maximum quantizing error must satisfy

$$\epsilon_{max} = \frac{\Delta}{2} = \frac{m_p}{q} \leq 0.01\, m_p$$

Hence, $q \geq 100$, and we choose $q = 128 = 2^7$. The number of bits per sample required is 7.

Since the Nyquist sampling rate is $2 f_m = 4000$ samples/s, the sampling rate for each signal is

$$f_s = 1.25(4000) = 5000\text{ samples/s}$$

There are eight time-division multiplexed signals, requiring a total of

$$8(5000) = 40{,}000\text{ samples/s}$$

Now, since each sample is encoded by 7 bits, therefore, the resultant bit rate will be

$$\frac{1}{T_b} = 7(40{,}000) = 280\text{ kb/s}$$

Further, we know that the minimum transmission bandwidth required is

$$f_B = \frac{1+0.2}{2}(280) = 168\text{ kHz} \quad \textbf{Ans.}$$

EXAMPLE 2.14. Let $X_\alpha(f)$ be the raised cosine spectrum with a roll-off factor α.

$$\textbf{Show that } \int_{-\infty}^{\infty} X_\alpha(f)\, df = 1$$

Solution: Earlier, we have derived the expression for $X_\alpha(f)$ i.e., P(f). The raised cosine spectrum for different value of α is shown in figure 2.38.

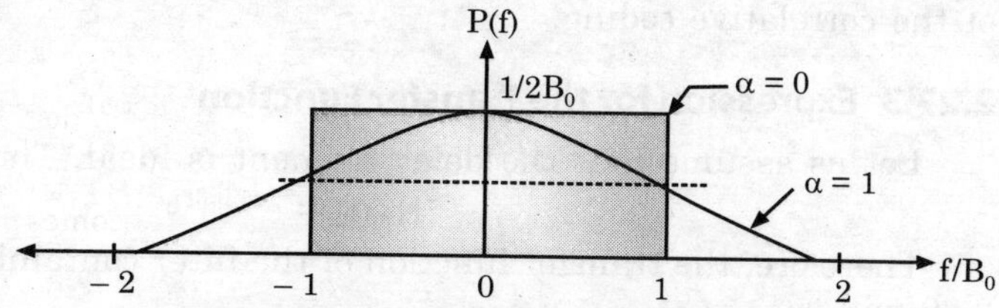

Fig. 2.38. Raised cosine spectrum

Let us consider the spectrum corresponding to α = 0. This means that

$$\int_{-\infty}^{\infty} P(f)\,df = \int_{-B_0}^{B_0} (1/2B_0)\,df = \frac{1}{2B_0}[f]_{-B_0}^{B_0} = \frac{1}{2B_0}[B_0 + B_0] = 1$$

The same can be proved for different values of α.

2.27 CORRELATIVE CODING

2.27.1 Correlative Coding or Partial Response

Till now we have treated the intersymbol interference (ISI) as an undesirable phenomenon which degrades the system performance. But by adding the intersymbol interference to the transmitted signal in a controlled manner it is possible to achieve a bit rate of 2B bits per second in a channel of bandwidth B Hz. Such schemes are called as correlative coding or partial response signaling schemes. Correlative and partial response are synonomous. These schemes are designed based on the thought that the intersymbol interference introduced into the transmitted signal is known, therefore, its effect at the receiver can be compensated. Thus, using the correlative coding it is possible to achieve the theoretical maximum signaling rate of 2B bits per second for a channel bandwidth of B Hz. This was predicted by Nyquist and we can achieve it by using realizable and perturbation tolerant filter.

2.27.2. Duobinary Signaling

The basic duobinary signaling scheme has been shown in figure 2.39.

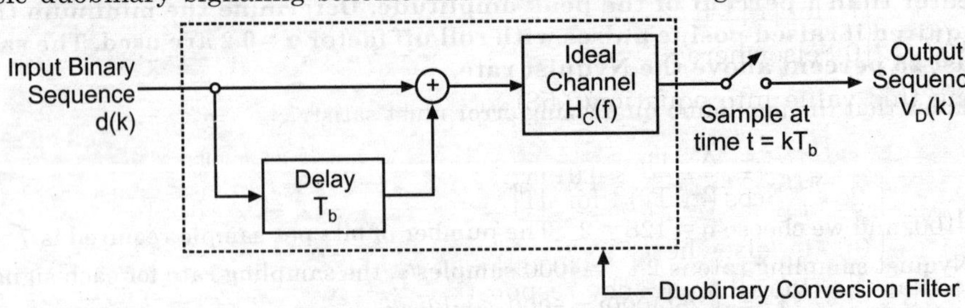

Fig. 2.39. Duobinary signaling scheme

The binary signal d(k) is first passed through a simple filter which consists of a single delay element. The present signal d(k) and its delayed version d(k – 1) are then added to get the duobinary signal at the output of the coder. This signal is mathematically expressed as:

$$V_D(k) = d(k) + d(k-1) \qquad ...(2.54)$$

One of the important effects of this transformation is that the sequence {d(k)} of the uncorrelated binary digits at the input is converted into a sequence $V_D(k)$ of correlated digits. This correlation between the adjacent transmitted pulses is equivalent to introducing intersymbol interference (ISI) in the transmitted signal. However this ISI is under designers control and this is the basis of the correlative coding.

2.27.3 Expression for the Transfer Function

Let us assume that the delay element is ideal. Therefore, its transfer function is given by

$$H_D(f) = e^{-j2\pi f T_b} \qquad ...(2.55)$$

Therefore, the transfer function of the filter containing the delay element and adder in figure 2.39 will be

$$1 + H_D(f) = 1 + e^{-j2\pi f T_b} \qquad ...(2.56)$$

Therefore, the overall transfer function of this basic filter cascaded with an ideal channel is given by,

$$H(f) = [1 + H_D(f)]\,[H_C(f)] = H_C(f)\,[1 + e^{-j2\pi fT_b}] \quad ...(2.57)$$

Substituting $1 = e^{+j\pi fT_b}.e^{-j\pi fT_b}$ and $e^{-j2\pi fT_b} = e^{-j\pi fT_b}.e^{-j\pi fT_b}$.

Therefore, $H(f) = H_C(f)\,[e^{j\pi fT_b}.e^{-j\pi fT_b} + e^{-j\pi fT_b}\,e^{-j\pi fT_b}]$

$$= 2\,H_C(f)\,e^{-j\pi fT_b}\,\frac{[e^{j\pi fT_b} + e^{-j\pi fT_b}]}{2}$$

But, $$\frac{e^{j\pi fT_b} + e^{-j\pi fT_b}}{2} = \cos(\pi fT_b)$$

Therefore, $$H(f) = 2H_C(f)e^{-j\pi fT_b}\cos(\pi fT_b) \quad ...(2.58)$$

This is required expression for the transfer function of the duobinary system.

2.27.4 Pulse Shape

For an ideal channel the relation between the channel bandwidth B and the bit rate R is given by,

Bandwidth $$B = \frac{1}{2}R$$

And the transfer function $H_C(f)$ is given by,

$$H_C(f) = \begin{cases} 1 & \text{for } |f| \le R/2 \\ 0 & \text{elsewhere} \end{cases} \quad ...(2.59)$$

Substituting this value into equation(2.58) for H(f), we get,

$$H(f) = \begin{cases} 2e^{-j\pi fT_b}\cos(\pi f\,T_b) & \text{for } |f| \le R/2 \\ 0 & \text{elsewhere} \end{cases} \quad ...(2.60)$$

This shows that the overall frequency response is of the form of a half cycle cosine function. The amplitude response and phase response are shown in figures 2.40(a) and (b) respectively.

The best thing about this frequency response is that it can be easily approximated.

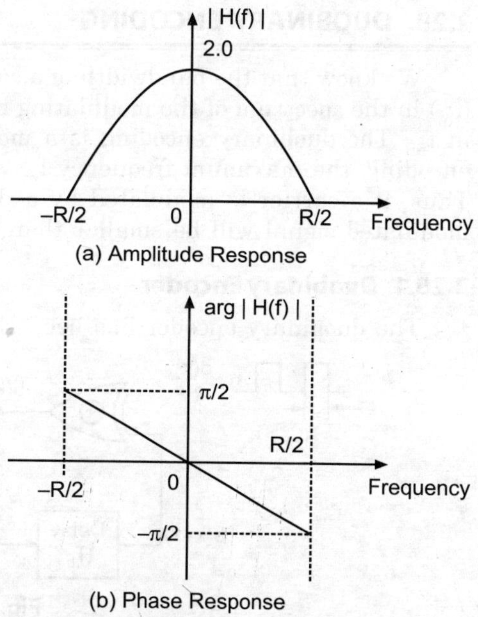

Fig. 2.40. Frequency response of duobinary conversion filter

2.27.5 Impulse Response

The impulse response of the duobinary filter h(t) can be obtained by taking the inverse Fourier transform of H(f).

Therefore, $h(t) = \text{IFT}[H(f)]$

or, $h(t) = \text{IFT}[2e^{-j\pi fT_b}\cos(\pi f\,T_b)] \quad |f| \le R/2$

We can imagine that the cosine function is being multiplied by a rectangular function extending from – R/2 to R/2.

Therefore, $h(t) = \text{IFT}[2e^{-j\pi fT_b}\cos(\pi f\,T_b).\text{rect}\,(f/R)]$

The corresponding value of impulse response is given by,

$$h(t) = \frac{\sin(\pi t/T_b)}{\pi t/T_b} + \frac{\sin[\pi(t-T_b)/T_b]}{\pi(t-T_b)/T_b}$$

or,
$$h(t) = \frac{\sin(\pi t/T_b)}{\pi t/T_b} - \frac{\sin(\pi t/T_b)}{\pi(t-T_b)/T_b} = \frac{T_b^2 \sin(\pi t/T_b)}{\pi t(T_b - t)} \quad ...(2.61)$$

So, the impulse response is a sinc function as depicted in figure 2.41.

We know that the bandwidth of a communication system depends on the maximum frequency (f_m) in the spectrum of the modulating (baseband) signal. The bandwidth will reduce with decrease in f_m. The **duobinary encoding** is a mode of encoding a binary bit stream. Due to this type of encoding the maxuimum frequency f_m will decrease as compared to that in the unencoded data. Thus, if a carrier is modulated by a duobinary encoded waveform., then the bandwidth of the modulated signal will be smaller than if the unencoded data were used.

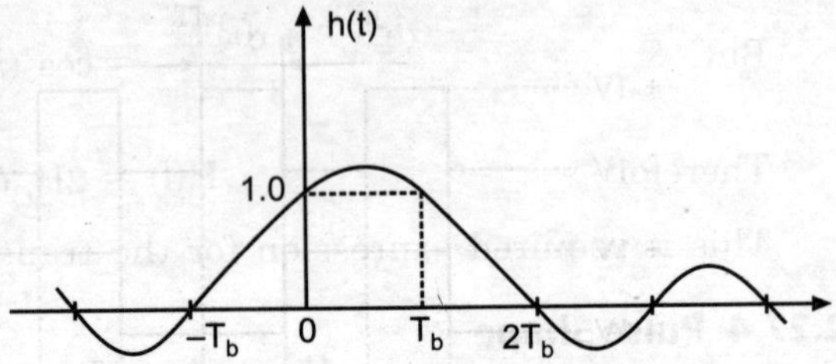

Fig. 2.41. Impulse response of a duobinary filter

2.28. DUOBINARY ENCODING

We know that the bandwidth of a communication system depends on the maximum frequency (f_m) in the spectrum of the modulating baseband signal. The bandwidth will reduce with decrease in f_m. The duobinary encoding is a mode of encoding a binary bit stream. Due to this type of encoding, the maximum frequency f_m will decrease as compared to that in the unencoded data. Thus, if a carrier is modulated by a duobinary encoded waveform, then the bandwidth of the modulated signal will be smaller than if the unencoded data were used.

2.28.1 Duobinary Encoder

The duobinary encoder has been shown in figure 2.42.

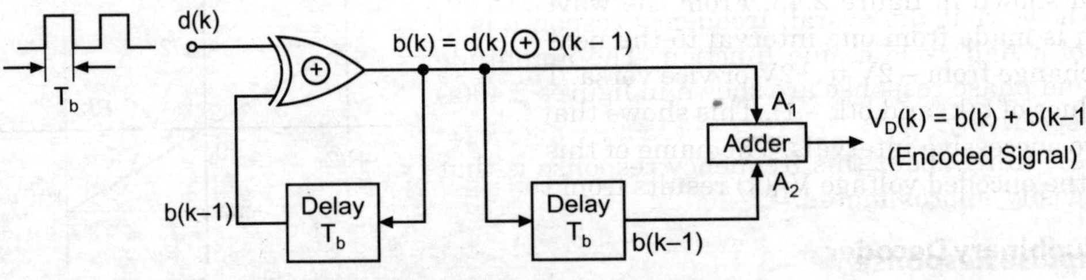

Fig. 2.42. Duobinary encoder

Working Operation

The input signal d(k) is the data bit stream with a bit duration equal to T_b. It changes its values between logic 0 and 1 levels. Let us assume that logic 0 level corresponds to – 1V and logic 1 level corresponds to +1V. The output of an EX-OR gate is feedback through a delay generating circuit to produce the signal b(k) at the output of the differential encoder as follows:

$$b(k) = d(k) \oplus b(k-1) \quad ...(2.62)$$

This is further delayed by one bit duration and b(k) and b(k – 1) are applied to an adder to produce the encoded signal $V_D(k)$.

$$V_D(k) = b(k) + b(k-1)$$

or, $$V_D(k) = [d(k) \oplus b(k-1)] + [d(k-1) \oplus b(k-2)] \quad ...(2.63)$$

The value of d(k) changes between + 1V and –1V. Hence, the value of b (k) at the output of the differential encoder also will change between +1V and –1V. Hence, b(k – 1) will change between $\pm 1V$. Therefore, the encoded signal $V_D(k)$ will change as follows:

$$V_D(k) = +2V \quad \text{if } b(k) = +1V \text{ and } b(k-1) = -1V$$
$$= 0V \quad \text{if } b(k) = b(k-1)$$
$$= -2V \quad \text{if } b(k) = -1V \text{ and } b(k-1) = +1V$$

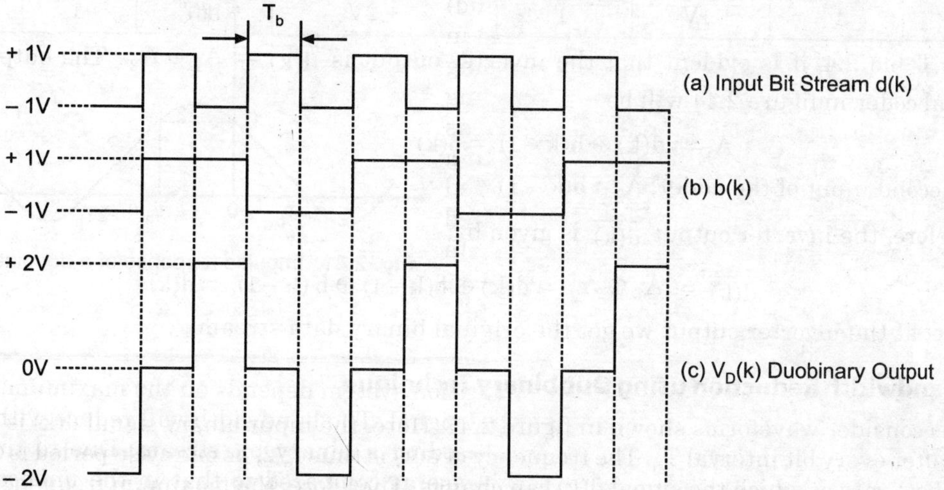

Fig. 2.43. Waveforms of d(k), b(k) and $V_D(k)$

Thus, the value of $V_D(k)$ in any interval k depends on both b(k) and b(k – 1). Therefore, we say that there is a correlation between the values of $V_D(k)$ in any two successive intervals. Therefore, the duobinary coding is also known as **correlative coding**. The waveforms for d(k), b(k) and $V_D(k)$ have been shown in figure 2.43. From the waveforms of figure 2.43, it is evident that when the transition is made from one interval to the next interval, it is not possible for the encoded signal $V_D(k)$ to change from – 2V to +2V or vice versa. Thus, value of $V_D(k)$ in any interval k is dependent on the values of b(k) and b(k – 1). This shows that there is a correlation between the values of $V_D(k)$ in any two successive intervals. The name of this coding system is appropriate because, in each bit interval, the encoded voltage $V_D(k)$ results from the combination of two bits.

2.28.2 Duobinary Decoder

The block diagram of a duobinary decoder is as shown in figure 2.44.

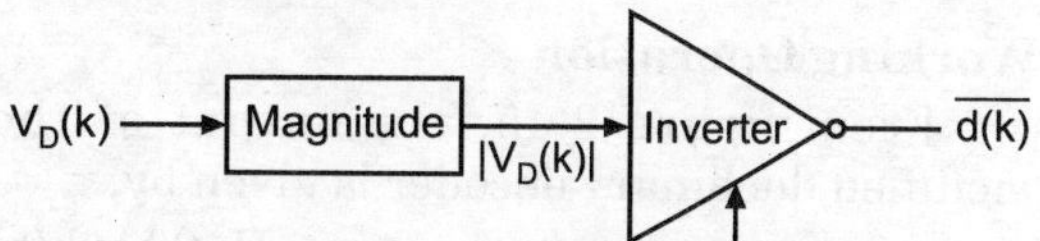

Fig. 2.44. Duobinary decoder

Working Operation

The duobinary decoder of figure 2.44 consists of a device, which provides the magnitude (absolute value) of its input. This means that if the input is – 2V or + 2V, output of the magnitude block will be 2V. The output of magnitude block is connected at the input of a logical inverter. For this inverter, we assume that logic 1 is +1V or greater and logic 0 is 0 volts. Table 2.4 shows that the decoded signal $\overline{d(k)}$ and the original data d(k) are exactly identical.

Table 2.4.

Adder input 1 (A_1)		Adder input A_2					
Voltage	**Logic**	**Voltage**	**Logic**	**Input voltage**	**Voltage**	**Logic**	**Logic**
– 1V	0	– 1V	0	– 1V	2V	1	0
– 1V	0	+ 1V	1	0V	0V	0	1
+ 1V	1	– 1V	0	0V	0V	0	1
+ 1V	1	+ 1V	1	+ 2V	2V	1	0

From Table 2.4, it is evident that the inverter output is $\overline{d(k)} = A_1 \oplus A_2$. The output of the differential coder in figure 2.44 will be

$$A_1 = d(k) \oplus b(k-1) = b(k) \qquad \text{...(2.64)}$$

The second input of the adder, $A_2 = b(k-1)$.

Therefore, the inverter output $\overline{d(k)}$ is given by,

$$\overline{d(k)} = A_1 \oplus A_2 = d(k) \oplus b(k-1) \oplus b(k-1) = d(k) \qquad \text{...(2.65)}$$

Hence, at the inverter output we get the original binary data stream.

2.28.3 Bandwidth Reduction using Duobinary Technique

Let us consider waveforms shown in figure 2.43. Here, the input binary signal d(k) is changing its value after every bit interval T_b. The frequency of d(k) is thus $f_b/2$, as one cycle period is 2 T_b. This is the highest rate at which the signal d(k) can change. Therefore, the maximum frequency f_m in its spectrum will have its highest possible value and hence the bandwidth will be maximum. Now, let us consider figure 2.43 which shows the duobinary encoded signal. The frequency of $V_D(k)$ is $f_b/4$ as the period of its one full cycle is 4 T_b. Thus, frequency of $V_D(k)$ is half that of b(k). Therefore, the maximum frequency f_m in its spectrum will reduce as compared to the encoded b(k) and the bandwidth requirement will reduce.

2.28.4 Modified Duobinary Encoder

The modified duobinary encoder is as shown in figure 2.45. In this encoder, the correlation between binary digits takes place over duration of 2 T_b instead of T_b.

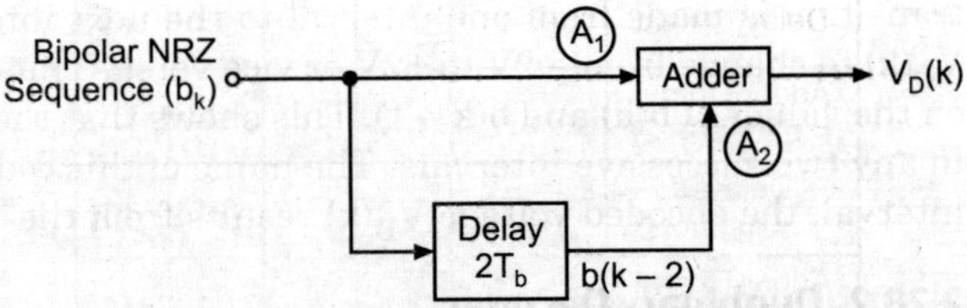

Fig. 2.45. Modified duobinary encoder

Working Operation

From figure 2.45, the output of the modified duobinary encoder is given by,

$$V_D(k) = b(k) - b(k-2) \qquad \text{...(2.66)}$$

In equation (2.66), the bipolar NRZ sequnce b(k) will have values ± 1. The output sequence $V_D(k)$ is a correlated sequence. As b(k) is a bipolar NRZ sequence. Equation (2.66) gives three possible values of $V_D(k)$ corresponding to various combinations of b(k) and b(k – 2). If $\hat{b}(k)$ represents the estimate of b(k) then its value is given by,

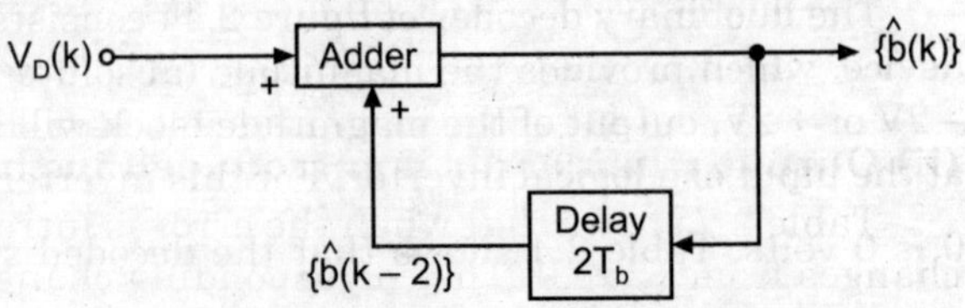

Fig. 2.46. Modified duobinary decoder

$$\hat{b}(k) = V_D(k) + \hat{b}(k-2) \qquad \text{...(2.67)}$$

This equation can be implemented by using the decoder of the modifid duobinary system as shown in figure 2.46.

Equation (2.67) indicates that if there is an error in $V_D(k)$, then this error will propagate in other values of $\hat{b}(k)$.

EXAMPLE 2.15. The binary data 011100101 is applied to the input of a modified duobinary system.

(i) Construct the modified duobinary coder output and corresponding receiver output without precoding.

(ii) Suppose that due to error during transmission the level produced by the third digit is reduced to zero. Construct the new receiver output.

Solution: The block diagram of the modified duobinary encoder and decoder have been shown in figure 2.47.

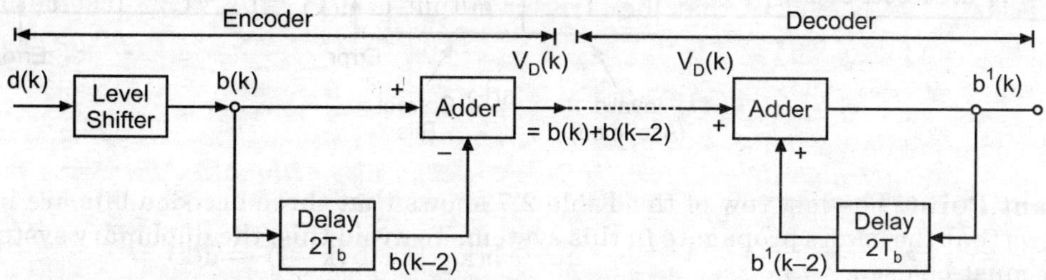

Fig. 2.47. Modified duobinary encoder and decoder

(i) Output without any error in the input.

The output is obtained as shown in table 2.6.

Table 2.6.

Assumed

Binary Input	d(k)			0	1	1	1	0	0	1	0	1
Level Shifter Output	b(k)	(+1)	(+1)	–1	+1	+1	+1	–1	–1	+1	–1	+1
Delay Output	b(k–2)			+1	+1	–1	+1	+1	+1	–1	–1	+1
Adder Output = b(k)–b(k–2)	$V_D(k)$			–2	0	+2	0	–2	–2	+2	0	0
Delayed Output	$b^1(k–2)$			+1	+1	–1	+1	+1	+1	–1	–1	+1
Adder Output $V_D(k)$ + b(k–2)	$b^1(k)$	+1	+1	–1	+1	+1	+1	–1	–1	+1	–1	+1
Binary Sequence	—	1	1	0	1	1	1	0	0	1	0	1

Encoder (rows 1–4); Decoder (rows 5–6)

Invalid (Assumed) — Valid Output

(ii) Output when error is introduced in the third digit.

Table 2.7 shows that when the error is introduced, the third digit of the received signal $V_D(k)$ changes from +2 to 0. The corresponding changes in the decoder are as shown in table 2.7.

Table 2.7.

Without error →	Transmitted Signal	b(k–2)			–2	0	(2)	0	–2	–2	+2	0	0
With error in the third digit →	Transmitted Signal	$V_D(k)$			–2	0	(0)	0	–2	–2	+2	0	0
	Delayed Output	$b^1(k-2)$			+1	+1	–1	+1	–1	+1	–3	–1	–1
	Adder Output $V_D(k) + b^1(k-2)$	$b^1(k)$	+1	+1	–1	+1	–1	+1	–3	–1	–1	–1	–1
	Binary Sequence at decoder Output	—	1	1	0	1	(0)	1	0	0	0	(0)	(0)

Invalid (Assumed) — Correct Bits — Error — Error Error

Important Point: The last row of the Table 2.7 shows that three decoded bits are in error. This shows that the errors propagate in this system. To avoid this, the duobinary system with precoder must be used.

2.28.5 Duobinary-Encoder with Precoder (Differential Encoder)

The modified duobinary system discussed in the previous section has an important limitation, which is that if there is an error in the received signal then it will propagate in the other values of $\hat{b}(k)$. The duobinary system discussed in this section is free from such a problem. The block diagram of duobinary encoder with precoder has been shown in figure 2.48.

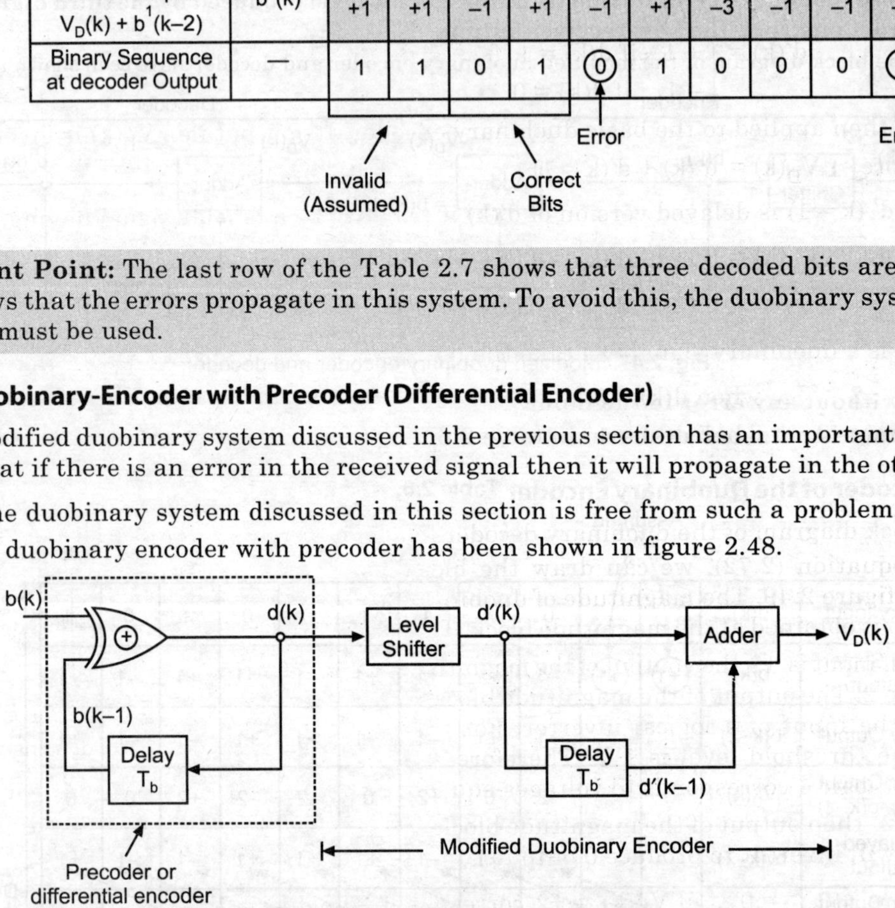

Fig. 2.48. Block diagram of duobinary encoder with precoder

The precoder is added to the modified duobinary encoder to obtain the duobinary encoder with precoder as shown in figure 2.48. The precoder consists of an Ex-OR gate along with a delay unit that introduces a delay of one bit duration T_b.

The precoder output d(k) is given by,

$$d(k) = b(k) \oplus d(k-1) \qquad ...(2.68)$$

where, b(k) is the unipolar stream of binary digits.

Equation (2.68) shows that precoder output is equal to zero, if both its inputs are alike (both 0 or 1) and it is 1 if any one of the input is 1.

Table 2.8. Operation of precoder

b(k) input	d(k – 1)	Precoder output d(k)	d′(k)
0	0	0	– 1
0	1	1	+ 1
1	0	1	+ 1
1	1	0	– 1

The precoder output is then applied to a level shifter which converts it into a bipolar signal having values + 1 or – 1 follows:

$$d'(k) = \begin{cases} +1 & \text{if } d(k) = 1 \\ -1 & \text{if } d(k) = 0 \end{cases} \quad ...(2.69)$$

d′(k) is then applied to the basic duobinary encoder. Its output i.e., $V_D(k)$ is given by,

Therefore, $$V_D(k) = d'(k) + d'(k-1) \quad ...(2.70)$$

As the d′ (k – 1) is delayed version of d′(k) it will also be a bipolar signal having value ± 1.

Therefore, $$V_D(k) = \begin{cases} \pm 2 & \text{if } d'(k) = d'(k-1) \\ 0 & \text{if } d'(k) \neq d'(k-1) \end{cases} \quad ...(2.71)$$

As this is a duobinary signal, we take the help of figure 2.48 to get

$V_D(k) = \pm 2$ when d(k) = 0

and $V_D(k) = 0$ when d(k) = 1 ...(2.72)

2.28.6 Decoder of the Duobinary Encoder

The block diagram of the duobinary decoder has been shown in figure 2.49.

From equation (2.72), we can draw the block diagram of figure 2.49. The magnitude of duobinary signal $V_D(k)$ is obtained at the magnitude block. This means that if input is ± 2 then output of the magnitude block will be 2. The output of the magnitude block is applied to the input of a logical inverter. For this inverter, the threshold level is 1 V. Therefore, we assume that logic 1 corresponds to voltages equal to or above + 1 V and logic 0 is 0 Volts. Thus, if $V_D(k) = \pm 2$, then output of the magnitude block will be 2, therefore, the inverter will treat it as logic 1 and will invert it to produce 0 output.

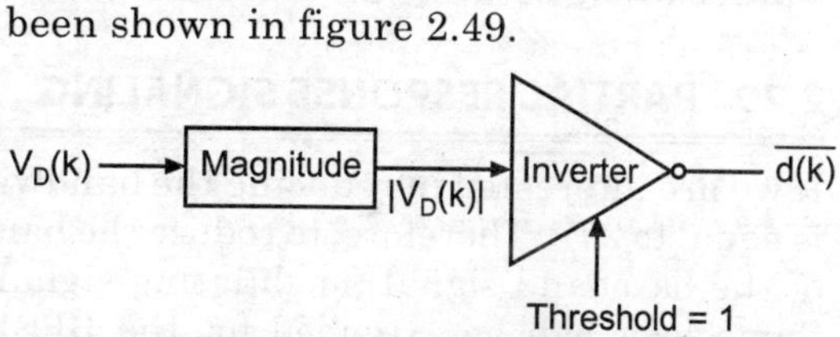

Fig. 2.49. Block diagram of duobinary decoder

Therefore, $\overline{d(k)} = 0$ if $V_D(k) = \pm 2$ corresponding to d(k) = 0

Alos, if $V_D(k) = 0$, then, inverter input is 0 and its output is 1.

Hence, $\overline{d(k)} = 1$ if $V_D(k) = 0$ corresponding to d(k) = 1

The receiver output is, therefore, given by,

output $\overline{d(k)} = 0$ if $V_D(k) = \pm 2$ which corresponds to d(k) = 0

output $\overline{d(k)} = 1$ if $V_D(k) = 0$ which corresponds to d(k) = 1.

Thus, the decoder recovers the original signal.

To prove that the error does not propagate

Let us consider table 2.9 which summarises the operation of the duobinary encoder with precoder. Table 2.9 also shows the encircled bits which represent the bits corresponding to the second bit in error.

Table 2.9. Summary of operation of the duobinary encoder, decoder with precoder

Data Bits at the Input	b(k)		0	Error 1 (0)	(0)	(0)	(1)	
Delayed Precoder Output	d(k–1)		0	0	1 (0)	1 (0)	1 (0)	Encoder
Precoder Output	d(k)	0	0	1 (0)	1 (0)	1 (0)	0 (1)	Encoder
Level Shifter Output	d′(k)		–1	1 (–1)	1 (–1)	1 (–1)	–1 (1)	Encoder
Delayed Shifter Output	d′(k–1)			–1	1 (–1)	1 (–1)	1 (–1)	Encoder
Adder Output	$V_D(k)$			0 (–2)	2 (–2)	2 (–2)	0 (0)	Encoder
Output of Magnitude Block	$\lvert V_D(k)\rvert$			0 (2)	2 (2)	2 (2)	0 (0)	Decoder
Inverter Output (Decoder Output)	$\overline{d(k)}$			1 (0)	0 (0)	0 (0)	1 (1)	Decoder

Only one error in the decoded output
Note that the error has not propagated

Important Point: The encircled bits of decoder output of Table 2.9 shows that there is only one error corresponding to the second bit only. The other bits have not changed. Thus it is proved that with the precoder included, the errors do not propagate.

2.29. PARTIAL RESPONSE SIGNALING

If we are interested in reducing the bandwidth of a modulated signal say AM, then its bandwidth is equal to 2 f_m. Therefore, to reduce the bandwidth of AM signal, we have to reduce the bandwidth of the baseband signal (modulating signal).

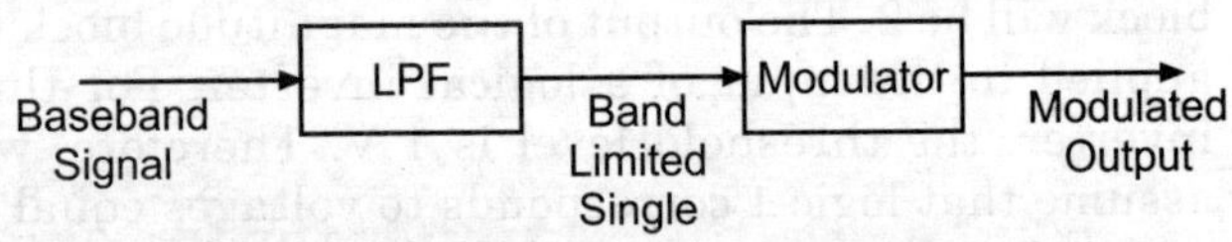

Fig. 2.50. Use of LPF to reduce the bandwidth

Same logic can be extended for the BPSK, QPSK or MSK signals, i.e., by reducing the bandwidth of their baseband signals it is possible to control the bandwidth. The question is how to reduce the bandwidth of the baseband signal? The answer is, by passing the baseband signal through a low pass filter.

But when we bandlimit a signal, it gets extended in the time domain. This will result in the intersymbol interference between the neighboring bits. Thus, the unfiltered baseband signal causes the interchannel interference whereas the filtered baseband leads to intersymbol interference. This is a great problem in reducing the bandwidth of the modulated signal. The solution to this problem is as follows: We know that if we apply an impulse of unit amplitude to a **brickwall filter** i.e., an ideal low pass filter with a transfer function H(f) given by,

$$H(f) = \begin{cases} 1 & \text{for } -f_c \le f \le f_c \\ 0 & \text{elsewhere} \end{cases} \qquad ...(2.73)$$

As shown in figure 2.51(a), then, at the response of this filter is given by,

$$h(t) = 2f_c \frac{\sin(2\pi f_c t)}{(2\pi f_c t)} \qquad ...(2.74)$$

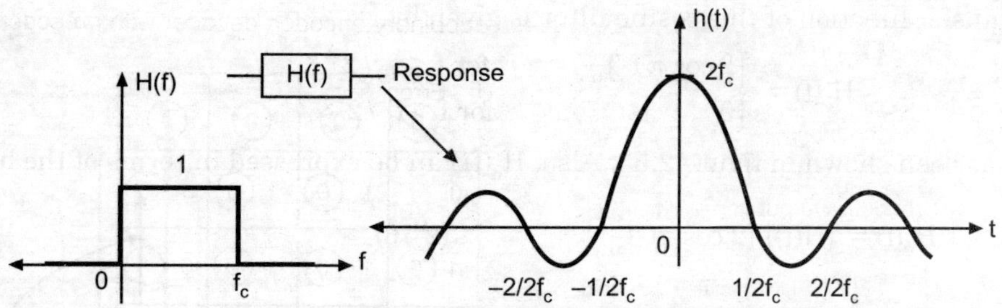

(a) Transfer function of a brickwall filter (b) Impulse response of the brickwall filter

Fig. 2.51.

The impulse response shown in figure 2.51 (b) is a sinc function that passes through zero at $t = \pm(1/2\,f_c), +(2/2\,f_c) +$ ... etc. and it has a maximum at $t = 0$. Now, let us assume that the unit amplitude impulses are generated at instants $k/2f_c$ (k is an integer) as shown in figure 2.52 (a). Then, the response of the brickwall filter has been shown in figure 2.52(b). This shows that at the instants $\pm 1/2\,f_c, \pm 2/2f_c + \ldots$, the responses due to all the other individual impulses pass through zero except for the response corresponding to the impulse at that particular instant.

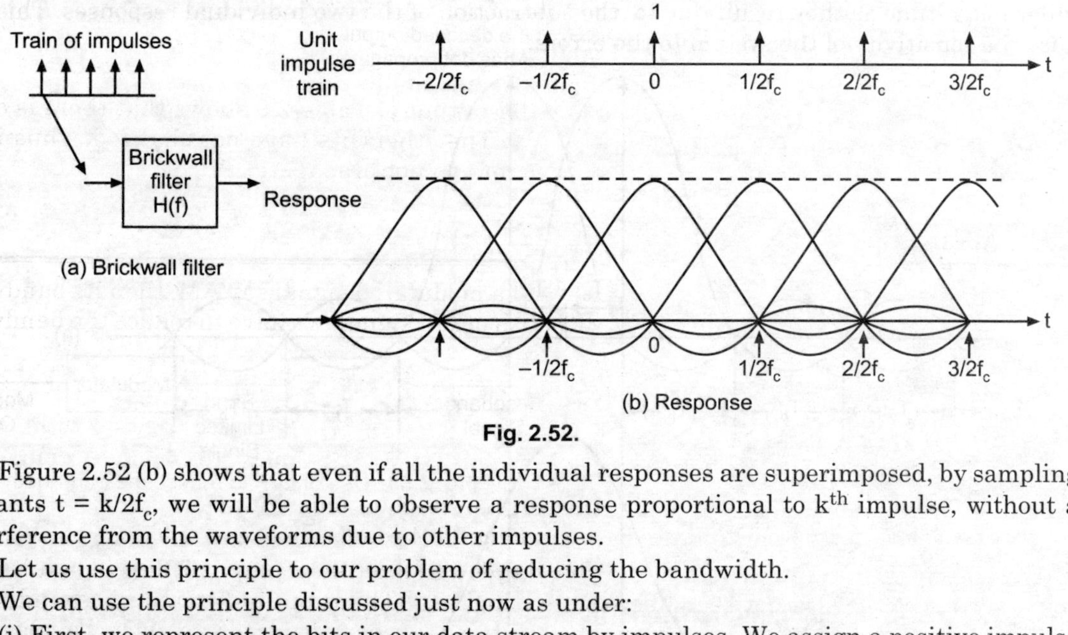

(a) Brickwall filter (b) Response

Fig. 2.52.

Figure 2.52 (b) shows that even if all the individual responses are superimposed, by sampling at instants $t = k/2f_c$, we will be able to observe a response proportional to k^{th} impulse, without any interference from the waveforms due to other impulses.

Let us use this principle to our problem of reducing the bandwidth.

We can use the principle discussed just now as under:

(i) First, we represent the bits in our data stream by impulses. We assign a positive impulse to a logic 1 and negative impulse to a logic 0 and let the impulses be separated from each other by T_b seconds.

(ii) Now, we pass this impulse train through a brickwall filter, with a cut off frequency:

$$f_c = 1/2\ Tb = f_b 2$$

(iii) Due to this bandlimiting process, we are able to reduce the bandwidth without introducing any intersymbol interference, due to sampling at instants $t = k/2\,f_c = k/f_b$.

Use of cosine filter

In practice, it is not possible to implement the brickwall filter. Hence, an alternative to this filter is used. It is a **cosine filter**.

The transfer function of this cosine filter is given by,

$$H_c(f) = \begin{cases} 2\cos \pi f T_b & \text{for } f \le f_b/2 \\ 0 & \text{for } f > f_b/2 \end{cases} \quad ...(2.75)$$

This has been shown in figure 2.53. Also, $H_c(f)$ can be expressed in terms of the brickwall filter as under:

$$H_c(f) = [H(f)]\,[2\cos \pi f T_b] \quad ...(2.76)$$

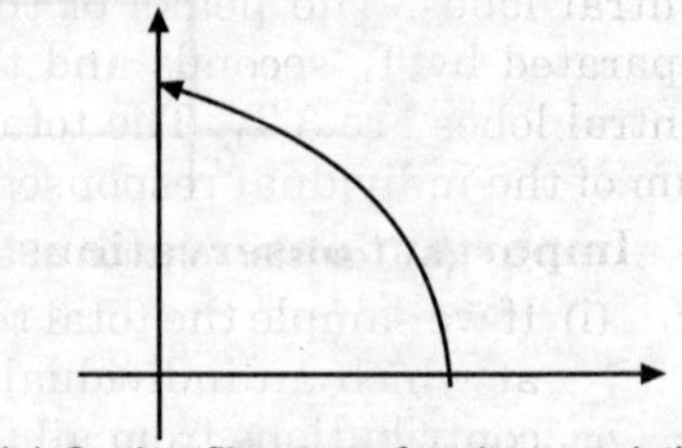

(a) Cosine filter transfer characteristics

Fig. 2.53.

Substituting $\cos \pi f T_b = \dfrac{e^{j\pi f T_b} + e^{-j\pi f T_b}}{2}$ we get

$$H_c(f) = [H(f)][e^{j\pi f T_b} + e^{-j\pi f T_b}] \quad ...(2.77)$$

Hence, $$H(c) = H(f)\,e^{j\pi f T_b} + H(f)\,e^{-j\pi T_b} \quad ...(2.78)$$

Taking the IFT (using the time shifting property of FT) of equation (2.78), we get

$$h_c(t) = \text{IFT}\,[H_c(f)] = h(t + T_b/2) + h(t - T_b/2) \quad ...(2.79)$$

These two parts have been shown in figure 2.54(a) and their resultant has been shown in figure 2.54 (b). The amplitudes of the sideslobes in the resultant response of figure 2.54(b) are of much smaller magnitude as they result due to the subtraction of the two individual responses. This will reduce the sensitivity of the system to the errors.

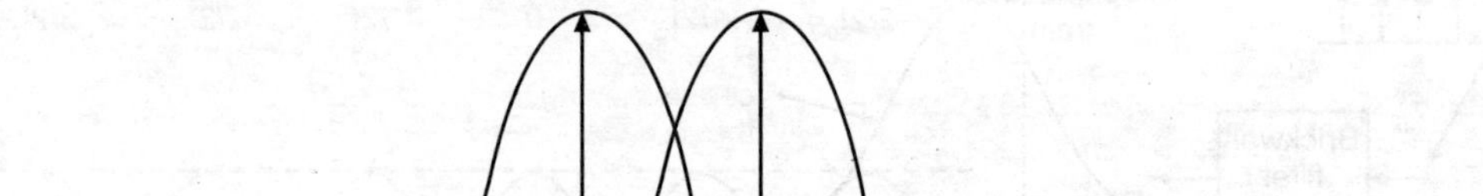

(a) Impulse response of cosine filter (b) Overall impulse response

Fig. 2.54.

2.29.1. Use of Cosine Filter

Now, let us consider that a bit stream is to be applied to the cosine filter. Hence, we shall convert it first into a stream of positive and negative impulses and then applied to the cosine filter.

The response of such a cosine filter is as shown in figure 2.55. This filter response has been drawn individually for five successive positive impulses. Remember that for the sake of simplicity we have drawn only the main lobes of central lobes. The peaks of the responses are separated by T_b seconds and the widths of the central lobes are 3 T_b. The total response is the sum of the individual responses.

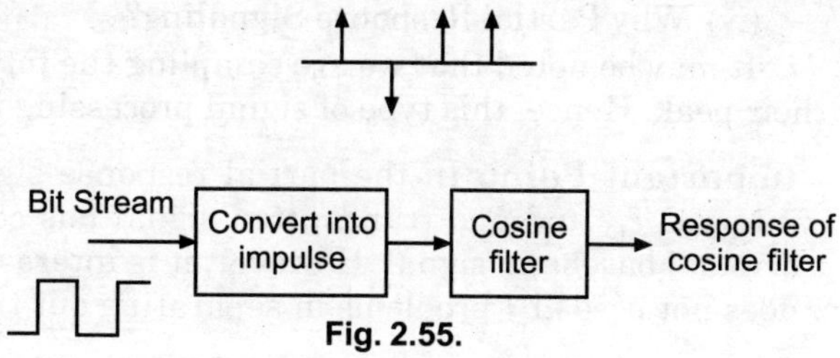

Fig. 2.55.

Important observations

(i) If we sample the total response at the time instants A, B, C ... etc in figure 2.56, the instants at which an individual response is at its maximum, then the resulting sample will have contributions from all the individual responses.

(ii) It is not possible to sample the total response at an instant of time where the sample corresponds to only a single individual response. This means that the sample of the total response will always be corresponding to all the individual responses.

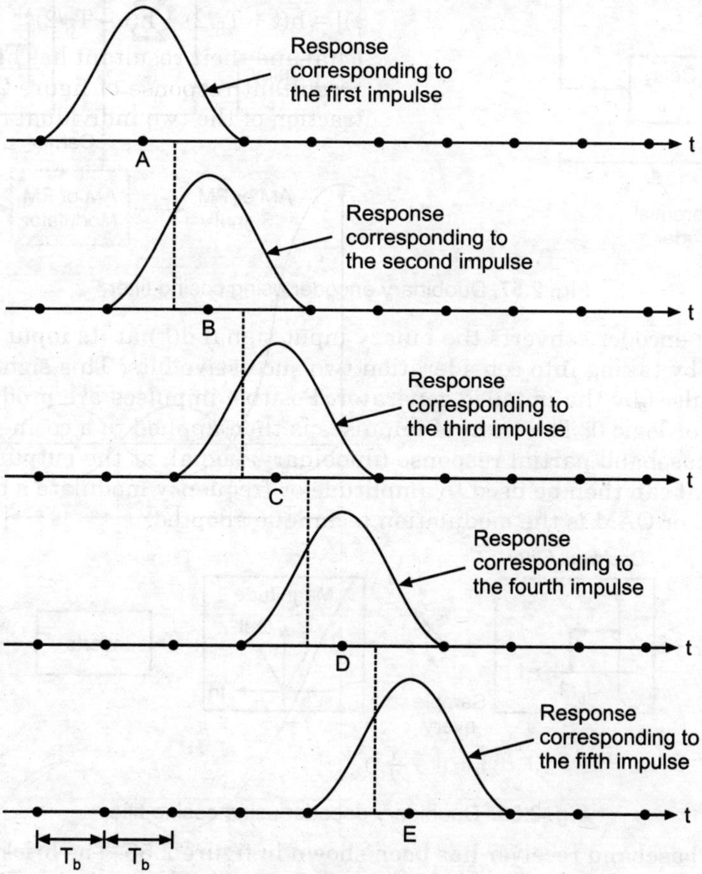

Fig. 2.56. Filter response to five separate impulses

(iii) If the total response is sampled at the instants which are at the centre of the peak points A, B, C ... i.e., at instants t = (2k – 1) $T_b/2$. Then the corresponding sample value will have equal contributions from only two individual responses. These sampling instants have been indicated by the dotted vertical lines in figure 2.56.

(iv) Why Partial Response Signaling?

It may be noted that we are sampling the total signal when the individual responses are not at their peak. Hence, this type of signal processing is known as the **partial response signaling.**

Important Point: In the partial response signaling, a signal is transmitted during each bit interval T_b, and and transmitted signal has contributions from the two successive bits of the original baseband signal. However, it is interesting to note that the superposition of waveforms does not pose any problems in separating out the original baseband signals.

2.29.2 Amplitude Modulation of the Partial Response Signal

The block diagram of a duobinary encoder using the cosine filter has been shown in figure 2.57.

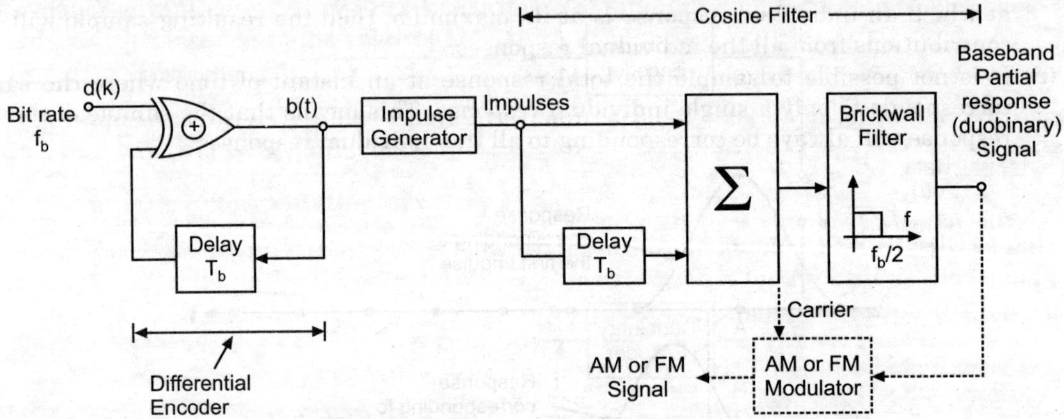

Fig. 2.57. Duobinary encoder using cosind filter

The differential encoder converts the binary input signal d(k) at its input into a differentially encoded signal b(t) by taking into consideration two successive bits. This signal is then converted into a train of impulses by the impulse generator. Positive impulses are produced for logic 1 and negative impulses for logic 0. The train of impulses is then applied to a cosine filter. As discussed earlier we get the baseband partial response (duobinary) signal, at the output of the cosine filter. This baseband signal can then be used to amplitude or frequency modulate a carrier. When AM is used either DSB/SC or QAM is the modulation technique adopted.

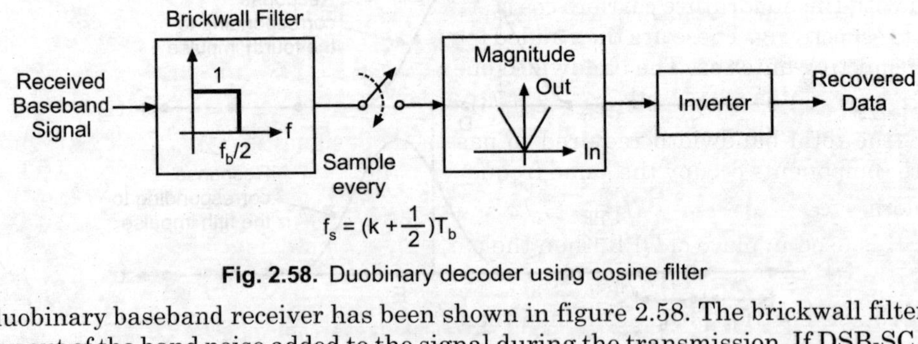

Fig. 2.58. Duobinary decoder using cosine filter

The duobinary baseband receiver has been shown in figure 2.58. The brickwall filter is used to remove any out of the band noise added to the signal during the transmission. If DSB-SC is adopted, then the duobinary baseband signal V_T in figure 2.58 is multiplied by the carrier $\sqrt{2}$ cos ω_0 t. Therefore, the output signal of the modulator is given by,

$$V_{DSB}(t) = \sqrt{2}\, V_T \cos \omega_0 t \qquad ...(2.80)$$

The bandwidth required to transmit this DSBSC signal is f_b which is twice the bandwidth of the baseband duobinary signal (In dSBSC BW = 2 f_m, here $f_m = f_b/2$)

Therefore, $$BW_{DSB} = 2 \times (f_b/2) = f_b \quad ...(2.81)$$

2.29.3 Quadrature Partial Response (QPR) Encoder

If the duobinary baseband partial response signal V_T is to be used to amplitude modulate two carriers in quadrature then the encoder shown in figure 2.59 is used and it is knows as the **quadrature partial response** (QPR) encoder.

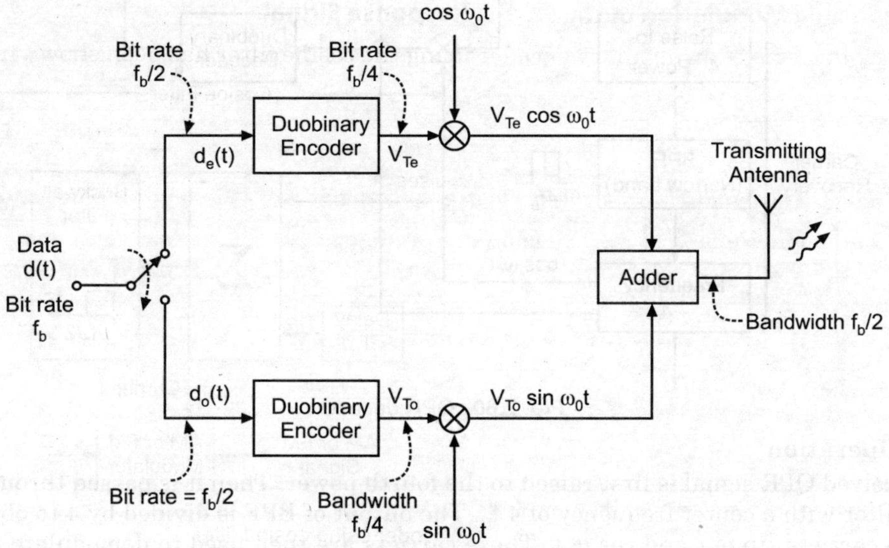

Fig. 2.59. QPR Encoder

Working Operation

The data d(t) at a bit rate f_b is applied at the input of the encoder. This data bit stream is first separated into even and odd bit streams $d_e(t)$ and $d_o(t)$. The bit rate of each of these bit streams is $f_b/2$. The odd and even bit streams are then applied to duobinary encoders which produce the baseband duobinary signals V_{Te} and V_{To} corresponding to the even and odd bit streams. It may be noted that the bit rate at the outputs of duobinary encoder is $f_b/4$. The even and odd duobinary signal are multiplied with the quadrature carriers cos ω_0 t and sin ω_0 t to produce the signals V_{Te} cos ω_0 t and V_{To} sin ω_0 t respectively. These are then added together to generate the QPR signal. It is transmitted by the transmitting antenna. The bandwidth of each QPR modulated signal is,

$$BW_{QPR} = 2\,(f_b/4) = f_b/2 \quad ...(2.82)$$

Hence, the total bandwidth required to pass a QPR signal is $BW_{QPR} = f_b/2$ because the two quadrature components occupy the same frequency band.

Therefore, $$BW_{QPR} = f_b/2 \quad ...(2.83)$$

If QPSK is used in place of QPR, then the bandwidth would have been,

$$BW_{QPSK} = f_b \quad ...(2.84)$$

But, if 16 QAM or 16 PSK is used, then the bandwidth would be

$$BW_{16\ QAM} = BW_{16\ PSK} = f_b/2 \quad ...(2.85)$$

Thus, the bandwidth of QPR is same as that of 16 QAM or 16 PSK.. The biggest advantage of using the QPR signal is that it does not display any (or in practice very small) sidelobes. Therefore, it is to be used when the spectrum width is a major problem. A disadvantage of QPR is that the envelope of the transmitted signal does not remain constant but it varies with time.

2.29.4 Quadrature Partial Response (QPR) Decoder

The block diagram of the QPR decoder has been shown in figure 2.60.

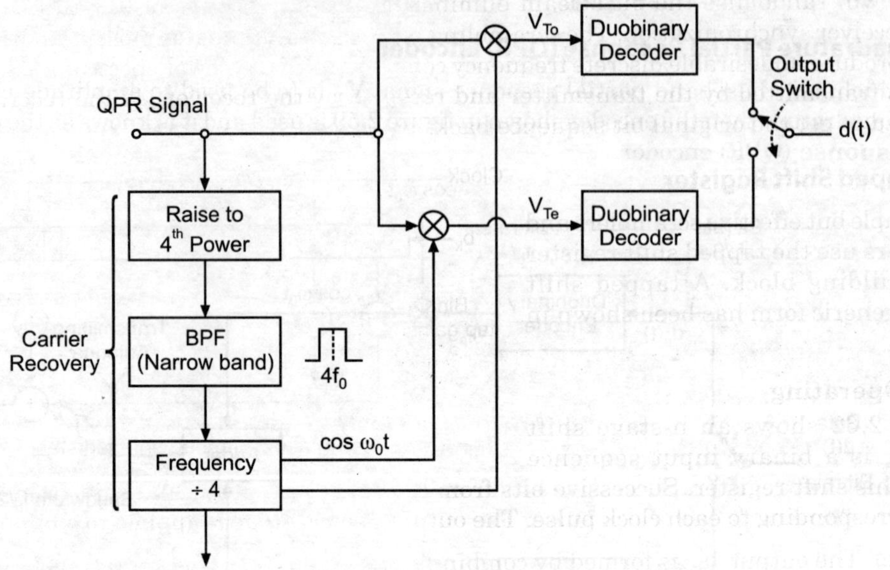

Fig. 2.60. QPR decoder

Working Operation

The received QPR signal is first raised to the fourth power. Then it is passed through a narrow bandpass filter with a center frequency of 4 f_0. The output of BPF is divided by 4 to obtain the two quadrature carriers sin $\omega_0 t$ and cos $\omega_0 t$. These carriers are then used to demodulate the received QPR signal and obtain the two baseband duobinary signals V_{Te} and V_{To}. Then from these signals, the duobinary decoder will generate the even and odd bit streams $d_o(t)$ and $d_e(t)$. The switch on the output side will then interleave $d_o(t)$ and $d_e(t)$ to generate the original data stream d(t). The duobinary decoders of figure 2.60 operate at a rate $f_b/2$ rather than f_b.

2.29.5 Partial Response Signaling using Transversal Filter

The duobinary uses successive bits whereas the modified duobinary system combines two successive bits. The partial response signaling system using transversal filter combines upto N successive bits. Such a system is shown in figure 2.61.

W_0, W_1, ... W_{N-1} are the tap weights for the tapped delay line used as transversal filter in figure 2.61. By changing the values of these tap gain we can change the type of partial response scheme. The spectral shape of the output is dependent on the values of the tap weights.

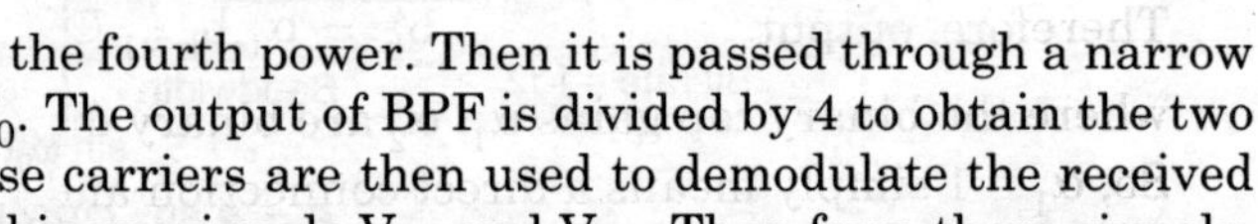

Fig. 2.61. Partial response signaling using tapped delay line or transversal filter

2.30 SCRAMBLERS

In communication applications, sometimes, it is essential to transmit the message in such a form that only the desired receiver should be able to understand it. For this it is necessary to perform coding. Scrambling is a coding operation applied to the message at the transmitter. The process of scrambling will randomise the bit stream eliminating the long strings of 0s and 1s which might affect the receiver synchronization. The scrambling process also eliminates most periodic bit patterns which can produce undesirable discrete frequency components in the power spectrum. The scrambling sequence is transmitted by the transmitter and received by the receiver. At the receiver end, it is unscrambled to get the original bit sequence back.

2.30.1 Tapped Shift Register

It is simple but effective scramblers and unscramblers use the tapped shift register as basic building block. A tapped shift register of generic form has been shown in figure 2.62.

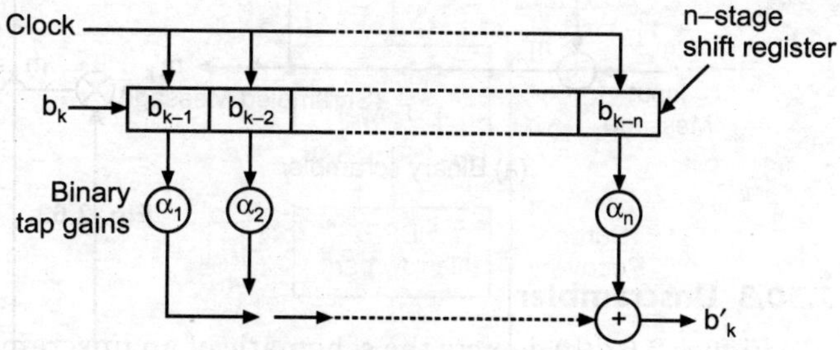

Fig. 2.62. Tapped shift register

Working Operating

Figure 2.62 shows an n-stage shift register. b_k is a binary input sequence applied to this shift register. Successive bits from b_k enter the register and shift from one stage to the next corresponding to each clock pulse. The output of each stage is applied to a binary tap again (α_1, α_2 ... etc). The output b'_k is formed by combining the bits in the register through a set of binary tap gains and modulo 2 adders.

Therefore, output $$b'_k = \alpha_1 b_{k-1} \oplus \alpha_2 b_{k-2} \oplus ... \oplus \alpha_n b_{k-n} \quad ...(2.86)$$

where the binary tap gains α_1, α_2 are binary digits i.e., 0 or 1.

So, $\alpha_1 = 1$ simply means a direct connection and $\alpha_1 = 0$ means no connection.

The symbol $\oplus$ represents a modulo-2 addition. The properties of modulo-2 addition are as under:

(i) $$b_1 \oplus b_2 = \begin{cases} 0 & \text{for } b_1 = b_2 \\ 1 & \text{for } b_1 \neq b_2 \end{cases} \quad ...(2.87)$$

(ii) $$b_1 \oplus b_2 \oplus b_3 = (b_1 \oplus b_2) \oplus b_3 = b_1 \oplus (b_2 \oplus b_3) \quad ...(2.88)$$

where, b_1, b_2 and b_3 are arbitary binary digits. We can implement the modulo -2 addition wtih exclusive - OR gates.

2.30.2 Scrambler

Figure 2.63 (a) shows the schematic of a scrambler. It uses a 4-stage shift register with tap gains $\alpha_1 = \alpha_2 = 0$ and $\alpha_3 = \alpha_4 = 1$. The clock line has not been shown for our convenience but it is very much present. m_k is the binary message sequence at the input to the scrambler. It is added to the register output, m''_k to obtain the scrambled message m'_k. The scrambled message m'_k is also fedback to the shift register input. Looking at figure 2.63 (a), we can write that,

$$m''_k = m'_{k-3} \oplus m'_{k-4} \quad ...(2.89)$$

and $$m'_k = m_k \oplus m''_k \quad ...(2.90)$$

Here, m'_k represents a scrambled version of m_k. The scrambled version is completely different from the original message m_k and, therefore, the message m_k cannot be recovered unless we have a proper unscrambled at the receiver. The scrambled signal m'_k is then transmitted over a suitable communication medium.

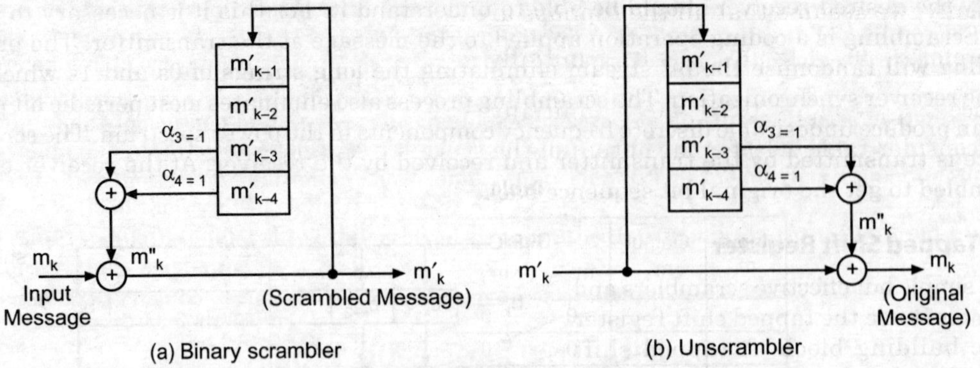

Fig. 2.63.

2.30.3 Unscrambler

Figure 2.63 (b) depicts the schematic of an unscrambler. It uses a 4-stage shift register with tap gains $\alpha_1 = \alpha_2 = 0.$ and $\alpha_3 = \alpha_4 = 1$. The unscrambler is used to recover the original message signal m_k from the scrambled signal m'_k. The unscrambler structure is exactly reverse as compared to the structure of the scrambler. It reproduces the original message signal m_k at its output. This is due to:

$$m'_k \oplus m''_k = (m_k \oplus m''_k) \oplus m''_k \qquad \text{since, } m'_k = m_k \oplus m''_k$$

$$m_k' \oplus m_k'' = m_k \oplus (m''_k \oplus m''_k)$$

or $$m_k' \oplus m_k'' = m_k \oplus 0 \qquad \text{since, } m''_k \oplus m''_k = 0$$

or $$m_k' \oplus m_k'' = m_k$$

Thus, by adding (modulo-2 addition), the scrambled signal m'_k with m''_k it is possible to obtain the original signal m_k. It may be noted that due to the identical structures of shift registers and same input m'_k for scrambler a well as unscrambler, we will get the same sequence m''_k for the scrambler as well as unscrambler.

Disadvantages

The most serious problem in the unscrambler is that one erroneous bit in m'_k will cause several output bit errors. And these errors will keep propagating. The error propagation will stop only when the unscrambler register is full of correct bits. There can be some specific input message sequenes (m_k) to the scrambler which will result in long strings of 0s or 1s in the scrambled sequence m'_k. This will result in loss of synchronization.

EXAMPLE 2.16. The data stream 10101010001 is an input to a scrambler shown in figure 2.64. Obtain the scrambled output assuming initial content of all registers to be zero.

Solution: Let us consider figure 2.64.

We have, $m''_k = m'_{k-3} \oplus m'_{k-2}$

and $m''_k = \text{Output} = m_k \oplus m''_k$

Therefore, output $m'_{k-3} \oplus m'_{k-2} \oplus m_k$

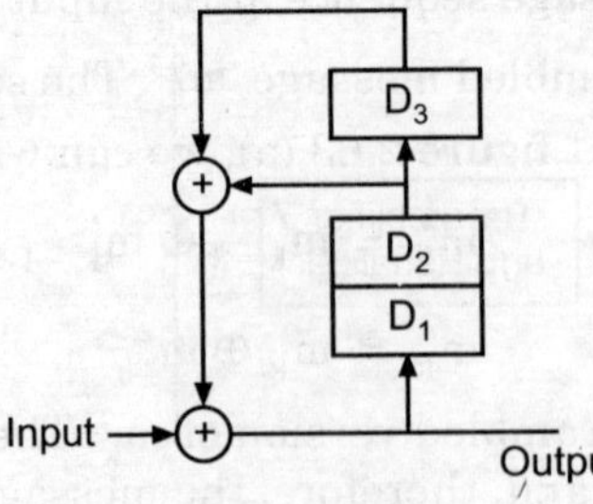

Fig. 2. 64.

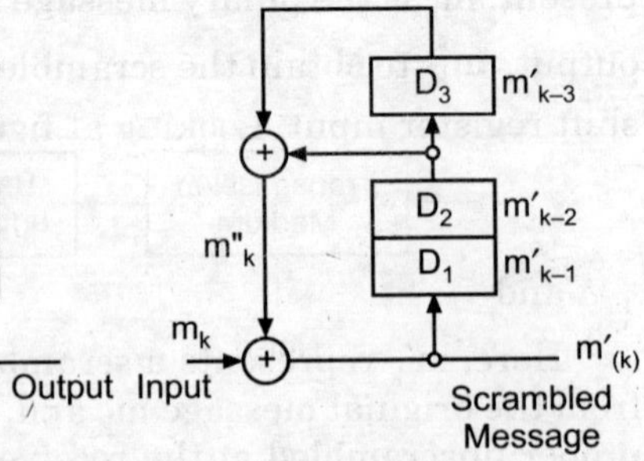

Fig. 2.65.

Initially, we assume that all the flipflops are in the reset condition.

Therefore, $m'_{k-2} = m'_{k-3} = 0$ initially

Hence, $m''_k = 0 \oplus 0 = 0$ initially

We consider table 2.10 for the output of the scrambler for the given input sequence.

Table 2.10.

Input m_k	Output m'_k	F/F Outputs m'_{k-1}	m'_{k-2}	m'_{k-3}	m''_k	
–	0	0	0	0	0	Initially
1	1	0	0	0	0	
0	0	1	0	0	0	
1	1	0	1	0	1	
0	1	1	0	1	1	
1	0	1	1	0	1	
0	1	0	1	1	0	
1	1	0	0	1	1	
0	1	1	0	0	0	
0	0	1	1	0	1	
0	1	0	1	0	1	
1	0	1	0	1	1	

Hence, the scrambled output will be

$$m'_k = 010110111010 \quad \textbf{Ans.}$$

EXAMPLE 2.17. A scrambler is shown in figure 2.66. Design the corresponding descrambler. If a sequence S = 101010100000111 is applied to the input of this scrambler, determine the output sequence T. Verify that if this T is applied to the input of the descrambler, the output is the sequence S.

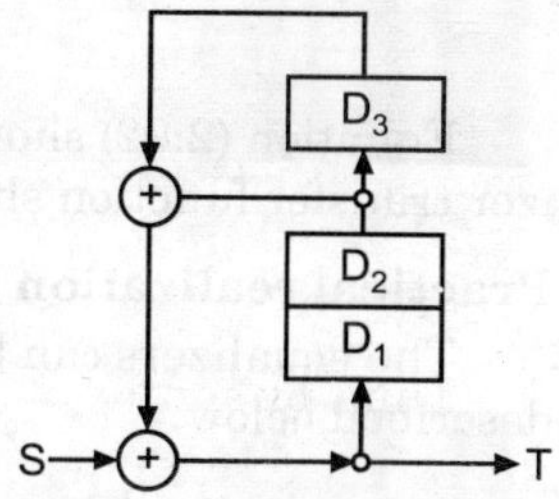

Fig. 2.66.

Solution: Solve yourself.

2.31 DIGITAL RECEIVERS AND REGENERATIVE REPEATERS

The functions of a receiver or regenerative repeater are as under:

(i) To reshape the incoming pulses using an equalizer
(ii) To extract the timing information
(iii) To make the decision about the symbol detection

Figure 2.67 shows a repeater which consists of a receiver and a regenerator.

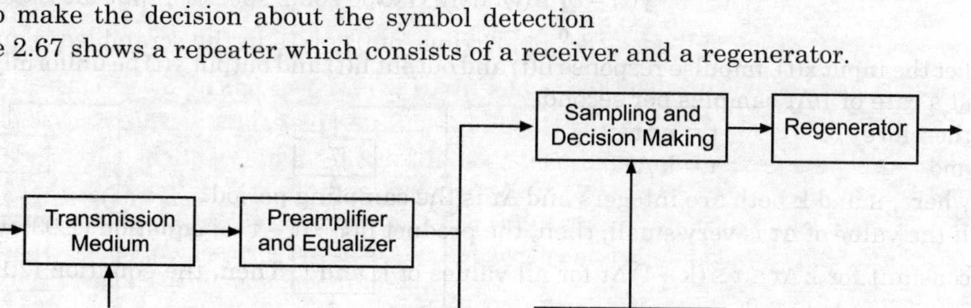

Fig. 2.67. Regenerative repeater

2.31.1 Equalization

Whenever a signal is passed through a communication channel, distortion will be introduced.

To compensate for the linear distortion, we can use a network called equalizer connected in cascade with the channel or system as shown in figure 2.68.

Principle of equalizer

The equalizer is designed in such a way that inside the frequency band of interest, the overall amplitude and phase responses of the cascade system shown in figure 2.68 are approximately equal to amplitude and phase responses for the distortionless transmission. Let us consider a communication channel with a transfer function $H_c(f)$. Let the transfer function of an equalizer be $H_{eq}(f)$. Then, the overall transfer function of the cascade connection is given by $H_c(f)\, H_{eq}(f)$. For the overall transmission through the cascaded connection of figure 2.68 to be distortionless, the overall transfer function should satisfy the following expression.

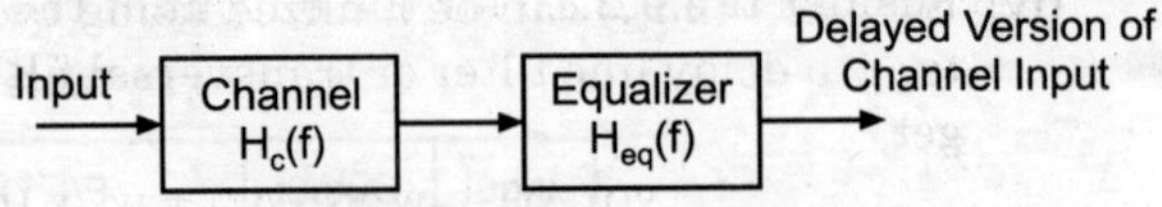

Fig. 2.68. Block diagram of equalization

$$H_c(f)\, H_{eq}(f) = k\, e^{-j2\pi f t_o} \quad \text{...(2.91)}$$

where k = Scaling factor and t_o = constant time delay

Therefore, the transfer function of the equalizer is given by,

$$H_{eq}(f) = \frac{k\, e^{-j2\pi f t_o}}{H_c(f)} \quad \text{...(2.92)}$$

Equation (2.92) shows the ideal value of the equalizer transfer function. Practically, the equalizer transfer function should be a close as possibe to the expression given in equation (2.92).

Practical realization of an equalizer

The equalizers can be practically realized using the structure of the **tapped-delay-line filter,** described below.

2.31.2 Tapped Delay Line Filter

Let us consider a time-invariant filter with an impulse response h(t). We assume that, h(t) = 0 for t < 0, i.e., the filter is causal. The impulse response of the filter is of finite duration i.e, h(t) = 0 for $t \geq T_f$.

(i) We can express the filter output y(t) produced in response of input x(t) as under:

$$y(t) = \int_0^{T_f} h(t).x(t-\tau)\, d\tau \quad \text{...(2.93)}$$

(ii) Let the input x(t), impulse response h(t) and output h(t) and output y(t) be uniformly sampled, at a rate of $1/\Delta\tau$ samples per second.

therefore $t = n\Delta\tau$ (2.94)

and $\tau = k\,\Delta\tau$ ...(2.95)

where, n and k both are integers and $\Delta\tau$ is the sampling period.

(iii) If the value of $\Delta\tau$ is very small, then, the product $h(t)\, x(t-\tau)$ of equation (2.93) will remain constant for $k\,\Delta\tau \leq \tau \leq (k+1)\,\Delta\tau$ for all values of k and t. Then, the equation (2.93) may be approximated by the convolution sum as under :

$$y(n\,\Delta\,\tau) = \sum_{k=0}^{N-1} h(k\,\Delta\tau).\, x(n\Delta\tau - k\,\Delta\,\tau)\,\Delta\tau \quad \text{...(2.96)}$$

when, $N\,\Delta\tau = T_f$

Substituting $h(k\,\Delta\tau)\,.\,\Delta\tau = w_k$ into equation (2.96), we get,

$$y(n\,\Delta T) = \sum_{k=0}^{N-1} w_k\,.\,x(n\,\Delta\tau - k\,\Delta\tau) \quad ...(2.97)$$

(iv) Equation (2.97) can be realized using the circuit shown in figure 2.68 which is known as a tapped delay line filter or transversal filter. Because, if we expand equation (2.97), then, we get,

$$y(n\,\Delta\tau) = w_0 x(n\,\Delta\tau) + w_1\,x\,(n\,\Delta\tau - \Delta\tau) + w_2\,x(n\,\Delta\tau - 2\,\Delta\tau) + ... + W_{N-1}\,x\,[n\,\Delta\tau - (N-1)\,\Delta\tau] \quad ...(2.98)$$

This expression is realized as shown in figure 2.69.

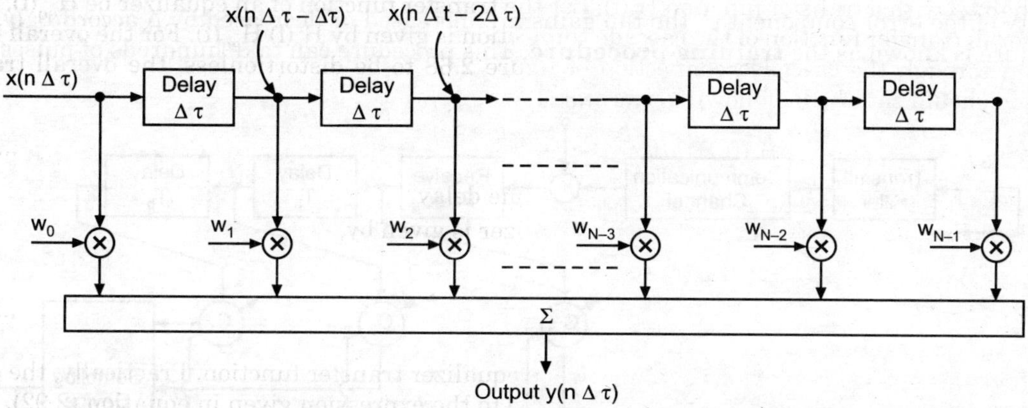

Fig. 2.69. Tapped delay line filter or transversal filter

2.31.3 Automatic Equalizers

In order to solve a set of simultaneous equations, the tap gains of the zero forcing equalizer described in the previous section are adjusted. This is called as the **trimming** of the equalizers. The process of equalizer trimming involves the following steps.

Trimming Procedure

(i) First, we send a pulse through the system

(ii) Then, we measure the output at the receiving filter output at the proper sampling instants.

(iii) Next, we calculate the values of the tap gains.

(iv) Accordingly, we set the gains on the taps of the equalizer.

In the automatic equalizers, the tap gains are adjusted automatically with the help of accurate and simple automatic system. These automatic systems are classified into two types.

(a) Preset type equalizers

(b) Adaptive equalizers

The preset type equalizers used a special sequence of pulses before to or during the breaks in the data transmission. The adaptive type equalizers adjusts itself continuously during the transmission of data by operating on the data signal itself.

The automatic equalizers adjsut the tap gains precisely at the optimum values by using the iterative technique.

2.31.4. Preset Equalizer

The block diagram of the preset equalizer has been shown in figure 2.70. Here, the components of the error vector are measured by transmitting a sequence of widely separated pulses through the system.

The output of the equalizer is measured at the proper instants of time. The tap gains are adjusted with the help of fixed iterations of the step size Δ. The sampling of the filter output at proper instants is done by using a timing circuit. This timing circuit is triggered by the peak detector. The center sample is compared with + 1 (or sliced at + 1) and the polarity of the error component ε_0^k is obtained. The polarities of the remaining error components are obtained by using the value of the filter output at $t = \pm j T_s$. The gate is opened at the end of the kth test pulse. Depending on the polarity of the error component ε^k the tap gains are increased or decreased by Δ accordingly. This procedure is known as the **training procedure**. This procedure can take hundreds of pulses.

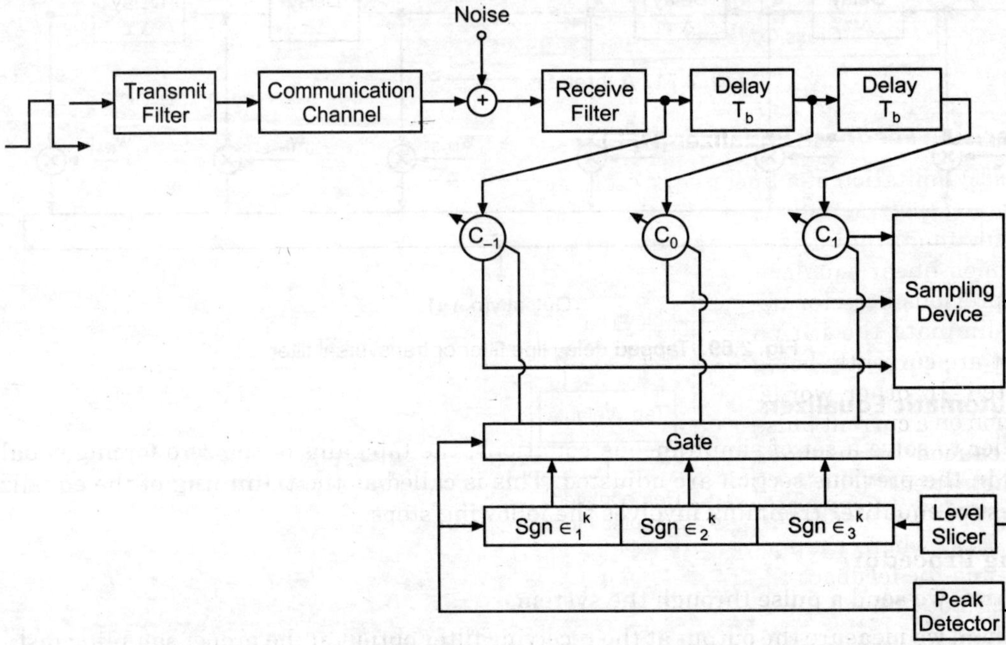

Fig. 2.70. A three tap preset equalizer

2.31.5 Adaptive Equalizer

The block diagram of an adaptive equalizer has been shown in figure 2.71. In the three tap adaptive equalizer of figure 2.71 the error vector ε^k is estimated continuously when the normal data transmission is going on. This is called as the **adaptive equalizer** because this scheme has the ability to change during the data transmission which eliminates the need for long training sessions, which is, therefore, the preset equalizer. The advantages of the adaptive equalizers are that they are more versatile, more accurate and cheaper than the preset equalizers.

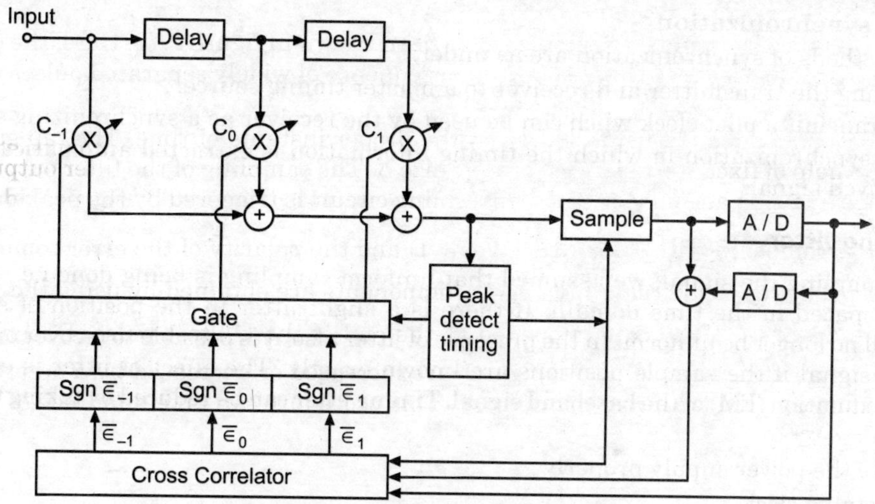

Fig. 2.71. A three tap adaptive equalizer

2.31.6 Decision Feedback Equalizer (DFE)

The basic limitation of a linear equalizer, such as the transversal filter, is the poor perform on channel having spectral nulls. A decision feedback equalizer (DFE) is a non-linear equalizer that uses previous detector decision to eliminate the ISI on pulses that are currently being demodulated. In other words, the distortion on a current pulse that was caused by previous pulses is subtracted. Figure 2.72 shows a simplified block diagram of a DFE where the forward filter and the feedback filter can each be a linear filter, such as transversal filter.

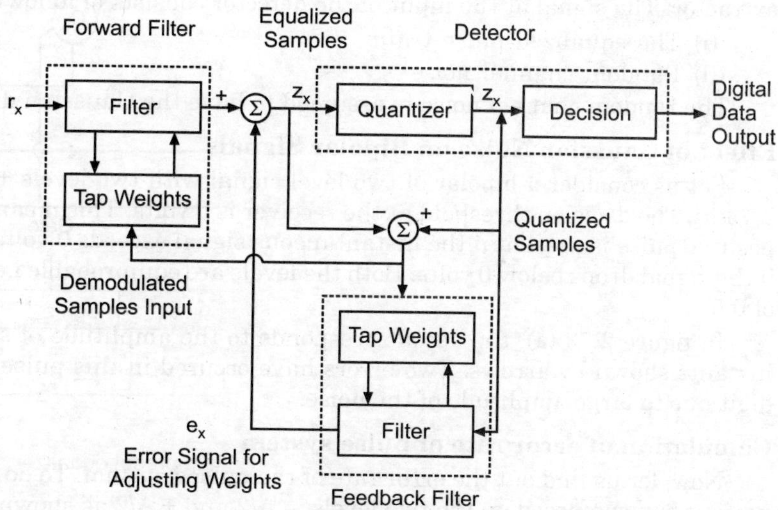

Fig. 2.72.

The non-linearity of the DFE stems from the non-linear characteristics of the detector that provides an input to be feedback filter. The basic idea of a DFE is that if the values of the symbols previously detected are known, then ISI contributed by these symbols can be cancelled out exactly at the output of the forward filter by subtracting past symbol values with appropriate weighting. The forward and feedback tap weights can be adjusted simultaneously to fulfill a criterion such as minimizing the MSE. The advantage of a DFE implementation is the feedback filter, which is additionally working to remove ISI, operates on noiseless quantized levels, and thus its output is free from channel noise.

2.31.7 Timing Extraction

The received digital signal needs to be sampled at precise instants. For this, we have to synchronize the clock signal at the receiver with the clock signal at the transmitter (symbol or bit synchronization).

Methods of synchronization

Three methods of synchronization are as under:

(i) Slaving the transmitter and receiver to a master timing source.

(ii) To transmit a pilot clock whch can be used by the receiver as a synchronizing signal.

(iii) Self synchronization in which the timing information is extracted automatically from the received signal.

2.31.8 Timing Jitter

While sampling the signal, we assumed that uniform sampling is being done i.e., the samples are equally spaced in the time domain. If there is a slight jitter in the position of samples, the sampling will no longer be uniform. In the presence of jitter also it is possible to recover or reconstruct the original signal if the sample positions are known exactly. The effect of jitter is equivalent to frequency modulation (FM) of the baseband signal. Timing jitter can be reduced by taking the following measures:

(i) Isolate the power supply properly

(ii) Use stable clock

2.31.9 Detection Error Probability

Once the transmission has passed through the equalizer, next step the detection at the detector. The detector samples the received signal on the basis of the clock signal provided by the timing extractor. The signal at the input of the detector consists of following two parts:

(i) The equalized pulse train

(ii) Random channel noise

The random channel noise is assumed to have the Gaussian distribution.

Effect of Gaussian Noise on Bipolar Signals

Let us consider a bipolar of two level signal with two levels + A/2 or – A/2 as shown in figure 2.73(*a*). The decision threshold at the receiver is 0 volts. This means that the decoder decides that a positive pulse is present if the instantaneous signal exceeds 0 volts, and a negative pulse is present if the signal drops below 0 volts. Both the levels are equiprobable i.e., they have the same probability of 0.5.

In figure 2.73 (a), the decoder responds to the amplitude of signal plus noise at the sampling instants shown by arrows. Two errors have occured in this pulse train in the second and the last digit due to large amplitude of the noise.

Calculation of error rate of pulse system

Now, let us find out the error rate of the pulsed system. To do so, we assume that the Gaussian noise is superimposed on the two levels, – A/2 and + A/2 as shown in figure 2.73(d).

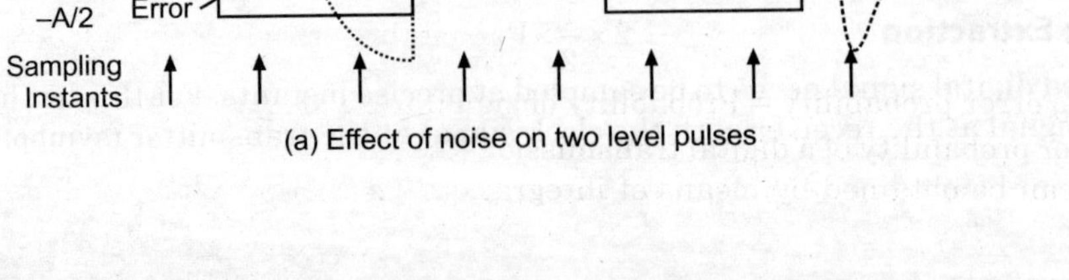

(a) Effect of noise on two level pulses

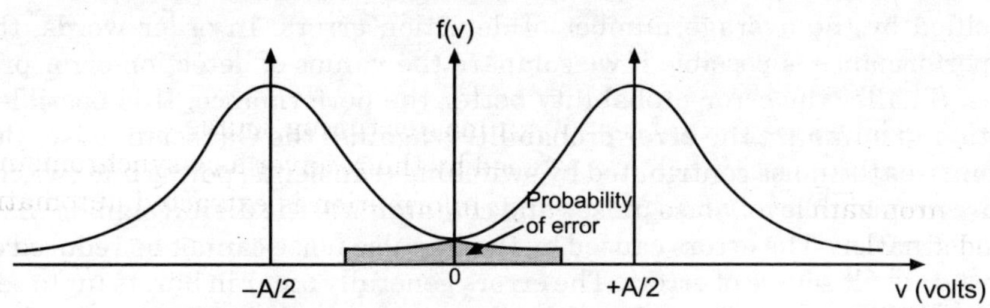

(b) Gaussian noise on a two level system

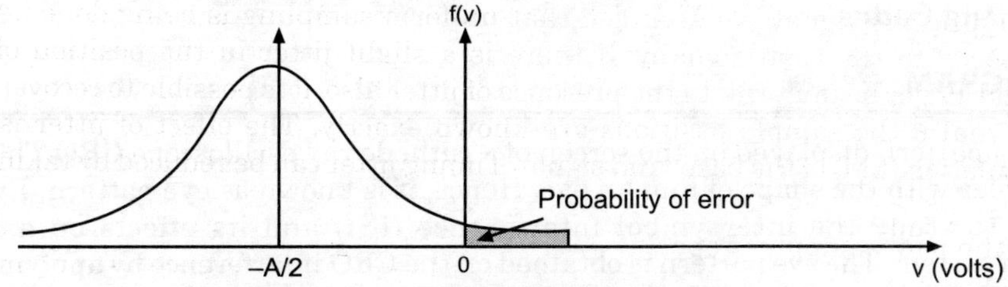

(c) Probability of error when signal is at –A/2 volt

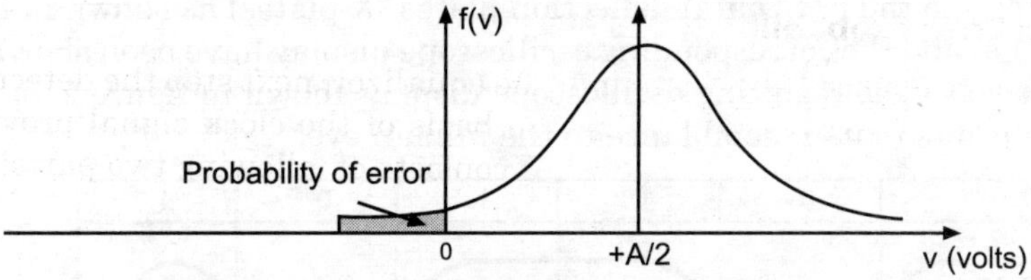

(d) Probability of error when signal is at +A/2 volt

Fig. 2.73.

Let us, find now the probability of error in the decision making at the decoder. When will the error occur. One situation is that the transmitted signal is – A/2 volts but due to noise the received voltage exceeds the decision threshold level of 0 volts. The probability of error is represented by the shaded area in figure 2.73 (c). The other situation in which an error can occur is when the transmitted voltage is + A/2 volts but the noise makes the received voltage less than 0 volts. (below the decision threshold). The probability of error is represented by the shaded area in figure 2.73 (d).

Error Probability

Both the signals i.e., – A/2 and + A/2 representing binary 0 and 1 respectively are equally likely i.e,. they have equal probabilities equal to 1/2 and the Gaussian distribution also is symmetrical. Therefore, the shaded areaa shown in figure 2.73 (c) and figure 2.73 (d) are equal. The probability of error is given by,

Error probability = [(Probability of signal at – A./2) × (Probability of signal in the shaded area for v ≥ 0)] + [(probability of signal at – A/2) × (Probability of signal in the shaded area for v ≤ 0)].

$$\text{Therefore, error probability} = \left[\frac{1}{2} \times \text{Probability of signal in shaded area for } v \geq 0\right]$$

$$+ \left[\frac{1}{2} \times \text{Probability of signal in shaded area for } v \leq 0\right]$$

$$= 2 \times \frac{1}{2} \times \text{Probability of signal in shaded area} \quad \ldots(2.99)$$

Therefore, error probability = Probability of signal in shaded area.

Thus, error probability of a digital transmission can be obtained by getting the shaded area. The shaded area can be obtained by means of integration. The performance of digital communication system is specified by the average number of detection errors. In order words, the meaningful comparison of performance is possible if we compare the values of detection error probabilitiesn of various systems. Smaller the error probability better the performance. It is possible to design the optimum defection to minimize the error probabilities against the Gaussian noise. But the impulse noise which comprises the noise contributed by switching transients, power line switching, lightning strikes also cause very high level noise pulses and contaminate the digital signals that are traveling from source to destination. The errors caused by the impulse noise cannot be reduced effectively and they are the most difficult source of errors. The errors generally occur in bursts (upto several hundred at a time). The burst errors can be corrected by means of using special type of codes called **Burst Error Correcting Codes.**

2.32 EYE DIAGRAM

Eye pattern is a pattern displayed on the screen of a cathode ray oscilloscope CRO.The shape of this pattern resembles with the shape of human eye. Hence, it is known as eye pattern. Eye pattern is a practical way to study the intersymbol interference (ISI) and its effects on a PCM or data communication system. The eye pattern is obtained on the CRO interference by applying the received signal to vertical deflection plates (Y-plates) of the CRO and a sawtooth wave at the transmission symbol rate i.e., ($1/T_b$) to the horizontal deflection plates (X-plates) as shown in figure 2.74 (c). The received digital signal and the corresponding oscilloscope display have been shown in figures 2.74(a) and 2.74(c) respectively. The resulting oscilloscope display shown in figure 2.74 (c) is called as the **eye pattern**. This is due to its resemblance to the human eye.

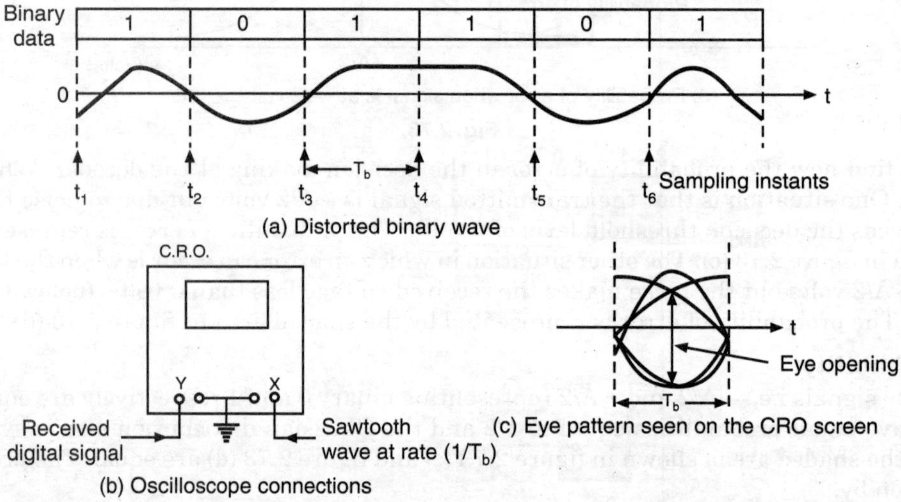

Fig. 2.74. Eye pattern

The interior region of the eye pattern is known as the eye opening. The eye pattern provides a great deal of information about the performance of the system.

Information obtained from eye pattern

(i) The width of the eye opening defines the time-interval over which the received wave can be sampled, without an error due to ISI. The best time for sampling is when the eye is open widest.

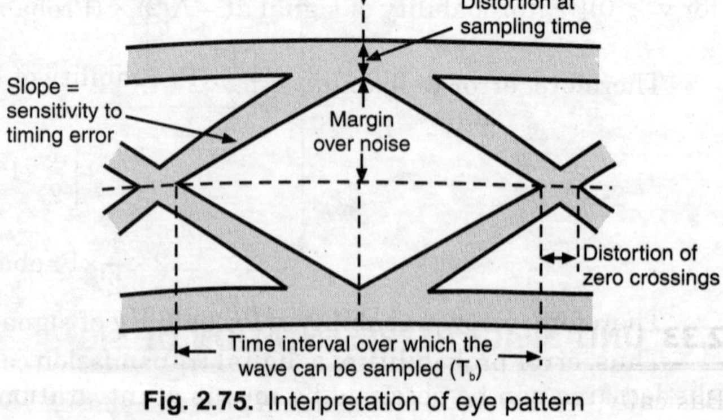

Fig. 2.75. Interpretation of eye pattern

(ii) The sensitivity of the system to the timing error is determined by the rate of closure of the eye as the sampling rate is varied. The height of eye opening at a specified sampling time defines the margin over noise.

(iii) When the effect of ISI is severe, the eye is completely closed and it is impossible to avoid errors due to the combined presence of ISI and noise in the system.

EXAMPLE 2.18. Consider that the bit sequence given below is to be transmitted. Bit sequence = 10110010. Draw the resulting waveform, if the sequence is transmitted using:

(i) Unipolar RZ (ii) Polar RZ (iii) AMI (iv) Split phase manchester (v) M-ary, where, M = 4. (Polar quaternary).

Solution: The required waveforms have been shown in figure 2.76.

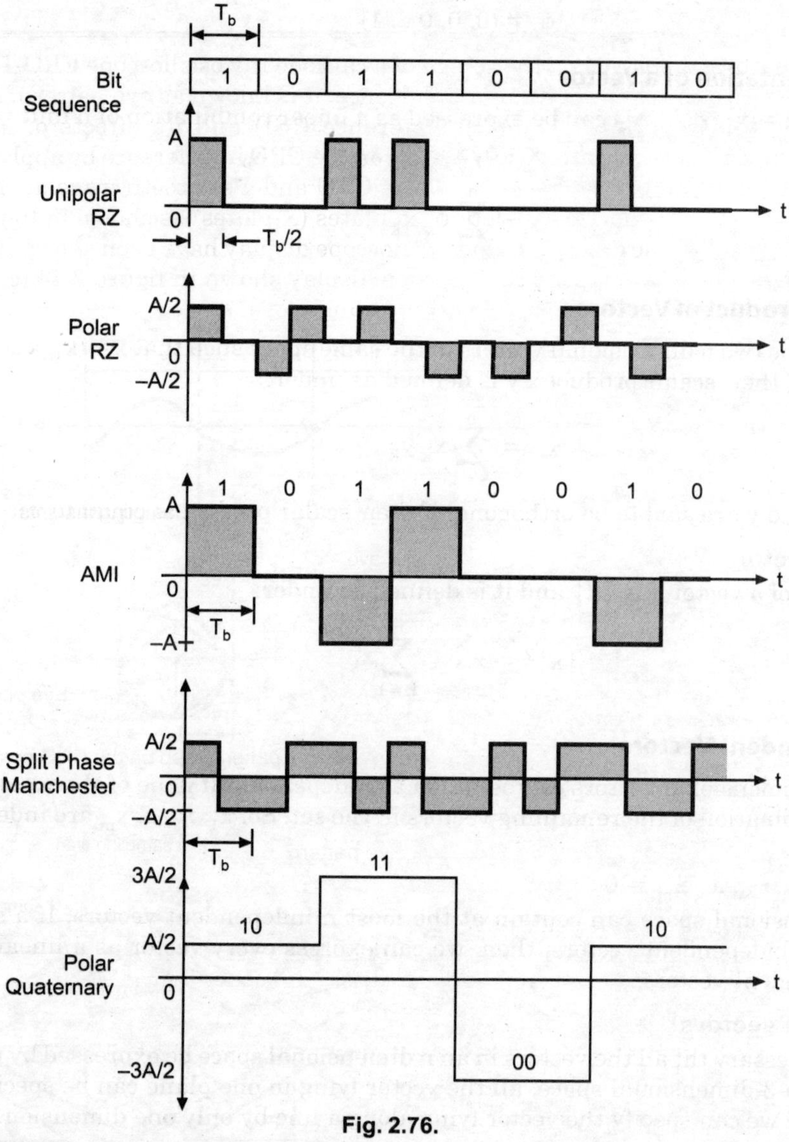

Fig. 2.76.

2.33 UNIT SPACE REPRESENTATION OF SIGNALS

Basically, we use the 3-dimensional physical space and 3-dimensional vectors in this space. But, it is possible to extend this concept to an n-dimensional space. A 3-dimensional vector is an entity specified by three numbers (x_1, x_2, x_3) in a certain order. Therefore, an n-dimensional vector is an entity which is specified by n numbers in a certain order say $(x_1, x_2, x_3....x_n)$.

2.33.1 Unit Vectors

Let the n unit vectors be denoted by $\phi_1, \phi_2, ... \phi_n$ and be defined as under:

$$\phi_1 = (1, 0, 0 ... 0)$$
$$\phi_2 = (0, 1, 0, ... 0)$$
$$\vdots$$
$$\vdots$$
$$\phi_n = (0, 0, 0 ... 1)$$

2.33.2 Representation of a Vector

Any vector $x = (x_1, x_2, ... x_n)$ can be expressed as a linear combination of n unit vectors as under:

$$x = \phi_1 x_1 + \phi_2 x_2 + ... \phi_n x_n$$

Therefore,
$$x = \sum_{k=1}^{n} \phi_x x_k \quad ...(2.100)$$

2.33.3 Scalar Product of Vectors

Let x and y be two n dimensional vectors in the same space such that $x = (x_1, x_2, ...x_n)$ and $y = (y_1, y_2, ..., y_n)$. Then, their scalar product x.y is defined as under:

$$x.y = \sum_{k=1}^{n} x_k y_k \quad ..(2.101)$$

Vectors x and y are said to be orthogonal if their scalar product is equal to zero i.e., if $x.y = 0$.

Length of a vector

The length of a vector x is $|x|$ and it is defined as under:

$$|x|^2 = x.x = \sum_{k=1}^{n} x_k^2 \quad ...(2.102)$$

2.33.4 Independent Vectors

A set of n-dimensional vectors can be called as independent if none of them can be represented as a linear combination of the remaining vectors in the set. So. $x_1, x_2, ... x_m$ are independent vectors if

$$a_1 x_1 + a_2 x_2 + ... a_m x_m \neq 0$$

An n dimensional space can contain at the most n independent vectors. If a space contains a maximum of n independent vectors, then, we can express every vector as a linear combination of these n independent vectors.

Dimensions of vectors

It is not necessary tht all the vectors in an n dimensional space be expressed by n numbers. This means that in a 3-dimensional space, all the vector lying in one plane can be specified by only two dimensions and we can specify the vector lying along a line by only one dimension.

2.33.5 Basis Vectors

A set of n independent vectors in an n-dimensional space are known as basis vectors. Every basis vector can be expressed as a linear combinations of n independent vectors. These n-independent vectors form the co-ordinates axis but they are not unique. The n-unit vectors $\phi_1, \phi_2..., \phi_n$ can be used as basis vectors. They are also denoted by $u_1, u_2...$ These vectors have an important property. They are all mutually orthogonal. It means,

$$\phi_j \cdot \phi_k = 0 \qquad \text{... if } j \neq k$$
$$= 1 \qquad \text{... if } j = k$$

The n-basis vectors which are mutually independent and mutually orthogonal are known as **orthonormal vectors** because they are orthogonal and their length is unity. The vector $x = (x_1, x_2 ... x_n)$ can be represented as under:

$$x = x_1\phi_1 + x_2\,\phi_2 + ...\, x_n\,\phi_n$$

If x is known, then we can obtain its component x_k by taking the scalar product of both sides with ϕ_k as under:

$$x.\phi_k = \phi_k \cdot (x_1\phi_k + x_2\,\phi_2 + ... + x_n\,\phi_n)$$
$$x.\phi_k = \phi_k \cdot x_1\,\phi_k + x_2\phi_2 + ...\, \phi_k \cdot x_n\,\phi_n$$
$$x.\phi_k = 0 + 0 + ... + x_k + 0$$

Therefore, $$x.\,\phi_k = x_k$$

Important Point: An n-dimensional vector can be represented completely if and only if we use n-basis vectors. A set of n-basis vectors is called as a complete set.

2.34. SIGNAL AS A VECTOR

An entity x which can be represented by an n-tuple i.e., $(x_1, x_2, ... x_n)$ is called as an n-dimensional vector. Therefore, if a signal x(t) can be specified using an n-tuple it can also be called as vector. Using the sampling thorem, we can express a bandlimited signal x(t) by the values of its samples. If the number of samples is n, then, signal is being represented by n values. Hence, it is also a vector.

2.34.1 Concept of Signal Space

The n-signals $\phi_1(t), \phi_2(t), ... \phi_n(t)$ are defined as independent signals if none of them can be represented as the linear combination of the remaining (n – 1) signals. Let us assume that the signal x(t) can be represented by a linear combination of n-independent signals as under:

$$x(t) = x_1\phi_1(t) + x_2\,\phi_2(t) + ... + x_n\,\phi_n(t)$$

$$x(t) = \sum_{k=1}^{n} x_k\,\phi_k(t) \qquad ...(2.103)$$

If every signal in a certain signal space can be represented by a linear combination of n independent signals, then, we will have a n-dimensional signal space. The n-independent signals $\phi_1(t), \phi_2(t)$ ... will be assumed to be orthogonal. It means,

$$\int_{\infty}^{\infty} \phi_1(t)\,\phi_k(t)\,dt = 0 \qquad \text{if, } j \neq k \qquad ...(2.104$$

$$= k_j \qquad \text{if, } j = k \qquad ...(2.105)$$

If, $k_j = 1$ for all j, then, the above mentioned set is orthonormal as well.

Important Point: Any band limited signal x(t) can be represented by a point $\{... x_{-k}, ... x_{-2}, x_{-1}, x_0, x_1, x_2 ... x_k, ...\}$, where x_1, x_2 ... represent the samples of signal x(t) taken at the Nyquist rate.

2.34.2 Vectors of Orthogonal Signals

If signals x(t) and y(t) are mutually orthogonal then their corresponding vectors x and y will also be orthogonal.

EXAMPLE 2.19. A signal space consists of the four signals $s_1(t)$, $s_2(t)$, $s_3(t)$ and $s_4(t)$ which are shown in figure 2.77. Represent these signals geometrically in the vector space representation. Also determine a suitable set of basis vectors and dimensionality of the signals.

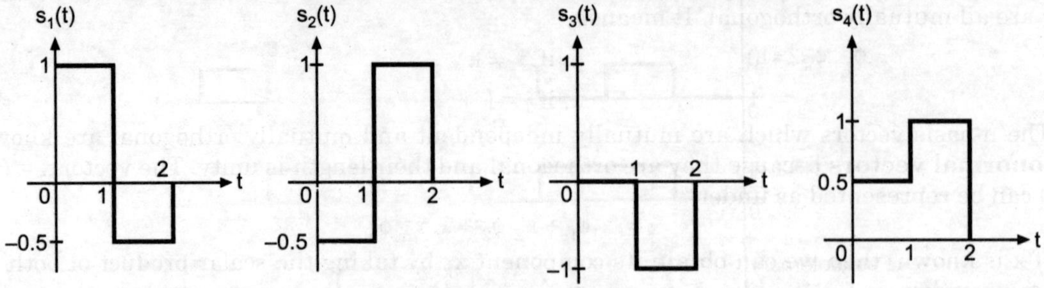

Fig. 2.77. Four given signals

Solution: (i) Basis signals

The two basis signals have been shown in figure 2.78. These are rectangular pulses of unit amplitude. They exist over the intervals 0 to 1 and 1 to 2.

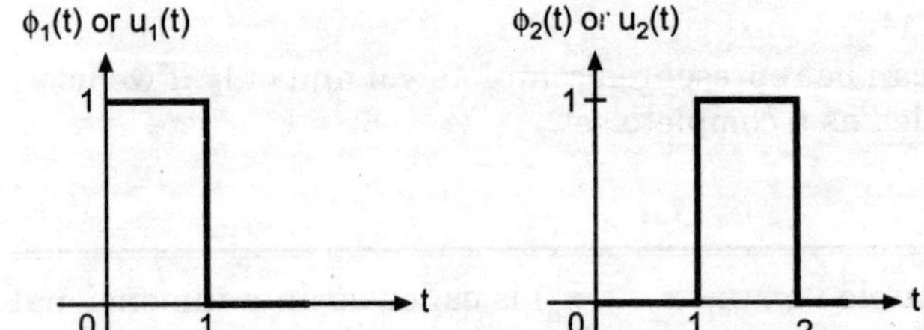

Fig. 2.78. Set of basis signals

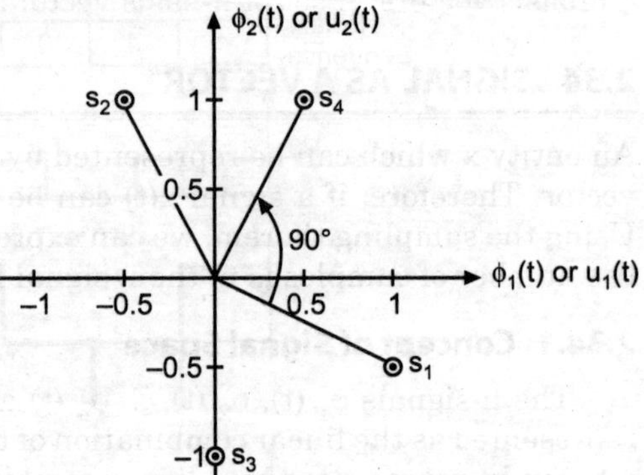

Fig. 2.79.

(ii) Obtain the vectors

In terms of the basis signals $\phi_1(t)$ and $\phi_2(t)$, the signal vectors s_1, s_2, s_3, s_4 corresponding to signals $s_1(t)$, $s_2(t)$, $s_3(t)$ and $s_4(t)$ are $s_1 = (1, -0.5)$, $s_2 = (-0.5, 1)$, $s_3(0, -1)$ and $s_4 = (0.5, 1)$. These points have been plotted in figure 2.79. The signals $s_1(t)$ and $s_4(t)$ are orthogonal.

EXAMPLE 2.20. Let $X_\alpha(f)$ be the raised cosine sepctrum with a roll-off factor α

Show that $\int_{\infty}^{\infty} X_\alpha(f)\,df$ ·

Solution: The raised cosine spectrum for different value of α is shown in figure 2.80.

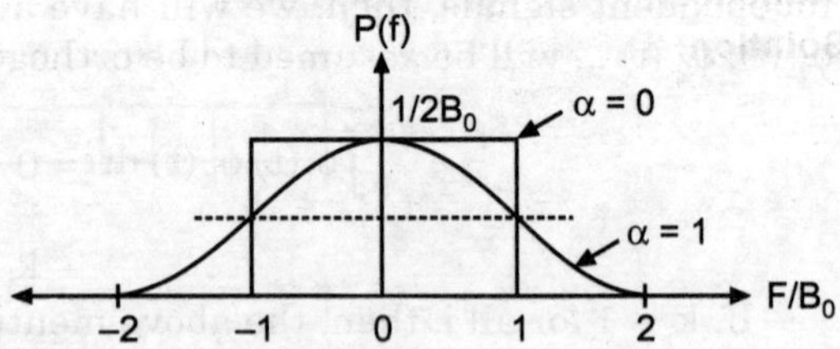

Fig. 2.80. Raised cosine spectrum

Now, let us consider the spectrum corresponding to $\alpha = 0$.

We have,
$$\int_{-\infty}^{\infty} P(f)\,df = \int_{-B_0}^{B_0} (1/2\,B_0)\,df$$

$$= \frac{1}{2B_0}[f]_{-B_0}^{B_0} = \frac{1}{2B_0}[B_0 + B_0] = 1$$

The same can be proved for different values of α.

EXAMPLE 2.21. Given a bit sequence of 01011001. Draw lines codes in NRZ–L, NRZ–I and pseudoternary formats. Compare them on the basis of their bandwidth and clocking capability.

Solution: Let us consider figure 2.81 for line formats.

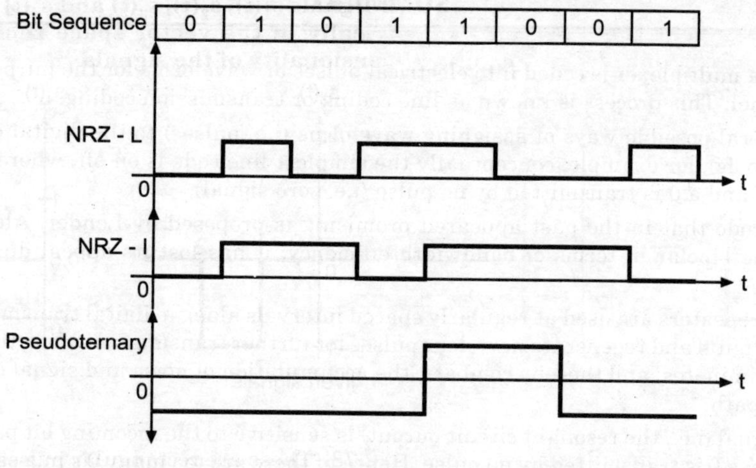

Fig. 2.81.

EXAMPLE 2.22. For the binary sequence 1101011101, construct RZ, AMI and Manchester format.

Solution: We consider figure 2.82 for the solution.

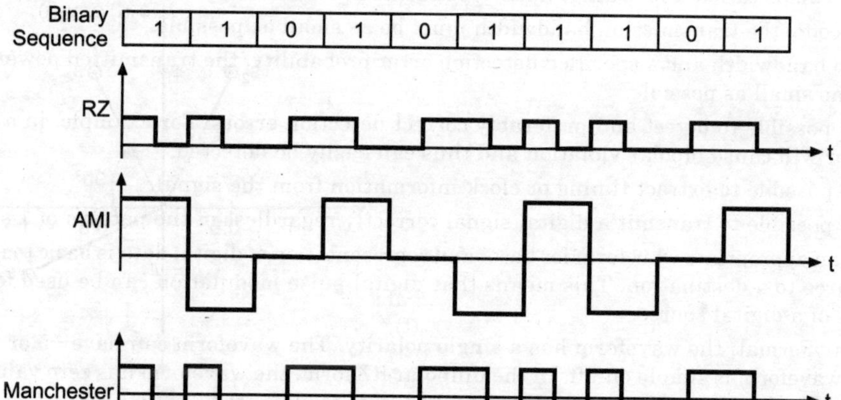

Fig. 2.82.

EXAMPLE 2.23. Sketch appropriate waveforms fo a sequence 111000110 using unipolar, and Manchester line coding schemes.

Solution: The required waveforms have been shown in figure 2.83.

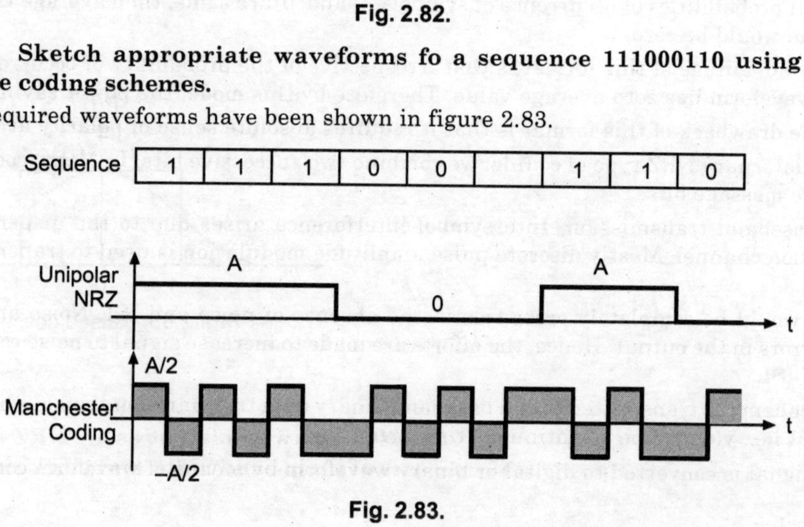

Fig. 2.83.

SUMMARY

- The output of a multiplexer is coded into electrical pulses or waveforms for the purpose of transmission over the channel. This process is known as line coding or transmission coding.
- There are several possible ways of assigning waveforms (i.e. pulses) to the digital data. In the binary case (two symbols), for example, conceptually the simplest line code is on-off, where a 1 is transmitted by a pulse p(t) and a 0 is transmitted by no pulse (i.e. zero signal).
- Another line code that in the past appeared promising is proposed by Lender. Although this code is better than the bipolar in terms of bandwidth efficiency, it has lost its appeal due to some practical problems.
- Regenerative repeaters are used at regularly spaced intervals along a digital transmission line to detect the incoming digital and regenerate new clear pulses for further transmission along the line. This process periodically eliminates, and thereby combats, the accumulation of noise and signal distortion along the transmission path.
- The timing signal (i.e., the resonant circuit output) is sensitive to the incoming bit pattern. In the on-off or bipolar case, a 0 is transmitted by no pulse. Hence, if there are too many 0's in a sequence (no pulses), there is no signal at the input of the resonant circuit and the sinusoidal output of the resonant circuit starts decaying, thus causing error in the timing information.
- The digital data may be transmitted by various transmission or line codes such as on-off, polar, bipolar and so on. This is called line-coding. Each type of line-code has its advantages and disadvantages.
- For a line-code, the transmission bandwidth must be as small as possible.
- For a given bandwidth and a specified detection error probability, the transmitted power for a line code should be as small as possible.
- It must be possible to detect and preferably correct detection errors. For example, in a bipolar case, a signal error will cause bipolar violation and thus can easily be detected.
- It must be possible to extract timing or clock information from the signal.
- It must be possible to transmit a digital signal correctly regardless of the pattern of 1 s and 0 s.
- The use of an appropriate waveform for baseband representation of digital data is basic to its transmission from a source to a destination. This means that digital pulse modulation can be used for transmitting the output of a digital source.
- In Unipolar Format, the waveform has a single polarity. The waveform can have +5 or +12 volts when high. The waveform is simple on-off. In the unipolar RZ form, the waveform has zero value when symbol '0' is transmitted and waveform has 'A' volts when '1' is transmitted.
- Since polar RZ and NRZ formats are bipolar, therefore, the average DC value is minimum in these waveforms. If probabilities of occurrence of symbols '1' and '0' are same, then average DC components of the waveform would be zero.
- The primary advantage of this format is that irrespective of the probability of occurrence of symbol '1' and '0' the waveform has zero average value. Therefore by this mode, the power saving is quite more. However, the drawback of this format is that it requires absolute sense of polarity at the receiver end.
- In a polar quaternary NRZ type of coding, we combine two successive bits. In M-ary coding, we combine 'k' successive message bits.
- In digital baseband transmission, Intersymbol interference arises due to the dispersive nature of a communication channel. Mostly discrete pulse amplitude modulation is used to transmit the baseband pulse signal.
- The output would be completely error free in the absence of noise and ISI. Noise and ISI combinely introduce errors in the output. Hence, the efforts are made to increase signal to noise ratio and minimize the effect of ISI.
- When the sequence is transmitted over a baseband binary data transmission system, the signal obtained at the output i.e., y(t) will be a continuous time signal.
- An analog signal is converted to digital or binary waveform by means of waveform coding techniques.

GLOSSARY

1. **Line code :** A system for translating logic ones and zeros into voltage or current levels for transmission.
2. **Negative logic :** A logic system in which a low level represents logic one and a high level represents logic zero.
3. **NRZ (non-return-to-zero) code :** A data line code in which the voltage or current does not necessarily return to zero between bits.
4. **Positive logic :** A logic system in which a high level represents logic one and a low level represents logic zero.
5. **RZ (return-to-zero) code :** A line code in which the voltage or current returns to zero at the end of each bit period.
6. **Unipolar code :** A line code in which the polarity of the voltage or the direction of the current remains the same at all times.
7. **Line code:** A system for translating logic ones and zeros into voltage or current levels for transmission.
8. **Vocoder:** Circuit for digitizing voice at a low data rate by using knowledge of the way in which voice sounds are produced.

SHORT QUESTIONS WITH ANSWERS

Q.1. What is Baseband binary data transmission system?

Ans. The data transmission system which makes use of a baseband channel for transmitting a binary data is known as baseband data transmission i.e., the system using a channel whose bandwidth is equal to the bandwidth of the message signal, called a baseband channel is referred to as baseband data transmission system.

Q.2. What is a Baseband binary PAM system? Explain.

Ans. Pulse amplitude modulation scheme is a system in which the amplitude of discrete pulses (carrier) vary along with the message signal. Thus, baseband binary PAM system is the scheme in which the amplitude transmitted pulses take any one of the two amplitude levels i.e., either 1 (or) 0.

Q.3. What is the drawback of binary PAM system ?

Ans. Following are the drawbacks:

(i) The bandwidth required for baseband binary PAM is approximately $\frac{r_b}{2}$ Hz, in order to transmit data at a rate of r_b bits/sec, with zero ISI.

(ii) If the bandwith available is exactly $\frac{r_b}{2}$ Hz, then the binary PAM system transmits data at a rate r_b bits/sec only if it uses an ideal low pass filter at the transmitter and receiver.

(iii) Thus, such filters are physically unrealizable and if any systems are approximated to these filters then those systems are extremely sensitive to the perturbations (or) disturbances in rate, timing (or) channel characteristics.

This is the drawback baseband binary PAM system.

Q.4. What do you mean an Eye Diagram ?

Ans. The performance of baseband PAM system depends on the amount of ISI and channel noise. The distribution of ISI and channel noise in the system can be readily observed by displaying the received waveform Y(t) on an oscilloscope using a sweep rate that is a fraction of the symbol rate r_s. The resulting display shape resembles a human eye and is widely known as the Eye pattern of the system.

Q.5. What is Equalization?

Ans. The process of correcting channel induced distortion is called equalization.

Q.6. What is the effect of ISI? Explain.

Ans. Following are the effects of ISI

(i) In the absence of ISI and noise, the transmitted bit can be decoded correctly at the receiver.

(ii) The presence of ISI will introduce errors in the decision at the receiver output.

(iii) Hence, the receiver can make an error in deciding whether it has received a logic 1 or a logic 0.

Q.7. What is Remedy to Reduce ISI?

Ans. Remedy to Reduce ISI

(i) It has been proved that the function which produces a zero intersymbol interference is a sinc function. Hence, instead of a rectangular pulse if we transmit a sinc pulse then the ISI can be reduced to zero.

(ii) This is known as Nyquist Pulse Shaping. The sinc pulse transmitted to have a zero ISI shown in figure 2.84 (a).

(iii) Further, we know that Fourier transform of a sinc pulse is a rectangular function. Hence, to preserve all the frequency components, the frequency response of the filter must be exactly flat in the pass band and zero in the attenuation band as shown in figure 2.84 (b).

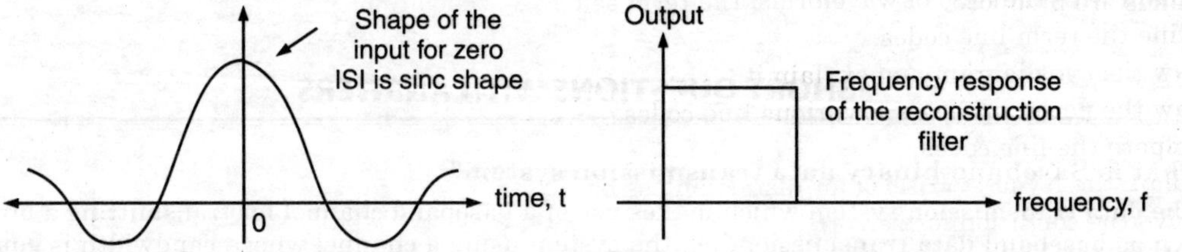

(a) Ideal pulse shape for zero ISI (b) Frequency response of the filter

Fig. 2.84.

Q.8. What is Nyquist's criterion for distortionless transmission? Explain.

Ans. We have observed that in absence of the ISI, we have

$$y(t_i) = \mu a_i$$

This expression shows that under these conditions, the ith transmitted bit can be decoded correctly. In order to minimize the effects of ISI, we have to design the transmitting and receiving filters properly. The transfer function of the channel and the shape of transmitted pulse are generally specified. Therefore, it becomes the first step towards design of filters. From this information we have to determine the transfer functions of the transmitting and receiving filters, to reconstruct the transmitted data sequence $\{b_k\}$. This is achieved by first extracting and then decoding the corresponding sequence of weights from the output y(t).

REVIEW QUESTIONS

1. Explain the difference between bandpass transmission and passband transmission.
2. Explain various PAM digital formats.
3. Explain the followings:
   (*i*) Intersymbol Interference, (*ii*) Eye pattern
4. What is Line coding ? Explain.
5. What are the properties of Line coding?
6. Draw the block diagram of a baseband digital communications of various blocks.
7. What is Nyquist criterion of zero ISI ? Explain.
8. Why do we need to use the discrete PAM formats?
9. State the important properties of line codes.
10. What is the meaning of word "RZ".
11. State true or false
    (a) The dc value of unipolar RZ format is zero.
    (b) The unipolar NRZ signal used two different amplitudes
    (c) Polar NRZ format has a zero dc value

12. In the bipolar NRZ or AMI line codes, the binary 0 is represented by
    (a) alternate zeros and ones. (b) alternate + A and – A amplituds.
    (c) zero amplitudes
13. Why do we need to use the line code formats?
14. How to improve the synchronization capacity of a code?
15. What is the difference betwen line coding and source coding?
16. What is ISI?
17. What is the cause of ISI?
18. What is the remedy to ISI?
19. What is the shape of transmitted pulse to avoid ISI?
20. Explain how the synch pulse can avoid ISI.
21. How to measure ISI on CRO?
22. What is the significance of eye opening?
23. What is the effect of ISI?
24. State various types of line codes?
25. What is the importance of dc component in line codes?
26. Explain with the help of waveforms, the term self synchronization.
27. Define the term line codes.
28. Draw the eye diagram and explain it.
29. Draw the power spectras of various line codes.
30. Compare the line codes.
31. Explain the Nyquist criteria for distortionless baseband binary transmission.
32. Derive the condition for zero $|S|$.
33. Explain the ideal solution to the problem of ISI and state various problems associated with it.
34. What is minimum bandwidth solution for pulse shape with zero ISI? Assume symbol duration is T.
    (U.P. Tech., Sem. Exam., 2004-2005)(5 Marks)
35. What is ISI? Give nyquist criterion for zero ISI raised cosine spectrum.
    (U.P. Tech., Sem. Exam., 2005-2006)(10 Marks)
36. What are different Nyquist criteria for zero Inter Symbol Interference (ISI).
    (U.P. Tech., Sem. Exam., 2006-2007)(5 Marks)
37. Define the following types of line codes for the electrical representations of binary data.
    (i) Unipolar NRZ signaling.
    (ii) Bipolar RZ signaling
    (iii) Split-phase Manchester. **(U.P. Tech., Sem. Exam., 2007-2008)(5 Marks)**
38. Write a short note on Raised Cosine Spectrum. **(U.P. Tech., Sem. Exam., 2006-2007)(5 Marks)**
39. Explain following line coding schemes with at least one example of each.
    1. Manchester2. Bipolar
40. Line coding used in T_1 multiplexing schemes is AMI with BBZS. Explain this line coding with its advantages over other line coding schemes.
41. Find out the power spectral density of NRZ and AMI signal format.
    (U.P. Tech., Sem. Exam., 2009-2010, 5 Marks)
42. For the binary sequence 011010110 construct NRZ, RZ, AMI and Manchester format.
    (U.P. Tech., Sem. Exam., 2006-2007, 5 Marks)

NUMERICAL PROBLEMS

1. Represent the data 10110100 using the following digital data formats with the help of neat figures:
   (i) Unipolar RZ (ii) Unipolar NRZ (iii) Split phase Manchester.
2. Given the binary sequence 1101110, draw the transmitted pulse waveform for
   (i) AMI RZ signaling format, and (ii) Split-phase (Manchester) signaling format.

3. In a binary PCM system on/off signaling is used. The matched filter receiver is used for detection of the signal. Calculate the probability of error if signaling rate is doubled.
[**Hint:** If signaling rate is doubled, bit period 'T' is reduced by half, hence energy is also reduced by half.]

$$\left[\textbf{Ans. } P_e = \frac{1}{2}\,\text{erfc}\sqrt{\frac{E}{8N_0}}\right]$$

OBJECTIVE TYPE QUESTIONS

1. Optimum system is the one
   (a) which is physically realisable
   (b) which must lead to unique solution
   (c) all of the two
2. If in a particular digital communication system application, the receiver designed is an optimum coherent receiver, the channel noise is white and bits are transmitted using rectangular pulses then which one of the following statements would be inconsistent? The receiver is
   (a) a correlation receiver
   (b) matched filter receiver
   (c) an integrate and dump receiver
   (d) a sample, hold and dump receiver
3. In manchester code, the symbol rate r_s and data rate r_s are related as
   (a) $r_b = 2r$ (b) $T_s = 2T_b$
   (c) $T_s = T_b/2$ (d) $r_s = r_b$
4. Transversal equaliser uses tapped delay lines to
   (a) reduce & SI
   (b) reduce BER
   (c) increase bit rate
   (d) increase band widths

Answers

1. (a) 2. (a) 3. (a) 4. (a)

❑❑❑

CHAPTER 3

DIGITAL MODULATION TECHNIQUES

SYLLABUS

Digital carrier system, Method of generation and detection of coherent & non-coherent binary ASK, FSK & PSK, Differential phase shift keying, quadrature modulation techniques (QPSK and MSK), M-ary Digital carrier Modulation.

3.1 INTRODUCTION

As discussed earlier, Modulation is defined as the process by which some characteristics of a carrier is varied in accordance with a modulating signal. In digital communications, the modulating signal consists of binary data or an M-ary encoded version of it. This data is used to modulate a carrier wave (usually sinusoidal) with fixed frequency. In fact, the input data may represent the digital computer outputs or PCM waves generated by digitizing voice or video signals. The channel may be a telephone channel, microwave radio link, satellite channel or an optical fiber. In digital communication, the modulation process involves switching or keying the amplitude, frequency or phase of the carrier in accordance with the input data*.

Thus, there are three basic modulation techniques for the transmission of digital data. They are known as amplitude-shift keying (ASK), frequency shift keying (FSK) and phase-shift keying (PSK) which can be viewed as special cases of amplitude modulation, frequency modulation and phase modulation respectively.

DO YOU KNOW?

Digital signals have become very important in both wired and wireless communication.

Chapter Outline

*Digital Signals have become very important in both wired and wireless commmunication. Some of the information to the transmitted is already in digital form.

The present chapter is devoted to detailed discussion of digital modulation techniques: their noise performance, spectral properties, their merits and limitations, applications and other related aspects.

3.2 DIGITAL MODULATION FORMATS

When we have to transmit a digital signal over a long distance, we need continuous-wave (CW) modulation. For this purpose, the transmission medium can be in form of radio, cable or other type of channel. Also, a carrier signal having some frequency f_c is used for modulation. Then the modulating digital signal modulates some parameter like frequency, phase or amplitude of the carrier. Due to this process, there is some deviation in carrier frequency f_c. This deviation is known as the bandwidth of the channel. This means that the channel has to transmit some range or band of frequencies. Such type of transmission is known as bandpass transmission and the communication channel is known as bandpass channel.

Here, the word bandpass is used since the range of frequencies does not start from zero Hz to f_m Hz. In fact, the range of frequencies from zero Hz to f_m Hz is known as **low-pass signal** and such channel is known as **low-pass channel.**

Now, when it is required to transmit digital signals on a bandpass channel, the amplitude, frequency or phase of the sinusoidal carrier is varied in accordance with the incoming digital data. Since the digital data is in discrete steps, the modulation of the bandpass sinusoidal carrier is also done in discrete steps. Due to this reason, this type of modulation (i.e., Digital modulation) is also known as switching or signaling. Now, if an amplitude of the carrier is switched depending on the input digital signal, then it is called Amplitude shift keying (ASK).

This process is quite similar to analog amplitude modulation. If the frequency of the sinusoidal carrier is switched depending upon the input digital signal, then it is known as the frequency shift keying (FSK). This is very much similar to the analog frequency modulation. If the phase of the carrier is switched depending upon the input digital signal, then it is called phase shift keying (PSK). This is similar to phase modulation. Since the phase and frequency modulation has constant amplitude envelope, therefore FSK and PSK also has a constant amplitude envelope. Because of constant amplitude of FSK and PSK, the effect of non-linearities, noise interference is minimum on signal detection. However, these effects are more pronounced on ASK. Therefore, FSK and PSK are preferred over ASK.

Figure 3.1 shows the waveforms for amplitude-shift keying, phase-shift keying and frequency shift keying. In these waveforms, a single feature of the carrier (i.e., amplitude, phase or frequency) undergoes modulation.

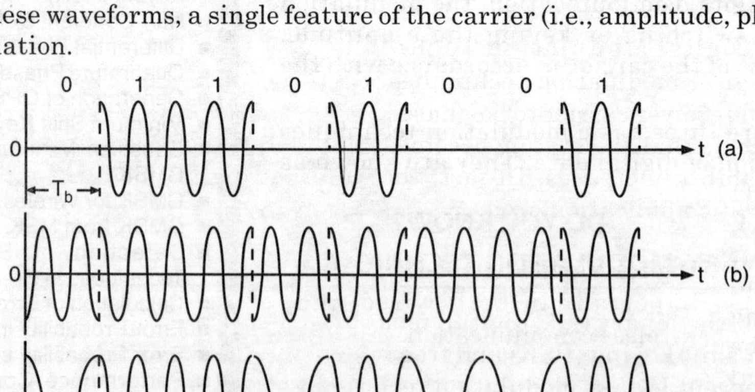

Fig. 3.1. The three basic forms of signaling binary information, (a) Amplitude-shift keying, (b) Phase-shift keying, (c) Frequency shift keying with continuous phase

In digital modulations, instead of transmitting one bit at a time, we transmit two or more bits simultaneously. This is known as M-ary transmission. This type of transmission results in reduced channel bandwidth. However, sometimes, we use two quadrature carriers for modulation. This process is known as **Quadrature modulation.**

Thus, we see that there are a number of modulation schemes available to the designer of a digital communication system required for data transmission over a bandpass channel.

Every scheme offers system trade-offs of its own. However, the final choice made by the designer is determined by the way in which the available primary communication resources such as transmitted power and channel bandwidth are best exploited. In particular, the choice is made in favour of a scheme which possesses as many of the following design characteristics as possible:

(i) Maximum data rate,
(ii) Minimum probability of symbol error,
(iii) Minimum transmitted power,
(iv) Maximum channel bandwidth,
(v) Maximum resistance to interfering signals,
(vi) Minimum circuit complexity.

3.3 TYPES OF DIGITAL MODULATION TECHNIQUES

(U.P.Tech., Sem. Examination, 2003-2004)

Basically, digital modulation techniques may be classified into coherent or non-coherent techniques, depending on whether the receiver is equipped with a phase-recovery circuit or not. The phase-recovery circuit ensures that the oscillator supplying the locally generated carrier wave receiver is synchronized* to the oscillator supplying the carrier wave used to originally modulate the incoming data stream in the transmitter.

(i) Coherent Digital Modulation Techniques

Coherent digital modulation techniques are those techniques which employ coherent detection. In coherent detection, the local carrier generated at the receiver is phase locked with the carrier at the transmitter. Thus, the detection is done by correlating received noisy signal and locally generated carrier. The coherent detection is a synchronous detection.

DO YOU KNOW?

Digital transmission uses frequency, phase, and amplitude variations, just as does analog transmission.

(ii) Non-coherent Digital Modulation Techniques

Non-coherent digital modulation techniques are those techniques in which the detection process does not need receiver carrier to be phase locked with transmitter carrier. The advantage of such type of system is that the system becomes simple. But the drawback of such a system is that the error probability increases. In fact, the different digital modulation techniques are used for various specific application areas.

3.4 COHERENT BINARY MODULATION TECHNIQUES

As mentioned earlier, the binary (i.e., digital) modulation has three basic forms amplitude-shift keying (ASK), phase-shift keying (PSK) and frequency-shift keying (FSK). In this section, let us discuss different coherent binary modulation techniques.**

---

*In both frequency and phase.

** To send data by radio, it is necessary to use a higher frequency carrier wave, just as for analog communication.

3.5 COHERENT BINARY AMPLITUDE SHIFT KEYING OR ON-OFF KEYING

(i) Definition

Amplitude shift keying (ASK) or ON-OFF keying (OOK) is the simplest digital modulation technique. In this method, there is only one unit energy carrier and it is switched on or off depending upon the input binary sequence.

Expression and Waveforms

The ASK waveform may be represented as,

$$s(t) = \sqrt{2P_s} \cos (2\pi f_c t) \text{ (To transmit '1')} \quad \ldots(3.1)$$

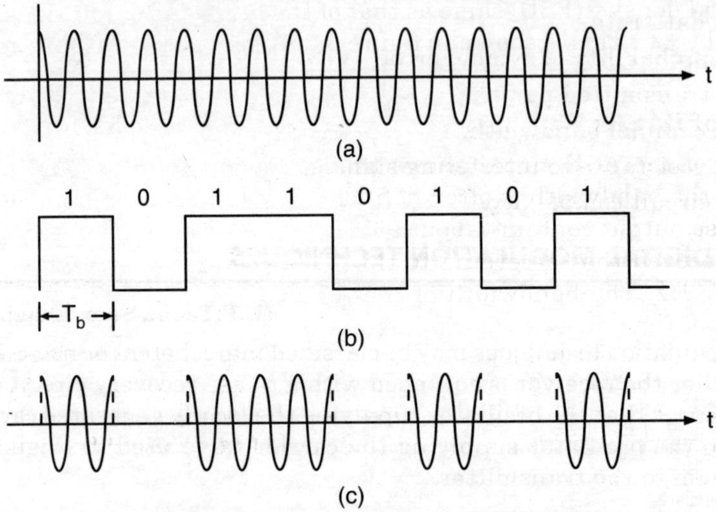

Fig. 3.2. Amplitude-shift keying waveforms, (a) Unmodulated carrier, (b) NRZ Unipolar bit sequence, (c) ASK waveform.

To transmit symbol '0', the signal s(t) = 0 i.e., no signal is transmitted. Signal s(t) contains some complete cycles of carrier frequency 'f_c'.

Hence, the ASK waveform looks like an ON-OFF of the signal. Therefore, it is also known as the **ON-OFF keying** (OOK). Figure 3.2 shows the ASK waveform.

3.5.1. Signal Space Diagram of ASK

The ASK waveform of equation (3.1) for symbol '1' can be represented as,

$$s(t) = \sqrt{P_s T_b} \cdot \sqrt{2/T_b} \cos (2\pi f_c t) = \sqrt{P_s T_b}\, \phi_1(t) \quad \ldots(3.2)$$

This means that there is only one carrier function $\phi_1(t)$. The signal space diagram will have two points on ϕ_1 (t). One will be at zero and other will be at $\sqrt{P_s T_b}$. Figure 3.3 shows this aspect.

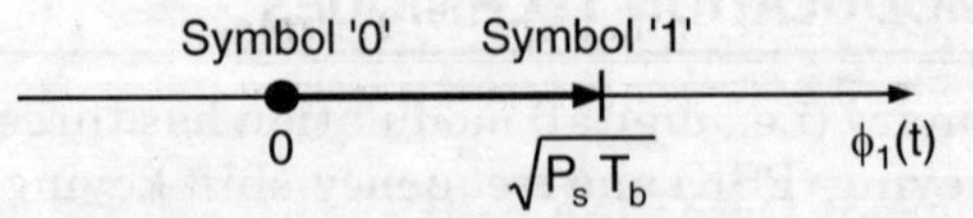

Fig. 3.3. Signal space diagram of ASK.

Thus, the distance between the two signal points is,

$$d = \sqrt{P_s T_b} = \sqrt{E_b} \quad \ldots(3.3)$$

3.5.2. Generation of ASK Signal

(i) Description and Working Operation

ASK signal may be generated by simply applying the incoming binary data (represented in unipolar form) and the sinusoidal carrier to the two inputs of a product modulator (i.e., balanced modulator). The resulting output will be the ASK waveform. This is shown in figure 3.4. Modulation causes a shift of the baseband signal spectrum.

(ii) Power Spectral Density (psd)

The ASK signal, which is basically the product of the binary sequence and the carrier signal, has a power spectral density (PSD) same as that of the baseband on-off signal but shifted in the frequency domain by $\pm f_c$. This is shown in figure 3.5. It may be noted that two impulses occur at $\pm f_c$.

(iii) Bandwidth of BASK

The spectrum of the ASK signal shows that it has an infinite bandwidth. However, for practical purpose, the bandwidth is often defined as the bandwidth of an ideal bandpass filter centered at f_c whose output contains about 95% of the total average power content of the ASK signal. It may be proved that according to this criterion the bandwidth of the ASK signal is approximately $3/T_b$ Hz. The bandwidth of the ASK signal can however, be reduced by using smoothed versions of the pulse waveform instead of rectangular pulse waveforms.

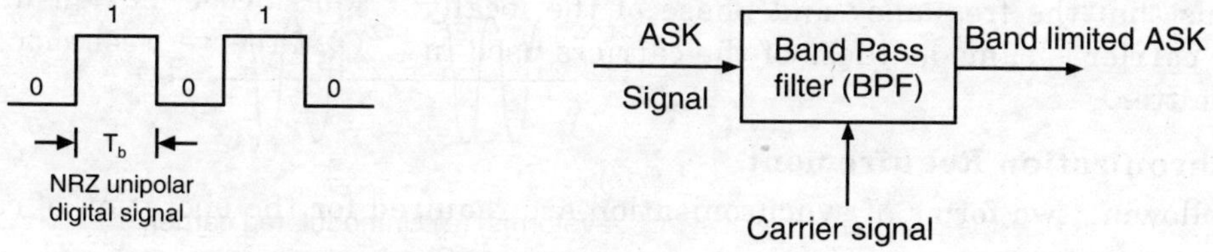

Fig. 3.4. Generation of binary ASK waveform

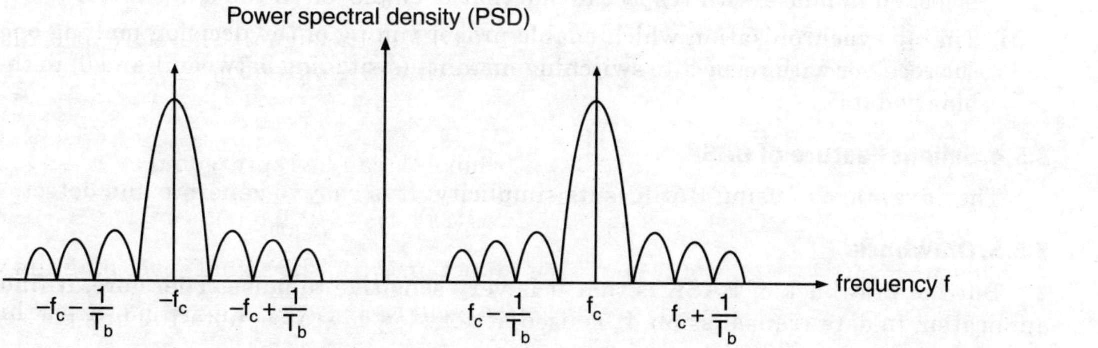

Fig. 3.5. Power spectral density of ASK signal.

3.5.3. BASK Reception: Coherent Detection or Demodulation of Binary ASK Signal

(i) Working Operation

The demodulation of binary ASK waveform can be achieved with the help of coherent detector as shown in figure 3.6. It consists of a product modulator which is followed by an integrator and a decision-making device. The incoming ASK signal is applied to one input of the product modulator.

The other input of the product modulator is supplied with a sinusoidal carrier which is generated with the help of a local oscillator. The output of the product modulator goes to input of the integrator. The integrator operates on the output of the multiplier for successive bit intervals and essentially performs a low-pass filtering action. The output of the integrator goes to the input of a decision-making device.*

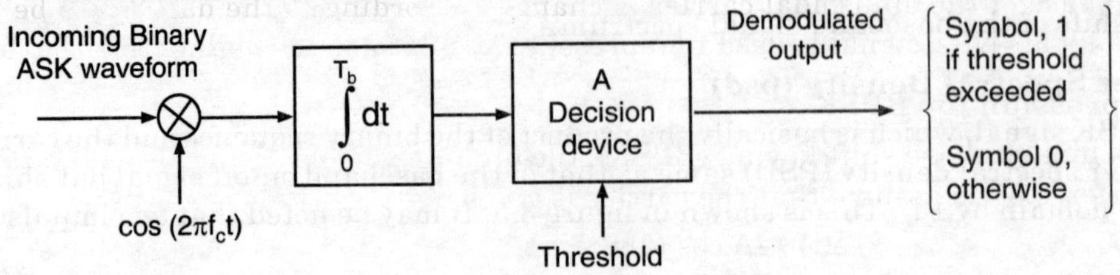

Fig. 3.6 Coherent detection of binary ASK signals.

Now, the decision-making device compares the output of the integrator with a present threshold. It makes a decision in favour of symbol 1 when the threshold is exceeded and in favour of symbol 0 otherwise. The coherent detection makes the use of linear operation. In this method we have assumed that the local carrier is in perfect synchronisation with the carriers used in the transmitter. This means that the frequency and phase of the locally generated carrier is same as those of the carriers used in the transmitter.

DO YOU KNOW?

Straight forward amplitude-shift keying (ASK) is rare in digital communication unless we count Morse code, but quadrature AM (QAM) is very common.

(ii) Synchronization Requirement

The following two forms of synchronisation are required for the operation of coherent (or synchronous detector):

(i) Phase synchronisation which ensures that carrier wave generated locally in the receiver is locked in phase with respect to one that is employed in the transmitter.

(ii) Timing synchronisation which enable proper timing of the decision making operation in the receiver with respect to switching instants (switching between 1 and 0) in the original binary data.

3.5.4. Salient Feature of BASK

The advantage of using BASK is its simplicity. It is easy to generate and detect.

3.5.5. Drawback

But the drawback of BASK is that it is very sensitive to noise, Therefore, it finds limited application in data transmission. It is used at very low bit rates, upto 100 bits per sec.

3.5.6. Bit Error Rate (BER) or Probability of Error

As a matter of fact, bit error rate (BER) or probability of error is a very important parameter. This parameter is used to judge the performance of a digital communication system. It is represented by P_e. P_e must be as small as possible.

---

*Since the high-frequency carrier is a sine wave for digital as well as analog signals, the same three basic parameters are available for modulation; amplitude, frequency and phase. All three, singly and in combination, are used in digital systems.

3.6 BINARY PHASE SHIFT KEYING (BPSK)*

(U.P.Tech., Sem. Exam., 2006-07)

(i) Definition

Binary phase shift keying (BPSK) is the most efficient of the three digital modulation, i.e., ASK, FSK and PSK. Hence, binary phase shift keying (BPSK) is used for high bit rates. In BPSK, phase of the sinusoidal carrier is changed according to the data bit to be transmitted. Also, a bipolar NRZ signal is used to represent the digital data coming from the digital source.

(ii) Expression for BPSK

In a binary phase shift keying (BPSK), the binary symbols '1' and '0' modulate the phase of the carrier. Let us assume that the carrier is given as,

$$s(t) = A \cos (2\pi f_c t) \quad ...(3.4)$$

Here 'A' represents peak value of sinusoidal carrier. For the standard 1 Ω load resistor, the power dissipated would be,

$$P = \frac{1}{2}A^2$$

or

$$A = \sqrt{2P} \quad ...(3.5)$$

Now, when the symbol is changed, then the phase of the carrier will also be changed by an amount of 180 degrees (i.e., π radians). Let us consider, for example,

For symbol '1', we have

$$s_1(t) = \sqrt{2P} \cos(2\pi f_c t) \quad ...(3.6)$$

If next symbol is '0', then we have

For symbol '0', we have

$$s_2(t) = \sqrt{2P} \cos (2\pi f_c t + \pi) \quad ...(3.7)$$

Now, because cos (θ + π) = – cos θ, therefore, the last equation can be written as,

$$s_2(t) = -\sqrt{2P} \cos (2\pi f_c t) \quad ...(3.8)$$

With the above equation, we can define BPSK signal combinely as,

$$s(t) = b(t) \sqrt{2P} \cos (2\pi f_c t) \quad ...(3.9)$$

where b(t) = + 1 when binary '1' is to be transmitted.

– 1 when binary '0' is to be transmitted

(iii) Binary Sequence and its Equivalent Signal b(t)

Figure 3.7 illustrates binary sequence and its equivalent signal b(t).

Important Point: *It may be observed from figure 3.7(b) that the signal b(t) is a* ***NRZ bipolar signal****. In fact, this signal directly modulates the carrier signal cos ($2\pi f_c t$).*

---

*When somewhat higher data rates are required in a band-limited channel than can be achieved with FSK, phase shift keying (PSK) is often used.

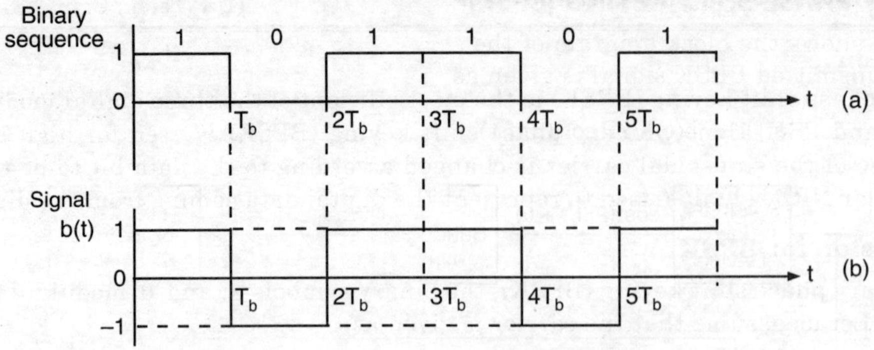

Fig. 3.7. (a) Binary sequence, (b) The corresponding bipolar signal b(t) .

3.6.1. Generation of BPSK Signal **(Important)**

BPSK signal may be generated by applying carrier signal to a balanced modulator. The binary data signal (0s and 1s) is converted into a NRZ bipolar signal by an NRZ encoder. Here, the bipolar signal b(t) is applied as a modulating signal to the balanced modulator.

Figure 3.8 shows the block diagram of a BPSK signal generator.

A NRZ level encoder converts the binary data sequence into bipolar NRZ signal.*

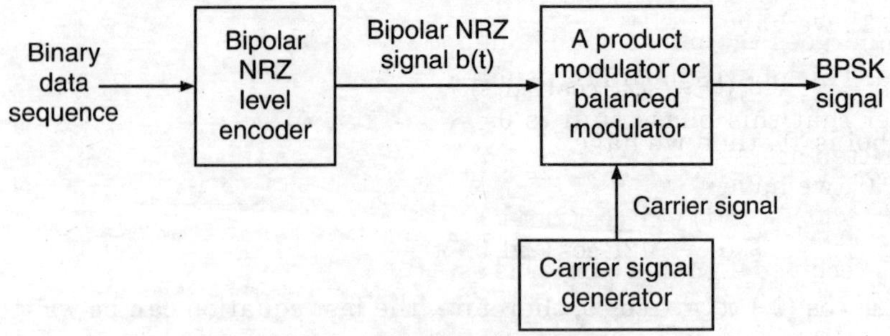

Fig. 3.8. Generation of BPSK

Table 3.1. Shows input digital and corresponding bipolar NRZ signal.

S.No.	Input digital signal	Bipolar NRZ signal b(t)	BPSK output signal
1.	Binary 0	b(t) = – 1	$-\sqrt{2P}\cos\omega_c t$
2.	Binary 1	b(t) = + 1	$+\sqrt{2P}\cos\omega_c t$

In above table,

(i) $P = \dfrac{E_b}{T_b}$, where, E_b is the signal energy and T_b is the bit duration

(ii) Also, $\omega_c = 2\pi f_c$

*Measuring phase requires a reference phase, which would be hard to maintain accurately. Usually, the phase of each symbol is compared with that of the previous symbol, rather than with a constant reference.

3.6.2. Reception of BPSK Signal: Coherent Detection

Figure 3.9 shows the block diagram of the scheme to recover baseband signal from BPSK signal. The transmitted BPSK signal is given as

$$s(t) = b(t)\sqrt{2P}\cos(2\pi f_c t)$$

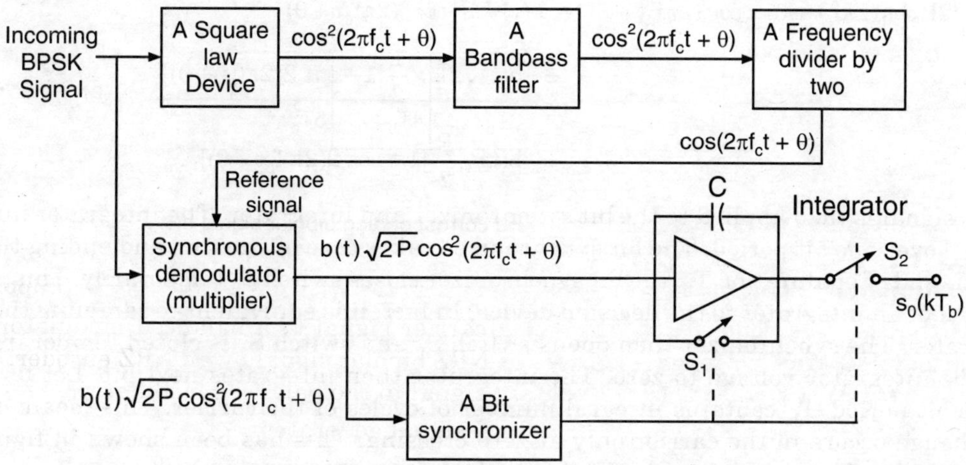

Fig. 3.9. Reception of baseband signal in BPSK signal.

This signal undergoes the phase change depending upon the time delay from transmitter end to receiver end. This phase change is, usually, a fixed phase shift in the transmitted signal.

Let us consider that this phase shift is θ. Because of this, the signal at the input of the receiver can be written as

$$s(t) = b(t)\sqrt{2P}\cos(2\pi f_c t + \theta) \qquad ...(3.10)$$

Now, from this received signal, a carrier is separated because this is coherent detection. As shown in the figure 3.9, the received signal is allowed to pass through a square law device. At the output of the square law device, we get a signal which is given as

$$\cos^2(2\pi f_c t + \theta)$$

Here, it may be noted that we have neglected the amplitude, since we are only interested in the carrier of the signal.

Again, we know that

$$\cos^2\theta = \frac{1+\cos 2\theta}{2}$$

Therefore, we have

$$\cos^2(2\pi f_c t + \theta) = \frac{1+\cos 2(2\pi f_c t + \theta)}{2} = \frac{1}{2} + \frac{1}{2}\cos 2(2\pi f_c t + \theta)$$

Here, $\frac{1}{2}$ represents a DC level. This signal is then allowed to pass through a bandpass filter (BPF) whose passband is centred around $2f_c$. Bandpass filter removes the DC level of $\frac{1}{2}$ and at the output, we obtain,

$$\cos 2(2\pi f_c t + \theta)$$

This signal is having frequency equal to 2 f_c. Hence, it is passed through a frequency divider by two. Thus, at the output of frequency divider, we get a carrier signal whose frequency is f_c i.e., $\cos(2\pi f_c t + \theta)$.

The synchronous (i.e., coherent) demodulator multiplies the input signal and the recovered carrier. Hence, at the output of multiplier, we get

$$b(t)\sqrt{2P}\cos(2\pi f_c t+\theta)\times\cos(2\pi f_c t+\theta) = b(t)\sqrt{2P}\cos^2(2\pi f_c t+\theta)$$

$$= b(t)\sqrt{2P}\times\frac{1}{2}[1+\cos 2(2\pi f_c t+\theta)]$$

$$= b(t)\sqrt{\frac{P}{2}}[1+\cos 2(2\pi f_c t+\theta)] \quad ...(3.11)$$

This signal is then applied to the bit synchronizer and integrator. The integrator integrates the signal over one bit period. The bit synchronized takes care of starting and ending times of a bit. At the end of bit duration T_b, the bit synchronizer closes switch S_2 temporarily. This connects the output of an integrator to the decision device. In fact, it is equivalent to sampling the output of integrator. The synchronizer then opens switch S_2 and switch S_1 is closed temporarily. This resets the integrator voltage to zero. The integrator then integrates next bit. Let us assume that one bit period 'T_b' contains integral number of cycles of the carrier. This means that the phase change occurs in the carrier only at zero crossing. This has been shown in figure 3.10. This BPSK waveform has full cycles of sinusoidal carrier.

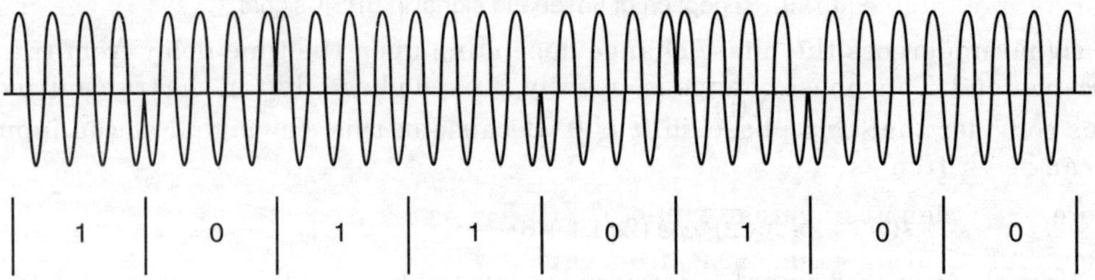

Fig. 3.10. The BPSK waveform

Also, in the kth bit interval, we can write output signal as under:

$$s_0(kT_b) = b(kT_b)\sqrt{\frac{P}{2}}\int_{(k-1)T_b}^{kT_b}[1+\cos(2\pi f_c t+\theta)]\,dt$$

This equation gives the output of an interval for k^{th} bit. Hence, integration is performed from $(k-1)T_b$ to kT_b. Here, T_b is the one bit period. We can write the above equation as under :

$$s_0(kT_b) = b(kT_b)\sqrt{\frac{P}{2}}\left[\int_{(k-1)T_b}^{kT_b} 1dt + \int_{(k-1)T_b}^{kT_b}\cos(2\pi f_c t+\theta)dt\right]$$

where $\int_{(k-1)T_b}^{kT_b}\cos 2(2\pi f_c t+\theta)\,dt = 0$, since average value of sinusoidal waveform is zero if integration is done over full cycles.

Hence, we can write above equation as,

$$s_0(kT_b) = b(kT_b)\sqrt{\frac{P}{2}}\int_{(k-1)T_b}^{kT_b} 1dt = b(kT_b)\sqrt{\frac{P}{2}}[t]_{(k-1)T_b}^{kT_b}$$

or $$s_0(kT_b) = b(kT_b)\sqrt{\frac{P}{2}}\{kT_b-(k-1)T_b\} = b(kT_b)\sqrt{\frac{P}{2}}\,T_b \quad ...(3.12)$$

Equation (3.12) shows that the output of the receiver depends on input.

Thus, $$s_0(kT_b) \propto b(kT_b)$$

Depending upon the value of $b(kT_b)$, the output $s_0(kT_b)$ is generated in receiver.

This signal is then applied to a decision device which decides whether transmitted symbol was zero or one.

3.6.3. The Spectrum of BPSK Signals

We know that the waveform b(t) is a NRZ binary waveform. In this waveform, there are rectangular pulses of amplitude $\pm V_b$. If we assume that each pulse is $\pm\frac{T_b}{2}$ around its centre, then it becomes easy to find Fourier transform of such pulse. The Fourier transform of this type of pulse is given as,

DO YOU KNOW?

Most phase-shift keying (PSK) systems use four phase angles for some what higher data rates than are achievable with FSK.

$$X(f) = V_b T_b \frac{\sin(\pi f T_b)^*}{(\pi f T_b)} \quad ...(3.13)$$

For a large number of such positive and negative pulses, the power spectral density S(f) is expressed as

$$S(f) = \frac{\overline{|X(f)|^2}}{T_s} \quad ...(3.14)$$

Here, $\overline{X(f)}$ denotes average value of X(f) due to all the pulses in b(t). And T_s is symbol duration. Substituting value of X(f) from equation (3.13) in equation (3.14), we get

$$S(f) = \frac{V_b^2 T_b^2}{T_s}\left[\frac{\sin(\pi f T_b)}{\pi f T_b}\right]^2$$

For BPSK, because only one bit is transmitted at a time, therefore, symbol and bit durations are same i.e., $T_b = T_s$. Then the last equation becomes,

$$S(f) = V_b^2 T_b \left[\frac{\sin(\pi f T_b)}{\pi f T_b}\right]^2 \quad ...(3.15)$$

This equation gives the power spectral density (psd) of baseband signal b(t). The BPSK signal is generated by modulating a carrier by the baseband signal b(t). Due to modulation of the carrier of frequency f_c, the spectral components are translated from f to $f_c + f$ and $f_c - f$. The magnitude of these components is divided by half.

Therefore, from equation (3.15) we can write the power spectral density of BPSK signal as under:

$$S_{BPSK}(f) = V_b^2 T_b \left\{\frac{1}{2}\left[\frac{\sin\pi(f_c-f)T_b}{\pi(f_c-f)T_b}\right] + \frac{1}{2}\left[\frac{\sin\pi(f_c+f)T_b}{\pi(f_c+f)T_b}\right]^2\right\}$$

*Here, $f_b = \frac{1}{T_b}$.

It may be noted that this equation consists of two half magnitude spectral components of same frequency 'f' above and below f_c. Let us assume that the value of $\pm V_b = \pm\sqrt{P}$. This means that the NRZ signal is having amplitudes of $+\sqrt{P}$ and $-\sqrt{P}$. Then the last equation becomes,

$$S_{BPSK}(f) = \frac{PT_b}{2}\left\{\left[\frac{\sin\pi(f-f_c)T_b}{\pi(f-f_c)T_b}\right]^2 + \frac{1}{2}\left[\frac{\sin\pi(f_c+f)T_b}{\pi(f_c+f)T_b}\right]^2\right\} \quad ...(3.16)$$

This equation gives power spectral density (psd) of BPSK signal for modulating signal b(t) having amplitudes equal to $\pm\sqrt{P}$.

Further, we know that the modulated signal is given as

$$s(t) = \pm\sqrt{2P}\cos(2\pi f_c t) \qquad [\text{Since, } A = \sqrt{2P}\,]$$

If $b(t) = \pm\sqrt{P}$, then the carrier becomes,

$$s(t) = \sqrt{2}\, b(t)\cos(2\pi f_c t) \quad ...(3.17)$$

Equation (3.15) describes power spectral density (psd) of the NRZ waveform. For one rectangular pulse, the shape of S(f) will be a sinc pulse as shown in figure 3.11.

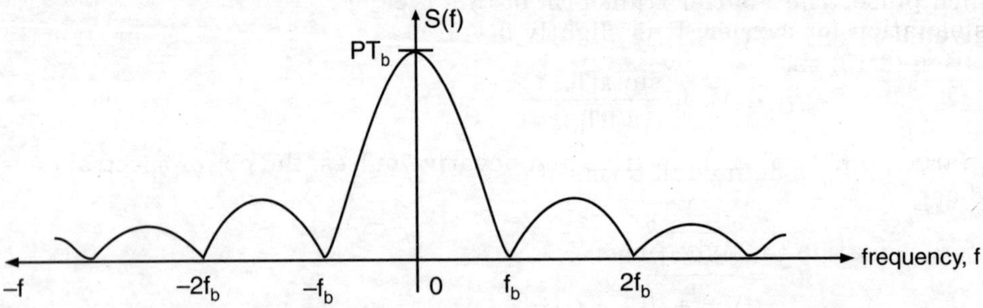

Fig. 3.11. Plot of power spectral density (psd) of NRZ baseband signal

It may be observed from this figure that the main lobe ranges from $-f_b$ to $+f_b$.* Because we have taken $\pm V_b = \pm\sqrt{P}$ in equation (3.15), therefore, the peak value of the main lobe is PT_b. Now let us consider the power spectral density (psd) of BPSK signal expressed by equation (3.16).

Figure 3.12 shows the plot of this equation. This figure clearly shows that there are two lobes, one at f_c and other at $-f_c$. The same spectrum of figure 3.11 has been placed at $+f_c$ and $-f_c$. However, the amplitudes of main lobes are $\frac{PT_b}{2}$ in figure 3.12.

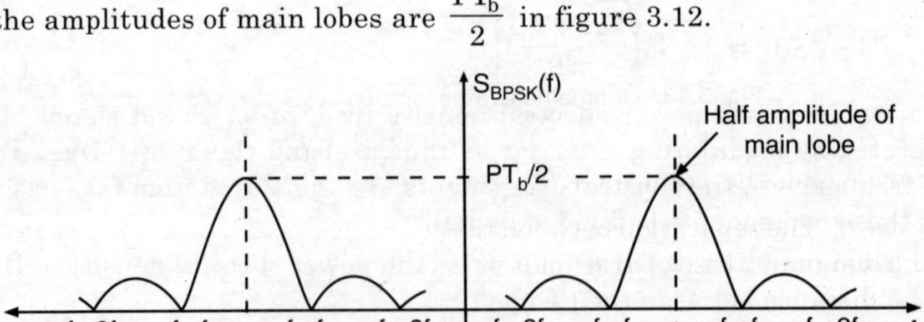

Fig. 3.12. Plot of power spectral density of BPSK signal

*Most phase-shift keying (PSK) systems use four phase angles for somewhat higher data rates than are achievable witb FSK.

Hence, they are reduced to half. The spectrum of S(f) as well as S_{BPSK} (f) extends overall the frequencies.

3.6.4. A Geometrical Representation for BPSK Signals

We know that BPSK signal carries the information about two symbols. These symbols are symbol '1' and symbol '0'. We can represent BPSK signal geometrically to show these two symbols. From equation (3.9), we know that BPSK signal is expressed as,

$$s(t) = b(t)\sqrt{2P}\cos(2\pi f_c t) \qquad ...(3.18)$$

Let us rearrange the last equation as,

$$s(t) = b(t)\sqrt{PT_b}\cdot\sqrt{\frac{2}{T_b}}\cos(2\pi f_c t) \qquad ...(3.19)$$

Now, let $\phi_1(t) = \sqrt{\frac{2}{T_b}}\cos(2\pi f_c t)$ represents an orthonormal carrier signal. Equation (3.17) also gives equation for carrier. It is slightly different than ϕ_1 (t) defined here. Then, we may write equation (3.19) as,

$$s(t) = b(t)\sqrt{PT_b}\,\phi_1(t) \qquad ...(3.20)$$

The bit energy E_b is defined in terms of power 'P' and bit duration T_b as,

$$E_b = PT_b \qquad ...(3.21)$$

Therefore, equation (3.20) becomes,

$$s(t) = \pm\sqrt{E_b}\,\phi_1(t) \qquad ...(3.22)$$

Here, b(t) is simply ± 1. Thus, on the single axis of $\phi_1(t)$, there will be two points. One point will be located at $+\sqrt{PT_b}$ or $+\sqrt{E_b}$ and other point will be located at $-\sqrt{PT_b}$ or $-\sqrt{E_b}$. This has been shown in figure 3.13.

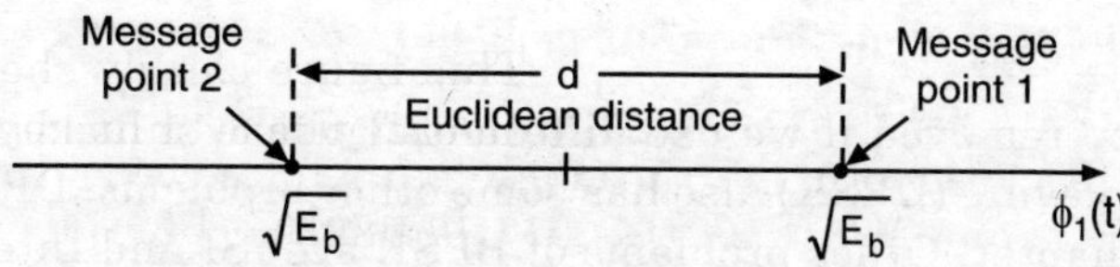

Fig. 3.13. Geometrical representation of BPSK signal

At the receiver end, the point at $+\sqrt{E_b}$ on $\phi_1(t)$ represents symbol '1' and point at $-\sqrt{E_b}$ represents symbol '0'. The separation between these two points represents the isolation in symbols '1' and '0' in BPSK signal. This separation is generally called distance 'd'. From figure 3.13, it is obvious that the distance between the two points is,

$$d = +\sqrt{E_b} - \left(-\sqrt{E_b}\right) = 2\sqrt{E_b} \qquad ...(3.23)$$

As this distance 'd' increases, the isolation between the symbols in BPSK signal is more. Thus, probability of error reduces.

3.6.5. Bandwidth for BPSK Signal

As discussed earlier, the spectrum of the BPSK signal is centred around the carrier frequency f_c.

If $f_b = \frac{1}{T_b}$, then for BPSK, the maximum frequency in the baseband signal will be f_b as shown in figure 3.12. In this figure, the main lobe is centred around carrier frequency f_c and extends from $f_c - f_b$ to $f_c + f_b$.

Therefore Bandwidth of BPSK signal will be,

BW = Highest frequency – Lowest frequency in the main lobe

$$BW = f_c + f_b - (f_c - f_b)$$

or

$$BW = 2f_b \quad ...(3.24)$$

Hence, the minimum bandwidth of BPSK signal is equal to twice of the highest frequency contained in baseband signal.

3.6.6. Salient Features of BPSK

(i) BPSK has a bandwidth which is lower than that of a BFSK signal.

(ii) BPSK has the best performance of all the three digital modulation techniques in presence of noise. It yields the minimum value of probability of error.

(iii) Binary phase shift keying (BPSK) has a very good noise immunity.

3.6.7. Drawbacks of BPSK

Figure 3.9 shows the block diagram of BPSK receiver. To regenerate the carrier in the receiver, we start by squaring $b(t)\sqrt{2P}\cos(2\pi f_c t + \theta)$. If the received signal is $-b(t)\sqrt{2P}\cos(2\pi f_c t + \theta)$, then the squared signal remains same as before. Hence, the recovered carrier is unchanged even if the input signal has changed its sign. Therefore, it is not possible to determine whether the received signal is equal to b(t) or – b(t). Infact, this results in ambiguity in the output signal.

Remedy

This problem can be removed if we use differential phase shift keying (DPSK). However, differential phase shift keying (DPSK) also has some other problems. DPSK will be discussed in detail later on in this chapter. Other problems of BPSK are ISI and Interchannel interference. However, these problems can be reduced to some extent by making use of filters.

3.6.8. Bit Error Rate (BER) or Probability of Error

The expression for probability of error, P_e, of a BPSK system is given by

$$P_e = \frac{1}{2}\,\text{erfc}\left[\sqrt{\frac{E}{N_o}}\right]^*$$

The above expression shows that the probability of error depends only on the energy contents of the signal i.e., E. Also, as the energy increases, the value of complementary error function (erfc) decreases and the value of P_e reduces.

*The derivation of this expression has been given sepatately in the end of this chapter.

3.7 COHERENT BINARY FREQUENCY SHIFT KEYING (BFSK)

(U.P.Tech., Sem. Exam; 2008-09)

In binary frequency shift keying (BFSK), the frequency of a sinusoidal carrier is shifted according to the binary symbol. In other words, the frequency of a sinusoidal carrier is shifted between two discrete values. However, the phase of the carrier is unaffected. This means that we have two different frequency signals according to binary symbols. Let there be a frequency shift by Ω. Then we can write following equations.

If $b(t) = \text{'1'}$, then $s_H(t) = \sqrt{2P_s}\cos(2\pi f_c + \Omega)\,t$ ...(3.25)

If $b(t) = \text{'0'}$, then $s_L(t) = \sqrt{2P_s}\cos(2\pi f_c - \Omega)\,t$ ...(3.26)

Hence, there is increase or decrease in frequency by Ω. Let us use the following conversion table to combine above two FSK equations:

Table 3.2. Conversion table for BFSK representation

b(t) Input	d(t)	$P_H(t)$	$P_L(t)$
1	+ 1V	+ 1V	0V
0	− 1V	0V	+ 1V

The equations (3.25) and (3.26) combinely may be written as

$$s(t) = \sqrt{2P_s}\cos\left[(2\pi f_c + d(t)\Omega)\,t\right] \quad \text{...(3.27)}$$

Hence, if symbol '1' is to be transmitted, the carrier frequency will be $f_c + \left(\frac{\Omega}{2\pi}\right)$ and is represented by f_H. If symbol '0' is to be transmitted, then the carrier frequency will be $f_c - \left(\frac{\Omega}{2\pi}\right)$ and is represented by f_L.

Therefore, we have

Thus, $f_H = f_c + \frac{\Omega}{2\pi}$ for symbol '1' ...(3.28)

$f_L = f_c - \frac{\Omega}{2\pi}$ for symbol '0' ...(3.29)

3.7.1. Generation of BFSK

(U.P.Tech., Sem. Exam; 2009-10) (10 marks)

It may be observed from Table 3.2 that $P_H(t)$ is same as b(t) and also $P_L(t)$ is inverted version of b(t). The block diagram for BFSK generation is shown in figure 3.14.

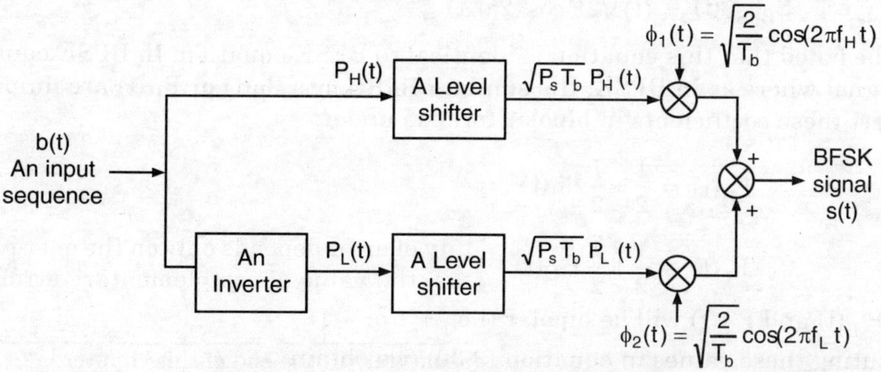

Fig. 3.14. Block diagram for BFSK generation

We know that input sequence b(t) is same as $P_H(t)$. An inverter is added after b(t) to get $P_L(t)$. The level shifter $P_H(t)$ and $P_L(t)$ are unipolar signals. The level shifter converts the '+1' level to $\sqrt{P_s T_b}$. Zero level is unaffected. Thus, the output of the level shifters will be either $\sqrt{P_s T_b}$ (if '+1') or zero (if input is zero). In other words, when a binary '0' is to be transmitted, $P_L(t) = 1$ and $P_H(t) = 0$, and for a binary '1' to be transmitted, $P_H(t) = 1$ and $P_L(t) = 0$. Hence, the transmitted signal will have a frequency of either f_H or f_L. Further, there are product modulators after level shifter. The two carrier signals $\phi_1(t)$ and $\phi_2(t)$ are used. $\phi_1(t)$ and ϕ_2 (t) are orthogonal to each other. In one bit period of input signal (i.e., T_b), ϕ_1 (t) or $\phi_2(t)$ have integral number of cycles.

Thus, the modulated signal is having continuous phase. Figure 3.15 shows such type of BFSK signal. The adder then adds the two signals.

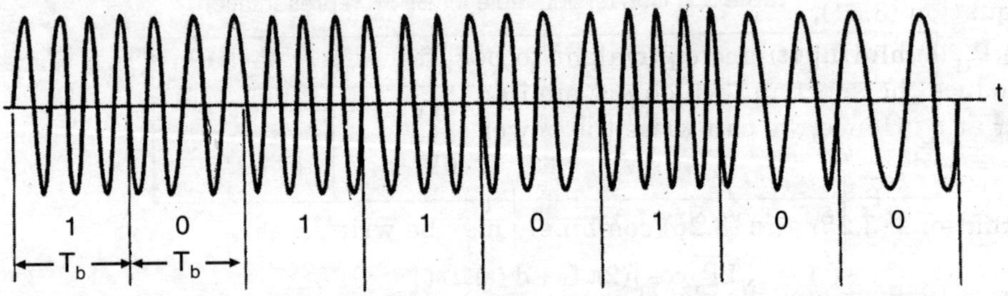

Fig. 3.15. The BFSK signal

Important Point: *Here it may be noted that outputs from both the multipliers are not possible at a time. This is because $P_H(t)$ and $P_L(t)$ are complementary to each other. Therefore, if $P_H(t) = 1$, then output will be only due to upper modulator and lower modulator output will be zero [since $P_L(t) = 0$)].*

3.7.2. The Spectrum of BFSK Signal

In figure 3.14, the BFSK signal s(t) may be written as,

$$s(t) = \sqrt{2P_s} P_H(t) \cos(2\pi f_H t) + \sqrt{2P_s} P_L(t) \cos(2\pi f_L t) \quad ...(3.30)$$

This is the expression for BFSK signal. Let us compare this equation with BPSK equation which is written below:

$$S_{BPSK}(t) = (t) \sqrt{2P} \cos(2\pi f_c t) \quad ...(3.31)$$

It may be noted that this equation is identical to BFSK equation. In BPSK equation, b(t) is a bipolar signal where as in BFSK, the similar coefficients $P_H(t)$ or P_L (t) are unipolar. Hence, let us convert these coefficients in bipolar form as under:

$$P_H(t) = \frac{1}{2} + \frac{1}{2} P'_H(t) \quad ...(3.32)$$

and

$$P_L(t) = \frac{1}{2} + \frac{1}{2} P'_L(t) \quad ...(3.33)$$

where $P'_H(t)$ and $P'_L(t)$ will be bipolar (i.e., + 1 or – 1).

Substituting these values in equation (3.30), we obtain

$$s(t) = \sqrt{2P_s}\left[\frac{1}{2}+\frac{1}{2}P_H^{'}(t)\right]\cos(2\pi f_H t) + \sqrt{2P_s}\left[\frac{1}{2}+\frac{1}{2}P_L^{'}(t)\right]\cos(2\pi f_L t)$$

or
$$s(t) = \sqrt{\frac{P_s}{2}}\cos(2\pi f_H t) + \sqrt{\frac{P_s}{2}}\cos(2\pi f_L t) + \sqrt{\frac{P_s}{2}}P_H^{'}(t)\cos(2\pi f_H t)$$

$$+\sqrt{\frac{P_s}{2}}P_L^{'}(t)\cos(2\pi f_L t) \quad ...(3.34)$$

DO YOU KNOW?

FSK is also used extensively in high-frequency radio systems for radio-teletype transmission.

In this equation, the first term represents the single frequency impulse situated at frequency f_H. The second term represents the impulse at f_L. These are constant amplitude pulses. The last two terms are identical to BPSK equation of equation (3.31).

Here $P_H'(t)$ and $P'_L(t)$ are equivalent to b(t). Therefore, these last two terms in equation (3.34) produce the spectrum which are similar to that of BPSK. One spectrum is located at f_H and other at f_L. Hence, we can write the power spectral density of BFSK as under:

$$S(f) = \sqrt{\frac{P_s}{2}}\left[\delta(f-f_H)+\delta(f+f_L)+\frac{P_sT_b}{2}\left\{\frac{\sin(\pi f_H T_b)}{\pi f_H T_b}\right\}^2+\frac{P_sT_b}{2}\left\{\frac{\sin(\pi f_L T_b)}{\pi f_L T_b}\right\}^2\right] \quad ...(3.35)$$

Figure 3.16 illustrates the plot of power spectral density of BFSK signal expressed by equation (3.35).

Also, f_H and f_L are selected such that,

$$f_H - f_L = 2f_b \quad ...(3.36)$$

With such types of selection, it is obvious from the spectrums in the above figure that the two frequencies f_H and f_L may be identified properly. The interference between the spectrums is not much with the above assumption.

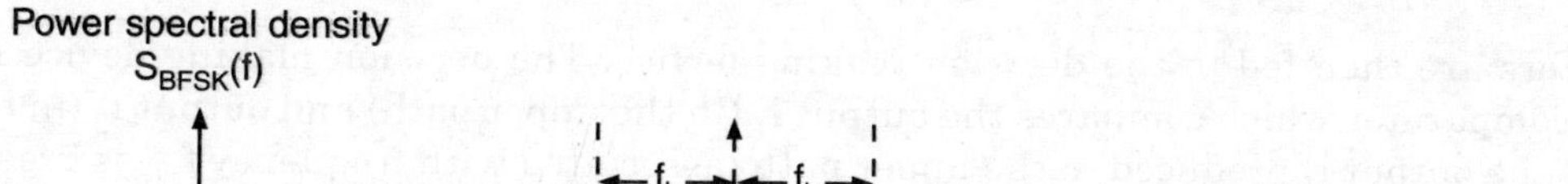

Fig. 3.16. Illustration of Power spectral density (psd) of a BFSK signal.

3.7.3. Bandwidth of BFSK Signal

From figure 3.16, it is obvious that the width of one lobe is $2f_b$. The two main lobes due to f_H and f_L are placed such that the total width due to both main lobes is $4f_b$.

Therefore, we have

Bandwidth of BFSK = $2f_b + 2f_b$

or $$BW = 4f_b \quad ...(3.37)$$

Now, if we compare this bandwidth with that of BPSK, we note that,

$$BW\ (BFSK) = 2 \times BW(BPSK) \quad ...(3.38)$$

3.7.4. BFSK Receiver: Coherent Detection of BFSK*

Figure 3.17 shows the block diagram of a scheme for demodulation of BFSK wave using coherent detection technique. The detector consists of two correlators that are individually tuned to two different carrier frequencies to represent symbols '1' and '0'. A correlator consists of a multiplier followed by an integrator. Then, the received binary FSK signal is applied to the multipliers of both the correlators. To the other input of the multipliers, carriers with frequency f_{c1} and f_{c2} are applied as shown in figure 3.17. The multiplied output of each multiplier is subsequently passed through integrators generating output l_1 and l_2 in the two paths. The output of the two

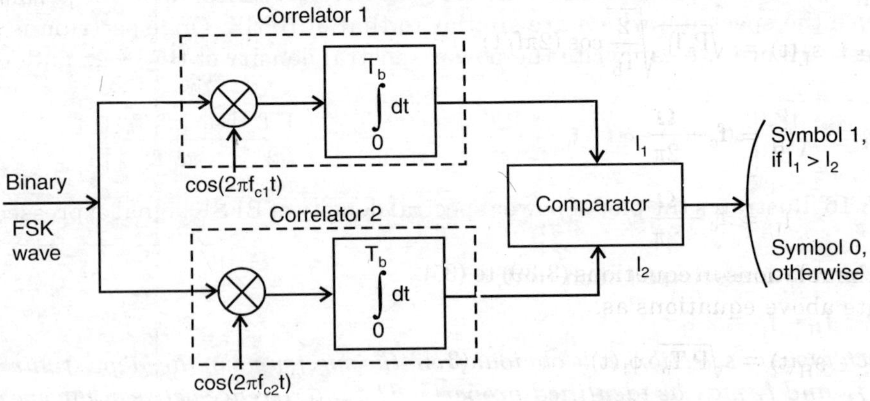

Fig. 3.17. Block diagram of BFSK receiver (detection of BFSK)

integrators are then fed to the decision making device. The decision making device is essentially a comparator which compares the output l_1 (in the upper path) and output l_2 (in the lower path). If the output l_1 produced in the upper path (associated with frequency f_{c1}) is greater than the output l_2 produced in the lower path (associated with frequency f_{c2}), the detector makes a decision in favour of symbol 1. If the output l_1 is less than l_2, then the decision making device decides in favour of symbol 0 (say). This type of digital communication receivers are also called correlation receivers. As discussed earlier, the detector based upon coherent detection requires phase and timing synchronisation.

3.7.5. Geometrical Representation of Orthogonal BFSK

As a matter of fact, orthogonal carriers are used for M-ary PSK and QASK. The different signal points are represented geometrically in $\phi_1\ \phi_2$-plane. For geometrical representation of BFSK signals, such orthogonal carriers are required. From figure 3.14, we know that two carriers $\phi_1(t)$ and $\phi_2(t)$ of two different frequencies f_H and f_L are used for modulation. To make $\phi_1(t)$ and $\phi_2(t)$ orthogonal, the frequencies f_H and f_L must be some integer multiple of band frequency 'f_b'.

Thus, $$f_H = m\ f_b \quad ...(3.39)$$

*Frequency-shift keying (FSK) uses two (and occasionally more than two) transmitted frequencies to achieve modest data rates with good performance in noisy channels.

and $$f_L = n\, f_b \quad ...(3.40)$$

Here; $f_b = \frac{1}{T_b}$, then the carriers would be

$$\phi_1(t) = \sqrt{\frac{2}{T_b}} \cos(2\pi m\, f_b\, t) \quad ...(3.41)$$

and $$\phi_2(t) = \sqrt{\frac{2}{T_b}} \cos(2\pi\, m f_b\, t) \quad ...(3.42)$$

The carriers $\phi_1(t)$ and $\phi_2(t)$ are orthogonal over the period T_b. We can write equation (3.25) and equation (3.26) as,

$$s_H(t) = \sqrt{P_s T_b} \sqrt{\frac{2}{T_b}} \cos(2\pi\, f_H t)$$

and $$s_L(t) = \sqrt{P_s T_b} \sqrt{\frac{2}{T_b}} \cos(2\pi f_L t)$$

Here $$f_H = f_c + \frac{\Omega}{2\pi}$$

and $$f_L = f_c - \frac{\Omega}{2\pi}$$

Using the relations in equations (3.39) to (3.42), we can write above equations as,

$$s_H(t) = \sqrt{P_s T_b} \cdot \phi_1(t) \quad ...(3.43)$$

and $$s_L(t) = \sqrt{P_s T_b} \cdot \phi_2(t) \quad ...(3.44)$$

Thus, based on the above two equations, we can draw the signal space diagram as shown in figure 3.18.

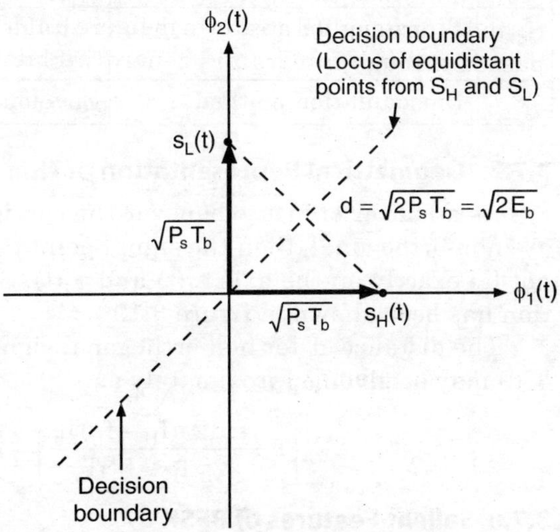

Fig. 3.18. Illustration of signal space representation of orthogonal BFSK signal.

3.7.5.1. Distance Between Signal Points

It may be noted that there are two signal points in the signal space. The distance between these two points may be evaluated as under:

$$d^2 = \left(\sqrt{P_s T_b}\right)^2 + \left(\sqrt{P_s T_b}\right)^2$$

or $$d^2 = 2P_s\, T_b$$

or $$d = \sqrt{2P_s T_b} \quad ...(3.45)$$

Since $P_s\, T_b = E_b$, we can write above relation (i.e., equation (3.45)) as under:

$$d = \sqrt{2E_b} \quad ...(3.46)$$

As compared to the distance of BPSK, we may observe that this distance is smaller than BPSK.

3.7.6. Performance Comparison of Three Basic Digital Modulation Techniques

We can compare three basic digital modulation techniques in the form of table 3.3.

Table 3.3.

S.No.	Parameter of comparison	Binary ASK	Binary FSK	Binary PSK
1.	Variable characteristic	Amplitude	Frequency	Phase
2.	Bandwidth (Hz) (spectral efficiency)	$2 f_b$	$4 f_b$	$2 f_b$
3.	Noise immunity	low	high	high
4.	Probability of error	high	low	low
5.	Performance in presence of noise	poor	Better than ASK	Best of three
6.	System complexity	Simple	Moderately complex	Very complex
7.	Bit rate or data rate	Suitable upto 100 bits/sec.	Suitable upto about 1200 bits/sec.	Suitable for high bit rates
8.	Demodulation method	Envelope detection	Envelope detection	Coherent detection

3.7.7. Geometrical Representation of Non-Orthogonal BFSK Signals

As a matter of fact, whenever the carriers $\phi_1(t)$ and $\phi_2(t)$ are non-orthogonal, then the signal point $S_H(t)$ or $S_L(t)$ would not lie exactly on the axis $\phi_1(t)$ and $\phi_2(t)$. Such a representation has been shown in figure 3.19.

The distance 'd' for non-orthogonal signal shown in figure 3.19 may be given approximately as,

$$d^2 = 2E_b\left[1 - \frac{\sin 2\pi(f_H - f_L)T_b}{2\pi(f_H - f_L)T_b}\right] \quad ...(3.47)$$

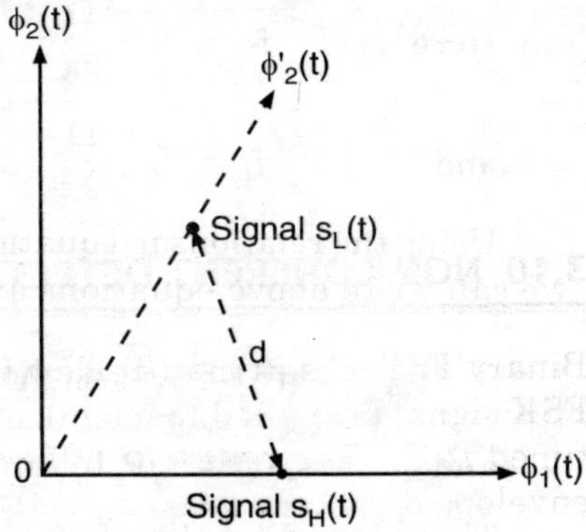

Fig. 3.19. Geometrical representation of non-orthogonal BFSK signal

3.7.8. Salient Features of BFSK

(i) BFSK is relatively easy to implement.

(ii) It has better noise immunity than ASK. Hence, the probability of error free reception of data is high.

3.7.9. Drawback of BFSK

The major drawback is its high bandwidth requirement*. Therefore, FSK is extensively used in low speed modems having bit rates below 1200 bits/sec.

3.8 NON-COHERENT BINARY MODULATION TECHNIQUES

As discussed earlier, coherent detection exploits knowledge of the carrier wave's phase reference, and thus providing the optimum error performance attainable with a digital modulation format of interest. However, when it is impractical to have knowledge of the carrier phase at the receiver, we make use of **non-coherent detection**. Thus, in this section, we shall study non-coherent binary modulation techniques i.e., we shall study non-coherent detection

DO YOU KNOW?

Frequency shift keying (FSK) uses two (and occasionally more than two) transmitted frequencies to achieve modest data rates with good performance in noisy channel.

*The maximum data rate for a channel is a function of bandwidth, modulation scheme, and signal-to-noise ratio.

of ASK and FSK. In the case of phase-shift keying (PSK), we cannot have **non-coherent PSK** since non-coherent means doing without phase information. However, there is a 'pseudo PSK' technique known as differential phase-shift keying (DPSK) which can be viewed as the non-coherent form of PSK.

3.9 NON-COHERENT BINARY AMPLITUDE SHIFT KEYING (ASK) (Important)

In the binary ASK case, the transmitted signal is defined as

$$s(t) = \sqrt{2P_s}\cos(2\pi f_c t)$$

Binary ASK signal can also be demodulated non-coherently using envelope detector. This greatly simplifies the design consideration required in synchronous detection. Non-coherent detection schemes do not require a phase-coherent local oscillator. This method involves some form of rectification and low pass filtering at the receiver. The block diagram of a non-coherent receiver for ASK signal has been shown in figure 3.20.

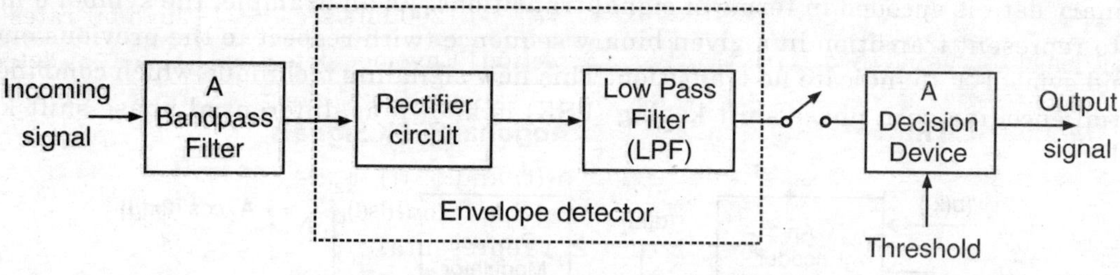

Fig. 3.20. Non-coherent ASK detector

3.10 NON-COHERENT DETECTION OF FSK

Binary FSK waves may be demodulated non-coherently using envelope detector. The received FSK signal is applied to a bank of two bandpass filters, one tuned to frequency f_{c1} and the other tuned to f_{c2}. Each filter is followed by an envelope detector. The resulting outputs of the two envelope detectors are sampled and then compared with each other. The arrangement for non-coherent detection of FSK signal has been shown in figure 3.21.

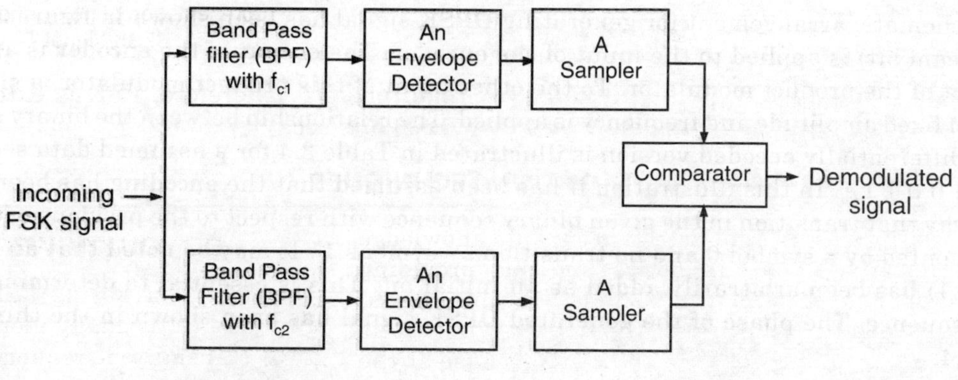

Fig. 3.21. Non-coherent detection of FSK binary signals

A decision is made in favour of symbol '1' if the envelope detector output derived from the filter tuned to frequency f_{c1} is larger than that derived from the second filter. Otherwise, a decision is made in favour of the symbol 0.

3.11 DIFFERENTIAL PHASE SHIFT KEYING (DPSK)

(i) Definition

We can view differential phase-shift keying as the non-coherent version of the PSK. Differential phase shift keying (DPSK) is differentially coherent modulation method. DPSK does not need a synchronous (coherent) carrier at the demodulator. The input sequence of binary bits is modified such that the next bit depends upon the previous bit. Therefore, in the receiver, the previous received bits are used to detect the present bit.

3.11.1 Generation of DPSK

(i) Encoding Technique

Thus, in order to eliminate the need for phase synchronisation of coherent receiver with PSK, a differential encoding system can be used with PSK. The digital information content of the binary data is encoded in terms of signal transitions. As an example, the symbol 0 may be used to represent transition in a given binary sequence (with respect to the previous encoded bit) and symbol '1' to indicate no transition. This new signaling technique which combines differential encoding with phase-shift keying (PSK) is known as differential phase-shift keying (DPSK).

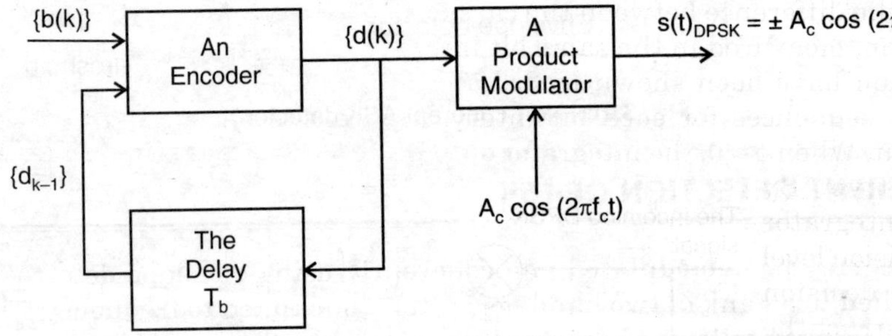

Fig. 3.22. Illustration of the scheme to generate DPSK signals.

(ii) Schematic Diagram

A schematic arrangement for generating DPSK signal has been shown in figure 3.23. The data stream b(t) is applied to the input of the encoder. The output of the encoder is applied to one input of the product modulator. To the other input of this product modulator, a sinusoidal carrier of fixed amplitude and frequency is applied. The relationship between the binary sequence and its differentially encoded version is illustrated in Table 3.4 for a assumed data sequence 0 0 1 0 0 1 0 0 1 1 1. In this illustration it has been assumed that the encoding has been done in such a way that transition in the given binary sequence with respect to the previous encoded bit is represented by a symbol 0 and no transition by symbol '1'. It may be noted that an extra bit (symbol 1) has been arbitrarily added as an initial bit. This is essential to determine the encoded sequence. The phase of the generated DPSK signal has been shown in the third row of Table 3.4.

Table 3.4. Differentially encoded sequences with phase.

Binary data {b(k)}		0	0	1	0	0	1	0	0	1	1
Differentially encoded data {d(k)}	1*	0	1	1	0	1	1	0	1	1	1
Phase of DPSK	0	π	0	0	π	0	0	π	0	0	0
Shifted differentially encoded data $\{d_{k-1}\}$		1	0	1	1	0	1	1	0	1	1
Phase of shifted DPSK		0	π	0	0	π	0	0	π	0	0
Phase comparison output		–	–	+	–	–	+	–	–	+	+
Detected binary sequence		0	0	1	0	0	1	0	0	1	1

* Arbitrary starting reference bit.

3.11.2. Detection of DPSK

For detection of the differentially encoded PSK (i.e., DPSK), we can use the receiver arrangement as shown in figure 3.23. The received DPSK signal is applied to one input of the multiplier. To the other input of the multiplier, a delayed version of the received DPSK signal by the time interval T_b is applied. The delayed version of the received DPSK signal (in the absence of channel noise) has been shown in the 4th row of the table. The output of the difference is proportional to cos (ϕ), here ϕ is the difference between the carrier phase angle of the received DPSK signal and its delayed version, measured in the same bit interval. The phase angle of the DPSK signal and its delayed version have been shown in 3rd and 5th rows respectively. The phase difference between the two sequences for each bit interval is used to determine the sign of the phase comparator output. When $\phi = 0$, the integrator output is positive whereas when $\phi = \pi$, the integrator output is negative. By comparing the integrator output with a decision level of zero volt, the decision device can reconstruct the binary sequence by assigning a symbol '0' for negative output and a symbol '1' for positive output. The reconstructed binary data is shown in the last row of the table. It is thus seen that in the absence of noise, the receiver can reconstruct the transmitted binary data exactly. *DPSK may be viewed as a non-coherent version of PSK. It may also be noted that the reconstruction is invariant with the choice of the initial bit in the encoded data. This has been illustrated in the example 3.1 given below.*

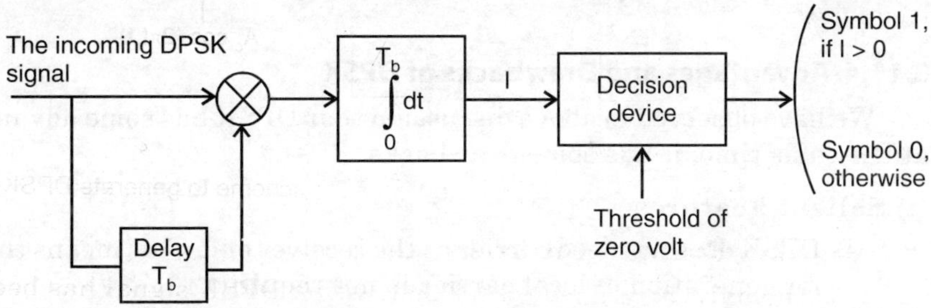

Fig. 3.23. Receiver for the detection of DPSK signals

EXAMPLE 3.1. A binary data stream 0 0 1 0 0 1 0 0 1 1 needs to be transmitted using DPSK technique. Prove that the reconstruction of the DPSK signal by the technique discussed in the previous article is independent of the choice of the extra bit.

Solution: In the last article, we have observed that DPSK signal can be detected accurately (in the absence of channel noise) without having a local oscillator for generation of synchronous carrier. The initial bit in the differentially encoded data was assumed to be '1'. In this example, we use the initial bit to be symbol '0' and verify that the reconstruction is invariant with the choice of the initial bit. The results obtained for this case are given in Table 3.5. It can be easily verified that the extra chosen bit 0 changes the phase of the DPSK sequence but the detected sequence remains invariant.

Table 3.5. Differentially encoded sequences with phase.

Binary data {b(k)}		0	0	1	0	0	1	0	0	1	1
Differentially encoded data {d(k)}	0*	1	0	0	1	0	0	1	0	0	0
Phase of DPSK	π	0	π	π	0	π	π	0	π	π	π
Shifted differentially encoded data $\{d_{k-1}\}$		0	1	0	0	1	0	0	1	0	0
Phase of shifted DPSK		π	0	π	π	0	π	π	0	π	π
Phase comparison output		–	–	+	–	–	+	–	–	+	+
Detected binary sequence		0	0	1	0	0	1	0	0	1	1
* Starting reference bit.											

3.11.3. Evaluation of Bandwidth of DPSK Signal

As discussed earlier that one previous bit is used to decide the phase shift of next bit. Thus, change in b(t) occurs only if input bit is at level '1'. No change happens if input bit is at level '0'. Because, one previous bit is always used to define the phase shift in next bit, therefore, the symbol can be said to have two bits. Hence, one symbol duration (T) is equivalent to two bits duration ($2T_b$) i.e.,

Symbol duration $\quad T = 2T_b$ ...(3.48)

Bandwidth is expressed as

$$BW = \frac{2}{T} = \frac{1}{T_b} = f_b \quad ...(3.49)$$

Hence, the minimum bandwidth in DPSK is equal to f_b, i.e., maximum baseband signal frequency.

3.11.4. Advantages and Drawbacks of DPSK

We have observed in above discussion that DPSK has some advantages over BPSK, however, at the same time, it has some drawbacks.

(i) Salient Features

(i) DPSK does not need carrier at the receiver end. This means that the complicated circuitry for generation of local carrier is not required.

(ii) The bandwidth requirement of DPSK is reduced as compared to that of BPSK.

(ii) Drawbacks

(i) The probability of error (i.e., bit error rate) of DPSK is higher than that of BPSK.

(ii) Because DPSK uses two successive bits for its reception, error in the first bit creates error in the second bit. Therefore, error propagation in DPSK is more. On the other hand, in BPSK single bit can go in error since detection of each bit is independent.

(iii) Noise interference in DPSK is more.

Important Point: *In DPSK, previous bit is used to detect next bit. Hence, if error is present in previous bit, detection of next bit can also be wrong. Hence, error is created in next bit also. Therefore, there is tendency of appearing errors in pairs in DPSK.*

3.11.5. Performance comparison of BPSK and DPSK

We can carry out the performance comparison of BPSK and DPSK in the form of table 3.6.

Table 3.6.

S.No.	Parameters of Comparison	BPSK	DPSK
1.	Variable characteristics	phase	phase
2.	Bandwidth	f_b	f_b
3.	Probability of error	Low	Higher than BPSK
4.	System complexity	Lower than DPSK	Higher than BPSK
5.	Demodulation technique	Synchronous	Synchronous
6.	Noise effect	Low	Higher than BPSK
7.	Requirement of synchronous carrier	is required	not required
8.	Bit determination at the receiver	Based on single bit interval	Based on signal received in two successive bit intervals

3.12 QUADRATURE PHASE SHIFT KEYING (QPSK) (U.P.Tech., Sem. Exam., 2005-06)

As a matter of fact, in communication systems, we have two main resources. These are the transmission power and the channel bandwidth. The channel bandwidth depends upon the bit rate or signaling rate f_b. In digital bandpass transmission, we use a carrier for transmission. This carrier is transmitted over a channel. If two or more bits are combined in some symbols, then the signaling rate will be reduced. Thus, the frequency of the carrier needed is also reduced. This reduces the transmission channel bandwidth. Hence, because of grouping of bits in symbols, the transmission channel bandwidth can be reduced. In quadrature phase shift keying (QPSK), two successive bits in the data sequence are grouped together. This reduces the bit rate or signaling rate (i.e., f_b) and thus reduces the bandwidth of the channel.*

In case of BPSK, we know that when symbol changes the level, the phase of the carrier is changed by 180°. Because, there were only two symbols in BPSK, the phase shift occurs in two levels only. However, in QPSK, two successive bits are combined. Infact, this combination of two bits forms four distinct symbols. When the symbol is changed to next symbol, then the phase of the carrier is changed by 45° ($\pi/4$ radians). Table 3.7 shows these symbols and their phase shifts.

Table 3.7. Symbol and corresponding phase shifts in QPSK

S.No.	Input successive bits		Symbol	Phase shift in carrier
1.	1(1 V)	0 (–1 V)	S_1	$\pi/4$
2.	0(–1 V)	0(–1 V)	S_2	$3\pi/4$
3.	0(–1 V)	1(1 V)	S_3	$5\pi/4$
4.	1(1 V)	1(1 V)	S_4	$7\pi/4$

Hence as shown in Table 3.7, there are four symbols and the phase is shifted by $\pi/4$ for each symbol.

*Because the digital input to a QPSK modulator is a binary (base 2) signal, to produce four different input combinations, the modulator requires more than a single input bit to determine the output conditions.

3.13 GENERATION OF QPSK

Figure 3.24 shows the block diagram of QPSK transmitter. Here, the input binary sequence is first converted to a bipolar NRZ type of signal. This signal is denoted by b(t). It represents binary '1' by + 1 V and binary '0' by – 1 V. This signal has been shown in figure 3.25(a). The demultiplexer divides b(t) into two separate bit streams of the odd numbered and even numbered bits. Here, $b_e(t)$ represents even numbered sequence and $b_0(t)$ represents odd numbered sequence. The symbol duration of both of these odd and even numbered sequences is $2T_b$. Hence, each symbol consists of two bits. Figure 3.25(b) and (c) illustrate the waveform of $b_e(t)$ and $b_0(t)$.

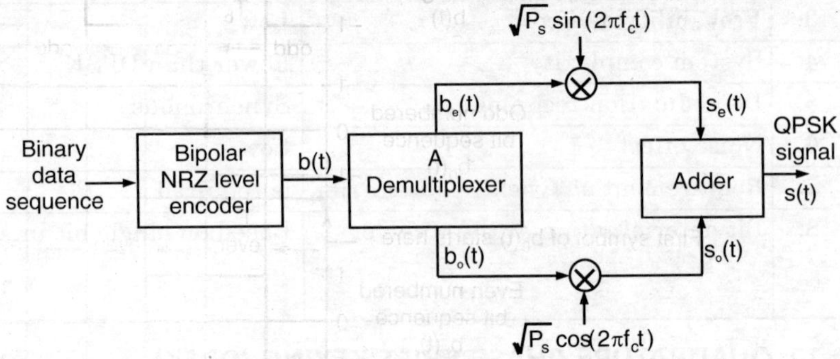

Fig. 3.24 Generation of QPSK.

It may be observed that the first even bit occurs after the first odd bit. Hence, even numbered bit sequence $b_e(t)$ starts with the delay of one bit period due to first odd bit. Thus, first symbol of $b_e(t)$ is delayed by one bit period 'T_b' with respect to first symbol of $b_0(t)$. This delay of T_b is known as offset. This shows that the change in levels of $b_e(t)$ and $b_0(t)$ cannot occur at the same time due to offset or staggering.

Also, the bit steam $b_e(t)$ modulates carrier $\sqrt{P_s}\cos(2\pi f_c t)$ and $b_0(t)$ modulates $\sqrt{P_s}\sin(2\pi f_c t)$. These modulators are the balanced modulators. The two carriers $\sqrt{P_s}\cos(2\pi f_c t)$ and $\sqrt{P_s}\sin(2\pi f_c t)$ have been shown in figure 3.25(d) and (e). There carriers are also known as **quadrature carriers.**

> **DO YOU KNOW?**
>
> In QPSK, each symbol represents two bits and the bit rate is twic the baud rate. This is called a dibit system.

The two modulated signals can be written as,

$$s_e(t) = b_e(t)\sqrt{P_s}\sin(2\pi f_c t) \quad ...(3.50)$$

and

$$s_0(t) = b_0(t)\sqrt{P_s}\cos(2\pi f_c t) \quad ...(3.51)$$

Hence, $s_e(t)$ and $s_0(t)$ are basically BPSK signals. The only difference is that $T = 2T_b$ here. The value of $b_e(t)$ and $b_0(t)$ would be + 1V or – 1V. Figure 3.25 (f) and (g) shows the waveforms of $s_e(t)$ and $s_0(t)$. The adder in figure 3.24 adds these two signals $b_e(t)$ and $b_0(t)$.

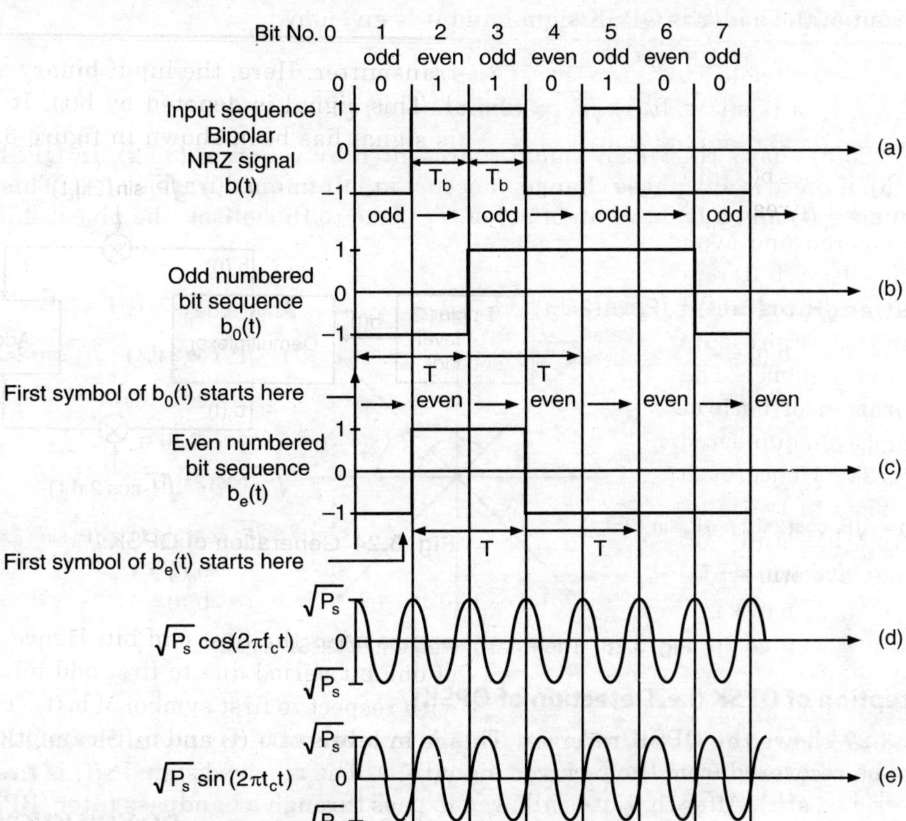

Fig. 3.25. QPSK waveforms, (a) Input sequence and its corresponding NRZ waveform, (b) Odd numbered bit sequence and its corresponding waveform (c) Even numbered bit sequence and its NRZ waveform (d) Basis function $f_1(t)$ (e) Basis function $f_2(t)$ (f) Binary PSK waveform for odd numbered channel (g) Binary PSK waveform for even numbered channel (h) Final QPSK waveform

The output of the adder is QPSK signal and it is given by,

$$s(t) = s_0(t) + s_e(t)$$

or
$$s(t) = b_0(t)\sqrt{P_s}\cos(2\pi f_c t) + b_e(t)\sqrt{P_s}\sin(2\pi f_c t) \qquad ...(3.52)$$

Figure 3.25(h) shows the QPSK signal represented by equation (3.52). In QPSK signal in figure 3.25(h), if there is any phase change, it occurs at minimum duration of T_b. This is because the two signals $s_e(t)$ and $s_0(t)$ have an offset of 'T_b'. Due to this offset, the phase shift in QPSK signal is $\frac{\pi}{2}$.

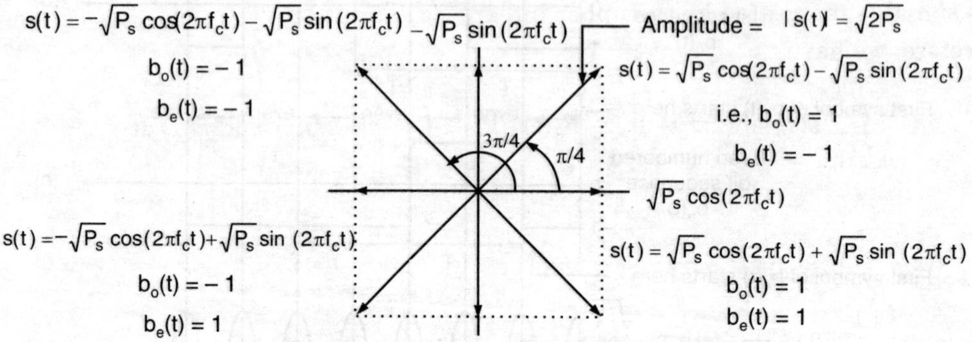

Fig. 3.26. The Phasor diagram of QPSK signal

3.13.1. Reception of QPSK (i.e. Detection of QPSK)

Figure 3.27 shows the QPSK receiver. This is synchronous reception. Hence, the coherent carrier is to be recovered from the received signal s(t). The received signal s(t) is first raised to its 4th power, i.e., $s^4(t)$. After that, it is allowed to pass through a bandpass filter (BPF) which is centred around $4f_c$. The output of the bandpass filter is a coherent carrier of frequency $4f_c$. This is divided by 4 and it provides two coherent quadrature carriers, i.e., cos ($2\pi f_c t$) and sin ($2\pi f_c t$). These coherent carriers are applied to two synchronous demodulators. These synchronous demodulators consist of multiplier and an integrator.*

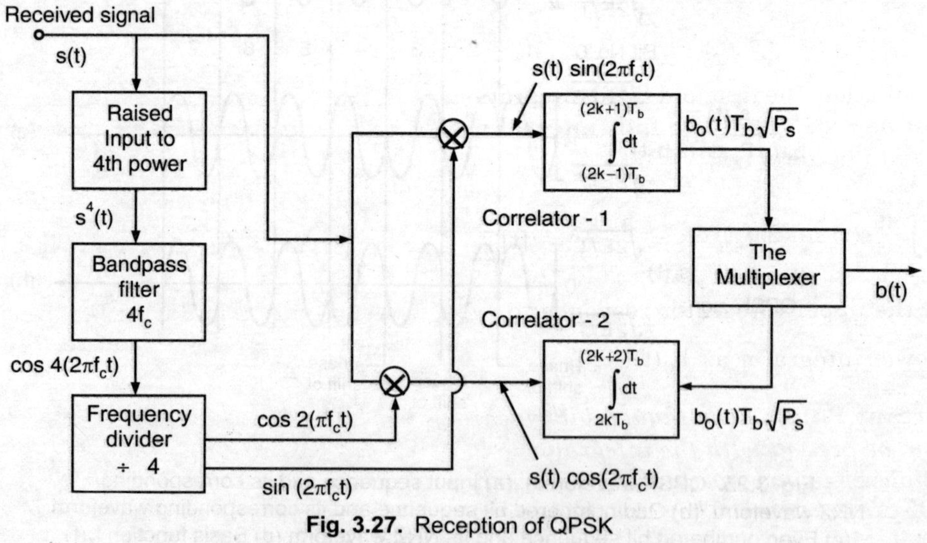

Fig. 3.27. Reception of QPSK

*The carrier recovery circuit reproduces the original transmit carrier oscillator signal. The recovered carrier must be frequency and phase coherent with the transmit reference carrier.

The incoming signal is applied to both the multipliers. Here, the integrator integrates the product signal over two bit interval (i.e., $T_s = 2T_b$). At the end of this period, the output of integrator is sampled. The outputs of the two integrators are sampled at the offset of one bit period, T_b. Hence, the output of the multiplexer is the signal b(t). This means that the odd and even sequences are combined by multiplexer.

Now, let us consider the product signal at the output of upper multiplier, i.e.,

$$s(t)\sin(2\pi f_c t) = b_0(t)\sqrt{P_s}\cos(2\pi f_c t)\sin(2\pi f_c t) + b_e(t)\sqrt{P_s}\sin^2(2\pi f_c t)$$

This signal is integrated by the upper integrator in figure 3.27.

Therefore, we have

$$\int_{(2k-1)T_b}^{(2k+1)T_b} s(t)\sin(2\pi f_c t)\,dt = b_0(t)\sqrt{P_s}\int_{(2k-1)T_b}^{(2k+1)T_b}\cos(2\pi f_c t)\sin(2\pi f_c t)\,dt$$

$$+\,b_e(t)\sqrt{P_s}\int_{(2k-1)T_b}^{(2k+1)T_b}\sin^2(2\pi f_c t)\,dt \qquad ...(3.53)$$

Now, since $\frac{1}{2}\sin(2x) = \sin x\cdot\cos x$

and $\sin^2(x) = \frac{1}{2}[1-\cos(2x)]$

Therefore, using these two trigonometric identities in equation (3.53), we get

$$\int_{(2k-1)T_b}^{(2k+1)T_b} s(t)\sin(2\pi f_c t)\,dt = \frac{b_o(t)\sqrt{P_s}}{2}\int_{(2k-1)T_b}^{(2k+1)T_b}\sin 4\pi f_c t\,dt + \frac{b_e(t)\sqrt{P_s}}{2}\int_{(2k-1)T_b}^{(2k+1)T_b} 1\cdot dt$$

$$-\frac{b_e(t)\sqrt{P_s}}{2}\int_{(2k-1)T_b}^{(2k+1)T_b}\cos 4\pi f_c t\,dt$$

In this equation, the first and third integration terms involve integration of sinusoidal carriers over two bit period. They have full (integral number of) cycles over two bit periods and thus integration will be zero, i.e.,

$$\int_{(2k-1)T_b}^{(2k+1)T_b} s(t)\sin(2\pi f_c t)\,dt = \frac{b_e(t)\sqrt{P_s}}{2}[t]_{(2k-1)T_b}^{(2k+1)T_b} = \frac{b_e(t)\sqrt{P_s}}{2}\times 2T_b = b_e(t)\sqrt{P_s}\,T_b \qquad ...(3.54)$$

Hence, the upper integrator responds to even sequence only. Similarly, we can obtain the output of lower integrator as $b_0(t)\sqrt{P_s}\,T_b$.

> **Important Point:** *Even though bit synchronizer has not been shown in figure 3.27, it is assumed to be present with the integrator to locate starting and ending times of integration. The multiplexer is also operated by bit synchronizer. The amplitudes of voltage marked in figure 3.27 are arbitrary. They can change depending upon the gains of the integrator.*

3.13.2. Concept of Carrier Synchronization in QPSK

Both the carriers are to be synchronized properly in coherent detection in QPSK. Figure 3.28 shows the PLL system for carrier synchronization in QPSK.

The fourth power of the input signal consists of discrete frequency component at $4f_c$.

We know that,

$$\cos 4(2\pi f_c t) = \cos(8\pi f_c t + 2\pi N)$$

where 'N' is the number of cycles over the bit period. It is always an integer value. When the frequency division by four takes place, the RHS of this equation becomes $\cos\left(2\pi f_c t + \frac{N\pi}{2}\right)$. This indicates that the output has a fixed phase error of $\frac{N\pi}{2}$. Differential encoding can be used to nullify the phase error events. The PLL remains locked with the phase of '$4f_c$' and then output of PLL is divided by 4. This provides a coherent carrier. A 90° phase shift is added to this carrier to produce a quadrature carrier.

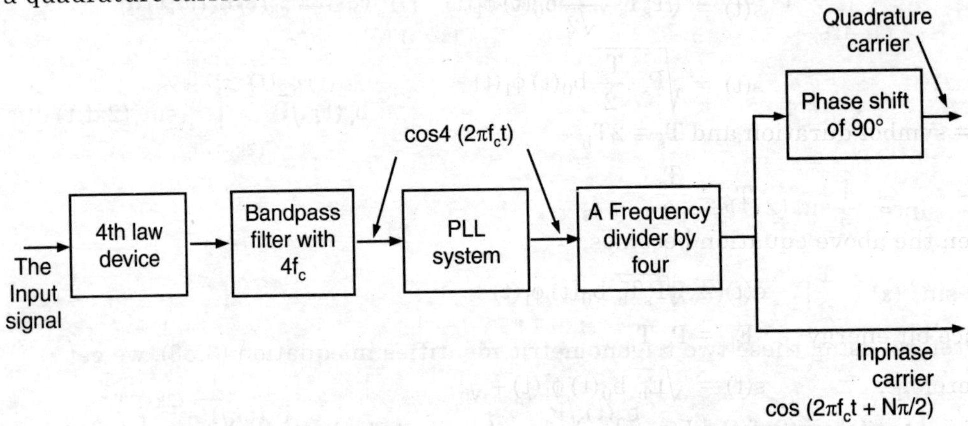

Fig. 3.28 PLL system used for carrier synchronization in QPSK

3.13.3. Signal Space Representation in QPSK Signals

Figure 3.29 shows the phasor diagram of QPSK signal. Depending upon the combination of two successive bits, the phase shift occurs in carrier. This means that the QPSK signal can be written as,

$$s(t) = \sqrt{2P_s}\cos\left[2\pi f_c t + (2m+1)\frac{\pi}{4}\right] \qquad m = 0, 1, 2, 3 \qquad ...(3.55)$$

Here, above expression takes four values and they represent the phasors of figure 3.29. This equation can be expanded as under:

$$s(t) = \sqrt{2P_s}\cos(2\pi f_c t)\cos\left[(2m+1)\frac{\pi}{4}\right] - \sqrt{2P_s}\sin(2\pi f_c t)\sin\left[(2m+1)\frac{\pi}{4}\right]$$

Let us rearrange the above equation as under:

$$s(t) = \left\{\sqrt{P_s T_s}\cos\left[(2m+1)\frac{\pi}{4}\right]\right\}\sqrt{\frac{2}{T_s}}\cos(2\pi f_c t)$$

$$-\left\{\sqrt{P_s T_s}\sin\left[(2m+1)\frac{\pi}{4}\right]\right\}\sqrt{\frac{2}{T_s}}\sin(2\pi f_c t) \qquad ...(3.56)$$

Again, let $\qquad \phi_1(t) = \sqrt{\frac{2}{T_s}}\cos(2\pi f_c t) \qquad ...(3.57)$

and $$\phi_2(t) = \sqrt{\frac{2}{T_s}} \sin(2\pi f_c t) \quad ...(3.58)$$

These two signals are known as orthogonal signals and they are used as carriers in QPSK modulator.

Let $$b_0(t) = \sqrt{2}\cos\left[(2m+1)\frac{\pi}{4}\right] \quad ...(3.59)$$

and $$b_e(t) = -\sqrt{2}\sin\left[(2m+1)\frac{\pi}{4}\right] \quad ...(3.60)$$

With the use of equations (3.57) to (3.60) we can write equation (3.56) as under:

$$s(t) = \sqrt{P_s T_s}\cdot\frac{1}{\sqrt{2}} b_0(t)\,\phi_1(t) + \sqrt{P_s T_s}\cdot\frac{1}{\sqrt{2}} b_e(t)\,\phi_2(t)$$

or $$s(t) = \sqrt{P_s\cdot\frac{T_s}{2}}\, b_0(t)\,\phi_1(t) + \sqrt{P_s\frac{T_s}{2}}\, b_e(t)\,\phi_2(t)$$

T_s = symbol duration and $T_s = 2T_b$

or $$T_b = \frac{T_s}{2} \quad ...(3.61)$$

Then the above equation becomes,

$$s(t) = \sqrt{P_s T_b}\, b_0(t)\,\phi_1(t) + \sqrt{P_s T_b}\, b_e(t)\,\phi_2(t) \quad ...(3.62)$$

Since bit energy $E_b = P_s\, T_b$

Therefore, $$s(t) = \sqrt{E_b}\, b_0(t)\,\phi_1(t) + \sqrt{E_b}\, b_e(t)\,\phi_2(t) \quad ...(3.63)$$

This equation gives signal space representation of QPSK signal. The two orthogonal signals $\phi_1(t)$ and $\phi_2(t)$ form the two axis of the signal space. Figure 3.29 shows the signal space representation. The possible 4 signal points have been shown by small circles on $\phi_1\ \phi_2$-plane.

From each signal point, we obtain two bits. For example, from point 'A', we obtain two bits as (1, 1) and from 'B' we obtain bits as (– 1, 1). The distance of any signal point from origin 'O', given as,

$$\text{Distance 'OB'} = \sqrt{P_s T_b + P_s T_b} = \sqrt{2P_s T_b} = \sqrt{P_s T_s} \qquad [\text{Since, } 2T_b = T_s] \quad ...(3.64)$$

or $$\text{'OB'} = \sqrt{E_s} \qquad [\because P_s\, T_s = E_s] \quad ...(3.65)$$

Hence, the length of each signal point from origin is $\sqrt{E_s}$. We know that $b_e(t)$ and $b_0(t)$ represent two successive bits. There is an offset of 'T_b' between $b_e(t)$ and $b_o(t)$. Therefore, $b_e(t)$ and $b_o(t)$ both cannot change their levels simultaneously. Hence, either $b_e(t)$ or $b_o(t)$ can change at a time. Let us say that $b_e(t) = b_o(t) = 1$ is representing signal point 'A' in figure 3.29. In the next bit interval, if $b_o(t) = -1$, then signal point will be 'D'. Otherwise, if $b_e(t)$ changes its level [i.e., $b_e(t) = -1$], then next signal point will be 'B'. Hence, from signal point 'A', then next signal points will be either 'D' or 'B'.

3.13.3.1. Distance Between Signal Points

As a matter of fact, the ability to determine a bit without error is measured by the distance between two nearest possible signal points in the signal space. Such points differed in signal bit. For example, signal points 'A' and 'B' are two nearest points since they differ by a signal bit $b_e(t)$. As 'A' and 'B' become closer to each other, the possibility of error increases. Therefore, this distance must be as large as possible. This distance is denoted by 'd'. In figure 3.29, the distance between signal points 'A' and 'B' can be given by,

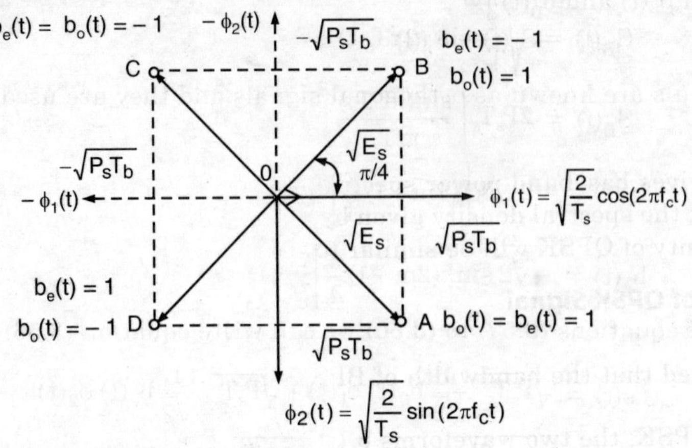

Fig. 3.29. The signal space representation for QPSK signals

$$d^2 = \left(\sqrt{E_s}\right)^2 + \left(\sqrt{E_s}\right)^2 = \sqrt{2E_s} \quad ...(3.66)$$

or

$$d = 2\sqrt{P_s T_b} = 2\sqrt{E_b} \quad ...(3.67)$$

✍ **Important Point:** *If we compare this distance with the distance of BPSK signals, then this shows that the distance for QPSK is the same as that for BPSK. Because, this distance represents noise immunity of the system, it shows that noise immunities of BPSK and QPSK are same.*

3.13.4. Spectrum of QPSK Signal

The input sequence b(t) is of bit duration T_b. Also, it is a NRZ bipolar waveform. Recall, the power spectral density of such waveform can be given as,

$$S(f) = V_b^2 T_b \left[\frac{\sin(\pi f T_b)}{\pi f T_b}\right]^2$$

Also, $V_b = \sqrt{P_s}$, then this equation becomes,

$$S(f) = P_s T_b \left[\frac{\sin(\pi f T_b)}{\pi f T_b}\right]^2 \quad ...(3.68)$$

This equation gives power spectral density (psd) of signal b(t). This signal is divided into $b_e(t)$ and $b_0(t)$ each of bit period $2T_b$. If we consider that symbols 1 and 0 are equally likely, then we can write power spectral densities (psds) of $b_e(t)$ and $b_0(t)$ as,

$$S_e(f) = P_s T_s \left[\frac{\sin(\pi f T_s)}{\pi f T_s}\right]^2 \quad ...(3.69)$$

and

$$S_0(f) = P_s T_s \left[\frac{\sin(\pi f T_s)}{\pi f T_s}\right]^2 \quad ...(3.70)$$

In these two equations, we have just replaced T_b by T_s and T_s is the period of bit in $b_e(t)$ and $b_0(t)$. Because, inphase and quadrature components [$b_e(t)$ and $b_0(t)$] are statistically independent, the baseband power spectral density of QPSK signal equals the sum of the individual power spectral densities of $b_e(t)$ and $b_0(t)$ i.e.,

$$S_B(f) = S_e(f) + S_0(f)$$

or
$$S_B(f) = 2P_sT_s\left[\frac{\sin(\pi f\, T_s)}{\pi f T_s}\right]^2 \qquad ...(3.71)$$

This equation gives baseband power spectral density of QPSK signal. Upon modulation of carrier frequency f_c, the spectral density given by above equation is shifted at $\pm f_c$. Thus plots of power spectral density of QPSK will be similar to that BPSK.

3.13.5. Bandwidth of QPSK Signal

We have observed that the bandwidth of BPSK signal is equal of $2f_b$. Here, $T_b = \frac{1}{f_b}$ is the one bit period. In QPSK, the two waveforms $b_e(t)$ and $b_0(t)$ form the baseband signals. One bit period for both of these signals is equal to $2T_b$. Therefore, bandwidth of QPSK signal will be

$$BW = 2\times\frac{1}{2T_b} = f_b \qquad ...(3.72)$$

Hence, the bandwidth of QPSK signal is half of the bandwidth of BPSK signal. Earlier, we have observed that noise immunity of QPSK and BPSK is same. This shows that inspite of the reduction in bandwidth in QPSK, the noise immunity remains same as compared to BPSK. BW of QPSK can also be obtained from figure 3.30.

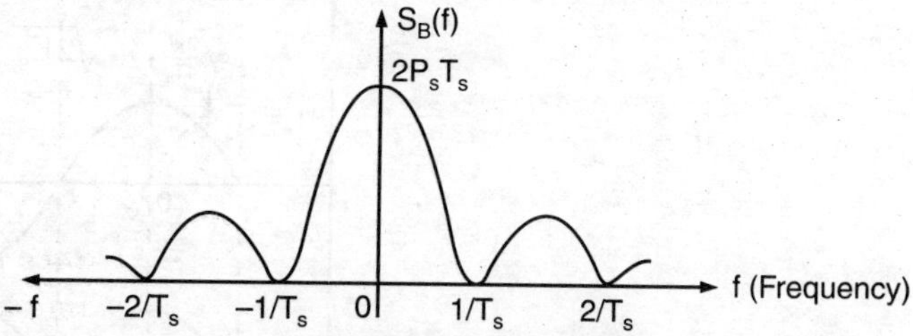

Fig. 3.30. Plot of power spectral density (psd) of QPSK signal

BW = Highest frequency – Lowest frequency in main lobe

$$BW = \frac{1}{T_s} - \left(-\frac{1}{T_s}\right) \text{ since carrier frequency } f_c \text{ cancels out}$$

$$BW = \frac{2}{T_s}$$

We know that
$$T_s = 2T_b$$

or
$$BW = \frac{2}{2T_b} = \frac{1}{T_b} = f_b \qquad ...(3.73)$$

3.13.6. Advantages of QPSK

QPSK has some certain advantages as compared to BPSK and DPSK as under:

(i) For the same bit error rate, the bandwidth required by QPSK is reduced to half as compared to BPSK.

(ii) Because of reduced bandwidth, the information transmission rate of QPSK is higher.

3.14 MINIMUM SHIFT KEYING (MSK)

We have discussed QPSK technique in last article. The bandwidth requirement of QPSK is high. Filters or other methods can overcome these problems, but they have other side effects. For example filters alter the amplitude of the waveform.

MSK overcomes these problems. In MSK, the output waveform is continuous in phase hence there are no abrupt changes in amplitude. The sidelobes of MSK are very small hence bandpass filtering is not required to avoid interchannel interference. Figure 3.31 shows the waveform of MSK. The binary bit sequence is shown at the top. Figure 3.31(a) shows the corresponding NRZ

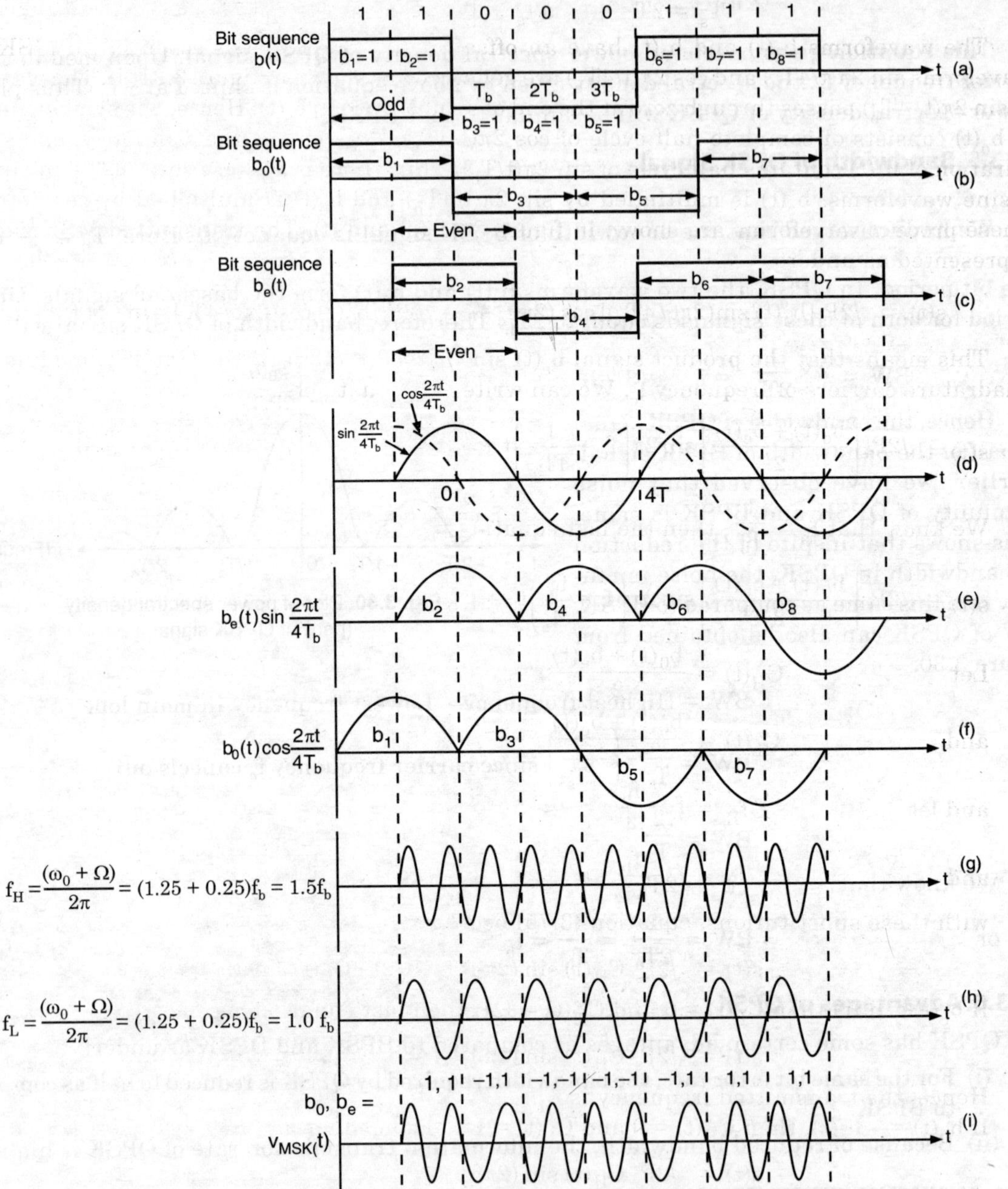

Fig. 3.31. (a) Bipolar NRZ waveform representing bit sequence (b) Odd bit sequence waveforms $b_0(t)$ (c) Even bit sequence waveform $b_e(t)$ (d) Waveforms of frequency $f_b/4$ used for smoothing of $b_e(t)$ and $b_o(t)$ (e) Modulating waveform of even sequence (f) Modulating waveform of odd sequence (g) Waveform of frequency f_H (h) Waveform of frequency f_L (i) MSK waveform.

waveform b(t). From b(t), two waveforms are generated for odd and even bits. $b_0(t)$ represents odd bits and $b_e(t)$ represents even bits. Figure 3.31(b) and (c) shows the waveform of $b_o(t)$ and $b_e(t)$. As shown in those waveforms b_1, b_3, b_5 etc. are represented by odd waveform i.e., $b_o(t)$.

The duration of each bit in $b_o(t)$ or $b_e(t)$ is $2T_b$, whereas it is T_b in b(t) i.e.,

$$T_s = 2T_b$$

The waveforms $b_o(t)$ and $b_e(t)$ have an offset of T_b. This offset is essential in MSK. Two waveforms sin $2\pi(t/4T_b)$ and cos $2\pi(t/4T_b)$ are generated as shown in figure 3.31(d). The waveform of sin $2\pi(t, 4T_b)$ passes through zero at the end of symbol time in $b_o(t)$. Hence, one symbol duration of $b_o(t)$ consists of complete half cycle of cos $2\pi(t/4T_b)$. This means that similarly, one symbol duration of $b_e(t)$ contains half cycle of sin $2\pi(t/T_b)$. Thus there is a phase shift of 'T_b' in sine and cosine waveforms. $b_e(t)$ is multiplied by sin $2\pi(t/4T_b)$ and $b_o(t)$ is multiplied by cos $2\pi(t/4T_b)$. These product waveforms are shown in figure 3.31 (e) and (f). The transmitted MSK signal is represented as under:

$$s(t) = \sqrt{2P_s}\,[b_e(t)\sin(2\pi t/4T_b)]\cos(2\pi f_c t) + \sqrt{2P_s}\,[b_0(t)\cos(2\pi t/4T_b)]\sin(2\pi f_c t)$$

This means that the product signal $b_e(t)$ sin $(2\pi t/4T_b)$ and $b_0(t)$ cos $(2\pi t/4T_b)$ modulate the quadrature carriers of frequency f_c. We can write last equation as,

$$s(t) = \sqrt{2P_s}\left[\frac{b_o(t)+b_e(t)}{2}\right]\sin 2\pi\left(f_c + \frac{1}{4T_b}\right)t + \sqrt{2P_s}\left[\frac{b_o(t)-b_e(t)}{2}\right]\sin 2\pi\left(f_c - \frac{1}{4T_b}\right)t \quad \text{...(3.74)}$$

We know that $f_b = \frac{1}{T_b}$, then the last equation (3.74) becomes,

$$s(t) = \sqrt{2P_s}\left[\frac{b_0(t)+b_e(t)}{2}\right]\sin 2\pi\left(f_c + \frac{f_b}{4}\right)t + \sqrt{2P_s}\left[\frac{b_0(t)-b_e(t)}{2}\right]\sin 2\pi\left(f_c - \frac{f_b}{4}\right)t \quad \text{...(3.75)}$$

Let $$C_H(t) = \frac{b_0(t)+b_e(t)}{2}$$

and $$C_L(t) = \frac{b_0(t)+b_e(t)}{2}$$

and let $$f_H = f_c + \frac{f_b}{4} \quad \text{...(3.76)}$$

and $$f_L = f_c - \frac{f_b}{4} \quad \text{...(3.77)}$$

with these substitutions, equation (3.75) becomes,

$$s(t) = \sqrt{2P_s}\,C_H(t)\sin(2\pi f_H t) + \sqrt{2P_s}\,C_L(t)\sin(2\pi f_L t) \quad \text{...(3.78)}$$

If $b_0(t) = b_e(t)$ then $C_L(t) = 0$ and $C_H(t) = \pm 1$, then last equation becomes,

$$s(t) = \sqrt{2P_s}\,C_H(t)\sin(2\pi f_H t) \quad \text{...(3.79)}$$

Hence, the transmitted frequency is f_H.
If $b_0(t) = -b_e(t)$, then $C_H(t) = 0$ and $C_L(t) = \pm 1$. Then equation (3.79) becomes,

$$s(t) = \sqrt{2P_s}\,C_H(t)\sin(2\pi f_L t) \quad \text{...(3.80)}$$

Hence, the transmitted frequency is f_L.

The frequencies f_H and f_L are chosen such that cos$(2\pi f_H t)$ and sin $(2\pi f_L t)$ are orthogonal over the interval T_b. For orthogonality following relation must be satisfied i.e.,

$$\int_0^{T_b} \sin(2\pi f_H t)\sin(2\pi f_L t)\,dt = 0 \qquad ...(3.81)$$

This relation will be satisfied if we have integers 'm' and 'n' such that,

$$2\pi (f_H - f_L)\, T_b = n\pi \qquad ...(3.82)$$

and

$$2\pi(f_H + f_L)\, T_b = m\pi \qquad ...(3.83)$$

Let us substitute values of f_H and f_L from equations (3.76) and (3.77) in above relations. From equation (3.82), we get

$$2\pi\left(f_c + \frac{f_b}{4} - f_c + \frac{f_b}{4}\right) T_b = n\pi$$

or

$$f_b\, T_b = n$$

or

$$f_b \times \frac{1}{f_b} = n \quad \Rightarrow \quad n = 1 \qquad ...(3.84)$$

Similarly from equation (3.83), we get

$$2\pi\left(f_c + \frac{f_b}{4} + f_c - \frac{f_b}{4}\right) T_b = m\,\pi$$

or

$$4f_c\, T_b = m$$

or

$$4f_c \times \frac{1}{f_b} = m \quad \Rightarrow \quad f_c = \frac{m}{4} f_b \qquad ...(3.85)$$

with n = 1 in equation (3.82), we get

$$2\pi(f_H - f_L)\, T_b = 1 \times \pi$$

or

$$(f_H - f_L) = \frac{1}{2T_b} = \frac{f_b}{2} \qquad ...(3.86)$$

Here, n = 1 means the difference between f_H and f_L is minimum and at the same time, (MSK) they are orthogonal. Therefore, this technique is called **minimum shift keying (MSK)**. This minimum difference is given by equation (3.86) above. From equation (3.85), we know that $f_c = \frac{m}{4} f_b$. This shows that carrier frequency 'f_c' is integer multiple of $\frac{f_b}{4}$. Substituting, this value of f_c in equation (3.76), we get

$$f_H = f_c + \frac{f_b}{4} = m\frac{f_b}{4} + \frac{f_b}{4} \quad \text{or} \quad f_H = (m+1)\frac{f_b}{4} \qquad ...(3.87)$$

Similarly, substituting $f_c = m\frac{f_b}{4}$ in equation (3.77), we get

$$f_L = (m-1)\frac{f_b}{4} \qquad ...(3.88)$$

Figure 3.31(g) and (h) shows the waveforms of $\sin(2\pi f_H t)$ and $\sin(2\pi f_L t)$. For these waveforms m = 5. Using equations (3.87) and (3.88), f_H and f_L are calculated with m = 5. Figure 3.31(i) shows the final MSK waveform. From equation (3.79), we know that if $b_0(t) = b_e(t)$, then transmitted waveform is of frequency f_H. And if $b_0(t) = -b_e(t)$ then the transmitted waveform is given by equation (3.75), which has frequency of f_L. This shows that MSK is basically FSK with reduced bandwidth and continuous phase.

3.14.1. Signal Space Representation of MSK and Distance between the Signal Points (i.e., Geometrical Representation of MSK)

Let us rearrange equation (3.74) as follows

$$s(t) = C_H(t)\sqrt{P_sT_s}\cdot\sqrt{\frac{2}{T_s}}\sin(2\pi f_H t) + C_L(t)\sqrt{P_sT_s}\cdot\sqrt{\frac{2}{T_s}}\sin(2\pi f_L t) \qquad ...(3.89)$$

Here let, $\phi_H(t) = \sqrt{2/T_s}\sin(2\pi f_H t)$ ...(3.90)

$\phi_L(t) = \sqrt{2/T_s}\sin(2\pi f_L t)$ ...(3.91)

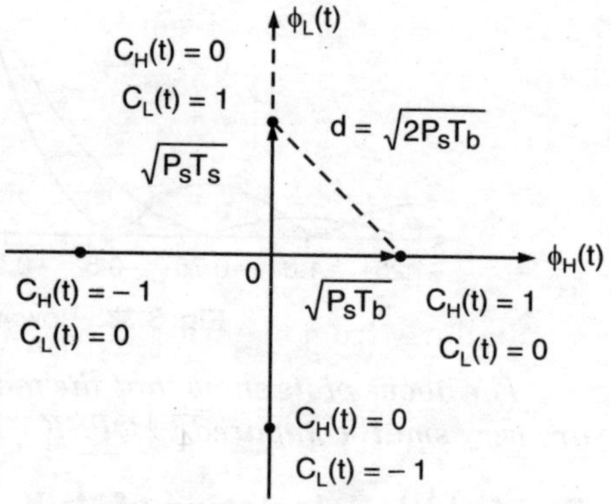

Fig. 3.32. Geometrical (Signal Space) representation of MSK signals

The carriers $\phi_H(t)$ and $\phi_L(t)$ are in quadrature. They are in quadrature because their frequencies are in quadrature. In QPSK the carriers are in quadrature because of phase shift. Depending on the values of $C_H(t)$ and $C_L(t)$, there will be four signal points in $\phi_H\ \phi_L$ plane. This has been illustrated in figure 3.32. The distance of each signal point from the origin is $\sqrt{P_sT_s}$.

Distance Between Signal Points

Since the points are symmetric, the distance between any two nearest points is same, i.e.,

$$d^2 = \left(\sqrt{P_sT_s}\right)^2 + \left(\sqrt{P_sT_s}\right)^2$$

or $$d = \sqrt{2P_sT_s} \qquad ...(3.92)$$

or $$d = \sqrt{2E_s} \quad (\text{since } P_s\,T_s = E_s) \qquad ...(3.93)$$

or $$d = \sqrt{4E_b} \quad (\text{since } E_s = 2E_b) = 2\sqrt{E_b} \qquad ...(3.94)$$

These relations give distance between signal points in MSK. This distance is same as in QPSK.

3.14.2. Power Spectral Density (psd) and Bandwidth of MSK

Let us consider the baseband signal of equation (3.78). The waveform which modulates $\sin(2\pi f_c t)$ is,

$$p(t) = \sqrt{2P_s}\,[b_0(t)\cos(2\pi t/4T_b)] \qquad ...(3.95)$$

or $$p(t) = \sqrt{2P_s}\;b_0(t)\cos(\pi f_b t/2) \qquad ...(3.96)$$

The power spectral density (psd) of above waveform is expressed as,

$$S_p(f) = \frac{32E_b}{\pi^2}\left[\frac{\cos(2\pi fT_b)}{1-(4fT_b)^2}\right]^2 \qquad ...(3.97)$$

when this signal modulates the carrier 'f_c' then the total power spectral density (psd) of baseband signal is divided by '4' and is placed at $\pm f_c$, i.e.,

$$S(f) = \frac{8E_b}{\pi^2}\left\{\frac{\cos 2\pi(f-f_c)T_b}{1-[4(f-f_c)T_b]}\right\}^2 + \frac{8E_b}{\pi^2}\left\{\frac{\cos 2\pi(f+f_c)T_b}{1-[4(f+f_c)T_b]^2}\right\}^2 \qquad ...(3.98)$$

The above equation gives power spectral density (psd) of MSK signal. Figure 3.33 shows the normalized spectral densities of MSK and QPSK. Normalization means maximum amplitudes of signals are scaled with respect to '1'.

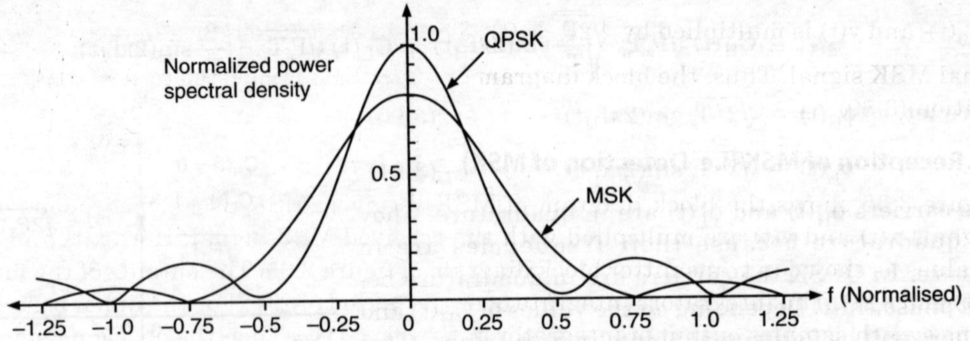

Fig. 3.33. Power spectral densities (psd) of MSK and QPSK.

The above plots show that the main lobe in MSK is wider than QPSK. The sides lobes in MSK are very small compared to QPSK.

Bandwidth Calculation of MSK

From figure 3.33, we observe that the width of main lobe in MSK is ± 0.75 i.e.,

$$fT_b = \pm 0.75$$

or $$f = \pm 0.75\, f_b$$

Hence, bandwidth will be equal to width of the main lobe i.e.,

$$BW = 0.75\, f_b - (-0.75\, f_b) = 1.5\, f_b \quad \text{...(3.99)}$$

Thus, the BW of MSK is higher than that of QPSK.

3.14.3. Generation of MSK

Figure 3.34 shows the block diagram of MSK transmitter. The two sinusoidal signals $\sin(2\pi f_c t)$ and $\cos(2\pi t/4T_b)$ are mixed (i.e., multiplied). The bandpass filters then pass only sum and

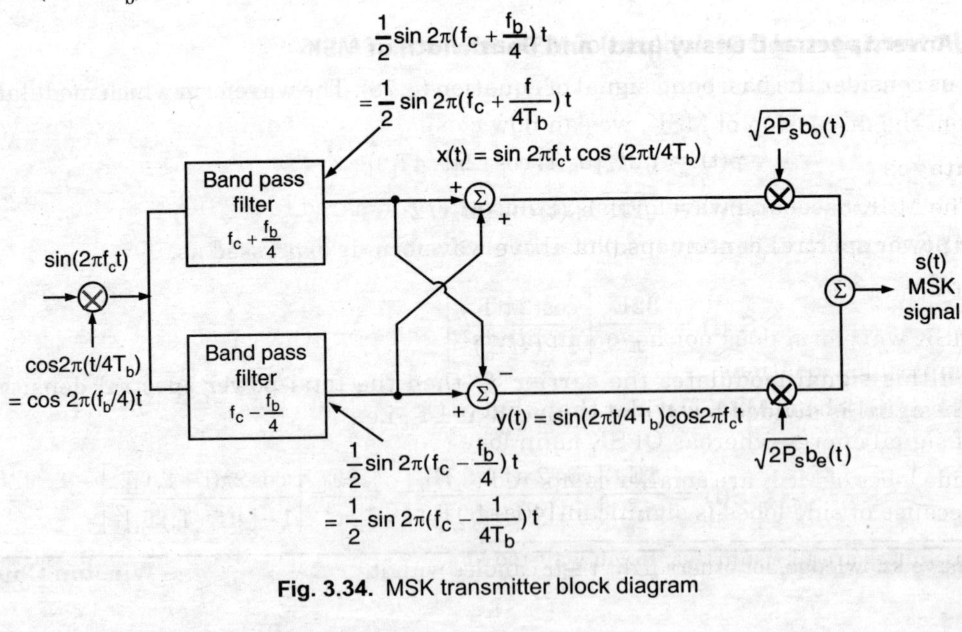

Fig. 3.34. MSK transmitter block diagram

difference components $f_c + \frac{f_b}{4}$ and $f_c - \frac{f_b}{4}$. The outputs of bandpass filters (BPFs) are then added and subtracted such that two signals x(t) and y(t) are generated. Signal x(t) is multiplied by $\sqrt{2P_s}\, b_0(t)$ and y(t) is multiplied by $\sqrt{2P_s}\, b_e(t)$. The outputs of the multipliers are then added to give final MSK signal. Thus, the block diagram of figure 3.34 is the step to step implementation of equation (3.89).

3.14.4. Reception of MSK (i.e. Detection of MSK)

Figure 3.35 shows the block diagram of MSK receiver. MSK uses synchronous detection. The signals x(t) and y(t) are multiplied with the received MSK signal. Here x(t) and y(t) have same values as shown in transmitter block diagram of figure 3.35. The outputs of the multipliers are $b_e(t)$ and $b_0(t)$. The integrators integrate over the period of $2T_b$. For the upper correlator, the sampling switch samples output of integrator at $t = (2k + 1)T_b$. Then the decision device decides whether $b_0(t)$ is + 1 or – 1. Similarly, lower correlator output is $b_e(t)$. The outputs of two decision devices are staggered by T_b. The switch S_3 operates at $t = kT_b$ and simply multiplexes the two correlator outputs.

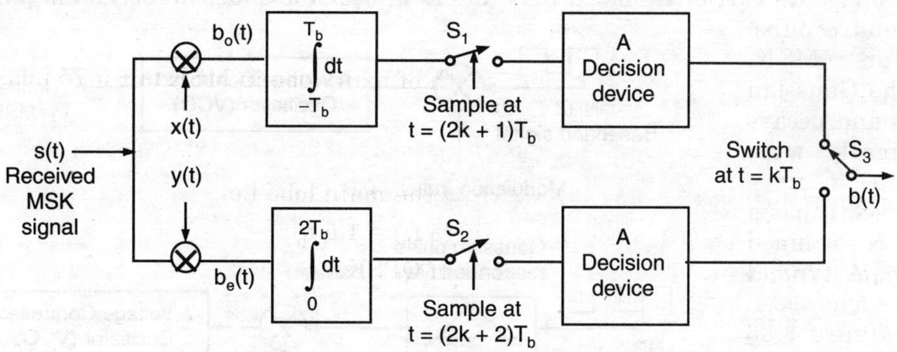

Fig. 3.35. MSK receiver block diagram.

3.14.5. Advantages and Drawbacks of MSK as Compared to QPSK

(U.P.Tech., Sem. Exam., 2008-09)(05 Marks)

From the discussion of MSK, we can now compare the advantages of MSK over QPSK.

Advantages

1. The MSK baseband waveforms are smoother compared to QPSK.
2. MSK signal have continuous phase in all the cases, whereas QPSK has abrupt phase shift of $\frac{\pi}{2}$ or π.
3. MSK waveform does not have amplitude variations, whereas QPSK signals have abrupt amplitude variations.
4. The main lobe of MSK is wider than that of QPSK. Main lobe of MSK contains around 99% of signal energy whereas QPSK main lobe contains around 90% signal energy.
5. Side lobes of MSK are smaller compared to that of QPSK. Hence, interchannel interference because of side lobes is significantly large in QPSK.

6. To avoid interchannel interference due to sidelobes, QPSK needs bandpass filtering, whereas it is not required in MSK.
7. Bandpass filtering changes the amplitude waveform of QPSK because of abrupt changes in phase. This problem does not exist in MSK.

The distance between signal points is same in QPSK as well as in MSK. Hence, the probability of error is also same. However, there are some drawback of MSK.

Drawbacks

1. The bandwidth requirement of MSK is 1.5 f_b, whereas it is f_b in QPSK. Actually, this cannot be said serious drawback of MSK. Because power to bandwidth ratio of MSK is more. In fact, 99% of signal power can be transmitted within the bandwidth of 1.2 f_b in MSK. While QPSK needs around 8 f_b to transmit the same power.
2. The generation and detection of MSK is slightly complex. Because of incorrect synchronization, phase jitter can be present in MSK. This degrades the performance of MSK.

3.15 GAUSSIAN MINIMUM SHIFT KEYING (i.e., GMSK)

Like Minimum shift keying (MSK), Gaussian MSK (GMSK) yields a constant amplitude and continuous phase RF carrier signal. It only differs in use of a Gaussian baseband pulse shape in place of square pulse shape for MSK. Because, the Gaussian pulse rises and decays asymptotically with respect to a zero response level, it has a much more constrained bandwidth. A typical GMSK system has been shown in figure 3.36 along with an unfiltered MSK system. The unfiltered MSK is generated by direct FSK modulation of a carrier with a baseband signal which is scaled in amplitude to produce a modulation index of 0.5. This value of modulation index produces a difference of 180° phase shift for the two data values. However, in GMSK, there is ISI (Inter symbol interference) which is a bandwidth limiting factor. GMSK is employed in GSM digital cellular radios and cellular digital packet data (CDPD) applications.

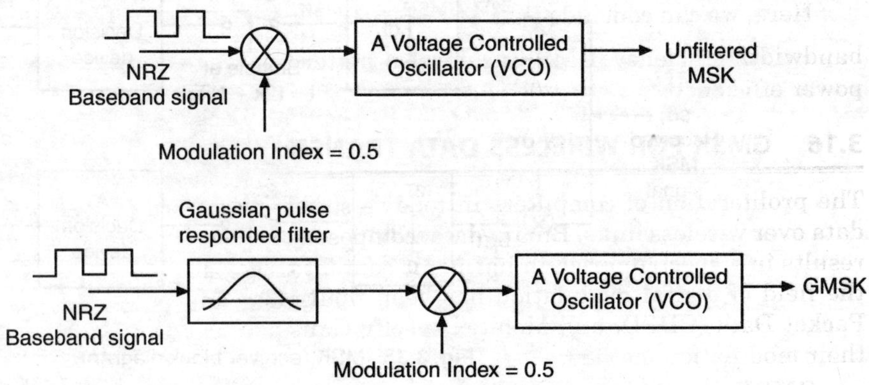

Fig. 3.36. GMSK as unfiltered MSK.

Gaussian Minimum Shift Keying (GMSK) is a modification of MSK. A filter used to reduce the bandwidth of a baseband pulse train prior to modulation is called a pre-modulation filter. The Gaussian pre-modulation filter smooths the phase trajectory of the MSK signal and hence limiting the instantaneous frequency variations. The result is an FM modulated signal with a much narrower bandwidth. This bandwidth reduction does not come for free since the pre-modulation filter smears the individual pulses in pulse train. As a consequence of this smearing in time, adjacent pulses interfere with each other generating what is commonly called inter-symbol interference or ISI. In the applications, where GMSK is used, the trade-off between power efficiency and bandwidth efficiency is well worth the cost.

Bit Error Rate (BER) for GMSK is given by

$$P_e = Q\left(\sqrt{\frac{2\alpha E_b}{N_0}}\right) \quad ...(3.100)$$

where α is a constant related to BT_b.

Table. 3.8. GMSK Parameter α Related to BT_b.

S.No.	The value of BT_b	The values of α
1.	0.25	0.68
2.	∞	0.85

It may be noted that the case where $BT_b \to \infty$ corresponds to MSK (i.e. the filter is all pass for a fixed symbol interval T_s).

Recall that the probability of error for plain MSK is given by

$$P_e \cong Q\left(\sqrt{\frac{2\alpha E_b}{N_0}}\right) \quad ...(3.101)$$

Here, we can conclude that $P_e^{GMSK} > P_e^{MSK}$. This arises from the trade off between power and bandwidth efficiency. GMSK achieves a better bandwidth efficiency than MSK at the expense of power efficiency.

3.16 GMSK FOR WIRELESS DATA TRANSMISSION

The proliferation of computers in today's society has increased the demand for transmission of data over wireless links. Binary data, composed of sharp "one to zero" and "zero to one" transitions, results in a spectrum rich in harmonic content that is not well suited to RF transmission. Hence, the field of digital modulation has been flourishing. Recent standards such as Cellular Digital Packet Data (CDPD) and Mobitex specify Gaussian Filtered Minimum Shift keying (GMSK) for their modulation method.

GMSK is a simple yet effective approach to digital modulation for wireless data transmission. To provide a good understanding of GMSK, we shall review the basics of MSK and GMSK, as well as how GMSK is implemented in CDPD and Mobitex systems.

DO YOU KNOW?

Gaussian minimum-shift keying is a special case of FSK that achieves the minimum bandwidth possible for a two-frequency FSK system at a given data rate.

GMSK modems reduce system complexity, and in turn lower system cost. There are, however, some important implementation details to be considered.

If we look at Fourier series expansion of a data signal, we observe harmonics extending to infinity. When these harmonics are summed, they give the data signal and its sharp transitions. Hence, an unfiltered NRZ data stream used to modulate an RF carrier will produce an RF spectrum of considerable bandwidth. The FCC has strict regulations about spectrum usage and such a system is generally considered impractical. But if we start to remove the high frequency harmonics from the Fourier series (i.e. pass the data signal through a lowpass filter), the transitions in the data will become progressively less sharp. This suggests that premodulation filtering is an effective method for reducing the occupied spectrum for wireless data transmission. In addition to a compact

Opportunity is rare, and wise men will never let it go by him. – **Bayard Taylor**

spectrum, a wireless data modulation scheme must have good bit error rate (BER) performance under noisy conditions. Its performance should also be independent of power amplifier linearity to allow the use of class C power amplifiers.

The academic field of 'Data Transmission' is full of modulation strategies that attempt to meet the above criteria. Most of them involve translation of data bits or patterns into a particular combination of phase, frequency or amplitude. Some of the more important techniques are listed in Table 3.9.

Table 3.9. Modulation Formats

S.No.	Modulation technique	Common Acronym
1.	Frequency Shift Keying	FSK
2.	Multi-level Frequency Shift Keying	MFSK
3.	Continuous Phase Frequency Shift Keying	CPFSK
4.	Minimum Shift Keying	MSK
5.	Gaussian Minimum Shift Keying	GMSK
6.	Phase Shift Keying	PSK
7.	Quadrature Phase Shift Keying	QPSK
8.	Differential Quadrature Phase Shift Keying	DQPSK
9.	$\pi/4$ Differential Quadrature Phase Shift Keying	$\pi/4$ DQPSK
10.	Quadrature Amplitude Modulation	QAM

Each of the modulation formats listed in Table 3.9 is suited to specific applications. In general, schemes that rely on more than two levels (e.g. QAM, QPSK) require better signal to noise ratios (SNR) than two-level schemes for similar BER performance. Additionally, in a wireless environment, multi-level schemes generally require greater power amplifier linearity than two-level schemes. The fact that GMSK uses a two-level continuous phase modulation (CPM) format has contributed to its popularity. Another point in its favour is that it allows the use of class C power amplifiers (relatively non-linear) and data rates approaching the channel BW (dependent on filter bandwidth and channel spacing).

3.17 GMSK FROM MSK

Prior to discussing GMSK in detail, we need to review MSK, from which GMSK is derived. MSK is a continuous phase modulation scheme where the modulated carrier contains no phase discontinuities and frequency changes occur at the carrier zero crossings. MSK is unique due to the relationship between the frequency of a logical zero and one the difference between the frequency of a logical zero and a logical one is always equal to half the data rate. In other words, the

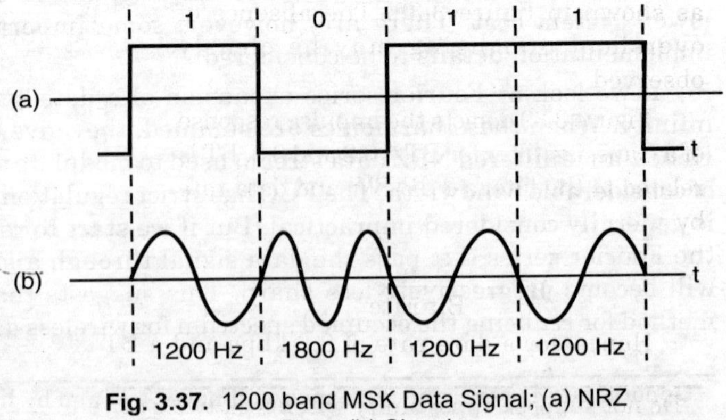

Fig. 3.37. 1200 band MSK Data Signal; (a) NRZ Data, (b) Signal

modulation index is 0.5 for MSK, and is defined as

$$m = \Delta f \times T$$

where $\Delta f = |f_{logic\ 1} - f_{logic\ 0}|$

$T = 1/\text{bit rate}$

For example, a 1200 bit per second baseband MSK data signal could be composed of 1200 Hz and 1800 Hz frequencies for a logical one and zero respectively as shown in figure 3.37.

Baseband MSK, as shown in figure 3.37, is a robust means of transmitting data in wireless systems where the data rate is relatively low compared to the channel BW.

An alternative method for generating MSK modulation can be realized by directly injecting NRZ data into a frequency modulator with its modulation index set for 0.5 as shown in figure 3.38. This approach is essentially equivalent to baseband MSK. However, in the direct approach, the VCO is part of the RF/IF section, whereas in baseband MSK, the voltage to frequency conversion takes place at baseband.

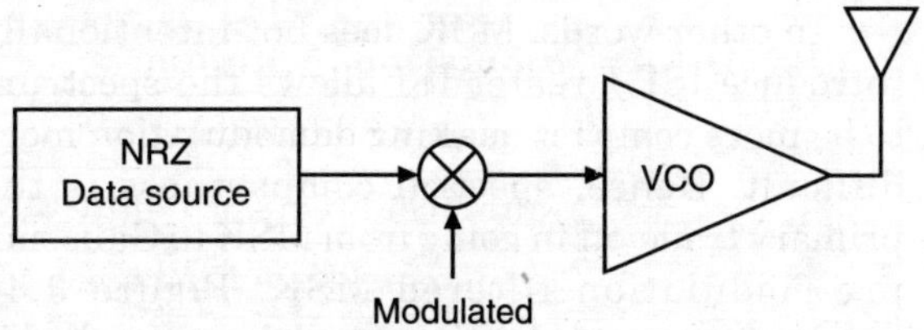

Fig. 3.38. Direct MSK modulation

The fundamental problem with MSK is that the spectrum is not compact enough to realize data rates approaching the RF channel BW. A plot of the spectrum for MSK reveals sidelobes extending well above the data rate. For wireless data transmission systems, which require more efficient use of the RF channel BW, it is necessary to reduce the energy of the MSK upper sidelobes. Earlier, we stated that straight forward means of reducing this energy is low pass filtering the data stream prior to presenting it to the modulator (pre-modulation filtering). The pre-modulation low pass filter must have a narrow BW with a sharp cut off frequency and very little overshoot in its impulse response. This is where the Gaussian filter characteristic comes in. It has an impulse response characterized by a classical Gaussian distribution (bell shaped curve), as shown in figure 3.39. The absence of overshoot or ringing may be clearly observed.

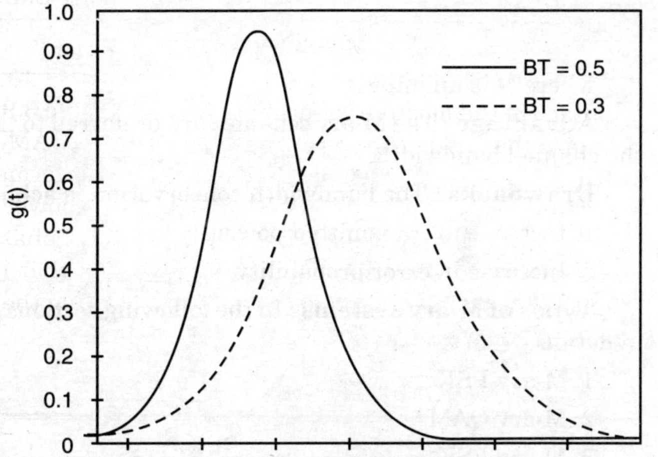

Fig. 3.39. Gaussian Filter Impulse Response for BT= 0.3 and BT = 0.5.

Figure 3.39 depicts the impulse response of a Gaussian filter for BT = 0.3 and 0.5. BT is related to the filter's 3 dB BW and data rate by

$$BT = \frac{f_{3dB}}{\text{Bit rate}}$$

Hence, for a data rate of 9.6 kbps and a BT of 0.3, the filter's – 3 dB cut off frequency is 2880

Hz.

Again from figure 3.39, it may be noted that a bit is spread over approximately 3 bit periods for BT = 0.3 and two bit periods for BT = 0.5. This gives rise to a phenomena called inter-symbol interference (ISI). For BT = 0.3 adjacent symbols or bits will interfere with each other more than for BT = 0.5. GMSK with BT = ∞ is equivalent to MSK.

In other words, MSK does not intentionally introduce ISI. Greater ISI allows the spectrum to be more compact, making demodulation more difficult. Hence, spectral compactness is the primary trade-off in going from MSK to Gaussian pre-modulation filtered MSK. Figure 3.40 displays the normalized spectral densities for MSK and GMSK. It may be observed that the sidelobe energy for GMSK is reduced. Ultimately, this means that channel spacing can be tighter for GMSK when compared to MSK for the same adjacent channel interference.

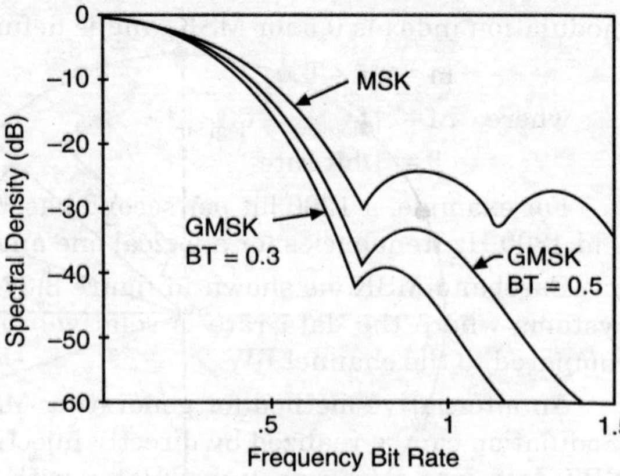

Fig. 3.40. Spectral Density for MSK and GMSK.

3.18 M-ARY MODULATION TECHNIQUES

In an M-ary signaling scheme, we can send one of the M possible signals such as $s_1(t)$, $s_2(t)$, ... $s_M(t)$, during each signaling interval of duration T seconds. The number of possible signals i.e., M is given by,

$$M = 2^N \tag{3.102}$$

where N is an integer

Advantage : The M-ary schemes are preferred to the binary schemes when we want to conserve the channel bandwidth.

Drawbacks : The bandwidth conservation is achieved at the cost of following disadvantages:

1. Increase in transmitted power.
2. Increase in error probability.

Types of M-ary systems : In the following sections, we shall discuss the following M-ary systems in details.

1. M-ary PSK
2. M-ary QAM
3. M-ary FSK

3.19 M-ARY PHASE SHIFT KEYING (PSK)

In QPSK, we have grouped together two bits to form messages. Depending on the two bit message (00, 01, 10 or 11), a sinusoidal signal of duration equal to 2 T_b and having a particular phase shift is transmitted. The QPSK signals differ by 90° or $\pi/2$ radians in phase with respect to each other.

The same principle can be further extended to obtain the M-ary PSK system. The M-ary PSK signals are obtained as under :

(i) Group N bits together to form N-bit symbols.

(ii) These symbols will extend over a period of NT_b where T_b is the duration of one bit.

(iii) Due to the grouping of N bits per symbol, we can have 2^N = M possible symbols.

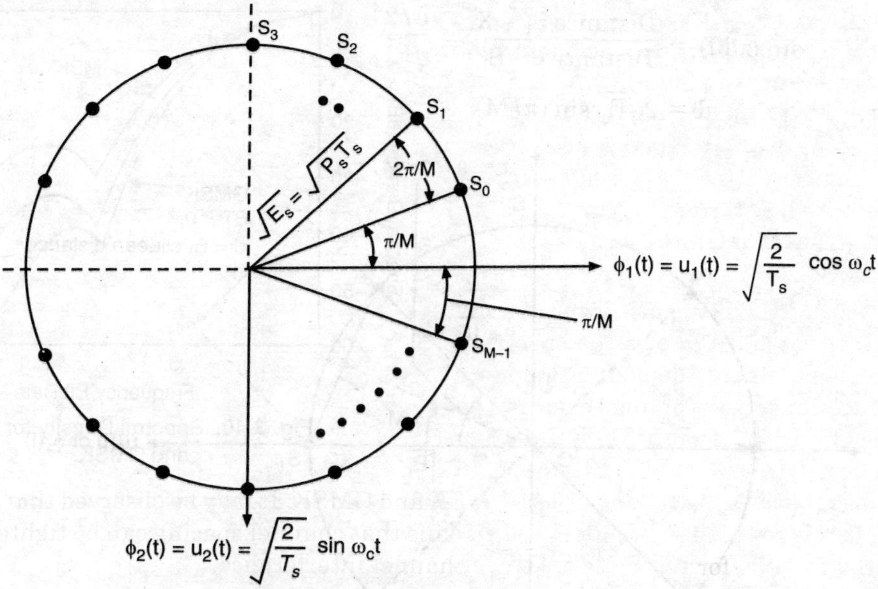

Fig. 3.41. Geometric representation of M-ary PSK signals

(iv) These M symbols are represented by sinusoidal signals of duration $T_s = NT_b$ which differ from one another by a phase $2\pi/M$ radians. Thus, the M-ary PSK waveform can be mathematically represented as,

$$V_{\text{M-ary PSK}} = \sqrt{2 P_s} \cos (\omega_c t + \phi_m) \qquad ...(3.103)$$

where m = 0, 1, ... M – 1

(v) In equation (3.103), ϕ_m represents the symbol phase angle, which is given by

$$\phi_m = (2m+1)\frac{\pi}{M} \qquad ...(3.104)$$

(vi) The M-ary PSK waveforms corresponding to equation (3.103) are represented by dots in figure 3.41. This is a signal space diagram in which the co-ordinate axes are formed by the orthonormal waveforms $u_1(t)$ and $u_2(t)$ given by the following equations :

$$u_1(t) = \sqrt{(2/T_s)} \cos \omega_c t$$

and $$u_2(t) = \sqrt{(2/T_s)} \sin \omega_c t$$

These are same orthonormal signals used for the signal space representation of a QPSK signal.

(vii) The distance of each M-ary PSK waveform from the origin is given by

$$\sqrt{E_s} = \sqrt{P_s T_s}$$

3.19.1 Distance between Signal Points (d)

The distance between signal points (d) is also called as the Euclidean distance. This distance is important in deciding the error probability of a communication sytem. The error probability goes on decreasing with increase in the Euclidean distane d. In order to obtain the expression for the distance d, let us consider figure 3.42.

(i) As shown in figure 3.42, the angle θ is equal to ($2\pi/M$) because 2 π corresponds to the complete circle and M is the number of messages.

(ii) Now, let us consider the right angle triangle $0 - S_1 - X$. Using the definition of sin θ, we get

$$\sin(\pi/M) = \frac{\text{Distance } S_1 - X}{\text{Distance } 0 - S_1} = \frac{d/2}{\sqrt{E_s}} \quad ...(3.105)$$

Therefore, $$d = 2\sqrt{E_s}\,\sin(\pi/M) \quad ...(3.106)$$

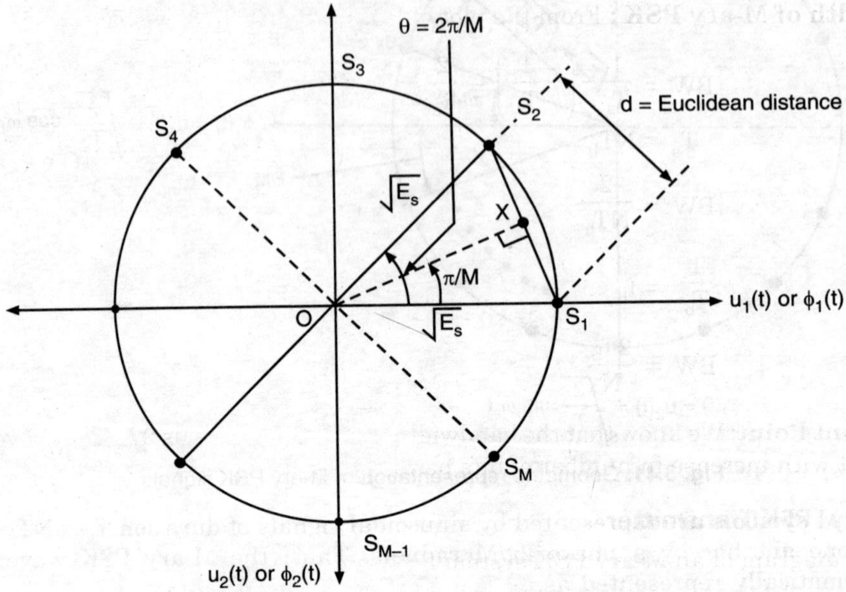

Fig. 3.42. Signal space diagram to calculate distance d

Equation (3.106) is a general expression. We can substitute any value of M to get the Euclidean distance d.

3.19.2 Power Spectral Density (PSD) and Bandwidth of M-ary PSK

(i) In the preceding section, we have obtained the expression for power spectral density of the baseband QPSK signal as under :

$$S_{B\,(QPSK)}(f) = 2\,P_s T_s \left[\frac{\sin(\pi f\,T_s)}{\pi f\,T_s}\right]^2$$

(ii) However, QPSK is a special case of M-ary PSK with M = 4. Therefore, we can use the above expression to write the expression for PSD of the baseband M-ary PSK system. The only modification required is, that we have to substitute $T_s = NT_b$ in the above expression.

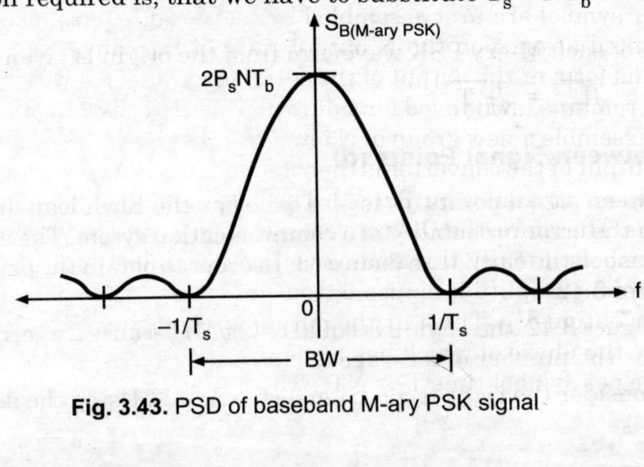

Fig. 3.43. PSD of baseband M-ary PSK signal

Therefore, $S_{B(M\text{-ary PSK})}(f) = 2P_sNT_b\left[\frac{\sin(\pi f NT_b)}{\pi f NT_b}\right]^2$ ...(3.107)

Using equation (3.107), we can plot the PSD of the baseband M-ary PSK as shown in figure 3.42.

Bandwidth of M-ary PSK : From the plot of PSD in figure 3.43, we have

$$BW = \frac{1}{T_s} - \left(-\frac{1}{T_s}\right) = \frac{2}{T_s}$$

But $T_s = NT_b$

Therefore, $BW = \frac{2}{NT_b}$

But, $\frac{1}{T_b} = f_b$

Hence, $BW = \frac{2f_b}{N}$ ...(3.108)

> **Important Point:** We know that the bandwidth of a BPSK system is $2f_b$. The above expression tells us that with increase in number of bits per message, the bandwidth reduces.

3.19.3 M-ary PSK Transmitter

The block diagram of an M-ary PSK system has been shown in figure 3.44.

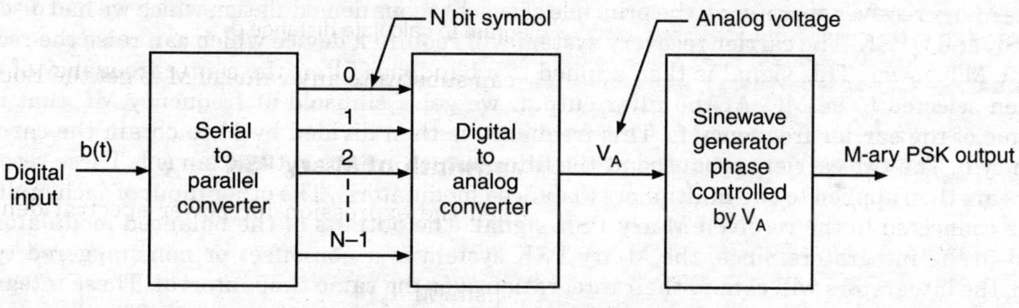

Fig. 3.44. M-ary PSK transmitter

Working Operation

The bit stream b (t) is applied to a serial to parallel converter. This block can store the N bits of a symbol. These N bits per symbol have been presented serially, in time sequence one after the other. The N bits per symbol are first assembled by the serial to parallel converter block. Then, all these bits are presented at once on the N output lines of the converter. Thus, the N-bit message appears in the parallel form at the output of the serial to parallel converter. The output of the serial to parallel converter remains unchanged for a duration of NT_b of a symbol which is the time required for the converter to assemble a new group of N bits. After every NT_b seconds, the converter output is updated. The N bit output of the converter is then applied to a D/A converter. Depending on the N bit digital input, it produces an analog output V_A. The N bit digital input can have $2^N = M$ number of possible combinations. Therefore, the D/A converter output V_A will have M number of distinct values, depending on the symbols. Finally this analog voltage is applied to a sinusoidal signal generator, which produces a constant amplitude sinusoidal output voltage, the phase ϕ_m of which is determined by the D/A converter output V_A. Thus, at the output of the transmitter, we get a fixed amplitude sinusoidal waveform, the phase of which has a one to one correspondence to the N bit symbols. The will change only once per symbol time $T_s = NT_b$. Thus, the M-ary PSK is generated.

3.19.4 M-ary PSK Receiver

The block diagram of M-ary PSK receiver has been shown in figure 3.45. This is same as the non-offset QPSK receiver.

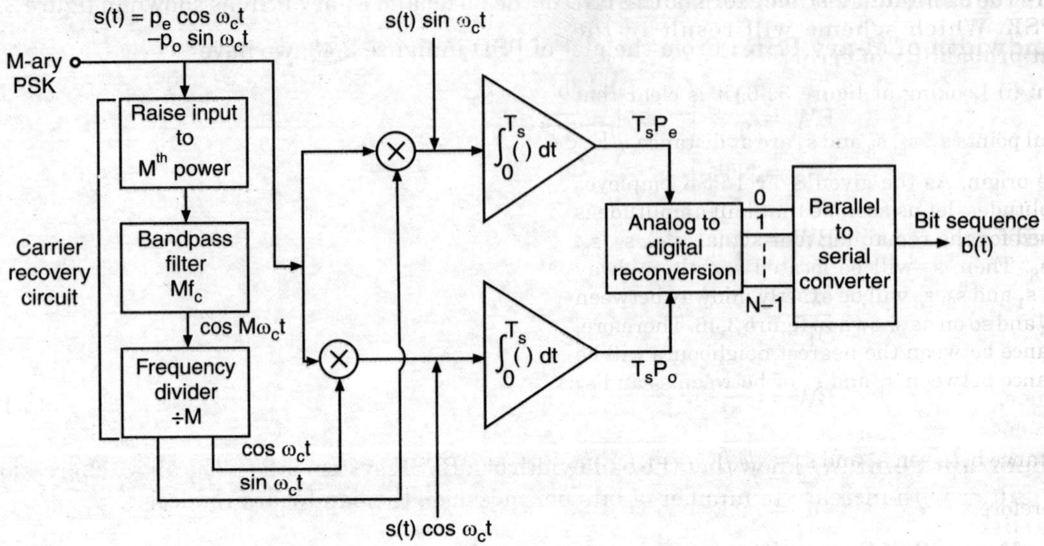

Fig. 3.45. M-ary PSK receiver

Working Operation

The M-ary receiver operates on the principle of synchronous demodulation which we had discussed for BPSK and QPSK. The carrier recovery system will require a device which can raise the received signal to M^{th} power. This signal is then applied to a bandpass filter, the center frequency of which has been selected to be Mf_c. At the filter output, we get a sinusoid at frequency Mf_c that is M^{th} harmonic of the carrier frequency f_c. This frequency is then divided by M to obtain the carrier at frequency f_c. The two carriers produced at the filter output are $\cos \omega_c t$ and $\sin \omega_c t$. These recovered carriers are then applied to two multipliers (balanced modulators). The other input of each multiplier block is connected to the received M-ary PSK signal. The outputs of the balanced modulators are applied to the integrators.Since, the M-ary PSK system is a non-offset or non-staggered type of system, the integrators will extend their integration over the same time interval. These integrators will work alongwith a bit synchronizer, which has not been shown in figure 3.45. The outputs of the integrators are proportional to $T_s p_e$ and $T_s p_o$ respectively and they (outputs) change at the symbol rate. These outputs are then applied to an A to D converter which yields the N bit transmitted signal. The signal is converted into b(t) by using a parallel to serial converter. Now, the operating systems with N = 4 bits and $M = 2^4 = 16$ are common. The bandwidth of such a system will be given by,

$$BW = \frac{2 f_b}{N} = \frac{2 f_b}{4} = \frac{f_b}{2} \qquad ...(3.109)$$

Advantages of M-ary PSK

(i) Bandwidth reduces with increase in the number of bits per symbol (N).

(ii) As the information is transmitted through signal phase and not through signal amplitude, this system operates very well in the situations where the signal amplitude varies due to the characteristics of the transmission medium.

Disadvantages of M-ary PSK :

(i) The probability of error increases with increase in the number of bits N per symbol, as the distance d decreases with increase in N.

(ii) The transmitter and receiver of M-ary PSK is very complex.

EXAMPLE 3.2. Given an 8-level PSK signal that employes two different amplitudes as shown in figure 3.46. Compute the distance between the nearest neighbours in terms of signal energy. Compare the minimum distance for standard 8-level PSK. Which scheme will result in the smallest probability of error?

Solution: (i) Looking at figure 3.46, it is clear that the signal points s_1, s_2, s_3 and s_4 are at distance $\sqrt{E_s}$ from the origin. As the given 8-level PSK employes two amplitudes, let us assume that half amplitude is being used for the remaining four signals i.e., s_5, s_6, s_7 and s_8. Then, s_5 will be located exactly midway between s_1 and s_2, s_6 will be exactly midway between s_2 and s_3 and so on as shown in figure 3.46. Therefore, the distance between the nearest neighbours will be the distance between s_1 and s_5 or between s_2 and s_5 etc.

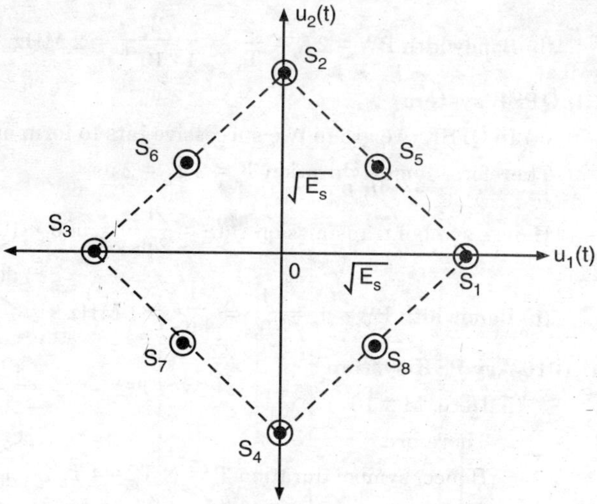

Fig. 3.46

Distance between s_1 and $s_2 = \sqrt{(\sqrt{E_s})^2 + (\sqrt{E_s})^2}$

Therefore, $$d = \sqrt{2\,E_s}$$

Therefore, distance between s_2 and $s_5 = \dfrac{d}{2} = \dfrac{\sqrt{2E_s}}{2} = \dfrac{\sqrt{E_s}}{2} = 0.707\sqrt{E_s}$ ...(i)

(ii) Now, let us consider the standard 8-level PSK system (M = 8).

The expression for minimum distance is given by

$$d_{min} = 2\sqrt{E_s}\,\sin(\pi/M)$$

Substituting M = 8 in the above equation, we get

$$d_{min} = 2\sqrt{E_s}\,\sin(\pi/8)$$

Therefore, $$d_{min} = 0.7653\sqrt{E_s} \quad ...(ii)$$

(iii) Comparing equations (i) and (ii), it is obvious that the distance is more for the standard 8-level PSK system. Therefore, error probability will be less for the standard 8-level PSK system.

EXAMPLE 3.3. In a digital continuous-time communication system, the bit rate of NRZ data stream is 1 Mbps and carrier frequcy of transmission is 100 MHz. Find the symbol rate of transmission and bandwidth requirement of the channel in following cases of different techniques used:

(i) BPSK system

(ii) QPSK system

(iii) 16-ary PSK system

Solution: The bit rate of NRZ data stream is 1 MHz. Therefore, the bit period will be

$$T_b = \frac{1}{f_b} = \frac{1}{1\text{MHz}} = 1\ \mu s \quad ...(i)$$

(i) BPSK system

(a) Symbol rate of transmission : In BPSK, each binary bit is a symbol. Therefore, symbol duration is equal to bit duration.

Therefore, $$T_s = T_b = 1\ \mu s$$

Hence, symbol rate of transmission $= \dfrac{1}{T_s} = \dfrac{1}{1\ \mu s} = 10^6$ symbols/sec **Ans.**

(b) Bandwidth $BW = 2\, f_b = \frac{2}{T_b} = \frac{2}{1 \times 10^{-6}} = 2$ MHz. **Ans.**

(ii) QPSK system

(a) In QPSK, we group two successive bits to form one symbol.

Therefore, symbol duration $T_s = 2\, T_b = 2\ \mu s$.

Hence, symbol transmission rate $= \frac{1}{2\ \mu s} = 500 \times 10^3$ symbols/sec **Ans.**

(b) Bandwidth $BW = f_b = \frac{1}{T_b} = \frac{1}{1\ \mu s} = 1$ MHz **Ans.**

(iii) 16-Ary PSK system :

(i) Here, $M = 16$

Therefore, $N = 4$

Hence, symbol duration $T_s = N\, T_b = 4\, T_b$

or $T_s = 4\ \mu s$

Therefore, symbol transmission rate $= \frac{1}{4\ \mu s} = 250 \times 10^3$ symbols/sec **Ans.**

(ii) Bandwidth $BW = \frac{2\, f_b}{N} = \frac{2 \times 1\ \text{MHz}}{4} = 500$ kHz. **Ans.**

EXAMPLE 3.4. Assume that you are required to transmit $f_b = 90$ Mb/s data in an authorised bandwidth of 20 MHz. Which modulation techniques would you consider? Explain why?

Solution: Given that $f_b = 90$ Mb/s and BW = 20 MHz. Hence, the technique that can be used here is M-ary PSK. The bandwidth of an M-ary PSK system is given by

$$BW = \frac{2\, f_b}{N}$$

where N = Number of bits/symbol

Therefore,
$$20\ \text{Mb/s} = \frac{2 \times 90\ \text{MHz}}{N}$$

or
$$N = \frac{180}{20} = 9 \quad \textbf{Ans.}$$

Important Point: If the number of bits N = 9 then the value of M is 2^N i.e., $2^9 = 512$. Hence, we will have to use a 512-ary PSK to transmit the 90 MHz signal in the available bandwidth of 20 MHz.

EXAMPLE 3.5. Derive an expression for the spectral spread 16-ary PSK system.

Solution: This is nothing but derivation of the power spectral density (psd) of 16-ary PSK. We have derived the expression for the PSD of a baseband signal of M-ary PSK.

We have
$$S_{B(\text{M-ary PSK})}(f) = 2P_s\, NT_b \left[\frac{\sin(\pi f N T_b)}{\pi f\, N T_b}\right]$$

In this equation, substitute the value of M = 16 i.e., N = 4 to obtain the PSD of the baseband signal as follows :

$$S_{B(\text{16-ary PSK})}(f) = 2P_s \times 4 \times T_b \left[\frac{\sin(\pi f \times 4T_b)}{\pi f \times 4T_b}\right] = 8P_s\, T_b \left[\frac{\sin(4\, \pi\, f\, T_b)}{4\, \pi f\, T_b}\right]$$

The corresponding spectrum is shown in figure 3.47.

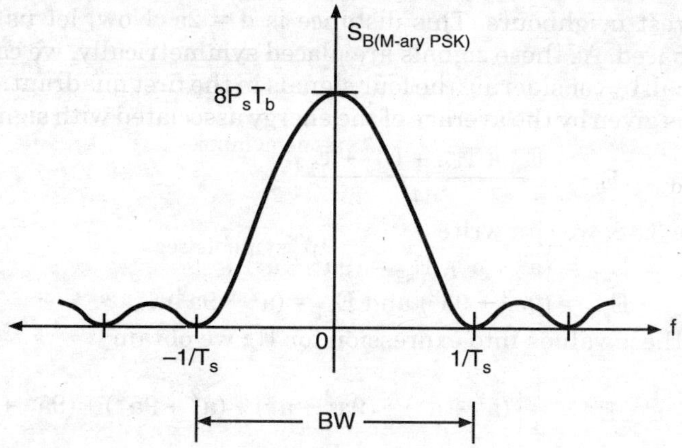

Fig. 3.47. Spectrum of the baseband signal of 16-ary PSK

EXAMPLE 3.6. Show that, in an AWGN channel, the detection performance of equal energy signals depends only on the 'distance' between the two pertinent message points in the signal space.

Solution: Even if the energy of the two transmitted signals is same, the receiver will make a correct decision if and only if the two signals are at a sufficient distance from each other in the constellation diagram. If the distance between them is small, then, the possibility of the receiver making mistake in making a decision about the received signal increases. This fact becomes evident from the expressions of error probabilities and the distance d. For the M-ary PSK system :

$$d = 2\sqrt{E_s}\,\sin(\pi/M) \qquad \text{...(i)}$$

and

$$P_e = \text{erfc}\sqrt{E_s/2N_0} \qquad \text{...(ii)}$$

As we increase the number of symbols M in equation (i), the value of sin (π/M) decreases and the value of d also decreases hence the value of E_s will decrease. The complementary error function of equation (ii) is a monotonically decreasing function. Hence, as E_s decreases, the value of P_e increases.

Important Point: This shows that the error probability increases with decrease in the distance d between the message points.

3.20 QUADRATURE AMPLITUDE SHIFT KEYING (QASK) OR QAM

In all the PSK methods discussed till now, one symbol is distinguished from the other in phase, but all the symbols transmitted using BPSK, QPSK or M-ary PSK are of same amplitude. The ability of a receiver to distinguish between one signal vector from another in presence of noise, depends on the distance between the vector end points. This suggests that the noise immunity will improve if the signal vectors differ not only in phase, but also in amplitude. Such a system is called as amplitude and phase shift keying system. In this system, the direct modulation of carriers in quadrature (i.e., cos $\omega_c t$ and sin $\omega_c t$) is involved, therefore, this system is called as the quadrature amplitude phase shift keying i.e., QAPSK or simply QASK. It is also known as **quadrature amplitude modulation (QAM)**.

3.20.1 Geometrical Representation of QASK

The geometrical representation is also called as signal space representation.

Let us assume that using QASK, we want to transmit a symbol consisting of 4 bits. This means that N = 4 and there are $2^4 = 16$ different possible symbols. Hence, the QASK system should be able to generate 16 different distinguishable signals. A possible geometric representation of 16 signals is shown in figure 3.48. In the geometric representation of figure 3.48, each signal point is equally distant from its nearest neighbours. This distance is d = 2a. Now, let us assume that all the 16 signals are equally spaced. As these signals are placed symmetrically, we can determine the energy associated with a signal, by considering the four signals in the first quadrant. The average normalized energy of each signal is given by the average of the energy associated with signals in the first quadrant.

Therefore, $$E_s = \frac{E_{s1} + E_{s2} + E_{s3} + E_{s4}}{4}$$

Looking at figure 3.48, we can write

$$E_{s1} = (a^2 + a^2),\ E_{s2} = (9a^2 + a^2)$$

$$E_{s3} = (9a^2 + 9a^2) \text{ and } E_{s4} = (a^2 + 9a^2)$$

Substituting all these values into expression for E_s, we obtain

$$E_s = \frac{1}{4}\left[(a^2 + a^2) + (9a^2 + a^2) + (a^2 + 9a^2) + (9a^2 + 9a^2)\right]$$

$$= 10\, a^2 \quad \text{...(3.110)}$$

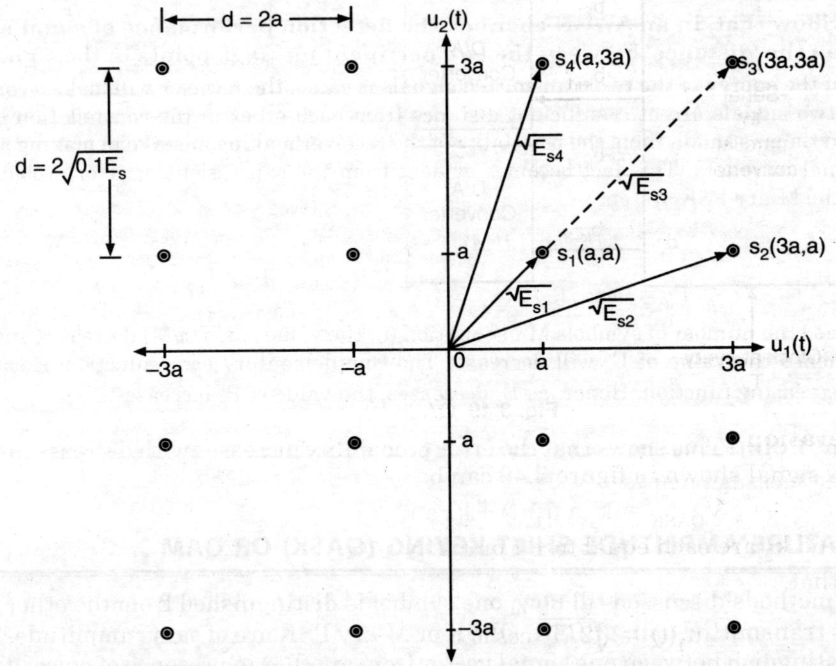

Fig. 3.48 Geometric representation of 16 signals in a QASK system (16-QAM)

Therefore, $$a = \sqrt{0.1\, E_s} \quad \text{...(3.111)}$$

and $$d = 2a = 2\sqrt{0.1\, E_s} \quad \text{...(3.112)}$$

where E_s = Normalized symbol energy. In this system, because each symbol consists of 4 bits, the normalized symbol energy is given by,

$$E_s = 4\, E_b \quad \text{...(3.113)}$$

where E_b is the normalized energy per bit.

Substituting equation (3.113) into equations (3.111) and (3.112), we get

$$a = \sqrt{0.4\, E_b} \text{ and } d = 2\sqrt{0.4\, E_b} \quad \text{...(3.114)}$$

Equation (3.114) shows that the distance d for QASK system is significantly less than the distance between the adjacent QPSK signals where $d = 2\sqrt{E_b}$. But, this distance d is greater than the 16-ary PSK, where,

or $$d = 2\sqrt{16\, E_b \times \sin^2 \frac{\pi}{16}} = 2\sqrt{0.15\, E_b} = \sqrt{0.6\, E_b} \quad ...(3.115)$$

Important Point: Thus, the 16 QASK system will have a low error rate as compared to 16-ary PSK but a higher error rate as compared to a QPSK sytem.

3.20.2 QASK Transmitter

The block diagram of a QASK transmitter is shown in figure 3.49.

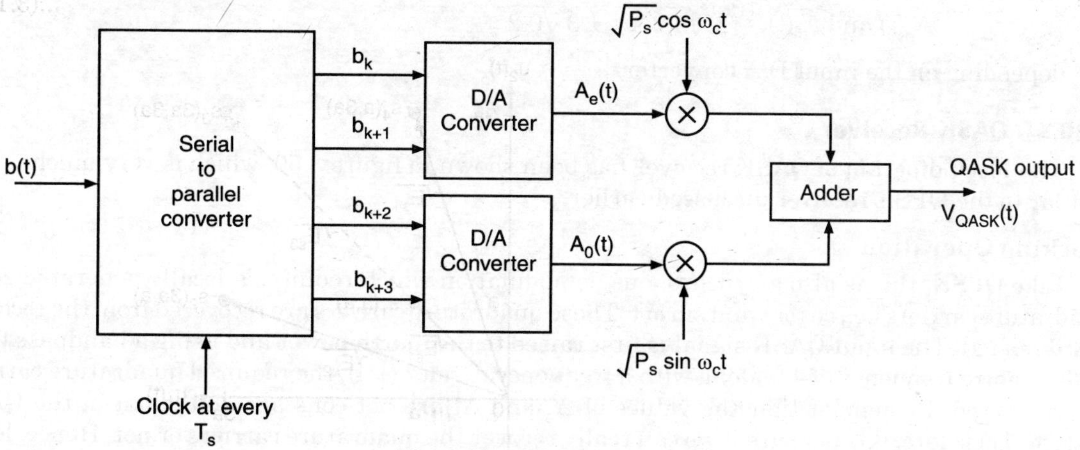

Fig. 3.49 QASK transmitter

Working Operation

The QASK signal shown in figure 3.49 can be mathematically represented as,

$$V_{QASK} = k_1\, a\, u_1(t) + k_2\, a\, u_2(t) \quad ...(3.116)$$

where k_1 and k_2 are each equal to ± 1 or ± 3.

We know that,

$$u_1(t) = \sqrt{2/T_s}\, \cos \omega_c t$$

and $$u_2(t) = \sqrt{2/T_s}\, \sin \omega_c t$$

and $$a = \sqrt{0.1\, E_s}$$

We can substitute these expressions into equation (3.116), to get

$$V_{QASK} = k_1 \times \sqrt{(0.2\, E_s / T_s)}\, \cos \omega_c t + k_2 \times \sqrt{(0.2\, E_s / T_s)}\, \sin \omega_c t \quad ...(3.117)$$

But, $\frac{E_s}{T_s} = P_s$.

Hence, equation for QASK is given by,

$$V_{QASK} = k_1 \sqrt{0.2\, P_s}\, \cos \omega_c t + k_2 \times \sqrt{0.2\, P_s}\, \sin \omega_c t \quad ...(3.118)$$

The QASK generator has been shown in figure 3.49. The bit stream b(t) is applied to a serial to parallel converter operating on a clock which has a period of T_s sec, which is the symbol duration.

The bits b(t) are stored by the converter and then presented in the parallel form. The four bit symbol is $b_{k+3}\, b_{k+2}\, b_{k+1}\, b_k$. Out of these four bits, the first two bits are applied to a D to A converter and the other two bits are applied to the second D to A converter.

The output of the first D/A converter is $A_e(t)$ which is used to modulate the carrier $\sqrt{P_s}\ \cos \omega_c t$, whereas the output of the second D/A converter i.e., $A_0(t)$ is used to modulate the carrier $\sqrt{P_s}\ \sin \omega_c t$ in the balanced modulators. The balance modulator outputs are added together to get the QASK output signal, which is expressed as follows :

$$V_{QASK}(t) = A_e(t) \times \sqrt{P_s}\ \cos \omega_c t + A_o(t) \times \sqrt{P_s}\ \sin \omega_c t \quad ...(3.119)$$

Comparing equations (3.119) and (3.118) we get,

$$A_e(t) \text{ and } A_o(t) = \pm\sqrt{0.2} \text{ or } \pm 3\sqrt{0.2} \quad ...(3.120)$$

depending on the input D/A converter.

3.20.3 QASK Receiver

The block diagram of QASK receiver has been shown in figure 3.50, which is very much similar to the QPSK receiver discussed earlier.

Working Operation

Like QPSK, this is also a synchronous demodulation which requires a locally generated set of quadrature carriers i.e., $\cos \omega_c t$ and $\sin \omega_c t$. These quadrature carriers are recovered from the received QASK signal. The input QASK signal is first raised to the fourth power and using a bandpass filter with a centre frequency of $4 f_c$ along with a frequency divider (÷ 4), the required quadrature carriers are recovered. Remember that the values of A_e and A_o are not constant and equal in the QASK system. Therefore, it is not sure if we can really recover the quadrature carriers or not. Hence, let us check whether we can really recover the carriers correctly. The input signal $V_{QASK}(t)$ is raised to fourth power as,

$$V^4_{QASK}(t) = P_s^{\,2}\,[A_e(t) \cos \omega_c t + A_o(t) \sin \omega_c t]^4$$

This signal is then passed through a bandpass filter with a centre frequency $4 f_c$, therefore, we neglect all the other terms except for those at frequency $4 f_c$.

Therefore,
$$V^4_{QASK}(t) = \frac{P_s}{8}[A_e^4(t) + A_o^4(t) - 6\,A_e^2(t)\,A_o^2(t)] \cos 4\,\omega_c t$$
$$+ \frac{P_s}{2}[A_e(t)\,A_o(t)\,[A_e^2(t) - A_o^2(t)] \sin 4\,\omega_c t \quad ...(3.121)$$

The average value of the coefficient of $\cos 4\,\omega_c t$ is not zero but the average value of the coefficient of $\sin 4\,\omega_c t$ will be zero. Thus, at the output of the bandpass filter and frequency divider combination, we get the quadrature carrier components $\cos \omega_c t$ and $\sin \omega_c t$. Then, two balanced modulators (multipliers) are used together with two integrators to recover the signals $A_e(t)$ and $A_o(t)$. Both the integrators integrate over one symbol interval T_s. The symbol time synchronizer which is not shown in figure 3.50 is actually used alongwith each integrator.

Finally, the original bits are obtained from $A_e(t)$ and $A_o(t)$ by using two A to D converters. The outputs of the two A to D converters are then applied to a serial to parallel converter to obtain the sequence b(t).

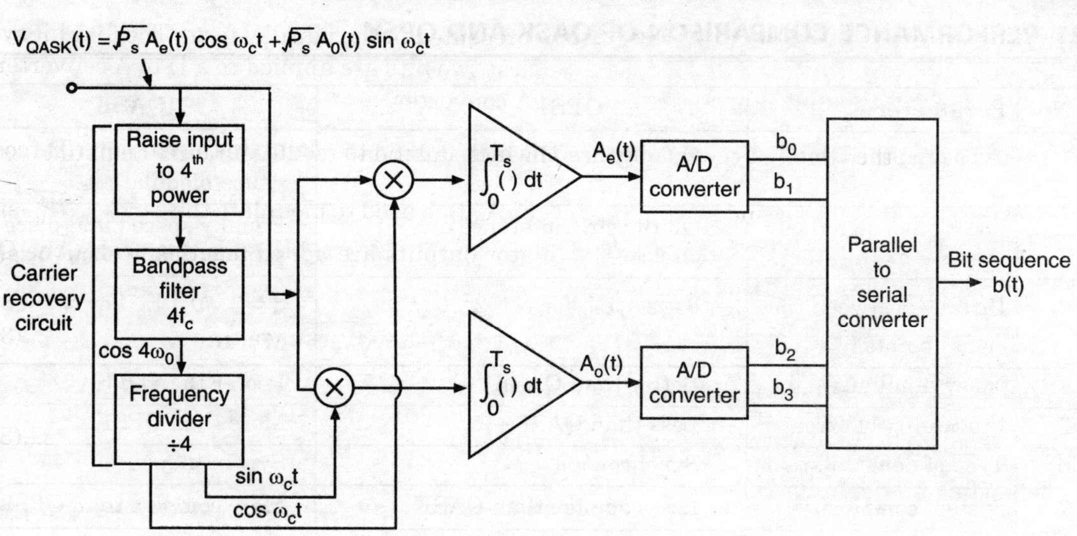

Fig. 3.50 The QASK receiver

3.20.4 Bandwidth of QASK System

The expression for the QASK output is very similar to that of a M-ary PSK. Therefore, the plot of power spectral density (psd) also will be same as that for the M-ary PSK. The spectrum of QASK is shown in figure 3.51, which is quite similar to that of a M-ary PSK.

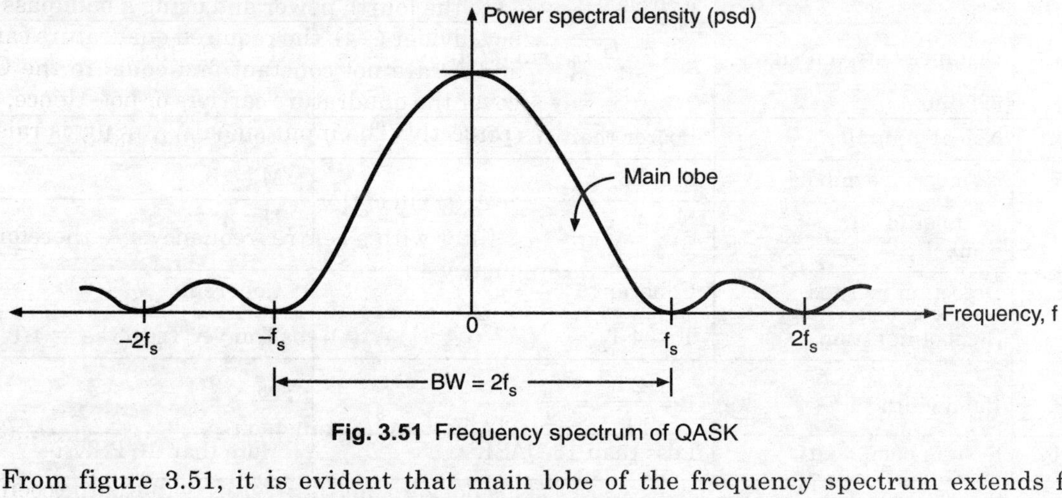

Fig. 3.51 Frequency spectrum of QASK

From figure 3.51, it is evident that main lobe of the frequency spectrum extends from $-f_s$ to $+f_s$. Therefore, the bandwidth of QASK is given by,

$$BW = f_s - (-f_s) = 2f_s \quad ...(3.122)$$

$$= \frac{2}{T_s} = \frac{2}{N\,T_b} \quad \text{(Since } T_s = N\,T_b\text{)}$$

Therefore,

$$BW = \frac{2f_b}{N} \quad \text{since } f_b = \frac{1}{T_b} \quad ...(3.123)$$

Thus, the bandwidth of QASK system is same as that of an M-ary PSK system.

3.21 PERFORMANCE COMPARISON OF QASK AND QPSK

S.No.	Parameter	QPSK	QASK
1.	Type of modulation	Quadrature phase modulation	Quadrature amplitude and phase modulation
2.	Location of signal points	On the circumference of a circle	Equally spaced and placed symmetrically about origin
3.	Distance between the signal points	$d = 2\sqrt{E_b}$ for N = 2	$d = 2\sqrt{0.4\, E_b}$ for N = 4 or M = 16
4.	Noisy immunity	Better than QASK	Poorer than QPSK
5.	Probability of error	Less than QASK	More than QPSK
6.	Type of demodulation	Synchronous	Synchronous
7.	System complexity	Less complex than QASK	More complex than QPSK

3.22. PERFORMANCE COMPARISON OF 16 PSK WITH 16 QASK

S.No.	Parameter	16 PSK	16 QASK
1.	Type of modulation	M-ary PSK with M = 16	M-ary QAM with M = 16
2.	Location of signal points	On the circumference of a circle	4 points in each quadrant
3.	Distance between signal points	$d = 2\sqrt{4\, E_b}\, \sin(\pi/M)$	$d = 2\sqrt{0.15\, E_b}$
4.	Noise immunity	Poorer than 16 QASK	Better than 16 PSK
5.	Number of symbols	M = 16	M = 16
6.	Number of bits per symbol	N = 4	N = 4
7.	Detection method	Coherent	Coherent
8.	Symbol duration	$T_s = 4\, T_b$	T_s ü ÿçÿ Tçe 63 se′ = 4 T_b
9.	Bandwidth	$B = \dfrac{2\, f_b}{N} = \dfrac{f_b}{2}$	$\dfrac{f_b}{2}$
10.	System complexity	Less than 16 QASK	More than 16 PSK

3.23. M-ARY FSK SYSTEM

An M-ary FSK system is the logical extension of a binary FSK system. In an M-ary FSK system, the symbols are not one bit long like BFSK but each symbol is N bit long with a symbol duration of $T_s = NT_b$. The block diagram of an N bit M-ary FSK system has been shown in figure 3.52.

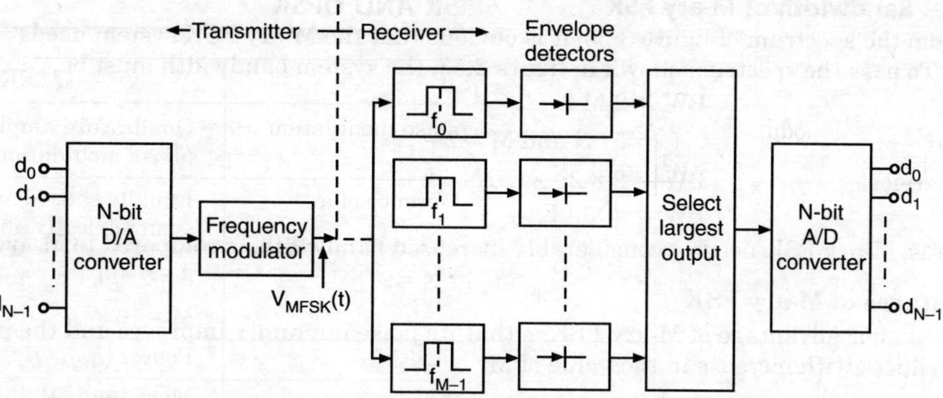

Fig. 3.52 An M-ary FSK communication system

Working Operation

At the transmitter an N bit symbol is presented after a period of T_s to an N bit D/A convertex. The output of the D/A converter is applied to a frequency modulator. The frequency modulator generates a carrier waveform whose frequency is determined by the analog modulating signal at the output of D/A converter. The transmitted signal, for the duration of the symbol interval has the frequency f_0, f_1 ... f_{M-1} where $M = 2^N$. For example, if N = 2 then $M = 2^2 = 4$. Then, there will be four different frequencies corresponding to the four different possible messages i.e., 00, 01, 11 and 10. The M-ary FSK signal is demodulated using envelope detectors as shown in figure 3.52. In receiver the incoming signal is applied to M parallel bandpass filters each followed by an envelope detector. These bandpass filters have center frequencies f_0, f_1, ... f_{M-1}. Therefore, these filters will pass only the signals of their center frequencies Remember that at any given instant of time only one frequency is being received. The outputs of bandpass filters are then applied to envelope detectors. They produce the voltage proportional to their ac input voltage. Outputs of all the envelope detectors are applied to a device which determines which of the detectors is the largest of all at the given time. The output of this device is applied to an N bit A to D converter which reproduces the originally transmitted N bit message.

3.23.1 Spectrum of M-ary FSK

As will be proved later, the probability of error is minimized if we select the frequencies f_0, f_1 ..., f_{M-1} such that the M signals are mutually orthogonal. To achieve this, the carrier frequency is selected to be equal to successive even harmonics of the symbol frequency $f_s = 1/T_s$. That means $f_0 = k\,f_s$, $f_1 = (k+2)\,f_s$, $f_2 = (k+4)\,f_s$ etc. The spectrum of such an M-ary FSK is as shown in figure 3.53, which is just an extension of the BFSK spectrum.

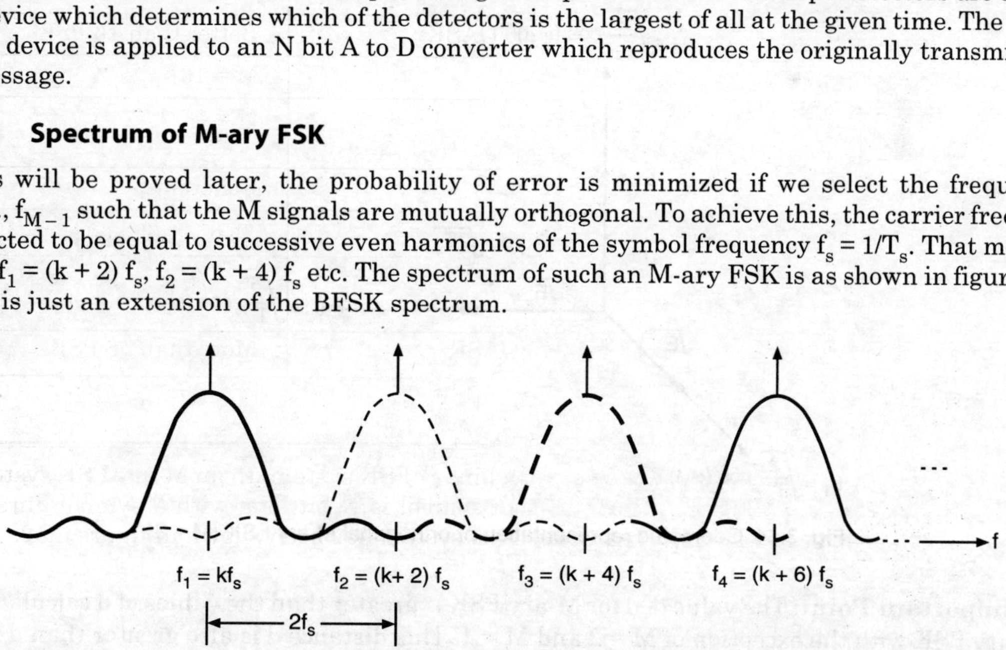

Fig. 3.53 Spectrum of an M-ary FSK

3.23.2 Bandwidth of M-ary FSK

From the spectrum of figure 3.53, it is obvious that the M-ary FSK system needs a large bandwidth. To pass the spectrum shown in figure 3.53, the system bandwidth must be,

$$BW = 2\,M\,f_s \quad ...(3.124)$$

But $f_s = f_b/N$ and $M = 2^N$

Therefore,

$$BW = 2 \times 2^N \times f_b/N \quad ...(3.125)$$

$$BW = 2^{N+1} F_{b/N}$$

Thus, M-ary FSK needs a considerably increased bandwidth as compared to M-ary PSK.

Advantages of M-ary FSK

The major advantage of M-ary FSK is that its noise immunity improves and the probability of error reduces with increase in the value of M.

Disadvantages

(i) A number of correctly tuned filters are required for the reception of M-ary FSK signal.

(ii) Large bandwidth requirements.

3.23.3 Geometric Representation of M-ary FSK

Figure 3.54 shows a signal space representation for orthogonal binary FSK (BFSK) signal. The signal space representation for M-ary FSK is the logical extension of binary FSK as shown in figure 3.54. The signal space representation of M-ary FSK is simply a co-ordinate system with M mutually orthogonal co-ordinate axis. The signal vectors will then be parallel to these co-ordinate axis. The geometric representation of M-ary FSK for M = 3 is shown in figure 3.54. As shown in this figure, the square of the length of the signal vector is the normalized signal energy. The distance between signal points is given by,

$$d = \sqrt{2\,E_s} = \sqrt{2\,N\,E_b} \quad ...(3.126)$$

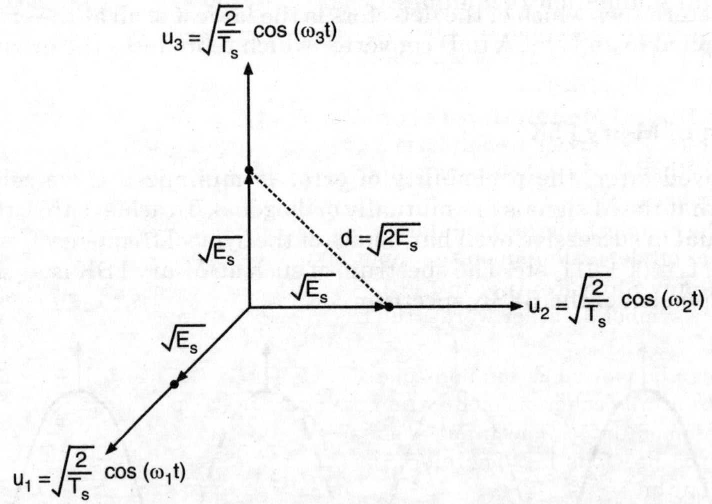

Fig. 3.54 Geometric representation of orthogonal M-ary FSK (M = 3)

Important Point: The value of d for M-ary FSK is greater than the values of d calculated for M-ary PSK with the exception of M = 2 and M = 4. This distance d is also greater than d in the 16-QASK system.

3.24. PERFORMANCE COMPARISON OF M-ARY PSK AND M-ARY FSK

S.No	Parameter	M-ary PSK	M-ary FSK
1.	Expression of transmitted signal	$\sqrt{2P_s}\cos(\omega_c t + \phi_m)$ $\phi_m = (2m+1)\dfrac{\pi}{4}$, $m = 0, 1, 2 \ldots M-1$	$\sqrt{2P_s}\cos(\omega_c t)$ with $c = 1, 2, \ldots M-1$
2.	Number of bits per symbol	$N \times (M = 2^N)$	$N \times (M = 2^N)$
3.	Symbol duration	$T_s = N\,T_b$	$T_s = N\,T_b$
4.	Information is transmitted through	Change in phase of carrier	Change in frequency of carrier
5.	Demodulation method	Coherent (synchronous)	Non-coherent
6.	Bandwidth (BW)	$2f_b / N$	$2^{N+1} f_b/N$
7.	Minimum distance between signal points	$d = 2\sqrt{E_s}\sin(\pi/M)$	$\sqrt{2NE_b}$
8.	Probability of error	More than that in M-ary FSK	Less than that in the M-ary PSK

GLOSSARY

1. **Coherent Digital Modulation Techniques:** Coherent digital modulation techniques are those techniques which employ coherent detection. In coherent detection, the local carrier generated at the receiver is phase locked with the carrier at the transmitter. Thus, the detection is done by correlating received noisy signal and locally generated carrier. The coherent detection is a synchronous detection.
2. **Non-coherent Digital Modulation Techniques:** Those techniques in which the detection process does not need receiver carrier to be phase locked with transmitter carrier. The advantage of such type of system is that the system becomes simple. But the drawback of such a system is that the error probability increases.
3. **ASK:** ASK is the simplest digital modulation technique. There is only one unit energy carrier and it is switched on or off depending upon the input binary sequence.
4. **Binary Frequency Shift Keying:** In BFSK, the frequency of a sinusoidal carrier is shifted according to the binary symbol. In other words, the frequency of a sinusoidal carrier is shifted between two discrete values.
5. **DPSK:** DPSK can be viewed as the non-coherent version of the PSK. Differential phase shift keying (DPSK) is differentially coherent modulation method. DPSK does not need a synchronous (coherent) carrier at the demodulator. The input sequence of binary bits is modified such that the next bit depends upon the previous bit. Therefore, in the receiver, the previous received bits are used to detect the present bit.
6. **Gaussian Minimum shift keying:** Gaussian MSK (GMSK) yields a constant amplitude and continuous phase RF carrier signal. It only differs in use of a Gaussian baseband pulse shape in place of square pulse shape for MSK. Because, the Gaussian pulse rises and decays asymptotically with respect to a zero response level, it has a much more constrained bandwidth.
7. **Non-coherent Digital Modulation Techniques:** Those techniques in which the detection process does not need receiver carrier to be phase locked with transmitter carrier. The advantage of such type of system is that the system becomes simple. But the drawback of such a system is that the error probability increases.

8. **ASK:** ASK is the simplest digital modulation technique. There is only one unit energy carrier and it is switched on or off depending upon the input binary sequence.
9. **Binary Frequency Shift Keying:** In BFSK, the frequency of a sinusoidal carrier is shifted according to the binary symbol. In other words, the frequency of a sinusoidal carrier is shifted between two discrete values. However, the phase of the carrier is unaffected. This means that we have two different frequency signals according to binary symbols.
10. **DPSK:** DPSK can be viewed as the non-coherent version of the PSK. Differential phase shift keying (DPSK) is differentially coherent modulation method. DPSK does not need a synchronous (coherent) carrier at the demodulator. The input sequence of binary bits is modified such that the next bit depends upon the previous bit. Therefore, in the receiver, the previous received bits are used to detect the present bit.
11. **Gaussian Minimum shift keying:** Gaussian MSK (GMSK) yields a constant amplitude and continuous phase RF carrier signal. It only differs in use of a Gaussian baseband pulse shape in place of square pulse shape for MSK. Because, the Gaussian pulse rises and decays asymptotically with respect to a zero response level, it has a much more constrained bandwidth.

SUMMARY

- Modulation is defined as the process by which some characteristics of a carrier is varied in accordance with a modulating signal.
- In digital communications, the modulating signal consists of binary data or an M-ary encoded version of it.
- The channel may be a telephone channel, microwave radio link, satellite channel or an optical fiber. In digital communication, the modulation process involves switching or keying the amplitude, frequency or phase of the carrier in accordance with the input data.
- There are three basic modulation techniques for the transmission of digital data. They are known as amplitude-shift keying (ASK), frequency shift keying (FSK) and phase-shift keying (PSK) which can be viewed as special cases of amplitude modulation, frequency modulation and phase modulation respectively.
- When we have to transmit a digital signal over a long distance, we need continuous-wave (CW) modulation. For this purpose, the transmission medium can be in form of radio, cable or other type of channel. Also, a carrier signal having some frequency f_c is used for modulation. Then the modulating digital signal modulates some parameter like frequency, phase or amplitude of the carrier.
- There is some deviation in carrier frequency f_c. This deviation is known as the bandwidth of the channel. This means that the channel has to transmit some range or band of frequencies. Such type of transmission is known as bandpass transmission and the communication channel is known as bandpass channel.
- When it is required to transmit digital signals on a bandpass channel, the amplitude, frequency or phase of the sinusoidal carrier is varied in accordance with the incoming digital data. Since the digital data is in discrete steps, the modulation of the bandpass sinusoidal carrier is also done in discrete steps. Due to this reason, this type of modulation (i.e., Digital modulation) is also known as switching or signaling.
- Because of constant amplitude of FSK and PSK, the effect of non-linearities, noise interference is minimum on signal detection. However, these effects are more pronounced on ASK. Therefore, FSK and PSK are preferred over ASK.
- In digital modulations, instead of transmitting one bit at a time, we transmit two or more bits simultaneously. This is known as M-ary transmission. This type of transmission results in reduced channel bandwidth. However, sometimes, we use two quadrature carriers for modulation. This process is known as Quadrature modulation.
- There are a number of modulation schemes available to the designer of a digital communication system required for data transmission over a bandpass channel.

- Every scheme offers system trade-offs of its own. However, the final choice made by the designer is determined by the way in which the available primary communication resources such as transmitted power and channel bandwidth.
- Coherent digital modulation techniques are those techniques which employ coherent detection. In coherent detection, the local carrier generated at the receiver is phase locked with the carrier at the transmitter. Thus, the detection is done by correlating received noisy signal and locally generated carrier. The coherent detection is a synchronous detection.
- Non-coherent digital modulation techniques are those techniques in which the detection process does not need receiver carrier to be phase locked with transmitter carrier.
- The binary (i.e., digital) modulation has three basic forms amplitude-shift keying (ASK), phase-shift keying (PSK) and frequency-shift keying (FSK).
- ASK signal may be generated by simply applying the incoming binary data (represented in unipolar form) and the sinusoidal carrier to the two inputs of a product modulator (i.e., balanced modulator).
- The demodulation of binary ASK waveform can be achieved with the help of coherent detector. It consists of a product modulator which is followed by an integrator and a decision making device.
- In a binary phase shift keying (BPSK), binary symbol '1' and '0' modulate the phase of the carrier.
- The BPSK signal may be generated by applying carrier signal to a balanced modulator. Here, the baseband signal b(t) is applied as a modulating signal to the balanced modulator.
- It is not possible to determine whether the received signal is equal to b(t) or – b(t). Infact, this results in ambiguity in the output signal.
- In binary frequency shift keying (BFSK), the frequency of the carrier is shifted according to the binary symbol. However, the phase of the carrier is unaffected. That is we have two different frequency signals according to binary symbols.
- Orthogonal carriers are used for M-ary PSK and QASK.
- The generation of BFSK is quite easy, even that it has many disadvantages compared to BPSK signal. Firstly, its bandwidth is greater than $4f_b$, which is almost double the bandwidth of BPSK. Also, the distance between the signal points is less in case of BFSK. Therefore, the error rate of BFSK is more as compared to BPSK.
- Coherent detection exploits knowledge of the carrier wave's phase reference, and thus providing the optimum error performance attainable with a digital modulation format of interest. However, when it is impractical to have knowledge of the carrier phase at the receiver, we make use of **non-coherent detection.**
- In the case of phase-shift keying (PSK), we cannot have "non-coherent PSK" since non-coherent means doing without phase information. However, there is a 'pseudo PSK' technique known as differential phase-shift keying (DPSK) which can be viewed as the non-coherent form of PSK.
- We can view differential phase-shift keying as the non-coherent version of the PSK. Differential phase shift keying (DPSK) is differentially coherent modulation method. DPSK does not need a synchronous (coherent) carrier at the demodulator.
- DPSK does not need carrier at the receiver end. This means that the complicated circuitry for generation of local carrier is not required.
- The bandwidth requirement of DPSK is reduced as compared to that of BPSK.
- The probability or error (i.e., bit error rate) of DPSK is higher than that of BPSK.
- Because DPSK uses two successive bits for its reception, error in the first bit creates error in the second bit. Therefore, error propagation in DPSK is more. On the other hand, in BPSK single bit can go in error since detection of each bit is independent.
- In communication systems, we have two main resources. These are the transmission power and the channel bandwidth. The channel bandwidth depends upon the bit rate or signaling rate f_b. In digital bandpass transmission, we use a carrier for transmission. This carrier is transmitted over a channel. If two or more bits are combined in some symbols, then the signaling rate will be reduced. Thus, the frequency of the carrier needed is also reduced. This reduces the transmission channel bandwidth. Hence, because of grouping of bits in symbols, the transmission channel bandwidth can be reduced. In quadrature phase shift keying (QPSK), two successive bits in the data sequence are grouped together. This reduces the bits rate or signaling rate (i.e., f_b) and thus reduces the bandwidth of the channel.

- In the bandpass transmission system, the digital signal modulates high frequency sinusoidal carrier. These are known as digital carrier modulation techniques. With the help of such techniques, it is possible to transmit data over long distances. However, in baseband transmission, the data is transmitted without modulation.
- During the transmission of data over the channel, it is corrupted by noise. Hence, at the receiver, the noisy signal is received. Therefore correct detection of the transmitted signal is necessary.
- The detection method must attenuate noise and amplify signal. This means that it must improve signal to noise ratio of the received signal.
- The detection method must check the received signal at the time instant in the bit interval when signal to noise ratio is maximum.
- The detection must be performed with minimum error probability.

SHORT QUESTIONS WITH ANSWERS

Q.1. What are the types of Digital Modulation Techniques? Explain.

Ans. Basically, digital modulation techniques may be classified into coherent or non-coherent techniques, depending on whether the receiver is equipped with a phase-recovery circuit or not. The phase-recovery circuit ensures that the oscillator supplying the locally generated carrier wave receiver is synchronized to the oscillator supplying the carrier wave used to originally modulate the incoming data stream in the transmitter.

(i) Coherent Digital Modulation Techniques

Coherent digital modulation techniques are those techniques which employ coherent detection. In coherent detection, the local carrier generated at the receiver is phase locked with the carrier at the transmitter. Thus, the detection is done by correlating received noisy signal and locally generated carrier. The coherent detection is a synchronous detection.

(ii) Non-coherent Digital Modulation Techniques

Non-coherent digital modulation techniques are those techniques in which the detection process does not need receiver carrier to be phase locked with transmitter carrier.

The advantage of such type of system is that the system becomes simple. But the drawback of such a system is that the error probability increases.

Q.2. Explain the principle of Binary Phase Shift Keying (BPSK).

Ans. In a binary phase shift keying (BPSK), the binary symbols '1' and '0' modulate the phase of the carrier. Let us assume that the carrier is given as,

$$s(t) = A\cos(2\pi f_c t)$$

Here 'A' represents peak value of sinusoidal carrier. For the standard 1 Ω load resistor, the power dissipated would be,

$$P = \frac{1}{2}A^2$$

or

$$A = \sqrt{2P}$$

Now, when the symbol is changed, then the phase of the carrier will also be changed by an amount of 180 degrees (i.e., π radians).

Let us consider, for example,

For symbol '1', we have

$$s_1(t) = \sqrt{2P}\cos(2\pi f_c t)$$

If next symbol is '0', then we have

For symbol '0', we have

$$s_2(t) = \sqrt{2P}\cos(2\pi f_c t + \pi)$$

Now, because $\cos(\theta + \pi) = -\cos\theta$, therefore, the last equation can be written as,

$$s_2(t) = -\sqrt{2P}\cos(2\pi f_c t)$$

With the above equation, we can define BPSK signal combinely as,

$$s(t) = b(t)\sqrt{2P}\cos(2\pi f_c t)$$

where $b(t) = +1$ when binary '1' is to be transmitted.

$= -1$ when binary '0' is to be transmitted.

Q.3. Explain the concept of Non-coherent Binary Amplitude Shift Keying (ASK).

Ans. In the binary ASK case, the transmitted signal is defined as

$$s(t) = \sqrt{2P_s}\cos(2\pi f_c t)$$

Binary ASK signal can also be demodulated non-coherently using envelope detector. This greatly simplifies the design consideration required in synchronous detection. Non-coherent detection schemes do not require a phase-coherent local oscillator. This method involves some form of rectification and low pass filtering at the receiver.

Q.4. Explain the principle of Differential Phase Shift Keying (DPSK).

Ans. We can view differential phase-shift keying as the non-coherent version of the PSK. Differential phase shift keying (DPSK) is differentially coherent modulation method. DPSK does not need a synchronous (coherent) carrier at the demodulator. The input sequence of binary bits is modified such that the next bit depends upon the previous bit.

REVIEW QUESTIONS

1. Explain coherent and non-coherent binary modulation techniques.
2. Explain coherent Binary ASK.
3. Describe coherent Binary FSK.
4. Explain the difference between different modulation techniques.
5. Draw the block diagram of QPSK system and explain its working.
6. Compare the bandwidth of QPSK system with that of BPSK system.
7. Represent QPSK signals in the signal space and find distance between them. What is the significance of each?
8. Write the power spectral density of BPSK and QPSK signals and draw the power spectrum of each.
9. Explain M-ary FSK system with the help of transmitters and receivers.
10. Determine the bandwidth required for M-ary FSK system. Draw the geometrical representation of M-ary FSK signals and find out distance between the signals.
11. Give the block diagram of a coherent signal reception system. Explain the effects of phase error and timing error on the error probability in a BPSK system employing coherent reception giving necessary derivations.
12. Explain the geometrical representation of Non-orthogonal BFSK.
13. Derive the expression for the spectrum of BPSK and sketch the same.
14. How is phase continuity achieved in MSK system?
15. Explain operation of differentially encoded PSK system. Explain why errors always occur in pairs in this system?
16. Explain the condition of orthogonally of the two BFSK signals and the signal space diagram for the orthogonal signal and hence derive an expression for the probability of error in synchronous detection of the BFSK signals using matched filters.
17. Derive an expression for the spectrum of BFSK and sketch it for the condition of orthogonality and minimum bandwidth.
18. What is the advantage of gray coding of the QPSK system? What is the advantage of differential encoding of the input to the QPSK system?

19. What type of synchronization is used in QPSK system?
20. Draw the signal space representation of QPSK and BPSK signals. Show that for the same data rate and same bit energy they have the same bit error rate probability on the same AWGN channel, although one of the signals occupy half the bandwidth.
21. What is the ambiguity in the decoded output in case of PSK systems? How is it corrected?
22. Draw the block diagram of DPSK modulator and explain how synchronization problem is avoided for its detection.
23. Explain the ASK system and derive the relation for error probability of binary ASK.

NUMERICAL PROBLEMS

1. Sketch the QPSK waveform for the sequence 1 1 0 1 0 1 0 0 1 0, assuming the carrier frequency to be equal to the bit rate.
2. Suggest a suitable synchronising circuit arrangement for extracting the carrier at the receiver end for coherent detection of binary PSK.
3. Sketch the waveform of the in-phase and quadrature components of a QPSK signals for the binary sequence 1 1 0 1 0 1 1 0 1 0.
4. For the sequence given in sketch the QPSK waveform.
5. Sketch the waveform of the MSK signal for the input binary sequence 1 1 0 0 1 0 0 0 1 0.
6. Sketch the QPSK waveform for the sequence 1 1 0 1 0 1 0 0 1 0, assuming the carrier frequency to be equal to the bit rate.
7. Suggest a suitable synchronising circuit arrangement for extracting the carrier at the receiver end for coherent detection of binary PSK.
8. Sketch the waveform of the in-phase and quadrature components of a QPSK signals for the binary sequence 1 1 0 1 0 1 1 0 1 0.
9. For the sequence given in Q8 sketch the QPSK waveform.
10. Sketch the waveform of the MSK signal for the input binary sequence 1 1 0 0 1 0 0 0 1 0.

□□□

UNIT– II

Concept of Probability, Random variable, Statistical averages, Correlation, Sum of Random Variables, Central Limit Theorem, Random Process, Classification of Random Processes, Power spectral density, Multiple random processes.

Above Syllabus has been covered in following chapter:

CHAPTER—4 : Probability and Random Variables

CHAPTER 4

PROBABILITY AND RANDOM VARIABLES

SYLLABUS

Concept of Probability, Random variable, Statistical averages, Correlation, Sum of Random Variables, Central Limit Theorem, Random Process, Classification of Random Processes, Power spectral density, Multiple random processes.

Inside this Chapter

4.1 INTRODUCTION

In previous chapter, we discussed how to apply Fourier transform to analyze some signals. All these signals were described by some fixed mathematical equations. Such type of signals are called **deterministic signals**. The behaviour of such type of signals as well as processing through linear time invariant (LTI) systems can be determined with the help of mathematical models. These mathematical models represent the complete behaviour of the signal at every instant of time. Hence, for such deterministic signals, there is no uncertainty about the value at any instant of time. This means that these signals are predictable.

There is one other class of signals, the behaviour of which cannot be predicted. Such type of signals are called **random signals**. These signals are called random signals because the precise value of these signals cannot be predicted in advance before they actually occur. The examples of random signals are the noise interferences in communication systems. This means that the noise interference during transmission is totally unpredictable. In the same way, the noise generated by the receiver itself is random. Even some other signals which are not noise signals are also random signals. These signals cannot be modelled mathematically. Actually the electromagnetic interference is the major source of random noise.

In the receiver, the thermal noise is caused by the random motion of the electrons. Although the random signals are not predictable in advance precisely, they can be described in terms of its statistical properties. It is possible to analyze the random signals statistically with the help of probability theory.

In fact, the probability theory is very essential mathematical tool in the design of digital communication systems.

4.2. BASIC DEFINITIONS RELATED TO PROBABILITY

Before we discuss the concept of probability and its applications in the field of communication, we must discuss few basic terms which are related to probability theory.

4.2.1. Experiment

An experiment is defined as the process which is conducted to get some results. If the same experiment is performed repeatedly under the same conditions, similar results are expected. But there are few experiments which do not produce the same result as we have stated above. The theory of probability mainly deals with such type of experiments. In fact, the theory of probability is applied to such type of experiments to predict the possibility of a particular output. An experiment is sometimes called **trial**. As an example, throw of a coin is an experiment or trial. This trial results in two outcomes namely **Head** and **Tail**. Similarly, drawing a card from a well shuffled pack is the trial or experiment and it results in 52 possible outcomes (i.e. cards). Hence, each experiment or a trial has an outcome and the possibility of this outcome can be predicted with the help of probability theory.

The outcomes of an event are called **equally** likely, if any one of them cannot be expected to occur in preference to another. As an example, the tossing of a coin results in two outcomes, Head and Tail. Both have same possibility of 50%. Such type of outcomes are called **equally** likely outcomes. On the other hand, if the box contains two black and four white balls, then getting a white or black ball is not equally likely. This is possible that we may get white balls sequentially.

4.2.2. Sample Space

A set of all possible outcomes of an experiment or trial is called the **sample space** of that experiment. It is generally denoted by 'S'. The total number of outcomes in a sample space is denoted by n(s). As an example, the tossing of a coin has two outcomes. Hence, we may write its sample space as:

$$S = \{H, T\}$$

where, $H \rightarrow$ Head

and $T \rightarrow$ Tail

Now, we consider another example. If three coins are tossed simultaneously, then each experiment has two outcomes namely H and T. Hence, the maximum possible number of outcomes will be eight.

$$S = \{HHH, HHT, HTH, THH, HTT, THT, TTH, TTT\}$$

So that $n(s) = 8$

4.2.3. Event

The expected subset of the sample space or happening is called an event. As an example, let us consider an experiment of throwing a cubic die. In this case, the sample space S will be as

$$S = \{1, 2, 3, 4, 5, 6\}$$

Now, if we want the number '3' to be an outcome or an even number, i.e., {2, 4, 6}, then this subset is called an **event**. This is denoted by letter 'E'. Hence event E is a subset of the sample space 'S'. If event E has only one outcome, then it is called an elementary event. On the other hand, if event E does not contain any out come, then it is called a null event. If E = S, then an event contains all the outcomes. Such as event is called a **certain event**. It always occurs, no matter what so ever is the outcome.

Two events A and B are called independent events if happening of event A has nothing to do

Pioneer in Communications : Albert Michelson

Albert Michelson (of Michelson-Morley fame) was an intense, practical man who developed ingenious physical instruments of extraordinary precision, mostly in the field of optics, harmonic analyzer, developed in 1898, could compute the first 80 coefficients of the Fourier series of a signal x(t) specified by any graphical description. The instrument could also be as a harmonic synthesizer, which could plot a function x(t) generated by summing the first harmonics (Fourier components) of arbitrary amplitudes and phases. This analyzer, then, had the ability of self-checking its operation by analyzing a signal x(t) and the adding resulting 80 components to see whether the sum yielded a close approximation of x(t).

with happening of *B*. As an example, let us consider an experiment of tossing a coin two times. In this case, occurrence of head in the first throw has nothing to do with the occurrence of head in the second throw. Hence, these two events are independent events since their outcomes are independent. On the other hand, if the outcome of one event is affected by other, then these events are called **dependent events.**

4.3 PROBABILITY

Probability may be defined as the study of random experiments. In any random experiment, there is always an uncertainty that a particular event will occur or not. As a measure of probability of occurrence of an event, a number between 0 to 1 is assigned. If it is sure that an event will occur, then we can say that its probability is 100% or 1. If it is not sure that an event will occur, then we can say that its probability is 0% or 0. If it is not sure whether the event will occur or not, then its probability is between 0 and 1.

Let us consider an example. The probability of occurrence of 28th February in a year is 1 since it is certain to occur every year. On the other hand, the probability of occurrence of 30th February in a year is 0 since it never comes. Again, the probability of occurrence of 29th February in a year is neither 0 nor 1. It is always between 0 and 1. Actually, it is 1/4 since, it occurs every leap year i.e., once in four years.

Therefore, from above discussion, we can write a mathematical expression for probability as Probability of an event A,

$$P(A) = \frac{\text{Number of possible favourable outcomes}}{\text{Total number of outcomes}}$$

For example, let us consider the probability of getting an even number in tossing of a die. In tossing of a die, an even number can occur in 3 ways out of 6 equally likely ways.

Therefore, $$P(A) = \frac{3}{6} = \frac{1}{2} = 0.5 = 50\%$$

4.4 PROPERTIES OF PROBABILITY

As discussed earlier, the sample space S contains all possible outcomes of an experiment. Also, an event is the subset of the sample space. Two events are said to be mutually exclusive if the occurrence of one of them precludes the occurrence of other. For example, in tossing of a coin, events Head and Tail are mutually exclusive. In throw of a die, the occurrence of number '4' will automatically exclude the occurrence of numbers 1, 2, 3 ,5 and 6. If an event contains all the outcomes then it is called certain event. Then, probability of this event is unity, i.e.,

$$P(A) = P(s) = 1$$

The properties of probability may be listed as under :

Property 1: The probability of a certain event is unity i.e.,

$$P(A) = 1 \quad ...(4.1)$$

Property 2: The probability of an event is always less than or equal to 1 and non-negative. Mathematically,

$$0 \le P(A) \le 1 \quad ...(4.2)$$

Property 3: If A and B are two mutually exclusive events, then

$$P(A + B) = P(A) + P(B) \quad ...(4.3)$$

Property 4: If A is any event, then the probability of not happening of A is

$$P(\bar{A}) = 1 - P(A) \quad ...(4.4)$$

where $\bar{A}$ represents the complement of event A.

Property 5: If A and B are any two events (not mutually exclusive events), then

$$P(A + B) = P(A) + P(B) - P(AB) \quad ...(4.5)$$

where P (AB) is called the probability of events A and B both occurring simultaneously. Such an event is called joint event of A and B, and the probability P (AB) is called the joint probability.

Now, if events A and B are mutually exclusive, then the joint probability, P(AB) = 0.

4.5 CONDITIONAL PROBABILITY

The concept of conditional probability is used in conditional occurrences of the events. Let us consider an experiment which involves two events A and B. Now the probability of event B, given that event A has occurred, is represented by P (B/A). Similarly, P (A/B) represents probability of event A given that event B has already occurred. Therefore, P (B/A) and P (A/B) are called **conditional probabilities**. These conditional probabilities may be defined in terms of their independent and joint probabilities as under:

The conditional probability of event B given that event A has already happened.

$$P(B/A) = \frac{P(AB)}{P(A)} \quad ...(4.6)$$

where P (AB) is the joint probability of A and B.

Similarly, the conditional probability of event A given that event B has already happened.

$$P(A/B) = \frac{P(AB)}{P(B)} \quad ...(4.7)$$

At this point, it may be noted that the joint probability has commutative property which states that

$$P(AB) = P(BA) \quad ...(4.8)$$

4.6 PROBABILITY OF STATISTICALLY INDEPENDENT EVENTS

If A and B are two events in an experiment, and possibility of occurrence of event B does not depend upon occurrence of event A, then these two events A and B are known as statistically independent events. The probability of event B, given that event A has already happened is expressed as

$$P(B/A) = \frac{P(AB)}{P(A)} \quad ...(4.9)$$

Again, since occurrence of event B does not depend on the occurrence of event A, then the probability of event B will be same as conditional probability P (B/A). Mathematically,

$$P(B/A) = P(B) \quad ...(4.10)$$

Now, putting the value of P(B/A) from equation (4.10) in equation (4.9), we have

$$P(AB) = P(A)\,P(B) \quad ...(4.11)$$

The probability of event A, given that event B has already happened, is expressed as:

$$P(A/B) = \frac{P(AB)}{P(B)} \quad ...(4.12)$$

Again, since events A and B are statistically independent, then the probability of event A will be same as conditional probability P (A/B). Mathematically,

$$P(A/B) = P(A) \quad ...(4.13)$$

Using equations (4.12) and 4.13), we have

$$P(AB) = P(A)\,P(B) \quad ...(4.14)$$

which is same as equation (4.11).

EXAMPLE 4.1. A box contains 3 red, 4 white and 5 black balls. One ball is drawn at random. Find the probability that it is (a) red (b) not Black (c) black or white.

Solution: Let R, W and B be the events of drawing a red, a white and a black ball respectively.

(a)
$$P(R) = \frac{\text{Ways of choosing a red ball}}{\text{Total no. of ways of choosing a ball}}$$

$$P(R) = \frac{3}{3+4+5} = \frac{3}{12} = \frac{1}{4}$$

(b) Probability of not getting a black ball is $P(\bar{B})$.

$$P(\bar{B}) = 1 - P(B)$$

But
$$P(B) = \frac{5}{3+4+5} = \frac{5}{12}$$

Therefore,
$$P(\bar{B}) = 1 - \frac{5}{12} = \frac{7}{12}$$

(c) Probability of getting a black or white ball will be P(B + W).

Therefore, $P(B + W) = P(B) + P(W)$ [Since B and W are two mutually exclusive events]

$$P(B+W) = \frac{5}{12} + \frac{4}{12}$$

$$P(B+W) = \frac{9}{12} = \frac{3}{4} \quad \textbf{Ans.}$$

EXAMPLE 4.2. Two dies are thrown simultaneously. Find the probability of getting a 5.

Solution: We know that each die has six faces.

Therefore, the total number of possible outcomes 6 × 6 = 36.

Each outcome is having an equal probability of occurrence. Therefore, probability of any outcome is 1/36.

We have 4 different combinations which give a 5. They are (4, 1), (3, 2), (2, 3), (1, 4).

Hence the probability of the overall event may be given as

$$P(A) = P(5) = \frac{1}{36} + \frac{1}{36} + \frac{1}{36} + \frac{1}{36} = \frac{4}{36} = \frac{1}{9} \quad \textbf{Ans.}$$

EXAMPLE 4.3. Two cards are drawn at random from an ordinary deck of 52 playing cards. Obtain the probability that one card is a Diamond and the other card is a Heart.

Solution: Since two cards are drawn, there are $\frac{52 \times 51}{1 \times 2} = 1326$ ways of drawing 2 cards from a deck of 52 cards.

Again, since there are 13 diamonds and 13 Hearts, there are 13 × 13 = 169 ways of drawing a Diamond and a Heart. Therefore, the probability of drawing that one card is diamond and other card is heart given by

$$P(A) = \frac{\text{The number of ways of drawing a diamond and a heart}}{\text{The total number of ways of drawing two cards}}$$

$$P(A) = \frac{169}{1326} = \frac{13}{102} \quad \textbf{Ans.}$$

EXAMPLE 4.4. A box contain 4 white and 3 black balls. Three balls are drawn from the box in succession. Find the probability that the first two balls are white and the third is black.

Solution: Let A_1 be the event that the first ball is a white ball.

Therefore, $$P(A_1) = \frac{4}{7}$$

Since there are four white balls in the box out of total, 7 balls.

Let A_2 be the event that the second ball is also a white ball.

Therefore, $$p\left(\frac{A_2}{A_1}\right) = \frac{3}{6}$$

Since, after the first drawing, there remain 3 white balls out of a total of 6 balls.

Let A_3 be the event that the third ball is black.

Therefore, $$P\left(\frac{A_3}{A_1\, A_2}\right) = \frac{3}{5}$$

Since after the second drawing, there are 3 black balls out of a total of 5 balls.

Therefore the required probability is

$$P = P(A_1 + A_2\, A_3) = P(A_1)\, P(A_2/A_1)\, P(A_3/A_1A_2) = \frac{4}{7} \times \frac{3}{6} \times \frac{3}{5} = \frac{6}{35} \quad \textbf{Ans.}$$

EXAMPLE 4.5. A box contains 5 white, 3 red and 2 black balls. Three balls are drawn in succession. Find the probability that the balls will be of different colours.

Solution: Let W, R, B be the events of drawing white, red and black balls, respectively. It may be observed that there are six possible ways of getting the balls of different colours as follow:

WRB, WBR, RWB, RBW, BWR, BRW

Now, the probability of first combination WRB will be

$$P(WRB) = \frac{5}{10}, \frac{3}{9}, \frac{2}{8}$$

Here, it may be noted that

$\frac{5}{10}$ is the probability of getting first ball white.

$\frac{3}{9}$ is the probability of getting second ball red.

$\frac{2}{8}$ is the probability of getting third ball black.

Similarly, probabilities for the other combinations may be calculated.

Since all the six combinations (events) are mutually exclusive, the required probability is

$$P = P(WRB + WBR + RWB + RBW + BWR + BRW)$$
$$= P(WRB) + P(WBR) + P(RWB) + P(RBW) + P(BWR) + P(BRW)$$

$$P = \left(\frac{5}{10}\times\frac{3}{9}\times\frac{2}{8}\right)+\left(\frac{5}{10}\times\frac{2}{9}\times\frac{3}{8}\right)+\left(\frac{3}{10}\times\frac{5}{9}\times\frac{2}{8}\right)+\left(\frac{3}{10}\times\frac{2}{9}\times\frac{5}{8}\right)+\left(\frac{2}{10}\times\frac{5}{9}\times\frac{3}{8}\right)+\left(\frac{2}{10}\times\frac{3}{9}\times\frac{5}{8}\right)$$

$$P = \frac{1}{4} \quad \textbf{Ans.}$$

EXAMPLE 4.6. Three students A, B and C are given a problem in Maths. The probabilities of their solving the problem are $\frac{3}{4}$, $\frac{2}{3}$ and $\frac{1}{4}$ respectively. Determine the probability that the problem is solved if all of them try to solve the problem. **(PTU, Jalandhar, Sem. Exam; 2003-04)**

Solution: The probabilities that A, B and C will not be able to solve the problem are $\left(1-\frac{3}{4}\right)$, $\left(1-\frac{2}{3}\right)$ and $\left(1-\frac{1}{4}\right)$

or $\frac{1}{4}$, $\frac{1}{3}$ and $\frac{3}{4}$ respectively.

Therefore, the probability that no one will solve the problem

$$= \frac{1}{4}\times\frac{1}{3}\times\frac{3}{4} = \frac{3}{48}$$

Hence the probability that the problem will be solved

$$= 1-\frac{3}{48} = \frac{45}{48} = \frac{15}{16} \quad \textbf{Ans.}$$

EXAMPLE 4.7. A box contains three white balls W_1, W_2, W_3 and two red balls R_1, R_2. Two balls are drawn in succession at random. Find the probability that the first drawn ball is white and the second ball is red.

Solution: Probability of event W_1 i.e. first ball is white will be

$$P(W_1) = \frac{\text{No. of white balls}}{\text{Total no. of balls}} = \frac{3}{5}$$

Now, if a white ball is drawn, there remain two white and two red balls.

Now, the conditional probability of the event of second ball red with first white

$$= P\left(\frac{R_2}{W_1}\right) = \frac{2}{4}$$

Hence, the probability of the event that the second ball is red with first white, will be

$$= P[W_1, R_2] = P\text{ (first ball white, second red)}$$

$$= P\left(\frac{R_2}{W_1}\right)\cdot P(W_1) = \frac{2}{4}\times\frac{3}{5} = \frac{6}{20} = \frac{3}{10} \quad \textbf{Ans.}$$

EXAMPLE 4.8. In a coin-tossing experiment, if the coin has head, one die is thrown and the result is recorded. But, if the coin has tail, two dice are thrown and their sum is recorded. Find the probability that the recorded number is 2.

Solution: We know that in a coin-tossing experiment, the probability of getting head = $\frac{1}{2}$.

Similarly, the probability of getting tail = $1-\frac{1}{2} = \frac{1}{2}$

Now, the probability of getting head with recorded number 2.

$$= \left(\frac{1}{2}\right)\left(\frac{1}{6}\right) = \frac{1}{12} \quad \text{...(i)}$$

Similarly, the probability of getting tail with recorded number 2, will be

$$= \left(\frac{1}{2}\right)\left(\frac{1}{36}\right) = \frac{1}{72} \quad \text{...(ii)}$$

Therefore, the required probability of getting recorded number 2 will be the sum of the equations (i) & (ii)

$$= \frac{1}{12}+\frac{1}{72} = \frac{6+1}{72} = \frac{7}{72} \quad \textbf{Ans.}$$

EXAMPLE 4.9. Two weak students in programming write a program. Their chance of writing a program correctly are $\frac{1}{8}$ and $\frac{1}{12}$. Now, if the probability of making a common error is $\frac{1}{10001}$ and they get the same answer, then find the probability that their program is correct.

Solution: It is given that the probability of two students A and B writing the program correctly are $\frac{1}{8}$ and $\frac{1}{12}$ respectively i.e., $P(A) = \frac{1}{8}$ and $P(B) = \frac{1}{12}$

Therefore, the probability of making common error $= \frac{1}{10001}$

Hence the required probability = Probability (Program is correct)

$$= \frac{\left(\frac{1}{8}\right)\left(\frac{1}{12}\right)}{\left(\frac{1}{8}\right)\left(\frac{1}{12}\right)+\left(1-\frac{1}{8}\right)\left(1-\frac{1}{12}\right)\left(\frac{1}{10001}\right)} = 0.9924 \quad \textbf{Ans.}$$

4.7 RANDOM VARIABLES

As discussed earlier, an event is the possible outcome of an experiment. The range of all the possible outcomes of an experiment is called the **sample space** 'S'. When a trial or experiment is performed, any one sample point in the sample space is the outcome of the trial. This means that a sample point always lies in the sample space 'S'. On the other hand, an event may correspond to a single sample point or set of sample points.

As an example, in an experiment of tossing a die, the sample space contains six sample points. Every time, the trial is performed, the outcome is any one sample point (number 1 to 6) in the sample space. In every trial, the outcome (sample point) will occur randomly. This has no fixed output. Hence, the outcome of a trial or experiment is a variable which can take values over the set of sample points.

Therefore, from above discussion, we can define a random variable as under:

A function which can take on any value from the sample space and its range is some set of real numbers is called a **random variable** of the experiment. Random variables are denoted by upper case letters such as X, Y etc. and the values taken by them are denoted by lower case letters with subscripts such as x_1, x_2, y_1, y_2 etc.

Random variables may be classified as under:

1. Discrete random variables
2. Continuous random variables.

4.7.1. Discrete Random Variables

A discrete random variable may be defined as the random variable which can take on only finite number of values in a finite observation interval. This means that the discrete random variable has countable number of distinct values.

For example, let us consider an experiment of tossing three coins simultaneously. In this case, there are eight possible outcomes. This will constitute the sample space S. Let the number of heads be the random variable X. Sample space S and random variable X may be written as

$$
\begin{array}{llllllllll}
S = \{ & HHH & HHT & HTH & THH & HTT & THT & TTH & TTT\} \\
X = \{ & x_1 & x_2 & x_3 & x_4 & x_5 & x_6 & x_7 & x_8 \} \\
\{ & 3 & 2 & 2 & 2 & 1 & 1 & 1 & 0 \}
\end{array}
$$

Hence, from above example, the concept of random variable is quite clear. From this example, it is also clear that random variable X takes on only a finite number of values i.e., 8. Therefore, it is a discrete random variable.

4.7.2. Continuous Random Variables

A random variable that takes on an infinite number of values is called a continuous random variable. Actually, there are several physical system (experiments) that generate continuous outputs or outcomes. Such systems generate infinte number of outputs or outcomes within the finite period. Continuous random variables may be used to define the outputs of such systems.

As an example, the noise voltage generated by an electronic amplifier has a continuous amplitude. This means that sample space S of the noise voltage amplitude is continuous. Therefore, in this case, the random variable X has a continuous range of values.

4.8 PROBABILITY FUNCTION OR PROBABILITY DISTRIBUTION OF A DISCRETE RANDOM VARIABLE

Let X be a discrete random variable and also let $x_1, x_2, x_3, \ldots$ be the values that X can take.

Then $$P(X = x_j) = f(x_j) \quad \ldots(4.15)$$

where $j = 1, 2, 3, \ldots$

will be the probability of x_j.

This $f(x_j)$ or simply f(x) is called the probability function or probability distribution of the discrete random variable.

EXAMPLE 4.10. In an experiment, three coins are tossed simultaneously. If the number of heads is the random variable, find the probability function for this random variable.

Solution: In the three tosses of coins, there are eight possible outcomes. This is the sample space S. Let the random variable, number of heads, be X.

Since, in the given experiment, the random variable X, number of heads, takes only finite values, this is a discrete random variable.

Writing the sample space S and values of discrete random variable X as

$$S = \{ \text{HHH} \quad \text{HHT} \quad \text{HTH} \quad \text{THH} \quad \text{HTT} \quad \text{THT} \quad \text{TTH} \quad \text{TTT} \}$$

$$X = \{ x_1 \quad x_2 \quad x_3 \quad x_4 \quad x_5 \quad x_6 \quad x_7 \quad x_8 \}$$

$$\{ 3 \quad 2 \quad 2 \quad 2 \quad 1 \quad 1 \quad 1 \quad 0 \}$$

Since the coin is fair, the probability of each of 8 possible outcomes, will be 1/8.

Now, $$P(X = 0) = P(x_8) = \frac{1}{8}$$

$$P(X = 1) = P(x_5) + P(x_6) + P(x_7)$$

or $$P(X = 1) = \frac{1}{8} + \frac{1}{8} + \frac{1}{8} = \frac{3}{8}$$

$$P(X = 2) = P(x_2) + P(x_3) + P(x_4)$$

or $$P(X = 1) = \frac{1}{8} + \frac{1}{8} + \frac{1}{8} = \frac{3}{8}$$

$$P(X = 3) = P(x_1) = \frac{1}{8}$$

Therefore, the probability function for discrete random variable X is as given below:

X	0	1	2	3
f(x)	$\frac{1}{8}$	$\frac{3}{8}$	$\frac{3}{8}$	$\frac{1}{8}$

4.9 CUMULATIVE DISTRIBUTION FUNCTION (CDF)

The Cumulative Distribution Function (CDF) of a random variable 'X' may be defined as the probability that a random variable 'X' takes a value less than or equal to x. Here x is the dummy variable. In other words, the cumulative distribution function (CDF) provides probabilistic description of a random variable.

Let us consider the probability of the event $X \leq x$. The probability of this event may be denoted as $P(X \leq x)$.

Now, according to the definition, the cumulative distribution function (CDF) may be written as

$$\text{CDF: } F_X(x) = P(X \leq x) \qquad ...(4.16)$$

where $F_X(x)$ is called cumulative distribution function (CDF) of a random variable X.

From equation (4.16), it may be observed that $F_X(x)$ is a function of dummy variable 'x' i.e. the value taken by random variable X.

It may also be noted that for any value of 'x', the CDF represents a probability. The CDF may be defined for discrete as well as continuous random variables. Because cumulative distribution function (CDF) basically represents the probability of random variable 'X' for event $X \leq x$, it is also called probability distribution function of the random variable or simply distribution function of the random variable. The CDF is sometimes also called cumulative probability distribution function.

4.9.1. Properties of Cumulative Distribution Function (CDF)

The properties of CDF may be listed as under:

Property 1: Since cumulative distribution function (CDF) is the probability distribution function i.e. it is defined as the probability of event $(X \leq x)$, its value is always between 0 and 1. This means that CDF is bounded between 0 and 1. Mathematically,

$$0 \leq F_X(x) \leq 1 \qquad ...(4.17)$$

Property 2: $$F_X(-\infty) = 0 \qquad ...(4.18)$$

and $$F_X(\infty) = 1 \qquad ...(4.19)$$

For first case, $x = -\infty$ means no possible event. Due to this fact, $P(X \leq -\infty)$ will always be zero. Therefore $F_X(-\infty) = 0$.

For second case, $x = \infty$ means $P(X \leq \infty)$. Since $P(X \leq \infty)$ includes probability of all possible events and the probability of a certain event is '1' therefore

$$F_X(\infty) = 1$$

Property 3: $$F_X(x_1) \leq F_X(x_2) \qquad \text{if } x_1 \leq x_2 \qquad ...(4.20)$$

The above property states that the CDF, $F_X(x)$ is a monotone non-decreasing function of x.

4.9.2. Cumulative Distribution Function (CDF) for Discrete Random Variables

If 'X' is a discrete random variable, then it takes on values at discrete points.

The Cumulative Distribution Function (CDF) is expressed as

$$F_X(x) = P(X \leq x) \qquad ...(4.21)$$

Now, if X is a discrete random variable, then it can take on values as $x_1, x_2, x_3 ..., x_n$. This means that x may be expressed in terms of $x_1, x_2, ..., x_n$.

Figure 4.1 shows how random variable can be assigned values.

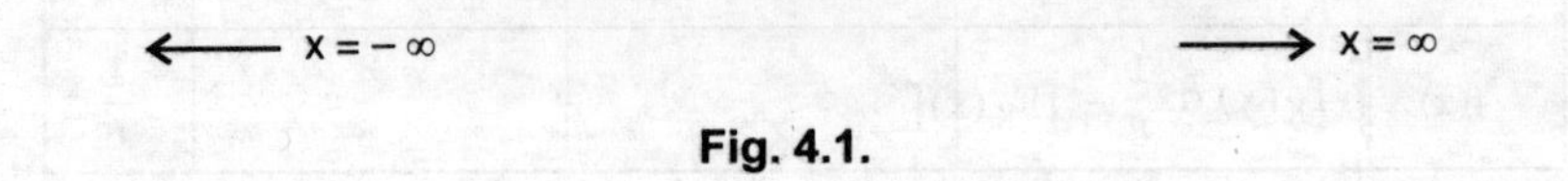

Fig. 4.1.

For a discrete random variable, the Cumulative Distribution Function (CDF) for a complete range of x may be expressed as

$$F_X(x) = 0 \quad \text{for } -\infty \le x < x_1$$
$$= \sum_{j=1}^{n} P(X = x_j) \quad \text{for } x_1 \le x \le x_n \qquad ...(4.22)$$
$$= 1 \quad \text{for } x_n < x < \infty$$

From above equation, it is clear that CDF of a discrete variable at any certain event is equal to the summation of the probabilities of random variable upto that event.

As x goes from $-\infty$ to ∞, $F_X(x)$ looks like a stair case with upward steps of height $P(X = x_j)$ at each $x = x_j$.

The CDF remains constant between the two events or steps.

4.10 PROBABILITY DENSITY FUNCTION (PDF)

The derivative of cumulative distribution function (CDF) with respect to some dummy variable is known as Probability Density Function (PDF). Probabiiity density function (PDF) is generally denoted by $f_X(x)$. Mathematically, PDF may be expressed as

$$\text{PDF: } f_X(x) = \frac{d}{dx} F_x(x) \qquad ...(4.23)$$

where x is a dummy variable.

Probability density function (PDF) is the more convenient representation for **continuous random variable**.

4.10.1. Properties of Probability Density Function (PDF)

In this article, we shall discuss a number of properties of PDF.

Property 1: Probability Density Function (PDF) is always non zero for all values of x.

Mathematically,

$$f_X(x) \ge 0 \quad \text{for all values of x} \qquad ...(4.24)$$

Proof:

We know that cumulative distribution function (CDF) increases monotonically. Therefore, the derivative of cumulative distribution function (CDF) will always be positive. Since PDF is obtained by taking derivative of CDF, therefore PDF will always be positive.

Property 2: The area under the PDF curve is always equal to unity. Mathematically,

$$\int_{-\infty}^{\infty} f_X(x)\,dx = 1 \qquad ...(4.25)$$

Proof: The expression for PDF is given as

$$f_X(x) = \frac{d}{dx} F_X(x)$$

Integrating both sides of above equation, we have

$$\int_{-\infty}^{\infty} f_X(x)\,dx = \int_{-\infty}^{\infty} \left[\frac{d}{dx} F_X(x)\right] dx$$

$$\int_{-\infty}^{\infty} f_X(x)\,dx = [F_X(x)]_{-\infty}^{\infty}$$

$$\int_{-\infty}^{\infty} f_X(x)\,dx = [F_X(\infty) - F_X(-\infty)] = 1 - 0 = 1$$

Property 3: The cumulative distribution function (CDF) may be obtained by integrating probability density function (PDF). Mathematically,

$$F_X(x) = \int_{-\infty}^{x} f_X(x)\,dx \qquad ...(4.26)$$

Proof: The expression for Probability Density Function (PDF) is given as

$$f_X(x) = \frac{d}{dx} F_x(x)$$

Integrating both sides of the above equation, we have

$$\int_{-\infty}^{x} f_X(x)\,dx = \int_{-\infty}^{x} \left[\frac{d}{dx} F_X(x)\right] dx$$

or
$$\int_{-\infty}^{x} f_X(x)\,dx = [F_X(x)]_{-\infty}^{x} = [F_X(x) - F_X(-\infty)]$$

or
$$= F_X(x) - 0 = F_X(x)$$

Hence, we have
$$F_X(x) = \int_{-\infty}^{x} f_X(x)\,dx$$

Hence Proved.

Property 4: Probability of the event $\{x_1 < X \le x_2\}$ is simply given by the area under the Probability Density Function (PDF) curve in the range $x_1 < X \le x_2$.

Mathematically,

$$P(x_1 < X \le x_2) = \int_{x_1}^{x_2} f_X(x)\,dx \qquad ...(4.27)$$

EXAMPLE 4.11. Probability density function (PDF) is given by the expression $f_X(x) = a\,e^{-b|x|}$. Here X is a random variable whose values lie in the range $x = -\infty$ to $x = +\infty$. Determine the following:

(i) The relationship between a and b.

(ii) Cumulative Distribution Function (CDF)

(iii) The probability that outcome lies between 1 and 2. (A.M.I.E., Examination, 1997)

Solution: (i) We know that the area under the PDF curve is always equal to 1.

$$\int_{-\infty}^{\infty} f_X(x)\,dx = 1,$$

Given that
$$f_X(x) = a\,e^{-b|x|}$$

Therefore,
$$\int_{-\infty}^{\infty} a\,e^{-b|x|}\,dx = 1$$

or
$$\int_{-\infty}^{0} a\,e^{-b(-x)}\,dx + \int_{0}^{\infty} a\,e^{-bx}\,dx = 1$$

or
$$\int_{-\infty}^{0} a\,e^{-b(-x)}\,dx + \int_{0}^{\infty} a\,e^{-bx}\,dx = 1$$

or $$\frac{a}{b}\left[e^{bx}\right]_{-\infty}^{0} - \frac{a}{b}\left[e^{-bx}\right]_{0}^{\infty} = 1$$

or $$\frac{a}{b}\,[e^0 - e^{-\infty} - e^{-\infty} + e^0] = 1$$

or $$\frac{a}{b}\,[1 - 0 - 0 + 1] = 1$$

or $$\frac{a}{b}\,[1 + 1] = 1$$

or $$\frac{2a}{b} = 1$$

or $$b = 2a$$

This is required relationship between a and b.

(ii) We know that cumulative distribution function (CDF) may be obtained by integrating PDF.

$$F_X(x) = \int_{-\infty}^{x} f_X(x)\,dx = \int_{-\infty}^{x} a\,e^{-b|x|}\,dx$$

Now, we shall break the above expression for x < 0, x = 0, x > 0.

$$\text{Therefore, } F_X(x) = \underbrace{\int_{-\infty}^{x} a\,e^{-b(-x)}\,dx}_{\text{This factor for } x<0} + \underbrace{\int_{x=0} a\,e^{-bx}dx + \int_{0^+}^{x} a\,e^{-bx}\,dx}_{\text{These two factors for } x \geq 0}$$

Hence, for x < 0, we have

$$\int_{-\infty}^{x} a\,e^{bx}\,dx = \frac{a}{b}\left[e^{bx}\right]_{-\infty}^{x} = \frac{a}{b}\,[e^{bx} - e^{-\infty}] = \frac{a}{b}\,e^{bx} = \frac{1}{2}\,e^{bx}$$

Similary, for x ≥ 0, we have

$$\int_{x=0} a\,e^{b}\,dx + \int_{0^+}^{x} a\,e^{-bx}\,dx$$

$$= \frac{a}{b}[e^0] - \frac{a}{b}\left[e^{-bx}\right]_{0^+}^{x}$$

$$= \frac{a}{b} \times 1 - \frac{a}{b}\,[e^{-bx} - e^{0^+}]$$

$$= \frac{a}{b} - \frac{a}{b}\,[e^{-bx} - 1] = \frac{a}{b}\,[1 - e^{-bx} + 1] \qquad [\text{Since } e^{0^+} \cong 1]$$

$$= \frac{a}{b}\,[2 - e^{-bx}] = \frac{a}{2a}\,[2 - e^{-bx}] \qquad (\text{Since } b = 2a)$$

$$= \frac{1}{2}\,(2 - e^{-bx}) = 1 - \frac{1}{2}\,e^{-bx}$$

Therefore, the expression for CDF will be given by

$$F_X(x) = \begin{cases} \frac{1}{2}\,e^{bx} & \text{for } x < 0 \\ 1 - \frac{1}{2}\,e^{-bx} & \text{for } x \geq 0 \end{cases}$$

Plot of cumulative distribution function (CDF) with respect to x:

We know from previous discussion that for a CDF $F_X(x)$

$$F_X(x) \text{ at } x = -\infty = 0$$

and $$F_X(x) \text{ at } x = \infty = 1$$

Also, $F_X(x)$ at $x = 0 = 1 - \frac{1}{2} e^{-bx}$

$$F_X(x) = 1 - \frac{1}{2} e^{-b.0} = 1 - \frac{1}{2} e^{o} = \frac{1}{2}$$

Hence, from the above information we can have an approximate plot of $F_X(x)$ as:

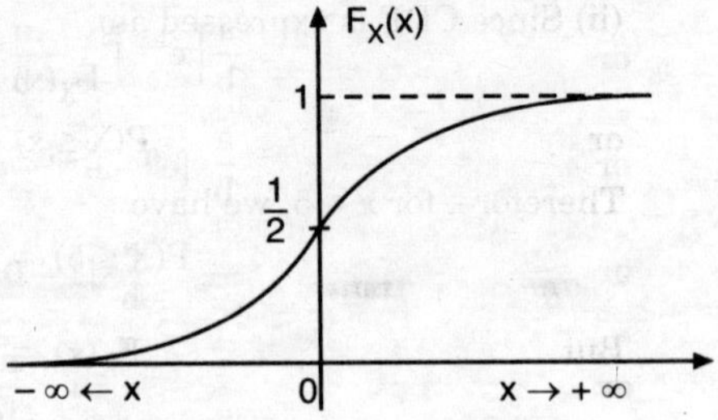

Fig. 4.2. Plot of CDF

(iii) We know that the probability that outcome lies between x_1 and x_2, is given by the expression

$$P(x_1 < X \le x_2) = \int_{x_1}^{x_2} f_X(x)\, dx$$

Given that $x_1 = 1$ and $x_2 = 2$

Therefore, $$P(1 < X \le 2) = \int_1^2 a e^{-b|x|} dx$$

For the interval 1 to 2, $e^{-b|x|} = e^{-bx}$

Hence, $$P(1 < X \le 2) = \int_1^2 a e^{-bx} dx = -\frac{a}{b}\left[e^{-bx}\right]_1^2 = -\frac{a}{b}[e^{-2b} - e^{-b}]$$

or $$P(1 < X \le 2) = \frac{a}{b}[e^{-b} - e^{-2b}]$$

But since $2a = b$,

So, $$\frac{a}{b} = \frac{1}{2}$$

Therefore, $$P(1 < X \le 2) = \frac{1}{2}[e^{-b} - e^{-2b}]$$ **Ans.**

EXAMPLE 4.12. The CDF for a certain random variable is given as

$$F_X(x) = \begin{cases} 0 & -\infty < x \le 0 \\ kx^2 & 0 < x \le 10 \\ 100\,k & 10 < x < \infty \end{cases}$$

(i) Find the value of k;
(ii) Find the value of $P(X \le 5)$;
(iii) Find the value of $P(5 < X \le 7)$;
(iv) Find the expression for PDF.

Solution: (i) The given CDF is

$$F_X(x) = \begin{cases} 0 & -\infty < x \le 0 \\ kx^2 & 0 < x \le 10 \\ 100k & 10 < x < \infty \end{cases} \quad \text{...(i)}$$

We know that the expression for CDF is given as

$$F_X(x) = \begin{cases} 0 & -\infty < x \le x_1 \\ \sum_{j=1}^{n} P(X = x_j) & x_1 < x \le x_n \\ 1 & x_n < x < \infty \end{cases} \quad \text{...(ii)}$$

Comparing the two equations (i) & (ii) we have

$$F_X(x) = 100\,k = 1 \quad \text{for } 10 < x < \infty$$

or $$k = \frac{1}{100}$$ **Ans.**

(ii) Since CDF is expressed as

$$F_X(x) = P(X \le x)$$

or

$$P(X \le x) = F_X(x)$$

Therefore, for x = 5, we have

$$P(X \le 5) = F_X(5)$$

But

$$F_X(x) = \frac{1}{100} \times x^2$$

Therefore,

$$P(X \le 5) = \frac{1}{100}(5)^2$$

$$P(X \le 5) = \frac{1}{100} \times 25 = \frac{1}{4} \quad \textbf{Ans.}$$

(iii) The PDF is expressed as

$$f_X(x) = \frac{d}{dx} F_X(x)$$

Since

$$F_X(x) = \begin{cases} 0 & -\infty < x \le 0 \\ \dfrac{x^2}{100} & \text{for } 0 < x \le 10 \\ 1 & \text{for } 10 < x < \infty \end{cases}$$

Now on differentiating $F_X(x)$ with respect to x, we have

PDF, $\quad f_X(x) = 0 \quad$ for $-\infty < x \le 0,\ 10 < x < \infty$

and

$$f_X(x) = \begin{cases} \dfrac{d}{dx}\left[\dfrac{x^2}{100}\right] & \text{for } 0 < x \le 10 \\ \dfrac{x}{50} & \text{for } 0 < x \le 10 \end{cases}$$

(iv) We know that $P(x_1 < X \le x_2) = \int_{x_1}^{x_2} f_X(x)\, dx$

Here, $\quad x_1 = 5,\ x_2 = 7$

$$P(5 < X \le 7) = \int_5^7 \frac{x}{50}\, dx \qquad \left[\text{Since } f_X(x) = \frac{x}{50}, \text{ for } 0 < x \le 10\right]$$

or

$$P(5 < X \le 7) = \frac{1}{50}\left[\frac{x^2}{2}\right]_5^7 = \frac{1}{100}[49 - 25]$$

$$P(5 < X \le 7) = \frac{1}{100} \times 24 = \frac{24}{100} \quad \textbf{Ans.}$$

EXAMPLE 4.13. A continuous random variable has a probability density function (PDF) expressed as

$$\mathbf{f_X(x) = 2\, e^{-2x} \qquad \text{for } x \ge 0}$$

Determine the probability that it will take a value between 1 and 3.

Solution: We know that the relation between probability and PDF is expressed as

$$P(x_1 < X \le x_2) = \int_{x_1}^{x_2} f_X(x)\, dx$$

Now the probability that a random variable will be between 1 and 3 is expressed as

$$P(1 < X \leq 3) = \int_1^3 f_X(x)\,dx = \int_1^3 2e^{-2x}\,dx$$

or $$P(1 < X \leq 3) = 2\left[\frac{e^{-2x}}{-2}\right]_1^3 = 2\left[-\frac{1}{2}(e^{-6} - e^{-2})\right] = -[-0.1328] = 0.1328 \qquad \textbf{Ans.}$$

EXAMPLE 4.14. A random process provides measurements x between the values 0 and 1 with a probability density function (PDF) given as

$$f_X(x) = 12x^3 - 21x^2 + 10x \quad \text{for } 0 \leq x \leq 1 = 0, \text{ otherwise}$$

Determine the followings:

(i) $P\left[X \leq \frac{1}{2}\right]$ **and** $P\left[X > \frac{1}{2}\right]$

(ii) Obtain a number k such that $P[X \leq k] = \frac{1}{2}$. **(U.P.Tech., Sem. Exam; 2008-09)(10 Marks)**

Solution: (i) We know that

$$P(x_1 < X \leq x_2) = \int_{x_1}^{x_2} f_X(x)\,dx$$

Therefore,

$$P(X \leq \frac{1}{2}) = \int_0^{1/2} f_X(x)\,dx = \int_0^{1/2} (12x^3 - 21x^2 + 10x)\,dx$$

or $$P(X \leq \frac{1}{2}) = \left[\frac{12x^4}{4} - \frac{21x^3}{3} + \frac{10x^2}{2}\right]_0^{1/2} = \left[3x^4 - 7x^3 + 5x^2\right]_0^{1/2}$$

or $$P(X \leq \frac{1}{2}) = \left[3\left(\frac{1}{2}\right)^4 - 7\cdot\left(\frac{1}{2}\right)^3 + 5\left(\frac{1}{2}\right)^2 - 0\right] = \frac{9}{16}$$

Also,

$$P(X > \frac{1}{2}) = 1 - P(X \leq \frac{1}{2}) = 1 - \frac{9}{16} = \frac{7}{16} \qquad \textbf{Ans.}$$

(ii) Since

$$P(x_1 < X \leq x_2) = \int_{x_1}^{x_2} f_X(x)\,dx$$

Therefore, $$P(X \leq k) = \int_0^k f_X(x)\,dx$$

But given that $$P(X \leq k) = \frac{1}{2}$$

Hence, $$\frac{1}{2} = \int_0^k (12x^3 - 21x^2 + 10x)\,dx = \left[\frac{12x^4}{4} - \frac{21x^3}{3} + \frac{10x^2}{2}\right]_0^k$$

or $$\frac{1}{2} = \left[3x^4 - 7x^3 + 5x^2\right]_0^k = 3k^4 - 7k^3 + 5k^2$$

or $6k^4 - 14k^3 + 10k^2 - 1 = 0$ ...(i)

The required answer will be the root of equation (i) which lies between 0 and 1.
Therefore, solving equation (i) for a root between 0 and 1, we get

$$k = 0.453 \quad \textbf{Ans.}$$

EXAMPLE 4.15. Determine whether the function given by expression

$$f_X(x) = \begin{cases} 0 & \text{for } x < 2 \\ \dfrac{1}{18}(3+2x) & \text{for } 2 \le x \le 4 \\ 0 & \text{for } x > 4 \end{cases}$$

is a density function? (U.P.Tech., Sem. Exam; 2007-08)(10 Marks)

Solution: If any given function is a probability density function (PDF), then it must satisfy following two basic conditions:

(i) $f_X(x) \ge 0$ for every x

(ii) $\int_{-\infty}^{\infty} f_X(x)\,dx = 1$

The given function is = 0 for x < 2

$$f_X(x) = \frac{1}{18}(3+2x) \text{ for } 2 \le x \le 4 = 0 \text{ for } x > 4$$

From this expression, it may be observed that first condition is satisfied.
For second condition, we have

$$\int_{-\infty}^{\infty} f_X(x)\,dx = \int_{-\infty}^{2} 0\,dx + \int_{2}^{4}\frac{1}{18}(3+2x)\,dx + \int_{4}^{\infty} 0\,dx$$

or
$$\int_{-\infty}^{\infty} f_X(x)\,dx = 0 + \frac{1}{18}\left[3x + x^2\right]_2^4 + 0 = \frac{1}{18}\left[\{3(4)+16\} - \{3(2)+4\}\right]$$

$$\int_{-\infty}^{\infty} f_X(x)\,dx = \frac{1}{18}[12+16-6-4] = \frac{1}{18}[28-10] = \frac{1}{18}\times 18 = 1$$

Therefore, second condition is also satisfied.
This means that the given function is a probability function (PDF). **Ans.**

EXAMPLE 4.16. Determine the constant k such that the function $f_X(x)$ given by the expression

$$f_X(x) = \begin{cases} \dfrac{1}{k} & \text{for } a \le x < b \\ 0 & \text{elsewhere} \end{cases}$$

is a probability density function (PDF). Also, find the cumulative distribution function (CDF) of the random variable X if k satisfies the conditions for $f_X(x)$ to be a probability density function (PDF).

Solution: Since it is given that $f_X(x)$ is a density function, therefore, it will satisfy the following requirement:

$$\int_{-\infty}^{\infty} f_X(x)\,dx = 1$$

Putting the value of I(x) in the given limit, we get

$$\int_a^b \frac{1}{k}\,dx = 1 \quad \text{or} \quad \frac{1}{k}\int_a^b dx = 1$$

or $\frac{1}{k}[x]_a^b = 1$ or $k = [b - a]$ **Ans.**

We know that CDF is given as

$$F_X(x) = \int_{-\infty}^{x} f_X(x)\,dx = \int_{a}^{x} f_X(x)\,dx = \int_{a}^{x} \frac{1}{k}\,dx = \int_{a}^{x} \frac{1}{b-a}\,dx$$

or $$F_X(x) = \frac{1}{b-a}\int_{a}^{x} dx = \frac{1}{b-a}[x]_a^x = \frac{1}{b-a}\cdot x - a = \frac{x-a}{b-a} \quad \textbf{Ans.}$$

EXAMPLE 4.17. Determine whether the following function is cumulative distribution function (CDF):

$$F_X(x) = \begin{cases} 0 \text{ for } x < a \\ \frac{1}{2}\left(\frac{x}{a}+1\right) \text{ for } -a \le x \le a \\ 1 \text{ for } x > a \end{cases}$$

(U.P.Tech., Sem. Exam; 2009-10)(10 Marks)

Solution: We know that the probability density function (PDF) is given as

$$f_X(x) = \frac{d}{dx}F_X(x)$$

$$f_X(x) = \begin{cases} \frac{1}{2a} \text{ for } -a \le x \le a \\ 0 \text{ elsewhere} \end{cases}$$

The required condition for a function $f_X(x)$ to be PDF is

$$\int_{-\infty}^{\infty} f_X(x)\,dx = 1, \qquad \text{...(i)}$$

Therefore,

$$\int_{-\infty}^{\infty} f_X(x)\,dx = \int_{-\infty}^{-a} f_X(x)\,dx + \int_{-a}^{a} f_X(x)\,dx + \int_{a}^{\infty} f_X(x)\,dx$$

$$\int_{-\infty}^{\infty} f_X(x)\,dx = \int_{-\infty}^{-a} 0\cdot dx + \int_{-a}^{a} \frac{1}{2a}\,dx + \int_{a}^{\infty} 0\cdot dx = 0 + \frac{1}{2a}[x]_{-a}^{a} + 0 = \frac{1}{2a}[a-(-a)] = 1$$

Hence, the function $f_X(x)$ satisfies the condition of a PDF. This means that $F_X(x)$ is a cumulative distribution function (CDF). **Ans.**

4.11 JOINT CUMULATIVE DISTRIBUTION FUNCTION

We may also define the outcome of an experiment by two random variables. In earlier article, we discussed the, cumulative distribution function (CDF) for a single random variable. In this article, we shall discuss the CDF for two random variables X and Y which is known as combined CDF or simply joint Distribution Function.

The joint Distribution Function or joint CDF $F_{XY}(x, y)$ of two random variables X and Y is defined as the probability that the random variable X is less than or equal to a specified value x and that the random variable Y is less than or equal to a specified value y.

The joint Cumulative Distribution Function $F_{XY}(x, y)$ may be defined as the probability that the outcome of an experiment will result in a sample point lying inside the range $(-\infty < X \le x, -\infty < Y \le y)$ of the joint sample space. The joint sample space is the combined sample space of X and Y. Mathematically,

$$F_{XY}(x, y) = P(X \le x, Y \le y) \qquad \text{...(4.28)}$$

4.11.1. Properties of Joint Cumulative Distribution Function

The Joint Cumulative Distribution Function has the following properties:

Property 1: The Joint Cumulative Distribution Function is a non-negative function.

Mathematically,

$$F_{XY}(x, y) \geq 0 \quad ...(4.29)$$

The Joint Cumulative Distribution Function is basically defined as the probability in the Joint sample space of random variables. We know that the probability always lies between 0 and 1. Therefore, the joint Cumulative Distribution Function also lies between 0 and 1 and hence non-negative.

Property 2: The Joint Cumulative Distribution Function is a monotone non-decreasing function of both x and y.

Property 3: The Joint Cumulative Distribution Function is always continuous everywhere in the xy-plane.

4.12 THE JOINT PROBABILITY DENSITY FUNCTION

The Joint Probability Density Function or simply Joint PDF is the PDF of two or more random variables.

The joint PDF of any two random variables X and Y may be defined as the partial derivative of the joint cumulative distribution function $F_{XY}(x, y)$ with respect to the dummy variables x and y. Mathematically,

$$f_{XY}(x, y) = \frac{\partial^2 F_{XY}(x, y)}{\partial x \, \partial y} \quad ...(4.30)$$

Here, we take partial derivative since the differentiation is with respect to two variables x and y.

4.12.1. Properties of Joint PDF

The Joint PDF has the following properties:

Property 1: The Joint PDF is non-negative. Mathematically,

$$f_{XY}(x, y) \geq 0 \quad ...(4.31)$$

Because the joint PDF is a derivative of a non-negative function (Joint CDF), therefore its value is always positive.

Property 2: The total volume under the surface of joint PDF is equal to unity.

Mathematically,

$$\int_{-\infty}^{\infty} \int_{-\infty}^{\infty} f_{XY}(x, y) \, dx \, dy = 1 \quad ...(4.32)$$

Because $f_{XY}(x, y)$ is a two-dimensional function, it represents a surface in the x, y plane. Hence, its integration is called as volume under the surface.

Property 3: The Joint PDF is continuous everywhere because joint CDF is continuous.

4.12.2. The Relationship Between Joint PDF and Probability

The relationship between probability and PDF over a certain interval is expressed as

$$P(x_1 < X \leq x_2) = \int_{x_1}^{x_2} f_X(x) \, dx \quad ...(4.33)$$

The above relationship may be extended for two random variables X and Y. Mathematically,

$$P(x_1 < X \leq x_2, y_1 < Y \leq y_2) = \int_{y_1}^{y_2}\int_{x_1}^{x_2} f_{XY}(x, y)\, dx\, dy \quad ...(4.34)$$

In above expression, the double integral represents the volume between x-y plane and the surface $f_{XY}(x, y)$.

Now, if the two random variables X and Y are statistically independent, then joint PDF of these two random variables becomes a product of two separate PDFs. Mathematically,

$$f_{XY}(x, y) = f_X(x)\, f_Y(y) \quad ...(4.35)$$

Putting this value of $f_{XY}(x, y)$ from equation in equation, we have

$$P(x_1 < X \leq x_2, y_1 < Y \leq y_2) = \int_{y_1}^{y_2}\int_{x_1}^{x_2} f_X(x)\, f_Y(y)\, dx\, dy \quad ...(4.36)$$

$$= \int_{y_1}^{y_2} f_Y(y)\, dy \int_{x_1}^{x_2} f_X(x)\, dx \quad ...(4.37)$$

The above equation provides the relationship between probability PDF of statistically independent random variables X and Y.

EXAMPLE 4.18. The joint PDF of two random variables is expressed as $F_{XY}(x, y)$. If the random variables X and Y are statistically independent, then show that

$$f_{XY}(x, y) = F_X(x)\, F_Y(y)$$

Here $f_X(x)$ is the PDF of random variable X and $f_Y(y)$ is the PDF of random variable Y.

Solution: We know that the joint PDF may be defined in terms of joint CDF as

$$f_{XY}(x, y) = \frac{\partial^2 F_{XY}(x, y)}{\partial x\, \partial y}$$

Now, if random variables X and Y are statistically independent, then

$$f_{XY}(x, y) = \frac{d}{dx} F_{XY}(x, y) \frac{d}{dy} F_{XY}(x, y) = f_X(x)\, f_Y(y)$$

EXAMPLE 4.19. A joint probability density function of two random variables X and Y is given as

$$f_{XY}(x, y) = \begin{cases} e^{-(x+y)} & \text{for } x \geq 0, y \geq 0 \\ 0 & \text{otherwise} \end{cases}$$

Determine the followings:

(i) $P(X < 1)$

(ii) $P(X > Y)$

(iii) $P(X + Y < 1)$

Solution: Given that joint probability density function is given as

$$f_{XY}(x, y) = e^{-(x+y)} = e^{-x} \cdot e^{-y} = f_X(x) \cdot f_Y(y)$$

Hence, X and Y are independent random variables.

Therefore, $f_X(x) = e^{-x}$ for $x \geq 0$

and $f_Y(y) = e^{-y}$ for $y \geq 0$

Now,

(i) $$P(X < 1) = \int_0^1 e^{-x}\, dx = \left[\frac{e^{-x}}{-1}\right]_0^1 = -\left[e^{-x}\right]_0^1$$

or $$P(X < 1) = -(e^{-1} - e^{-0}) = -\left(\frac{1}{e} - 1\right) = 1 - \frac{1}{e} \quad \textbf{Ans.}$$

(ii) $$P(X > Y) = 1 - P(X < Y)$$

or $$P(X < Y) = \int_0^{\infty}\left[\int_0^{y} f_{XY}(x,y)\, dx\right] dy = \int_0^{\infty} e^{-y}\left[\frac{e^{-x}}{-1}\right]_0^{y} dy$$

or $$P(X < Y) = -\int_0^{\infty} e^{-y}\cdot(e^{-y} - 1)\, dy = -\left(\frac{e^{-2y}}{-2} + e^{-y}\right)_0^{\infty} = \left(\frac{1}{2}e^{-2y} - e^{-y}\right)_0^{\infty}$$

$$= \left[\left(\frac{1}{2}e^{-\infty} - e^{-\infty}\right) - \left(\frac{1}{2}e^{-0} - e^{-0}\right)\right] = \frac{1}{2}\cdot 0 - 0 - \frac{1}{2}\cdot 1 + 1$$

or $$P(X < Y) = 0 - 0 - \frac{1}{2} + 1 = \frac{1}{2}$$

Hence, $$P(X > Y) = 1 - P(X < Y) = 1 - \frac{1}{2} = \frac{1}{2} \quad \textbf{Ans.}$$

(iii) $$P(X + Y < 1) = \int_0^1\left[\int_0^{1-x} e^{-(x+y)}\, dy\right] dx$$

or $$P(X + Y < 1) = \int_0^1 e^{-x}\left[\frac{e^{-y}}{-1}\right]_0^{1-x} dx = -\int_0^1 e^{-x}\cdot[e^{-(1-x)} - 1)]\, dx$$

or $$P(X + Y < 1) = -\int_0^1 [e^{-1} - e^{-x}]\, dx = 1 - \frac{2}{e} \quad \textbf{Ans.}$$

EXAMPLE 4.20. The joint probability density function of two random variables X and Y is given as

$$f_{XY}(x, y) = C(2x + y) \quad \text{for } 0 \le x \le 2,\ 0 \le y \le 3$$
$$= 0 \quad \text{elsewhere}$$

Determine the value of constant C.

Solution: For a joint PDF, we know that

$$\int_{-\infty}^{\infty}\int_{-\infty}^{\infty} f_{XY}(x,y)\, dxdy = 1$$

Putting the value of $f_{XY}(x, y)$, we get

$$\int_0^3\int_0^2 C(2x + y)\, dxdy = 1$$

or $$C\int_0^3\left[\frac{2x^2}{2} + yx\right]_0^2 dy = 1 \quad C\int_0^3 (2y + 4)\, dy = 1$$

or $$C\left(\frac{2y^2}{2} + 4y\right)_0^3 = 1 \quad C(y^2 + 4y)_0^3 = 1$$

or $$C(9 + 12 - 0 - 0) = 1 \quad 21\cdot C = 1 = \frac{1}{21}$$

or, $$C = \frac{1}{21} \quad \textbf{Ans.}$$

4.13 MARGINAL DENSITIES

When the probability density functions $f_X(x)$ and $f_Y(y)$ for any single random variable are obtained from joint PDF, then $f_X(x)$ and $f_Y(y)$ are called marginal PDF or simply marginal densities.

We can obtain the marginal PDFs as follows:

Since, the CDF for a single random variable may be obtained from its PDF as

$$F_X(x) = \int_{-\infty}^{x} f_X(x)\, dx \quad \text{...(4.38)}$$

The above expression may be extended for any two random variable, X and Y as

$$F_{XY}(x, y) = P(X \le x, Y \le y) \quad \text{...(4.39)}$$

$$F_{XY}(x, y) = \int_{-\infty}^{y} \int_{-\infty}^{x} f_{XY}(x, y)\, dx\, dy \quad \text{...(4.40)}$$

Now, if we have to find CDF of random variable X, then the value of Y does not matter. Therefore,

$$F_X(x) = P(X \le x) = P(X \le x, -\infty < y \le \infty) \quad \text{...(4.41)}$$

$$= \int_{-\infty}^{\infty} \int_{-\infty}^{x} f_{XY}(x, y)\, dx\, dy \quad \text{...(4.42)}$$

For a single random variable, the PDF may be obtained by differentiating CDF.

$$f_X(x) = \frac{d}{dx} F_X(x) \quad \text{...(4.43)}$$

In this equation, differentiation is done with respect to x because y varies from $-\infty$ to $+\infty$.

$$\text{or} \qquad f_X(x) = \frac{d}{dx}\left[\int_{-\infty}^{\infty} \int_{-\infty}^{x} f_{XY}(x, y)\, dx\, dy\right] \quad \text{...(4.44)}$$

$$\text{or} \qquad f_X(x) = \int_{-\infty}^{\infty} f_{XY}(x, y)\, dx\, dy \quad \text{...(4.45)}$$

Here, dx is cancelled due to differentiation. Therefore, the above equation shows that the PDF $f_X(x)$ may be obtained by integrating joint PDF $f_{XY}(x, y)$ with respect to undesired random variable y from $-\infty$ to ∞.

This $f_X(x)$ is then called marginal PDF. In the same way, the marginal PDF for random variable y may be obtained. It is expressed as

$$f_Y(y) = \int_{-\infty}^{\infty} f_{XY}(x, y)\, dx \quad \text{...(4.46)}$$

Therefore the marginal densities are:

$$f_X(x) = \int_{-\infty}^{\infty} f_{XY}(x, y)\, dy \quad \text{...(4.47)}$$

$$\text{and} \qquad f_Y(y) = \int_{-\infty}^{\infty} f_{XY}(x, y)\, dx \quad \text{...(4.48)}$$

4.14 CONDITIONAL PROBABILITY DENSITY FUNCTION

Out of the two random variables, one variable may take a fixed value. In this case, the PDF is called conditional. As an example, out of the two continuous random variables X and Y, let X = x.

Then we may find the conditional, PDF of Y given that X = x as,

$$f_Y(y/x) = \frac{f_{XY}(x, y)}{f_X(x)} \quad ...(4.49)$$

where $f_X(x)$ is the marginal density of random variable X.

Similarly, we may find the conditional PDF of X given that Y = y. Then

$$f_X(x/y) = \frac{f_{XY}(x, y)}{f_Y(y)} \quad ...(4.50)$$

4.14.1. Properties of Conditional PDF

The conditional PDF has following properties:

Property 1: A conditional PDF is always non-negative function. Mathematically,

$$f_X(x/y) \geq 0 \quad ...(4.51)$$

and

$$f_Y(y/x) \geq 0 \quad ...(4.52)$$

The conditional PDF is the ratio of two PDFs which are non-negative. Therefore, conditional PDF is also non-negative.

Property 2: The area under the conditional PDF is equal to unity. Mathematically,

$$\int_{-\infty}^{\infty} f_X(x/y)\, dx = 1 \quad ...(4.53)$$

and

$$\int_{-\infty}^{\infty} f_Y(y/x)\, dy = 1 \quad ...(4.54)$$

Property 3: If two random variables X and Y are statistically independent, then the conditional PDF reduces to marginal density.

Mathematically,

$$f_Y(y/x) = f_Y(y) \quad ...(4.55)$$

and

$$f_X(x/y) = f_X(x) \quad ...(4.56)$$

4.15 STATISTICAL AVERAGES OF RANDOM VARIABLES

We know that the Probability Density Function (PDF) gives some kind of information about the random variable. However, the interpretation of this information is quite complex. There are some other measures or numbers which give more useful and quick information about the random variable. Collectively these characteristic numbers or measures are known as statistical averages. Some useful and convenient informations like mean or average, moments, standard deviation, variance etc. may be obtained for random variables. These factors are special characteristics of the Probability Density Function (PDF) and actually they depend on the particular type of PDF.

Now we shall discuss these statistical averages.

4.15.1. Mean or Average

The mean or average of any random variable is expressed by the summation of the values of random variables X weighted by their probabilities. Mean value of a random variable is denoted by m_x.

Mean value is also known as expected value of random variable X.

$$m_x = E(X) \quad ...(4.57)$$

where E[ ] represents expection operator.

Therefore both m_x and E (X) have the same meaning.

Also, in general term, mean or average is given by

$$\text{Mean or average of } X = \frac{\text{Arithmetic sum of all values of X}}{\text{Total number of values of X}}$$

4.15.2. Mean Value of Discrete Random Variable

Let the discrete random variable X be take the following values

$$X = \{x_1, x_2, x_3 \ldots x_n\} \quad ...(4.58)$$

Then the mean or average value m_x is expressed as

$$m_x = x_1P(x_1) + x_2P(x_2) + x_3P(x_3) + \ldots + x_nP(x_n) \quad ...(4.59)$$

Therefore, the mean value of a discrete random variable X is equal to the summation of values of X weighted by probability of that particular value.

Hence, mean value of discrete random variable

$$m_x = E(X) = \bar{X} = \sum_{i=1}^{n} x_i.P(x_i) \quad ...(4.60)$$

Here, $\bar{X}$ is also the notation for mean values.

4.15.3. Mean Values of Continuous Random Variables

If random variable X becomes continuous, the sample points $x_1, x_2, x_3 \ldots, x_n$ becomes quite close to each other such that $(x_2 - x_1 \cong 0)$. Therefore, summation in equation (4.60) for discrete random variable converts to an integration over the complete range of $x(-\infty < x < \infty)$.

For continuous random variable X, the mean or average value is expressed as

$$m_x = \int_{-\infty}^{\infty} x f_X(x)\, dx \quad ...(4.61)$$

where $f_X(x)$ = Probability density function (PDF)

Also, if a function g(x) transforms random variable X into another random variable, then mean or average of g(x) may be expressed as

$$m = \overline{g(x)} = E[g(x)] \quad ...(4.62)$$

$$m = \int_{-\infty}^{\infty} g(x)\, f_X(x)\, dx \quad ...(4.63)$$

EXAMPLE 4.21. The random variable Z is the function of another random variable X in such a way that Z = cos (X) and X is uniformly distributed in the interval (– π, π) i.e.,

$$f_X(x) = \begin{cases} \frac{1}{2\pi} & \text{for } -\pi < x < \pi \\ 0 & \text{otherwise} \end{cases}$$

Determine the mean value of Z.

Solution: The given random variable is

$$Z = \cos(X)$$

Now, assuming g(x) = cos (X). The above equation may be written as

$$Z = g(x)$$

We know that the mean value of random variable Z = g(x) is expressed as

$$m = \int_{-\infty}^{\infty} g(x)\, f_X(x)\, dx$$

Putting the values of g(x) and $f_X(x)$, we have

$$m = \int_{-\pi}^{\pi} \cos x \left(\frac{1}{2\pi}\right) dx = \frac{1}{2\pi}\int_{-\pi}^{\pi} \cos x\, dx$$

or $$m = \frac{1}{2\pi}[-\sin x]_{-\pi}^{\pi} = \frac{1}{2\pi}[0] = 0 \quad \textbf{Ans.}$$

Thus, it may be observed from above that since the given function is Z = cos (X), it has zero mean value over the period of one cycle.

4.15.4. Moments and Variance

The nth moment of any random variable X may be defined as the mean value of X^n.

i.e. $$g(x) = X^n \quad ...(4.64)$$

then the mean is written as

$$\text{mean} = \overline{g(x)} \quad ...(4.65)$$

$$\overline{X^n} = \int_{-\infty}^{\infty} x^n f_X(x)\, dx = E(X^n) \quad ...(4.66)$$

$$\overline{X^n} = \int_{-\infty}^{\infty} x^n f_X(x)\, dx$$

Putting n = 1, the above equation becomes

$$\overline{X} = E(X) = \int_{-\infty}^{\infty} x\, f_X(x)\, dx \quad ...(4.67)$$

or $$\overline{X} = m_x \quad ...(4.68)$$

Hence, the first moment of random variable X will be same as its mean value.

If n = 2, then equation (4.66) becomes

$$\overline{X^2} = E(X^2) = \int_{-\infty}^{\infty} x^2 f_X(x)\, dx \quad ...(4.69)$$

where $\overline{X^2}$ is known as mean square value of random variable X.

Similarly, the central moments are the moments of the difference between random variable X and its mean m_x. Therefore, the nth central moment may be given as

$$E[(X - m_x)^n] = \int_{-\infty}^{\infty} (x - m_x)^n f_X(x)\, dx \quad ...(4.70)$$

The second central moment for n = 2, is known as *variance* of random variable X i.e.,

$$\text{Variance}[X] = E[(X - m_x)^2] \quad ...(4.71)$$

$$= \int_{-\infty}^{\infty} (x - m_x)^2 f_X(x)\, dx \quad ...(4.72)$$

Hence, it is clear that the variance provides an indication about randomness of the random variable.

Variance is generally represented by σ_x^2 i.e.,

$$\sigma_x^2 = \text{Variance (X)} \quad ...(4.73)$$

$$\sigma_x^2 = \text{Var (X)}$$

or $$\sigma_x^2 = E[(X - m_x)^2] \quad ...(4.74)$$

or $$\sigma_x^2 = \int_{-\infty}^{\infty} (x - m_x)^2 f_X(x)\, dx \quad ...(4.75)$$

Expanding equation (4.74), we get

$$\sigma_x^2 = E[X^2 - 2\, m_x X + m_x^2] \quad ...(4.76)$$

Since, $(X - m_x)^2 = X^2 + m_x^2 - 2\, m_x X$

The E [ ] operator represents mean or average value, therefore it is linear. Thus,

$$\sigma_x^2 = E(X^2) - 2\, m_x E(X) + m_x^2 \quad ...(4.77)$$

or $$\sigma_x^2 = E(X^2) - 2\, m_x m_x + m_x^2$$

or $$\sigma_x^2 = E(X^2) - m_x^2 \quad ...(4.78)$$

or $$\sigma_x^2 = \text{mean square value} - \text{square of mean value}$$

The above equation may also be written as

$$\text{Variance} = \sigma_x^2 \quad ...(4.79)$$

or $$\text{Variance} = \overline{X^2} - m_x^2 \quad ...(4.80)$$

The square root of variance is known as **standard deviation** of a random variable X. Standard deviation gives the measure of spread observed over the values of X related to mean value.

Therefore,

$$\text{Standard deviation} = \sqrt{\text{Variance}} \quad ...(4.81)$$

$$= \sqrt{\sigma_x^2} = \sigma_x \quad ...(4.82)$$

Thus, standard deviation $= \sigma_x$

$$= \sqrt{E(X^2) - m_x^2} \quad ...(4.83)$$

EXAMPLE 4.22. The probability density function (PDF) of a continuous random variable is of the form

$$f_X(x) = \frac{1}{2} e^{-|x|} \quad \text{for } -\infty < x < \infty$$

Determine the mean of the random variables.

Solution: We know that for a continuous random variable mean is expressed as

$$m_x = \int_{-\infty}^{\infty} x f_X(x)\, dx = \int_{-\infty}^{\infty} x \cdot \frac{1}{2} e^{-|x|}\, dx = \int_{-\infty}^{\infty} \frac{x}{2} e^{-|x|}\, dx$$

Let $$\frac{x}{2} e^{-|x|} = f(x)$$

Then $$m_x = \int_{-\infty}^{\infty} f(x)\,dx$$

But f(x) is an odd function since

$$f(-x) = -\frac{x}{2}e^{-|x|} = -f(x)$$

Therefore $$\int_{-\infty}^{\infty} f(x)\,dx = 0$$

Thus $$m_x = 0 \qquad \textbf{Ans.}$$

EXAMPLE 4.23. The probability density function (PDF) of a continuous random variable X in the range (–3, 3) is defined as follow:

$$f_X(x) = \begin{cases} = \dfrac{1}{16}(3+x)^2 & \text{for } -3 \le x \le -1 \\ \dfrac{1}{16}(2-6x)^2 & \text{for } -1 \le x \le 1 \\ = \dfrac{1}{16}(3-x)^2 & \text{for } 1 \le x \le 3 \end{cases}$$

Verify that the area under the curve is unity. Also prove that the mean is zero.

(A.M.I.E., Examination, 1998)

Solution: We know that the area under the PDF curve is expressed as

$$\text{area} = \int_{-\infty}^{\infty} f_X(x)\,dx = \int_{-3}^{3} f_X(x)\,dx$$

$$\text{area} = \int_{-3}^{-1} f_X(x)\,dx \int_{-1}^{1} f_X(x)\,dx \int_{1}^{2} f_X(x)\,dx$$

or $$\text{area} = \int_{-3}^{1} \frac{1}{16}(3+x)^2 dx + \int_{-1}^{1} \frac{1}{16}(2-6x^2)\,dx + \int_{1}^{3} \frac{1}{16}(3-x)^2\,dx$$

$$\text{area} = \frac{1}{48}[8+32+8] = \frac{1}{48} \times 48 = 1$$

Therefore, the area under the PDF curve is unity.

We know that for a continuous random variable, the mean is expressed as

$$m_x = \int_{-\infty}^{\infty} x f_X(x)\,dx$$

where $f_X(x)$ = probability density function (PDF)

Now

$$m_x = \int_{-3}^{3} x \cdot f_X(x)\,dx = \int_{-3}^{1} \frac{1}{16} \cdot x(3+x)^2 dx + \int_{-1}^{1} \frac{1}{16} \cdot x(6-2x^2)\,dx + \int_{1}^{3} \frac{1}{16} \cdot x(3-x)^2 dx$$

$$m_x = \frac{1}{16}\int_{-3}^{-1} (9x+6x^2+x^3)\,dx + \frac{1}{16}\int_{-1}^{1} (6x-2x^3)dx + \frac{1}{16}\int_{1}^{3} (9x-6x^2+x^3)dx = 0$$

Hence, mean value is zero. **Ans.**

EXAMPLE 4.24. A random variable X has the uniform distribution given by

$$f_X(x) = \begin{cases} \dfrac{1}{2\pi} & \text{for } 0 \le x \le 2\pi \\ 0 & \text{otherwise} \end{cases}$$

Determine m_x, $\overline{X^2}$ and σ_x. **(U.P.Tech., Sem. Exam; 2007-08)(05 Marks)**

Solution: We know that the mean value of a continuous random variable is expressed as

$$m_x = \int_{-\infty}^{\infty} x f_X(x)\,dx = \int_0^{2\pi} x \cdot \frac{1}{2\pi}\,dx$$

$$m_x = \frac{1}{2\pi}\int_0^{2\pi} x\,dx = \frac{1}{2\pi}\left[\frac{x^2}{2}\right]_0^{2\pi} = \frac{1}{4\pi}\left[x^2\right]_0^{2\pi} = \frac{1}{4\pi}[4\pi^2 - 0] = \pi \quad \textbf{Ans.}$$

We know that the mean square value of a continuous variable is given as

$$\overline{X^2} = E(X^2) = \int_{-\infty}^{\infty} x^2 f_X(x)\,dx = \int_0^{2\pi} x^2 \frac{1}{2\pi}\,dx = \frac{1}{2\pi}\int_0^{2\pi} x^2\,dx$$

or

$$\overline{X^2} = \frac{1}{2\pi}\left[\frac{x^3}{3}\right]_0^{2\pi} = \frac{1}{6\pi}\left[x^3\right]_0^{2\pi} = \frac{1}{6\pi}[8\pi^3 - 0] = \frac{1}{6\pi}\cdot 8\,\pi^3 = \frac{4}{3}\pi^2 \quad \textbf{Ans.}$$

Variance of a continuous random variable is expressed as

$$\sigma_x^2 = E[X^2] - m_x^2 = \frac{4}{3}\pi^2 - \pi^2 = \frac{1}{3}\pi^2$$

Also the standard deviation σ_x is given as the square root of variance i.e.,

$$\sigma_x = \sqrt{\sigma_x^2} = \sqrt{\frac{1}{3}\pi^2} = \frac{\pi}{\sqrt{3}} \quad \textbf{Ans.}$$

EXAMPLE 4.25. The probability density function (PDF) of a random variable X is given as $f_X(x)$. A random variable Y is defined as Y = aX + b where a < 0. Obtain PDF of Y in terms of the PDF of X.

Solution: We know that the Probability Density Function (PDF) may be expressed as

$$P(x < X \le x + dx) = f_X(x)\,dx$$

Similarly for random variable Y

$$P(y < Y \le y + dy) = f_Y(y)\,dy$$

Because the probabilities of random variables X and Y are same, we may equate the above two equations. Thus we get

$$f_Y(y)\,dy = f_X(x)\,dx = f_X(x)\cdot\frac{dx}{dy} \quad \text{...(i)}$$

Also given that $y = ax + b$

thus $\dfrac{dy}{dx} = a = \dfrac{1}{a}$ and $x = \dfrac{y-b}{a}$

Putting all these values in equation (i), we get

$$f_Y(y) = \frac{1}{a} f_X\left(\frac{y-b}{a}\right)$$

This is the required expression of PDF of Y in terms of PDF of X.

4.16 UNIFORM DISTRIBUTION

The Probability Density Function (PDF) of an uniform distribution is expressed as

$$\text{Uniform PDF: } f_X(x) = 0 \quad \text{for } x < m - \frac{A}{2} \ \& \ x > m + \frac{A}{2}$$

$$= \frac{1}{A} \quad \text{for } \left(m - \frac{A}{2}\right) \leq x \leq \left(m + \frac{A}{2}\right) \qquad ...(4.84)$$

Figure 4.3 shows the PDF of an uniformly distributed random variable.

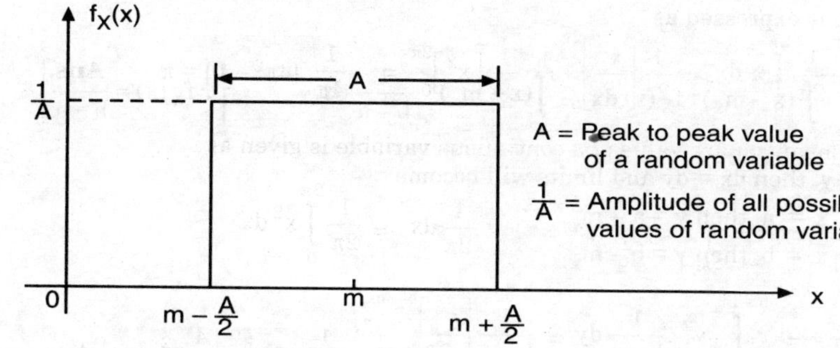

Fig. 4.3. PDF of an uniformly distributed random variable

In above figure, the peak to peak value is A and amplitude is uniform i.e. A. From figure 4.3, it is also clear that the random variable and its PDF are continuous. Therefore uniform distribution is utilised for continuous random variable.

Also, the value of PDF, $f_X(x)$ is same for all possible values of a random variable. Thus this distribution is known as **Uniform Distribution.**

EXAMPLE 4.26. Prove that the mean and variance of a random variable X having an uniform distribution in the interval [a, b] are

$$m_x = \frac{a+b}{2} \quad \text{and } \sigma_x^2 = \frac{(a-b)^2}{12}$$

Solution: An uniform distribution in the interval [a, b] is sketched in the figure 4.4.

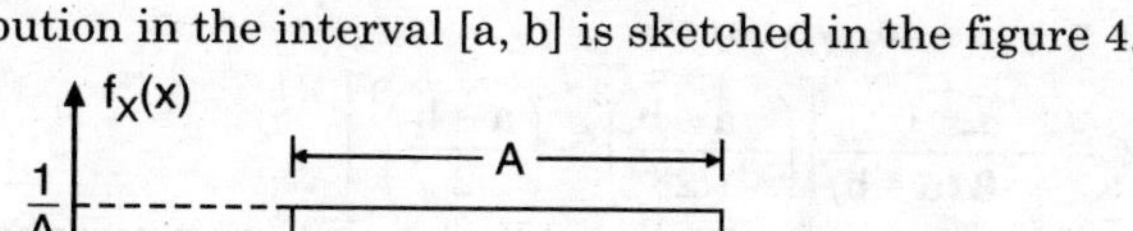

Fig. 4.4 Uniform distribution having interval [a, b]

We know that the mean value of a continuous random variable is expressed as

$$m_x = \int_{-\infty}^{\infty} x \, f_X(x) \, dx$$

$$m_x = \int_a^b x \cdot \frac{1}{A} \, dx$$

From figure 4.4, it may be observed that the value of A will be the difference between upper and lower limits of interval i.e.,

$$A = b - a$$

With this value of A, the mean value will become

$$m_x = \int_a^b x \cdot \frac{1}{b-a}\, dx = \frac{1}{b-a}\int_a^b x\, dx = \frac{1}{2(b-a)}\left[\frac{x^2}{2}\right]_a^b = \frac{1}{2(b-a)}\left[x^2\right]_a^b$$

or $$m_x = \frac{1}{2(b-a)}[b^2 - a^2] = \frac{1}{2(b-a)}\cdot(b-a)(b+a) = \frac{b+a}{2} = \frac{a+b}{2} \quad \text{...(i)}$$

Also, the variance is expressed as

$$\sigma_x^2 = \int_{-\infty}^{\infty}(x - m_x)^2 f_X(x)\, dx = \int_a^b (x - m_x)^2 \cdot \frac{1}{b-a}\, dx \qquad \left[\because f_X(x) = \frac{1}{b-a}\right]$$

Now, let $x - m_x = y$, then $dx = dy$ and limits will become

if $x = a$, then $y = a - m_x$

and if $x = b$, then $y = b - m_x$

So that, $$\sigma_x^2 = \int_{a-m_x}^{b-m_x} y^2 \cdot \frac{1}{b-a}\, dy = \frac{1}{b-a}\left[\frac{y^3}{3}\right]_{a-m_x}^{b-m_x}$$

or $$\sigma_x^2 = \frac{1}{3(b-a)}\left[(b-m_x)^3 - (a-m_x)^3\right]$$

Putting the value of $m_x = \frac{a+b}{2}$ in above equation, we get

$$\sigma_x^2 = \frac{1}{3(b-a)}\left[\left(b - \frac{a+b}{2}\right)^3 - \left(a - \frac{a+b}{2}\right)^3\right]$$

or $$\sigma_x^2 = \frac{1}{3(b-a)}\left[\left(\frac{b-a}{2}\right)^3 - \left(\frac{a-b}{2}\right)^3\right]$$

writing $(b-a) = -(a-b)$, we have

$$\sigma_x 2 = -\frac{1}{3(a-b)}\left[\left(-\frac{a-b}{2}\right)^3 - \left(\frac{a-b}{2}\right)^3\right]$$

$$\sigma_x^2 = -\frac{1}{3(a-b)}\cdot -\frac{(a-b)^3}{4} = \frac{(a-b)^2}{12} \quad \text{...(ii)}$$

Because $(a-b)^2 = (b-a)^2$, the above equation may be written as

$$\sigma_x^2 = \frac{(a-b)^2}{12}$$

But $A = b - a,$

Therefore, $$\sigma_x^2 = \frac{A^2}{12} \quad \text{...(iii)}$$

From figure 4.4, we have

$$a = m - \frac{A}{2}$$

and $$b = m + \frac{A}{2}$$

thus $$a + b = m - \frac{A}{2} + m + \frac{A}{2} = 2m$$

or $$\frac{a+b}{2} = m$$

Putting this value in equation (i), we get
Mean value of uniform distribution

$$m_x = \frac{a+b}{2} = m$$

Using equations (ii) and (iii), we get
Variance of Uniform Distribution

$$\sigma_x^{\ 2} = \frac{(a-b)^2}{12} = \frac{(b-a)^2}{12} = \frac{A^2}{12} \quad \textbf{Ans.}$$

4.17 GAUSSIAN OR NORMAL DISTRIBUTION

Gaussian Distribution is also known as Normal Distribution. It is defined for a continuous random variable. The PDF for a Gaussian random variable is expressed as

Gaussian PDF

$$f_X(x) = \frac{1}{\sigma\sqrt{2\pi}} e^{-(x-m)^2/2\sigma^2} \quad ...(4.85)$$

In this equation

m = mean value of the random variable

σ^2 = variance of the random variable

Figure 4.5 shows the plot of Gaussian PDF $f_X(x)$ with respect to x.

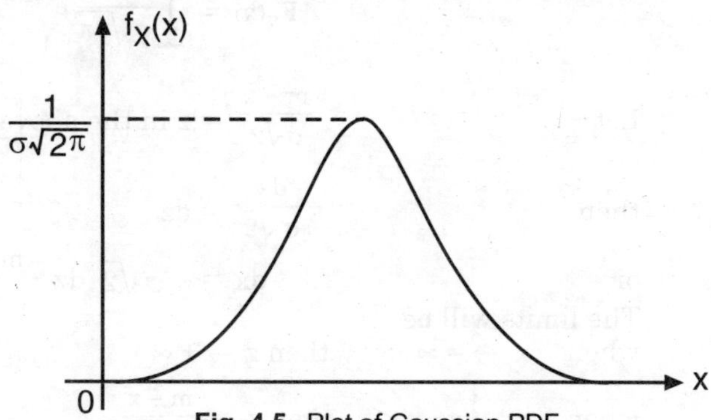

Fig. 4.5. Plot of Gaussian PDF

4.17.1. Properties of Gaussian PDF

The several properties of Gaussian PDF are as under:

Property 1: The peak value occurs at x = m i.e. mean value.
Mathematically,

$$f_X(x) = \frac{1}{\sigma\sqrt{2\pi}} \text{ at } x = m \text{ (mean value)} \quad ...(4.86)$$

This value may be obtained by putting x = m in equation (4.85).

Property 2: The plot of Gaussian PDF exhibit even symmetry around mean value.
Mathematically,

$$f_X(m - \sigma) = f_X(m + \sigma) \quad ...(4.87)$$

Property 3: The area under the PDF curve is 1/2 for all values of x below mean value and 1/2 for all values of x above mean value. Mathematically,

$$P(X \le m) = P(X > m) = \frac{1}{2} \quad ...(4.88)$$

Property 4: As $\sigma \to 0$, the Gaussian function approaches to δ (impulse) function located at x = m. This is because the area under the PDF curve is always one (unity). Also the area of impulse function is also unity.

Importance

The Gaussian distribution is used for continuous random variables. We know that the random motion of the thermally agitated electrons produces thermal noise. This thermal noise has Gaussian distribution. The random errors in the experimental measurements create the measured values to have Gaussian distribution about the true value. The Gaussian distribution is very important in the analysis of communication and statistical systems.

EXAMPLE 4.27. Find out the Cumulative Distribution Function (CDF) of the Gaussian Random variable.

Solution: We know that the Cumulative Distribution Function (CDF) may be obtained by integrating PDF as

CDF: $$F_X(x) = \int_{-\infty}^{x} f_X(x)\,dx \quad \text{...(i)}$$

But Gaussian PDF is expressed as

$$f_X(x) = \frac{1}{\sigma\sqrt{2\pi}} e^{-(x-m)^2/2\sigma^2}$$

Putting this value of $f_X(x)$ in equation (i)

$$F_X(x) = \int_{-\infty}^{x} \frac{1}{\sigma\sqrt{2\pi}} e^{-(x-m)^2/2\sigma^2}\,dx \quad \text{...(ii)}$$

Let $\frac{m-x}{\sigma\sqrt{2}} = z$ in the above equation

then $$-\frac{dx}{\sigma\sqrt{2}} = dz$$

or $$dx = -\sigma\sqrt{2}\,dz$$

The limits will be

when $x \to -\infty$; then $z \to +\infty$

when $x \to x$; then $z \to \frac{m-x}{\sigma\sqrt{2}}$

Putting all these values in equation (ii)

$$F_X(x) = \int_{\infty}^{\frac{m-x}{\sigma\sqrt{2}}} \frac{1}{\sigma\sqrt{2\pi}} \cdot e^{-z^2}\left(-\sigma\sqrt{2}\,dz\right) = -\frac{1}{\sqrt{\pi}} \int_{\infty}^{\frac{m-x}{\sigma\sqrt{2}}} e^{-z^2}\,dz$$

Interchanging the limits, we get

$$= \frac{1}{\sqrt{\pi}} \int_{\frac{m-x}{\sigma\sqrt{2}}}^{\infty} e^{-z^2}\,dz$$

Rearranging, we get

$$F_X(x) = \frac{1}{2} \cdot \frac{2}{\sqrt{\pi}} \int_{\frac{m-x}{\sigma\sqrt{2}}}^{\infty} e^{-z^2}\,dz \quad \text{...(iii)}$$

We know that the error function is given by

$$\text{erfc}(u) = \frac{2}{\sqrt{\pi}} \int_{u}^{\infty} e^{-z^2}\,dz \quad \text{...(iv)}$$

Here $$u = \frac{m-x}{\sigma\sqrt{2}}$$

Using equations (iii) and (iv), we get

Gaussian CDF: $$F_X(x) = \frac{1}{2}\ \text{erfc}\left(\frac{m-x}{\sigma\sqrt{2}}\right)$$

4.18 RAYLEIGH DISTRIBUTION

Rayleigh Distribution is used for continuous random variables. It is produced from two Gaussian random variables. Let X and Y be independent Gaussian random variables having mean value zero and variance σ^2. Mathematically,

$$m_x = m_y = 0 \quad \text{...(4.89)}$$

and $$\sigma_x^{\ 2} = \sigma^2 \quad \text{...(4.90)}$$

Figure 4.6 shows the rectangular to polar conversion.

From figure 4.6, it may be observed that independent Gaussian random variables X and Y are related to Rayleigh Distribution random variables R and ϕ such that

$$R = \sqrt{X^2 + Y^2} \quad \text{...(4.91)}$$

and $$\phi = \tan^{-1}\left(\frac{Y}{X}\right) \quad \text{...(4.92)}$$

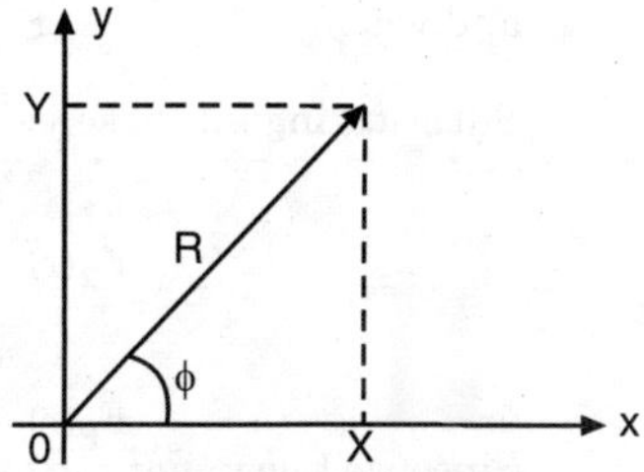

Fig. 4.6. Rectangular to polar conversion

From these equations, it is evident that R will always be positive because the variables X and Y are squared.

Also, the range of ϕ will be from 0 to 2π because it repeats after 2π radians. This means that ϕ completes one circle in figure 4.6 when it goes from 0 to 2π. After transformation of Gaussian random variables X and Y into R, it is found that R is having Rayleigh Probability Density Function (PDF) which is expressed as

$$f_R(r) = \frac{r}{\sigma^2}\, e^{-r^2/2\sigma^2} \quad \text{...(4.93)}$$

Figure 4.7 shows the Rayleigh PDF defined by equation (4.93).

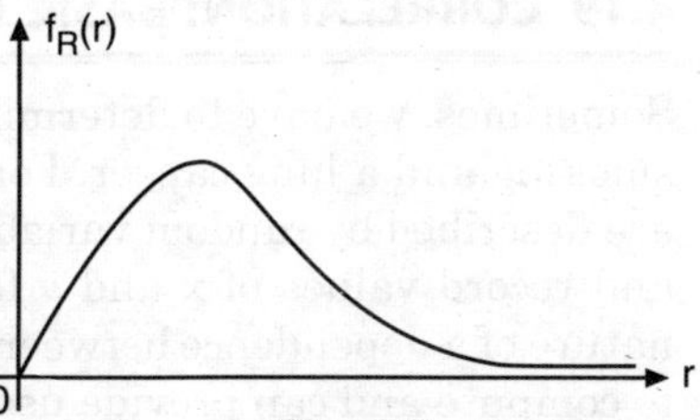

Fig. 4.7.

Therefore, from PDF curve, it is clear that

$$f_R(r) = 0 \qquad \text{for } r < 0 \quad \text{...(4.94)}$$

$$f_R(r) = \frac{r}{\sigma^2}\, e^{-r^2/2\sigma^2} \qquad \text{for } r \geq 0 \quad \text{...(4.95)}$$

This means that R has always positive value and never goes for negative value.

Importance

Rayleigh Distribution is always used for modelling of statistics of signals transmitted through radio channels such as cellular radio.

EXAMPLE 4.28. If Probability distribution function (PDF) of Rayleigh distribution is given by the expression

$$f_R(r) = \frac{r}{\sigma^2}\, e^{-r^2/2\sigma^2}$$

then above that the cumulative distribution function (CDF) of Rayleigh Distribution will be

$$F_R(r) = P(R \leq r) = \left(1 - e^{-r^2/2\sigma^2}\right) u(r)$$

Solution: We know that CDF may be obtained from PDF as

$$F_X(x) = \int_{-\infty}^{x} f_X(x)\,dx$$

where $$F_R(r) = \int_{-\infty}^{x} f_R(r)\,dr$$

Now since r never becomes negative, therefore integration limits will be from 0 to r.

Thus, $$F_R(r) = \int_0^r \frac{r}{\sigma^2} e^{-r^2/2\sigma^2}\,dr \quad \text{...(i)}$$

Substituting $$\frac{r^2}{2\sigma^2} = t$$

So that $$\frac{2r\,dr}{2\sigma^2} = dt \quad \text{or} \quad \frac{r\,dr}{\sigma^2} = dt$$

when $r = 0$ then $t = 0$

and when $r = r$ then $t = \dfrac{r^2}{2\sigma^2}$

Substituting all these values in equation (i), we get

$$F_R(r) = \int_0^{r^2/2\sigma^2} e^{-t}\,dt = \left[-e^{-t}\right]_0^{r^2/2\sigma^2} = \left[-e^{-r^2/2\sigma^2} + e^0\right] = \left[-e^{-r^2/2\sigma^2} + 1\right]$$

or $$F_R(r) = 1 - e^{-r^2/2\sigma^2}$$

Since we know that $r \geq 0$ always, therefore the above equation may be multiplied by unit step function u(r).

Hence Rayleigh CDF:

$$F_R(r) = \left(1 - e^{-r^2/2\sigma^2}\right) u(r).$$

4.19 CORRELATION: BASIC CONCEPTS

Sometimes, we have to determine the nature or degree of dependence between two entities, such as smoking and a lung cancer. For this, let us consider a random experiment with two outcomes that are described by random variables (RVs) x and y. Also, we conduct various trials of this experiment and record values of x and y for each trial. From this data, it is quite possible to determine the nature of a dependence between x and y. The covariance of RVs x and y is one measure that is simple to compute and can provide useful information about the dependence between x and y.

The covariance σ_{xy} of two Random Variable (RVs) may defined as under:

$$\sigma_{xy} = \overline{(x - \bar{x})(y - \bar{y})} \quad \text{...(4.96)}$$

Here, it may be noted that the concept of covariance is a natural extension of the concept of variance, which may be defined as under:

$$\sigma_x^2 = \overline{(x - \bar{x})(x - \bar{x})}$$

Let us consider a case of two variables of x and y that are dependent such that they tend to vary in harmony. This means that if x increases, y increases, and if x decreases, y also decreases. As an example, x may be the average daily temperature of a city and y the volume of soft drink sales that day in the city. It is reasonable for the two quantities to vary in harmony for most of the cases. We consider the following experiment.

We choose pick a random day and record the average temperature of that particular day by the value of x and the soft drink sales volume on that particular day by the value of y. We carryout this measurement for several days (trials) and record the data x and y for each trial. We know plot points (x, y) for all the trials. This plot is known as the scatter diagram. It appears as shown in figure 4.8(a). This plot indicates that when x is large, y is likely to be large. It may be noted that in a few cases, a low average temperature will also be paired with highar soft drink sales because of some typical situation. This is very much clear from the scatter diagram in figure 4.8 (a).

In this example, the variable $(x - \bar{x})$ represents the difference between actual and average values of x, and $y - \bar{y}$ represents the difference between actual and average values of y. Therefore, let us plot $(y - \bar{y})$ vs. $(x - \bar{x})$. This is the same as the scatter diagram in figure 4.8 (a) with the origin shifted to $(\bar{x}, \bar{y})$, as in figure 4.8(b).

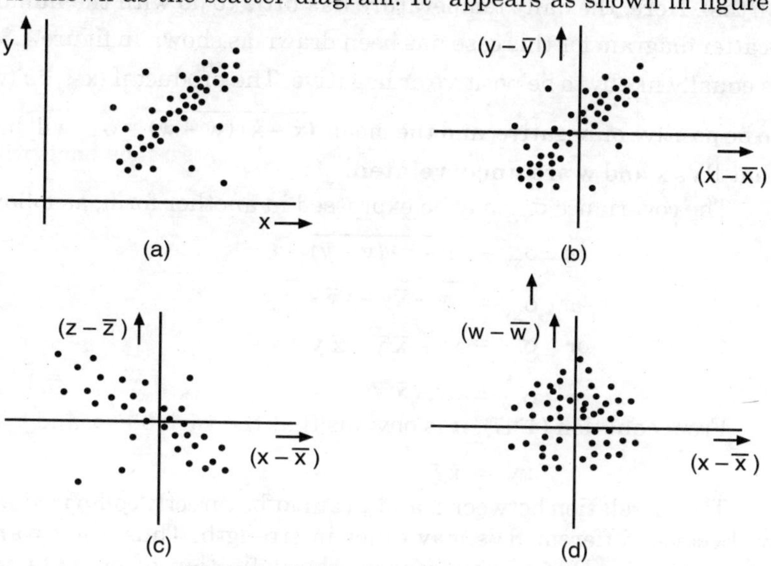

Fig. 4.8. Illustrate scatter diagrams

Important Point: This shows that a day with an above-average temperature is likely to produce above-average soft drink sales, and a day with a below-average temperature is likely to produce below-average soft drink sales.

Thus, we conclude if $x - \bar{x}$ is positive, $y - \bar{y}$ is likely to be positive, and if $x - \bar{x}$ is negative, $y - \bar{y}$ is more likely to be negative. Therefore, the quantity $(x - \bar{x})(y - \bar{y})$ will be positive for most of the trials. We compute this product for every pair, add these products, and then divide by the number of trials. The result is the mean value of $(x - \bar{x})(y - \bar{y})$, i.e., the covariance $\sigma_{xy} = \overline{(x - \bar{x})(y - \bar{y})}$. The covariance is positive in the example under consideration.

Important Point: In such cases, we say that a positive correlation exists between variables x and y. We can conclude that a positive correlation implies variation of two variables in harmony (in the same direction, up or down).

After this, let us consider the case of the two variables: x, the average daily temperature, and z, the sales volume of sweaters that day. Now, as x (daily average temperature) increases, z (the sweater sales volume) tends to decrease. A particular scatter diagram for this experiment has been shown in figure 4.8(c). Thus, if $x - \bar{x}$ is positive (above-average temperature), $z - \bar{z}$ is likely to be negative (below-average sweater sales). Similarly, when $x - \bar{x}$ is negative, $z - \bar{z}$ is likely to be positive. The product $(x - \bar{x})(z - \bar{z})$ will be negative for most of the trials, and the mean $\overline{(x - \bar{x})(z - \bar{z})} = \sigma_{xz}$ will be negative.

Important Point: In such a case, we say that negative correlation exists between x, and y are unrelated. It means that they are dependent, but when one increases, the other decreases, and vice versa.

Lastly, let us consider the variables x (the average daily temperature) and w (the number of births). Here, the daily temperature has little to do with the number of children born. A particular scatter diagram for this case has been drawn as shown in figure 4.8(d). If $(x - \bar{x})$ is positive, $(w - \bar{w})$ is equally likelly to be positive or negative. The product if $(x - \bar{x})(w - \bar{w})$ is therefore equally likely to be positive or negative, and the mean $\overline{(x - \bar{x})(w - \bar{w})} = \sigma_{xw}$ will be zero. In such a case, we conclude that RVs x and w are **uncorrelated**.

The covariance σ_{xy} may be expressed in another form, as follows. By definition, we have

or $\sigma_{xy} = \overline{(x - \bar{x})(y - \bar{y})}$

or $\sigma_{xy} = \overline{xy} - \overline{\bar{x}y} - \overline{x\bar{y}} + \overline{\bar{x}\,\bar{y}}$

or $\sigma_{xy} = \overline{xy} - \bar{x}\,\bar{y} - \bar{x}\,\bar{y} + \bar{x}\,\bar{y}$

or $\sigma_{xy} = \overline{xy} - \bar{x}\,\bar{y}$ ...(4.97)

From equation (4.97), it is obvious that the variables x and y are uncorrelated ($\sigma_{xy} = 0$) if,

$$\overline{xy} = \bar{x}\,\bar{y} \quad \text{... (4.98)}$$

The correlation between x and y cannot be directly compared with the correlation between z and w. Because different RVs may differ in strength. Thus, the covariance value must be normalized appropriately. Due to this reason, the definition of **correlation coefficient** is quite useful. **Correlation coefficient** ρ_{xy} is σ_{xy} normalized by $\sigma_x\,\sigma_y$,

$$\rho_{xy} = \frac{\sigma_{xy}}{\sigma_x \sigma_y} \quad \text{...(4.99)}$$

Therefore, if x and y are uncorrelated, then, $\rho_{xy} = 0$. Also, it can be illustrated that

$$-1 \le \rho_{xy} \le 1 \quad \text{...(4.100)}$$

Independence vs. Uncorrelatedness

We know that for independent RVs, we have

$$\overline{xy} = \bar{x}\,\bar{y} \text{ and } \sigma_{xy} = 0$$

Hence, independent random variables (RVs) are uncorrelated. It may be noted that whereas independent variables are uncorrelated, the converse is not necessarily true i.e., uncorrelated variables are generally not independent. Basically, independence is a stronger and more restrictive condition than uncorrelatedness. For independent variables, when the expectations exist, we know that

$$\overline{g_1(x)\,g_2(y)} = \overline{g_1(x)}\,\overline{g_2(y)}$$

for any functions $g_1(.)$ and $g_2(.)$, whereas for uncorrelatedness, the only requirement is that

$$\overline{xy} = \bar{x}\,\bar{y}$$

There is only one special case for which independence and uncorrelatedness are equivalent i.e., when random variables x and y are jointly Gaussian. It may be noted that when x and y are jointly Gaussian, individually, x and y are also Gaussian.

Mean Square of the Sum of Uncorrelated Variables

Now, if x and y are uncorrelated, then for $z = x + y$, we have

$$\sigma_z^2 = \sigma_x^2 + \sigma_y^2 \quad \text{...(4.101)}$$

This means that the variance of the sum is the sum of variances for uncorrelated RVs. We have

$$\sigma_z^2 = \overline{[(x - \bar{x}) + (y - \bar{y})]^2}$$

$$\text{or} \quad \sigma_z^2 = \overline{(x-\bar{x})} + \overline{(y-\bar{y})^2} + 2\overline{(x-\bar{x})(y-\bar{y})}$$

$$\text{or} \quad \sigma_z^2 = \sigma_x^2 + \sigma_y^2 + 2\sigma_{xy}$$

Since x and y are uncorrelated, $\sigma_{xy} = 0$, and equation (4.101) follows. If x and y have zero means, then z also has a zero mean, and the mean square values of these variables become equal to their variances. Thus, we write

$$\overline{(x+y)^2} = \overline{x^2} + \overline{y^2} \qquad \text{...(4.102)}$$

if x and y are uncorrelated and have zero means. Therefore, equations (4.101) and (4.102) are valid not only when x and y are independent, however, also under the less restrictive condition that x and y be uncorrelated.

4.20 SUM OF RANDOM VARIABLES

In several applications, it is quite useful to characterize the random variable (RV)z that is the sum of two random variable (RVs) x and y:

$$z = x + y$$

Since z = x + y, y = z – x irrespective of the value of x. Therefore, the event $z \le z$ is the joint event [$y \le z - x$ and x to have any value in the range $(-\infty, \infty)$]. Hence,

$$F_z(z) = P(z \le z) = P(x \le \infty, y \le z - x)$$

$$\text{or} \quad F_z(z) = \int_{-\infty}^{\infty}\int_{-\infty}^{z-x} p_{xy}(x, y)\,dy\,dx = \int_{-\infty}^{\infty} dx \int_{-\infty}^{z-x} p_{xy}(x, y)\,dy$$

will continue to be Gaussian. The particular fact that linear filter output to a Gaussian signal input is a Gaussian signal is highly significant and is one of the most useful results in communication analysis.

4.21 CENTRAL LIMIT THEOREM

Under certain conditions, the sum of a large number of independent random variable (RVs) becomes a Gaussian random variable, independent of the probability densities of the variable that is added. The very important statement of this trend is popularly known as the **central limit theorem.**

A particular trend toward a Gaussian distribution when a large number of functions are convolved has been shown in figure 4.9. Basically, we assume all PDFs are identical, that is, a gate function 0.5 Π (x/2). Figure 4.9 shows the successive convolutions of gate functions. The trend toward a bell-shaped density is quite obvious.

This particular result that the distribution of the sum of n independent Bernoulli random variables, when properly normalized, converges toward Gaussian distribution was established first by A. de Moivre in the early 1700s. Here, it may noted that the normalized sum is the sample average (or sample mean) of n random variables.

Central Limit Theorem (for the sample mean)

Let us consider that $x_1, \ldots x_n$ are the independent random samples from a given distribution with mean μ and variance σ^2 with $0 < \sigma^2 < \infty$. Then, for any value x, we have

$$\lim_{n \to \infty} P\left[\frac{1}{\sqrt{n}}\sum_{i=1}^{n}\frac{x_i - \mu}{\sigma} \le x\right] = \int_{-\infty}^{x}\frac{1}{\sqrt{2\pi}}e^{-v^2/2}dv \qquad \text{...(4.103)}$$

or equivalently,

$$\lim_{n \to \infty} P\left[\frac{\tilde{x}_n - \mu}{\sigma/\sqrt{n}} > x\right] = Q(x) \qquad \text{...(4.104)}$$

Here, it may be noted that

$$\tilde{x}_n = \frac{x_1 + \ldots + x_n}{n}$$

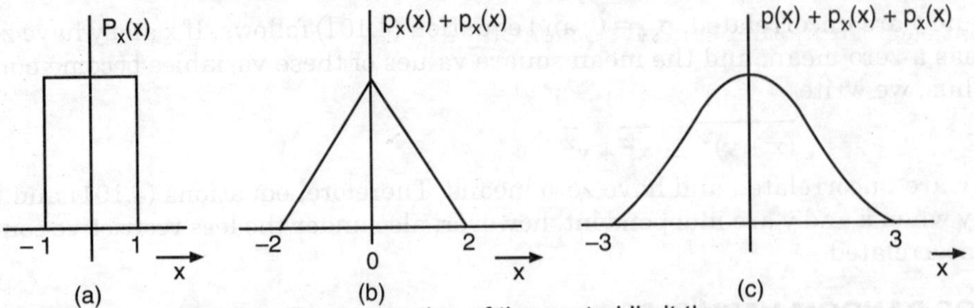

Fig. 4.9. Demonstration of the central limit theorem.

is called as the sample mean. The meaning is that the sample mean of variation with non-zero finite variance converges to gaussian distribution with a fixed decreasing variance σ^2/n. In other words, irrespective of the true distribution of sample can be approximated by gaussian distribution with mean $n\mu$ and variance $n\sigma^2$.

EXAMPLE 4.29. Given a communication system which transmits a data packet of 1024 bits. Each bit can be in error with probability of 10^{-2}. Determine the (approximate) probability that more than 30 of the 1024 bits are in error.

Solution: Let us define a random variable x_i such that $x_i = 1$ if the ith bit is in error and $x_i = 0$ if not. Therefore, we have

$$v = \sum_{i=1}^{1024} x_i$$

is the number of errors in the data packet. Now, we have to determine $P(v > 30)$.

Because $P(x_i = 1) = 10^{-2}$ and $P(x_i = 0) = 1 - 10^{-2}$, we will have to find :

$$P(v > 30) = \sum_{m=31}^{1024} \binom{1024}{m} (10^{-2})^m (1 - 10^{-2})^{1024 - m}$$

Which may be very lengthy. Let us now apply the central limit theorem to solve this problem in an approximate manner.

First, let us

$$\overline{x_i} = 10^{-2} \times (1) + (1 - 10^{-2}) \times (0) = 10^{-2}$$

$$\overline{x_i^2} = 10^{-2} \times (1)^2 + (1 - 10^{-2}) \times (0) = 10^{-2}$$

As a result, we have

$$\sigma_i^2 = \overline{x_i^2} - (\overline{x_i})^2 = 0.0099$$

Based upon the central limit theorem, $v = \Sigma_i x_i$ is approximately Gaussian with mean of $1024 \times 10^{-2} = 10.24$ and variance $1024 \times 0.0099 = 10.1376$. Because,

$$y = \frac{v - 10.24}{\sqrt{10.1376}}$$

is a standard Gausian with zero mean and unit variance.

$$P(v > 30) = P\left(y > \frac{30 - 10.24}{\sqrt{10.1376}}\right)$$

or $\quad P(v > 30) = P(y > 6.20611)$

or $\quad P(v > 30) = Q\,(6.20611)$

or $\quad P(v > 30) \cong 1.925 \times 10^{-10}$ **Ans.**

Central Limit Theorem (for the sum of independent random variables):

Let us consider that the random vriables $x_1, \ldots x_n$ are independent but not necessarily identically distributed. Each of the random variable x_i has mean μ_i and non-zero variance $\sigma_i^2 < \infty$. Further, let us assume that each third-order central moment

$$\overline{|x_i - \mu_i|^3} < \infty, \quad i = 1, \ldots, n$$

and also, we consider $\lim\limits_{n \to \infty} \sum\limits_{i=1}^{n} \overline{|x_i - \mu_i|^3} \left(\sum\limits_{i=1}^{n} \sigma_i^2 \right)^{\frac{3}{2}} = 0$

Then, random variable $$y(n) = \frac{\sum\limits_{i=1}^{n} x_i - \sum\limits_{i=1}^{n} \mu_i}{\sqrt{\sum\limits_{i=1}^{n} \sigma_i^2}}$$

converges to a standard Gaussian density as $n \to \infty$, i.e.,

$$\lim_{n \to \infty} P[y(n) > x] = Q(x) \qquad \ldots(4.105)$$

Important Point: The central limit theorem yields an explanation for the well-known fact that many random variables in practical experiments are approximately Gaussian. As an example, communication channel noise is the sum effect of several different random disturbance sources (*e.g.*, sparks lightning, static electricity). Based upon the central limit theorem, noise as the sum of all these random disturbances must be approximately Gaussian.

4.22 RANDOM PROCESS

Let there be a random experiment E having outcome λ from the sample space S. This means that $\lambda \in S$. Thus, everytime an experiment is conducted, the outcome λ will be one of the sample point in sample space. If this outcome λ is associated with time, then a function of λ and time t is formed i.e., $X(\lambda, t)$. Then the function $X(\lambda, t)$ is known as **random process**. Hence, when any random experiment E is given a time dimension, then each outcome appears at some certain time and the random experiment will be converted to Random Process. A random process is the function of two variables λ and t.

Here, λ belongs to sample space S and $-\infty < t < \infty$. At some specific time say $t = t_0$, $X(\lambda, t_0)$ is a random variable whose value depends upon λ. Also, for a specific outcome λ_i, there is a single time function $X(\lambda_i, t_0)$. This function is known as sample function whereas the collection of all the sample functions is known an **ensemble**.

We can represent random process as X(t) i.e. here λ is dropped.

Therefore, a random process X(t) may represent several functions of t and λ, a single time function, a random variable or a single number.

In practical life, there are various physical phenomenon which may be represented with the help of random process.

As an example, let us consider a set of voltage waveforms generated by thermal electron motion in a large number of identical resistors.

We can represent the voltage produced at each instant by the resistor as v. Then the random process X(v, t) denotes the output voltages from the set of resistors.

Figure 4.10 shows the waveforms produced by these resistors randomly and different parameters of random process.

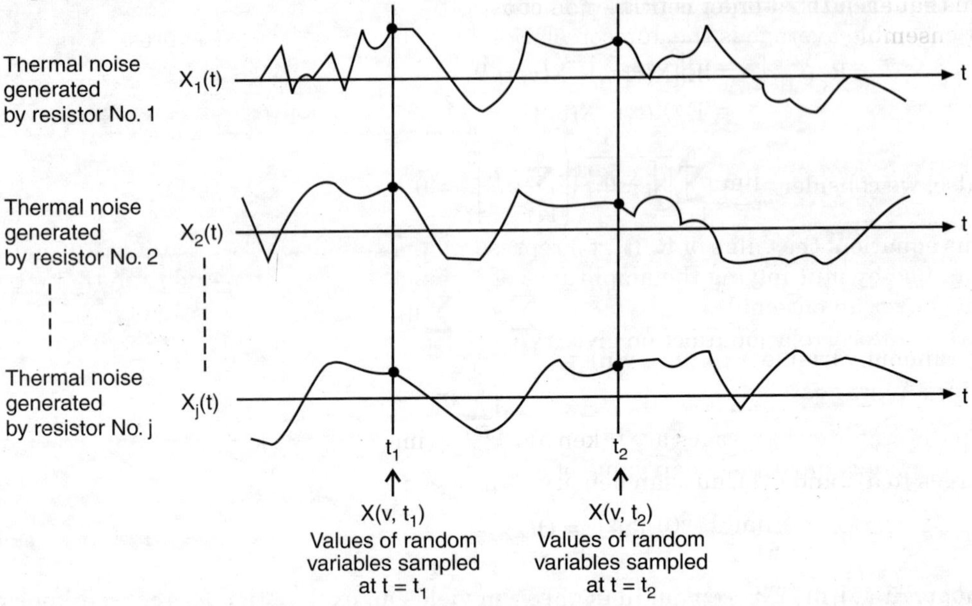

Fig. 4.10.

4.22.1. Ensemble and Time Averages

We know that the random variables may be specified with the help of different probability distributions. Random processes cannot be completely specified with the help of PDFs. Generally ensemble and time statistics are used to specify the random processes. Most generally, statistical averages such as mean $m_x(t)$ and autocorrelation function R_τ (t_1, t_2) are used to describe random process.

4.22.2. Ensemble Averages

In case of ensemble averages, the average is taken over the ensemble of waveforms, keeping the time constant or fixed.

As an example, if we desire ensemble mean value in random process of figure 4.10, then we will have to take mean of $v_1(t_1)$, $v_2(t_1)$, $v_3(t_1)$ ..., $v_i(t_1)$.

This ensemble mean is taken at time $t = t_1$.

In the same way, we can take ensemble mean at time $t = t_2$ also. The ensemble mean at $t = t_1$ and $t = t_2$ need not be same. Similarly, ensemble mean at other values of t may also be evaluated. Thus ensemble mean becomes function of time t. Therefore it is represented as $m_x(t)$.

We know that the mean value for a continuous random variable is expressed as

$$m_x = \int_{-\infty}^{\infty} x f_X(x)\, dx \qquad ...(4.106)$$

Using this expression, the ensemble mean may be defined as

$$m_x(t) = E[X(v, t] \qquad ...(4.107)$$

Hence, Ensemble mean value

$$m_x(t) = \int_{-\infty}^{\infty} x f_X(x, t)\, dx \qquad ...(4.108)$$

In this equation, the time t is treated as constant and the integration is done with respect to x. The next ensemble average is the autocorrelation function which may be expressed as

$$R_\tau(t_1, t_2) = E[X(v, t_1) \cdot X(v, t_2)] \qquad ...(4.109)$$

$$= E[X(t_1) \cdot X(t_2)]$$

$$= \int_{-\infty}^{\infty}\int_{-\infty}^{\infty} x_1 x_2 f x_1 f x_2 (x_1, x_2)\, dx_1\, dx_2 \qquad ...(4.110)$$

In this equation, the value of $R_\tau(t_1, t_2)$ represents the similarity of amplitudes at time t_1 and t_2 and is obtained by multiplying the amplitudes of sample functions at t_1 and t_2 and taking mean of this product over an ensemble.

Thus the autocorrelation function gives the information about frequency content of the process.

4.22.3. Time Averages

When the statistical averages are taken along the time; they are known as time averages. As an example, we may define time mean value of a sample function x(t) as

$$\text{Time mean value: } \langle m_x \rangle = \lim_{T \to \infty} \frac{1}{T} \int_{-T/2}^{T/2} X(t)\, dt \qquad ...(4.111)$$

The equation (4.111) is a standard equation to find average or mean value of a function.

Let us consider the random process shown in figure (4.10). We can obtain time mean of waveform $X_1(t)$ by integrating it over a long period. The time mean of $X_1(t)$ and $X_2(t)$ need not be same. Thus the collection of time means becomes a function of the sample function $X_i(t)$.

The autocorrelation function may be expressed using time averaging as

$$\langle R_\tau(\varepsilon) \rangle = \lim_{T \to \infty} \frac{1}{T} \int_{-T/2}^{T/2} x(t) \cdot x(t + \tau)\, dt \qquad ...(4.112)$$

4.23. CLASSIFICATION OF RANDOM PROCESS

Random processes may be classified into the following broad categories:

4.23.1. Stationary and Non-Stationary Random Process

A random process X(t) is called stationary if its statistics are not affected by any shift in the time origin. We may define stationary process in terms of ensemble averages as:

(i) The ensemble mean is independent of time. Mathematically,

$$m_x(t) = m_x(t_1) = m_x(t_2) = m_x(t_3) = ... = \text{constant at all the time instants} \qquad ...(4.113)$$

(ii) The autocorrelation function $R_\tau(t_1, t_2)$ depends only upon the time difference $t_2 - t_1$.

$$R_\tau(t_1, t_2) = R_\tau(t_1 + t, t_2 + t) \qquad ...(4.114)$$

Therefore, the autocorrelation function of any stationary process is a function of time-difference and is given as

$$R_\tau(t_1, t_2) = R_\tau(t_2 - t_1) \qquad ...(4.115)$$

$$= R_\tau(\rho)$$

where $\rho = t_2 - t_1$

Thus, the system of resistors producing thermal noise voltages represents a stationary process due to the fact that the noise generated by resistor does not depend on time.

4.23.2. Wide Sense Stationary (i.e., Weakly Stationary) Process

The process may not be stationary in strict sense, still the mean and autocorrelation functions are independent of shift of time region. Such type of process is known as **wide sense stationary process.**

All the processes in practice, are non-stationary since every process has some start and end. This means that the statistics of such process are dependent upon time. A true stationary process should start at $t = -\infty$ and should not stop till $t = \infty$. Such type of process is not possible practically. The process may appear stationary over a certain period of time. Then this will be wide sense stationary process.

4.23.3. Ergodic Process

A random process is known as ergodic process if the time-averages are equal to ensemble averages. Hence for a ergodic process, we have

$$m_x = \langle m_x \rangle \quad \text{...(4.116)}$$

$$R_\tau (t_1, t_2) = \langle R_\tau (\varepsilon) \rangle \quad \text{...(4.117)}$$

For any ergodic process, all ensemble averages will be equal to corresponding time averages for any particular sample function. The time averages are not a function of time. When time and ensemble averages are the same, it implies that ensemble averages also are not a function of time.

Thus an ergodic process is always stationary but converse in not true.

Ergodicity of the process may be defined in terms of some statistical averages like mean and autocorrelation.

The random process is ergodic in the mean if

$$m_x = \langle m_x \rangle$$

and variance of $\langle m_x \rangle \to 0$ as $T \to \infty$

Similarly, the random process is ergodic in the autocorrelation if

$$R_\tau (t_1, t_2) = \langle R_\tau (\varepsilon) \rangle \quad \text{...(4.118)}$$

and variance of $\langle R_\tau (\varepsilon) \rangle \to 0$ as $T \to \infty$

Hence, the Ergodicity of any random process may be determined by evaluating statistical averages of single sample function. This means that a single sample function represents entire random process.

4.23.4. Gaussian Process

Let us consider a random process denoted by X(t) for an interval which starts at a time $t = 0$ and lasts until $t = T$.

Now, let us weigh the random process X(t) by any function g(t) and then integrate the product X(t) g(t) over this observation time-interval $(0 \le t \le T)$.

Thus, Gaussian we obtain a random variable Y defined as

$$Y = \int_0^T X(t)\, g(t)\, dt \quad \text{...(4.119)}$$

where Y is a linear function of X(t).

Also, the random variable Y depends upon the argument function X(t) g(t) over the observation time interval 0 to T. Here, we assume that the weighing function g(t) is such that the mean square value of the random variable

$$Y = \int_0^T X(t)\, g(t)\, dt \quad \text{is finite.} \qquad ...(4.120)$$

Now, if the random variable Y is a Gaussian distributed random variable for each value of function g(t), then the process X(t) is known as a **Gaussian process**. Also, a random variable Y has a Gaussian distribution if its probability density function has the following form:

$$f_Y(y) = \frac{1}{\sqrt{2\pi}\, \sigma_y} e^{\{[-(y - mY)^2]/2\sigma Y\}^2} \qquad ...(4.121)$$

Here, m_Y is the mean and σ_Y^2 is the variance of the random variable Y. A Gaussian distributed random variable is completely characterized by specifying its mean and the variance.

Figure 4.11 exhibits a plot of Gaussian distributed probability density function (pdf) for $m_Y = 0$ and $\sigma_Y^2 = 1$.

It may be noted that a Gaussian process has two main features or advantages:

(i) Firstly, the Gaussian process has many properties which make analytic results possible.

(ii) Secondly, the random process produced by physical phenomena are often such that a Gaussian model is an appropriate one.

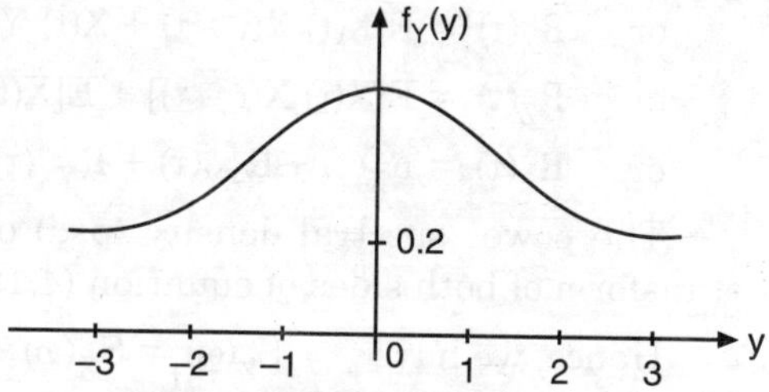

Fig. 4.11. Normalized Gaussian distributed pdf of a random variably

Also the use of a Gaussian model to describe physical phenomena is generally confirmed by the experiments. Thus, the Gaussian process is very important in the study of communication systems.

A Gaussian process has the following important properties:

(i) If a Gaussian process X(t) is made to apply to a stable linear filter, then the random process Y(t) produced at the output of the filter is also Gaussian.

(ii) Let us consider the samples of random process $X(t_1), X(t_2), ..., X(t_n)$, obtained by observing a random process X(t) at time $t = t_1, t = t_2, ..., t = t_n$.

Now, if the random process X(t) is Gaussian, then this set of random variables are jointly Gaussian for any value of n, with their n-fold joint pdf being completely determined by specifying the set of means.

Here, means used for more than one mean (or average), i.e.

$$m_X(t_i) = E[X(t_i)],\ i = 1, 2, ..., n$$

and the set of autocorrelation function

$$C_X(t_j, t_i) = E[\{X(t_j) - m_X(t_j)\}\{X(t_i) - m_X(t_i)\}] \qquad ...(4.122)$$

for $\quad i = j = 1, 2, 3 ..., n$

(iii) Also, if a Gaussian process is wide-sense stationary (WSS), then the process is also stationary in the strict sense.

(iv) If the set of random variables $X(t_1), X(t_2) ..., X(t_n)$, obtained by sampling a Gaussian process X(t) at times $t = t_1, t = t_2, ..., t = t_n$, are not correlated i.e.,

$$E[\{X(t_j) - m_X(t_j)\}\{X(t_i) - m_X(t_i)\}] = 0 \qquad ...(4.123)$$

for $i \neq j$

then this set of random variables are statistically independent.

4.24 SUM OF RANDOM PROCESSES

Now, we shall discuss the sum of two random processes.

Let us consider two WSS processes X(t) and Y(t) with zero means. The sum of random processes is another random process denoted as Z(t). Also, the statistical parameters of Z(t) may be determined in terms of random processes X(t) and Y(t).

Thus, the random process Z(t) may be expressed as

$$Z(t) = X(t) + Y(t) \quad ...(4.124)$$

and the autocorrelation function of Z(t) be found as

$$R_Z(\tau) = E[Z(t)\, Z(t-\tau)]$$

$$R_Z(\tau) = E[\{X(t) + Y(t)\}\, \{X(t-\tau)\, Y(t-\tau)\}]$$

or, $R_Z(\tau) = E[X(t)\, X(t-\tau) + X(t)\, Y(t-\tau) + X(t-\tau)\, Y(t) + Y(t)\, Y(t-\tau)]$

or $R_Z(\tau) = E[X(t)\, X(t-\tau)] + E[X(t)\, Y(t-\tau)] + E[X(t-\tau)\, Y(t)] + E[Y(t)\, Y(t-\tau)]$

or $R_Z(\tau) = R_X(\tau) + R_{XY}(\tau) + R_{YX}(\tau) + R_Y(\tau)$ ....(4.125)

The power spectral density (psd) of random process Z(t) may be obtained by taking Fourier transform of both sides of equation (4.125)

Hence, we have $S_Z(\omega) = S_X(\omega) + S_{XY}(\omega) + S_{YX}(\omega) + S_Y(\omega)$ ...(4.126)

The power spectral density (psd) of the sum of two WSS random processes is equal to the sum of the individual power spectral densities (psds) plus the sum of the cross power spectral densities (psds) $S_{XY}(\omega)$ and $S_{YX}(\omega)$.

If random processes X(t) and Y(t) are non-correlated then the cross power spectral densities $S_{XY}(\omega)$ and $S_{YX}(\omega)$ are zero. Now equation (4.126) can be written as

$$S_Z(\omega) = S_X(\omega) + S_Y(\omega)$$

Hence, the power spectral density (psd) of the sum of two uncorrelated WSS random processes is equal to the sum of their individual power spectral densities.

4.25 CORRELATION FUNCTION: DETAILED ASPECTS

The correlation function provides a measure of similarity or coherence between a given signal (process) and a replica of the same signal or other signal (process) by a variable amount.

4.25.1. Autocorrelation Function

Autocorrelation function may be defined as a measure of similarity between a signal or process and its replica by a variable amount.

The autocorrelation function of a stationary process X(t) may be defined as

$$R_{X(t_j - t_i)} = E[X(t_j)\, X(t_i)] \text{ for any } t_j \text{ and } t_i \quad ...(4.127)$$

where $X(t_j)$ and $X(t_i)$ are the random variables obtained by observing the process X(t) at times t_j and t_i, respectively.

Also, the autocorrelation function depends only on the time difference $(t_j - t_i)$.

Using $\tau = t_j - t_i$ in equation (4.127), we get

$$R_X(\tau) = E[X(t)\, X(t-\tau)] \quad ...(4.128)$$

The expressions for X(t) and X(t – τ) are viewed as random variables. The variable τ is known as **time-lag or time-delay parameter**. The autocorrelation function for stationary random process is independent of a shift of time origin.

4.25.2. Properties of Autocorrelation Function of Random Process

The autocorrelation function $R_X(\tau)$ of a wide-sense stationary (WSS) process has the following important properties:

(i) The autocorrelation function of WSS process X(t) is always an even function of time-lag. Autocorrelation function $R_X(\tau)$ satistifies the following mathematical relationship, i.e.

$$R_X(\tau) = R_X(-\tau) \quad ...(4.129)$$

Proof: According to the definition of autocorrelation function of WSS random process, we have

$$R_X(\tau) = E[X(t)\, X(t-\tau)] \quad ...(4.130)$$

Substituting $\tau = -\tau$, in equation (4.130), we get

$$R_X(-\tau) = E[X(t)\, X(t+\tau)]$$

or
$$R_X(-\tau) = E[X(t+\tau)\, X(t)] \quad ...(4.131)$$

For WSS process, we know that

$$X(t)\, X(t-\tau) = X(t+\tau)\, X(t) \quad ...(4.132)$$

From last three equations, we conclude that

$$R_X(\tau) = R_X(-\tau)$$

(ii) The mean square value of a WSS process is equal to the autocorrelation function of the random process for zero time-lag. It may be written as

$$R_X(0) = R_X(\tau)\big|_{\tau=0} = E[X(t)\, X(t-\tau)]\big|_{\tau=0}$$

or
$$R_X(0) = E[X(t)\, X(t-0)] = E[X^2(t)]$$

or
$$R_X(0) = E[X^2(t)] \quad ...(4.133)$$

(iii) The autocorrelation function of a WSS random process has the maximum magnitude at zero time-lag ($\tau = 0$). This property states that

$$|R_X(\tau)| = \text{magnitude of } R_X(\tau) = R_X(0) \quad ...(4.134)$$

Proof: We know that the mean square value of the difference between X(t) and $X(t-\tau)$ is always non-negative, i.e.

$$E[\{X(t) - X(t-\tau)\}^2] \geq 0$$

or
$$E[X^2(t) - 2X(t)\, X(t-\tau) + X^2(t-\tau)] \geq 0 \quad ...(4.135)$$

Since expectation operation is a linear operation, we can write above equation as

$$E[X^2(t)] - 2E[X(t)\, X(t-\tau)] + E[X^2(t-\tau)] \geq 0 \quad ...(4.136)$$

Also, for a WSS random process, we have

$$E[X^2(t)] = E[X^2(t-\tau)] = R_X(0) \quad ...(4.137)$$

and
$$E[X(t)\, X(t-\tau)] = R_X(\tau) \quad ...(4.138)$$

Substituting equations (4.133) and (4.137) in equation (4.138), we get

$$|R_X(\tau)| \leq R_X(0) \quad ...(4.139)$$

Thus, the autocorrelation functions $R_X(\tau)$ provide means of decreasing the independance of two random variables obtained by observing a random process X(t) τ seconds apart. The autocorrelation function is a measure of the rate of the fluctuation of random process. Various types of random processes have been illustrated in figure 4.12.

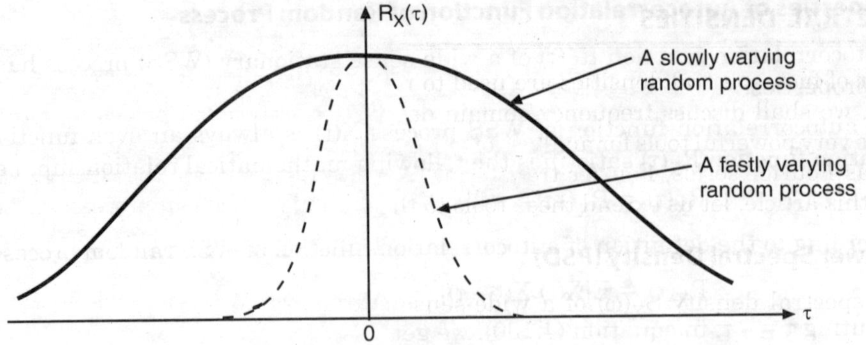

Fig. 4.12. Illustration of autocorrelation functions of slowly and rapidly varying random processes

4.25.3. Cross-Correlation Functions

An autocorrelation function is determined for single random process but cross correlation function is determined for two random processes. Let us consider two random processes X(t) and Y(t) with autocorrelation functions $R_X(t, u)$ and $R_Y(t , u)$, respectively. There will be two cross-correlation functions between two random process X(t) and Y(t). These cross-correlations may be defined as

$$R_{XY}(t, u) = E[X(t)\,Y(u)] \quad \text{...(4.140)}$$

and

$$R_{YX}(t, u) = E[Y(t)\,X(u)] \quad \text{...(4.141)}$$

where t and u are the two values of time at which the processes are observed. The cross-correlation function of two random processes X(t) and Y(t) may be expressed conveniently in matrix form as

$$R(t, u) = \begin{bmatrix} R_X(t,u) & R_{XY}(t,u) \\ R_{YX}(t,u) & R_Y(t,u) \end{bmatrix} \quad \text{...(4.142)}$$

R(t, u) is called the correlation matrix of random processes X(t) and Y(t).

Equation (4.142) may also be written as

$$R(t, u) = \begin{bmatrix} R_X(t-u) & R_{XY}(t-u) \\ R_{YX}(t-u) & R_Y(t-u) \end{bmatrix} \quad \text{...(4.143)}$$

If both random processes X(t) and Y(t) are WSS processes, then equation (4.142) can be written as equation (4.143). In addition, both random processes X(t) and Y(t) are said to be jointly wide-sense stationary.

It may be noted that in general, the cross-correlation function is not an even function of τ like an autocorrelation function. It is having following symmetry relation

$$R_{XY}(\tau) = R_{YX}(-\tau) \quad \text{...(4.144)}$$

Moreover, cross-correlation does not have a maximum at the origin like the autocorrelation function. The two random processes X(t) and Y(t) are said to be incoherent or orthogonal if cross-correlation function of X(t) and Y(t) is zero.

$$R_{XY}(\tau) = 0 \quad \text{...(4.145)}$$

The two random processes are said to be non-correlated, if their cross-correlation functions $R_{XY}(\tau)$ are equal to the multiplication of mean values.

$$R_{XY}(\tau) = E[X(t)]\,E[Y(t)] \quad \text{...(4.146)}$$

The incoherent or orthogonal processes are non-correlated processes with E[X(t)] and or E[Y(t)] = 0.

4.26 SPECTRAL DENSITIES

As a matter of fact, spectral densities are used to represent random process in frequency-domain. In this article, we shall discuss frequency-domain description of random processes. Frequency-domain methods are very powerful tools for analysis of various types of signals. We already know some powerful tools such as Fourier series, Fourier transform, Laplace transform for the analysis of detetministic signals. In this article, let us extend these tools to the stationary random processes.

4.26.1. Power Spectral Density (PSD)

Power spectral density $S_X(\omega)$ of a wide-sense stationary (WSS) random process X(t) may be defined as

$$S_X(\omega) = \text{Fourier transform } \{R_X(\tau)\}$$

$$= \int_{-\infty}^{\infty} R_X(\tau)\, e^{-j\omega\tau} d\tau \qquad ...(4.147)$$

where $R_X(\tau)$ = Autocorrelation function of random process X(t).

Thus, we can say that the Fourier transform of autocorrelation function $R_X(\tau)$ of a random process X(t) is called power spectral density (psd) of random process. Equation (4.147) was initially developed by Albert Einstein. Later it was developed by N. Weiner and Khenchine. Hence, this equation is known as Einstein-Weiner Khenchine (EWK) equation. Power spectral density (psd) is used to measure the energy of the signal in the frequency band $[\omega, \omega + d\omega]$.

Cross Power Spectral Density (CPSD).

The cross power spectral density of two jointly WSS random processes X(t) and Y(t) may be defined as

$$S_{XY}(\omega) = \text{Fourier transform } \{R_{XY}(\tau)\}$$

$$\boxed{S_{XY}(\omega) = \int_{-\infty}^{\infty} R_{XY}(\tau)\, e^{-j\omega\tau} d\tau} \qquad ...(4.148)$$

where $R_{XY}(\tau)$ = cross-correlation function of random processes X(t) and Y(t).

Properties of Power Spectral Density

Power spectral density (psd) of a random process has the following main properties.

(i) Power spectral density of a random process is a real function of frequency ω.

Proof: We know that the autocorrelation function of a random process is an even function, i.e.

$$R_X(\tau) = R_X(-\tau)$$

Therefore,
$$S_X(\omega) = \int_{-\infty}^{\infty} R_X(\tau)\, e^{-j\omega\tau} d\tau$$

or
$$S_X(\omega) = \int_{-\infty}^{\infty} R_X(\tau)\, [\cos \omega\tau - j \sin \omega\tau]\, d\tau$$

$$S_X(\omega) = \int_{-\infty}^{\infty} R_X(\tau) \cos \omega\tau\, d\tau - j \int_{-\infty}^{\infty} R_X(\tau) \sin \omega\tau\, d\tau \qquad ...(4.149)$$

Now let $\int_{-\infty}^{\infty} R_X(\tau) \cos \omega\tau\, d\tau = I$ ...(4.150)

Substituting $t = -\tau$ in equation (4.150), we get

$$\int_{-\infty}^{\infty} R_X(-t)\sin(-\omega t)(-dt) = I$$

or $$-\int_{-\infty}^{\infty} R_X(-t)\sin\omega t\, dt = I \quad ...(4.151)$$

Using the even property of autocorrelation function, we obtain

$$-\int_{-\infty}^{\infty} R_X(\tau)\sin\omega\tau\, d\tau = I \quad ...(4.152)$$

After adding equations (4.151) and (4.152), we get

$$I = 0 \quad ...(4.153)$$

Substituting $\int_{-\infty}^{\infty} R_X(\tau)\sin\omega\tau\, d\tau = 0$ in equation (4.153), we get

$$S_X(\omega) = \int_{-\infty}^{\infty} R_X(\tau)\cos\omega\tau\, d\tau \quad ...(4.154)$$

Here $S_X(\omega)$ is a real function of ω.

(ii) Power spectral density (psd) of a random process X(t) is an even function of frequency ω, i.e.

$$S_X(\omega) = S_X(-\omega)$$

(iii) Power spectral density (psd) of a random process X(t) is a non-negative function of ω, i.e.,

$$S_X(\omega) \geq 0 \text{ for all } \omega$$

4.26.2. Energy Spectral Density (ESD)

The energy spectral density (psd) $\psi_X(\omega)$ may be defined as a measure of density of the energy contained in random process X(t) in Joules per Hertz. It may be noted that since the amplitude spectrum of a real-valued random process X(t) is an even function of ω, the energy spectral density of such a signal is symmetrical about the vertical axis passing through the origin.

Thus, total energy of the random process X(t) is defined as

$$E = \frac{1}{2\pi}\int_{-\infty}^{\infty} \psi_X(\omega)\, d\omega \quad ...(4.155)$$

Multiple Random Processes

For two real random processes x(t) and y(t), we define the cross-correlation function $R_{xy}(t_1, t_2)$ as

$$R_{xy}(t_1, t_2) = \overline{x(t_1)\, y(t_2)} \quad ...(4.156)$$

The two processes are said to be **jointly stationary** (in the wide sense) if each of the processes is individually wide-sense stationary and if

$$R_{xy}(t_1, t_2) = R_{xy}(t_2 - t_1)$$

$$R_{xy}(t_1, t_2) = R_{xy}(\tau) \quad ...(4.157)$$

Uncorrelated, Orthogonal (Incoherent), and Independent Processes

Two processes x(t) and y(t) are said to be **uncorrelated** if their cross-correlation function is equal to the product of their means; that is,

$$R_{xy}(\tau) = \overline{x(t)\, y\,(t+\tau)} = \overline{x}\;\overline{y} \quad ...(4.158)$$

This implies that RVs x(t) and y(t + τ) are uncorrelated for all t and τ.

Processes x(t) and y(t) are said to be incoherent, or orthogonal, if

$$R_{xy}(\tau) = 0 \quad ...(4.159)$$

Incoherent, or orthogonal processes are uncorrelated processes with $\overline{x}$ and or $\overline{y} = 0$.

Processes x(t) and y(t) are **independent** random processes if the random variables $x(t_1)$ and $y(t_2)$ are independent for all possible choices of t_1 and t_2.

Cross-Power Spectral Density:

We define the cross-power spectral density S_{xy} (f) for two random processes x(t) and y(t) as

$$S_{xy}(f) = \lim_{T\to\infty} \frac{\overline{X_T(f)\, Y_T(f)}}{T} \quad ...(4.160)$$

where $X_T(f)$ and $Y_T(f)$ are the Fourier transforms of the truncated processes x(t) Π (t/T) and y(t) Π (t/T), respectively. It can be shown that,

$$R_{xy}(\tau) \Leftrightarrow S_{xy}(f) \quad ...(4.161)$$

It can be seen from equations (4.156) and (4.157) that for real random processes x(t) and y(t),

$$R_{xy}(\tau) = R_{yx}(-\tau) \quad ...(4.162)$$

Therefore,

$$S_{xy}(f) = S_{yx}(-f) \quad ...(4.163)$$

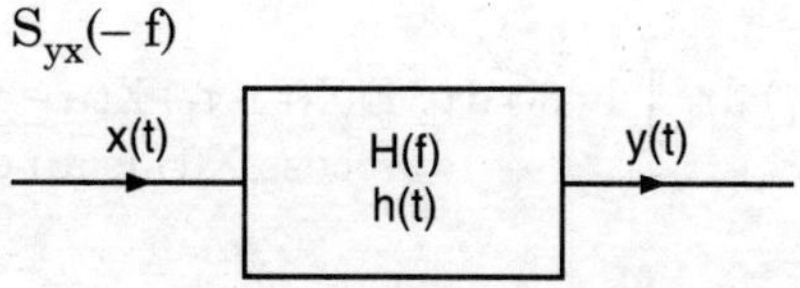

Fig. 4.13. Transmission of a random process through a linear time-invariant system

4.27 RESPONSE OF LINEAR SYSTEMS TO RANDOM INPUTS

In this article, we shall discuss about response of linear systems to random inputs. This means that we shall study transmission of random process through linear systems or linear filters.

Suppose that a random process X(t) is applied to the input of a linear time-invariant (LTI) system having an impulse response h(t). Let the input random process X(t) produces a random process Y(t) at the system output. The block diagram of this LTI system has been shown in figure 4.14. Here, we shall determine the mean and the auto-correlation function of the output random process Y(t) in terms of input random process X(t). Also, we are assuming that X(t) is a wide-sense stationary (WSS) process.

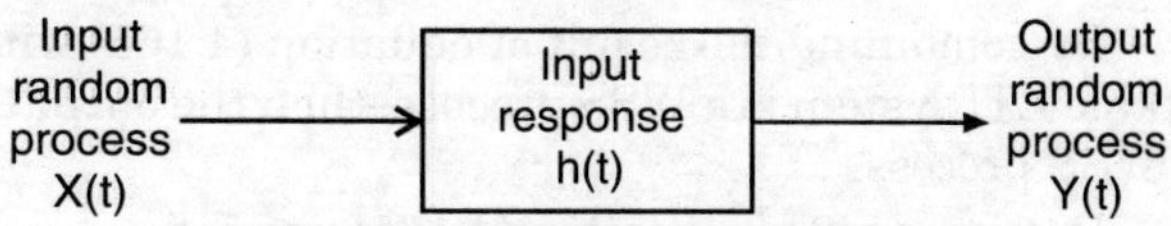

Fig. 4.14. Determination of response of a linear system to a random input.

The mean of the output random process is expressed as

$$m_Y(t) = E\,[Y(t)] = E\left[\int_{-\infty}^{\infty} h(\tau)\, X(t-\tau)\, d\tau\right]$$

or

$$m_Y(t) = \int_{-\infty}^{\infty} h(\tau)\, E[X(t-\tau)]d\tau = \int_{-\infty}^{\infty} h(\tau)\, m_X(t-\tau)]\, d\tau \quad ...(4.164)$$

when the input random process X(t) is wide-sense stationary, the mean $m_X(t)$ is a constant m_X. Therefore, equation (4.164) may be written as

$$m_Y = \int_{-\infty}^{\infty} h(\tau)\, m_X\, d\tau = m_X \int_{-\infty}^{\infty} h(\tau)\, d\tau = m_X\, H(0) \quad ...(4.165)$$

where H(0) = zero-frequency response of the system.

In other words, we can say that the mean of the output random process of a stable linear time-invariant (LTI) system is equal to the mean of the input random process multiplied by the zero-frequency response of the system.

We shall now determine the autocorrelation function of the output random process Y(t). Let t and u be two values of time at which output process is observed. Autocorrelation function of output random process is defined as

$$R_Y(t, u) = E[Y(t)\, Y(u)] \quad ...(4.166)$$

By using convolution integral, equation (4.166) can be written as

$$R_Y(t, u) = E\left[\int_{-\infty}^{\infty} h(\tau_1)\, X(t-\tau_1)\, d\tau_1 \int_{-\infty}^{\infty} h(\tau_2)\, X(u-\tau_2)\, d\tau_2\right] \quad ...(4.167)$$

Assuming $E[X^2(t)]$ is finite for all t and the system is stable, we can interchange the order of the expectation operator and integration w.r.t. τ_1 and τ_2 to get

$$R_Y(t, u) = \int_{-\infty}^{\infty} h(\tau_1)\, d\tau_1 \int_{-\infty}^{\infty} h(\tau_2)\, d\tau_2\, E[X(t-\tau_1)\, X(u-\tau_2)]$$

or,

$$R_Y(t, u) = \int_{-\infty}^{\infty} h(\tau_1)\, d\tau_1 \int_{-\infty}^{\infty} h(\tau_2)\, d\tau_2\, R_X(t-\tau_1, u-\tau_2) \quad ...(4.168)$$

When the input X(t) is a wide-sense stationary process, the autocorrelation function of X(t) is only a function of the difference between the observation time $t - \tau_1$ and $u - \tau_2$. Substituting $\tau = t - u$ in equation (4.168), we get

$$R_Y(\tau) = \int_{-\infty}^{\infty}\int_{-\infty}^{\infty} h(\tau_1)\, h(\tau_2)\, R_X(\tau - \tau_1 + \tau_2)\, d\tau_1\, d\tau_2 \quad ...(4.169)$$

$R_Y(\tau)$ is the function of time lag τ.

By combining the result of equation (4.165) with equation (4.169), we see that if the input to a stable LTI system is a WSS process, then the output of the linear time-invariant (LTI) system is also a WSS process.

Putting $\tau = 0$ in equation (4.165), we get

$$R_Y(0) = \int_{-\infty}^{\infty}\int_{-\infty}^{\infty} h(\tau_1)\, h(\tau_2)\, R_X(\tau_2 - \tau_1)\, d\tau_1\, d\tau_2 \quad ...(4.170)$$

Also we know that

$$R_Y(0) = E[Y^2(t)] \quad ...(4.171)$$

EXAMPLE 4.30. A random variable X follows Gaussian probability density function with parameters μ and σ. Find its probability distribution function.

Solution: We know that the relationship between probability density function (pdf) and the probability distribution function of a random variable X is expressed as

$$F_X(z) = \int_{-\infty}^{z} f_x(x)\,dx \quad \text{...(i)}$$

Since we have given that the random variable X follows Gaussian probability density function then

$$F_X(z) = \int_{-\infty}^{z} \frac{1}{\sqrt{2\pi\sigma}} e^{-(x-\mu)/2\sigma^2} dx \quad \text{...(ii)}$$

Substituting $\frac{x-\mu}{\sigma} = y$ in equation (ii), we get

$$F_x(z) = \int_{-\infty}^{[(Z-\mu)/\sigma]} \frac{1}{\sqrt{2\pi}\,\sigma} e^{-y^2/2}(\sigma dy) = \frac{1}{\sqrt{2\pi}} \int_{-\infty}^{[(z-\mu)/\sigma]} e^{-y^2/2} dy$$

or
$$F_x(z) = \frac{1}{\sqrt{2}}\left[\int_{-\infty}^{0} e^{-y^2/2} dy + \int_{0}^{(z-\mu)/2} e^{-y^2/2} dy\right] = \left[-\int_{0}^{\infty} e^{-y^2/2} dy + \int_{0}^{(z-\mu)} e^{-y^2/2} dy\right] \quad \text{...(iii)}$$

Also error function is defined as

$$\text{erf}(x) = \frac{1}{\sqrt{2\pi}} \int_{0}^{x} e^{-y^2/2} dy \quad \text{...(iv)}$$

and
$$\text{erf}(\infty) = \frac{1}{\sqrt{2\pi}} \int_{0}^{\infty} e^{-y^2/2} dy = \frac{1}{2} \quad \text{...(v)}$$

Putting equations (iv) and (v) in equation (iii), we get

$$F_x(z) = \frac{1}{2} + \text{erf}\left(\frac{z-\mu}{\sigma}\right) \quad \textbf{Ans.}$$

EXAMPLE 4.31. Prove that for a Gaussian random variable X,

$$P[x_1 \le X \le x_2] = \text{erf}\left(\frac{x_2-\mu}{\sigma}\right) - \text{erf}\left(\frac{x_2-\mu}{\sigma}\right)$$

here erf (x) = Error function

μ and σ = parameters of Gaussian probability density function and x_1, x_2 = any real numbers.

Solution: We know that

$$P[x_1 \le X \le x_2] = \int_{x_1}^{x_2} f_X(x)\,dx \quad \text{...(i)}$$

here
$$f_X(x) = \frac{1}{\sqrt{2\pi\sigma}} e^{(x-\mu)^2/2\sigma^2} \quad \text{...(ii)}$$

Substituting equation (ii) in equation (i), we get

$$P[x_1 \le X \le x_2] = \int_{x_1}^{x_2} \frac{1}{\sqrt{2\pi}\,\sigma} e^{(x-\mu)^2/2\sigma^2} dx = \frac{1}{\sqrt{2\pi}\,\sigma} \int_{x_1}^{x_2} e^{(x-\mu)^2/2\sigma^2} dx \quad \text{...(iii)}$$

Putting $\left(\frac{x-\mu}{\sigma}\right) = y$ in equation (iii), we get

$$P[x_1 \le X \le x_2] = \frac{1}{\sqrt{2\pi}\,\sigma} \int_{[(x_1-\mu)/\sigma)}^{[(x_2-\mu)/\sigma]} e^{y^2/2} \sigma\, dy = \frac{1}{\sqrt{2\pi}} \int_{[(x_1-\mu)/\sigma]}^{[(x_2-\mu)/\sigma]} -e^{y^2/2}\, dy$$

$$= \frac{1}{\sqrt{2\pi}}\left[-\int_{[(x_1-\mu)/\sigma]}^{0} e^{y^2/2}dy + \int_{0}^{[(x_2-\mu)/\sigma]} e^{y^2/2}dy\right]$$

$$= \frac{1}{\sqrt{2\pi}}\left[-\int_{0}^{(x_1-\mu)/\sigma} e^{y^2/2}dy + \int_{0}^{(x_2-\mu)/\sigma} e^{y^2/2}dy\right]$$

$$P[x_1 \le X \le x_2] = -\operatorname{erf}\left(\frac{x_1-\mu}{\sigma}\right) + \operatorname{erf}\left(\frac{x^2-\mu}{\sigma}\right) \quad \textbf{Ans.}$$

EXAMPLE 4.32. The marginal probability density functions of two random variables X and Y are given below

$$f_X(x) = \begin{cases} 3(1-x)^2, & \text{for } 0 \le x \le 1 \\ 0, & \text{for } x < 0, x > 1 \end{cases}$$

and

$$f_Y(y) = \begin{cases} 3(1-y)^2, & \text{for } 0 \le y \le 1 \\ 0, & \text{for } y < 0, y > 1 \end{cases}$$

Determine:

(i) Mean of random variables X and Y

(ii) Variance of random variables X and Y. **(U.P.Tech., Sem. Exam.; 2007-08)**

Solution: (i) Mean of $X = \mu_X = E(X) = \int_{-\infty}^{\infty} x f_X(x)\, dx$

$$E(X) = \int_0^1 x\, 3(1-x)^2\, dx = \frac{1}{4}$$

and the mean of $Y = \mu_Y$

$$E(Y) = \int_{-\infty}^{\infty} y f_Y(y)\, dy = \int_0^1 y\, 3(1-y)^2\, dy = \frac{1}{4}$$

(b) Variance of $X = \sigma_X^2 = E(X^2) - \mu_X^2$ ...(i)

Now, $E(X^2)$ may be found as

$$E(X^2) = \int_{-\infty}^{\infty} x^2 f_X(x)\, dx = \int_0^1 x^2\, 3(1-x)^2\, dx = \frac{1}{10}$$

Similarly, we can find $E(Y^2)$ as

$$E(Y^2) = \int_{-\infty}^{\infty} y^2 f_Y(y)\, dy = \int_0^1 y^2\, 3(1-y)^2\, dy = \frac{1}{10}$$

Substituting the values $E[X^2]$ and μ_X in equation, we get

$$\sigma_X^2 = E(X^2) - \mu_X^2 = \frac{1}{10} - \left(\frac{1}{4}\right)^2 = \frac{8-5}{80} = \frac{3}{80}$$

Similarly, variance of Y is determined as

$$\sigma_Y^2 = E(Y^2) - \mu_Y^2 \quad \text{...(ii)}$$

Putting the values of $E(Y^2)$ and μ_Y in equation (ii), we get

$$\sigma_Y^2 = \frac{1}{10} - \left(\frac{1}{4}\right)^2 = \frac{3}{80} \quad \textbf{Ans.}$$

EXAMPLE 4.33. Given two random processes X(t) and Y(t) as

$$X(t) = Z_1(t) + 3Z_2(t - \tau)$$
$$Y(t) = Z_2(t + \tau) + 3Z_1(t - \tau)$$

here $Z_1(t)$ and $Z_2(t)$ are independent white noise processes each with a variance equal to 0.5. Determine,

(i) autocorrelation functions of X(t) and Y(t),

(ii) cross-correlation function of X(t) and Y(t). (U.P.S.C., I.E.S., Examination, 1997)

Solution: (a) Autocorrelation function of X(t) is expressed as

$$R_X(t_1, t_0) = E[X(t_1)\, X(t_0)] = E[\{Z_1(t_1) + 3Z_2(t_1 - \tau)\}\, \{Z_1(t_0)\, 3Z_2(t_0 - \tau)\}]$$
$$= E[Z_1(t_1)\, Z_1(t_0)] + 3E[Z_1(t_1) - Z_2(t_0 - \tau)] + 3E[Z_2(t_1 - \tau)\, Z_1(t_0)] + 9[Z_2(t_1 - \tau)\, Z_2(t_0 - \tau)] \quad ...(i)$$

But we know that

$$E[Z_1(t_1)\, Z_2(t_0 - \tau] = 0$$

Since, $Z_1(t)$ and $Z_2(t)$ are independent.

Similarly, $E[Z_2(t_1 - \tau)\, Z_1(t_0)] = 0$

For $t_1 \neq t_0$, $E[Z_1(t_1)\, Z_1(t_0)] = 0$

If $t_1 = t_0$ $\quad E[Z_1(t_1)\, Z_1(t_0)] = E\,[Z_1{}^2(t_1)] = \sigma^2 + \mu^2 = 0.5\, \delta(t_1 - t_0)$ ...(ii)

Similarly $\quad E[Z_2(t_1 - \tau)\, Z_2(t_0 - \tau)] = 0 \text{ for } t_1 \neq t_0 = 0.5 \text{ for } t_1 = t_0 = 0.5\, \delta(t_1 - t_0)$ ...(iii)

Substituting equations (ii) and (iii) in equation (i), we get

$$R_X(t_1, t_0) = 0.5\, \delta(t_1 - t_0) + 4.5\, \delta(t_1 - t_0) \quad ...(iv)$$

Similarly, $\quad R_X(t_1, t_0) = 0.5\, \delta(t_1 - t_0) + 4.5\, \delta(t_1 - t_0)$ ...(v)

From equations (iv) and (v), we see that the random processes X(t) and Y(t) are wide-sense stationary (WSS).

(ii) Cross-correlation function of random processes X(t) and Y(t) is expressed as

$$R_{XY}(t_1, t_0) = E[X(t_1)\, Y(t_0)] = E[\{Z_1(t_1) + 3Z_2(t_1)\}\{Z_2(t_0 + \tau) + 3Z_1(t_0 - \tau)\}]$$
$$= E[Z_1(t_1)\, Z_2(t_0 + \tau)] + 3E[Z_1(t_1)\, Z_1(t_0 - \tau)]$$
$$+ 3E[Z_2(t_1) Z_2(t_0 + \tau)] + 9E[Z_2(t_1)\, Z_1(t_0 - \tau) \quad ...(vi)$$

Now $\quad E[Z_1(t_1)\, Z_2(t_0)] = 0 \quad$ (Since $Z_1(t_1)$ and $Z_2(t_0)$ are independent random variables)

Therefore, $\quad E[Z_1(t_1) Z_1(t_0 - \tau)] = \begin{cases} 0, \text{ if } t_1 \neq t_0 - \tau \\ 0.5, \text{if } t_1 = t_0 - \tau \end{cases}$

$$E[Z_2(t_1 - \tau) Z_2(t_0)] = \begin{cases} 0, \text{ if } t_0 \neq t_1 - \tau \\ 0.5, \text{if } t_0 = t_1 - \tau \end{cases}$$

Now, $\quad E[Z_2(t_1 - \tau)\, Z_1(t_0 - \tau)] = 0$

Substituting above values in equation (vi), we get

$$R_{XY}(t_1, t_0) = 1.5\delta(t_1 - t_0 + \tau) + 1.5\, \delta(t_0 - t_1 + \tau) \quad ...(vii)$$

Because cross-correlation depends upon the difference of t_1 and t_0 therefore, random processes X(t) and Y(t) are jointly stationary or simply stationary. **Ans.**

SUMMARY

- There is one other class of signals, the behaviour of which cannot be predicted. Such type of signals are called random signals.
- These signals are called random signals because the precise value of these signals cannot be predicted in advance before they actually occur. The examples of random signals are the noise interferences in communication systems.
- An experiment is defined as the process which is conducted to get some results. If the same experiment is performed repeatedly under the same conditions, similar results are expected.
- The outcomes of an event are called equally likely, if any one of them cannot be expected to occur in preference to another. As an example, the tossing of a coin results in two outcomes, Head and Tail. Both have same possibility of 50%. Such type of outcomes are called equally likely outcomes.
- Probability may be defined as the study of random experiments. In any random experiment, there is always an uncertainty that a particular event will occur or not.
- The concept of conditional probability is used in conditional occurrences of the events.
- The conditional probability of event B given that event A has already happened.

$$P(B/A) = \frac{P(AB)}{P(A)}$$

where P (AB) is the joint probability of A and B.
Similarly, the conditional probability of event A given that event B has already happened.

$$P(A/B) = \frac{P(AB)}{P(B)}$$

The joint probability has commutative property which states that

$$P(AB) = P(BA)$$

- If A and B are two events in an experiment, and possibility of occurrence of event B does not depend upon occurrence of event A, then these two events A and B are known as statistically independent events.
- A function which can take on any value from the sample space and its range is some set of real numbers is called a **random variable** of the experiment.
- Random variables may be classified as under:
  (i) Discrete random variables
  (ii) Continuous random variables.
- A discrete random variable may be defined as the random variable which can take on only finite number of values in a finite observation interval. This means that the discrete random variable has countable number of distinct values.
- A random variable that takes on an infinite number of values is called a continuous random variable. Actually, there are several physical system (experiments) that generate continuous outputs or outcomes. Such systems generate infinite number of outputs or outcomes within the finite period.
- The Cumulative Distribution Function (CDF) of a random variable 'X' may be defined as the probability that a random variable 'X' takes a value less than or equal to x.
- The derivative of cumulative distribution function (CDF) with respect to some dummy variable is known as Probability Density Function (PDF). Probability density function (PDF) is generally denoted by $f_X(x)$. Mathematically, PDF may be expressed as

$$\text{PDF: } f_X(x) = \frac{d}{dx} F_X(x)$$

where x is a dummy variable.
Probability density function (PDF) is the more convenient representation for **continuous random variable**.

The joint Cumulative Distribution Function $F_{XY}(x, y)$ may be defined as the probability that the outcome of an experiment will result in a sample point lying inside the range $(-\infty < X \le x, -\infty < Y \le y)$ of the joint sample space. The joint sample space is the combined sample space of X and Y.

Mathematically,

$$F_{XY}(x, y) = P(X \le x, Y \le y)$$

- The joint PDF of any two random variables X and Y may be defined as the partial derivative of the joint cumulative distribution function $F_{XY}(x, y)$ with respect to the dummy variables x and y.

  Mathematically,

$$f_{XY}(x, y) = \frac{\partial^2 F_{XY}(x, y)}{\partial x\, \partial y}$$

- The relationship between probability and PDF over a certain interval is expressed as

$$P(x_1 < X \le x_2) = \int_{x_1}^{x_2} f_X(x)\, dx$$

- In case, when the probability density functions $f_X(x)$ and $f_Y(y)$ for any single random variable are obtained from joint PDF, then $f_X(x)$ and $f_Y(y)$ are called marginal PDF or simply marginal densities.
- Out of the two random variables, one variable may take a fixed value. In this case, the PDF is called conditional.
- There are some other measures or numbers which give more useful and quick information about the random variable. Collectively these characteristic numbers or measures are known as **statistical averages.**
- The mean or average of any random variable is expressed by the summation of the values of random variables X weighted by their probabilities. Mean value of a random variable is denoted by m_x. Mean value is also known as expected value of random variable X.

$$m_x = E(X)$$

  where E[ ] represents expection operator.
- Also, in general term, mean or average is given by

$$\text{Mean or average of X} = \frac{\text{Arithmetic sum of all values of X}}{\text{Total number of values of X}}$$

- Mean value of Discrete random variable

$$m_x = E(X) = \bar{X} = \sum_{i=1}^{n} x_i . P(x_i)$$

  Here $\bar{X}$ is also the notation for mean values.
- Continuous random variable X, the mean or average value is expressed as

$$m_x = \int_{-\infty}^{\infty} x f_X(x)\, dx$$

  where $f_X(x)$ = Probability density function (PDF)
- The nth moment of any random variable X may be defined as the mean value of X^n.

  i.e. $\quad g(x) = X^n$
- Variance is generally represented by σ_x^2 i.e.,

$$\sigma_x^2 = \text{Variance (X)}$$

$$\sigma_x^2 = \text{Var (X)}$$

or

$$\sigma_x^2 = E[(X - m_x)^2]$$

or $$\sigma_x^2 = \int_{-\infty}^{\infty} (x - m_x)^2 f_X(x)\, dx\,.$$

- The square root of variance is known as **standard deviation** of a random variable X. Standard deviation gives the measure of speed observed over the values of X related to mean value. Therefore,

  $$\text{Standard deviation} = \sqrt{\text{Variance}} = \sqrt{\sigma_x^2} = \sigma_x\,.$$

- The Probability Density Function (PDF) of an uniform distribution is expressed as

  $$\text{Uniform PDF: } f_X(x) = \begin{cases} 0 \text{ for } x < m - \dfrac{A}{2} \text{ and } x > m + \dfrac{A}{2} \\ = \dfrac{1}{A} \quad \text{for } \left(m - \dfrac{A}{2}\right) \le x \le \left(m + \dfrac{A}{2}\right) \end{cases}$$

- The value of PDF, $f_X(x)$ is same for all possible values of a random variable. Thus, this distribution is known as **Uniform Distribution.**
- Gaussian Distribution is also known as Normal Distribution. It is defined for a continuous random variable. The PDF for a Gaussian random variable is expressed as
  Gaussian PDF

  $$f_X(x) = \frac{1}{\sigma\sqrt{2\pi}} e^{-(x-m)^2/2\sigma^2}$$

  In this equation m = mean value of the random variable
  σ^2 = variance of the random variable.
- Rayleigh Distribution is used for continuous random variables. It is produced from two Gaussian random variables. Let X and Y be independent Gaussian random variables having mean value zero and variance σ^2. Mathematically,

  $$m_x = m_y = 0$$

  and $$\sigma_x^2 = \sigma^2$$
- Rayleigh Distribution is always used for modelling of statistics of signals transmitted through radio channels such as cellular radio.
- Let there be a random experiment E having outcome λ from the sample space S. Thus, every-time an experiment is conducted, the outcome λ will be one of the sample point in sample space. If this outcome λ is associated with time, then a function of λ and time t is formed i.e., X(λ, t). Then the function X(λ, t) is known as **random process.**
- When the statistical averages are taken along the time they are known as time averages. As an example, we may define time mean value of a sample function x(t) as

  $$\text{Time mean value: } \langle m_x \rangle = \lim_{T \to \infty} \frac{1}{T} \int_{-T/2}^{T/2} X(t)\, dt$$

- The autocorrelation function may be expressed using time averaging as

  $$\langle R_\tau(\varepsilon) \rangle = \lim_{T \to \infty} \frac{1}{T} \int_{-T/2}^{T/2} x(t) \cdot x(t+\tau)\, dt$$

- A random process X(t) is called stationary if its statistics are not affected by any shift in the time origin.
- The process may not be stationary in strict sense, still the mean and autocorrelation functions are independent of shift of time region. Such type of process is known as wide sense stationary process.
- A random process is known as ergodic process if the time-averages are equal to ensemble averages. Hence for a ergodic process, we have

  $$m_x = \langle m_x \rangle$$

  $$R_\tau(t_1, t_2) = \langle R_\tau(\varepsilon) \rangle$$

- It may be noted that a Gaussian process has two main advantages:
  (i) Firstly, the Gaussian process has several properties which make analytic results possible.
  (ii) Secondly, the random process produced by physical phenomena are often such that a Gaussian model may appropriate.
- A Gaussian process has the following important properties:
  (i) If a Gaussian process X(t) is made to apply to a stable linear filter, then the random process Y(t) produced at the output of the filter is also Gaussian.
  (ii) Let us consider the samples of a random process $X(t_1), X(t_2), ..., X(t_n)$, obtained by observing a random process X(t) at time $t = t_1, t = t_2, ..., t = t_n$.
- Also, if a Gaussian process is wide-sense stationary (WSS), then the process is also stationary in the strict sense.
- The power spectral density of the sum of two uncorrelated WSS random processes is equal to the sum of their individual power spectral densities.
- The correlation function provides a measure of similarity or coherence between a given signal and a replica of the same signal or other signal by a variable amount.
- Autocorrelation function may be defined as a measure of similarity between a signal or process and its replica by a variable amount.
  The autocorrelation function of a stationary process X(t) may be defined as

$$R_{X(t_j - t_i)} = E[X(t_j)\, X(t_i)] \text{ for any } t_j \text{ and } t_i$$

  here $X(t_j)$ and $X(t_i)$ are the random variables obtained by ovserving the process X(t) at times t_j and t_i, respectively.
- The autocorrelation function of WSS process X(t) is always an even function of time-lag. Autocorrelation function $R_X(\tau)$ satistifies the following mathematical relationship, i.e.

$$R_X(\tau) = R_X(-\tau).$$

- The mean square value of a WSS process is equal to the autocorrelation function of the random process for zero time-lag. It may be written as

$$R_X(0) = R_X(\tau)\big|_{\tau=0} = E[X(t)\, X(t-\tau)]\big|_{\tau=0}$$

  or $$R_X(0) = E[X(t)\, X(t-0)] = E[X^2(t) = E[X^2(t)]$$

- The autocorrelation function of a WSS random process has the maximum magnitude at zero time-lag (τ = 0). This property states that

$$|R_X(\tau)| = \text{magnitude of } R_X(\tau) = R_X(0)$$

- Spectral densities are used to represent random process in frequency-domain. Frequency-domain methods are very powerful tools for analysis of various types of signals.
- Power spectral density $S_X(\omega)$ of a wide-sense stationary (WSS) random process X(t) may be defined as

$$S_X(\omega) = \text{Fourier transform } \{R_X(\tau)\} = \int_{-\infty}^{\infty} R_X(\tau)\, e^{-j\omega\tau} d\tau$$

  here $R_X(\tau)$ = Autocorrelation function of random process X(t).
- The cross power spectral density of two jointly WSS random processes X(t) and Y(t) may be defined as density of two jointly WSS random processes X(t) and Y(t) is defined as

$$S_{XY}(\omega) = \text{Fourier transform } \{R_{XY}(\tau)\} = \int_{-\infty}^{\infty} R_{XY}(\tau)\, e^{-j\omega\tau} d\tau$$

  here $R_{XY}(t)$ = Cross-correlation function of random processes X(t) and Y(t).
- The energy spectral density $\psi_X(\omega)$ may be defined as a measure of density of the energy contained in random process X(t) in Joules per Hertz. It may be noted that since the amplitude spectrum of a real-valued random process X(t) is an even function of ω, the energy spectral density of such a signal is symmetrical about the vertical axis passing through the origin.
  Thus, total energy of the random process X(t) is defined as

$$E = \frac{1}{2\pi} \int_{-\infty}^{\infty} \Psi_X(\omega)\, d\omega$$

GLOSSARY

1. **Experiment:** An experiment is defined as the process which is conducted to get some results. If the same experiment is performed repeatedly under the same conditions, similar results are expected. But there are few experiments which do not produce the same result as we have stated above. The theory of probability mainly deals with such type of experiments.
2. **Equally Likely Outcomes:** The outcomes of an event are called equally likely, if any one of them cannot be expected to occur in preference to another. As an example, the tossing of a coin results in two outcomes, Head and Tail. Both have same possibility of 50%. Such type of outcomes are called equally likely outcomes.
3. **Sample Space:** A set of all possible outcomes of an experiment or trial is called the sample space of that experiment.
4. **Event:** The expected subset of the sample space or happening is called an event.
5. **Probability:** Probability may be defined as the study of random experiments. In any random experiment, there is always an uncertainty that a particular event will occur or not.
6. **Random Variable:** A function which can take on any value from the sample space and its range is some set of real numbers is called a random variable of the experiment. Random variables are denoted by upper case letters such as X, Y etc. and the values taken by them are denoted by lower case letters with sub-scripts such as x_1, x_2, y_1, y_2 etc.
7. **Continuous Random:** A random variable that takes on an infinite number of values is called a continuous random variable. Actually, there are several physical system (experiments) that generate continuous out-puts or outcomes. Such systems generate infinte number of outputs or outcomes within the finite period. Continuous random variables may be used to define the outputs of such systems.
8. **Probability Density Function (PDF):** The derivative of cumulative distribution function (CDF) with respect to some dummy variable is known as Probability Density Function (PDF). Probability density function (PDF) is generally denoted by $f_X(x)$.

SHORT QUESTIONS WITH ANSWERS

Q.1. What are random signals? What is significance of random signals in probability theory?

Ans. There is one other class of signals, the behaviour of which cannot be predicted. Such type of signals are called random signals. These signals are called random signals because the precise value of these signals cannot be predicted in advance before they actually occur. The examples of random signals are the noise interferences in communication systems. This means that the noise interference during transmission is totally unpredictable. In the same way, the noise generated by the receiver itself is random. Even some other signals which are not noise signals are also random signals. These signals cannot be modelled mathematically. Actually the electro magnetic interference is the major source of random noise.

Q.2. What do you mean by Experiment in probability theory?

Ans. An experiment is defined as the process which is conducted to get some results. If the same experiment is performed repeatedly under the same conditions, similar results are expected. But there are few experiments which do not produce the same result as we have stated above. The theory of probability mainly deals with such type of experiments. In fact, the theory of probability is applied to such type of experiments to predict the possibility of a particular output. An experiment is sometimes called trial. As an example, throw of a coin is an experiment or trial. This trial results in two outcomes namely Head and Tail.

Q.3. What is sample space?

Ans. A set of all possible outcomes of an experiment or trial is called the sample space of that experiment. It is generally denoted by 'S'. The total number of outcomes in a sample space is denoted by n(s). As an example, the tossing of a coin has two outcomes. Hence, we may write its sample space as:

$$S = \{H, T\}$$

where, $H \rightarrow$ Head and $T \rightarrow$ Tail

Q.4. What is an probability theory?

Ans. The expected subset of the sample space or happening is called an event. As an example, let us consider an experiment of throwing a cubic die. In this case, the sample space S will be as

$$S = \{1, 2, 3, 4, 5, 6\}$$

Now, if we want the number '3' to be an outcome or an even number, i.e., {2, 4, 6}, then this subset is called an event. This is denoted by letter 'E'. Hence event E is a subset of the sample space 'S'. If event E has only one outcome, then it is called an elementary event. On the other hand, if event E does not contain any out come, then it is called a null event. If E = S, then an event contains all the outcomes. Such as event is called a certain event. It always occurs, no matter what so ever is the outcome.

Q.5. What do you mean by probability? Explain.

Ans. Probability may be defined as the study of random experiments. In any random experiment, there is always an uncertainty that a particular event will occur or not. As a measure of probability of occurrence of an event, a number between 0 to 1 is assigned. If it is sure that an event will occur, then we can say that its probability is 100% or 1. If it is not sure that an event will occur, then we can say that its probability is 0% or 0. If it is not sure whether the event will occur or not, then its probability is between 0 and 1.

Q.6. Write properties of probability? Explain.

Ans. Property 1: The probability of a certain event is unity i.e.,

$$P(A) = 1$$

Property 2: The probability of any event is always less than or equal to 1 and non-negative. Mathematically,

$$0 \le P(A) \le 1$$

Property 3: If A and B are two mutually exclusive events, then

$$P(A + B) = P(A) + P(B)$$

Property 4: If A is any event, then the probability of not happening of A is

$$P(\bar{A}) = 1 - P(A)$$

where $\bar{A}$ represents the complement of event A.

Property 5: If A and B are any two events (not mutually exclusive events), then

$$P(A + B) = P(A) + P(B) - P(AB)$$

where P (AB) is called the probability of events A and B both occurring simultaneously. Such an event is called joint event of A and B, and the probability P (AB) is called the joint probability.

Now, if events A and B are mutually exclusive, then the joint probability, P(AB) = 0.

Q.7. What is conditional probability?

Ans. The concept of conditional probability is used in conditional occurrences of the events. Let us consider an experiment which involves two events A and B. Now the probability of event B, given that event A has occurred, is represented by P (B/A). Similarly, P (A/B) represents probability of event A given that event B has already occurred. Therefore, P (B/A) and P (A/B) are called conditional probabilities.

Q.8. Explain random variable?

Ans. A function which can take on any value from the sample space and its range is some set of real numbers is called a random variable of the experiment. Random variables are denoted by upper case letters such as X, Y etc. and the values taken by them are denoted by lower case letters with subscripts such as x_1, x_2, y_1, y_2 etc.

Random variables may be classified as under:

1. Discrete random variables
2. Continuous random variables.

Q.9. What is a continuous random variable?

Ans. A random variable that takes on an infinite number of values is called a continuous random variable. Actually, there are several physical system (experiments) that generate continuous outputs or outcomes. Such systems generate infinte number of outputs or outcomes within the finite period. Continuous random variables may be used to define the outputs of such systems.

As an example, the noise voltage generated by an electronic amplifier has a continuous amplitude. This means that sample space S of the noise voltage amplitude is continuous. Therefore, in this case, the random variable X has a continuous range of values.

Q.10. What do you mean by probability density function (PDF)?

Ans. The derivative of cumulative distributive function (CDF) with respect to some dummy variable is known as Probability Density Function (PDF). Probability density function (PDF) is generally denoted by $f_X(x)$. Mathematically, PDF may be expressed as–

$$\text{PDF: } f_X(x) = \frac{d}{dx} F_X(x)$$

where x is a dummy variable.

Probability density function (PDF) is the more convenient representation for continuous random variable.

Q.11. What do you mean by conditional probability density function?

Ans. Out of the two random variables, one variable may take a fixed value. In this case, the PDF is called conditional. As an example, out of the two continuous random variables X and Y, let X = x.

Then we may find the conditional, PDF of Y given that X = x as,

$$f_Y(y/x) = \frac{f_{XY}(x, y)}{f_X(x)}$$

where $f_X(x)$ is the marginal density of random variable X.

Similarly, we may find the conditional PDF of X given that Y = y. Then

$$f_X(x/y) = \frac{f_{XY}(x, y)}{f_Y(y)}$$

Q.12. What is mean or average?

Ans. The mean or average of any random variable is expressed by the summation of the values of random variables X weighted by their probabilities. Mean value of a random variable is denoted by m_x.

Mean value is also known as expected value of random variable X.

$$m_x = E(X)$$

where E[ ] represents expection operator.

Therefore both m_x and E (X) have the same meaning.

Also, in general term, mean or average is given by

$$\text{Mean or average of} \quad X = \frac{\text{Arithmetic sum of all values of X}}{\text{Total number of values of X}}$$

Q.13. Explain uniform distribution?

Ans. The Probability Density Function (PDF) of an uniform distribution is expressed as

$$\text{Uniform PDF: } f_X(x) = \begin{cases} 0 & \text{for } x < m - \frac{A}{2} \ \& \ x > m + \frac{A}{2} \\ \frac{1}{A} & \text{for } \left(m - \frac{A}{2}\right) \le x \le \left(m + \frac{A}{2}\right) \end{cases}$$

Figure 4.15 shows the PDF of an uniformly distributed random variable.

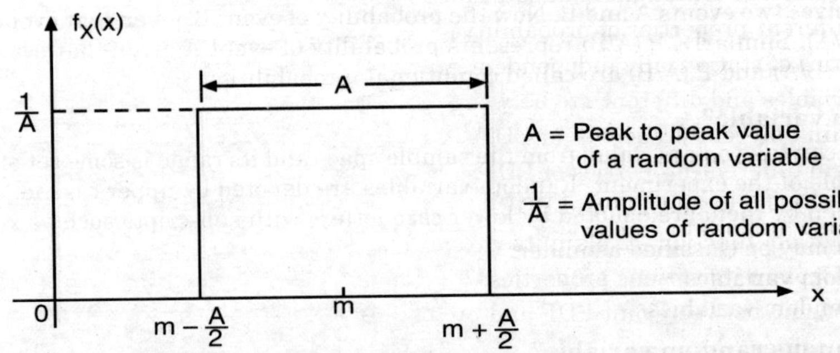

Fig. 4.15. PDF of an uniformly distributed random variable

In above figure, the peak to peak value is A and amplitude is uniform i.e. A. From figure 4.15, it is also clear that the random variable and its PDF are continuous. Therefore uniform distribution is utilised for continuous random variable.

Also, the value of PDF, $f_X(x)$ is same for all possible values of a random variable. Thus, this distribution is known as Uniform Distribution.

Q.14. What is importance of Gaussian distribution?

Ans. The Gaussian distribution is used for continuous random variables. We know that the random motion of the thermally agitated electrons produces thermal noise. This thermal noise has Gaussian distribution. The random errors in the experimental measurements create the measured values to have Gaussian distribution about the true value. The Gaussian distribution is very important in the analysis of communication and statistical systems.

Q.15. What do you mean by random process? Explain.

Ans. Let there be a random experiment E having outcome λ from the sample space S. Thus every time an experiment is conducted, the outcome λ will be one of the sample point in sample space. If this outcome λ is associated with time, then a function of λ and time t is formed i.e., $X(\lambda, t)$. Then the function $X(\lambda, t)$ is known as random process. Hence when any random experiment E is given a time dimension, then each outcome appears at some certain time and the random experiment will be converted to Random Process. A random process is the function of two variables λ and t.

Q.16. What is Ergodic process? Explain.

Ans. A random process is known as ergodic process if the time-averages are equal to ensemble averages. Hence for a ergodic process, we have

$$m_x = \langle m_x \rangle$$

$$R_\tau (t_1, t_2) = \langle R_\tau (\varepsilon) \rangle$$

For any ergodic process, all ensemble averages will be equal to corresponding time averages for any particular sample function. The time averages are not a function of time. When time and ensemble averages are the same, it implies that ensemble averages also are not a function of time.

Thus an ergodic process is always stationary but converse in not true.

Ergodicity of the process may be defined in terms of some statistical averages like mean and autocorrelation.

The random process is ergodic in the mean if

$$m_x = \langle m_x \rangle$$

and variance of $\langle m_x \rangle \to 0$ as $T \to \infty$

Similarly, the random process is ergodic in the autocorrelation if

$$R_\tau (t_1, t_2) = \langle R_\tau (\varepsilon) \rangle$$

and variance of $\langle R_\tau (\varepsilon) \rangle \to 0$ as $T \to \infty$

Hence, the Ergodicity of any random process may be determined by evaluating statistical averages of single sample function. This means that a single sample function represents entire random process.

REVIEW QUESTIONS

1. Define probability. Explain conditional and joint probabilities.
2. State and prove different properties of probability.
3. How the probabilities of statistically independent events are related with each other.
4. Define random variables and differentiate between discrete and continuous random variables.
5. Define CDF. Explain different properties of CDF.
6. Define PDF. Explain different properties of PDF.
7. Derive the relationship between CDF and PDF.
8. Define joint distribution function. Explain its various properties.
9. Define joint PDF. Explain its various properties.
10. Explain the relationship between joint PDF and probability.
11. Define marginal densities.
12. Explain conditional PDF and its properties.
13. Define mean or average or expected value of a random variable. Explain how it is calculated for continuous and discrete random variables.
14. Explain variance and standard deviation of a random variable.
15. Define random process and explain ensemble and sample function.
16. Explain ensemble and time averages.
17. Explain the stationary random process with the help of one example.
18. Define ergodic process. Explain the difference between ergodic process and stationary process.

NUMERICAL PROBLEMS

1. A wheel of chance is divided into three equal segments coloured Green (G), Red (R) and Yellow (Y). The wheel is spinned twice and the outcome is taken as resulting colour sequence. That is GR, RG etc. Let A be

the event which represents "neither colour is Yellow". Let B be the event which represents "matching colours". Calculate P(A), P(B), P(AB) and P(A + B).

[Ans. $P(A) = \frac{4}{9}$, $P(B) = \frac{1}{3}$, $P(AB) = \frac{2}{9}$ and $P(A + B) = \frac{5}{9}$]

2. Three honest coins are tossed simultaneously and the resulting sequence of heads and tails is observed. Let A be the event in which "first two tosses match." Let B be the event representing "exactly two heads". Find out P(A), P(B), P(B/A) and P(A/B).

[Ans. $P(A) = \frac{1}{2}$, $P(B) = \frac{3}{8}$, $P(B/A) = \frac{1}{4}$ and $P(A/B) = \frac{1}{3}$]

3. A long binary message contains 1428 binary 1's and 2668 binary 's. What is the probability of obtaining binary 1 in any received bit? **[Ans.** P(A = 1) = 0.5486]

4. Which of the following functions satisfy the properties of PDF and Why? Hence state whether the function is PDF or not.

(a) $f_X(x) = \frac{1}{\pi}\left(\frac{1}{1+x^2}\right)$

(b) $f_X(x) = |x|$ for $|x| < 1$
$= 0$ otherwise

(c) $f_X(x) = \frac{1}{6}(8 - x)$ for $4 \le x \le 10$
$= 0$ otherwise

[Hint: For all above function apply $\int_{-\infty}^{\infty} f_X(x)\,dx = 1$ If this property holds, then it is PDF].

[Ans. (a) This is valid PDF, (b) This is valid PDF, (c) This is not PDF]

5. The PDF of a random variable is given as

$f_X(x) = Ke^{-bx}$ for $x \ge 0$
$= 0$ for $x < 0$ and $K, b > 0$

(i) Find the value of K in terms of b.
(ii) Find m_x and σ^2_x.

6. The PDF of a random variable 'X' is given as

$f_X(x) = \frac{1}{A^2}(x + A)$ for $x < 0$

$= \frac{1}{A^2}(-x + A)$ for $x \ge 0$

Find out the mathematical expression for CDF.

[Ans. $f_X(x) = \frac{1}{2} + \frac{1}{A}x + \frac{1}{2A^2}x^2$ for $-A \ge x < 0$ $\frac{1}{2} + \frac{1}{A}x - \frac{1}{2A^2}x^2$ for $0 \le x < A$]

7. A class of modulated signals are modelled by

$x(t) = A X(t) \cos(\omega_c t + \theta)$

Here X(t) is the message signal and A cos ($\omega_c t + \theta$) is the carrier. The message signal X(t) is a zero mean stationary random process. The carrier amplitude A and frequency ω_c are constants. The carrier phase θ is a random variable uniformly distributed in the interval (–π, π) X(t) and θ are independent find out mean and autocorrelation function of x(t) and state whether the process is stationary or not.

[Ans. Mean of x(t) = 0 Auto correlation function $R_\tau(t_1, t_2) = \frac{A^2}{2} E[X(t_1).X(t_2)].\cos \omega_c(t_1 - t_2)$

$= \frac{A^2}{2} R_\tau(t_1 - t_2) . \cos \omega_c(t_1 - t_2)$]

Since mean and $R_\tau(t_1, t_2)$ does not depend on time 't' the process x(t) is stationary in wide sense

□□□

UNIT—III

Performance Analysis of Digital communication system; Optimum linear Detector for Binary polar signaling, General Binary Signaling, Coherent Receivers for Digital Carrier Modulations, Signal Space Analysis of Optimum Detection, vector Decomposition of White Noise, Random Processes, General Expression for Error Probability of optimum receivers.

Above Syllabus has been covered in following chapter:

CHAPTER—5 : Performance Analysis of Digital Communication System

CHAPTER 5

PERFORMANCE ANALYSIS OF DIGITAL COMMUNICATION SYSTEM

SYLLABUS

Performance Analysis of Digital communication system; Optimum linear Detector for Binary polar signaling, General Binary Signaling, Coherent Receivers for Digital Carrier Modulations, Signal Space Analyhsis of Optimum Detection, vector Decomposition of White Noise Random Processes, General Expression for Error Probability of optimum receivers.

Inside this Chapter

5.1 INTRODUCTION

In analog communications, the user objective is to achieve high fidelity in waveform reproduction; Hence, the suitable performance criterion is the output signal-to-noise ratio. The choice of this criterion indicates that the signal-to-noise ratio reflects the quality of the message and is related to the ability of a listener to interpret a message.

In digital communication system, the transmitter input is chosen from a finite set of possible symbols. The objective at the receiver is not to reproduce the waveform that carries the symbol with fidelity; instead, the receiver aims to accurately determine which particular symbol was transmitted among the set of possible ones. Because each symbol is represented by a particular waveform at the transmitter, our goal is to decide, from the noisy received signal, which particular waveform was originally transmitted. Logically, the appropriate figure of merit in a digital communication system is the probability of error in this decision at the receiver. In particular, the probability of bit error, also known as the bit error rate (BER), is a direct quality measure of the communication system. Not only is the BER important to digital signal sources, it is also directly related to the quality of signal reproduction for analog signal source.

The communication system which you have studied till now (analog or digital) have been compared, based on the points such as bandwidth requirement, ease of multiplexing, ease of implementation etc. But, we have not yet compared these systems based on their performance in presence of noise.

Definition of noise

When the signal voltage arrives at the demodulator of a receiver, it will be accompanied by a voltage waveform which varies with time in an entirely unpredictable manner. This unpredictable voltage waveform is a random process called "Noise". The signal accompanied by noise is said to be contaminated or corrupted by noise. Noise is an unwanted electrical disturbance which gives rise to audible or visual disturbances in communication system and errors in the digital communication.

5.1.1 Source of Noise

Noise can arise from different types of sources. The noise sources are:

(i) Natural sources
(ii) Man made sources
(iii) Fundamental or internal sources.

(i) Natural sources of noise

The natural phenomena that give rise to noise are solar flares, electronic storms and radiation in space. The noise received by the receiving antenna from the natural sources can be reduced by repositioning the antenna (proper orientation of the antenna). Noise originating from the sun and outer space is known as extraterrestrial noise. The extraterrestrial noise can be sub-divided into two groups as solar noise and cosmic noise. Our sun being a large body at a very high temperature radiates a lot of noise.

The noise radiation from sun varies with temperature changes on the surface of the sun. These temperature changes follow a cycle of 11 years hence, the periods of great electrical disturbances also repeat after every 11 years. The cosmic noise comes from the stars. This is identical to the noise radiated by Sun because stars also are large hot bodies. This noise is called as black body noise and it is distributed uniformly over the entire sky. We also receive noise from the center of our galaxy, other galaxies and other virtual point sources such as Quasars and Pulsars.

(ii) Man made sources of noise

The man made noise is generated due to the make and break process in a current carrying circuit. The examples are: electrical motors, welding machines, ignition system of the automobiles, high current circuit using SCRs, fluorescent lamps etc.

(iii) Fundamental or Internal sources of noise

- The fundamental sources of noise are within the electronic equipment. They are called as fundamental sources because they are the integral part of the physical nature of the material used for making electronic components.

- Error Probability of BPSK (with Coherent DEtection)
- Bit Error Probability of Coherently Detected BFSK
- Bit Error Probability for a Non Coherently Detected Binary Orthogonal FSK
- Error Probability of DPSK System
- Probability of Error of QPSK System
- Comparison of Error Performance for Different Types of Binary Systems
- Symbol Error Performance for M-ary System (M > 2)
- Probability of Symbol Error for MFSK (Coherent Detection)
- Bit Error Probability Versus Symbol Error Probability
- Effects of Intersymbol Interference (ISI)
- Error Probability of Binary PCM
- Use of Signal Space to Calculate P_e
- Error Probability of 16 QAM (16 QASK)
- Vector Decomposition of White Noise Random Processes

- This type of noise follows certain rules. Therefore, it can be eliminated by properly designing the electronic circuits and equipments.

Types of noise

The fundamental noise sources produce different types of noise. They are as follows:

(i) Thermal noise
(ii) Shot noise
(iii) Partition noise
(iv) Low frequency or flicker, noise
(v) High frequency or transit time, noise.

(i) Thermal noise or Johnson noise

The free electrons within a conductor are always in random motion. This random motion is due to the thermal energy received by them. The distribution of these free electrons within a conductor at a given instant of time is not uniform. It is possible that an excess number of electrons may appear at one end or the other of the conductor. The average voltage resulting from this non-uniform distribution is zero but the average power is not zero. As this power results from the thermal energy, it is called as the thermal noise power.

Therefore, $$E_n = \sqrt{4\,k\,TBR}$$

This is the expression for the rms value of the thermal noise voltage.

(ii) Shot noise

The shot noise is produced due to shot effect. The shot noise is produced in all the amplifying devices rather in all the active devices. The shot noise is produced due to the random variations in the arrival of electrons (or holes) at the output electrons of an amplifying device. Therefore, it appears as a randomly varying noise current superimposed on the output. The shot noise sounds like a shower of lead shots falling on a metal sheet.

(iii) Partition noise

Partition noise is generated when a current gets divided between two or more paths. It is generated due to the random fluctuations in the current division. Therefore, the partition noise in a transistor will be higher than that in a diode. The devices like gallium arsenide FET draw almost zero gate bias current, hence, keeping the partition noise to its minimum value.

(iv) Low frequency or flicker noise

The flicker noise will appear at frequencies below a few kilohertz. It is sometimes called as "1/f" noise. In the semiconductor devices, the flicker noise is generated due to the fluctuations in the carrier density. These fluctuations in the carrier density will cause the fluctuations in the conductivity of the material. This will produce a fluctuating voltage drop when a direct current flows through a device. This fluctuating voltage is called as flicker noise voltage. The mean square value of flicker noise voltage is proportional to the square of direct current flowing through the device.

(v) High frequency or transit noise

If the time taken by an electron to travel from the emitter to the collector of a transistor becomes comparable to the period of the signal which is being amplified then, the transit time effect takes place. This effect is observed at very high frequencies. Due to the transit time effect, some of the carriers may diffuse back to the emitter. This gives rise to an input admittance, the conductance component of which increases with frequency. The minute currents induced in the input of the device by the random fluctuations in the output current, will create random noise at high frequency. Once this noise appears, it goes on increasing with frequency at a rate of 6 dB per octave.

5.1.2 White Gaussian Noise

In this chapter, let us discuss the effect of white noise on the performance of different communication systems. White noise is the noise whose power spectral density is uniform over the entire frequency range of interest, as shown in figure 5.1.

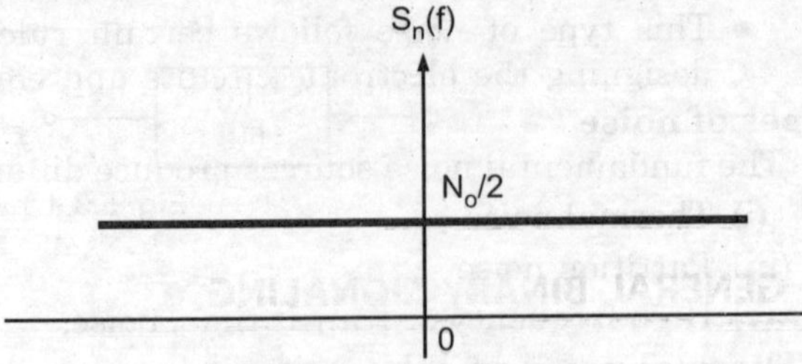

Fig. 5.1. Power spectral density of white noise

Why is it called as white noise?

The white noise contains all the frequency components in equal proportion. This is analogous with white light which is a superposition of all visible spectral components.

Why is it called as gaussian noise?

The white noise has a gaussian distribution. That means the PDF of white noise has the shape of gaussian PDF. Therefore it is called as gaussian noise.

Power spectral density of white noise:

As shown in figure 5.1, the power spectral density (psd) of a white noise is given by,

$$S_n(f) = \frac{N_0}{2} \qquad \text{... (5.1)}$$

This equation shows that the power spectral density of white noise is independent of frequency.

As N_0 is constant, the psd is uniform over the entire frequency range including the positive as well as the negative frequencies. N_0 in equation (5.1) is defined as:

$$N_o = K\, T_e \qquad \text{...(5.2)}$$

where, K = Boltzmann's constant and

T_e = Equivalent noise temperature of the system.

EXAMPLE 5.1. Derive and plot the auto correlation function of a white gaussian noise which has a power spectral density of $N_o/2$.

Solution: We know that the power spectral density and auto correlation function form Fourier transform pair. That means,

$$R(\tau) \xleftrightarrow{\text{F.T.}} S(f) \text{ i.e. FT } \{R(\tau)\} = S(f)$$

or $R(\tau) = \text{IFT}\{S(f)\}$

Therefore, for the white noise the autocorrelation function $R(\tau)$ is obtained as under:

$R(\tau) = \text{IFT}\{S_n(f)\} = \text{IFT}\{N_o/2\}$

Therefore, $R(\tau) = \frac{N_o}{2}\delta(t)$

This is the expression for the auto correlation function of white noise. The auto correlation function can be plotted as shown in figure 5.2.

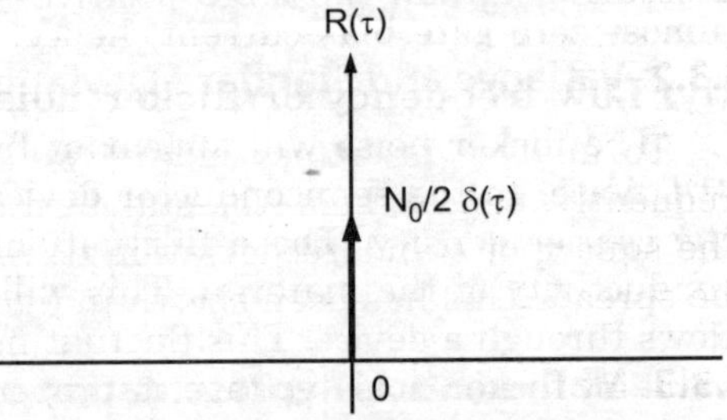

Fig. 5.2. Auto correlation function of white noise

> **Important Point:** As depicted in figure 5.2, the auto correlation function of white noise is a delta function weighted by a factor $(N_o/2)$ at $\tau = 0$. The meaning of this is that any two samples of white noise are totally uncorrected.

5.2 OPTIMUM LINEAR DETECTOR FOR BINARY POLAR SIGNALING

In the binary communication systems, the information is transmitted as either 0 or 1 in each time interval T. The binary polar signalling scheme is as shown in figure 5.3. The binary digits of 0 and 1 originating from the source are represented by $\pm x(t)$ respectively. These bits travel over a distortionless but noisy channel. So the received signal is represented as,

$$y(t) = \pm x(t) + n(t) \qquad \text{... } 0 \le t \le T$$

where n(t) = Gaussian channel noise.

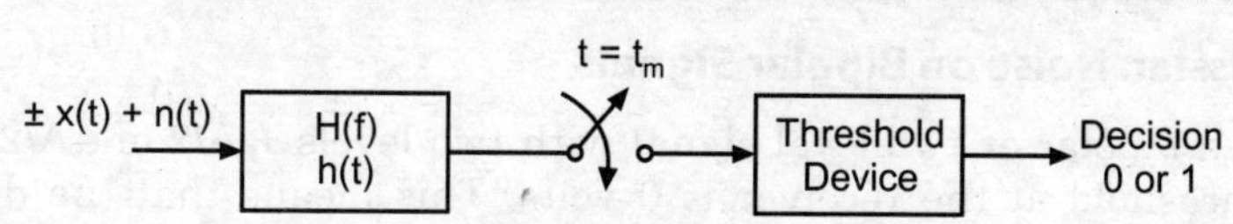

Fig. 5.3. Linear receiver

5.3 GENERAL BINARY SIGNALING

Noise entering into or contributed by a system results in errors in the detected signal. The number of errors or error rate is dependent on both the signal and noise amplitudes and the characteristics of the noise. Noise signal of large amplitude spikes will introduce errors in the detected signal. The noise getting added to the signal is assumed to be Gaussian noise.

5.3.1 Gaussian Noise and its PDF

In many applications, the noise is considered to be Gaussian eventhough in practice, it is not so. A noise is said to be Gaussian noise if it follows the Gaussian distribution. The familiar example of Gaussian noise is the thermal noise generated in a resistor. The probability density function (PDF) of Gaussian noise is as shown in figure 5.4

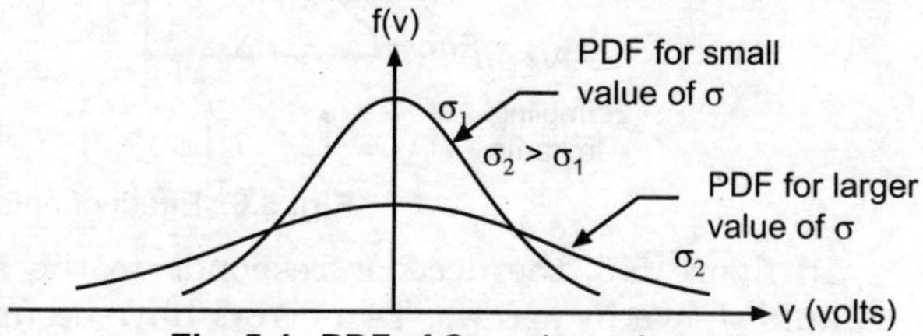

Fig. 5.4. PDF of Gaussian noise

The Gaussian noise PDF is of Bell shape as shown in figure 5.4 and the area under the curve shows a total probability of 1 or 100% The peak value is centered around zero value of v. This shows that the mean value or average value of the Gaussian noise is zero. The PDF shows that the larger values of v have less chance of occurrence, than the smaller values. This is because the PDF decreases with increase in the vlaues of v. The voltage near zero volt has a greater probability of occurrence than the large positive or negative voltages.

5.3.2 Variance and Standard Deviation

To analyze probable signals, we have to use the terms variance and standard deviation frequently. Out of them, the standard deviation σ is a measure of spread of the observed values. The square of standard deviation σ^2 is called as variance. The larger the variance, the larger is the spread. This has been shown in figure 5.4, where $\sigma_2 > \sigma_1$. Hence, spread of curve of σ_2 is more.

5.3.3 Mathematical Representation of PDF

The Gaussian PDF curves of figure 5.4 may be expressed mathematically as under:

$$f(v) = \frac{1}{\sqrt{2\pi}\sigma} e^{-v^2/2\sigma^2} \qquad \text{...(5.3)}$$

where $f(v)$ = Probability density of voltage v

σ = Standard deviation and

σ^2 = Variance

5.3.4 How to use the PDF curves?

To find the probability that the voltage v is greater than A volts, the shaded area of figure 5.5, must be computed. The area under the entire curve is 1 representing 100% probability. To get the shaded area, equation (5.3) must be integrated for v > A.

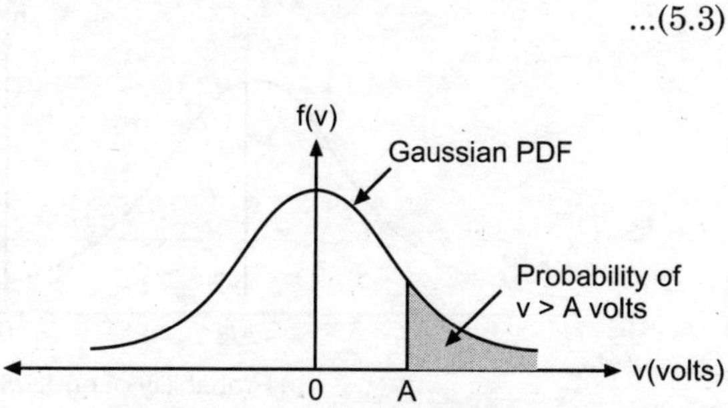

Fig. 5.5. Shaded area represents the probability that v > A

5.3.5 Effect of Gaussian Noise on Bipolar Signals

Let us consider a bipolar or two level signal with two levels + A/2 or –A/2 as shown in figure 5.6. The decision threshold at the receiver is 0 volts. This means that the decoder decides that a positive pulse is present if the instantaneous signal exceeds 0 volts, and a negative pulse is present if the signal drops below 0 volts. Both the levels are equiprobable i.e., they have the same probability of 0.5.

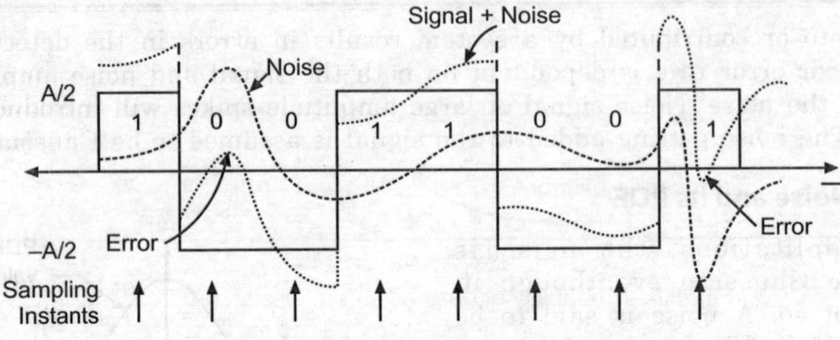

Fig. 5.6. Effect of noise on two level pulses

In figure 5.6, the decoder responds to the amplitude of signal plus noise at the sampling instants shown by arrows. Two errors have occured in this pulse train in the second and the last digit due to large amplitude of the noise.

5.3.6 Determination of Error Rate of Pulse System

Now, let us find out the error rate of the pulsed system, we have discussed so far. To do so, let us assume that the Gaussian noise is superimposed on the two levels, – A/2 and + A/2 as shown in figure 5.7(a).

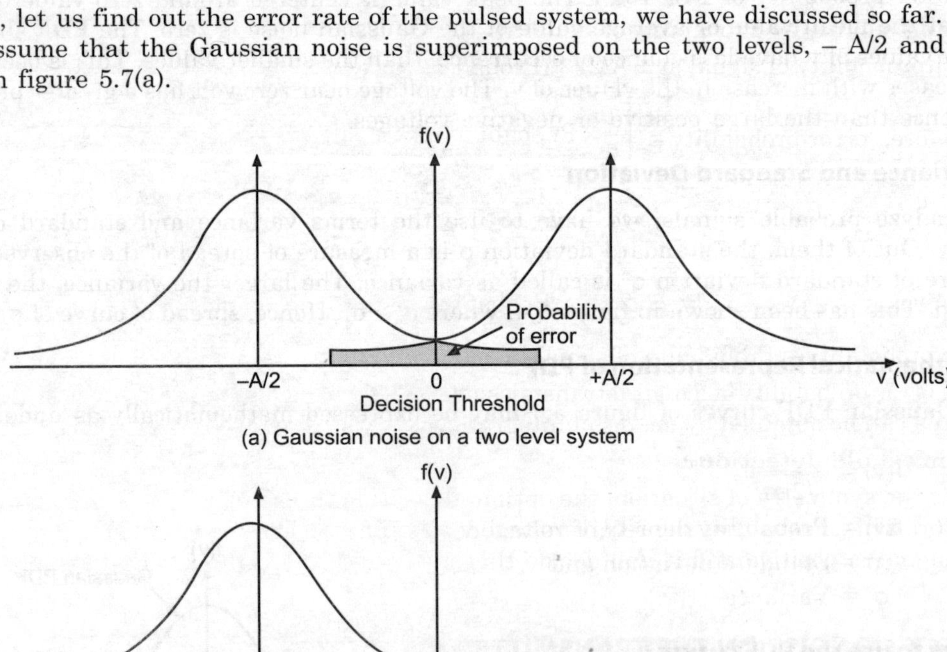

(a) Gaussian noise on a two level system

(b) Probability of error when signal is at –A/2 volt

Fig. 5.7.

Now, let us find the probability of error in the decision making at the decoder. When will the error occur? One situation is that the transmitted signal is A/2 volts but due to noise the received voltage exceeds the decision threshold level of 0 volts. The probability of error is represented by the shaded area in figure 5.7(b). The other situation in which an error can occur is when the transmitted voltage is +A/2 volts but the noise makes the received voltage less than 0 volts (below the decision threshold). The probability of error is represented by the shaded area in figure 5.8.

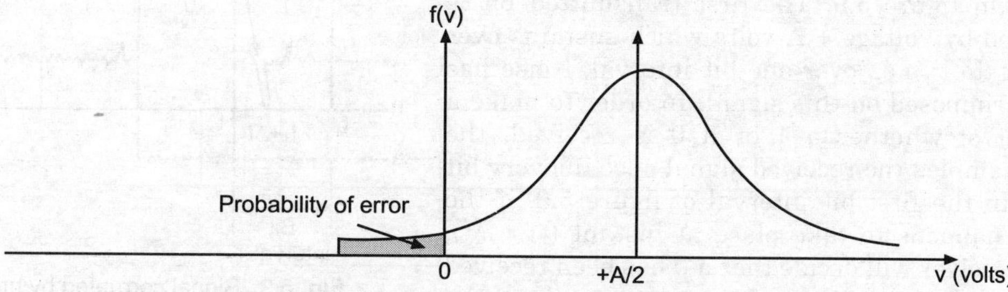

Fig. 5.8. Probability of error when signal is at + A/2 volt

5.3.7 Error probability

Both the signals i.e., –A/2 and + A/2 representing binary 0 and 1 respectively are equally likely i.e., they have equal probabilities equal to 1/2 and the Gaussian distribution also is symmetrical.

Therefore, the shaded areas shown in figure 5.7(b) and figure 5.8 are equal. The probability of error is given by:

Error probability = [(Probability of signal at –A/2)] × (Probability of signal in the shaded area for $v \geq 0$)] + [(probability of signal at + A/2) × (Probability of signal in the shaded area for $v \leq 0$)

$$\text{Therefore, error probability} = \left[\frac{1}{2} \times \text{Probability of signal in shaded area for } v \geq 0\right]$$

$$+ \left[\frac{1}{2} \times \text{Probability of signal in shaded area for } v \leq 0\right]$$

$$= 2 \times \frac{1}{2} \times \text{Probability of signal in shaded area} \qquad \text{...(5.4)}$$

Error probability = Probability of signal in shaded area.

Thus, error probability of a digital transmission can be obtained by getting the shaded area. The shaded area can be obtained by means of integration.

Binary threshold detection

Due to the symmetry of situation, the optimum decision threshold will be zt zero, as shown in figure 5.7(a). This means that the receivers will decide that a 1 is received if the sampled value of received signal is positive and it will decide that a 0 is received if the received sample signal is negative.

5.4 EFFECT OF NOISE ON THE TRANSMITTED SIGNAL

The data transmission systems used for the transmission of a sequence of binary digits, 0s and 1s as discussed earlier. These digits can be represented by different patterns such as unipolar RZ and NRZ, bipolar RZ and NRZ, AMI and Manchester. In general the binary digits are encoded in such a way that a 1 is represented by $x_1(t)$ and a 0 is represented by signal $x_2(t)$, where $x_1(t)$ and $x_2(t)$,where $x_1(t)$ and $x_2(t)$ each have a duration T. The resulting signal may be transmitted directly or used to modulate a carrier. The received signal is corrupted by noise. Hence, there is a probability that the receiver will make an error in deciding whether a 1 or a 0 was transmitted. In this chapter, let us make calculations of such error probabilities and discuss methods to minimize them. Consider a binary sequence 1 0 1 1 being transmitted. While travelling from the transmitter to receiver, noise gets added to it. Thus, the signal received by a receiver is corrupted by noise as shown in figure 5.9. The first transmitted bit is represented by voltage + A volts which sustains over the time t_1 to t_2, i.e., over one bit interval. Noise has been superimposed on this signal. In order to make a judgement of whether a 1 or a 0 is received, the receiver, samples the received signal once in every bit interval. In the first bit interval of figure 5.9. if the sampling happens to take place at instant $(t_1 + \Delta t)$, then the receiver will decide that a 0 has been received thus introducing an error. In order to reduce the probability of error the sampling instant in each interval should be selected in such a way that the signal amplitude is maximum at the instant of sampling.

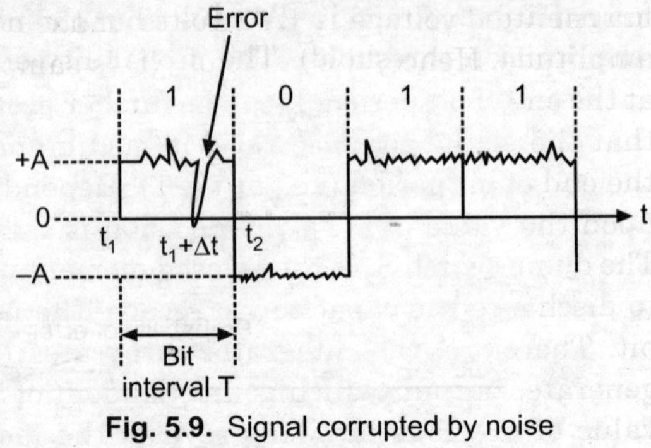

Fig. 5.9. Signal corrupted by noise

In the following sections, we shall discuss various types of detection methods to detect the baseband signals.

5.4.1 Requirements of Detection Technique

(i) A detection technique must have minimum probability of error.2.

(ii) It should sample (check) the received signal in every bit interval at the instant when the signal has maximum possible amplitude.

(iii) The detection method must maximize the signal to noise ratio, by amplifying the signal and attenuating the noise.

(iv) One of the detection techniques used is integrate and dump circuit.

5.4.2. Detection Techniques

(i) Integrate and dump receiver (filter) (ii) Optimum filter

(iii) Matched filter (iv) Correlator or coherent receiver

5.5 INTEGRATE AND DUMP FILTER (i.e., RECEIVER)

Let us consider a very simple and basic detector circuit for the detection of digital signals. Figure 5.10 illustrates the circuit of such type of detector which is called an integrate and dump filter.

Here, the digital signal x(t) is distorted by white noise n(t) during the transmission over channel. This noisy signal [x(t) + n(t)] is applied to the input of integrate and dump filter. The capacitor is discharged fully at the beginning of the bit interval. This is achieved by temporarily closing switch S_2 at the beginning of the bit interval. The integrator then integrates noisy input signal over one bit period. This integrated signal is shown as y(t) in the figure. For the square pulse input, the

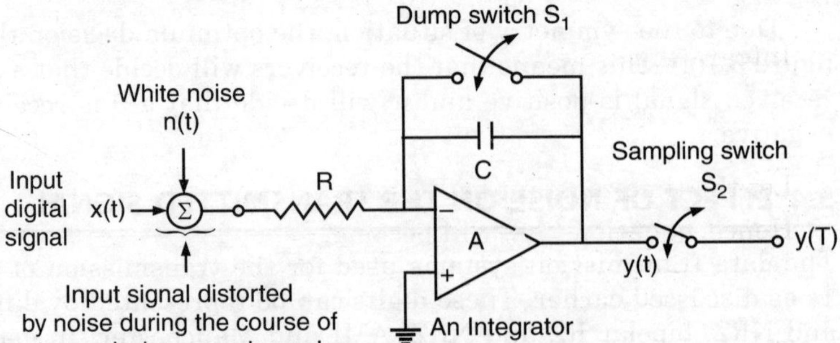

Fig. 5.10. Integrate and dump filter

output of the integrator would be a triangular pulse as shown in figure 5.10.

Figure 5.11(b) illustrates the waveform of y(t). At the end of bit period i.e., at t = T, the magnitude of y(t) attains its maximum amplitude. Hence, the value of y(t) is sampled at the end of bit period. It can be further proved that the signal to noise ratio is maximum at the end of bit period (i.e., at t = T). Depending upon the value of y(T), the decision is taken. The dump switch S_2 is then closed momentarily to discharge the capacitor to receive the next bit. Therefore, the integrator integrates (i.e. generates output) which is independent of the value of previous bit. This reveals the fact that the detection in integrate and dump filter is unaffected by the values of previous bits. In figure 5.11(b), the output of integrator would decrease after t > T.

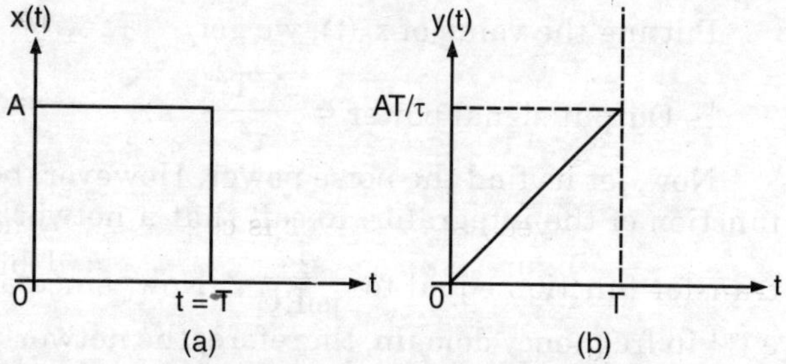

Fig. 5.11. (a) Input pulse to the integrator (Here, we have assumed that the noise is absent). This pulse represents binary '1'. The width of pulse is 'T'
(b) Output of integrator. The initial output is zero. At t = T, the output of integrator is y(t) = AT/τ.

5.5.1. Calculation of Signal to Noise Ratio for the Integrate and Dump Filter

We know that the output of the integrator may be written as,

$$y(t) = \frac{1}{RC}\int_0^T [x(t) + n(t)]\,dt \qquad \text{...(5.5)}$$

Here, the integration is performed over one bit period i.e., from 0 to T. The noisy signal [x(t) + n(t)] is the input to the integrator. Equation (5.5) may be modified as

$$y(t) = \frac{1}{RC}\int_0^T x(t)\,dt + \frac{1}{RC}\int_0^T n(t)\,dt \qquad \text{...(5.6)}$$

or $\qquad y(t) = x_0(t) + n_0(t)$

where $x_0(t)$ is the output signal voltage and $n_0(t)$ is the output noise voltage.
Now, let us consider output signal voltage which is

$$x_0(t) = \frac{1}{RC}\int_0^T x(t)\,dt \qquad \text{...(5.7)}$$

Because the value of x(t) = A from 0 to T, the above equation can be written as,

$$x_0(t) = \frac{1}{RC}\int_0^T A\,dt = \frac{A}{RC}\int_0^T 1\cdot dt$$

Integrating, we get

$$x(t) = \frac{A}{RC}[t]_0^T \quad \text{or} \quad x_0(t) = \frac{AT}{RC} \qquad \text{...(5.8)}$$

Let the time constant be RC = τ.

Hence, equation (5.8) will become,

$$x_0(t) = \frac{AT}{\tau} \qquad \text{...(5.9)}$$

The normalized signal power in standard resistance of 1 Ω would be,

$$\text{Output signal power} = \frac{x_0^2(t)}{1\Omega}$$

Putting the value of $x_0(t)$, we get

$$\text{Output signal power} = \frac{A^2T^2}{\tau^2} \qquad ...(5.10)$$

Now, let us find the noise power. However, before doing so, we have to calculate the transfer function of the integrator. Recall that a network which performs integration operation has the transfer function equal to $\frac{1}{j\omega RC}$.* Now, since a delay of t = T in time domain is equivalent to $e^{-j\omega T}$ in frequency domain, therefore, the network performing integration over the period of T can be represented by the transfer function as under:

$$H(f) = \frac{1 - e^{-j\omega T}}{j\omega RC} \qquad ...(5.11)$$

For $\omega = 2\pi f$ and $RC = \tau$, equation (5.11) takes the form

$$H(f) = \frac{1 - e^{-j2\pi fT}}{j2\pi f\tau}$$

Simplifying, we get

$$H(f) = \frac{1 - [\cos(2\pi fT) - j\sin(2\pi fT)]^{**}}{j2\pi f\tau}$$

Separating the real and imaginary parts, we have

$$H(f) = \frac{\sin(2\pi fT)}{2\pi f\tau} - j\frac{1 - \cos(2\pi fT)}{2\pi f\tau}$$

Then, the magnitude of above transfer function will become

$$|H(f)|^2 = \frac{\sin^2(2\pi fT) + 1 - 2\cos(2\pi fT) + \cos^2(2\pi fT)}{(2\pi f\tau)^2}$$

Simplifying the last equation, we obtain,

$$|H(f)|^2 = \frac{\sin^2(\pi fT)}{(\pi f\tau)^2} \qquad ...(5.12)$$

Now, the average power of the output noise signal $n_0(t)$ may be obtained by integrating its power density spectrum.

This means that

$$\text{Power, } P = \int_{-\infty}^{\infty} S(f)\,df$$

For a standard 1 Ω resistance, the noise power would be $\frac{\overline{n_0^2(t)}}{1\,\Omega} = \overline{n_0^2(t)}$.

Here, mean square value of noise signal is considered since it is a random signal. i.e.,

$$\text{Noise power, } \quad \overline{n_0^2(t)} = \int_{-\infty}^{\infty} S_{no}(f)\,df \qquad ...(5.13)$$

Again, we know that the input and output power spectral densities are related as,

$$S_{no}(f) = |H(f)|^2 S_{ni}(f) \qquad ...(5.14)$$

where H(f) = transfer function of filter,

* In fact, this can be very easily verified by taking the example of an circuit. The reason is that an RC low pass filter also basically performs integration operation.

**Using Euler's identity $e^{j\theta} = \cos\theta + j\sin\theta$.

$S_{no}(f)$ = power spectral density (psd) of output noise

$S_{ni}(f)$ = power spectral density (psd) of input noise.

Again, let us assume that white noise is present, therefore, the power spectral density (psd) of this noise would be,

$$S_{ni}(f) = \frac{N_0}{2} \quad ...(5.15)$$

Substituting this value in equation (5.14), we get

$$S_{no}(f) = |H(f)|^2 \cdot \frac{N_0}{2}$$

Substituting this value in equation (5.13), we have

$$\overline{n_0^2(t)} = \int_{-\infty}^{\infty} |H(f)|^2 \cdot \frac{N_0}{2} df$$

Substituting the value of $|H(f)|^2$ from equation (5.12) in above equation, we obtain

$$\overline{n_0^2(t)} = \int_{-\infty}^{\infty} \frac{\sin^2(\pi f T)}{(\pi f \tau)^2} \cdot \frac{N_0}{2} df = \frac{N_0}{2} \int_{-\infty}^{\infty} \frac{\sin^2(\pi f T)}{(\pi f \tau)^2} df \quad ...(5.16)$$

Now, let us substitute $\pi f \tau = x$

so that $\quad dx = \pi\tau\, df \quad$ or $\quad df = \frac{1}{\pi\tau} dx$

Further, since $\quad f = \frac{x}{\pi\tau}$

therefore, $\quad \pi f T = \frac{xT}{\tau}$

With all these substitutions, equation (5.16) becomes,

$$\overline{n_0^2(t)} = \frac{N_0}{2} \int_{-\infty}^{\infty} \frac{\sin^2\left(\frac{xT}{\tau}\right)}{x^2} \cdot \frac{1}{\pi\tau} dx$$

Rearranging above equation, we get

$$\overline{n_0^2(t)} = \frac{N_0}{2} \int_{-\infty}^{\infty} \frac{\sin^2\left(\frac{xT}{\tau}\right)}{\left(\frac{xT}{\tau}\right)} \cdot \left(\frac{T}{\tau}\right)^2 \cdot \frac{1}{\pi\tau} dx = \frac{N_0}{2} \cdot \frac{T^2}{\pi\tau^3} \int_{-\infty}^{\infty} \frac{\sin^2\left(\frac{xT}{\tau}\right)}{\left(\frac{xT}{\tau}\right)} \cdot dx$$

Now, let $\quad \frac{xT}{\tau} = u$

so that $dx = \frac{\tau}{T} du$ and limits will remain unchanged, therefore, above equation will become,

$$\overline{n_0^2(t)} = \frac{N_0}{2} \cdot \frac{T^2}{\pi\tau^3} \int_{-\infty}^{\infty} \frac{\sin^2 u}{u^2} \cdot \frac{\tau}{T} du = \frac{N_0 T}{2\pi\tau^2} \int_{-\infty}^{\infty} \left(\frac{\sin u}{u}\right)^2 du$$

Because, here the function $\frac{\sin u}{u}$ is squared, the above equation can be written as,

$$\overline{n_0^2(t)} = \frac{N_0 T}{2\pi\tau^2} \cdot 2\int_{0}^{\infty} \left(\frac{\sin u}{u}\right)^2 du = \frac{N_0 T}{2\pi\tau^2} \cdot 2 \cdot \frac{\pi}{2} = \frac{N_0 T}{2\tau^2} \quad ...(5.17)$$

This relation describes the noise power at the output. The signal to noise power ratio at output of integrator can be obtained as,

$$\text{Signal to noise ratio, } \left(\frac{S}{N}\right)_0 = \frac{\text{Signal power}}{\text{Noise power}}$$

Substituting the values of signal power from equation (5.10) and noise power from equation (5.17), we get

$$\left(\frac{S}{N}\right)_0 = \frac{\frac{A^2T^2}{\tau^2}}{\frac{N_0 T}{2\tau^2}} = \frac{2A^2T}{N_0} = \frac{A^2T}{N_0/2}$$

Hence, we conclude Signal to Noise ratio of Integrate and Dump receiver:

$$\left(\frac{S}{N}\right)_0 = \frac{A^2T}{N_0/2} \qquad ...(5.18)$$

This signal to noise ratio is also known as **figure of Merit.**

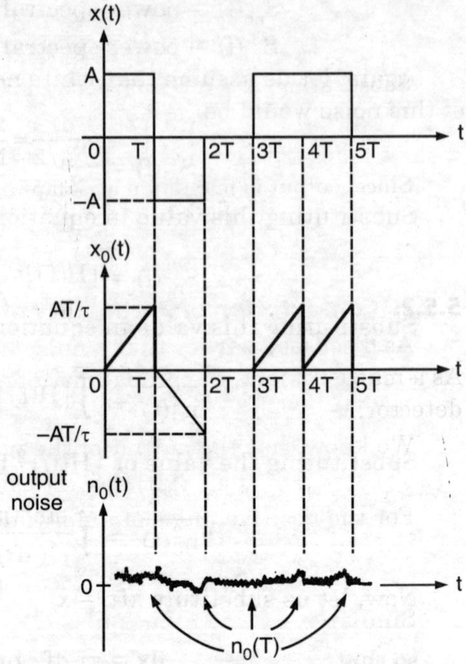

Fig. 5.12. (a) Input signal to the integrate and dump filter (b) Output signal of the integrate and dump filter (c) Output noise of the integrate and dump filter

Few points about Signal to noise ratio:

(i) The result in equation (5.18) shows that signal to noise ratio improves in proportion to sampling period T. It is also increased as signal amplitude 'A' is increased.

(ii) Because noise has gaussian distribution and zero mean value at any time, therefore, the output of integrator also increases by a very small margin at the end of bit interval. Figure 5.12 shows the waveform of $x_0(t)$ and $n_0(t)$ for the input signal x(t).

Figure 5.12 shows the waveforms of integrate and dump filter receiver for the rectangular pulses input.

It may observed that the output signal voltage reaches to the value of $\pm\frac{AT}{\tau}$ at the sampling instant. This is the maximum signal voltage. But the noise voltage $n_0(t)$ does not increase in same proportion. This is due to the fact that the noise has zero average value and Gaussian distribution.

EXAMPLE 5.2. Show that for an integrate and dump filter receiver, the maximum signal to noise ratio is expressed as,

$$\left(\frac{S}{N}\right)_0 = \frac{2E}{N_0}$$

Given that the input signal x(t) is rectangular pulses of amplitudes ± A and duration T.

Solution: We know that the energy of the signal x(t) is given by the relation

$$E = \int_{-\infty}^{\infty} x^2(t)\, dt \qquad ...(i)$$

Given that x(t) is a rectangular pulse of amplitude ± A and duration, T_1. Therefore, equation (i) can be written as

$$E = \int_0^T (\pm A)^2\, dt = A^2 \int_0^T dt = A^2[t]_0^T = A^2\, T \qquad ...(ii)$$

We know that the signal to noise ratio of integrate and dump filter receiver is given as,

$$\left(\frac{S}{N}\right)_0 = \frac{A^2T}{N_0/2} \qquad ...(iii)$$

Substituting, $A^2T = E$ from equation (ii) in equation (iii), we get

$$\left(\frac{S}{N}\right)_0 = \frac{E}{N_0/2} = \frac{2E}{N_0}$$

Since, output is maximum at sampling instant, T, therefore, the above value is maximum i.e.,

$$\left(\frac{S}{N}\right)_{0max} = \frac{2E}{N_0}$$

Hence Proved

5.5.2. Computation of Probability of Error for Integrate and Dump Filter Receiver

As discussed earlier that a noise interference may lead to wrong decision at the receiver end. As a matter of fact, the probability of error denoted by P_e is a good measure of performance of the detector.

We know that the output of the integrator is expressed as,

$$y(t) = x_0(t) + n_0(t)$$

For the positive pulse of amplitude A, $x_0(t)$ is given as,

$$x_0(t) = \frac{AT}{\tau} \quad \text{for} \quad x(t) = A$$

Similarly, for the input pulse of amplitude – A, $x_0(t)$ is given by,

$$x_0(t) = -\frac{AT}{\tau} \quad \text{for} \quad x(t) = -A \qquad ...(5.19)$$

Therefore, the output y(t) may be written as

$$y(t) = \frac{AT}{\tau} + n_0(t) \quad \text{for} \quad x(t) = A \qquad ...(5.20)$$

Similarly,

$$y(t) = -\frac{AT}{\tau} + n_0(t) \quad \text{for} \quad x(t) = -A \qquad ...(5.21)$$

Let us consider that x(t) = – A. Further, if noise $n_0(t)$ is greater than $\frac{AT}{\tau}$, then the output y(t) would be positive according to equation (5.20). After that the receiver will decide in favour of symbol + A, which is wrong decision. This means that an error is introduced. Similarly, let us consider that x(t) = + A. If noise $n_0(t) > -\frac{AT}{\tau}$ then, the output y(t) will be negative according to equation (5.21). This leads to decision in favour of – A, which is erroneous. Based on the above discussion, we can make conclusions about probability of error in the form of a Table 5.1.

Table 5.1. Probability of error in integrate and dump filter receiver

S. No.	Input x(t)	Value of $n_0(t)$ for error in the output	The probability of error P_e
1.	– A	An error will be introduced if $n_0(t) > \frac{AT}{\tau}$	In this case, the probability of error may be obtained by calculating probability that $n_0(t) > \frac{AT}{\tau}$
2.	+ A	An error will be introduced if $n_0(t) < -\frac{AT}{\tau}$	In this case, the probability of error may be obtained by calculating probability that, $n_0(t) < -\frac{AT}{\tau}$

As shown in Table 5.1, the error will be introduced depending upon probability that $n_0(t)$ takes a particular value. These probabilities may be obtained from PDF of $n_0(t)$. Recall that the probability density function (PDF) of the Gaussian distributed function is expressed by standard relation as under:

$$f_X(x) = \frac{1}{\sigma\sqrt{2\pi}} e^{-(x-m)^2/2\sigma^2} \quad ...(5.22)$$

where, $f_X(x)$ = the PDF of random function x.

m = the mean value

and σ = the standard deviation.

Here, because we have to evaluate PDF for white Gaussian noise, therefore, we have

$$x = n_0(t)$$

Since this noise has zero mean value i.e., m = 0, therefore equation (5.22) may be written as,

$$f_X\{n_0(t)\} = \frac{1}{\sigma\sqrt{2\pi}} e^{-[n_0(t)]^2/2\sigma^2} \quad ...(5.23)$$

The standard deviation σ is expressed as,

$$\sigma = [\text{mean square value} - \text{square of mean value}]^{1/2}$$

This means that $\sigma_x = \left[\overline{X^2} - m_x^2\right]^{1/2}$

Further, we have mean square value $\overline{X^2} = \overline{n_0^2(t)} = \frac{N_0 T}{2\tau^2}$

Also, mean value $m_x = 0$ for this noise.

Therefore, we have

$$\sigma = \left[\overline{n_0^2(t)}\right]^{1/2} = \sqrt{\frac{N_0 T}{2\tau^2}} \quad ...(5.24)$$

Hence equation (5.23), can be written as,

$$f_X\{n_0(t)\} = \frac{1}{\sqrt{\frac{N_0 T}{2\tau^2}}\sqrt{2\pi}} \cdot e^{-[n_0(t)]^2 \Big/ 2\left(\frac{N_0 T}{2\tau^2}\right)}$$

At this stage, it may be noted that $n_0(t)$ is the function like 'x'. It is a random variable and we are evaluating its PDF.

On simplifying above equation, we get

$$f_X\{n_0(t)\} = \frac{\tau}{\sqrt{\pi N_0 T}} e^{-[n_0(t)]^2 \Big/ \left(\frac{N_0 T}{\tau^2}\right)} \quad ...(5.25)$$

This equation describes PDF of white Gaussian noise.

Figure 5.13 shows the graphical representation of this PDF.

From the property of PDF, we know that,

$$P\left[n_0(t) > \frac{AT}{\tau}\right] = \int_{\frac{AT}{\tau}}^{\infty} f_X[n_0(t)]\, d[n_0(t)] \quad ...(5.26)$$

This equation gives the probability that $n_0(t)$ takes values greater than $\frac{AT}{\tau}$.

The above integration gives the area under the curve from $\frac{AT}{\tau}$ onwards. It has been illustrated by shaded region in figure. Similarly, the probability that $n_0(t)$ attains value less than $\frac{AT}{\tau}$ is given by area under the curve from $-\frac{AT}{\tau}$ onwards on left side. This portion has also been shown shaded in figure 5.13.

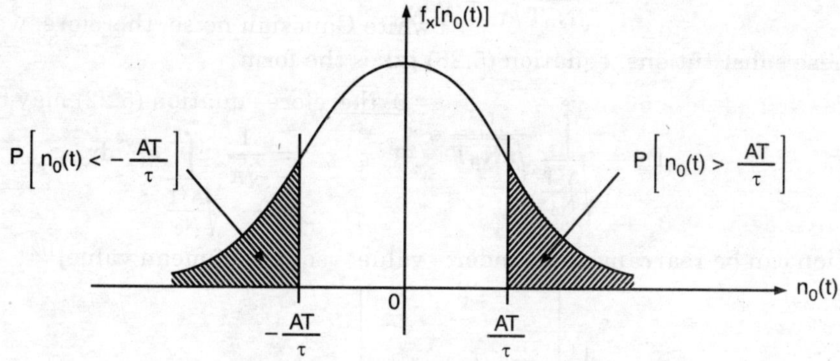

Fig. 5.13. Illustration of PDF of white Gaussian noise having zero mean

Since the PDF curve is symmetric, therefore, we can write.

$$P\left[n_0(t) > \frac{AT}{\tau}\right] = P\left[n_0(t) < -\frac{AT}{\tau}\right] \quad ...(5.27)$$

We know that the above probabilities represent error probability. Because, occurrence of – A or + A is mutually exclusive, therefore, the probability of error is given by either of the two in above equation i.e.,

$$P_e = P\left[n_0(t) > \frac{AT}{\tau}\right] = P\left[n_0(t) < -\frac{AT}{\tau}\right]$$

Substituting value of $P\left[n_0(t) > \frac{AT}{\tau}\right]$ in above equation from equation (5.26), we get

$$P_e = P\left[n_0(t) > \frac{AT}{\tau}\right] = \int_{\frac{AT}{\tau}}^{\infty} f_X[n_0(t)]\, d\,[n_0(t)]$$

Again, substituting value of $f_X[n_0(t)]$ from equation (5.25), in above equation, we get

$$P_e = \int_{\frac{AT}{\tau}}^{\infty} \frac{\tau}{\sqrt{\pi N_0 T}}\, e^{-[n_0(t)]^2 \Big/ \left(\frac{N_0 T}{\tau^2}\right)} d[n_0(t)] \quad ...(5.28)$$

Now, let us put

$$\frac{[n_0(t)]^2}{\frac{N_0 T}{\tau^2}} = y^2$$

so that $$\frac{n_0(t)}{\sqrt{N_0 T}/\tau} = y$$

Thus, $$d[n_0(t)] = \frac{\sqrt{N_0 T}}{\tau} dy$$

when $n_0(t) \rightarrow \infty$, then $y \rightarrow \infty$

when $$n_0(t) = \frac{AT}{\tau}$$

then $$y = \frac{AT/\tau}{\sqrt{N_0 T}/\tau} = \sqrt{\frac{A^2 T}{N_0}}$$

With all these substitutions, equation (5.28) takes the form

$$P_e = \int_{\sqrt{\frac{A^2T}{N_0}}}^{\infty} \frac{\tau}{\sqrt{\pi N_0 T}} e^{-y^2} \frac{\sqrt{N_0 T}}{\tau} dy = \frac{1}{\sqrt{\pi}} \int_{\sqrt{\frac{A^2T}{N_0}}}^{\infty} e^{-y^2} dy$$

This equation can be rearranged as under:

$$P_e = \frac{1}{2} \cdot \left[\frac{2}{\sqrt{\pi}} \int_{\sqrt{\frac{A^2T}{N_0}}}^{\infty} e^{-y^2} dy \right] \quad ...(5.29)$$

The integration inside brackets may be evaluated with the help of **complementary error function** i.e.,

$$\frac{2}{\sqrt{\pi}} \int_u^{\infty} e^{-y^2} dy = \text{erfc}(u)$$

Recall that this is a standard result and generally evaluated using numerical methods. Thus, with the help of this definition, equation (5.29) becomes,

$$P_e = \frac{1}{2} \text{erfc} \sqrt{\frac{A^2T}{N_0}} \quad ...(5.30)$$

This equation describes the probability of error P_e of the integrate and dump filter receiver. Now, since $A^2T = E$, i.e., energy of the bit, therefore, we have

$$P_e = \frac{1}{2} \text{erfc} \sqrt{\frac{E}{N_0}} \quad ...(5.31)$$

Important Point: *It may be noted that basically 'erfc' is the monotonically decreasing function. Therefore, P_e falls rapidly as the ratio $\frac{E}{N_0}$ increases. Hence, the maximum value of P_e is $\frac{1}{2}$ when $\frac{E}{N_0}$ is very very small. This means that even if the signal is lost entirely in the noise even then, the probability of error P_e will be $\frac{1}{2}$. Thus, the receiver would make incorrect decisions for half number of times.*

A man is better known by the company he keeps. – **Anonymous**

5.6 THE OPTIMUM FILTER (i.e. OPTIMUM RECEIVER)

(Important)

In the previous article, we discussed Integrate and Dump receiver. But, at this point, let us think whether Integrate and dump filter is an optimum* filter for the purpose of minimizing the probability of errors P_e. For this purpose, we shall discuss a generalized filter to receive binary coded signals. It is known as **optimum filter**. Therefore, let us consider the generalized Gaussian noise which is having zero mean. In the last section, we assumed 'white' Gaussian noise of zero mean.

Let the received signal be a binary waveform. Let us again assume that the polar NRZ signal is used to represent binary 1's and 0's. i.e.,

for binary '1',

$$x_1(t) = +A \quad \text{for one bit period T}$$

and for binary '0',

$$x_2(t) = -A \quad \text{for one bit period T}$$

Hence, the input signal x(t) will be either $x_1(t)$ or $x_2(t)$ depending upon the polarity of the NRZ signal. Figure 5.14 shows the block diagram of a receiver for such a binary coded signal.

As shown in this figure, the noise n(t) is added to the signal $x(t)$ in the channel during the transmission.

Thus, input to the optimum filter is [x(t) + n(t)] i.e.,

Input to the receiver = x(t) + n(t)

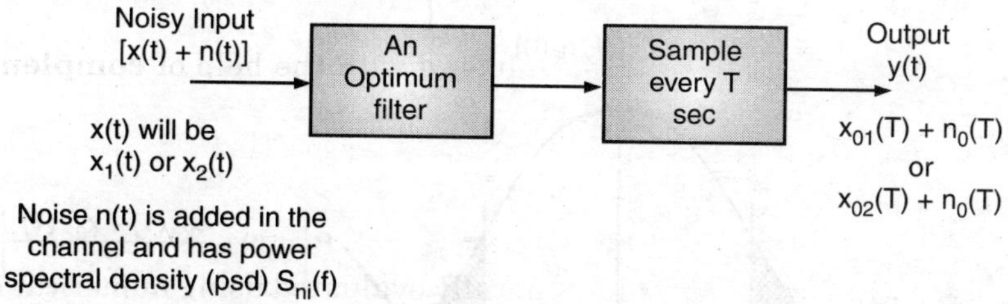

Fig. 5.14. Block diagram of receiver for binary coded signal

and output from the receiver = $x_{01}(T) + n_0(T)$ or $x_{02}(T) + n_0(T)$

Also, in the absence of the noise n(t), the output of the receiver will be,

$$y(T) = x_{01}(T) \quad \text{if} \quad x(t) = x_1(t)$$

and

$$y(T) = x_{02}(T) \quad \text{if} \quad x(t) = x_2(t)$$

Hence, in the absence of noise, decisions are taken clearly. However, if noise is present then, we select $x_1(t)$ if y(T) is closer to $x_{01}(T)$ than $x_{02}(T)$ and we select $x_2(t)$ if y(T) is closer to $x_{02}(T)$ than $x_{01}(T)$.

Therefore, the decision boundary will be in the middle of $x_{01}(T)$ and $x_{02}(T)$.

It is expressed as,

$$\text{Decision boundary} = \frac{x_{01}(T) + x_{02}(T)}{2} \quad \text{...(5.32)}$$

5.6.1. Calculation of Probability of Error (P_e) for Optimum Filter

The probability of error (P_e) may be evaluated on the same basis as in the last section. Here, we shall consider generalized Gaussian noise. Let us consider that $x_2(t)$ was transmitted, but $x_{01}(T)$ is greater than $x_{02}(T)$. If noise $n_0(T)$ is positive and larger in magnitude compared to the

* Here, the word optimum means to have maximum advantages.

voltage difference $\frac{1}{2}[x_{01}(T) + x_{02}(T)] - x_{02}(T)$, then, in this case, the incorrect decision will be taken. This means that the error will be generated if,

$$n_0(T) \geq \frac{x_{01}(T) + x_{02}(T)}{2} - x_{02}(T) \geq \frac{x_{01}(T) - x_{02}(T)}{2} \quad ...(5.33)$$

We have obtained the probability density function (PDF) for $n_0(t)$ in the previous section. It is given as

$$f_X[n_0(t)] = \frac{1}{\sigma\sqrt{2\pi}} e^{-[n_0(t)]^2/2\sigma^2} \quad ....(5.34)$$

Here, $n_0(t)$ is the random function whose PDF is given by above equation. Also, σ is its standard deviation and the function has zero mean value.

Hence, to evaluate the probability of error, we must integrate the area under the PDF curve from $n_0(T) \geq \frac{x_{01}(T) - x_{02}(T)}{2}$. This portion of the curve has been shown shaded in figure 5.15.

Therefore, we have

$$P_e = P\left[n_0(T) \geq \frac{x_{01}(T) - x_{02}(T)}{2}\right] = \int_{\frac{x_{01}(T) - x_{02}(T)}{2}}^{\infty} f_X[n_0(t)]\, d[n_0(t)]$$

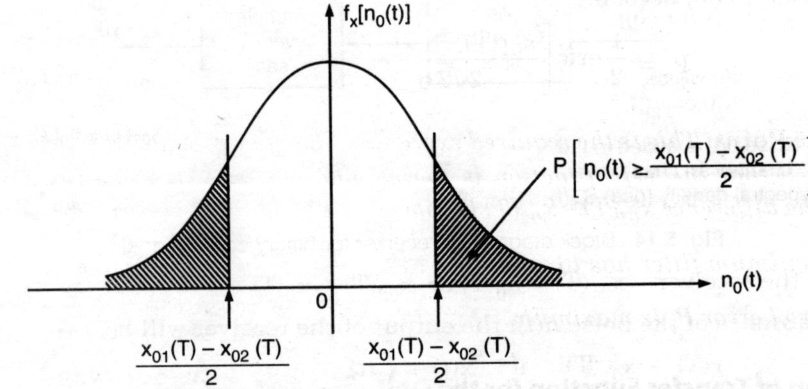

Fig. 5.15. Evaluation of P_e for optimum filter receiver.

Using equation (5.34), the last equation becomes,

$$P_e = \int_{\frac{x_{01}(T) - x_{02}(T)}{2}}^{\infty} \frac{1}{\sigma\sqrt{2\pi}} e^{-[n_0(t)]^2/2\sigma^2} \quad ...(5.35)$$

Let us substitute $\frac{[n_0(t)]^2}{2\sigma^2} = y^2$

so that $n_0(t) = \sigma\sqrt{2}\,y$

or $d[n_0(t)] = \sigma\sqrt{2}\,dy$

when $n_0(t) = \infty$, then $y = \infty$

and then $$n_o(t) = \frac{x_{01}(T) - x_{02}(T)}{2}$$

then, we have $$y = \frac{x_{01}(T) - x_{02}(T)}{2\sqrt{2}\,\sigma}$$

Substituting all these values in equation (5.34), we obtain,

$$P_e = \int_{\frac{x_{01}(T) - x_{02}(T)}{2\sqrt{2}\,\sigma}}^{\infty} \frac{1}{\sigma\sqrt{2\pi}} e^{-y^2} \cdot \sigma\sqrt{2}\,dy = \frac{1}{\sqrt{\pi}} \int_{\frac{x_{01}(T) - x_{02}(T)}{2\sqrt{2}\,\sigma}}^{\infty} e^{-y^2} dy$$

Let us rearrange this equation as under:

$$P_e = \frac{1}{2}\left[\frac{2}{\sqrt{\pi}} \int_{\frac{x_{01}(T) - x_{02}(T)}{2\sqrt{2}\,\sigma}}^{\infty} e^{-y^2} dy\right] \quad ...(5.36)$$

To solve this integration, let us use the following standard result.

$$\frac{2}{\sqrt{\pi}} \int_{u}^{\infty} e^{-y^2} dy = \text{erfc}\,(u)$$

Then, equation (5.36) becomes

$$P_e = \frac{1}{2}\,\text{erfc}\left[\frac{x_{01}(T) - x_{02}(T)}{2\sqrt{2}\,\sigma}\right] \quad ...(5.37)$$

Important Point: *This is the required expression for error probability P_e of optimum filter. It may be noted that the 'erfc' function is the monotonically decreasing function. Hence, P_e decreases as the difference $x_{01}(T) - x_{02}(T)$ becomes greater and the rms noise voltage σ becomes smaller. The optimum filter has to maximize the ratio $\frac{x_{01}(T) - x_{02}(T)}{\sigma}$ in such a manner that the probability of error P_e is maximum.*

5.6.2. Evaluation of Transfer Function for the Optimum Filter

Now, let us find the transfer function of the optimum filter in such a way that it will maximize the ratio $\frac{x_{01}(T) - x_{02}(T)}{\sigma}$. For this, let us represent the difference signal $x_{01}(T) - x_{02}(T)$ by $x_0(T)$. Thus, we have

$$x_0(T) = x_{01}(T) - x_{02}(T) \quad ...(5.38)$$

This means that the optimum filter has to maximize the ratio $\frac{x_0(T)}{\sigma}$. Now, let us derive the transfer function of the optimum filter such that the square of the ratio $\frac{x_0(T)}{\sigma}$ is maximized. The square of this ratio is,

$$\left(\frac{S}{N}\right)_0 = \frac{x_0^2(T)}{\sigma^2}$$

A pleasant companion reduces the length of the journey. – Syrus

where $\left(\frac{S}{N}\right)_0$ is known as the signal to noise power ratio. Thus, we are maximizing square of $\left[\frac{x_0(T)}{\sigma}\right]$ for mathematical convenience. In this equation, $x_0^2(T)$ is the normalized signal power in 1 Ω load.

Further, $\sigma^2 = \overline{n_0^2(T)} = E[n_0^2(T)]$ is normalized noise power because mean value of noise is zero.* Thus, the optimum filter has to maximize the ratio

$$\left(\frac{S}{N}\right)_0 = \frac{x_0^2(T)}{\overline{n_0^2(T)}} \text{ or } \frac{x_0^2(T)}{E[n_0^2(T)]} \text{ or } \frac{x_0^2(T)}{\sigma^2} \quad ...(5.39)$$

Again, let $X_0(f)$ be Fourier transform of $x_0(t)$. If $X(f)$ is the Fourier transform of input difference signal $x(t)$ [i.e., $x(t) = x_1(t) - x_2(t)$], then we have,

$$X_0(f) = H(f)\,X(f) \quad ...(5.40)$$

Here, $H(f)$ is the transfer function of optimum filter. Also, $x_0(T)$ is found by taking inverse Fourier transform of $X_0(f)$ i.e.,

$$x_0(T) = \text{IFT}[X_0(f)] = \int_{-\infty}^{\infty} X_0(f)\, e^{j2\pi fT} df$$

We know that $$x_0(T) = \int_{-\infty}^{\infty} H(f)\,X(f)\, e^{j2\pi fT}\, df \quad ...(5.41)$$

Also, the input noise to the optimum filter is $n(t)$. Let, its power spectral density (psd) be $S_{ni}(f)$. The output noise of optimum filter is $n_0(t)$. Let its power spectral density be $S_{n0}(f)$.

We know that the input and output power spectral densities of noise are related as,

$$S_{n0}(f) = |H(f)|^2 S_{ni}(f) \quad ...(5.42)$$

The normalized noise power can be obtained by integrating the power spectral density.

Thus, we have

$$\sigma^2 = \int_{-\infty}^{\infty} S_{n0}(f)\, df^{**} = \int_{-\infty}^{\infty} |H(f)|^2 S_{ni}(f)\, df \quad ...(5.43)$$

Substituting for $x_0(T)$ from equation (5.41) and σ^2 from equation (5.43) in equation (5.39), we get,

$$\left(\frac{S}{N}\right)_0 = \frac{x_0^2(T)}{\sigma^2} = \frac{\left|\int_{-\infty}^{\infty} H(f)\,X(f)\, e^{j2\pi fT} df\right|^2}{\int_{-\infty}^{\infty} |H(f)|^2 S_{ni}(f)\, df} \quad ...(5.44)$$

Here, let us use the well known Schwarz's inequality which states that

$$\left|\int_{-\infty}^{\infty} \theta_1(x)\,\theta_2(x)\, dx\right|^2 \le \int_{-\infty}^{\infty} |\theta_1(x)|^2 dx \int_{-\infty}^{\infty} |\theta_2(x)|^2 dx \quad ...(5.45)$$

---

* $\because \sigma^2 = \overline{n_0^2(T)} - (\text{mean value})^2$.

** By property of psd.

Let $\qquad \theta_1(f) = \sqrt{S_{ni}(f)}\, H(f) \quad \text{and} \quad \theta_2(f) = \frac{1}{\sqrt{S_{ni}(f)}} X(f) e^{j2\pi fT}$

With all these substitutions, the equation (5.44) becomes,

$$\left(\frac{S}{N}\right)_0 = \frac{\left|\int_{-\infty}^{\infty} \theta_1(f) \cdot \theta_2(f)\, df\right|^2}{\int_{-\infty}^{\infty} |\theta_1(f)|^2\, df}$$

Applying Schwarz's inequality of equation (5.45) to the numerator of above equation, we have

$$\left(\frac{S}{N}\right)_0 \leq \frac{\int_{-\infty}^{\infty} |\theta_1(f)|^2\, df \int_{-\infty}^{\infty} |\theta_2(f)|^2\, df}{\int_{-\infty}^{\infty} |\theta(f)|^2\, df}$$

$$\left(\frac{S}{N}\right)_0 \leq \int_{-\infty}^{\infty} |\theta_2(f)|^2\, df \leq \int_{-\infty}^{\infty} \frac{1}{S_{ni}(f)} |X(f) e^{j2\pi fT}|^2\, df$$

Now, substituting value of $\theta_2(f)$ in the above equation, we have

$$\left|X(f) e^{j2\pi fT}\right|^2 = |X(f)|^2 \qquad [\text{Since, } |e^{j2\pi fT}| = 1]$$

Thus, we have

$$\left(\frac{S}{N}\right)_0 \leq \int_{-\infty}^{\infty} \frac{|X(f)|^2}{S_{ni}(f)} df \qquad \text{...(5.46)}$$

The signal to noise power ratio expressed by equation (5.46) will be maximum when we consider equality.

In other words, we must have

$$\left(\frac{S}{N}\right)_{0\,max} = \int_{-\infty}^{\infty} \frac{|X(f)|^2}{S_{ni}(f)} df$$

This is possible only when equality applies in Schwarz's inequality i.e., in equation (5.45),

$$\left|\int_{-\infty}^{\infty} \theta_1(x)\, \theta_2(x)\, dx\right|^2 = \int_{-\infty}^{\infty} |\theta_1(x)|^2\, dx \int_{-\infty}^{\infty} |\theta_2(x)|^2\, dx$$

This equality is possible only if we have $\theta_1(f) = k\theta^*{}_2(f)$

Substituting the values of $\theta_1(f)$ and $\theta_2(f)$, we obtain,

$$\sqrt{S_{ni}(f)}\, H(f) = k \cdot \frac{1}{\sqrt{S_{ni}(f)}} X^*(f)\, e^{-j2\pi fT}$$

Therefore, when this equation is satisfied, the signal to noise ratio $\left(\frac{S}{N}\right)_0$ will be maximum. In this case, the transfer function H(f) can be obtained from above equation as under:

$$H(f) = k \cdot \frac{X^*(f)}{S_{ni}(f)} e^{-j2\pi fT}$$

It may be noted that at the beginning of this derivation, we stated that

$$\left(\frac{S}{N}\right)_0 = \frac{x_0^2(T)}{\sigma^2} = \left[\frac{x_{01}(T) - x_{02}(T)}{\sigma}\right]^2$$

Now we can have the following conclusion:

The optimum filter which minimizes the probability of error (P_e) has to maximize the ratio $\left[\frac{x_{01}(T) - x_{02}(T)}{\sigma}\right]^2$. *To maximize this ratio, the filter has the transfer function given by following expression,*

$$H(f) = k \cdot \frac{X^*(f)}{S_{ni}(f)} e^{-j2\pi fT} \qquad ...(5.47)$$

and the maximized ratio is expressed as,

$$\left(\frac{S}{N}\right)_{0\,max} = \left[\frac{x_0^2(T)}{\sigma^2}\right]_{max} = \left[\frac{x_{01}(T) - x_{02}(T)}{\sigma}\right]^2_{max}$$

or
$$\left(\frac{S}{N}\right)_{0\,max} = \int_{-\infty}^{\infty} \frac{|X(f)|^2}{S_{ni}(f)} df \qquad ...(5.48)$$

5.7 MATCHED FILTER

In previous section, we have discussed optimum filter. For this filter, we have considered generalized Gaussian noise. When this noise is white Gaussian noise, then the optimum filter is known as **matched filter.** Recall that for the white Gaussian noise, the power spectral density is given as,

$$S_{ni}(f) = \frac{N_0}{2} \qquad ...(5.49)$$

5.7.1. Calculation of Impulse Response for the Matched Filter

The transfer function of the optimum filter is expressed as,

$$H(f) = k \cdot \frac{X^*(f)}{S_{ni}(f)} e^{-j2\pi fT} \qquad ...(5.50)$$

In this equation, if we substitute $S_{ni}(f) = \frac{N_0}{2}$ i.e., psd of white noise, then H(f) becomes transfer function of matched filter.

This means that transfer function of matched filter, $H(f) = k \cdot \frac{X^*(f)}{\frac{N_0}{2}} e^{-j2\pi fT}$

or
$$H(f) = \frac{2k}{N_0} X^*(f)\, e^{-j2\pi fT} \qquad ...(5.51)$$

From the property of Fourier transform, we know that

$$X^*(f) = X(-f) \qquad ...(5.52)$$

Using this property, we can write equation (5.51) as under:

$$H(f) = \frac{2k}{N_0} X(-f)\, e^{-j2\pi fT}$$

The impulse response of a matched filter can be evaluated by taking inverse Fourier transform of above equation.

Thus, we have

$$h(t) = \text{IFT}\,[H(f)] = \text{IFT}\left[\frac{2k}{N_0} X(-f)\, e^{-j2\pi fT}\right] \quad \text{...(5.53)}$$

The inverse Fourier transform of $X(-f)$ is $x(-t)$ and $e^{-j2\pi fT}$ represents time shift of T seconds. Hence, we have

$$\text{FT}\,[x(-t)] = X(-f)$$

and $$\text{FT}\,[x(T-t)] = X(-f)\, e^{-j2\pi fT}$$

With the help of all these properties of Fourier transform, the equation (5.53) takes the form

$$h(t) = \frac{2k}{N_0} x(T-t) \quad \text{...(5.54)}$$

Note that we have considered $x(t) = x_1(t) - x_2(t)$, therefore, equation (5.54) will become,

$$h(t) = \frac{2k}{N_0} [x_1(T-t) - x_2(T-t)] \quad \text{...(5.55)}$$

Thus, two equations (5.54) and (5.55) give the required impulse response of the matched filter.

EXAMPLE 5.3. A polar NRZ waveform has to be received with the help of a matched filter. Here, binary 1 is represented by a rectangular positive pulse. Also, binary zero is represented by a rectangular negative pulse. Determine the impulse response of the matched filter. Also, sketch it.

Solution: Let $x_1(t)$ represent the positive rectangular pulse with duration T as shown in figure 5.16(a).

Let $x_2(t)$ represent a negative rectangular pulse with duration T as shown in figure 5.16(b)

Thus, we have

$$x_1(t) = +A \quad \text{for} \quad 0 \le t \le T$$

and $$x_2(t) = -A \quad \text{for} \quad 0 \le t \le T \quad \text{...(i)}$$

Let us calculate the difference signal x(t) i.e.,

$$x(t) = x_1(t) - x_2(t) \quad \text{for } 0 \le t \le T$$

or $$x(t) = A - (-A)$$

or $$x(t) = 2A \quad \text{for} \quad 0 \le t \le T \quad \text{...(ii)}$$

This difference signal has been shown in figure 5.16(c). The time reversed version of x(t) will be,

$$x(-t) = 2A \quad \text{for} \quad -T \le t \le 0 \quad \text{...(iii)}$$

It may be noted that that in the above equation, the time reference is inverted. This signal has been shown in figure 5.16(d).

Now let us delay $x(-t)$ by T seconds.

i.e., $$x(T-t) = 2A \quad \text{for} \quad 0 \le t \le T \quad \text{...(iv)}$$

The signal has been shown in figure 5.16 (e)

Impulse response of a matched filter is expressed as,

$$h(t) = \frac{2k}{N_0} x(T-t)$$

Substituting $x(T-t)$ from equation (iv), we get

$$h(t) = \frac{2k}{N_0} \cdot 2A \quad \text{for} \quad 0 \le t \le T$$

or $$h(t) = \frac{4Ak}{N_0} \quad \text{for} \quad 0 \le t \le T$$

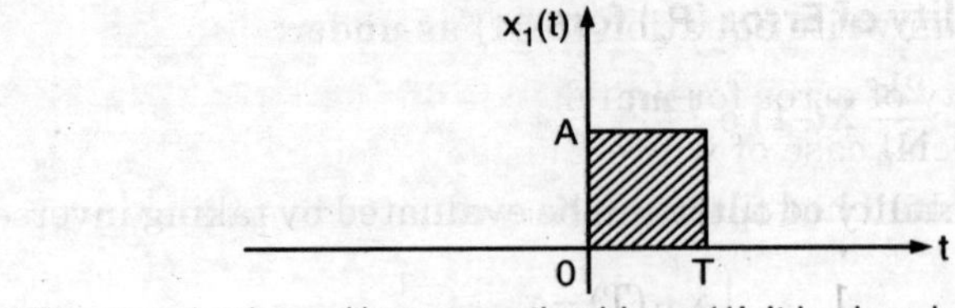

(a) The signal pulse $x_1(t)$ representing binary '1'. It is given by equation (i)

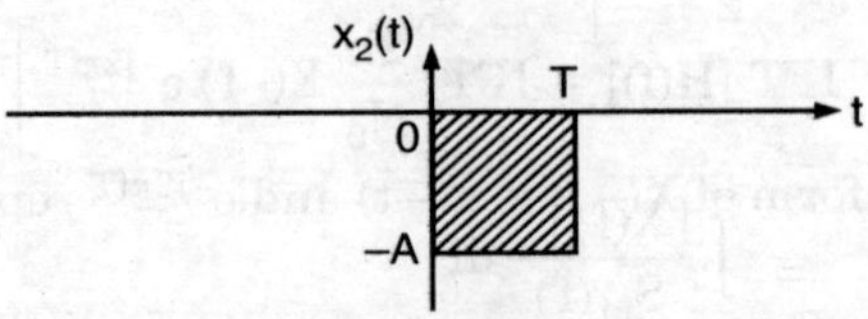

(b) The signal pulse $x_2(t)$ representing binary '0'. It is given by equation (i)

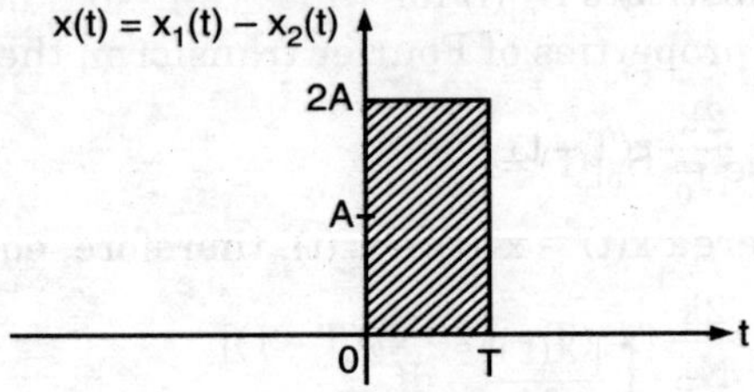

(c) The difference of $x_1(t)$ and $x_2(t)$. It has the amplitude of 2A. It is given by equation (ii)

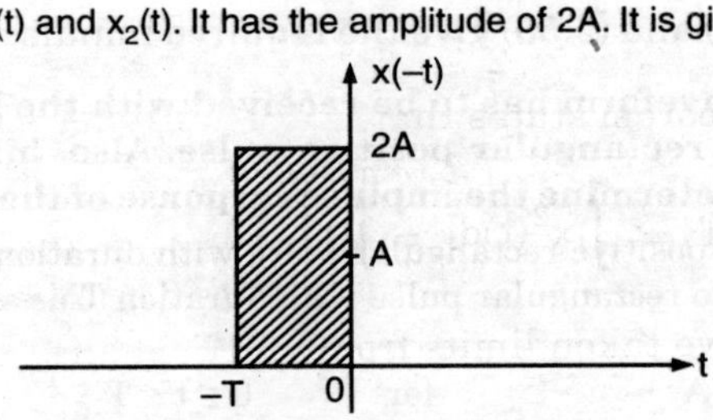

(d) The difference signal which has been folded in time. It also has amplitude equal to 2A. It is given by equation (ii)

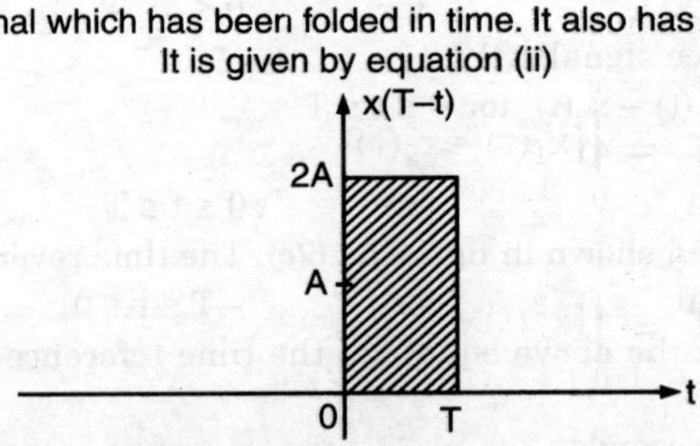

(e) The difference signal which has been folded in time. It also has amplitude equal to 2A. It is given by equation (iii)

Fig. 5.16.

Figure 5.17 shows the sketch of this impulse response.

It may be noted that with $\frac{2k}{N_0} = 1$, this figure will be same as figure 5.16(e) and figure 5.16(c). *This proves that the shape of the impulse response of the matched filter is similar (or matched) to the shape of the input signal x(t).* ***Hence, it is known as matched filter.***

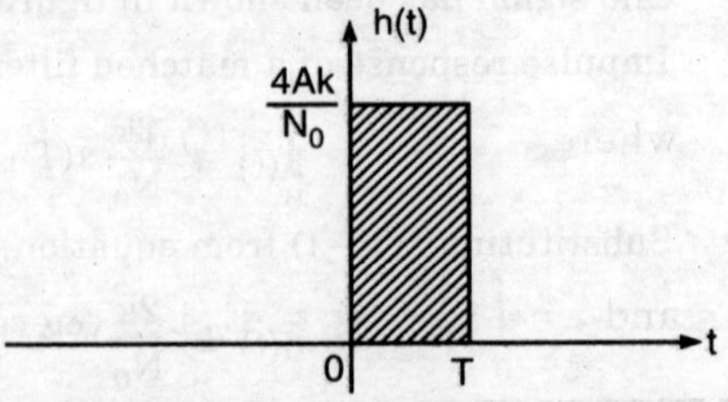

Fig. 5.17. Impulse response of a matched filter for a rectangular pulse input

5.7.2. Calculation of Probability of Error (P_e) for the Matched Filter

To evaluate the probability of error for matched filter, let us again start with optimum filter and we shall consider the special case of white Gaussian noise.

We know that error probability of optimum filter is expressed as,

$$P_e = \frac{1}{2}\,\text{erfc}\left[\frac{x_{01}(T) - x_{02}(T)}{2\sqrt{2}\,\sigma}\right] \quad \text{...(5.56)}$$

In this equation, we have

$$\left[\frac{x_{01}(T) - x_{02}(T)}{\sigma}\right]^2_{max} = \int_{-\infty}^{\infty} \frac{|X(f)|^2}{S_{ni}(f)}\,df$$

In this equation, let us substitute $S_{ni}(f)$ for white Gaussian noise.

Thus, we have

$$\text{psd of white noise} = S_{ni}(f) = \frac{N_0}{2}$$

Hence,

$$\left[\frac{x_{01}(T) - x_{02}(T)}{\sigma}\right]^2_{max} = \int_{-\infty}^{\infty} \frac{|X(f)|^2}{\frac{N_0}{2}}\,df = \frac{2}{N_0}\int_{-\infty}^{\infty} |X(f)|^2\,df \quad \text{...(5.57)}$$

Also, Parseval's power theorem states that,

$$\int_{-\infty}^{\infty} |X(f)|^2\,df = \int_{-\infty}^{\infty} x^2(t)dt = \int_0^T x^2(t)dt \quad \text{...(5.58)}$$

In the last integral we have taken limits from 0 to T because x(t) exists from 0 to T only. We know that $x(t) = x_1(t) - x_2(t)$.

Hence, above equation becomes,

$$\int_{-\infty}^{\infty} |X(f)|^2\,df = \int_0^T [x_1(t) - x_2(t)]^2 dt$$

or

$$\int_{-\infty}^{\infty} |X(f)|^2\,df = \int_0^T [x_1^2(t) + x_2^2(t) - 2x_1(t)\,x_2(t)]\,dt$$

or

$$\int_{-\infty}^{\infty} |X(f)|^2\,df = \int_0^T x_1^2(t)\,dt + \int_0^T x_2^2(t)\,dt - 2\int_0^T x_1(t)\,x_2(t)\,dt \quad \text{...(5.59)}$$

where, $\int_0^T x_1^2(t)\,dt = E_1$ i.e., energy of $x_1(t)$ by standard relations.

and $\int_0^T x_2^2(t)\,dt = E_2$ i.e., energy of $x_2(t)$ by standard relations.

He who establishes his argument by noise and command shows that his reason is weak

– Michel de Montaigne

and $\int_0^T x_1(t)\, x_2(t) = E_{12}$ represents energy due to autocorrelation between $x_1(t)$ and $x_2(t)$.

Now, if we choose $x_1(t) = -x_2(t)$, then these energies will be equal, i.e.,

$$E_1 = E_2 = -E_{12} = E \quad \text{...(5.60)}$$

Substituting all these values in equation (5.60), we get

$$\int_{-\infty}^{\infty} |X(f)|^2 df = [E + E - 2(-E)] = 4E \quad \text{...(5.61)}$$

Substituting this value of $\int_{-\infty}^{\infty} |X(f)|^2 df$ in equation (5.57), we get

$$\left[\frac{x_{01}(T) - x_{02}(T)}{\sigma}\right]^2_{max} = \frac{2}{N_0} \cdot 4E = \frac{8E}{N_0} \quad \text{...(5.62)}$$

Therefore,

$$\left[\frac{x_{01}(T) - x_{02}(T)}{\sigma}\right]_{max} = 2\sqrt{2} \cdot \sqrt{\frac{E}{N_0}}$$

Substituting this value of $\left[\frac{x_{01}(T) - x_{02}(T)}{\sigma}\right]$ in equation (5.56), we get probability of error of matched filter as under:

$$P_e = \frac{1}{2} \text{erfc} \sqrt{\frac{E}{N_0}} \quad \text{...(5.63)}$$

Important Point: *Here, erfc is the monotonically decreasing function. Therefore, when we substitute* $\left[\frac{x_{01}(T) - x_{02}(T)}{\sigma}\right]_{max}$ *as the argument of erfc, we obtain minimum value of probability of error* P_e. *This means that the above equation really provides minimum probability of error of matched filter.*

Thus, Minimum error probability of matched filter:

$$P_e = \frac{1}{2} \text{erfc} \sqrt{\frac{E}{N_0}} \quad \text{...(5.64)}$$

5.7.3. Few Points about Error Probability of Error, P_e of Matched Filter

From the equation (5.64), we can draw the following important conclusions:

(i) It is obvious from equation (5.64) that the error probability depends only upon the signal energy E. It does not depend on the shape of the signal.

(ii) We have obtained equation (5.64) by considering that $x_1(t) = -x_2(t)$. Also, in previous sections we have derived error probability for Integrate and Dump filter where we have taken $x_1(t) = +A$ and $x_2(t) = -A$.

In other words, $x_1(t) = -x_2(t) = A$ and we obtained error probability of integrator as,

$$P_e = \frac{1}{2} \text{erfc} \sqrt{\frac{E}{N_0}} \quad \text{...(5.65)}$$

The jealous man knows nothing, suspects a great deal and fears everything. – Curt Goetz

Hence, *Integrator is equivalent to matched filter when $x_1(t) = -x_2(t) = A$ or in other words, we can say that for a rectangular bipolar pulse input, the integrate and dump filter is same as a matched filter.*

EXAMPLE 5.4. Prove that the maximum signal to noise ratio for the matched filter is found to be,

$$\left(\frac{S}{N}\right)_{0\max} = \frac{2E}{N_0}$$

Solution: To derive this equation, let us consider the signal to noise power ratio of optimum filter which is given as,

$$\left(\frac{S}{N}\right)_{0\max} = \int_{-\infty}^{\infty} \frac{|X(f)|^2}{S_{ni}(f)} df \qquad ...(i)$$

We know that when white noise is present, then optimum filter is known as matched filter.

psd of white noise is $S_{ni}(f) = \frac{N_0}{2}$. Substituting this value in equation (i), we get

$$\left(\frac{S}{N}\right)_{0\max} = \int_{-\infty}^{\infty} \frac{|X(f)|^2}{\frac{N_0}{2}} df = \frac{2}{N_0}\int_{-\infty}^{\infty} |X(f)|^2 df \qquad ...(ii)$$

Also, Rayleigh's energy theorem states that,

$$\int_{-\infty}^{\infty} |X(f)|^2 df = \int_{-\infty}^{\infty} x^2(t)\, dt = E \text{ (energy)} \qquad ...(iii)$$

With this result, equation (ii) becomes

$$\left(\frac{S}{N}\right)_{0\max} = \frac{2}{N_0}\cdot E$$

Hence, maximum signal to noise power ratio of matched filter will be

$$\left(\frac{S}{N}\right)_{0\max} = \frac{2E}{N_0} \qquad ...(iv)$$

Now, this equation can be rearranged as under:

$$\left(\frac{S}{N}\right)_{0\max} = \frac{E}{\frac{N_0}{2}}$$

Here, E is energy of signal x(t) and $\frac{N_0}{2}$ is power spectral density (psd) of white noise.

Thus, we have

$$\left(\frac{S}{N}\right)_{0\max} = \frac{\text{Energy of the signal } x(t)}{\text{psd of white noise}}$$

Hence Proved

EXAMPLE 5.5. Prove that for a matched filter, the maximum signal component occurs at t = T (i.e., sampling instant) and has magnitude equal to E, i.e., energy of the signal x(t).

Solution: We know that the Fourier transform of the output signal $x_0(t)$ is expressed as,

$$X_0(f) = X(f)\, H(f) \qquad ...(i)$$

where X(f) = Fourier transform of input signal x(t)

and H(f) = Transfer function of matched filter

Substituting value of H(f) from equation (5.51), we get

$$X_0(f) = X(f)\cdot\frac{2k}{N_0} X^*(f)\, e^{-j2\pi fT} = \frac{2k}{N_0} X(f)\, X^*(f)\, e^{-j2\pi fT} \qquad ...(ii)$$

Now, since $X(f) X^*(f) = |X(f)|^2$, therefore, equation (ii) becomes,

$$X_0(f) = \frac{2k}{N_0} |X(f)|^2 e^{-j2\pi fT} \quad ...(iii)$$

From this expression, we can obtain $x_0(t)$ by taking inverse Fourier transform.

Thus, we have

$$x_0(t) = \int_{-\infty}^{\infty} X_0(f) e^{j2\pi ft} df * = \int_{-\infty}^{\infty} \frac{2k}{N_0} |X(f)|^2 e^{-j2\pi fT} . e^{j2\pi ft} df$$

or
$$x_0(t) = \frac{2k}{N_0} \int_{-\infty}^{\infty} |X(f)|^2 e^{j2\pi f(t-T)} df$$

At t = T, $x_0(t)$ will become,

$$x_0(T) = \frac{2k}{N_0} \int_{-\infty}^{\infty} |X(f)|^2 e^{j2\pi f(T-T)} df$$

or
$$x_0(T) = \frac{2k}{N_0} \int_{-\infty}^{\infty} |X(f)|^2 df \quad [\text{Since } e^{j2\pi f(T-T)} = e^0 = 1] \quad ...(iv)$$

By Rayleigh's energy theorem, we can write

$$\int_{-\infty}^{\infty} |X(f)|^2 df = \int_{-\infty}^{\infty} x^2(t) dt = E,$$

Hence, equation (iv) becomes

$$x_0(T) = \frac{2k}{N_0} \cdot E$$

Also, maximum value of $x_0(T)$ will result when $\frac{2k}{N_0} = 1$.

Thus, $x_0(T) = E$ when $t = T$ **Hence Proved**

EXAMPLE 5.6. Prove that the output signal of a matched filter is proportional to a shifted version of the autocorrelation function of the input signal to which the filter is matched.

Solution: For a matched filter, the output signal $x_0(t)$ may be obtained from its spectrum $X_0(f)$ by taking inverse Fourier transform.

This means that

$$x_0(t) = \int_{-\infty}^{\infty} X_0(f) e^{j2\pi ft} df$$

Substituting $X_0(f)$ from equation (iii) [last example], we get

$$x_0(t) = \int_{-\infty}^{\infty} \frac{2k}{N_0} |X(f)|^2 e^{-j2\pi fT} . e^{j2\pi ft} df$$

or
$$x_0(T) = \frac{2k}{N_0} \int_{-\infty}^{\infty} |X(f)|^2 e^{j2\pi f(t-T)} df \quad ...(i)$$

* By definition of IFT.

A man of character will make himself worthy of position he is given. – **Mahatma Gandhi**

Recall that the energy spectral density (esd) is given as

$$\psi(f) = |X(f)|^{2*}$$

Hence, equation (i) becomes,

$$x_0(t) = \frac{2k}{N_0}\int_{-\infty}^{\infty}\psi(f)e^{j2\pi f(t-T)}\,df \quad \text{...(ii)}$$

We know that one of the property of esd states that the autocorrelation function $R(\tau)$ and energy spectral density $\psi(f)$ form a Fourier transform pair.

This means that

$$R(\tau) \leftrightarrow \psi(f) \quad \text{...(iii)}$$

Thus, we have

$$R(\tau) = \int_{-\infty}^{\infty}\psi(f)\,e^{j2\pi f\tau}\,df \quad \text{...(iv)}$$

and

$$\psi(f) = \int_{-\infty}^{\infty}R(\tau)\,e^{-j2\pi f\tau}\,d\tau \quad \text{...(v)}$$

Applying equation (iv) to RHS of equation (ii), we get

$$x_0(t) = \frac{2k}{N_0}R(t-T)$$

This proves that $x_0(t)$ is proportional to autocorrelation function of $x(t)$ shifted to right by T. Here, $\frac{2k}{N_0}$ is a constant term and it depends upon psd of noise. **Hence Proved**

EXAMPLE 5.7. In a binary transmission, one of the messages is represented by a rectangular pulse x(t). An other message is transmitted by the absence of the pulse. Evaluate the signal to noise ratio at t = T. Assuming white noise with psd equal to $\frac{N_0}{2}$. Also sketch the impulse response of the matched filter and output of the matched filter.

Solution: Evaluation of Impulse response:

The rectangular pulse can be represented as

$$x(t) = \begin{cases} A & \text{for } 0 \le t \le T \\ 0 & \text{otherwise} \end{cases} \quad \text{...(i)}$$

Further, we know that impulse response of the matched filter is given by,

$$h(t) = \frac{2k}{N_0}x(T-t) \quad \text{...(ii)}$$

From equation (i), we can write,

$$x(-t) = \begin{cases} A & \text{for } -T \le t \le 0 \\ 0 & \text{otherwise} \end{cases} \quad \text{...(iii)}$$

Let us delay this signal by T to get

$$x(T-t) = \begin{cases} A & \text{for } 0 \le t \le T \\ 0 & \text{otherwise} \end{cases} \quad \text{...(iv)}$$

* According to the Standard definition of esd.

He that is good at making excuses, is seldom good for anything else. – Benjamin Franklin

Substituting this value in equation (ii), we obtain

$$h(t) = \begin{cases} \frac{2k}{N_0} A & \text{for } 0 \le t \le T \\ 0 & \text{otherwise} \end{cases} \qquad \text{...(v)}$$

This equation gives impulse response of the matched filter. Figure 5.18 illustrates the sketch of this response.

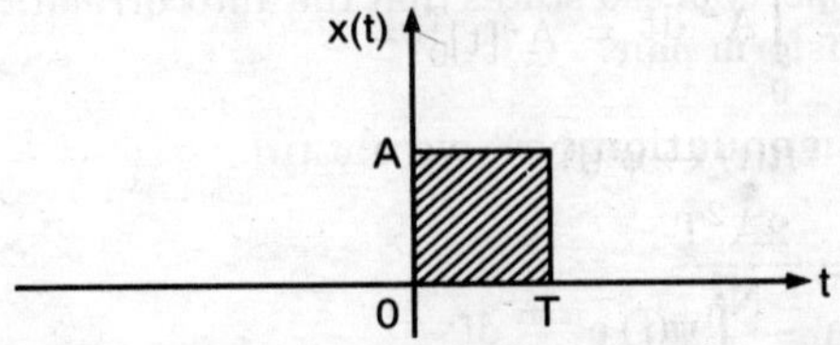

(a) Input signal x(t). It is rectangular pulse given by equation (i)

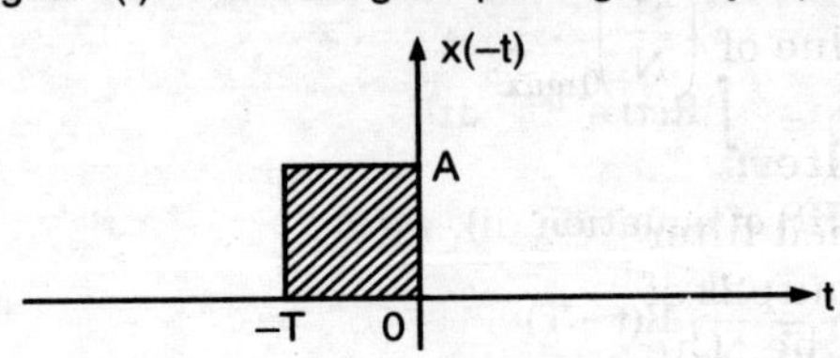

(b) Input signal pulse folded in time. This signal is represented by equation (iii)

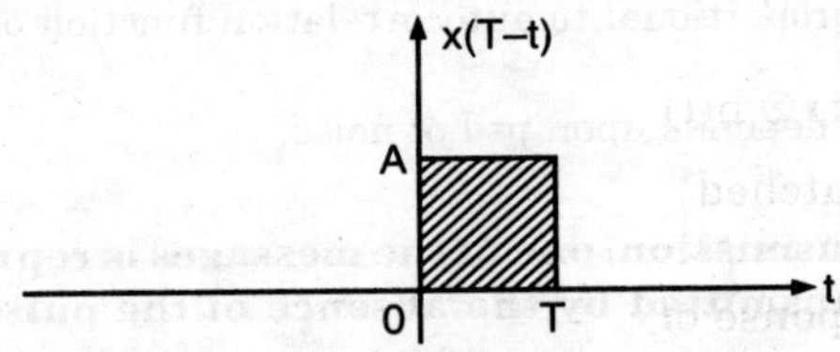

(c) Folded signal of Fig. (b) is delayed by T. This signal is represented by equation (iv)

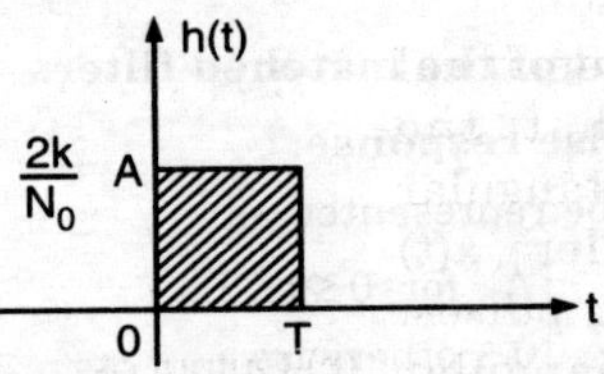

(d) Impulse response of the matched filter to the input signal pulse in figure (a)

Fig. 5.18.

Calculation of $\left(\frac{S}{N}\right)_{0\,max}$

We know that the maximum signal to noise ratio of the matched filter $\left(\frac{S}{N}\right)_{0\,max}$ is given by,

$$\left(\frac{S}{N}\right)_{0\,max} = \frac{2E}{N_0} = \frac{E}{N_0/2} \qquad \text{...(vi)}$$

Here, $\frac{N_0}{2}$ is given as psd of white noise.

Further, we can evaluate energy of signal x(t) by,

$$E = \int_{-\infty}^{\infty} x^2(t)\, dt$$

Substituting, x(t) = A and changing the limits from 0 to T as described in equation (i), the last equation becomes,

$$E = \int_{0}^{T} A^2\, dt \;=\; A^2[t]_0^T = A^2Ts$$

Substituting this value in equation (vi), we obtain,

$$\left(\frac{S}{N}\right)_{0\,max} = \frac{2A^2T}{N_0}$$

which is the required value of $\left(\frac{S}{N}\right)_{0\,max}$

Output of the Matched Filter:

We know that the matched filter is an LTI system. Thus, the output of the matched filter will be the convolution of input and impulse response.

Thus, we have $y(t) = x(t) \otimes h(t)$

where y(t) = output of matched filter,

h(t) = impulse response of matched filter,

and x(t) = input signal.

Without going into mathematical details, we can state that the convolution of two similar rectangular pulses is a triangular pulse. Here, x(t) and h(t) are two rectangular pulses. Therefore, y(t) will be a triangular pulse. This has been shown in figure 5.19.

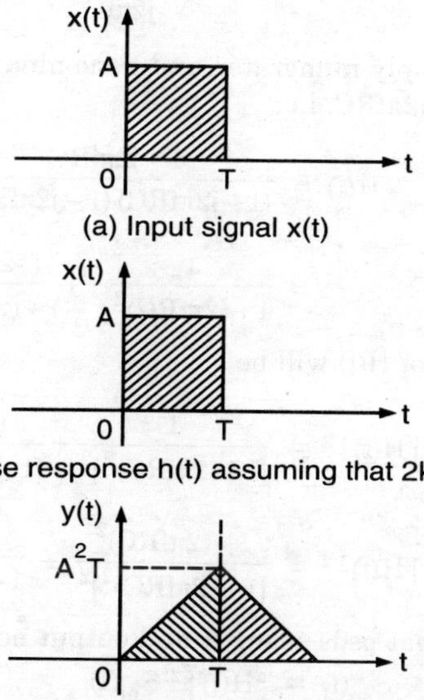

(a) Input signal x(t)

(b) Impulse response h(t) assuming that $2k/N_0 = 1$

(c) Output of the matched filter for rectangular pulse input. It may be observed that output is maximum at t = T.

Fig. 5.19.

Figure 5.19 shows the matched filter maximizes its output at t = T. Hence, y(t) has a value of A^2T at t = T.

EXAMPLE 5.8. If the rectangular pulse is applied to a RC filter shown in figure 5.20 instead of a matched filter, evaluate the maximum signal to noise ratio $\left(\frac{S}{N}\right)_{0\,max}$ which can be obtained by RC filter and compare it with that obtained by corresponding matched filter.

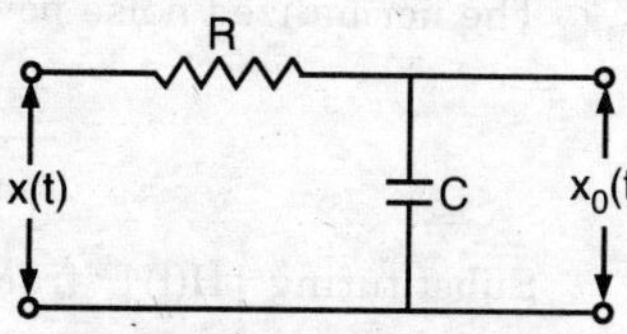

Fig. 5.20. An RC filter to receive rectangular pulse

Solution: When rectangular pulse is applied as input to the RC filter, its output $x_0(t)$ will be as shown in figure 5.21 (b).

The output $x_0(t)$ of the RC filter can be expressed in two parts as under:

$$x_0(t) = A(1 - e^{-t/RC}) \qquad \text{for } 0 \le t \le T$$

$$A(1 - e^{-t/RC})\, e^{-(t-T)/RC} \quad \text{for} \quad t > T \qquad \text{...(i)}$$

Here, the first part represents charging of the capacitor when pulse is applied for $0 \le t \le T$. The second part represents a discharging of the capacitor when the pulse $x(t)$ is removed after $t > T$. The maximum value of $x_0(t)$ occurs at $t = T$ as shown in figure 5.21(b). Therefore, substituting $t = T$ in first part of equation (i), we get

$$x(T) = A(1 - e^{-T/RC}) \qquad \text{...(ii)}$$

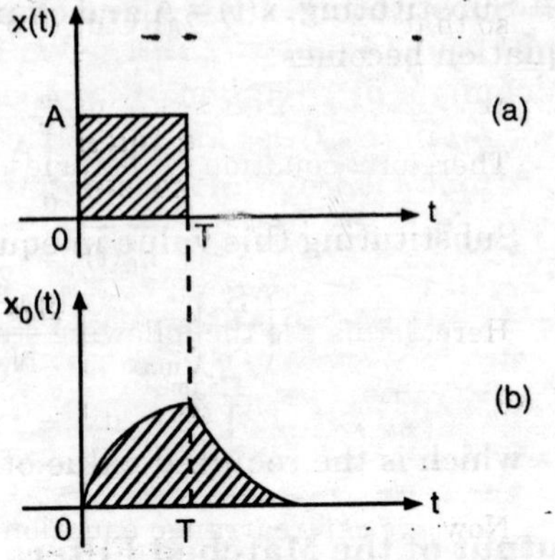

Fig. 5.21. (a) Rectangular pulse input to the RC filter (b) Output of the RC filter in response to the rectangular input pulse

Now, let us calculate the transfer function of the RC filter. Figure 5.22 shows the values of different parameters of RC filter in frequency domain.

From figure 5.22, we can write H(f) as,

$$H(f) = \frac{X_0(f)}{X(f)} = \frac{\dfrac{1}{j2\pi fC}}{R + \dfrac{1}{j2\pi fC}} = \frac{1}{1 + j2\pi fRC}$$

Let us multiply numerator and denominator of above equation by $1 - j2\pi fRC$, i.e.,

$$H(f) = \frac{1 - j2\pi fRC}{(1 + j2\pi fRC)(1 - j2\pi fRC)}$$

$$= \frac{1}{1 + (2\pi fRC)^2} + j\frac{(-2\pi fRC)}{1 + (2\pi fRC)^2}$$

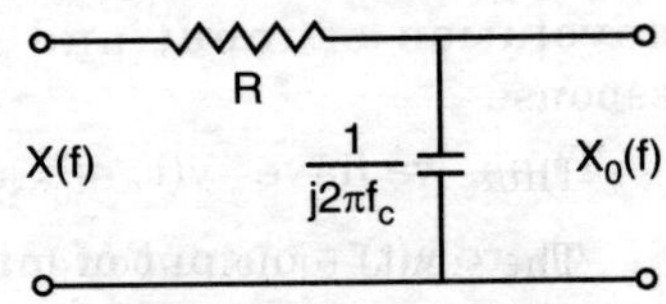

Fig. 5.22. Frequency domain representation of RC network

Magnitude of H(f) will be,

$$|H(f)|^2 = \frac{1}{[1 + (2\pi fRC)^2]^2} + \frac{(2\pi fRC)^2}{[1 + (2\pi fRC)^2]^2}$$

or

$$|H(f)|^2 = \frac{1 + (2\pi fRC)^2}{[1 + (2\pi fRC)^2]^2} = \frac{1}{1 + (2\pi fRC)^2} \qquad \text{...(iii)}$$

We know that psds of input and output noise are related as,

$$S_{no}(f) = |H(f)|^2 S_{ni}(f)$$

The psd of white Gaussian noise is, $S_{ni}(f) = \frac{N_0}{2}$; hence, above equation becomes,

$$S_{no}(f) = |H(f)|^2 \cdot \frac{N_0}{2}$$

The normalized noise power $\overline{n_0^2(t)}$ is obtained by integrating its power spectral density (psd), i.e.,

$$\overline{n_0^2(t)} = \int_{-\infty}^{\infty} |H(f)|^2 \cdot \frac{N_0}{2}\, df \; * = \frac{N_0}{2} \int_{-\infty}^{\infty} |H(f)|^2\, df \qquad \text{...(iv)}$$

Substituting $|H(f)|^2$ from equation (iii) in above equation,

$$\overline{n_0^2(t)} = \frac{N_0}{2} \int_{-\infty}^{\infty} \frac{1}{1 + (2\pi fRC)^2}\, df$$

Since 'f' is squared in above equation, therefore, we can change integration limits from $(-\infty, \infty)$ to $(0, \infty)$, i.e.,

* By property of psd.

$$\overline{n_0^2(t)} = \frac{N_0}{2}\cdot 2\int_0^{\infty}\frac{1}{1+(2\pi fRC)^2}\,df = N_0\int_0^{\infty}\frac{1}{1+(2\pi fRC)^2}\,df \quad ...(v)$$

Let $2\pi fRC = y$

so that $2\pi RC\, df = dy$

or $df = \dfrac{dy}{2\pi RC}$

Therefore, equation (v) becomes,

$$\overline{n_0^2(t)} = \frac{N_0}{2\pi RC}\int_0^{\infty}\frac{dy}{1+y^2} \quad ...(vi)$$

Here, let us use the following standard result of integration:

$$\int_0^{\infty}\frac{x^{m-1}}{1+x^n}\,dx = \frac{\pi/n}{\sin\left(\frac{m\pi}{n}\right)} \qquad n > m > 0 \quad ...(vii)$$

Now, we can rearrange equation (vi) as,

$$\overline{n_0^2(t)} = \frac{N_0}{2\pi RC}\int_0^{\infty}\frac{y^{1-1}}{1+y^2}\,dy \qquad \text{since } y^{1-1} = y^0 = 1$$

Comparing this equation with equation (vii), we get

$$m = 1 \text{ and } n = 2$$

Thus, $$\overline{n_0^2(t)} = \frac{N_0}{2\pi RC}\cdot\frac{\pi/2}{\sin(\pi/2)} \quad ...(viii)$$

The signal to noise power ratio is given as

$$\left(\frac{S}{N}\right)_0 = \frac{x_0^2(t)}{\overline{n_0^2(t)}}$$

Substituting $x_0^2(t)$ from equation (ii) and $\overline{n_0^2(t)}$ from equation (viii), we get

$$\left(\frac{S}{N}\right)_0 = \frac{A^2(1-e^{-T/RC})^2}{\frac{N_0}{4RC}} = \frac{4RC\,A^2(1-e^{-T/RC})^2}{N_0}$$

Again, let $v = \dfrac{T}{RC}$, then above equation becomes,

$$\left(\frac{S}{N}\right)_0 = \frac{4\cdot\frac{T}{v}A^2(1-e^{-v})^2}{N_0} = \frac{4A^2T}{N_0}\cdot\frac{(1-e^{-v})^2}{v} \quad ...(ix)$$

For maximum value of $\left(\frac{S}{N}\right)_0$, $\dfrac{d\left(\frac{S}{N}\right)_0}{dv}$ should be equal to zero, i.e.,

$$\left(\frac{S}{N}\right)_{0\max} = \frac{d\left(\frac{S}{N}\right)_0}{dv} = \frac{4A^2T}{N_0}\cdot\frac{2ve^{-v}(1-e^{-v})-(1-e^{-v})^2}{v^2} = 0$$

Therefore, $2v\,e^{-v}(1-e^{-v}) = (1-e^{-v})^2$

or $2v\,e^{-v} = 1 - e^{-v}$

or $2v\,e^{-v} + e^{-v} = 1$

or $e^{-v}(1+2v) = 1$

or $$e^v = 1 + 2v$$

Solving, we get $$v = 1.26$$

Since $v = \frac{T}{RC}$, we can write,

$$\frac{T}{RC} = 1.26$$

Maximum value of $\left(\frac{S}{N}\right)_0$ will be obtained by substituting, $v = \frac{T}{RC} = 1.26$ in equation (ix)

Thus, $$\left(\frac{S}{N}\right)_{0\,max} = \frac{4A^2T}{N_0}\frac{(1-e^{-1.26})^2}{1.26} = 0.814 \cdot \frac{2A^2T}{N_0} \quad ...(x)$$

In a last example we have obtained $\left(\frac{S}{N}\right)_{0\,max}$ for a rectangular pulse input to the matched filter.

Thus, we have

$$\left(\frac{S}{N}\right)_{0\,max} \text{(matched filter)} = \frac{2A^2T}{N_0}$$

With this result we can write equation (x) as under:

$$\left[\left(\frac{S}{N}\right)_{0\,max} \text{(RC filter)}\right] = 0.814 \times \left[\left(\frac{S}{N}\right)_{0\,max} \text{matched filter}\right]$$

Thus, because of RC filter, the maximum signal to noise ratio obtained is reduced to 0.814 of that of matched filter. **Hence Proved**

EXAMPLE 5.9. Find the impulse response of the matched filter for a Gaussian signal pulse given by,

$$\mathbf{x(t) = \frac{1}{\sigma\sqrt{2\pi}} e^{-t^2/2\sigma^2}}$$

The noise on the channel is a white noise with power density spectrum of $\frac{N}{2}$.

Calculate the maximum signal to noise ratio $\left(\frac{S}{N}\right)_{0\,max}$ achieved by this filter.

Solution: The given gaussian pulse is,

$$x(t) = \frac{1}{\sigma\sqrt{2\pi}} e^{-t^2/2\sigma^2} \quad ...(i)$$

We know that the impulse response of the matched filter is given as,

$$h(t) = \frac{2k}{N_0} x(T-t)$$

Let $\frac{2k}{N_0} = 1$, then above equation will be,

$$h(t) = x(T-t) = \frac{1}{\sigma\sqrt{2\pi}} e^{-(T-t)^2/2\sigma^2} \quad ...(ii)$$

This equation gives the required impulse response of the matched filter for Gaussian signal pulse.

A strong positive mental attitude will create more miracles than any wonder drug.

– Patricia Neal

We know that the maximum signal to noise ratio of matched filter is given as,

$$\left(\frac{S}{N}\right)_{0\max} = \frac{2E}{N_0} = \frac{2E}{N} \quad \left(\text{since } \frac{N_0}{2} = \frac{N}{2} \text{ given}\right) \quad ...(iii)$$

Here, E is the energy of the signal x(t). It can be calculated as,

$$E = \int_{-\infty}^{\infty} x^2(t)\, dt$$

Substituting value of x(t) from equation (i), we get

$$E = \int_{-\infty}^{\infty} \frac{1}{2\pi\sigma^2} e^{-t^2/\sigma^2} dt = \frac{1}{2\pi\sigma^2} \int_{-\infty}^{\infty} e^{-t^2/\sigma^2} dt \quad ...(iv)$$

Now, substituting

$$\frac{t^2}{\sigma^2} = y^2$$

so that $\frac{t}{\sigma} = y$

or $dt = \sigma\, dy$

however, the integration limits will remain same.

Now, substituting these values in equation (iv), we get

$$E = \frac{1}{2\pi\sigma^2} \int_{-\infty}^{\infty} e^{-y^2} dy = \frac{1}{2\pi\sigma} \int_{-\infty}^{\infty} e^{-y^2} dy \quad ...(v)$$

Let us use the following standard result of integration:

$$\int_{-\infty}^{\infty} e^{-a^2x^2+bx} dx = \frac{\sqrt{\pi}}{a} e^{b^2/(4a^2)}, a > 0$$

Here, putting b = 0 and a = 1, we get

$$\int_{-\infty}^{\infty} e^{-x^2} dx = \sqrt{\pi} \quad ...(vi)$$

Using the result of above equation, we can obtain equation (v) as,

$$E = \frac{1}{2\pi\sigma} \cdot \sqrt{\pi} = \frac{1}{2\sigma\sqrt{\pi}} \quad ...(viii)$$

Substituting this value of E in equation (iii), the maximum signal to noise ratio is given as,

$$\left(\frac{S}{N}\right)_{0\max} = \frac{2 \cdot \frac{1}{2\sigma\sqrt{\pi}}}{N} = \frac{1}{N\sigma\sqrt{\pi}} \quad \textbf{Ans.}$$

EXAMPLE 5.10. Two messages are transmitted by mark and space using a simple binary pulse shown in figure 5.23. Find error probability of the optimum receiver assuming that the probability of x(t) being present is 0.5. Assume that the channel noise is a white noise of power spectral density $\frac{N}{2}$, where $N = 10^{-4}$.

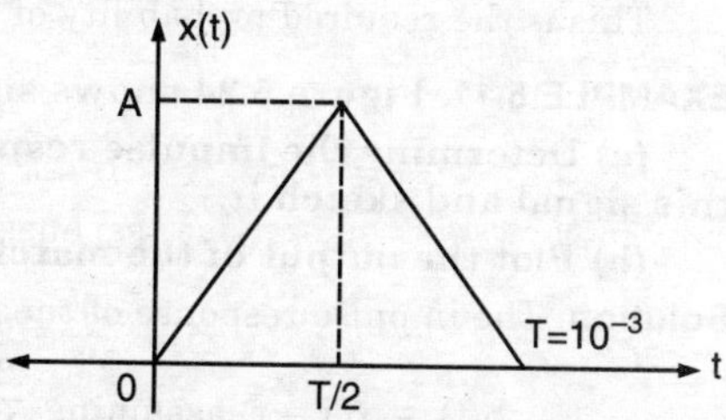

Fig. 5.23. Transmitted signal x(t)

Solution: The pulse of figure 5.23 can be represented mathematically as,

$$x(t) = \begin{cases} \frac{2A}{T} t & \text{for } 0 < t \le T/2 \\ 2A - \frac{2A}{T} t & \text{for } T/2 < t \le T \end{cases} \quad ...(i)$$

The energy of signal x(t) can be calculated as under:

$$E = \int_{-\infty}^{\infty} x^2(t)\,dt$$

Substituting value of x(t), we get

$$E = \int_0^{T/2}\left(\frac{2A}{T}t\right)^2 dt + \int_{T/2}^{T}\left(2A - \frac{2A}{T}t\right)^2 dt = \frac{4A^2}{T^2}\int_0^{T/2} t^2 dt + 4A^2\int_{T/2}^{T}\left(1-\frac{t}{T}\right)^2 dt$$

or $$E = \frac{4A^2}{T^2}\left[\frac{t^3}{3}\right]_0^{T/2} + 4A^2\int_{T/2}^{T}\left(1-\frac{2t}{T}+\frac{t^2}{T^2}\right)dt = \frac{4A^2}{T^2}\cdot\frac{T^3/8}{3} + 4A^2\left[t - \frac{2}{T}\cdot\frac{t^2}{2} + \frac{1}{T^2}\cdot\frac{t^3}{3}\right]_{T/2}^{T}$$

or $$E = \frac{4A^2}{T^2}\cdot\frac{T^3}{24} + 4A^2\left\{\left(T - \frac{T^2}{T} + \frac{1}{T^2}\cdot\frac{T^3}{3}\right) - \left(\frac{T}{2} - \frac{1}{T}\cdot\frac{T^2}{4} + \frac{1}{3T^2}\cdot\frac{T^3}{8}\right)\right\}$$

or $$E = \frac{A^2T}{6} + 4A^2\frac{T}{24} = \frac{A^2T}{3} \quad \text{...(ii)}$$

Since the channel noise is white noise, the optimum filter is same as matched filter. Hence, we can use the relations derived for matched filter. The probability of error of matched filter is given as,

$$P_e = \frac{1}{2}\,\text{erfc}\sqrt{\frac{E}{N_0}} \quad \text{...(iii)}$$

Here, it may be noted that this probability of error, P_e is obtained for the transmission and reception of two symbols $x_1(t)$ and $x_2(t)$. Even though not stated earlier, it is assumed that probability of occurrence of $x_1(t)$ and $x_2(t)$ is same. Since there are only two symbols, probability of occurrence of $x_1(t)$ and $x_2(t)$ both will be 0.5. Hence, this example probability of x(t) [i.e., one of the symbols] is given 0.5. Thus, the probability of other symbol which is represented by space will also be 0.5. Thus, the error probability given by equation (iii) above can be used to calculate P_e in this example.

Here $N_0 = N = 10^{-4}$ given

and $E = \frac{A^2T}{3} = \frac{A^2\times 10^{-3}}{3}$ since $T = 10^3$ given.

Substituting these values in equation (iii) we get

$$P_e = \frac{1}{2}\,\text{erfc}\sqrt{\frac{A^2\times10^{-3}}{3}\cdot\frac{1}{10^{-4}}} = \frac{1}{2}\,\text{erfc}\sqrt{\frac{10A^2}{3}}$$

This is the required probability of error. **Ans.**

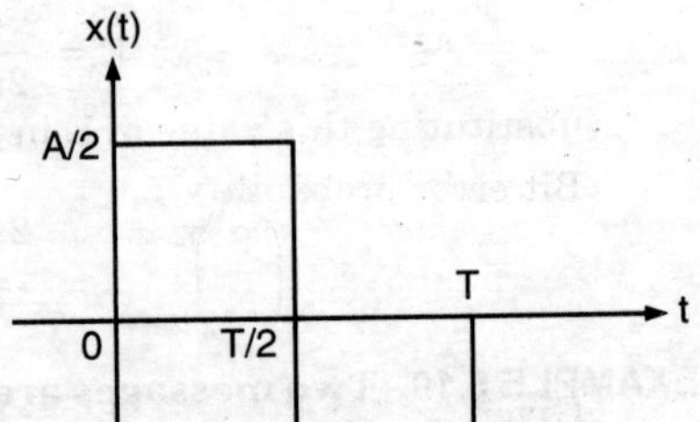

Fig. 5.24. Input signal

EXAMPLE 5.11. Figure 5.24 shows signal x(t)

(a) Determine the impulse response of the filter matched to this signal and sketch it.

(b) Plot the output of the matched filter.

Solution: The impulse response of the matched filter is given as

$$h(t) = x(T-t) \text{ assuming } \frac{2k}{N_0} = 1$$

The impulse response is sketched in figure 5.25(a), it can be described as,

$$h(t) = \begin{cases} \dfrac{-A}{2} & \text{for} \quad 0 \le t \le \dfrac{T}{2} \\ \dfrac{A}{2} & \text{for} \quad \dfrac{T}{2} \le t \le T \end{cases}$$

The output of the matched filter is obtained by convolution of x(t) and h(t), i.e.,

$$y(t) = x(t) \otimes h(t)$$

Figure 5.25(b) shows the output signal. The waveform has maximum value at t = T, it is $\frac{A^2T}{4}$.

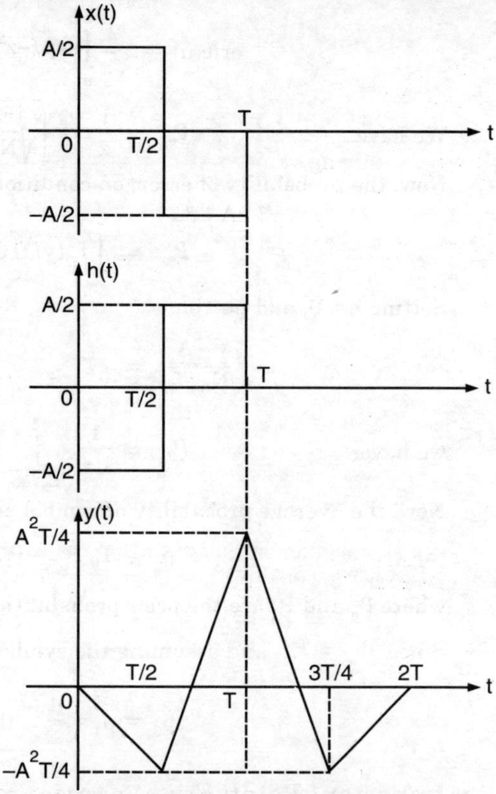

Fig. 5.25. (a) The signal pulse x(t), (b) Impulse response of the signal of (a), (c) Output waveform of the matched filter

EXAMPLE 5.12. A received binary NRZ signal assumes the voltage levels of 500 millivolts and 500 millivolts respectively for '1' and '0' transmission with a bit rate or r bits/second. The signal is corrupted by additive white Gaussian noise with a two-sided spectral density of 10^{-6} volts 2/Hz. The received signal is processed by an Integrate and Dump circuit in every bit interval and compared with a zero threshold to take a bit decision.

Assuming '1' and '0' transmission to be equally likely, the maximum value of r such that the bit error probability ≤ 10^{-5}.

Given: $$\frac{1}{\sqrt{2\pi}}\int_x^{\infty} e^{-z^2/2}\,dz = 10^{-5} \text{ at } x = 4.27$$

$$\text{erfc}(x) = \frac{2}{\sqrt{\pi}}\int_x^{\infty} e^{-z^2/2}\,dz = 2 \times 10^{-5} \text{ at } x = 3.02$$

Solution:

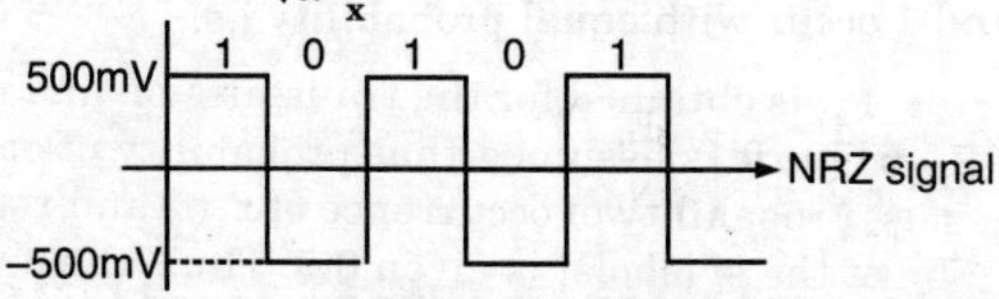

Fig. 5.26.

The transmitted signal is

$$s_i(t) = \begin{cases} s_1(t) = +A = +500 \text{ mV}, 0 \le t \le T \\ s_2(t) = -A = -500 \text{ mV}, 0 \le t \le T \end{cases}$$

then, energy of the difference signal at the filter input

$$E_d = \int_0^T [s_1(t) - s_2(t)]^2\,dt = \int_0^T (2A)^2\,dt = 4A^2T.$$

Bit error probability,

$$P_s = Q\left(\sqrt{\frac{E_d}{2N_0}}\right) = Q\left(\sqrt{\frac{2A^2T}{N_0}}\right)$$

Given $$P_s \le 10^{-5}$$

then, $$Q\left(\sqrt{\frac{2A^2T}{N_0}}\right) \le 10^{-5}$$

or $$\frac{2A^2T}{N_0} = 4.27$$

Now as $$\frac{N_o}{2} = 10^{-6} \text{ volts/Hz}$$

Therefore, $$T = \frac{2 \times 10^{-6} \times (4.27)^2}{2 \times (500 \times 10^{-3})^2} = 7.29 \times 10^{-5}$$

$$\text{erfc}(u) = \frac{2}{\sqrt{\pi}} \int_u^{\infty} \exp(-z^2)\, du$$

We have, $$P_{e_0} = \frac{1}{2} \text{erfc}\left(\sqrt{\frac{E_b}{N_0}}\right)$$

Now, the probability of error, on condition that sending symbol is 0,

$$P_{e_0} = \int_{-\infty}^{\infty} f_Y(y/1)\, dy = \frac{1}{\sqrt{\pi N_0/T_b}} \int_{-\infty}^{\lambda} \exp\left(-\frac{(y-A)^2}{N_0/T_b}\right) dy$$

Setting $\lambda = 0$, and putting

$$\frac{y-A}{\sqrt{N_0/T_b}} = -z$$

we have, $$P_{e_0} = \frac{1}{\sqrt{\pi}} \int_{\sqrt{E_b/N_0}}^{\infty} \exp(-z^2)\,.\,dz = \frac{1}{2} \text{erfc}\left(\sqrt{\frac{E_b}{N_0}}\right)$$

Now, the average probability of symbol error P_e in the receiver is given by,

$$P_e = P_0\, P_{e_0} + P_1, P_{e_1}$$

where P_e and P_1 are the prior probabilities of binary symbols 0 and 1, respectively,

Since $P_{e_0} = P_{e_1}$ and assuming the symbols 0 and 1 occur with equal probability i.e.,

$$p_0 = p_1 = \frac{1}{2} \quad \text{then,} \quad P_e = \frac{1}{2} \text{erfc}\left(\sqrt{\frac{E_b}{N_0}}\right)$$

Important Point: *Hence, average probability of symbol error in a binary coded PCM receiver depends solely on E_b/N_0, the ratio of transmitted signal energy per bit to noise spectral density.*

EXAMPLE 5.13. Prove that signal to noise ratio, $\left(\frac{S}{N}\right)_o \leq \frac{2E}{\eta}$

where E is the energy of the input signal s(t) and η/2 is the power spectral density of the input noise η(t).

Solution: Suppose the frequency response of the linear filter is H(ω).

Then, the matched filter output,

$$g_o(T) = k \int_{-\infty}^{\infty} H(\omega)\,.\,s(\omega) e^{j\omega T} d\omega$$

where $$s(\omega) = \int [s(t)]$$

Output noise power,

$$N_0 = E[\eta_o^{\,2}(t)] = \frac{\eta}{2} \cdot \frac{1}{2\pi} \int_{-\infty}^{\infty} |H(\omega)^2 d\omega$$

$$\left(\frac{S}{N}\right)_o = \frac{\left[\frac{1}{(2\pi)}\right]^2 \left|\int_{-\infty}^{\infty} H(\omega)s(\omega)\, e^{j\omega T}\, d\omega\right|^2}{\left(\frac{\eta}{2}\right)\left[\frac{1}{2\pi}\right] \int_{-\infty}^{\infty} |H(\omega)^2\, d\omega} \quad \text{...(i)}$$

Now, using Schwarz inequality, we have

$$\left|\int_{-\infty}^{\infty} H(\omega)s(\omega)\,e^{j\omega T}\,d\omega\right|^2 \le \int_{-\infty}^{\infty} |H(\omega)|^2\,d\omega \int_{-\infty}^{\infty} |s(\omega)|^2\,d\omega \qquad ...(ii)$$

Using equations (ii) and (i), we get

$$\left(\frac{S}{N}\right)_o \le \frac{2}{\eta}\cdot\frac{1}{2\pi}\int_{-\infty}^{\infty} |s(\omega)|^2\,d\omega = \frac{2E}{\eta}$$

where $E = \frac{1}{2\pi}\int_{-\infty}^{\infty} |s(\omega)|^2\,d\omega$, energy of input signal.

EXAMPLE 5.14. Compute the matched filter output over (0, T) to the pulse waveform

$$s(t) = \begin{cases} e^{-t} & 0 \le t \le T \\ 0 & \text{otherwise} \end{cases}$$

Solution: For the given s(t), the impulse response of the matched filter is,

$$h(t) = s(T - t).$$

or $$h(t) = e^{-(T-t)}$$

Now, output $$z(t) = s(t) \otimes h(t)$$

or $$z(t) = \int_{-\infty}^{\infty} e^{-\tau}\,.\,e^{-(T-1+\tau)}\,.\,d\tau$$

or $$z(t) = -\frac{e^{-T}}{2}\,.\left[e^{t-2\tau}\right]_0^t = e^{-T}\,.\,\frac{e^t - e^{-t}}{2} = e^{-T}\,\text{sinht} \qquad \textbf{Ans.}$$

5.8 THE CORRELATOR: COHERENT RECEPTION

In this article, we shall discuss a little different type of receiver which is known as **correlator.** Figure 5.27 shows the block diagram of this correlator.

In figure 5.27, f(t) represents input noisy signal,

i.e., $$f(t) = x(t) + n(t).$$

The signal f(t) is multiplied to the locally generated replica of input signal x(t). Then, result of multiplication f(t) · x(t) is integrated. The output of the integrator is sampled at t = T (i.e., end of one symbol period). Then based on this sampled value, decision is made. This is how the correlator works. It is known as correlator because it correlates the received signal f(t) with a stored replica of the known signal x(t). In the block diagram of figure 5.27, the product f(t) x(t) is integrated over one symbol period, i.e., T.

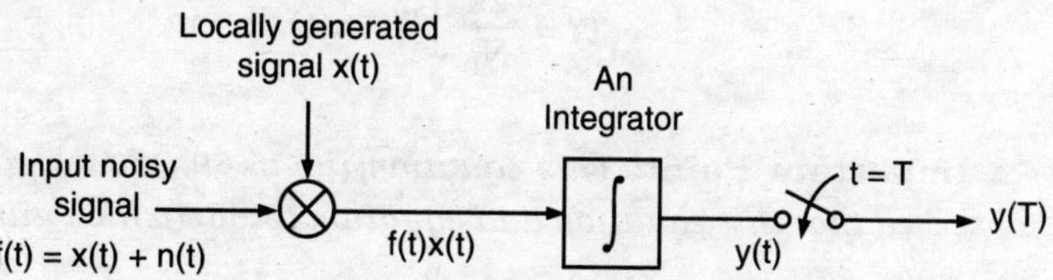

Fig. 5.27. Block diagram of a correlator

Thus, output y(t) will be,

$$y(t) = \int_0^T f(t)\,x(t)\,dt$$

output of correlator,

$$y(T) = \int_0^T f(t)\, x(t)\, dt \qquad ...(5.66)$$

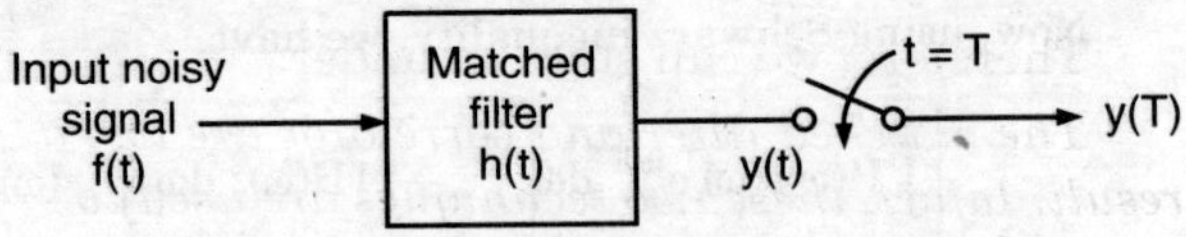

Fig. 5.28. Block diagram of a matched filter receiver

Now consider the matched filter as shown in figure 5.28.

In this block diagram, it may be observed that the matched filter does not need locally generated replica of input signal x(t). The output of the matched filter is obtained by convolution of input f(t) and its impulse response h(t).

This means that

$$y(t) = f(t) \otimes h(t) = \int_{-\infty}^{\infty} f(\tau) \cdot h(t-\tau)\, d\tau \qquad ...(5.67)$$

We know that the impulse response h(t) of the matched filter is given by,

$$h(t) = \frac{2k}{N_0} x(T-t) \qquad ...(5.68)$$

Therefore, $$h(t-\tau) = \frac{2k}{N_0} x(T-t+\tau)$$

Substituting this value of $h(t-\tau)$ in equation (5.67), we get

$$y(t) = \int_{-\infty}^{\infty} f(\tau) \frac{2k}{N_0} x(T-t+\tau)\, d\tau$$

Because the integration is performed over one bit period, therefore, we can change integration limits from 0 to T.

Hence, we write

$$y(t) = \frac{2k}{N_0} \int_0^T f(\tau)\, x(T-t+\tau)\, d\tau$$

At t = T, the last equation becomes,

$$y(T) = \frac{2k}{N_0} \int_0^T f(\tau)\, x(T-T+\tau)\, d\tau = \frac{2k}{N_0} \int_0^T f(\tau)\, x(\tau)\, d\tau$$

Now, let us substitute $\tau = t$ just for convenience for notation, then we have output of matched filter

$$y(T) = \frac{2k}{N_0} \int_0^T f(t)\, x(t)\, dt \qquad ...(5.69)$$

Important Point: *This equation (i.e. 5.69) gives the output of matched filter. It may be observed that this equation and equation (5.66) (which gives output of correlator) are identical.*

In equation (5.69) the constant $\frac{2k}{N_0}$ is present which can be normalized to 1. The similarity between equation (5.66) and equation (5.69) shows that the matched filter and correlator provides same output.

Education begins a gentleman, conversation completes him. – **Dr. Thomas Fuller**

Therefore, we can state as under:

The matched filter and correlator are two distinct, independent techniques which yield same result. Infact, these two techniques are used to synthesize the optimum filter.

5.9 COHERENT RECEIVERS FOR DIGITAL CARRIER MODULATION

We have already discussed the coherent detectors for digital carrier modulation (ASK, FSK, PSK, etc.) Now, let us obtain the error probability of all these systems one by one. Thus, in the following articles, we shall obtain the bit error probabilities of the binary systems such as binary ASK, binary PSK, differentially encoded binary PSK (DEPSK), binary FSK etc. and compare their error performance. The detection of these signals can be achieved by simply using the coherent detection.

5.10. DETECTION OF SIGNALS IN NOISE

A source is capable of transmitting M different signals $S_1(t)$, $S_2(t)$, ..., $S_M(t)$ etc. All these signals are equally likely and each one has a probability of (1/M). Let one of these M possible signals be transmitted in each time slot of duration T seconds. Then for an AWGN channel, a possible realization x(t) of the received random process X(t) is described as,

$$x(t) = s_i(t) + W(t) \quad .... \quad 0 \le t \le T$$
$$i = 1, 2, 3, ... M$$

Where W(t) is a sample function of the white noise process W(t). It has a zero mean value and power spectral density $N_0/2$. When the receiver receives the signal x(t), which is the sum of received signal and noise. The receiver has to observe the signal x(t) and make an estimate of the transmitted signal $S_i(t)$ to decide which symbol was transmitted.

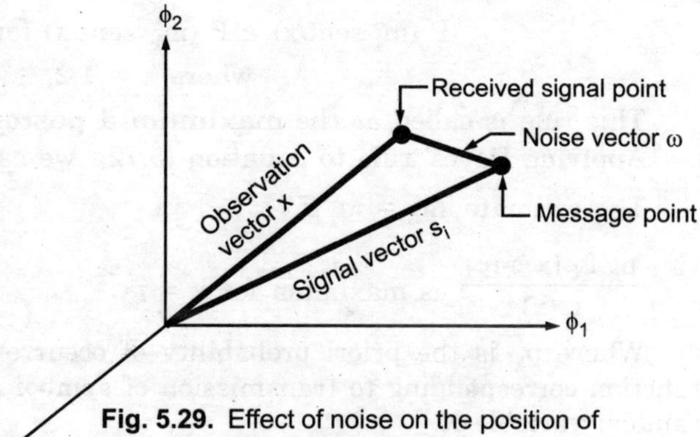

Fig. 5.29. Effect of noise on the position of received signal point

5.10.1. Signal Constellation

The transmitted signal $S_i(t)$ is applied to a bank of correctors. An appropriate set of N orthonormal basis functions is also applied. The resulting corrector outputs define the signal vector S_i. These vectors give us all the information about the transmitted signal $S_i(t)$. Hence we can represent $S_i(t)$ by a point in the Euclidean space of dimension $N \le M$. Such a point is called as the transmitted signal point or message point. All the message points in the N-dimensional Euclidean space is called as the signal constellation. The representation of the transmitted signal is simple but that of the received signal is not so simple because of the presence of the additive noise W(t). When the signal x(t) is applied to the bank of N correleators the new vector x produced by the corrector output is called as the observation or received vector. This vector is denoted by x and given by,

$$x = S_i + W \quad i = 1, 2, ...M \qquad ... (5.70)$$

Based on the observation vector x, we can represent the received signal x(t) by a point in the same Euclidean space in which we have represented the transmitted signal. This point is referred to as received signal point. Figure 5.29 illustrates relationship between the observation vector x, signal vector S_i and noise vector W for N = 3.

5.11. MAXMIMUM LIKELIHOOD DETECTOR

Let us now tackle the detection problem. If the observation vector x is given then we can perform a mapping from x to get an estimate $\hat{m}$ of the transmitted symbol m_i, in a certain way so as to minimize the average probability of symbol error in the decision. This is known as the maximum likelihood detection.

Maximum likelihood detection

Assume that if the observation vector has a value x, then the receiver will make the decision of $\hat{m} = m_i$.

The average probability of symbol error in such a decision is given by,

$$P_e(m_i, x) = P(m_i \text{ not sent}/x) = P(m_i \text{ sent}/x) \quad \text{...(5.71)}$$

Where $P(m_i \text{ sent}/x)$ is the conditional probability that m_i was sent provided x is received.

5.11.1 Optimum Decision Rule

We are interested in minimizing the probability of error in mapping each observation vector into a decision. Hence we have to deduce an optimum decision rule from equation (5.71). The optimum decision rule is as follows:

The estimate $\hat{m}_i = m_i$ if

$$P(m_i \text{ sent}/x) \geq P(m_k \text{ sent}/x) \text{ for all } k \neq i$$

$$\text{where } k = 1, 2, \ldots M \quad \text{...(5.72)}$$

This rule is called as the **maximum a posteriori probability.**

Applying Bayes' rule to equation (5.72), we can restate the decision rule as follows:

The estimate $\hat{m}_i = m_i$ if

$$\frac{p_k f_X(x/m_k)}{f_x(x)} \text{ is maximum for } k = i \quad \text{... (5.73)}$$

Where p_k is the priori probability of occurrence of symbol m_k, $f_X(x/m_k)$ is the likelihood function corresponding to transmission of symbol m_k, and $f_X(x)$ is the unconditional joint pdf of random variable X.

From equation (5.73), we may note the following points:

1. The denominator term $f_X(x)$ is independent of the transmitted signal.
2. Then, a priori probability $p_k = p_i$ because all the transmitted signals are equally likely.

So the decision rule may be further simplified as follows:

The estimate $\hat{m}_i = m_i$, if

$$f_X(x/m_k) \text{ is maximum for } k = i \quad \text{... (5.74)}$$

Generally it is more convenient to work with the natural logarithm of the likelihood functions than the likelihood functions itself.Such a natural logarithm of the likelihood functions is called as meric. The likelihood function $f_X(x/m_k)$ is always non negative. Also if $X > Y > 0$ then $\log_e X > \log_e Y$. Hence, the decision rule of equation (5.74) can be restated as follows:

The estimate $\hat{m}_i = m_i$, if

$$\log_e [f_X(x/m_k)] \text{ is maximum for } k = i \quad \text{... (5.75)}$$

The decision rule stated above is called as the maximum likelihood and device used for its implementations is called as maximum likelihood detector. Looking at equation (5.75), we conclude that the maximum likelihood detector computes the metric of each transmitted message, compares them and then makes a decision based on the maximum of them.

5.12 PROBABILITY OF ERROR (P_e)

The receiver or detector is supposed to make a correct decision about which symbol is transmitted by a transmitter. But it is not always possible for a practical receiver to make a correct decision. The wrong decision of receiver is called as an error and the probability of such error is called as error probability. The error probability is denoted by P_e. All the errors should be made to reduce the error probability to its minimum possible value. The error probability P_e is also called as the average probability of symbol error. A system with smaller value of P_e is better than the system having larger value of P_e.

5.12.1 Relation between P_e and Euclidean Distance

In previous chapter we have defined the term Euclidean distance (d). The error probability depends on the Euclidean distance. The error probability P_e decreases with increase in the Euclidean distance.

5.13. INTRODUCTION TO THE ERROR PERFORMANCE

The performance of digital communication system can be measured and compared on the basis of their error probability P_e. We are going to take help of the expressions derived for the probability of error of the matched filters or the optimum filter. Probability of error P_e is the probability of the detector making an incorrect decision. The error probability is of two types:

(i) Probability of bit error (P_B)

(ii) Probability of symbol error

Probability of bit error is used for the binary communication system whereas the probability of symbol error is used for the communication system with the M-ary communication systems. In the following sections, we shall obtain the bit error probabilities of the binary systems such as, binary ASK, binary PSK, differentially encoded binary PSK (DEPSK), binary FSK etc. and compare their error performance. The detection of these signal can be done by using the **coherent detection (using optimum or matched filters)** or using the non-coherent detection, using a bank of filters (for FSK).

5.14 ERROR PROBABILITY OF ASK

We know that the ASK (amplitude shift keying) signal is represented as:

Binary 1: $x_1(t) = \sqrt{2 P_s} \cos \omega_c t$

Binary 0: $x_2(t) = 0$

where P_s = Average normalized signal power

$$P_s = \frac{A^2}{2}$$

Now, let us obtain the expression for probability of error using the matched filter. We have already derived the expression of error probability with optimum filter as

$$P_e = \frac{1}{2} \operatorname{erfc}\left[\frac{x_{01}(T) - x_{02}(T)}{2\sqrt{2}\,\sigma}\right] \quad \text{... (5.76)}$$

We have also derived the expression for maximum output signal to noise ratio of an optimum filter as,

$$\left[\frac{x_{01}(T) - x_{02}(T)}{\sigma}\right]^2_{max} = \int_{-\infty}^{\infty} \frac{|X(f)|^2}{S_{ni}(f)} df \quad \text{... (5.77)}$$

The value of psd for a white noise input is $S_{ni}(f) = N_0/2$. Substituting this into equation (5.77), we get the expression for maximum signal to noise ratio of a matched filter as:

$$\left[\frac{x_{01}(T) - x_{02}(T)}{\sigma}\right]^2_{max} = \frac{2}{N_0}\int_{-\infty}^{\infty} |X(f)|^2 df \quad \text{... (5.78)}$$

According to the Rayleigh's energy theorem,

$$\int_{-\infty}^{\infty} |X(f)|^2 df = E = \int_{-\infty}^{\infty} x^2(t)\, dt \quad \text{... (5.79)}$$

Substituting this value into equation (5.78), we get

$$\left[\frac{x_{01}(T) - x_{02}(T)}{\sigma}\right]^2_{max} = \frac{2}{N_0}\int_{-\infty}^{\infty} x^2(t)\, dt \quad \text{... (5.80)}$$

But as the signal x(t) is present only over a bit interval T, the limits of integration will change from 0 to T.

Therefore, $$\left[\frac{x_{o1}(T) - x_{o2}(T)}{\sigma}\right]^2_{max} = \frac{2}{N_0}\int_0^T x^2(t)\, dt \quad \text{... (5.81)}$$

Where $$x(t) = x_1(t) - x_2(t)$$

But in case of ASK, $x_2(t) = 0$. Therefore $x(t) = x_1(t) = \sqrt{2 P_s}\cos\omega_c t$

We have $$\left[\frac{x_{o1}(T) - x_{o2}(T)}{\sigma}\right]^2_{max} = \frac{2}{N_0}\int_0^T 2 P_s \cos^2 \omega_c t\, dt \quad \text{... (5.82)}$$

or $$\left[\frac{x_{o1}(T) - x_{o2}(T)}{\sigma}\right]^2_{max} = \frac{4 P_s}{N_0}\int_0^T \cos^2 \omega_c t\, dt \quad \text{... (5.83)}$$

But, $$\cos^2 \omega_c t = \frac{1 + \cos 2\omega_c t}{2}$$

or $$\left[\frac{x_{o1}(T) - x_{o2}(T)}{\sigma}\right]^2_{max} = \frac{4 P_s}{N_0}\int_0^T \frac{1 + \cos 2\omega_c t}{2}$$

or $$\left[\frac{x_{o1}(T) - x_{o2}(T)}{\sigma}\right]^2_{max} = \frac{2 P_s}{N_0}\left\{\int_0^T 1\, dt + \int_0^T \cos 2\omega_c t\, dt\right\}$$

$$= \frac{2 P_s}{N_0}\left\{[t]_0^T + \frac{1}{2\omega_c}[\sin 2\omega_c t]_0^T\right\}$$

or $$\left[\frac{x_{o1}(T) - x_{o2}(T)}{\sigma}\right]^2_{max} = \frac{2 P_s}{N_0}\left\{T + \frac{\sin 2\omega_c T}{2\omega_c}\right\} \quad \text{... (5.84)}$$

Let us consider the second term on the right hand side of equation (5.83) i.e.,

$$\sin 2\omega_c T = \sin (2 \times 2 \times \pi f_c T)$$
$$= \sin 4 \pi f_c T \quad \text{... (5.85)}$$

We assume that the frequency of the carrier signal (f_c) is selected such that there are k number of complete cycles of the carrier during one bit duration T, as shown in figure 5.30.

In figure 5.30, we have k = 2.

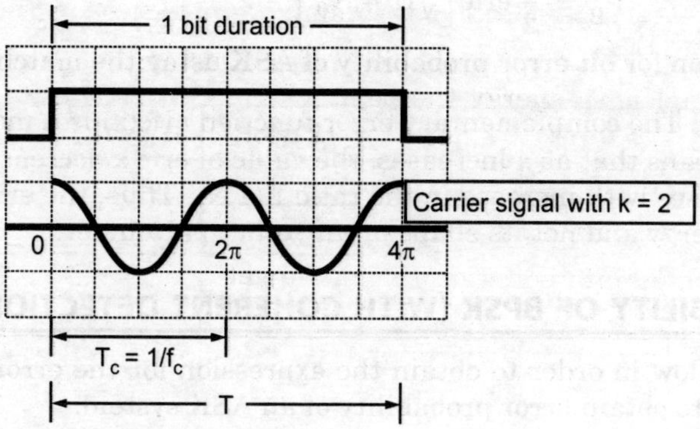

Fig. 5.30. Relation between bit period T and carrier frequency

From figure 5.30, it is obvious that,

$$\text{one bit period} \quad T = 2\,T_c = \frac{2}{f_c} \quad \text{as } T_c = \frac{1}{f_c}$$

or
$$f_c T = 2 \quad \text{... (5.86)}$$

Here, k = 2, therefore, equation (5.86) can be written in general as under:

$$f_c T = k \quad \text{... (5.87)}$$

Substituting this value into equation (5.85), we get,

$$\sin 2\omega_c T = \sin 4\pi k \quad \text{where } k = 1, 2 \quad \text{.... (5.88)}$$

or
$$\sin 2\,\omega_c T = \sin 4\pi, \sin 8\,\pi, \sin 12\,\pi \text{...etc. for } k = 1, 2, 3, \text{.....}$$

or
$$\sin 2\,\omega_c T = 0 \text{ for all values of k.}$$

Therefore, equation (5.84) gets modified as under:

$$\left[\frac{x_{o1}(T) - x_{o2}(T)}{\sigma}\right]^2_{max} = \frac{2P_s T}{N_0} \quad \text{... (5.89)}$$

Taking square root of both the sides, we get

$$\left[\frac{x_{o1}(T) - x_{o2}(T)}{\sigma}\right]_{max} = \sqrt{\frac{2\,P_s\,T}{N_0}} \quad \text{... (5.90)}$$

Now, let us substitute equation (5.90) into equation (5.76) to obtain the error probability for ASK as under:

$$P_e = \frac{1}{2} \text{erfc}\left[\frac{\sqrt{\frac{2\,P_s\,T}{N_0}}}{2\sqrt{2}}\right]$$

or
$$P_e = \frac{1}{2} \text{erfc}\left[\sqrt{\frac{P_s\,T}{4\,N_0}}\right] \quad \text{... (5.91)}$$

This is the expression for the bit error probability denoted by P_B.

$$P_B = \frac{1}{2} \operatorname{erfc}\left[\sqrt{\frac{P_s T}{4 N_0}}\right]$$

But, $P_s T = E$ = Energy of the signal. (Energy per bit)

Therefore, $$P_B = \frac{1}{2} \operatorname{erfc}\left[\sqrt{E/4 N_0}\right]$$

This is the expression for bit error probability of ASK using the matched filter.

> **Important Point:** The complementary error function erfc(x) is a monotonically decreasing function of x. This means that as x increases, the value of erfc x decreases. Hence, the bit error probability P_B decreases with increase in the ratio $E/4 N_0$. Thus, the error probability depends only on the signal energy and not its shape or any other parameter.

5.15. ERROR PROBABILITY OF BPSK (WITH COHERENT DETECTION)

The various steps to follow in order to obtain the expression for the error probability, are exactly same as those followed to obtain error probability of an ASK system.

We know that the BPSK signal is represented as follows:

Binary 1: $x_1(t) = -\sqrt{2 P_s} \cos \omega_c t$

Binary 2: $x_2(t) = \sqrt{2 P_s}$

Therefore $x_2(t) = -x_1(t)$

We have to use the matched filter for detection of BPSK signal. The expression for error probability of an optimum filter is given by

$$P_e = \frac{1}{2} \operatorname{erfc}\left[\frac{x_{o1}(T) - x_{o2}(T)}{2\sqrt{2}\,\sigma}\right] \quad \text{... (5.92)}$$

The expression for the signal to noise ratio of a matched filter is given by,

$$\left[\frac{x_{o1}(T) - x_{o2}(T)}{\sigma}\right] = \frac{2}{N_0} \int_{-\infty}^{\infty} |X(f)|^2 \, df \quad \text{... (5.93)}$$

Using the Rayleigh's energy theorem, we have

$$\int_{-\infty}^{\infty} |X(f)|^2 \, df = \int_{-\infty}^{+\infty} x^2(t) dt = \int_0^T x^2(t)\, dt \quad \text{... (5.94)}$$

The limits of integration of the last term in equation (5.94) are 0 to T because x(t) is present only over one bit interval T. Substituting equation (5.94) into equation (5.93), we get,

$$\left[\frac{x_{01}(\tau) - x_{o2}(T)}{\sigma}\right]^2_{max} = \frac{2}{N_0} \int_0^T x^2(t) dt \quad \text{...(5.95)}$$

But, $x(t) = x_1(t) - x_2(t)$

and for BPSK, $x_2(t) = -x_1(t)$

Therefore, $x(t) = 2x_1(t) = 2\sqrt{2 P_s} \cos \omega_c t$

Substituting this value of x(t) into equation (5.95), we get

$$\left[\frac{x_{o1}(T) - x_{o2}(T)}{\sigma}\right]^2_{max} = \frac{2}{N_0} \int_0^T 8 P_s \cos^2 \omega_c t \, dt$$

$$= \frac{16\,P_s}{N_0}\int_0^T \cos^2 \omega_c\, t\, dt \qquad \text{... (5.96)}$$

But $\cos^2 \omega_c t = \dfrac{1+\cos 2\omega_c t}{2}$

Substituting this value of $\cos^2 \omega_c t$ into equation (5.96), we obtain

$$\left[\frac{x_{o1}(T)-x_{o2}(T)}{\sigma}\right]^2_{max} = \frac{16\,P_s}{N_0}\int_0^T \frac{1+\cos 2\,\omega_c\, t}{2}$$

or $$\left[\frac{x_{o1}(T)-x_{o2}(T)}{\sigma}\right]^2_{max} = \frac{8\,P_s}{N_0}\left[\int_0^T 1\,dt + \int_0^T \cos 2\,\omega_c\, t\, dt\right]$$

or $$\left[\frac{x_{o1}(T)-x_{o2}(T)}{\sigma}\right]^2_{max} = \frac{8\,P_s}{N_0}\left\{[t]_0^T + \frac{1}{2\omega_c}(\sin 2\,\omega_c\, t)_0^T\right\}$$

or $$\left[\frac{x_{o1}(T)-x_{o2}(T)}{\sigma}\right]^2_{max} = \frac{8\,P_s}{N_0}\left\{T + \frac{\sin 2\,\omega_c\, T}{2\,\omega_c}\right\} \qquad \text{...(5.97)}$$

The value of second term in the RHS of equation (5.97) is zero,

Therefore, $$\left[\frac{x_{o1}(T)-x_{o2}(T)}{\sigma}\right]^2_{max} = \frac{8\,P_s T}{N_0} \qquad \text{... (5.98)}$$

But P_s T = Energy E.

Therefore, $$\left[\frac{x_{o1}(T)-x_{o2}(T)}{\sigma}\right]^2_{max} = \frac{8\,E}{N_0}$$

Taking the square root of both sides, we have

Therefore, $$\left[\frac{x_{o1}(T)-x_{o2}(T)}{\sigma}\right]_{max} = \sqrt{\frac{8E}{N_0}} \qquad \text{... (5.99)}$$

Substituting this expression into equation (5.99), we get the error probability for BPSK as under:

$$P_e = \frac{1}{2}\,\text{erfc}\left[\frac{1}{2\sqrt{2}}\cdot\sqrt{\frac{8E}{N_0}}\right]$$

Therefore, $$P_e = \frac{1}{2}\,\text{erfc}\left[\sqrt{\frac{E}{N_0}}\right] \qquad \text{... (5.100)}$$

This is the expression for error probability of BPSK with matched filter receiver.

This is the expression for bit error probability P_B.

It indicates that the probability of error depends on the energy contents of the signal E. It does not depend on the shape of the signal. As the energy increases, value of erfc function will decrease and the probability of error will also reduce.

This result can be expressed in terms of the Q functions as under:

$$P_B = Q\sqrt{\frac{2E}{N_0}} \qquad \text{... (5.101)}$$

This is because the relation between erfc and Q function is,

$$Q(x) = \frac{1}{2}\,\text{erfc}\,(x/\sqrt{2}) \text{ and erfc}\,(x) = 2\,Q\left(x\sqrt{2}\right)$$

Important Point: The erfc function is a monotonic decreasing function. Therefore the value of erfc $\left[\sqrt{E/N_0}\right]$ will be less than erfc $\left[\sqrt{E/4N_0}\right]$. Hence the error probability with BPSK technique is less than with the ASK technique. Hence BPSK system will be superior in performance as compared to the ASK system.

EXAMPLE 5.15. Obtain the error probability for a BPSK system having a bit rate of 1 Mbits/s. The receiver receives the waveforms $S_1(t) = A\cos\omega_c t$ and $S_2(t) = -A\cos\omega_c t$. The received signals are coherently detected using a matched filter. If A = 10 mV and single sided noise power spectral density is $N_0 = 10^{-11}$ W/Hz. Assume that the signal power and energy per bit are normalized.

Solution: We know that $$A = \sqrt{\frac{2E_b}{T}} = 10 \times 10^{-3}\ \text{V}$$

But $$T = \text{one bit period} = \frac{1}{1\times 10^6} = 1\ \mu s$$

Therefore, $$E_b = \frac{A^2 T}{2} = \frac{(10\times 10^{-3})^2 \times 1\times 10^{-6}}{2}$$

Therefore, $$E_b = 5 \times 10^{-11}\ \text{Joules}$$

This is the energy per bit.

Therefore, $$\sqrt{\frac{E_b}{N_0}} = \sqrt{\frac{5\times 10^{-11}}{1\times 10^{-11}}} = 2.24$$

Therefore, Bit error probability $$P_B = \frac{1}{2}\,\text{erfc}\left[\sqrt{\frac{E_b}{N_0}}\right] = \frac{1}{2}\,\text{erfc}\,|2.24| = \frac{1}{4}\times 0.00041$$

$$P_B = 2.05 \times 10^{-4}$$ **Ans.**

5.15.1 Probability of Bit Error for Coherently Detected Differentially Encoded BPSK (DEBPSK)

We have discussed DEBPSK system eariler. Sometimes a signal experiences an inversion while travelling through a network. (Such as the telephone network). The conventional BPSK system cannot identify the phase reversal but the DEBPSK can. The type of detection used for DEBPSK also is the coherent detection using the matched filter. The probability of bit error for coherently detected, differentially encoded DEBPSK is given by,

$$P_B = 2Q\left(\sqrt{\frac{2E}{N_0}}\right)\left[1 - Q\left(\sqrt{\frac{2E}{N_0}}\right)\right] \quad \text{... (5.102)}$$

But $$Q(x) = \frac{1}{2}\,\text{erfc}\left(x/\sqrt{2}\right)$$

Therefore, $$P_B = 2\times\frac{1}{2}\,\text{erfc}\left(\sqrt{\frac{2E}{2N_0}}\right)\left[1 - \frac{1}{2}\,\text{erfc}\left(\sqrt{\frac{2E}{2N_0}}\right)\right]$$

$$P_B = \text{erfc}\left(\sqrt{E/N_0}\right)\left[1 - \frac{1}{2}\,\text{erfc}\left(\sqrt{E/N_0}\right)\right] \quad \text{... (5.103)}$$

This expression shows that error performance has degraded as compared to that of coherently detected BPSK. This is because of the differential encoding. The differential encoding results in a pair of errors at the receiver, due to a single error introduced in the system.

5.16 BIT ERROR PROBABILITY OF COHERENTLY DETECTED BFSK

Here, also, let us assume that a matched filter is being used for the detection of BFSK signal. In BFSK (binary frequency shift keying), the received signal is as follows:

Binary 1: $x_1(t) = A \cos(\omega_c + \Omega) t$

Binary 2: $x_2(t) = A \cos(\omega_c - \Omega) t$

We know that one way of synthesizing a matched filter is to construct a correlation receiver system, because the correlation receiver will give exactly the same performance as a matched filter provided that the locally generated waveform is $[x_1(t) - x_2(t)]$

Therefore, local signal: $x_1(t) - x_2(t) = A \cos(\omega_c + \Omega) t - A \cos(\omega_c - \Omega)t$ ... (5.104)

But $A = \sqrt{2 P_s}$

Therefore, $x_1(t) - x_2(t) = \sqrt{2 P_s} [\cos(\omega_c + \Omega) t - \cos(\omega_c - \Omega)t]$ ... (5.105)

We know that the error probability of a matched filter is given by,

$$P_e = \frac{1}{2} \operatorname{erfc}\left[\frac{x_{o1}(T) - x_{o2}(T)}{2\sqrt{2}\,\sigma}\right] \quad ... (5.106)$$

And for a matched filter detection, the maximum signal to noise ratio is given by,

$$\left[\frac{x_{o1}(T) - x_{o2}(T)}{\sigma}\right]^2_{max} = \frac{2}{N_0}\int_0^T x^2(t)\, dt$$

But $x(t) = x_1(t) - x_2(t)$

or, $x(t) = \sqrt{2 P_s} [\cos(\omega_c + \Omega) t - \cos(\omega_c - \Omega)t]$ [From equation (5.105)]

Substituting this value of x(t) into equation (5.95), we get,

$$\left[\frac{x_{o1}(T) - x_{o2}(T)}{\sigma}\right]^2_{max} = \frac{2}{N_0}\int_0^T \left\{\sqrt{2 P_s}\,[\cos(\omega_c + \Omega)\, t - \cos(\omega_c - \Omega)t]\right\}^2 dt \quad ... (5.107)$$

But we do not have value of $x^2(t)$. i.e., the RHS of the above equation.

To obtain the value of $x^2(t)$, we have

$$x^2(t) = \left\{\sqrt{2 P_s}\,[\cos(\omega_c + \Omega)\, t - \cos(\omega_c - \Omega)t]\right\}^2$$

$$= 2 P_s [\cos(\omega_c + \Omega) t - \cos(\omega_c - \Omega)t]^2 \quad ... (5.108)$$

Let us use the following standard trigonometric identity which states that,

$\cos(A + B) - \cos(A - B) = -2 \sin A \sin B$

Applying this identity to equation (5.108), we get,

$$x^2(t) = 2 P_s [-2 \sin(\omega_c t) \sin(\Omega t)]^2$$

$$x^2(t) = 2 P_s [4 \sin^2(\omega_c t) \sin^2(\Omega t)]$$

Therefore, $x^2(t) = 2 P_s [2 \sin^2(\omega_c t) \times 2 \sin^2(\Omega t)]$ ... (5.109)

We know that $2 \sin^2 \theta = 1 - \cos 2\theta$

Therefore, $x^2(t) = 2 P_s [(1 - \cos 2\omega_c t)(1 - \cos 2\Omega t)]$

$$x^2(t) = 2 P_s [1 - \cos 2\Omega t - \cos 2\omega_c t + \cos 2\omega_c t \cos \Omega t] \quad ...(5.110)$$

In this equation, we use: $\cos A \cos B = \frac{1}{2}[\cos(A - B) + \cos(A + B)]$ to get,

$$x^2(t) = 2 P_s\left[1 - \cos 2\Omega t - \cos 2\omega_c t + \frac{1}{2}\cos 2(\omega_c - \Omega)t + \frac{1}{2}\cos 2(\omega_c + \Omega)t\right] \quad ...(5.111)$$

Thus, we have obtained the expression for $x^2(t)$

Now, we obtain the value of $\int_0^T x^2(t)\,dt$.

Taking integration of the sides of equation (5.111), we get,

$$\int_0^T x^2(t)\,dt = \int_0^T 2P_s\left\{1-\cos 2\,\Omega t-\cos 2\omega_c t+\frac{1}{2}[\cos 2(\omega_c-\Omega)t+\cos 2(\omega_c+\Omega)t]\right\}dt$$

$$\text{or} \int_0^T x^2(t)\,dt = 2P_s\left\{\int_0^T 1\,dt-\int_0^T \cos 2\,\Omega t\,dt-\int_0^T \cos 2\,\omega_c\,t\,dt+\frac{1}{2}\int_0^T \cos 2(\omega_c-\Omega)\,t\,dt+\frac{1}{2}\int_0^T \cos 2\,\omega_c+\Omega)t\,dt\right\}$$

$$\text{or} \int_0^T x^2(t)\,dt = 2P_s\left\{T-\frac{\sin 2\Omega\,T}{2\Omega}-\frac{\sin 2\,\omega_c\,T}{2\omega_c}+\frac{1}{2}\frac{\sin 2(\omega_c-\Omega)T}{2(\omega_c-\Omega)}+\frac{1}{2}\frac{\sin 2(\omega_c+\Omega)}{2\,(\omega_c+\Omega)}\right\} \quad \text{... (5.112)}$$

Therefore,

$$\int_0^T x^2(t)\,dt = 2\,P_s\,T\left\{1-\frac{\sin 2\,\Omega T}{2\Omega T}-\frac{\sin 2\,\omega_c T}{2\omega_c\,T}+\frac{1}{2}\frac{\sin 2(\omega_c-\Omega)T}{2(\omega_c-\Omega)T}+\frac{1}{2}\frac{\sin 2(\omega_c+\Omega)T}{2\,(\omega_c+\Omega)T}\right\} \quad \text{...(5.113)}$$

If we assume that the offset angular frequency Ω is very small as compared to the carrier angular frequency ω_c, then the last three terms in equation (5.113) each will have a form (sin 2 ω_cT)/2 ω_cT. This ratio approaches zero as the value of $\omega_c t$ increases.

Generally, $\omega_c\,T >> 1$ therefore the last three terms in RHS of equation (5.113) can be neglected. Therefore, equation (5.179) gets modified to.

$$\int_0^T x^2(t)\,dt = 2\,P_s\,T\left[1-\frac{\sin 2\Omega T}{2\,\Omega T}\right] \quad \text{... (5.114)}$$

Thus, we have obtained the expression for $\int_0^T x^2(t)$ dt. Now, substituting it into equation (5.95), we get

$$\left[\frac{x_{01}(T)-x_{02}(T)}{\sigma}\right]^2_{max} = \frac{2}{N_0}\times 2\,P_s T\left[1-\frac{\sin 2\Omega T}{2\,\Omega T}\right]$$

$$\text{or} \left[\frac{x_{01}(T)-x_{02}(T)}{\sigma}\right]^2_{max} = \frac{4\,P_s\,T}{N_0}\left[1-\frac{\sin 2\,\Omega T}{2\,\Omega T}\right] \quad \text{... (5.115)}$$

The value of the quantity $\left[\frac{x_{01}(T)-x_{02}(T)}{\sigma}\right]^2_{max}$ in equation (5.115) attains its largest value when Ω is selected so that $2\,\Omega T = \frac{3\,\pi}{2}$. Substituting this value of Ω into equation (5.115), we get,

$$\left[\frac{x_{01}(T)-x_{02}(T)}{\sigma}\right]^2_{max} = \frac{4\,P_s\,T}{N_0}\left[1-\frac{\sin(3\pi/2)}{(3\pi/2)}\right]$$

$$\text{or} \left[\frac{x_{01}(T)-x_{02}(T)}{\sigma}\right]^2_{max} = \frac{4\,P_s T}{N_0}\left[1-\frac{(-1)}{(3\pi/2)}\right] = 4.84\frac{P_s T}{N_0} \quad \text{... (5.116)}$$

Taking square root of both the sides, we get

$$\left[\frac{x_{o1}(T) - x_{o2}(T)}{\sigma}\right]_{max} = \sqrt{\frac{4.84\, P_s\, T}{N_0}} \quad \text{... (5.117)}$$

Substituting this value of $\left[\frac{x_{o1}(T) - x_{o2}(T)}{\sigma}\right]_{max}$ into equation (5.92), we obtain the error probability as under:

$$P_e = \frac{1}{2}\,\text{erfc}\left[\frac{1}{2\sqrt{2}} \cdot \sqrt{\frac{4.84\, P_s\, T}{N_0}}\right]$$

Therefore,

$$P_e = \frac{1}{2}\,\text{erfc}\left[\sqrt{\frac{0.6\, P_s\, T}{N_0}}\right]$$

But $P_s\, T$ = Energy E of one bit.

Therefore,

$$P_e = \frac{1}{2}\,\text{erfc}\left[\sqrt{\frac{0.6\, E}{N_0}}\right] \quad \text{.. (5.118)}$$

This is the required expression for bit error probability of BFSK when matched filter is used for detection (P_B). Let us compare this equation with the probability of error obtained for BPSK equation. As the erfc is a monotonic decreasing function, the error probability for BFSK system is higher than that of a BPSK system. This happens because in BPSK, $x_2(t) = -x_1(t)$ and in BFSK this conditions is not satisfied.

5.17 BIT ERROR PROBABILITY FOR A NON COHERENTLY DETECTED BINARY ORTHOGONAL FSK

Now, let us consider the BFSK system in which the non coherent detection using the bandpass filter is done. The bit error probability of such a system is given by,

$$P_B = \frac{1}{2}\exp\left[\frac{-A^2}{4\, N_0\, W_f}\right] \quad \text{... (5.119)}$$

Where
A = Peak signal amplitude
$N_0/2$ = PSD of white noise
W_f = Filter bandwidth

Equation (5.119) states that the error performance of the non-coherent BFSK system is dependent on the bandwidth of the filter W_f. The error probability decreases with decreases in the filter bandwidth W_f. This expression is valid only when the intersymbol interference (ISI) is negligible.

5.18 ERROR PROBABILITY OF DPSK SYSTEM

To obtain the error probability of a differential PSK (DPSK) system, we shall show the sub optimum nature of DPSK by, considering the system in terms of the phasor diagrams shown in figure 5.31.

Phasors when no noise is present

In figure 5.31(a), the phasor shows that when no noise is present the phase of received signal is either 0 or π. Therefore a decision boundary is located at angle $\pi/2$. The receiver decides that a 1 was transmitted if the phase difference between the consecutive bits differs by less than $\pi/2$ and that a 0 was transmitted if the phase difference between the consecutive bits differs by more than $\pi/2$.

Phase difference between consecutive bits	*Decision at the receiver*
Less than $\pi/2$	1 is transmitted
More than $\pi/2$	0 is transmitted.

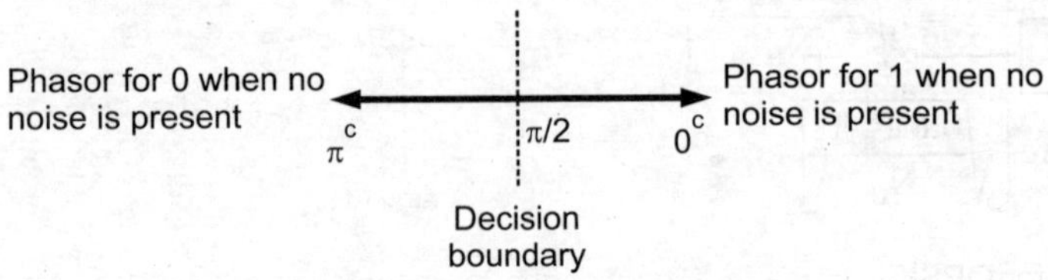

(a) Phasor when no noise is present

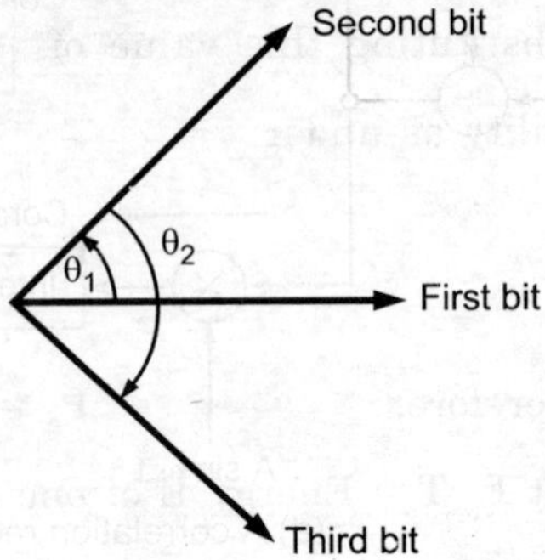

(b) Phasor when noise is present

Fig. 5.31. Reception of DPSK

Phasors when Noise is Present

Figure 5.31(b) shows the phasors for three consecutive received bits. Each bit was transmitted as 1, but because of noise, each phasor is perturbed from the x-axis as shown in figure 5.31(b). The DPSK receiver compares bit 2 with bit 1 and makes a decision that bit 2 was 1 because θ_1 is less than 90°. The DPSK receiver then compares bit 3 with bit 2 and decides that bit 3 was a 0 because the angle θ_2 is greater than 90°. Here, the receiver has made an error in the decision making. This error is due to the fact that the DPSK receiver uses only one previous bit as a reference. Instead if all the previous positive bits were somehow averaged and if this average is used as a stable reference, then the error described earlier would not have occurred at all, because then the system would have been BPSK and not DPSK. Thus, DPSK is sub-optimum and results in a higher probability of error than in BPSK where we have a stable reference phase. The bit error probability of DPSK is given by:

$$P_B = \frac{1}{2} e^{-E/N_0}$$

Where E =Energy per bit.

5.19 PROBABILITY OF ERROR OF QPSK SYSTEM

In order to understand the error probability, let us reproduce the QPSK receiver and the phasor diagram of QPSK, as shown in figure 5.32(a) and (b) respectively. From figure 5.32(a), it is obvious that two correlators are required and the locally generated reference waveforms are A cos $\omega_c t$ and sin $\omega_c t$. The received signal plus noise is passed through these correlators to generate the even and odd bit sequences $b_e(t)$ and $b_o(t)$ as explained earlier. These bit sequences are then added together to obtain the required message signal as shown in figure 5.32(b).

Let us observe figure 5.32(b) carefully. The reference waveform of correlator 1, i.e., A cos $\omega_c t$ is at an angle $\phi = 45°$ to the axes of orientation of all the four possible signals. Therefore, the axis A cos $\omega_c t$ can be treated as the decision boundary. Note that this decision boundary is at $\phi = 45°$ therefore, any correlator of the two present can make a mistake if a phase shift of 45° or $\pi/4$ radians occurs in the corresponding carrier. Therefore, the probability that correlator 1 or correlator 2 make an error is given by,

$$P_1'(e) = P_2'(e) = \frac{1}{2} \text{erfc} \sqrt{\frac{E}{N_0} \cos^2 \phi} \quad \text{... (5.120)}$$

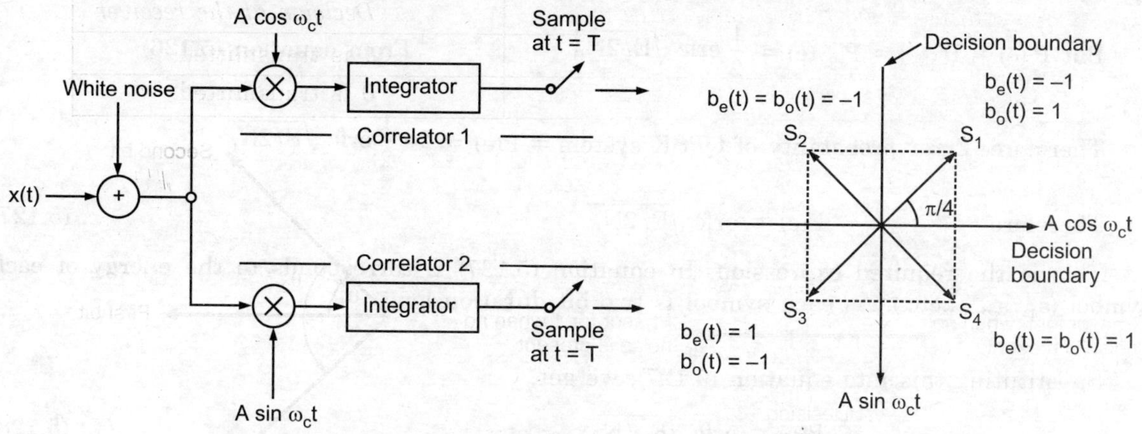

(a) A correlation receiver for QPSK (b) Signal space representation of QPSK

Fig. 5.32.

Substituting $\phi = 45°$ in equation (5.120), we obtain

$$P_1'(e) = P_2'(e) = \frac{1}{2}\,\text{erfc}\sqrt{\frac{E}{N_0}(\cos 45)^2}$$

But, $(\cos 45)^2 = 1/2 \therefore P_1(e) = P_2(e) = \frac{1}{2}\,\text{erfc}\sqrt{E/2N_0}$ ... (5.121)

The probability "P(c)" that the QPSK receiver will correctly identify the transmitted signal is equal to the product of the individual probabilities of correct identification of the two correlators.

Therefore, Probability of correct reception $P(c) = P_1'(c) \times P_2'(c)$ ... (5.122)

where $P_1'(c)$ = Probability that correlator 1 receives signal correctly.

$P_2'(c)$ = Probability that correlator 2 receives signal correctly.

But $P_1'(c) = 1 - P_1'(e)$ and $P_2'(c) = 1 - P_2'(e)$. Substituting these values into equation (5.122) we get.

$$P(c) = \left[1 - P_1'(e)\right] \times \left[1 - P_2'(c)\right]$$

Therefore, $P(c) = 1 - P'_1(e) - P'_2(e) + P'_1(e)\,P'_2(e)$ ... (5.123)

Substitute $P'_1(e) = P'_2(e) = P'(e)$ in equation (5.123), we get

$$P(c) = 1 - 2\,P'(e) + (P'(e))^2 \quad \text{... (5.124)}$$

Normally, P′(e) is very small. Therefore $P'_2(e)$ is still smaller. Hence, we can neglect the third term on RHS of equation (5.124) to get,

$$P(c) = 1 - 2\,P'(e) \quad \text{... (5.125)}$$

Hence, the error probability of a QPSK system is given by,

$$P(c) = 1 - P(c) = 1 - [1 - 2\,P'(e)]$$

Therefore, $P(c) = 2\,P'(e)$ ...(5.126)

But $P'(e) = P'_1(e) = P'_2(e) = \frac{1}{2}\,\text{erfc}\sqrt{E/2N_0}$ ... From equation (5.126)

Therefore, Error probability of QPSK system $= P(e) = 2\times\frac{1}{2}\,\text{erfc}\sqrt{E/2N_0}$

Therefore,
$$P(e) = \text{erfc}\sqrt{E/2N_0} \quad \text{... (5.127)}$$

This is the required expression. In equation (5.127) E corresponds to the energy of each symbol (s_1, s_2, ...etc.). As each symbol is two bit duration long.

$$E = 2\,E_b$$

Substituting this into equation (5.127), we get,

$$P(e) = \text{erfc}\sqrt{E_b/N_0} \quad \text{... (5.128)}$$

5.20 PERFORMANCE COMPARISON OF ERROR PERFORMANCE FOR DIFFERENT TYPES OF BINARY SYSTEMS

Figure 5.33 shows the bit error probability for various types of binary systems. The bit error probabilities of various binary systems have been listed in table 5.2 for the sake of comparison.

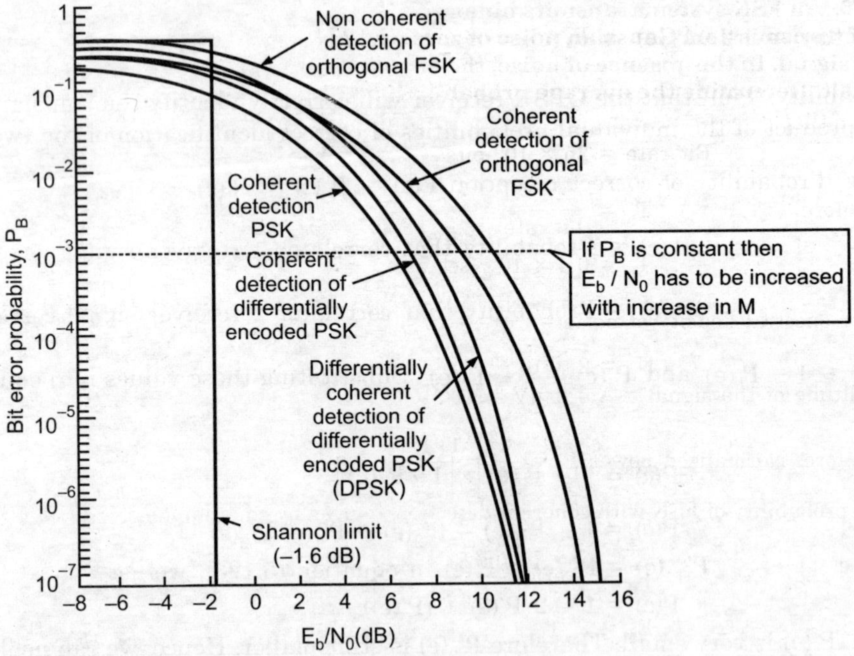

Fig. 5.33. Illustration of bit error probability for several types of binary systems

Table 5.2. Bit error probabilities of various binary system

S.No.	Name of modulation system	Expression for bit error probability P_B
1.	Amplitude shift keying (coherent detection)	$\frac{1}{2}\text{erfc}\left[\sqrt{E/4N_0}\right]$
2.	BPSK (coherent)	$\frac{1}{2}\text{erfc}\left[\sqrt{E/N_0}\right]$
3.	Differentially encoded PSK (coherent)	$\text{erfc}\left[\sqrt{E/N_0}\right]\left[1+\frac{1}{2}\text{erfc}\left(\sqrt{E/N_0}\right)\right]$
4.	BFSK (coherent)	$P_e=\frac{1}{2}\text{erfc}\left[\sqrt{\frac{0.6\,E}{N_0}}\right]$
5.	BFSK (non-coherent)	$\frac{1}{2}\exp\left[-\frac{A^2}{4\,N_0\,W_f}\right]$
6.	Differential PSK (coherent)	$\frac{1}{2}\exp\left[-E/N_0\right]$
7.	QPSK (coherent)	$\text{erfc}\left[\sqrt{E_b/N_0}\right]$

EXAMPLE 5.16. An FSK system transmits binary data at the rate of 2.5 × 10⁶ bits per second. During the course of transmission, Gaussian noise of zero mean of power spectral density 10⁻²⁰ Watts/Hz is added to the signal. In the absence of noise, the amplitude of received sinusoidal wave for digit 1 or 0 is 1 microvolt. Determine the average probability of symbol error, assuming coherent detection.

Solution: Given that

(i) Bit rate $= 25\times 10^6$ bits/sec.

Therefore,
$$T_b = \frac{1}{\text{bit rate}} = \frac{1}{2.5\times10^6}$$

or
$$T_b = 0.4\times10^{-6}\text{ sec or }0.4\ \mu\text{s}$$

(ii) Power spectral density of noise $= \frac{N_0}{2} = 10^{-20}$ Watt/Hz.

or
$$N_0 = 2\times10^{-20}$$

(iii) Amplitude of the signal $= A = 1\ \mu V = 1\times10^{-6}$

Therefore, Normalized power $P_s = \frac{A^2}{2} = \frac{1\times10^{-12}}{2}$

The error probability of FSK with coherent detection is given by equation (5.118) as under:

$$P_e = \frac{1}{2}\text{erfc}\left[\frac{0.6\,E_b}{N_0}\right]^{1/2}$$

But
$$E_b = P_s\,T_b$$

or
$$P_e = \frac{1}{2}\text{erfc}\left[\frac{0.6\,P_s\,T_b}{N_0}\right]^{1/2}$$

Substituting the values, we get

$$P_e = \frac{1}{2}\text{erfc}\left[\frac{0.6\times1\times10^{-12}\times1}{2\times2\times10^{-12}\times2.5\times10^{6}}\right]^{1/2} = \frac{1}{2}\text{erfc}\,[7.745]^{1/2}$$

Therefore, $P_e = \frac{1}{2}\text{erfc}\,[2.78]$

or $P_e \equiv 1.5 \times 10^{-4}$ **Ans.**

EXAMPLE 5.17. A received signal is either + 2V or – 2V hold for a time T. The signal is corrupted by white Gaussian noise of power spectral density 10^{-4} volt²/Hz. If the signal is processed by an integrator and dump receiver, what is the minimum time T during which a signal must be sustained if the probability of error is not exceed 10^{-4}?

Solution: The probability of error of integrate and dump receiver is given as:

$$P_e = \frac{1}{2}\text{erfc}\left[\frac{A^2T}{N_0}\right]^{1/2}$$

It has been given that $\frac{N_0}{2} = 10^{-4}$

Therefore, $N_0 = 2 \times 10^{-4}$ and $A = 2$

The error probability should not exceed 10^{-4}

Therefore, $P_e \le 10^{-4}$

Therefore, $$10^{-4} \le \frac{1}{2}\text{erfc}\left[\frac{A^2T}{N_0}\right]^{1/2}$$

or $$10^{-4} \le \frac{1}{2}\text{erfc}\left[\frac{2^2 \times T}{2\times10^{-4}}\right]^{1/2}$$

Using the sign of equality we get,

$$\text{erfc}\left[\frac{4\,T_{min}}{2\times10^{-4}}\right]^{1/2} = 2 \times 10^{-4}$$

Using the table of erfc we get,

$$\left[\frac{4\,T_{min}}{2\times10^{-4}}\right]^{1/2} \cong 2.6$$

Therefore, $$\frac{4\,T_{min}}{2\times10^{-4}} = 6.76$$

or $T_{min} = 0.338 \times 10^{-3}$ sec. or 0.338 msec. **Ans.**

Thus the minimum time for which the signal should be extended is 0.338 msec.

5.21 SYMBOL ERROR PERFORMANCE FOR M-ARY SYSTEM (M > 2)

In the preceding sections, we have discussed about the bit error probability of various binary communication systems. Now we will learn about the error performance of M-ary system. The symbol error probability is denoted by $P_e(M)$.

5.21.1 Probability of Symbol Error for MPSK

If the energy to noise ratio of an M-ary PSK signal is large, then, the sy nbol error probability using the coherent PSK is given by,

$$P_e(M) = 2Q\left[\sqrt{\frac{2E_s}{N_0}}\sin\frac{\pi}{M}\right] \quad ...(5.129)$$

But $$Q(x) = \frac{1}{2}\,\text{erfc}\left(x/\sqrt{2}\right)$$

Therefore, $$P_e(M) = 2\times\frac{1}{2}\,\text{erfc}\left[\frac{1}{\sqrt{2}}\times\sqrt{\frac{2\,E_s}{N_0}}\,\sin\frac{\pi}{M}\right]$$

or $$P_e(M) = \text{erfc}\left[\sqrt{E/N_0}\,\sin\frac{\pi}{M}\right] \quad \text{... (5.130)}$$

In equations (5.129) and (5.130), we get

$P_e(M)$ = Probability of symbol error

E = Energy per symbol = $E_b\,(\log_2 M)$

M = Number of messages being transmitted = 2^N

N = Number of bits per symbol

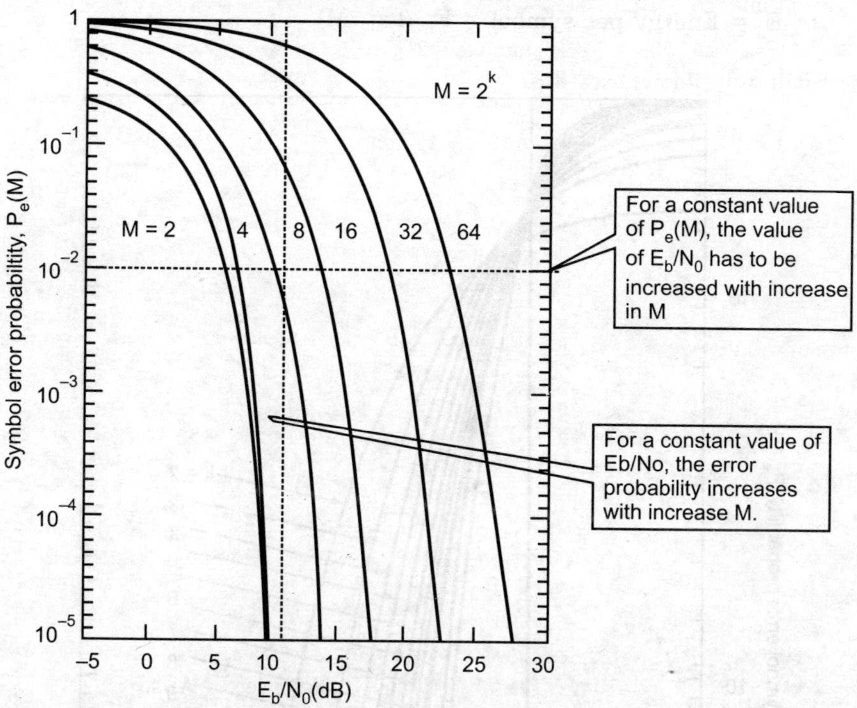

Fig. 5.34. Symbol error performance for M-ary PSK system using coherent detection

Equations (5.129) and (5.130) indicate that the probability of symbol error depends on the ratio $\sqrt{E_s/N_0}$ and M. The error probability decreases i.e. error performance improves with increases in the value of $\sqrt{E_s/N_0}$ but error probability increases with increase in the value of M. Figure 5.34 shows the graph of symbol error probability versus (E_s/N_0) with coherent detection.

✍ **Important Point:** *Figure 5.34 indicates that in order to obtain the same symbol error, we have to increase the value of signal energy to noise ratio* $\sqrt{E_s/N_0}$ *as the value of M goes on increasing. If the ratio* $\sqrt{E_s/N_0}$ *is maintained constant then with increases in the value of M, the error probability goes on increasing.*

Thus, the error performance of BPSK is superior to the M-ary PSK system with M > 2.

5.22 PROBABILITY OF SYMBOL ERROR FOR MFSK (COHERENT DETECTION)

In M-ary FSK signal, M orthogonal carrier waves are used to represent M different symbol or message. The symbol error probability $P_e(M)$ for M equally likely orthogonal signals have an upper bound given by:

$$P_e(M) = (M - 1)\, Q\left[\sqrt{E_s / N_0}\right] \qquad \text{... (5.131)}$$

But $$Q(x) = \frac{1}{2}\,\text{erfc}\left[x/\sqrt{2}\right]$$

Therefore, $$P_e(M) = (M - 1) \times \frac{1}{2}\,\text{erfc}\left[\sqrt{E_s / 2N_0}\right] \qquad \text{.. (5.132)}$$

Where M = Number of symbol or message

E_s = Energy per symbol = E_b ($\log_2$ M)

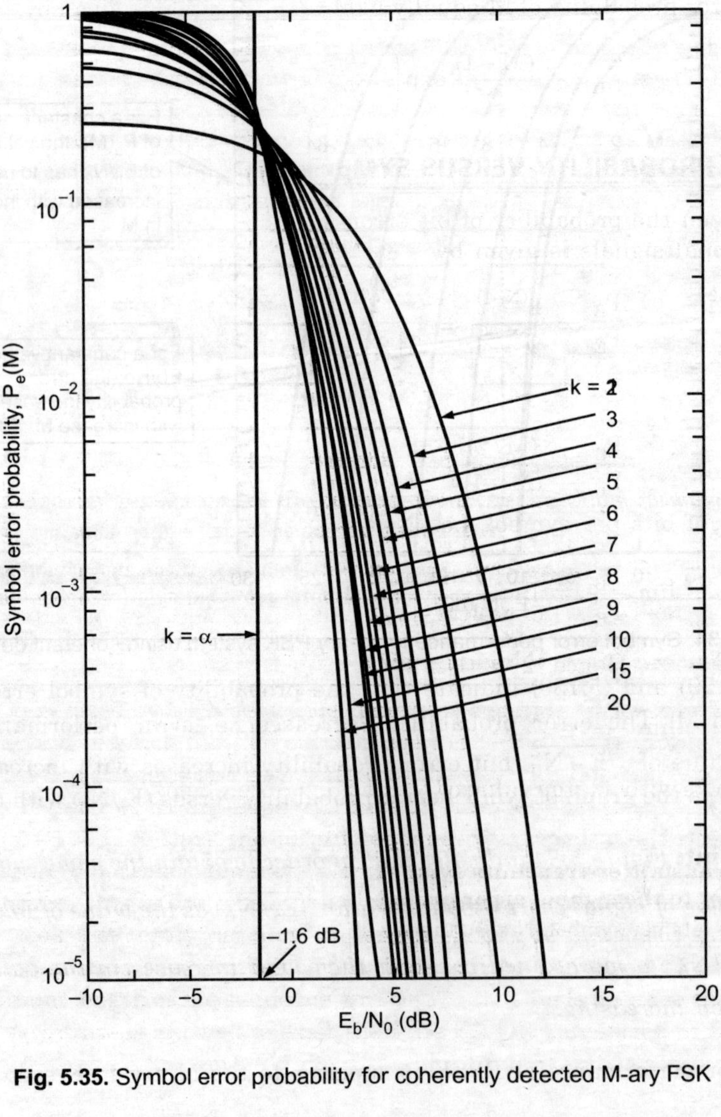

Fig. 5.35. Symbol error probability for coherently detected M-ary FSK

Equations (5.131) and (5.132) indicate that the symbol error probability increases (error performance becomes poor) with increase in the value of M. But, the error probability decreases (performance improves) with increase in the ratio $\sqrt{E_s/N_0}$. Figure 5.35 shows a graph of symbol error probability verus (E_s/N_0) using the coherent detection.

> **Important Point:** *Figure 5.35 indicates that to obtain the same symbol error we have to increase the value of signal energy to noise ratio E_s/N_0 as M is increased. If the ratio E_s/N_0 is maintained constant then with increase in the value of M, the error probability goes on increasing.*

5.22.1 Performance Comparison of MPSK and MFSK System

If the ratio E_s/N_0 and M are same for the MPSK and MFSK system then the symbol error probability of MPSK system is less than that of MFSK system.Hence, the error performance of MPSK system is better than that of MFSK system.

5.22.2 Symbol Error Probability of MFSK by Non Coherent Detection

The symbol error probability of M-equally likely MFSK signal is given by;

$$P_e(m) < \frac{(M-1)}{2}\exp\left(-\frac{E_s}{2N_0}\right) \quad \text{... (5.133)}$$

5.23 BIT ERROR PROBABILITY VERSUS SYMBOL ERROR PROBABILITY

The relation between the probability of bit error (P_B) and the probability of symbol error P_E for a set of M-orthogonal signals is given by

$$\frac{P_B}{P_E} = \frac{2^{(N-1)}}{2^N - 1} = \frac{2^N}{2\left[2^N - 1\right]}$$

But $2^N = M$

Therefore, $$\frac{P_B}{P_E} = \frac{(M/2)}{(M-1)} \quad \text{... (5.134)}$$

If the number of bits per symbol, i.e., N is very large (approaches ∞) then

$$\lim_{N\to\infty}\frac{P_B}{P_E} = \lim_{N\to\infty}\frac{M}{2(M-1)} = \lim_{N\to\infty}\frac{2^N}{2(2^N-1)}$$

as $N \to \infty$, $2^N \to \infty$. Hence, $2^N - 1 \cong 2^N$

or $$\lim_{N\to\infty}\frac{P_B}{P_E} = \frac{1}{2} \quad \text{... (5.135)}$$

Thus, for symbols with a large value of N, the bit error probability is twice as high as the symbol error probability.

EXAMPLE 5.18. A transmitter transmits symbols with 3-bits per symbol. Calculate the ratio of bit error probability P_B to the symbol error probability P_E.

Solution: With three bits per symbol (N = 3), there will be 8 possible transmitted symbols $M = 2^3 = 8$ as follows:

8–Possible transmitted symbols

	0	0	0
	0	0	1
	0	1	0
	0	1	1
	1	0	0
Transmitted symbol	1	0	1
	1	1	0
	1	1	1
			Bit Position

Fig. 5.36.

Bit error probability (P_B)

Let us consider the bit position box in the above. Let us obtain the number of ways to cause error to binary 1. In the bit position box of the given figure, there are 4-ones and 4-zeros. We can consider any other bit position, the same thing is observed. The number of ways to cause error to binary-1 in any bit position is 2^{N-1} i.e., $2^2 = 4$.

Symbol error probability (P_E)

The number of ways a symbol error can made is $2^N - 1 = 2^3 - 1 = 7$. Hence, the final relationship P_B/P_E for orthogonal signalling is obtained by calculating the value of the ratio 2^{N-1} and $2^N - 1$.

Therefore,
$$\frac{P_B}{P_E} = \frac{2^{N-1}}{2^N - 1} = \frac{4}{7}$$

5.24 EFFECTS OF INTERSYMBOL INTERFERENCE

Earlier we have discussed the concept of intersymbol interference. In practice ISI must be accounted because it is a second source of interference. The ISI is generated due to dispersed pulse (in the time domain) due to the use of bandlimiting filters. The ISI can be observed at various points in the system such as transmitter output, in the communication channel, or at the receiver input. The effect of this additional interference is the degradation of error performance (increase in the error probability) for coherent as well as the non-coherent reception.

5.25 ERROR PROBABILITY OF BINARY PCM

As we know that the PCM signal is a train of binary 0s and 1s. Therefore, let us assume that the binary 0 is represented by 0 volts and a binary 1 is represented by + A volts and let the bit duration be T sec.

$$x_1(t) = A \quad \text{for } 0 \le t \le T \quad \text{... For binary 1}$$
$$\text{and} \quad x_2(t) = 0 \quad \text{for } 0 \le t \le T \quad \text{... For binary 0} \qquad ...(5.136)$$

Assuming that an optimum filter is used for detection, we can write the expression for its error probability as,

$$P_e = \frac{1}{2}\,\text{erfc}\left[\frac{x_{o1}(T) - x_{o2}T)}{2\sqrt{2}\,\sigma}\right] \qquad ...(5.137)$$

$$\left[\frac{x_{o1}(T) - x_{o2}(T)}{2\sqrt{2}\,\sigma}\right]^2_{max} = \int_{-\infty}^{\infty} \frac{|X|(f)|^2}{S_{ni}(f)} \qquad ...(5.138)$$

Assuming that the input noise is white gaussian noise, substituting $S_{ni}(f) = N_0/2$, in equation (5.138), we get

$$\left[\frac{x_{o1}(T) - x_{o2}(T)}{2\sqrt{2}\,\sigma}\right]^2_{max} = \frac{2}{N_0}\int_{-\infty}^{\infty} |X(f)|^2\, df \qquad \dots (5.139)$$

But, according to Parseval's theorem, we have $\int_{-\infty}^{\infty} |X(f)|^2 = P = \int_{-\infty}^{\infty} x^2(t)\,dt$

Therefore,

$$\left[\frac{x_{o1}(T) - x_{o2}(T)}{2\sqrt{2}\,\sigma}\right]^2_{max} = \frac{2}{N_0}\int_{-\infty}^{\infty} x^2(t)\, dt \qquad \dots (5.140)$$

As the input signal $x(t) = x_1(t) - x_2(t) = x_1(t)$.... as $x_2(t) = 0$

$$\left[\frac{x_{o1}(T) - x_{o2}(T)}{2\sqrt{2}\,\sigma}\right]^2_{max} = \frac{2}{N_0}\int_{-\infty}^{\infty} x_1^2(t).\, dt \qquad \dots (5.141)$$

$$= \frac{2}{N_0}\int_{0}^{T} A^2\, dt \qquad \dots[\text{Since } x_1(t) = A, \text{ for } 0 \le t \le T]$$

$$= \frac{2A^2T}{N_0}$$

Taking square root of both the sides, we get

$$\left[\frac{x_{o1}(T) - x_{o2}(T)}{\sigma}\right]_{max} = \sqrt{\frac{2A^2T}{N_0}} \qquad \dots(5.142)$$

Substituting this into equation (5.137), we get

$$P_e = \frac{2}{1}\,\text{erfc}\left[\frac{1}{2\sqrt{2}} \times \sqrt{\frac{2A^2T}{N_0}}\right]$$

Therefore,

$$P_e = \frac{1}{2}\,\text{erfc}\left[\sqrt{\frac{A^2T}{4N_0}}\right] \qquad \dots (5.143)$$

This is the expression for error probability of PCM.

By substituting $A^2T = E$ in the above expression, we obtain,

$$P_e = \frac{1}{2}\,\text{erfc}\left[\sqrt{\frac{E}{4N_0}}\right] \qquad \dots (5.144)$$

EXAMPLE 5.19. A signal is either $s_1(t) = A\cos(2\pi f_0 t)$ or $s_2(t) = 0$ for an interval $T = \frac{n}{f_0}$ with 'n' an integer. The signal is corrupted by white noise with PSD = $\frac{N_0}{2}$. Find the transfer function of the matched filter for this signal. Write an expression for the probability of error P_e.

Solution: The given signal is an BASK signal. It has been given that,

$$x_1(t) = A\cos(2\pi f_0 t) = \sqrt{2P_s}\,\cos(2\pi f_0 t)$$

and

$$x_2(t) = 0$$

$$A = \sqrt{2P_s} \text{ or } P_s = \frac{A^2}{2}$$

We have already derived the expressions for the transfer function and error probability. They are as under:

$$H(f) = \frac{2k}{N_0} X^*(f) e^{-j2\pi fT}$$

and
$$P_e = \frac{1}{2} \text{erfc} \sqrt{\frac{E}{4 N_0}}$$

where
$$X(f) = \text{F.T. of } x(t) \text{ and } E = P_s T = \frac{A^2}{2} T$$

Expression for transfer function

$$H(f) = \frac{2k}{N_0} X^*(f) e^{-j2\pi fT}$$

$$x(t) = x_1(t) - x_2(t) = s_1(t) - s_2(t) = A \cos (2\pi f_0 t)$$

Therefore,
$$X(f) = FT[x(t)] = FT[A \cos 2\pi f_0 t]$$

$$X(f) = \frac{A}{2} \{\delta(f - f_0) + \delta(f + f_0)\}$$

As the imaginary part in the above expression is zero.

$$X^*(f) = X(f) = \left\{\frac{A}{2} \delta(f - f_0) + \delta(f + f_0)\right\}$$

Substituting, we get,
$$H(f) = \frac{2k}{N_0} \times \frac{A}{2} \{\delta(f - f_0) + \delta(f + f_0) e^{-j2\pi fT}\}$$

$$H(f) = \frac{kA}{N_0} \{\delta(f - f_0) + \delta(f + f_0)\} e^{-j2\pi fT}$$

This is the desired result.

Expression for P_e

$$P_e = \frac{1}{2} \text{erfc} \sqrt{\frac{E}{4 N_0}}$$

EXAMPLE 5.20. It is given that the binary data is transmitted at a rate of 10^6 bits/second over a channel having a bandwidth 3 MHz. Assume that the noise PSD at the receiver is $\frac{N_0}{2} = 10^{-10}$ watts/Hz. Find the average carrier required at the receiver input for coherent PSK and DPSK-signaling schemes to maintain a probability of error $P_e = 10^{-4}$.

Solution: We know that the average error probability of PSK system is given by

$$P_e = \frac{1}{2} \text{erfc} \sqrt{\frac{E}{N_0}}$$

But
$$P_e = 10^{-4} \text{ (given)}$$

Therefore,
$$10^{-4} = \frac{1}{2} \text{erfc} \sqrt{\frac{E}{N_0}}$$

or
$$2 \times 10^{-4} = \text{erfc} \sqrt{\frac{E}{N_0}}$$

But
$$\text{erf}(u) = 1 - \text{erfc}(u)$$

Therefore, $1 - 2 \times 10^{-4} = 1 - \text{erfc}\sqrt{\frac{E}{N_0}}$

or $0.998 = \text{erf}\sqrt{\frac{E}{N_0}}$

Further, we have

$$\text{erf}(2.6) \cong 0.998$$

Therefore, $\sqrt{\frac{E}{N_0}} = 2.6$

$$\frac{E}{N_0} = 6.76$$

or $E = 6.76\ N_0$

But $N_0 = 2 \times 10^{-10}$

Therefore, $E = 6.67 \times 2 \times 10^{-10} = 1.352 \times 10^{-9}$ Joules

But $E = PT$ and $T = \frac{1}{\text{Bit rate}} = \frac{1}{10^6}$

Therefore, $P = \frac{E}{T} = 1.352 \times 10^{-10} \times 10^6 = 1.352$ mW **Ans.**

The is the required carrier power.

Now, the error probability of DPSK system is given by,

$$P_e = \frac{1}{2} e^{-E_b/N_0}$$

Therefore, $10^{-4} = \frac{1}{2} e^{-E_b/N_0}$

or $\frac{-E_b}{N_0} = -8.5171$

or $\frac{E_b}{N_0} = 8.5171,$

or $E_b = 8.5171\ N_0$

or $E_b = 8.5171 \times 2 \times 10^{-10} = 1.7 \times 10^{-9}$ Joules

Hence, power will be $P = \frac{E_b}{T_b} = 1.7 \times 10^{-9} \times 10^6 = 1.7$ mW **Ans.**

Important Point:

For the same value of P_e, the DPSK needs to transmit more power than that transmitted by a PSK system.

EXAMPLE 5.21. An on-off binary system uses the pulse waveforms,

$$s_1(t) = \begin{cases} s_1(t) = A\sin\left(\frac{\pi t}{T}\right), & 0 \le t \le T \\ s_2(t) = 0, & \text{otherwise} \end{cases}$$

Let A = 0.2 mV and T = 2 μs. Additive white gaussian noise with power spectral density $\frac{\eta}{2} = 10^{-15}$ Watts/Hz is added to the signal. Determine the probability of error when $P(s_1) = P(s_2) = \frac{1}{2}$. Take Q $(\sqrt{10}) = 7.83 \times 10^{-4}$.

Solution: The given system is an ASK system.

Therefore, $s_1(t) = A \sin (\pi t/T)$

and $s_2(t) = 0$

We know that the error probability of an ASK system is given by

$$P_e = \frac{1}{2} \text{erfc}\left\{\frac{x_{01}(T) - x_{02}(T)}{2\sqrt{2}\,\sigma}\right\} \quad \text{... (i)}$$

But, $$\left\{\frac{x_{01}(T) - x_{02}(T)}{2\sqrt{2}\,\sigma}\right\}^2_{max} = \frac{2}{N_0}\int_0^T x^2(t)$$

where $x(t) = x_1(t) - x_2(t) = s_1(t) - s_2(t)$

$x(t) = A \sin (\pi t./T)$

Therefore, $$\left\{\frac{x_{01}(T) - x_{02}(T)}{\sigma}\right\}^2_{max} = \frac{2}{N_0}\int_0^T A^2 \sin^2(\pi t/T)$$

But $$\sin^2 x = \frac{1}{2}(1 - \cos 2x)$$

Therefore, $$\left\{\frac{x_{01}(T) - x_{02}(T)}{\sigma}\right\}^2_{max} = \frac{2}{N_0}\int_0^T \frac{A^2}{2}\left[1 - \cos\left(\frac{2\pi t}{T}\right)\right]dt$$

$$= \frac{2}{N_0}\int_0^T A^2 dt - \frac{2}{N_0}\int_0^T \frac{A^2}{2}\cos\frac{2\pi t}{T}dt$$

$$= \frac{A^2}{N_0}[t]_0^T - \frac{A^2}{N_0}\left[\sin\frac{2\pi t}{T}\right]_0^T = \frac{A^2 T}{N_0} - 0$$

or $$\left\{\frac{x_{01}(T) - x_{02}(T)}{\sigma}\right\}^2_{max} = \frac{A^2 T}{N_0} \quad \text{...(ii)}$$

Substituting equation (ii) into (i), we get,

$$P_e = \frac{1}{2} \text{erfc}\left\{\frac{x_{01}(T) - x_{02}(T)}{2\sqrt{2}\,\sigma}\right\} = \frac{1}{2}\text{erfc}\left\{\frac{1}{2\sqrt{2}}\sqrt{\frac{A^2 T}{N_0}}\right\} \quad \text{... (iii)}$$

We know that the relation between erfc and Q function is given by

$$Q(x) = \frac{1}{2}\text{erfc}\left(\frac{x}{\sqrt{2}}\right)$$

Hence, we rearrange equation (iii) as under:

$$P_e = \frac{1}{2}\text{erfc}\left\{\frac{\sqrt{A^2\,T/4\,N_0}}{\sqrt{2}}\right\}$$

Therefore, $$P_e = Q\left\{\sqrt{\frac{A^2\,T}{4\,N_0}}\right\} \quad \text{... (iv)}$$

But, $A = 0.2 \times 10^{-3}$ V, $T = 2 \times 10^{-6}$ sec, $\frac{N_0}{2} = 10^{-15}$ W/Hz.

Therefore, $$P_e = Q\left\{\sqrt{\frac{(0.2\times10^{-3})^2\times2\times10^{-6}}{4\times2\times10^{-15}}}\right\} = Q\left[\sqrt{10}\right]$$

or $P_e = 7.83 \times 10^{-4}$ **Ans.**

EXAMPLE 5.22. Binary data is transmitted over a microwave link of a rate of 10^6 bits/sec and the PSD of noise at the receiver input is 10^{-10} watts/Hz. Find the average carrier power required to maintain an average probability of error $P_e \leq 10^{-4}$ for coherent binary FSK. What is the required channel bandwidth?

Solution: Given: Bit rate = 10^6 bits/s

$$\frac{N_0}{2} = 10^{-10}\ \text{W/Hz}$$

Therefore, $N_0 = 2 \times 10^{-10}$ W/Hz

$P_e \leq 10^{-4}$; $P_s = ?$

The error probability of FSK with coherent detection is given by,

$$P_e = \frac{1}{2} \text{erf}_c \left[\frac{0.6\, E_b}{N_0}\right]^{1/2}$$

But $E_b = P_s T_b$

Therefore,

$$P_e = \frac{1}{2} \text{erf}_c \left[\frac{0.6\, P_s T_b}{N_0}\right]^{1/2}$$

But

$$T_b = \frac{1}{\text{bit rate}}$$

Therefore,

$$10^{-4} = \frac{1}{2} \text{erfc} \left[\frac{0.6\, P_s}{2 \times 10^{-10} \times 10^6}\right]^{1/2}$$

or

$$2 \times 10^{-4} = \text{erf}_c\, [3000\, P_s]^{1/2}$$

But, 1 – erfc (u) = erf(u)

Hence,

$$1 - 2 \times 10^{-4} = 1 - \text{erfc}\, [3000\, P_s]^{1/2}$$

or

$$0.9998 = \text{erf}\, [3000\, P_s]$$

We have,

$$0.9998 = \text{erf}\, [2.8]$$

or

$$3000\, P_s = 2.8$$

$$P_s = \frac{2.8}{3 \times 10^3} = 0.933\ \text{mW}$$ **Ans.**

EXAMPLE 5.23. Error probability of 10^{-5} is desired and a channel bandwidth of 20 kHz is available for 16 QAM or QPSK system. Calculate the value of E_b/η required for each of these systems.

Solution: Given that: $P_e = 10^{-5}$, B = 20 kHz, Bit Rate = 80 kbps

We know that the error probability of a QPSK system is given by

$$P_e = \text{erfc} \sqrt{E_b / N_0}$$

Therefore,

$$10^{-5} = \text{erfc} \sqrt{E_b / N_0}$$

But erf (u) = 1 – erfc (u)

Therefore,

$$1 - 10^{-5} = 1 - \text{erfc} \sqrt{E_b / N_0} = \text{erf} \sqrt{E_b / N_0}$$

or

$$0.99999 = \text{erf} \sqrt{E_b / N_0}$$

Then, we have

$$\sqrt{E_b / N_0} = 3$$

or

$$E_b/N_0 = 9$$ **Ans.**

5.26 SIGNAL SPACE ANALYSIS OF OPTIMUM DETECTION

Thus far, we have discussed only the simple linear threshold detection type receivers for the channels under the Gaussian noise. But the major constraint on such receivers is their linear structure. A truly optimum receiver is that receiver which can detect an M-ary signal with channel noise n(t) with a minimum error probability. The channel output in this case is given by,

$$y(t) = x_i(t) + n(t) \qquad 0 \le t \le T \quad i = 1, 2, \dots M$$

In order to analyze this, we recognize that the channel output y(t) is a random process. The receiver must make a decision by transforming y(t) into a finite dimensional space. For such an analysis, it is necessary to use the geometrical representation of signals and noises.

5.26.1 Vector Space Representation of Signals

In practice, we use the 3-dimensional physical space and 3-dimensional vectors in this space. But it is possible to extend this concept to an n-dimensional space. A 3-dimensional vector is an entity specified by three numbers (x_1, x_2, x_3) in a certain order. Hence, an n-dimensional vector is an entity which is specified by n numbers in a certain order say $(x_1, x_2, \dots x_n)$.

5.26.2 Unit Vectors

Let the n unit vectors be denoted by $\phi_1, \phi_2 \dots \phi_n$ and may be defined as under:

$$\phi_1 = (1, 0, 0, \dots 0)$$
$$\phi_2 = (0, 1, 0, \dots 0)$$
$$\vdots$$
$$\vdots$$
$$\phi_n = (0, 0, 0, \dots 1)$$

5.26.3 Representation of a Vector

Any vector $x = (x_1, x_2, \dots x_n)$ can be represented as a linear combination of n unit vectors as follows:

$$x = \phi_1\, x_1 + \phi_n x_2 + \dots \phi_n x_n \qquad \dots(5.145)$$

Therefore,

$$x = \sum_{k=1}^{n} \phi_k x_k \qquad \dots(5.146)$$

5.26.4 Scalar Product of Vectors

Let x and y be two n dimensional vectors in the same space such that $x = (x_1, x_2, \dots x_n)$ and $y = (y_1, y_2, \dots y_n)$. Then their scalar product x.y is defined as,

$$x \,.\, y = \sum_{k=1}^{n} x_k y_k \qquad \dots(5.147)$$

Vectors x and y are said to be orthogonal if their scalar product is equal to zero i.e., if x.y = 0.

Length of a vector

The length of a vector x is $|x|$ and it is defined as follows:

$$|x|^2 = x.x = \sum_{k=1}^{n} x_k^2 \qquad \dots(5.148)$$

5.26.5 Independent Vectors

A set of n-dimensional vectors can be called as independent if none of them can be represented as a linear combination of the remaining vectors in the set. So, $x_1, x_2, \dots x_m$ are independent vectors if $a_1\, x_1 + a_2\, x_2 + \dots + a_m\, x_m \ne 0$

An n-dimensional space can contain at the most n independent vectors. if a space contains a maximum of n independent vectors, then we can express every vector as a linear combination of these n independent vectors.

Dimensions of vectors

It is not necessary that all the vectors in an n dimensional space be expressed by n numbers. This means that in a 3-dimensional space, all the vector lying in one plane can be specified by only two dimensions and we can specify the vector lying a line by only one dimension.

5.26.6 Basis Vectors

A set of n independent vectors in an n-dimensional space are known as basis vectors. Every basis vectors can be expressed as a linear combination of n independent vectors. These n-independent vectors form the co-ordinate axis but they are not unique. The n-unit vector $\phi_1, \phi_2, \ldots \phi_n$ can be used as basis vectors. They are also denoted by $u_1, u_2 \ldots$ These vectors have an important property. They are all mutually orthogonal. That is,

$$\phi_j.\phi_k = 0 \quad \ldots \text{ if } j \neq k \qquad \ldots(5.149)$$

$$= 1 \quad \text{if } j = k \qquad \ldots(5.150)$$

The n-basis vectors which are mutually independent and mutually orthogonal are called as orthonormal vectors because they are orthogonal and their length is unity. The vector $x = (x_1, x_2, \ldots x_n)$ can be represented as under:

$$x = x_1\phi_1 + x_2\phi_2 + \ldots + x_n \phi_n)$$

If x is known, then we can obtain its component x_k by taking the scalar product of both sides with ϕ_k as follows:

$$x.\phi_k = \phi_k.(x_1\phi_1 + x_2\phi_2 + \ldots + x_n\phi_n)$$

$$= x_k \,.\, x_1\phi_1 + \phi_{k\,.}x_2 + \phi_k.x_n\phi_n$$

$$= 0 + 0 + \ldots xk + 0$$

Therefore, $\quad x.\phi_k = x_k$

An n-dimensional vector can be represented completely if and only if we use n-basis vectors. A set of n-basis vectors is called as a complete set.

5.26.7 Signal as a vector

An entity x which can be represented by an n-tuple i.e., $(x_1, x_2, \ldots x_n)$ is called as an n-dimensional vector. Therefore, if a signal x(t) can be specified using an n-tuple it can also be called as vector. Using the sampling theorem, we can express a bandlimited signal x(t) by the values of its samples. If the number of samples is n then signal is being represented by n values. Hence, it is also a vector.

5.26.8 Concept of Signal Space

The n-signals $\phi_1(t), \phi_2(t), \ldots \phi_n(t)$ are defined as independent signals if none of them can be represented as the linear combination of the remaining (n – 1) signals.

Let us assume that the signal x(t) can be represented by a linear combination of n-independent signals as follows:

$$x(t) = x_1\phi_1(t) + x_2\phi_2(t) + \ldots x_n\phi_n(t)$$

$$x(t) = \sum_{k=1}^{n} x_k \, \phi_k(t) \qquad \ldots(5.151)$$

If every signal in a certain signal space can be represented by a linear combination of "n" independent signals then we will have a n-dimensioinal space.

The n-independent signals $\phi_1(t), \phi_2(t), \ldots$ will we assumed to be orthogonal. That means,

$$\int_{-\infty}^{\infty} \phi_j(t)\,\phi_k(t)\,dt = 0 \;...\; \text{if } j \neq k \quad ...(5.152)$$

$$= k_j \;...\; \text{if } j = k \quad ...(5.153)$$

If $k_j = 1$ for all j, then the above mentioned set is orthonormal as well.

Important Point: Any band-limited signal x(t) can be represented by a point $(...x_{-k}, x_2, x_{-1}, x_0, x_1, x_2, ... x_k,...)$ where $x_1, x_2...$ represent the samples of signal x(t) taken at the Nyquist rate.

5.26.9 Vectors of Orthogonal Signals

If signals x(t) and y(t) are mutually orthogonal then their corresponding vectors x and y will also be orthogonal.

EXAMPLE 5.24. A signai space consists of the four signals $s_1(t)$, $s_2(t)$, $s_3(t)$ and $s_4(t)$ which are shown in figure 5.37. Represent these signals geometrically in the vector space representation. Also determine a suitable set of basis vectors and dimensionality of the signals.

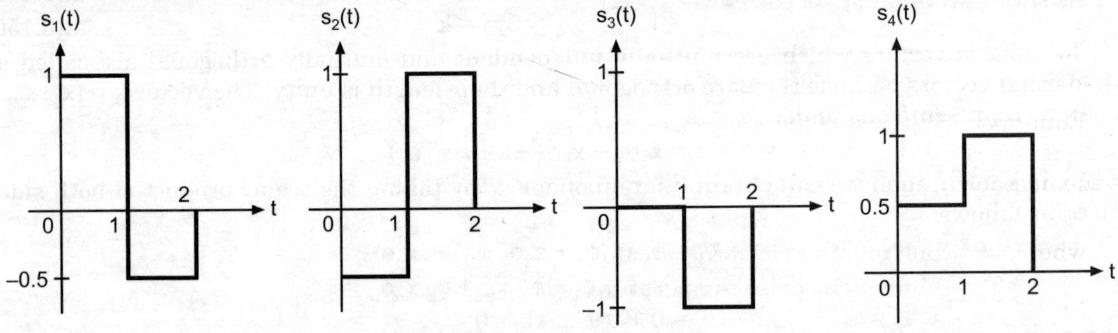

Fig. 5.37. Four given signals

Solution: (i) The two basis signals are as shown in figure 5.38. These are rectangular pulses of unit amplitude. They exist over the intervals 0 to 1 and 1 to 2.

(ii) Now, let us obtain the vectors. In terms of the basis signals $\phi_1(t)$ and $\phi_2(t)$, the signal vectors s_1, s_2, s_3, s_4 corresponding to signals $s_1(t)$, $s_2(t)$, $s_3(t)$ and $s_4(t)$ are $s_1 = (1, -0.5)$, $s_2 = (-0.5, 1)$, $s_3 = (0, -1)$ and $s_4 = (0.5, 1)$. These points are plotted in figure 5.39.

The signals $s_1(t)$ and $s_4(t)$ are orthogonal.

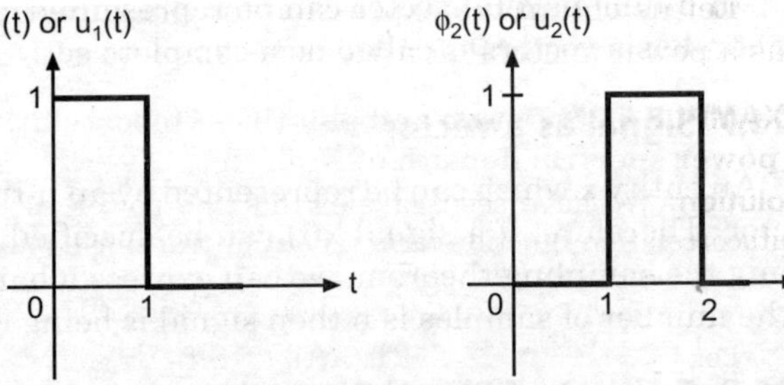

Fig. 5.38. Set of Basis signals

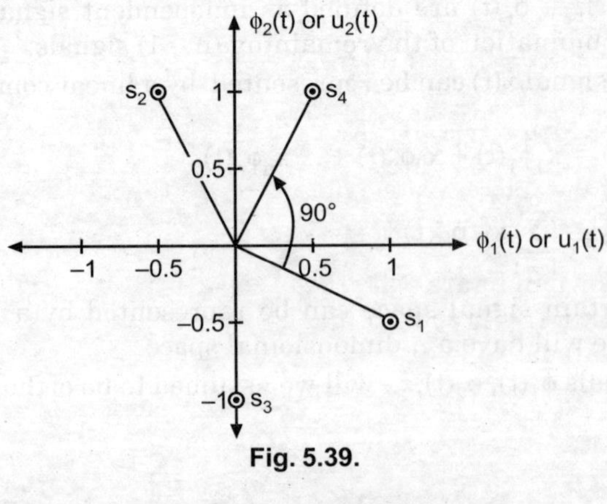

Fig. 5.39.

5.27 VECTOR DECOMPOSITION OF WHITE NOISE RANDOM PROCESS

5.27.1 White Noise Signal

White noise is the noise whose power spectral density is uniform over the entire frequency range of interest as shown in figure 5.40.

Why is it called as white noise?

The white noise contains all the frequency components in equal proportion. This is analogous with white light which is a superposition of all visible spectral components.

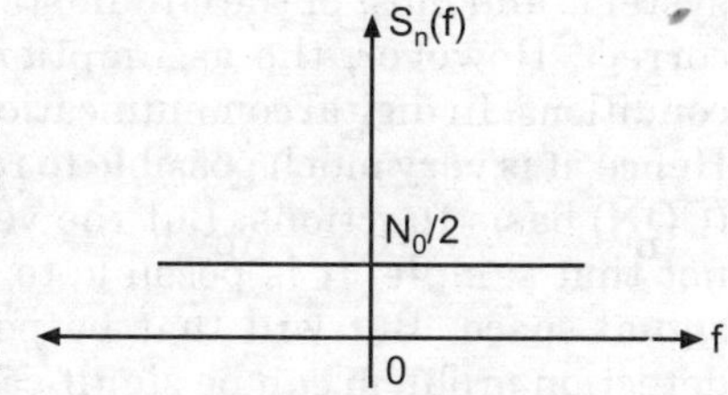

Fig. 5.40. Power spectral density of white noise

Why is it called as Gaussian noise?

The white noise has a Gaussian distribution. This means that the PDF of white noise has the shape of Gaussian PDF. Therefore, it is called as Gaussian noise.

Power spectral density of white noise

As shown in figure 5.40, the power spectral density (psd) of a white noise is given by,

$$S_n(f) = \frac{N_o}{2} \quad ...(5.154)$$

Equation (5.154) shows that the power spectral density of white noise is independent of frequency. As N_o is constant, the psd is uniform over the entire frequency range including the positive as well as the negative frequencies. N_o in equation (5.154) is defined as:

$$N_o = K\,T_e \quad ...(5.155)$$

where, K = Boltzmann's constant, and,

T_e = Equivalent noise temperature of the system

5.27.2 Auto Correlation Function of White Gaussian Noise

Let us obtain the expression for the autocorrelation function of a white Gaussian noise which has a power spectral density of $N_o/2$.

EXAMPLE 5.25. Derive and plot the autocorrelation function of the white Gaussian noise which has a power spectral densith of $N_o/2$.

Solution: We have seen earilier that the power spectral density and autocorrelation function form a Fourier transform pair. That means

$$R(\tau) \xleftrightarrow{FT} S(f)$$

i.e., $$FT\{R(\tau)\} = S(f)$$

or $$R(\tau) = IFT\{S(f)\}$$

Therefore for the white noise the autocorrelation function $R(\tau)$ is obtained as follows :

$$R(\tau) = IFT\{S_n(f)\} = IFT\{N_o/2\}$$

$$R(\tau) = \frac{N_o}{2}\,\delta(t) \quad \textbf{Ans.}$$

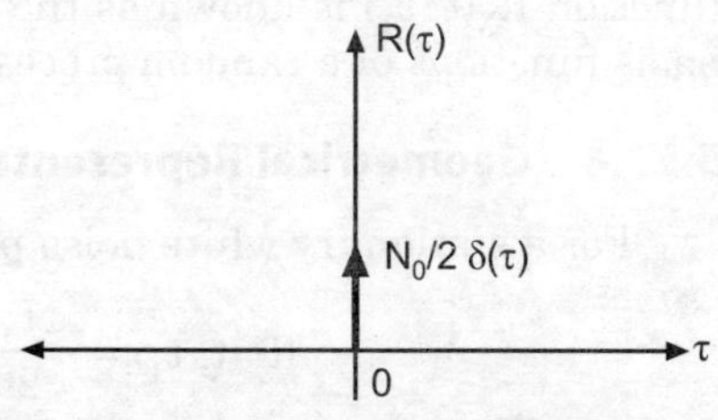

Fig. 5.41. Autocorrelation function of white density

This is the expression for the autocorrelation function of white noise. The autocorrelation function can be plotted as shown in figure 5.41.

Important Point: As shown in figure 5.41, the autocorrelation function of white noise is a delta function weighted by a factor ($N_o/2$) at $\tau = 0$. The meaning of this is that any two samples of white noise are totally uncorrelated.

Now, let us concentrate on the noise which can be described as an ergodic random process. Further, we shall assume that the probability density of the noise is Gaussian. In a lot of communication systems and most of the circumstances the assumption of Gaussian density proves to be perfectly correct. However, the assumption of Gaussian density can be wrong under certain operating conditions. In digital communications the message signal is always one of the M possible waveforms. Hence, it is very much possible to represent all M waveforms through a set of complete orthonormal (CON) basis functions. But the vector decomposition of the random noise n(t) at the receiver is not that sample. It is possible to represent a deterministic signal by one vector, a point in the signal space. But will that be possible for a random variable? If the answer is yes then the detection problem can be significantly simplified. Consider a complete orthonormal (CON) set of basis functions $\{\phi_4(t)\}$ for a signal defined over [0, T]. Then any deterministic signal s(t) in this signal space will satisfy the following condition.

$$\int_0^T | s(t) - \sum_k S_k \phi_k(t) |^2 dt = 0 \qquad ...(5.156)$$

This means that for $t \varepsilon [0, T]$ the following equality is true.

$$s(t) = \sum_k S_k \phi_k(t) \qquad ...(5.157)$$

But, this is not true for a random process defined over [0, T]. Some modifications are necessary.

5.27.3 Basis Function for a Random Process

For deterministic signals we can derive the basis signals via the Gram-Schmidt orthogonalization procedure. But this procedure is invalid for the random processes. A random process X(t) is an ensemble of signals. Therefore, the basis signals $\{f_k(t)\}$ should also depend on the characteristics of the random process. Here, we state that the orthonormal basis functions must be solutions of the following expression:

$$\lambda_i \phi_i(t) = \int_0^T R_X(t, t_1)\, \phi_i(t_1) dt_1 \qquad 0 \le t \le T \qquad ...(5.158)$$

The solution of this expression is known as Karhunel-Loeve expansion and the auto correlation function $R_X(t, t_1)$ is known as the Kernel function. The Karhunel-Loeve expansion shows that the basis functions of a random process X(t) depend on its autocorrelation function $R_x(t, t_1)$.

5.27.4 Geometrical Representation of White Noise Process

For a stationary white noise process X(t), the autocorrelation function is given by,

$$R_X(t, t_1) = \frac{N_o}{2} \delta(t - t_1) \qquad ...(5.159)$$

For this special kernel, the integration in equation (5.158) gets reduced to a simple form as follows:

$$\lambda_i \phi_i(t) = \int_0^T \frac{N_o}{2} \delta(t - t_1)\, \phi_i dt_1 = \frac{N_o}{2} \phi_i(t) \qquad t \in (0, T) \qquad ...(5.160)$$

The meaning of this result is that any CON set of basis functions can be used for representing stationary white noise process.

Also $\qquad \lambda_i = \dfrac{N_o}{2} \qquad$ ... for all i

For an M-ary transmission the M signals are represented by the set of orthonormal basis function $\{\phi_k(t)\}$ as follows:

$$S_i(t) = \sum_k S_{ik}\,\phi_k(t) \qquad i = 1, ..., M \qquad ...(5.161)$$

Based on equation (5.160), these basis function are also suitable for representing the white channel noise $n_w(t)$ such that,

$$n_w(t) \overset{m,s}{=} \sum_k n'_k \phi_k(t) \qquad i = 1, ... M \qquad ...(5.162)$$

The received signal y(t) is given by,

$$y(t) = S_i(t) + n_w(t) \qquad ...(5.163)$$

$$\overset{m,s}{=} \sum_k S_{ik}\phi_k(t) + \sum_k n_k \phi_k(t) \qquad ...(5.164)$$

$$\overset{m,s}{=} \sum_k y_k \phi_k(t) \qquad ...(5.165)$$

Let us define

$$y_k = \int_0^T y(t)\,\phi_k{}^*(t)dt = S_{ik} + n_k \qquad ...(5.166)$$

Thus, the channel output signal can be effectively represented by a sequence of random variables $\{y_k\}$ of equation (5.166). This means that the optimum receiver for white noise channels can be derived from the information corresponding to $\{y_k\}$. It may be noted that the white noise X(t) consists of an ensemble of sample functions. The coefficients are given by,

$$x_k = \int_0^T x(t)\,\phi_k{}^*(t)dt \qquad k = 1, 2 ... \qquad (5.167)$$

These coefficients in the decomposition of equation (5.162) will be different for each sample function. Hence, the coefficients are random variables. Each sample function will have a specific vector $(x_1, x_2, ... x_n)$ and will correspond to one point in the signal space as shown in figure 5.41.

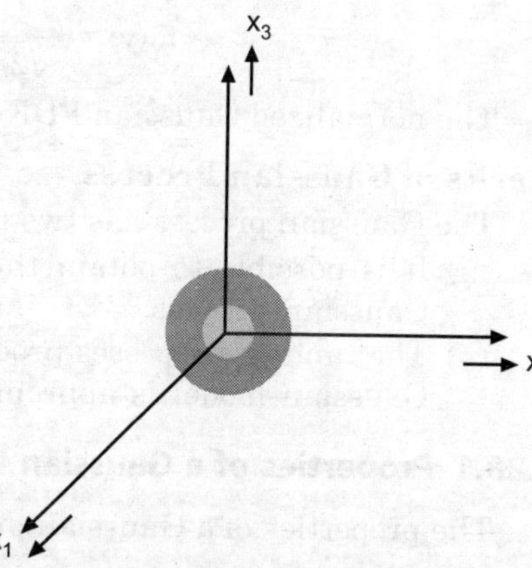

Fig. 5.42. Geometric representation of a Gaussian random process

5.28 GAUSSIAN PROCESS

In this section, we will discuss an important family of random processes known as Gaussian processes. Let the observation interval for a random process X(t) be t = 0 to t = T. Let us weight the random process X(t) by a function g(t), and obtain a random variable Y by integrating the product g(t) X(t) over the interval 0 to T. The random variable is given by,

$$Y = \int_0^T g(t)\,X(t)\,dt \qquad ...(5.168)$$

Y is the linear functional of X(t).

Definition

If in the expression for Y, the weighting function g(t) is such that the mean square value of the random variable Y is a Gaussian distributed random variable for every g(t), then the process X(t) is called as a Gaussian process. In simple words the process X(t) is a Gaussian process if every linear functional of X(t) is a Gaussian random variable. We have already defined a Gaussian variable and its properties. A random variable is called as Gaussian random variable if its probability density function (PDF) is of the following nature.

$$f_y(y) = \frac{1}{\sqrt{2\pi}\sigma_y} e^{-(y-m_y)^2/2\sigma_y^2} \quad ...(5.169)$$

where, m_y = Mean of random variable Y.

σ_y^2 = Variance

The plot of Gaussian PDF is shown in figure 5.43. It may be noted that the Gaussian random variable has been normalized to get the mean at $m_y = 0$ and variance $\sigma_y^2 = 1$.

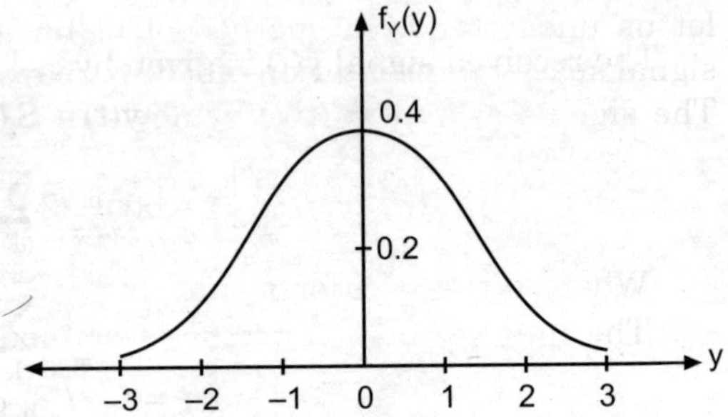

Fig. 5.43. Normalized PDF curve of a Gaussian variable

The normalized Gaussian PDF can be mathematically expressed by substituting $m_y = 0$ and $\sigma_y^2 = 1$ into equation (5.169) as follows:

$$f_Y(y) = \frac{1}{\sqrt{2\pi}} e^{-(y^2/2)} \quad ...(5.170)$$

The normalized Gaussian PDF is denoted as N(0, 1).

Merits of Gaussian Process

The Gaussian process has two main advantages:

(i) It is possible to obtain the analytical results because of many properties exhibited by Gaussian process.

(ii) The random processes produced by the physical phenomenon are generally such that the Gaussian model is appropriate to explain them.

5.28.1 Properties of a Gaussian Process

The properties of a Gaussian process are as follows:

Property 1

If a Gaussian process X(t) is applied to a stable linear filter then the process Y(t) developed at the filter output is also Gaussian.

Property 2

Consider the set of random variables or samples $X(t_1), X(t_2)... X(t_n)$ obtained by observing a random process X(t) at the time instants $t_1, t_2, ... t_n$. If the process X(t) is Gaussian, then these random variables are jointly Gaussian for any value of n. Their joint PDF is completely determined by specifying the set of means as follows:

$$m_x(t_i) = E[X(t_i)], i = 1, 2, ... n$$

and the set of autocovariance functions.

$$C_X(t_k, t_i) = [E(X(t_k) - m_{X(t_k)}) (X(t_i))]k, i = 1, 2, ... n$$

Property 3

If a Gaussian process is wide sense stationary, then the process is also stationary in strict sense.

Property 4

If the random variables $X(t_1), X(t_2), \ldots X(t_n)$ obtained by sampling a Gaussian process X(t) at times $t_1, t_2, \ldots t_n$ are uncorrelated, that means if,

$$E[X(t_k) - m_{X(t_k)})\, X(t_i) - m_{X(t_i)})] = 0 \qquad i \neq k$$

Then, these random variables are statistically independent.

5.29 USE OF SIGNAL SPACE TO CALCULATE (P_e)

We have discussed the procedure of plotting the signal space diagram of various systems. Now let us understand the method of using it for the calculation of error probability. Consider the signal space representation of BPSK shown in figure 5.43 (a). This signal space is one dimensional. The signals s_1 and s_2 (two symbols in BPSK) are given by,

$$\left.\begin{matrix} s_1 \\ s_2 \end{matrix}\right\} = b(t)\sqrt{2P_s}\ \cos\omega_c t \ \ldots.\ 0 < t \leq T_b \qquad \ldots (5.171)$$

Where b(t) = + 1 for s_1 and b(t) = – 1 for s_2. P_s represents the signal power.

The unit vector u(t) or $\phi(t)$ is defined as under:

$$u(t) = \sqrt{2/T_b}\ \cos\omega_c t$$

Therefore,
$$\left.\begin{matrix} s_1 \\ s_2 \end{matrix}\right\} = b(t)\sqrt{P_s T_b}\sqrt{2/T_b}\ \cos\omega_c t$$

$$= b(t)\sqrt{P_s T_b}\ u(t)\ \ 0 \leq t \leq T_b \qquad \ldots (5.172)$$

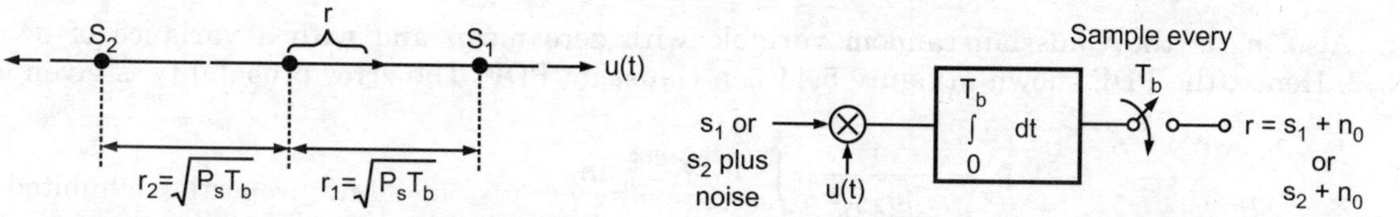

(a) Signal space representation of BPSK signal (b) Correlation receiver for BPSK

Fig. 5.43.

Figure 5.43(b) shows the correlator receiver for the BPSK signal. The input to this receiver is (s_1 + noise) or (s_2 + noise) where as its output is "r" which can be either r_1 or r_2 such that,

$$\left.\begin{matrix} r_1 = s_1 + n_0 \\ r_2 = s_2 + n_0 \end{matrix}\right\} \qquad \ldots (5.173)$$

Figure 5.43(a) shows the signal vectors each of length $\sqrt{P_s T_b}$ measured in terms of unit vector u(t).

The correlator receiver of figure 5.43(b) produces a response r as stated in equation (5.173). Imagine that in some interval, due to noise a response r is produced such that $r = r_1$, $r \neq r_2$. Then how to decide whether s_1 was transmitted or s_2 was transmitted? This decision is made by finding $|r - r_1|$ and $|r - r_2|$. If $|r - r_1| < |r - r_2|$ then, s_1 was transmitted and if $|r - r_2| < |r - r_1|$ then s_2 was transmitted. This indicates that if the response is r_k then, it corresponds to the signal s_k to which r is closest.

5.29.1. Error Probability of BPSK

Let the type of detection be synchronous detection. The correlator receiver of figure 5.43(b) can work as a synchronous detector for BPSK. The noise in the present case is given by,

$$n(t) = n_0 u(t) = n_0\sqrt{2/T_b}\ \cos\omega_c t \qquad \ldots (5.174)$$

It may be noted that the u(t) is being used alongwith the noise because BPSK is a single dimensional signal. n_0 is a Gaussian random variable with a variance $\sigma^2_0 = N_0/2$. Now, we assume that s_2 was transmitted. This signal reaches at the receiver input as $s_2 + n(t)$.

Therefore, receiver input = $s_2 + n(t)$

$$\text{Input} = -\sqrt{P_s T_b}\, u(t) + n_0\, u(t)$$

$$\text{(nput} = \left(n_0 - \sqrt{P_s T_b}\right) u(t) \quad \text{... (5.175)}$$

We know that $s_1 = \sqrt{P_s T_b}\, u(t)$. Therefore, any positive signal at the receiver output will be treated as s_1.

If the signal s_2 is to be mistaken as s_1, then from equation (5.175), it is clear than $n_0 > \sqrt{P_s T_b}$.

Hence, the probability of error is equal to the probability of n_0 being higher than $\sqrt{P_s T_b}$. This is graphically shown in figure 5.44. The shaded portion represents the probability of error P_e.

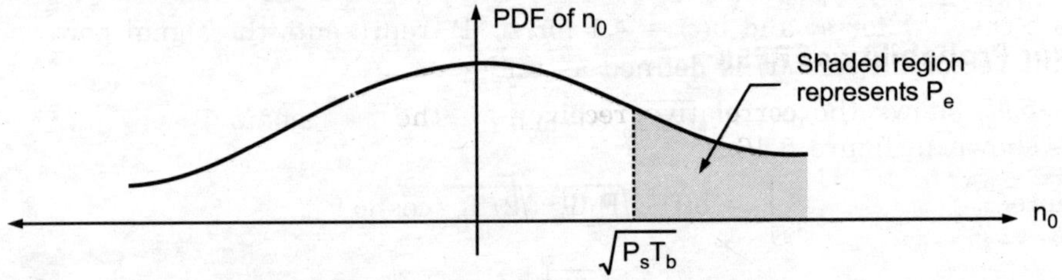

Fig. 5.44. Graphical representation of P_e.

Also, n_0 is the Gaussian random variable with zero mean and with a variance of $\sigma^2_0 = N_0/2$. Hence, the PDF shown in figure 5.44 is a Gaussian PDF. The error probability is given by,

$$P_e = \frac{1}{\sqrt{2\pi\sigma_0^2}} \int_{\sqrt{P_s T_b}}^{\infty} e^{-n_0^2/2\sigma_0^2}\, dn_0$$

$$P_e = \frac{1}{\sqrt{2\pi N_0/2}} \int_{\sqrt{P_s T_b}}^{\infty} e^{-n_0^2/2\sigma_0^2} dn_0 = \frac{1}{\sqrt{\pi N_0}} \int_{\sqrt{P_s T_b}}^{\infty} e^{-n_0^2/N_0} dn_0$$

Let $n^2_0/N_0 = y^2$

Therefore, $\quad \dfrac{n_0}{\sqrt{N_0}} = y \qquad dn_0 = \sqrt{N_0}\, dy$

For $\quad n_0 = \sqrt{P_s T_b} \qquad y = \sqrt{P_s T_b / N_0}$

and for $\quad n_0 = \infty \qquad y = \infty$

Hence, the expression of P_e gets transformed as follows:

$$P_e = \frac{1}{\sqrt{\pi}} \int_{\sqrt{P_s T_b/N_0}}^{\infty} e^{-y^2} dy$$

Rearranging, we get,

$$P_e = \frac{1}{2} \times \frac{2}{\sqrt{\pi}} \int_{\sqrt{P_s T_b/N_0}}^{\infty} e^{-y^2} dy = \frac{1}{2} \text{erfc} \sqrt{\frac{P_s T_b}{N_0}} \quad \text{... (5.176)}$$

But the distance between the two signals s_1 and s_2 from the signal space diagram of figure 5.44 is

$$d = 2\sqrt{P_s T_b}$$

Therefore, $\sqrt{P_s T_b} = d/2$ and $P_s T_b = d^2/4$

Substituting this value into equation (5.176), we obtain

$$P_e = \frac{1}{2} \text{erfc} \sqrt{\frac{d^2}{4 N_0}} \quad \text{... (5.177)}$$

Important Point: *The error probability goes on decreasing with increase in the distance "d" between the adjacent symbols in the signal space diagram.*

5.29.2 Error Probability of BFSK

Figure 5.45 shows the correlative receiver for the orthogonal BFSK and its signal space diagram is shown in figure 5.46.

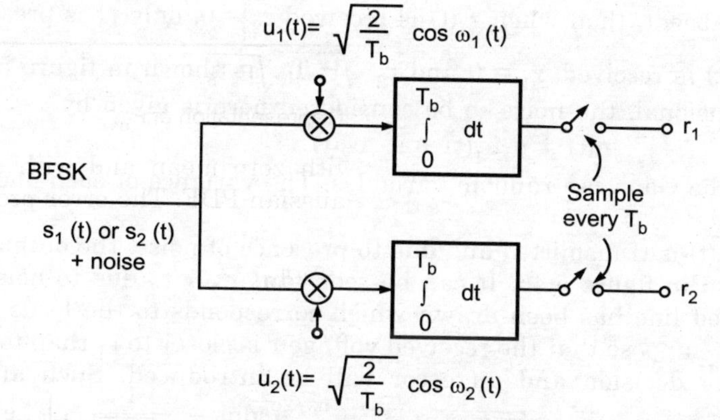

Fig. 5.45. Correlator receiver for BFSK

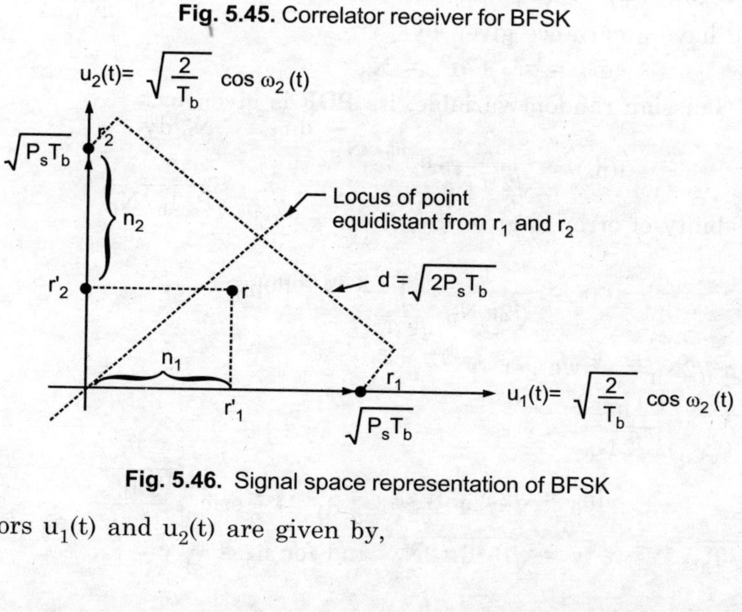

Fig. 5.46. Signal space representation of BFSK

The unit vectors $u_1(t)$ and $u_2(t)$ are given by,

$$u_1(t) = \sqrt{\frac{2}{T_b}} \cos \omega_1 t \quad \text{... (5.178)}$$

$$u_2(t) = \sqrt{\frac{2}{T_b}} \cos \omega_2 t \quad \text{... (5.179)}$$

The values of ω_1 and ω_2 are selected so as to ensure the orthogonality. The transmitted signals s_1 and s_2 are of power P_s and they are given by.

$$s_1(t) = \sqrt{2P_s} \cos \omega_1 t$$

And $$s_2(t) = \sqrt{2P_s} \cos \omega_2 t$$

Rearranging the equations, we get,

$$s_1(t) = \sqrt{P_s T_b}\sqrt{2/T_b} \cos \omega_1 t = \sqrt{P_s T_b}\, u_1(t) \quad \text{... (5.180)}$$

And $$s_2(t) = \sqrt{P_s T_b}\sqrt{2/T_b} \cos \omega_2 t = \sqrt{P_s T_b}\, u_2(t) \quad \text{... (5.181)}$$

The correlator receiver of figure 5.45 produces signals r_1 and r_2 respectively.

If the noise is absent, then when $s_1(t)$ is received $r_2 = 0$, only r_1 is present and $r_1 = \sqrt{P_s . T_b}$. Similarly when $s_2(t)$ is received, $r_1 = 0$ and r_2 $\sqrt{P_s T_b}$ as shown in figure 5.46. Since the BPSK signal is two dimensional, the noise to be considered here is given by

$$n(t) = n_1 u_1(t) + n_2 u_2(t) \quad \text{... (5.182)}$$

n_1 and n_2 are the Gaussian random variables. The variance of each one of them is $\sigma^2_1 = \sigma^2_2 = N_0/2$.

Assume that $s_2(t)$ is transmitted and due to presence of noise, the outputs of the receiver are r'_1 and r'_2 as shown in figure 5.46. It can be seen that $r'_2 \neq r_2$ due to noise n_2 and $r'_1 \neq 0$ due to noise n_1. A dotted line has been drawn which corresponds to the locus of points equidistant from r_1 and r_2. Now suppose that the received voltage r is closer to r_1 than to r_2. Then the receiver will make a wrong decision and an error will be introduced. Such an error will occur if $n_1 > (r_2 - n_2)$ or if $(n_1 + n_2) > \sqrt{P_s T_b}$ Since n_1 and n_2 are uncorrelated the random variable $n_0 = (n_1 + n_2)$ will have a variance given by,

$$\sigma^2_0 = \sigma^2_1 + \sigma^2_2 = N_0 \quad \text{... (5.183)}$$

Since n_0 is a Gaussian random variable, its PDF is given by,

$$f(n_0) = \frac{1}{\sqrt{2\pi N_0}} e^{-n_0^2/2N_0} \quad \text{... (5.184)}$$

And the probability of error is given by,

$$P_e = \frac{1}{\sqrt{2\pi N_0}} \int_{\sqrt{P_s T_b}}^{\infty} e^{-n_0^2/2N_0} dn_0 \quad \text{... (5.185)}$$

Substituting $n^2_0/2N_0 = y^2$ we get,

$$\frac{n_0}{\sqrt{2N_0}} = y$$

Therefore, $$dn_0 = \sqrt{2N_0}\, dy$$

For $n_0 = \sqrt{P_s T_b}$, $y = \sqrt{P_s T_b / 2N_0}$ and for $n_0 = \infty$, $y = \infty$

Therefore, $$P_e = \frac{1}{\sqrt{2\pi N_0}} \int_{\sqrt{P_s T_b / 2 N_0}}^{\infty} e^{-y^2} \sqrt{2 N_0}\, dy$$

$$= \frac{1}{\sqrt{\pi}} \int_{\sqrt{P_s T_b / 2 N_0}}^{\infty} e^{-y^2} dy = \frac{1}{2} \times \frac{2}{\sqrt{\pi}} \int_{\sqrt{P_s T_b / 2N_0}}^{\infty} e^{-y^2} dy$$

or $$P_e = \frac{1}{2} \operatorname{erfc} \sqrt{\frac{P_s T_s}{2 N_0}} \quad \text{... (5.186)}$$

From the signal space diagram of figure 5.46, we get,

$$d = \sqrt{2 P_s T_b}$$

Therefore, $$d^2 = 2 P_s T_b$$

or $$P_s T_b = d^2/2$$

Substituting this into equation (5.186), we obtain

$$P_e = \frac{1}{2} \operatorname{erfc} \sqrt{\frac{d^2}{4 N_0}} \quad \text{... (5.187)}$$

Important Point: *The error probability of BFSK is same as that of BPSK when P_e is expressed in terms of distance d. However the expression for P_e in terms of P_s T_b and N_0 shows the error probability of BFSK is higher than that of BPSK.*

5.29.3 Error Probability of QPSK

Figure 5.47 shows the signal space representation of a QPSK system. The two unit vectors are

$$u_1(t) = \sqrt{2/T_s} \cos \omega_c t \quad \text{... (5.188)}$$

$$u_2(t) = \sqrt{2/T_s} \sin \omega_c t \quad \text{... (5.189)}$$

where T_s = symbol duration $T_s = 2\ T_b$

Since the signal space is two dimensional, the relevant noise is given by,

$$n(t) = n_1 u_1(t) + n_2\, u_2(t) \quad \text{... (5.190)}$$

Where n_1 and n_2 are independent Gaussian random variables with variance $N_0/2$.

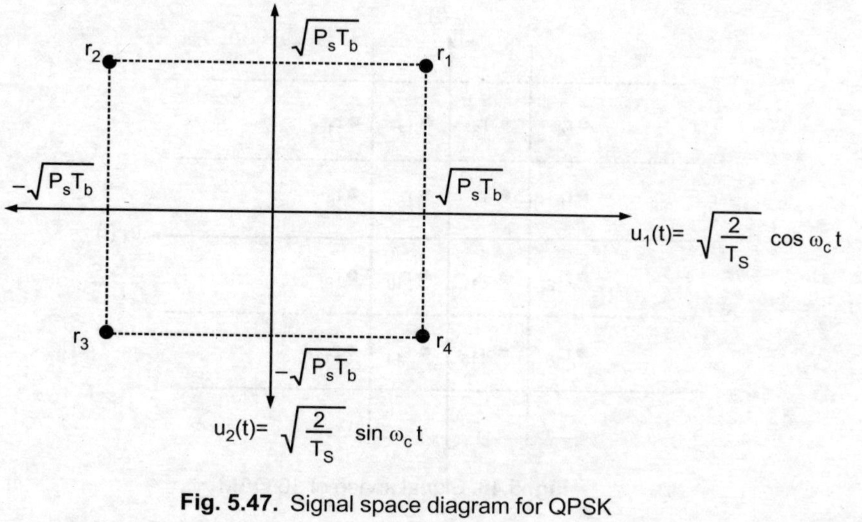

Fig. 5.47. Signal space diagram for QPSK

Let us now calculate the probability of correct reception and obtain the error probability. If s_1 is the transmitted, then the decision of the receiver is correct if and only if r_1 is in the first quadrant of figure 5.47. If the noise is sufficient, then the received signal r_1 will be driven out of the first quadrant. This will happen if and only if n_1 and n_2 both are in the range from $-\sqrt{P_s T_b}$ to infinity.

But, the distance "d" for QPSK is $2\sqrt{P_s T_b}$

Therefore, $\sqrt{P_s T_b} = d/2$

And, the probability of correct decision is given by,

$$P(C/s_1) = P\left(n_1 > -\frac{d}{2}, n_2 > -\frac{d}{2}\right) = \int_{-d/2}^{\infty} f(n_1)\,dn_1 \int_{-d/2}^{\infty} f(n_2)\,dn_2$$

But n_1 and n_2 are independent random variables.

Therefore,
$$P(C/s_1) = \int_{-d/2}^{\infty} \frac{e^{-n_1^2/N_0}}{\sqrt{\pi N_0}}\,dn_1 \int_{-d/2}^{\infty} \frac{e^{-n_2^2/N_0}}{\sqrt{\pi N_0}}\,dn_2 \quad ...(5.191)$$

But, $n_1 = n_2$

Therefore,
$$P(C/s_1) = \left[\frac{1}{\sqrt{\pi N_0}} \int_{-d/2}^{\infty} e^{-n_1^2/N_0}\,dn_1\right]^2 \quad ...(5.192)$$

It can be easily proved that,

$$P(C/s_1) = \left[1 - \frac{1}{2}\text{erfc}\sqrt{d^2/4N_0}\right]^2 \quad ...(5.193)$$

If each symbol s_1, s_2, s_3, s_4 are equally likely then the error probability is given by,

$$P_e = 1 - P(c) = 1 - \left[1 - \frac{1}{2}\,\text{erfc}\,\sqrt{d^2/4N_0}\right]^2 \quad ...(5.194)$$

5.30 ERROR PROBABILITY OF 16 QAM (16 QASK)

The signal space diagram of 16 QAM is shown in figure 5.48.

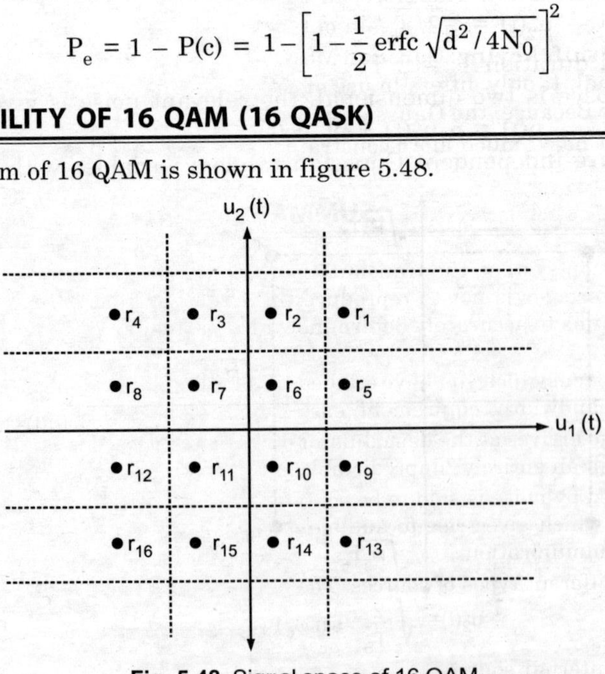

Fig. 5.48. Signal space of 16 QAM

Let us calculate the error probability for the symbol such as s_6 in figure 5.48 which is located at (a, a). This is signal has the largest probability of error. The signals s_7, s_{10} and s_{11} also will have the largest error probability. The minimum distance is given by,

$$d = \sqrt{0.4\,E_s} = \sqrt{1.6\,E_b}$$

And, the error probability is given by,

$$P_e \le 4\times\frac{1}{2}\,\text{erfc}\left[\frac{1.6\,E_b}{4\,N_0}\right]^{1/2} = 2\,\text{erfc}\left[0.4\frac{E_b}{N_0}\right]^{1/2}$$

GLOSSARY

1. **Coherent Digital Modulation Techniques:** Coherent digital modulation techniques are those techniques which employ coherent detection. In coherent detection, the local carrier generated at the receiver is phase locked with the carrier at the transmitter. Thus, the detection is done by correlating received noisy signal and locally generated carrier. The coherent detection is a synchronous detection.
2. **Non-coherent Digital Modulation Techniques:** Those techniques in which the detection process does not need receiver carrier to be phase locked with transmitter carrier. The advantage of such type of system is that the system becomes simple. But the drawback of such a system is that the error probability increases.
3. **ASK:** ASK is the simplest digital modulation technique. There is only one unit energy carrier and it is switched on or off depending upon the input binary sequence.
4. **Binary Frequency Shift Keying:** In BFSK, the frequency of a sinusoidal carrier is shifted according to the binary symbol. In other words, the frequency of a sinusoidal carrier is shifted between two discrete values. However, the phase of the carrier is unaffected. This means that we have two different frequency signals according to binary symbols.
5. **DPSK:** DPSK can be viewed as the non-coherent version of the PSK. Differential phase shift keying (DPSK) is differentially coherent modulation method. DPSK does not need a synchronous (coherent) carrier at the demodulator. The input sequence of binary bits is modified such that the next bit depends upon the previous bit. Therefore, in the receiver, the previous received bits are used to detect the present bit.
6. **Gaussian Minimum shift keying:** Gaussian MSK (GMSK) yields a constant amplitude and continuous phase RF carrier signal. It only differs in use of a Gaussian baseband pulse shape in place of square pulse shape for MSK. Because, the Gaussian pulse rises and decays asymptotically with respect to a zero response level, it has a much more constrained bandwidth.

SUMMARY

- In digital communication system, the transmitter input is chosen from a finite set of possible symbols. The objective at the receiver is not to reproduce the waveform that carries the symbol with fidelity; instead, the receiver aims to accurately determine which particular symbol was transmitted among the set of possible ones.
- The communication system which you have studied till now (analog or digital) have been compared, based on the points such as bandwidth requirement, ease of multiplexing, ease of implementation etc.
- When the signal voltage arrives at the demodulator of a receiver, it will be accompanied by a voltage waveform which varies with time in an entirely "unpredictable manner".
- The signal accompanied by noise is said to be contaminated or corrupted by noise. Noise is an unwanted electrical disturbance which gives rise to audible or visual disturbances in communication system and errors in the digital communication.
- Noise can arise from different types of sources. The noise sources are:
  (i) Natural sources
  (ii) Man made sources
  (iii) Fundamental or internal sources.

- The natural phenomena that give rise to noise are solar flares, electronic storms and radiation in space. The noise received by the receiving antenna from the natural sources can be reduced by repositioning the antenna (proper orientation of the antenna).
- The noise radiation from sun varies with temperature changes on the surface of the sun. These temperature changes follow a cycle of 11 years hence, the periods of great electrical disturbances also repeat after every 11 years. The cosmic noise comes from the stars.
- The man made noise is generated due to the make and break process in a current carrying circuit. The examples are: electrical motors, welding machines, ignition system of the automobiles, high current circuit using SCRs, fluorescent lamps etc.
- Partition noise is generated when a current gets divided between two or more paths. It is generated due to the random fluctuations in the current division. Therefore, the partition noise in a transistor will be higher than that in a diode. The devices like gallium arsenide FET draw almost zero gate bias current, hence, keeping the partition noise to its minimum value.
- The flicker noise will appear at frequencies below a few kilohertz. It is sometimes called as "1/f noise. In the semiconductor devices, the flicker noise is generated due to the fluctuations in the carrier density.
- The detection theory helps in designing a decision making device which looks at the input signals and makes a guesses about whether 0 or 1 was transmitted.
- The estimation theory has been developed in order to design a processor which uses the information in the received signal to extract the estimates of some physical parameters or waveforms. There is always a possibility of error getting introduced into the detection as well as estimation.
- A channel in digital communication system should have the following two characteristics.
  (i) The channel should be linear. It should have a sufficient bandwidth to ensure undistorted transmission of $s_i(t)$.
  (ii) The signal $s_i(t)$ is contaminated by white gaussian noise.
- When we have to transmit a digital signal over a long distance, we need continuous-wave (CW) modulation. For this purpose, the transmission medium can be in form of radio, cable or other type of channel.
- If the frequency of the sinusoidal carrier is switched depending upon the input digital signal, then it is known as the frequency shift keying (FSK). This is very much similar to the analog frequency modulation. If the phase of the carrier is switched depending upon the input digital signal, then it is called phase shift keying (PSK).
- In digital modulations, instead of transmitting one bit at a time, we transmit two or more bits simultaneously. This is known as M-ary transmission. This type of transmission results in reduced channel bandwidth. However, sometimes, we use two quadrature carriers for modulation. This process is known as Quadrature modulation.
- Every scheme offers system trade-offs of its own. However, the final choice made by the designer is determined by the way in which the available primary communication resources such as transmitted power and channel bandwidth are best exploited. In particular, the choice is made in favour of a scheme which possesses as many of the following design characteristics as possible:
  (i) Maximum data rate,
  (ii) Minimum probability of symbol error,
  (iii) Minimum transmitted power,
  (iv) Maximum channel bandwidth,
  (v) Maximum resistance to interfering signals,
  (vi) Minimum circuit complexity.
- ASK signal may be generated by simply applying the incoming binary data (represented in unipolar form) and the sinusoidal carrier to the two inputs of a product modulator (i.e., balanced modulator).
- The advantage of using BASK is its simplicity. It is easy to generate and detect.
- But the drawback of BASK is that it is very sensitive to noise, therefore, it finds limited application in data transmission. It is used at very low bit rates, upto 100 bits per sec.
- Binary phase shift keying (BPSK) is the most efficient of the three digital modulation, i.e., ASK, FSK and PSK. Hence, binary phase shift keying (BPSK) is used for high bit rates.
- The signal space is an N-dimensional Euclidean space which as N mutually perpendicular axes labelled $\phi_1, \phi_2, \dots \phi_N$.
- The transmitted signal $S_i(t)$ is applied to a bank of correctors. An appropriate set of N orthonormal basis functions is also applied. The resulting corrector outputs define the signal vector S_i. These vectors give us all the information about the transmitted signal $S_i(t)$.
- The receiver or detector is supposed to make a correct decision about which symbol is transmitted by

a transmitter. But it is not always possible for a practical receiver to make a correct decision. The wrong decision of receiver is called as an error and the probability of such error is called as error probability.

- The error probability is denoted by P_e. All the errors should be made to reduce the error probability to its minimum possible value.
- The error probability P_e is also called as the average probability of symbol error. A system with smaller value of P_e is better than the system having larger value of P_e.

SHORT QUESTIONS WITH ANSWERS

Q. 1. What is meant by noise?

Ans. When the signal voltage arrives at the demodulator of a receiver, it will be accompanied by a voltage waveform which varies with time in an entirely "unpredictable manner". This unpredictable voltage waveform is a random process called "Noise". The signal accompanied by noise is said to be "contaminated" or corrupted by noise. Noise is an unwanted electrical disturbance which gives rise to audible or visual disturbances in communication system and errors in the digital communication.

Q. 2. What are source of noise.

Ans. Noise can arise from different types of sources. The noise sources are:

(i) Natural sources
(ii) Man made sources
(iii) Fundamental or internal sources.

Q.3. What is meant by shot noise?

Ans. The shot noise is produced due to shot effect. The shot noise is produced in all the amplifying devices rather in all the active devices. The shot noise is produced due to the random variations in the arrival of electrons (or holes) at the output electrons of an amplifying device. Therefore, it appears as a randomly varying noise current superimposed on the output. The shot noise sounds like a shower of lead shots falling on a metal sheet.

Q.4. What is meant by detection and estimation?

Ans. Detection: The detection theory helps in designing a decision making device which looks at the input signals and makes a guesses about whether 0 or 1 was transmitted.

Estimation: The estimation theory has been developed in order to design a processor which uses the information in the received signal to extract the estimates of some physical parameters or waveforms. There is always a possibility of error getting introduced into the detection as well as estimation.

Q.5. Write the expression for minimum error probability of matched filter.

Ans. Minimum error probability of matched filter:

$$P_e = \frac{1}{2}\,\text{erfc}\sqrt{\frac{E}{N_0}}$$

Q.6. What is meant by quadrature modulation?

Ans. In digital modulations, instead of transmitting one bit at a time, we transmit two or more bits simultaneously. This is known as M-ary transmission. This type of transmission results in reduced channel bandwidth. However, sometimes, we use two quadrature carriers for modulation. This process is known as Quadrature modulation.

Q.7. What is signal space?

Ans. The signal space is an N-dimensional Euclidean space which as N mutually perpendicular axes labelled $\phi_1, \phi_2, \cdots \phi_N$.

Q.8. Define probability of error (P_e)?

Ans. The receiver or detector is supposed to make a correct decision about which symbol is transmitted by a transmitter. But it is not always possible for a practical receiver to make a correct decision. The wrong decision of receiver is called as an error and the probability of such error is called as error probability. The error probability is denoted by P_e. All the errors should be made to reduce the error probability to its minimum possible value. The error probability P_e is also called as the average probability of symbol error. A system with smaller value of P_e is better than the system having larger value of P_e.

REVIEW QUESTIONS

1. Explain the various techniques to detect the baseband digital signals.
2. Explain how integrator is used to detect baseband digital signals. Obtain an expression of signal to noise ratio for Integrate and Dump receiver.
3. Derive an expression for error probability for an integrator.
4. Derive an expression for error probability for optimum filter.
5. Derive an expression for transfer function for optimum filter.
6. Explain matched filter. How it differs from optimum filter. Derive an expression for impulse response of the matched filter.
7. Derive an expression for error probability of matched filter.
8. What is a correlator ? Explain is the difference and similarity between correlator and matched filter?
9. Derive expressions for the error probabilities of binary PCM, ASK, PSK and FSK signals which use matched filter detection in presence of white Gaussian noise.
10. Explain coherent and noncoherent binary modulation techniques.
11. Explain coherent Binary ASK.
12. Describe coherent Binary FSK.
13. Explain the condition of orthogonally of the two BFSK signals and the signal space diagram for the orthogonal signal and hence derive an expression for the probability of error in synchronous detection of the BFSK signals using matched filters.
14. Derive an expression for the spectrum of BFSK and sketch it for the condition of orthogonality and minimum bandwidth.
15. What is the advantage of gray coding of the QPSK system? What is the advantage of differential encoding of the input to the QPSK system?
16. What type of synchronization is used in QPSK system?
17. Draw the signal space representation of QPSK and BPSK signals. Show that for the same data rate and same bit energy they have the same bit error rate probability on the same AWGN channel, although one of the signals occupy half the bandwidth.
18. What is the ambiguity in the decoded output in case of PSK systems? How is it corrected?
19. Draw the block diagram of DPSK modulator and explain how synchronization problem is avoided for its detection.
20. Explain the ASK system and derive the relation for error probability of binary ASK.
21. Define error probability.
22. Derive the expression for probability of bit error for coherently detected DEPSK.
23. State the expression of bit error probability for non coherently detected BFSK.
24. What is the error probability of DPSK signal?
25. Explain the symbol error performance for M-ary systems.
26. Obtain the expression for symbol error for MPSK systems.
27. Compare the MPSK and MFSK systems.

NUMERICAL PROBLEMS

1. In a binary PCM system on/off signaling is used. The matched filter receiver is used for detection of the signal. Calculate the probability of error if signaling rate is doubled.

   [**Hint:** If signaling rate is doubled, bit period 'T' is reduced by half, hence energy is also reduced by half.] $\left[\textbf{Ans. } P_e = \frac{1}{2}\,\text{erfc}\sqrt{\frac{E}{8N_0}}\right]$

2. An FSK system transmits binary data at the rate of 2.5×10^6 bits per second. During the transmission, the white noise of zero mean and power spectral density 10^{-20} watts/Hz is added to the signal. In absence of noise, the amplitude of the received sinusoidal wave for digit 1 or 0 is 1 microvolt. Find the average probability of symbol error for current binary FSK system. $\left[\textbf{Ans. } P_e = \frac{1}{2}\,\text{erfc}\,\sqrt{6}\right]$

3. A binary PCM uses NRZ signaling has error probability of 10^{-6}. Assume that the signalling rate is doubled, then find new value of the average probability of error.
   [**Hint:** Here NRZ signaling is used. Hence energy equation and error probability equation will be different than those derived in sec. Since signalling rate is doubled, bit period and hence energy is reduced by half.
   [**Ans.** $P_e = 10^{-3}$]

OBJECTIVE TYPE QUESTIONS

1. Optimum system is the one
   (a) which is physically realisable
   (b) which must lead to unique solution
   (c) all of the two
2. If in a particular digital communication system application, the receiver designed is an optimum coherent receiver, the channel noise is white and bits are transmitted using rectangular pulses then which one of the following statements would be inconsistent? The receiver is
   (a) a correlation receiver
   (b) matched filter receiver
   (c) an integrate and dump receiver
   (d) a sample, hold and dump receiver
3. In manchester code, the symbol rate r_s and data rate r_s are related as
   (a) $r_b = 2r$ (b) $T_s = 2T_b$
   (c) $T_s = T_b/2$ (d) $r_s = r_b$
4. Transversal equalizer uses tapped delay lines to
   (a) reduce and SI
   (b) reduce BER
   (c) increase bit rate
   (d) increase bandwidths
5. If optimum coding/decoding is used at the transmitter and receiver, error free data will be recovered at the receiver output, provided E_b/N_o (E_b = Received signal energy/bit and $N_o/2$ power spectral density of the channel noise at receiver input) at the receiver input is larger than
   (a) 0 dB (b) – 1.59 dB
   (c) 1.33 dB (d) 2.15 dB
6. The bit rate of a digital communication system is 34 Mpbs. The modulation scheme is QPSK. The baud rate of the system is
   (a) 68 Mpbs (b) 34 Mbps
   (c) 17 Mbps (d) 85 Mbps
7. Comparison of MSK and QPSK schemes show that
   (a) MSK requires less power
   (b) QPSK requires less power
   (c) filtering in simple in MSK
   (d) no comparison
8. For a BPSK scheme, the bit error probability is given by
   (a) $\frac{1}{2}\text{erfc}\left(\sqrt{\frac{E_b}{2N_o}}\right)$
   (b) $\frac{1}{2}\text{erfc}\left(\frac{1}{2}\sqrt{\frac{E_b}{2N_o}}\right)$
   (c) $\frac{1}{2}\text{erfc}\left(\sqrt{\frac{E_b}{N_o}}\right)$
   (d) $\frac{1}{2}\text{erfc}\left(\frac{1}{2}\sqrt{\frac{E_b}{N_o}}\right)$
9. For a DPSK scheme, the bit error probability is given by
   (a) $\frac{1}{2}\text{erfc}\sqrt{\frac{E_b}{2000}}$ (b) $\frac{1}{2}\text{erfc}\sqrt{\frac{E_b}{2000}}$
   (c) $\frac{1}{2}\exp\left(\frac{-E_b}{N_o}\right)$ (d) $\frac{1}{2}\exp\left(\frac{-E_b}{2000}\right)$
10. For non-coherent FSK, the bit error rate is given by
   (a) $\frac{1}{2}\exp\left(\frac{-A^2T_b}{2000}\right)$ (b) $\frac{1}{2}\exp\left(\frac{-A^2T_b}{N_o}\right)$
   (c) $\frac{1}{2}\text{erfc}\left(\frac{-A^2T_b}{2000}\right)$ (d) $\frac{1}{2}\text{erfc}\left(\frac{-A^2T_b}{N_o}\right)$

11. For coherent FSK system, the bit error probability is given by

(a) $\frac{1}{2}\text{erfc}\left(\frac{\sqrt{E_b}}{4N_o}\right)$ (b) $\frac{1}{2}\text{erfc}\left(\frac{\sqrt{E_b}}{4N_o}\right)$

(c) $\frac{1}{2}\exp\left(\frac{\sqrt{E_b}}{2000}\right)$ (d) $\frac{1}{2}\exp\left(\frac{\sqrt{E_b}}{2000}\right)$

12. For coherent ASK scheme, the error probability is given by

(a) $P_e = \frac{1}{2}\text{erfc}\left(\frac{\sqrt{E_b}}{4N_o}\right)$ (b) $P_e = \frac{1}{2}\text{erfc}\left(\frac{\sqrt{E_b}}{2N_o}\right)$

(c) $P_e = \frac{1}{2}\text{erfc}\left(\frac{\sqrt{E_b}}{8N_o}\right)$ (d) $P_e = \frac{1}{2}\text{erfc}\left(\frac{\sqrt{E_b}}{N_o}\right)$

Answers

1. (a) 2. (a) 3. (a) 4. (a) 5. (b) 6. (a) 7. (a)
8. (c) 9. (c) 10. (a) 11. (b) 12. (c)

❑❑❑

UNIT– IV

Spread spectrum Communications: Frequency Hopping Spread Spectrum (FHSS) systems, Direct Sequence Spread Spectrum, Code Division Multiple Access of DSSS, Multiuser Detection, OFDM Communications.

Above Syllabus has been covered in following chapter:

CHAPTER—6 : Spread Spectrum Communications

CHAPTER 6

SPREAD SPECTRUM COMMUNICATIONS

SYLLABUS

Spread spectrum communications: Frequency Hopping Spread Spectrum (FHSS) Systems, Direct Sequence Spread Spectrum, Code division multiple access of DSSS, Multiuser detection, OFDM communications.

6.1. INTRODUCTION

Till now we have discussed a number of digital communication systems. The focus of our attention while discussing those systems was on two important factors, viz.

How to utilize the channel bandwidth efficiently?

How to minimize the amount of transmitted power?

However, the efficient utilization of bandwidth and minimizing the transmitted power are not the only problems faced by a communication system. Some other problems encountered by it are as under:

Problems encountered by a communication system

(i) In the areas such as military communication, the information has to be *secured*. This means that an unauthorized user is not expected to access the information. Also he should not be allowed to interfere the communication by any means.

(ii) Sometimes a hostile transmitter (say used by terrorists) can **jam** the transmission. To avoid this the channel should be **immune** to any external interference.

(iii) Even in the non-military communications an unintentional interference is caused by a user who is transmitting its information through a channel which is already being used.

Remedy

These problems can be successfully solved by using a

Inside this Chapter

technique called ***Spread Spectrum Modulation***. We shall discuss this modulation technique in detail in this chapter.

6.2. HOW IS THE SS SIGNAL DIFFERENT FROM THE NORMAL SIGNAL?

The spread spectrum signal is different from a normal signal, in the following aspects:

This signal occupies a larger bandwidth than that of a normal signal. (Therefore, the name spread spectrum). The spread spectrum signal invariably uses some kind of coding. The spectrum spreading at the transmitter and despreading (opposite to spreading) at the receiver is obtained with the help of this code word. The code word associated with an SS signal is independent of the information carried by the signal. The most important point is that the SS signal is *pseudorandom* in nature. This makes it appear like *random noise*. Therefore, the normal receiver cannot demodulate the SS signal. Only, a specially designed receiver can demodulate it to recover the original information.

6.3. APPLICATIONS OF SPREAD SPECTRUM MODULATION TECHNIQUE

The spread spectrum signals are used in the following applications:

(i) In combating the intentional interference (jamming).

(ii) **In rejecting the unintentional interference from some other user**

This is possible to achieve by assigning a different code for the signals from various users. This type of communication which allows multiple users to use simultaneously a common channel for transmission of information is called a Code Division Multiple Access (CDMA).

(iii) **To avoid the self interference due to multipath propagation**

A signal can take multiple paths while travelling over a communication channel. The signal components following different path lengths will result in a dispersed signal at the receiver. This can be considered as the self-interference. This type of interference also can be suppressed by using the SS modulation.

(iv) **In low probability of intercept (LPI) signals**

A message can be hidden in the background noise by spreading/its bandwidth using the code word and then transmitting the coded signal at a low power level. Due to these modifications, the probability of being intercepted (detected) is reduced to a great extent. Therefore, such a spread and coded signal is called as the low probability of-intercept (LPI) signal.

(v) **In obtaining the message privacy**

The message privacy can be obtained by superimposing a pseudorandom pattern on the transmitted message

6.4. CLASSIFICATION OF SPREAD SPECTRUM MODULATION TECHNIQUES

The SS modulation techniques are broadly categorized into two categories namely the averaging type systems and the avoidance type systems as shown in figure 6.1.

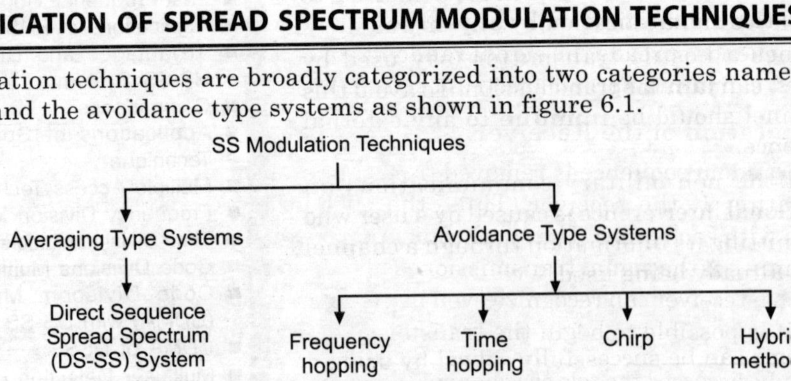

Fig. 6.1. Classification of spread spectrum technique

The averaging systems reduce the interference by averaging it over a long period. The direct sequence spread spectrum (DS-SS) system is an averaging system. The avoidance systems reduce the interference by making the signal avoid the interference over a large fraction of time. The avoidance systems are further classified depending on the type of modulation used. Some of the avoidance type systems using different modulation techniques are as under:

(i) Frequency hopping system
(ii) Time hopping system
(iii) Chirp
(iv) Hybrid modulation system

We will discuss all these systems in detail in this chapter one by one.

6.5. A MODEL OF SPREAD SPECTRUM DIGITAL COMMUNICATION SYSTEM

1. Block Diagram

The block diagram shown in figure 6.2 illustrates the basic elements of a spread spectrum digital communication system.

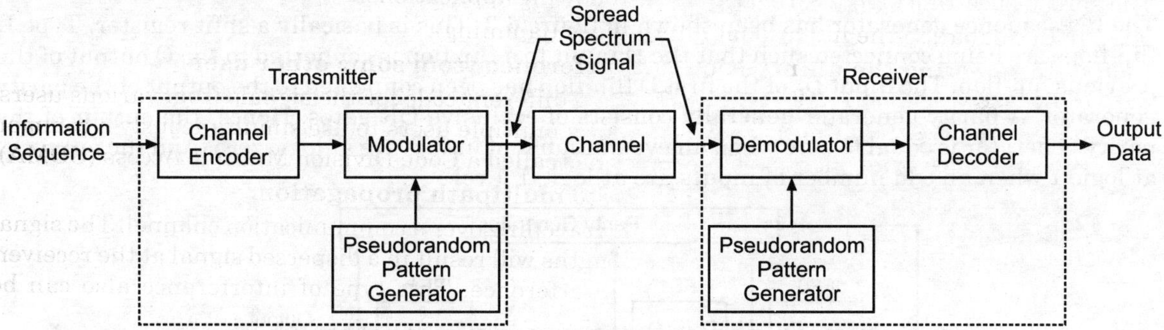

Fig. 6.2. Model of spread spectrum digital communication system

2. Working Operation of Transmitter

The information sequence at the input of the system is a binary information sequence. The same signal is recovered at the output of the system as output data signal*. In addition to these basic building blocks of a digital communication system, two additional blocks called *pseudo-random pattern generator* are used as shown in figure 6.2. One of them is connected to the modulator on the transmitter side whereas the other is connected to the demodulator on the receiving side. Both these generators are identical to each other. These generators generate a pseudorandom or pseudonoise (PN) binary sequence. It is impressed on the transmitted signal at the modulator. Thus, the modulated signal along with the pseudorandom sequence travels over the communication channel. This sequence spreads the signal randomly over a wide frequency band. Thus, the output of the modulated signal is a spread spectrum signal.

3. Working Operation of the Receiver

The pseudorandom sequence is removed from the received signal, by the other *pseudorandom generator* operating at the receiver. Thus, the pseudorandom pattern generators operate in synchronization with each other. The synchronization between these pattern generators is achieved before the beginning of the signal transmission. This is done by transmitting a fixed pseudorandom bit pattern which a receiver can recognize even in presence of interference. Once this synchronization is established, it is possible to begin the transmission. Thus in the spread spectrum receiver, the

* We have already discussed the role of a channel encoder, channel decoder, modulator and demodulator in chapter 1.

receiver can demodulate the transmitted signal if and only if a known pseudo-noise sequence has been transmitted along with the information signal. Two types of interferences are present in the Spread Spectrum digital communication system namely the narrow band or broadband interference.

4. Modulation Techniques

The modulation techniques used are under:

(i) Phase shift keying (PSK), and

(ii) Frequency shift keying (FSK)

If PSK is used, then the PN sequence generated at the modulator is used along with the PSK modulation to shift the phase of the PSK signal pseudorandomly. The resulting signal at the modulator output is called as *Direct-Sequence (DS) spread spectrum signal.* If binary or M-ary FSK is being used, then the frequency of the FSK signal is shifted pseudorandomly. The resulting signal at the output of the modulator is called as *Frquency Hopped (FH) spread spectrum signal.*

6.6. PSEUDO-NOISE (PN) SEQUENCES

The PN sequence generator has been shown in figure 6.3. This is basically a shfit register. Type D flip-flops are being connected such that the D input to a flip flop is connected to the Q output of the previous flip-flop. The input D_0 of the first D flip-flop has been connected to the output of the parity generator. A parity generator generally consists of exclusive-OR gates. Hence, the output of the parity generator is equal to zero when an even number of inputs are at logic zero. And the output is at logic 1 when an odd number of inputs are at logic 1 level.

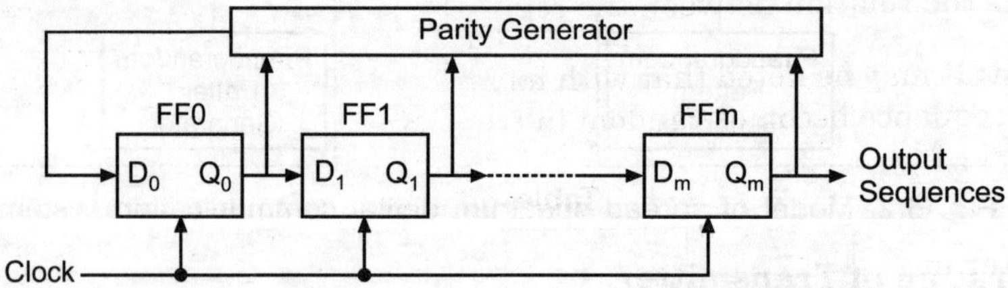

Fig. 6.3. A pseudo-random sequence generator

The inputs to the parity generator are the outputs from the flip-flops. However, it is not essential to connect all the Q outputs to the input of parity generator.

Output of Parity Generator	Status of Inputs
Logic '0'	Even number of inputs are at logic '0'
Logic '1'	Odd number of inputs are at logic '1'

The character generated by a PN sequence generator depends on the number of flip-flops used (m) and on the selection of which flip-flop outputs are connected to the inputs of parity generator. The state of each flip-flop changes and gets shifted to the next flip-flop corresponding to each pulse of the clock. A shift register of 'm' flip-flops will have 2^m number of states; *i.e.*, $Q_0\ Q_1 \ldots Q_{m-1} = 000 \ldots 0$ to $Q_0\ Q_1 \ldots Q_{m-1} = 111 \ldots 1$. Thus, the output sequence will repeat itself after every 2^m bits. The PN generator of figure 6.3 cannot generate a truly random sequence because this structure is a deterministic structure. This is the reason why, the sequence repeats itself. In order to make the random sequence "look like" truly random, its length should be increased sufficiently. With the MOS large scale integration the number of flip flops can be increased to about 2000 per shift register. The maximum length of the sequence will be $2^m - 1$. This is because the state 000...0 must be excluded. If all zero state is allowed to exist then the EX-OR gates used in the parity generator will produce a zero output all the time. To avoid this the all zero state should be excluded.

6.6.1. Definition of a PN Sequence

A pseudo-noise (PN) sequence may be defined as a coded sequence of 1s and 0s with certain auto-correlation properties.

6.6.2. The Maximum Length Sequence

The maximum length sequence is a type of cyclic code which represents a commonly used periodic PN sequence. Such a sequence has long periods and requires a linear feedback shift register. A PN sequence generated by an m-register linear feedback shift register is of length $2^m - 1$. A shift register of length m consists of m flip flops and all of them operate on the same clock. At each clock pulse, the state of each flip-flop is shifted to the next one. So as to prevent the shift register from getting emptied, we have to apply an input continuously to the first flip-flop. This input (called feedback) is computed by using a logical function (Boolean function) of the states of all the flip-flops. For a linear type feedback register, a feedback function is obtained by using modulo-2 addition (EX-OR gate) of the outputs of various flip-flops as shown in figure 6.4. For m = 3 as shown in figure 6.4 (3 flip-flops), the maximum length sequence at the generator output will always be periodic with a period of

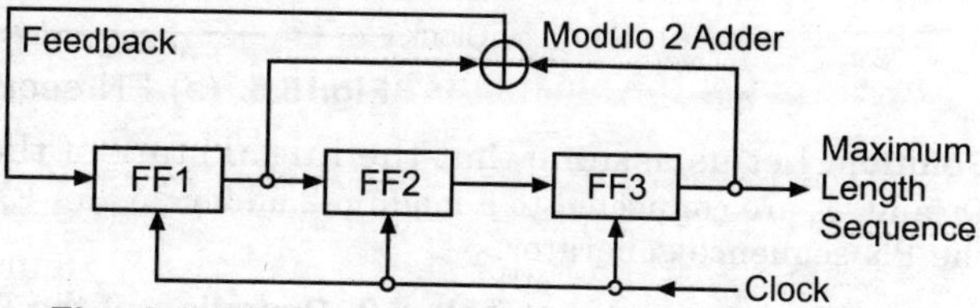

Fig. 6.4. Maximum length sequence gemerator

$$N = 2^m - 1 \quad ...(6.1)$$

where m = length of shift register

Table 6.1 shows the relation between the register length m and PN sequence length N.

Important Point: It may be noted that with increase in the value of m, the sequence length N increase and the sequence becomes random in true sense.

Table 6.1.

Length of register m	7	8	9	10	11	12	13
PN sequence length N	127	255	511	1023	2047	4095	8191

6.6.3. Properties of Maximum-Length Sequences

Maximum length sequences have many properties possessed by a truly rendom sequence. Some of the important properties are as under:

(i) Balance property

(ii) Run property

(iii) Correlation property

Balance Property

In each period of a maximum length sequence, the number of 1s is always one more than the number of 0s. This is called as the balance property.

Run Property

This property states that among the *runs* of 1s and 0s in each period of maximum lenth sequence, one half the runs of a each kind are of length one, one fourth are of length two, one eight are of length three, and so on. In this statement, the word *run* means a subsequence of identical symbols (1s or 0s) within one period of the sequence. And the length subsequence is the length of the run. The total number of runs will be $(m + 1)/2$ if an m stage feedback shfit register is being used.

Correlation Property

This property states that the auto-correlation of a maximum length sequence is periodic and binary valued.

EXAMPLE 6.1. For the PN sequence generator shown in figure 6.5(a), obtain and draw the PN sequence.

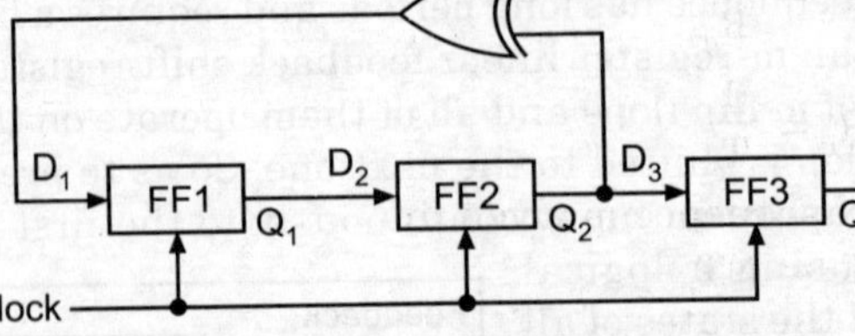

Fig. 6.5. (a) PN sequence generator

Solution: Let us assume that the initial state of the shift register is $Q_3\ Q_2\ Q_1 = 001$. The outputs Q_2 and Q_3 are connected to a modulo-2 adder i.e., an EX-OR gate. Table 6.2 summarizes the operation of the PN sequence generator.

Table 6.2. Operation of the PN sequence generator

Clock pulse Number	Shift register outputs Q_3	Q_2	Q_1	EX-OR gate output $Q_3 \oplus Q_2$	PN sequence Q_3
0	0	0	1	$0 \oplus 0 = 0$	0
1	0	1	0	$0 \oplus 1 = 1$	0
2	1	0	1	$1 \oplus 0 = 1$	1
3	0	1	1	$0 \oplus 1 = 1$	0
4	1	1	1	$1 \oplus 1 = 0$	1
5	1	1	0	$1 \oplus 1 = 1$	1
6	1	0	0	$1 \oplus 0 = 1$	1
7	0	0	0	$0 \oplus 0 = 0$	0
8	0	1	0	$0 \oplus 1 = 1$	0
9	1	0	1	$1 \oplus 0 = 1$	1
10	0	1	1	$0 \oplus 1 = 1$	0

The sequence repeats after this

The PN sequence obtained at the Q_3 output is as shown in figure 6.5. (b).

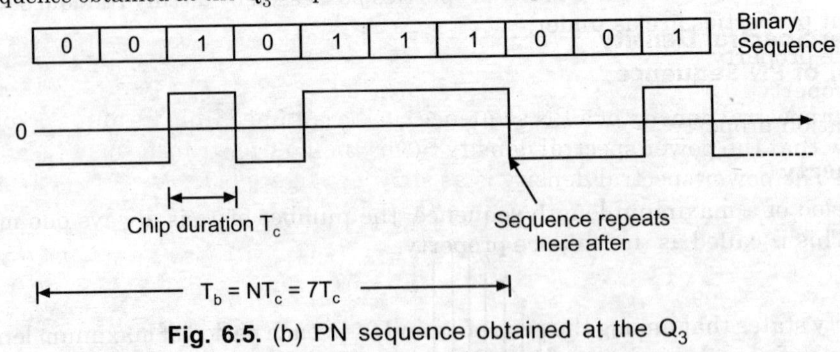

Fig. 6.5. (b) PN sequence obtained at the Q_3

Important Observations

Some of the important observations from figure 6.5(b) may be listed as under:

(i) The length of the PN sequence obtained at the Q_3 output is equal to $2^m - 1$. In this case, m = 3, hence, the maximum length of the sequence is $N = 2^3 - 1 = 7$

(ii) The PN sequence repeats itself after every 7 clock cycles.

(iii) The PN sequence is an NRZ type signal with logic 1 represented by + 1 and binary 0 is represented by − 1.

(iv) The duration of every bit is known as the chip duration T_c and the chip rate R_c is defined as the number of bits (chips) per second.

Therefore $$T_c = \frac{1}{R_c} \quad ...(6.1(a)$$

or $$R_c = \frac{1}{T_c} \quad ...(6.2)$$

(v) The period of the PN sequence is given by

$$T_b = NT_c$$

Important Point: It may be noted that the choice of 100 as the initial state is an arbitary choice. Any other state of the remaining six can be be treated as the initial state.

6.6.4. Auto-correlation Function of PN Sequences

The auto-correlation function R(τ) of the PN sequence c(t) is given by:

$$R(\tau) = \frac{1}{T_b}\int_{-T_b/2}^{T_b/2} c(t).c(t-\tau)\,dt \quad ...(6.3)$$

This is as per the basic definition of the auto-correlation of a periodic signal. By solving the above expression, we get,

$$R(\tau) = \begin{cases} 1 - \frac{N+1}{NT_c}|\tau| & \text{for } |\tau| < T_c \\ -\frac{1}{N} & \text{elsewhere} \end{cases} \quad ...(6.4)$$

The auto-correlation function R (τ) of equation (6.4) can be plotted as shown in figure 6.6. This shows that the auto-correlation R (τ) is a periodic function of time and it is a two valued function.

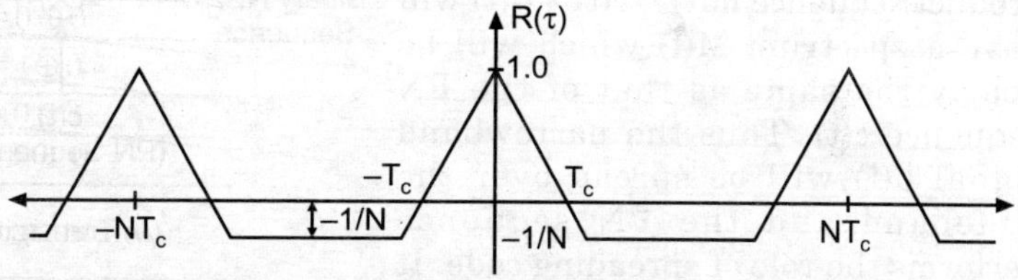

Fig. 6.6. Auto-correlation of a PN sequence

6.6.5. Power Spectral Density (PSD) of PN Sequence

The power spectral density of a PN sequence can be obtained from its auto-correlation function R(τ). We know that the power spectral density S(f) is the Fourier transform of the auto-correlation function R(τ). The power spectral density is as shown in figure 6.7.

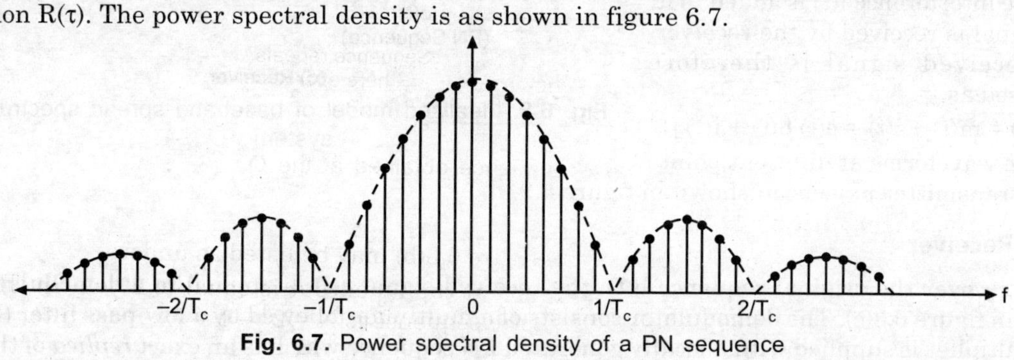

Fig. 6.7. Power spectral density of a PN sequence

6.7. A NOTION OF SPREAD SPECTRUM

The most important advantage of spread spectrum modulation is that it provides protection against externally generated interfering signals. Such signals are called as a jamming signals. The information bearing signal is made to occupy a bandwidth which is much larger than the minimum bandwidth required for its transmission. This will make the signal to appear like a noise and blends into the background. Thus, the spread spectrum is a method of camouflaging the message signal. One of the methods to used for widening the bandwidth of the data sequence is to use the modulation. Let us consider figure 6.8 which shows the transmitter, channel and receiver of an idealized model of baseband spread spectrum system.

Definition of Jamming

An externally generated interfering signal is called as a jamming signal and such an intentional interference is called as jamming.

6.7.1. Working Operation of the Encoder (i.e., Transmitter)

The input data sequence is denoted by d(t). This data sequence is first converted into an NRZ sequence b(t) by the NRZ encoder. The NRZ signal b (t) and the pseudonoise signal c(t) are applied to the two inputs of a product modulator. At the output of the product modulator, we obtain the spread spectrum signal. The spectrum of this signal is quite spread out as compared to the spectrum b (t) which is a narrow band signal. Thus, a data sequence b(t) is used to modulate a wideband pseudo-noise sequence c(t) by applying these two sequences to the product modulator or multiplier. Both sequences b(t) and c(t) are in polar form. According to Fourier transform theory, the multiplication of signals in the time domain results in convolution of their frequency domain *i.e.*,

$$c(t) \times b(t) \xleftrightarrow{F} C(f) * B(f) \qquad \text{...(6.5)}$$

Therefore, if the data sequence b(t) is narrowband and the PN sequence c(t) is a wideband sequence, then the product sequence m(t) = c(t) × b(t) will have a spectrum M(f) which will be nearly the same as that of the PN sequence c(t). Thus the narrowband signal b(t) will be spread over the wideband and the PN sequence performs the role of spreading code. It may be noted that the transmitted signal m(t) is a baseband signal. The spread spectrum signal m(t) is transmitted over the channel where an additive interference i(t) is added to it. The signal is received by the receiver. The received signal is therefore expressed as,

$$r(t) = m(t) + i(t) = c(t)\, b(t) + i(t)$$

The waveforms at different points in the transmitter have been shown in figure 6.9.

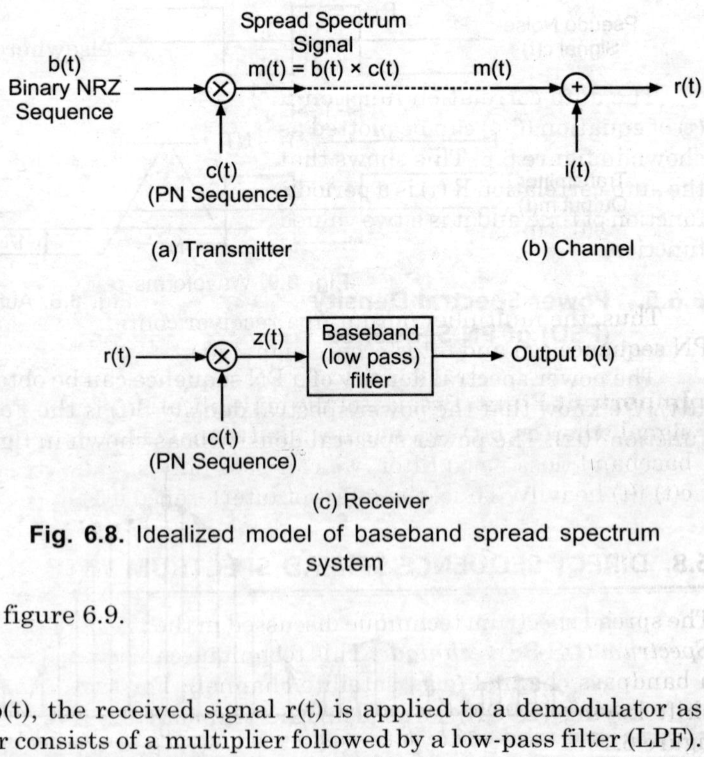

Fig. 6.8. Idealized model of baseband spread spectrum system

6.7.2. Receiver

To recover the original sequence b(t), the received signal r(t) is applied to a demodulator as shown in figure 6.8(c). The demodulator consists of a multiplier followed by a low-pass filter (LPF). The multiplier is supplied with a locally generated PN sequence which is an exact *replica* of the PN sequence used at the transmitter. The receiver needs to operate in perfect *synchronization* with the transmitter. The demodulated signal is given by,

$$z(t) = c(t) \times r(t) \quad ...(6.6)$$

Substituting the expression for r(t) i.e.,

$$r(t) = c(t)\, b(t) + i(t)$$

We get, $z(t) = c(t)\, [c(t)\, b(t) + i(t)]$

Since, $z(t) = c^2(t)\, b(t) + c(t)\, i(t) \quad ...(6.7)$

In equation (6.7), it may be noted that the desired signal b(t) has been multiplied by $c^2(t)$.

We know that, $c(t) = \pm 1$

Therefore, $c^2(t) = +1$ for all the values of t.

Hence, equation (6.7) gets converted to,

$$z(t) = b(t) + c(t)i(t) \quad ...(6.8)$$

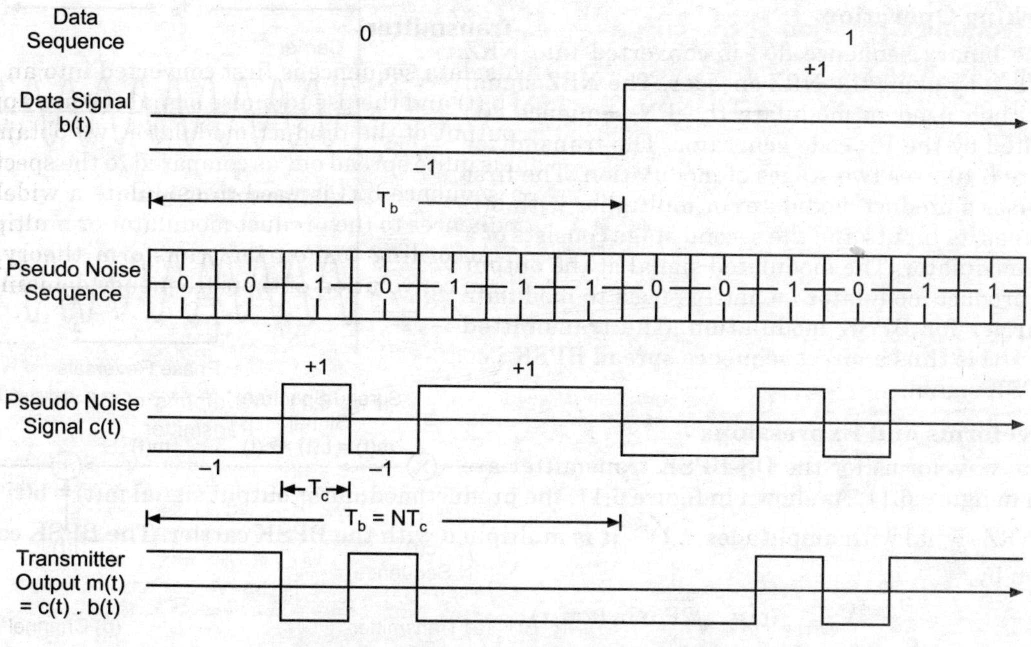

Fig. 6.9. Waveforms of DS-SS transmitter

Thus, the multiplier output in a receiver contains the desired signal b(t) and the product of the PN sequence c(t) and interference signal i(t).

Important Point: Because of the multiplication, the product signal c(t).i(t) becomes a wideband signal whereas b(t) is a narrow band signal. Therefore, by applying the multiplier output to a baseband (loss pass) filter, we can pass only signal b(t) and attenuate the interference signal c(t) i(t) heavily. Thus, the effect of interference is reduced to a great extent.

6.8. DIRECT SEQUENCE SPREAD SPECTRUM WITH COHERENT BPSK

The spread spectrum technique discussed in the previous section is called as *Direct Sequence Spread Spectrum* (DS-SS) *technique*. This technique can be used in practice for transmission of signal over a bandpass channel (e.g., satellite channel). For such an application, the coherent binary phase shift keying (BPSK) is used in the transmitter and receiver. The transmitter as shown in figure 6.10.

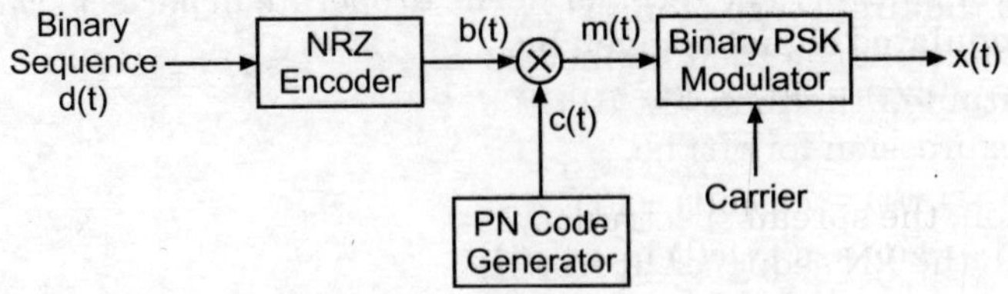

Fig. 6.10. Direct sequence spread spectrum coherent PSK transmitter

6.8.1. DS-BPSK transmitter

1. Block Diagram

The DS-BPSK transmitter has been shown in figure 6.10.

2. Working Operation

The binary sequence d(t) is converted into NRZ signal b(t) by using the NRZ encoder. The NRZ signal b(t) is then used to modulate the PN sequence c(t) generated by the PN code generator. The transmitter of figure 6.10 uses two stages of modulation. The first stage uses a product modulator or multiplier with b(t) and c(t) as its inputs and the second stage consists of a BPSK modulator. The modulated signal at the output of the product modulator i.e. m(t) is used to modulate the carrier for BPSK modulation. The transmitted signal x(t) is thus a direct sequence spread BPSK i.e., DS-BPSK signal.

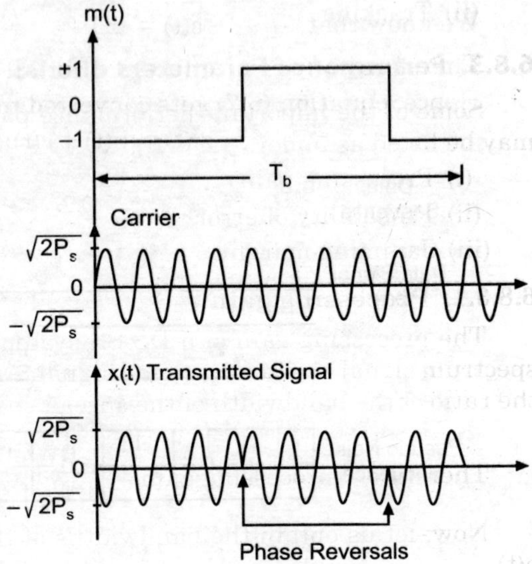

Fig. 6.11. Waveforms of DS-BPSK transmitter

3. Waveforms and Expressions

The waveforms for the DS-BPSK transmitter are shown in figure 6.11. As shown in figure 6.11. the product modulator output signal m(t) = b(t) × c(t) is an NRZ signal with amplitudes ±1V . It is multiplied with the BPSK carrier. The BPSK carrier is given by,

$$V_{carrier}(t) = \sqrt{2P_s} \sin(2\pi f_c t) \quad \text{...(6.9)}$$

The output of BPSK modulator i.e., x(t) is transmitted, x(t) is given mathematically as under:

$$x(t) = m(t) \times V_{carrier}(t) = m(t) \times \sqrt{2P_s} \sin(2\pi f_c t)$$

But, $m(t) = \pm 1$

Therefore, $$x(t) = \pm\sqrt{2P_s} \sin(2\pi f_c t) \quad \text{...(6.10)}$$

Thus, the phase shift of x(t) is 0° corresponding to a positive m(t) and it is 180° corresponding to a negative m(t).

6.8.2. DS-BPSK Receiver

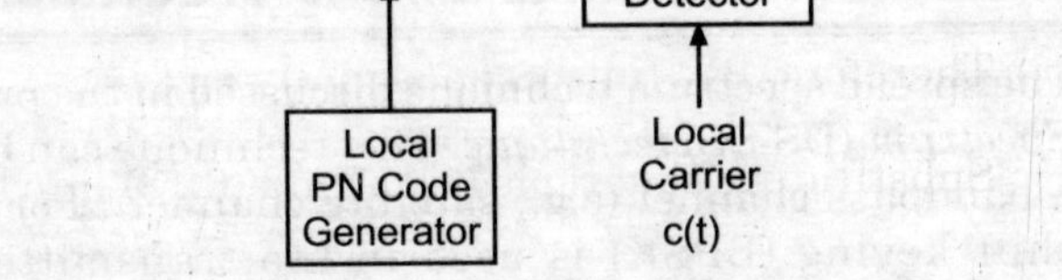

Fig. 6.12. The DS-BPSK receiver

1. Block diagram

The DS-BPSK receiver has been shown in figure 6.12.

2. Working Operation

The received signal y(t) and the locally generated replica of the PN-sequence are applied to a multiplier. This is the first stage of multiplication. The multiplier performs the de-spreading operation. Output of multiplier is then applied to a coherent BPSK detector with a local carrier applied to it. At the output of the coherent BPSK detector we get back the original data sequence.

3. Synchronization

For proper operation, the spread spectrum system requires a local PN sequence at the receiver to be synchronized with the PN sequence at transmitter. The synchronization is carried out in two parts:

(i) Acquision, and

(ii) Tracking

6.8.3. Performance Parameters of a DS-SS System

Some of the important performance parameters of a direct sequence spread spectrum system may be listed as under:

(i) Processing gain

(ii) Probability of error

(iii) Jamming margin

6.8.3.1. Processing gain

The processing gain of a DS-SS system represents the gain achieved by processing a spread spectrum signal over an unspread signal. Another way to define the processing gain PG is that it is the ratio of the bandwidth of the spread spectrum signal to the bandwidth of the unspread signal.

Therefore, $$\text{processing gain} = \frac{\text{BW of spread signal}}{\text{BW of unspread signal}} \quad ...(6.11)$$

Now, let us obtain the bandwidths of the spread signal and that of the unspread data signal b(t).

Bandwidth of Spread Signal m(t)

As shown in figure 6.13, the spread signal m(t) is obtained by multiplying the NRZ signals b(t) and PN sequence c(t).

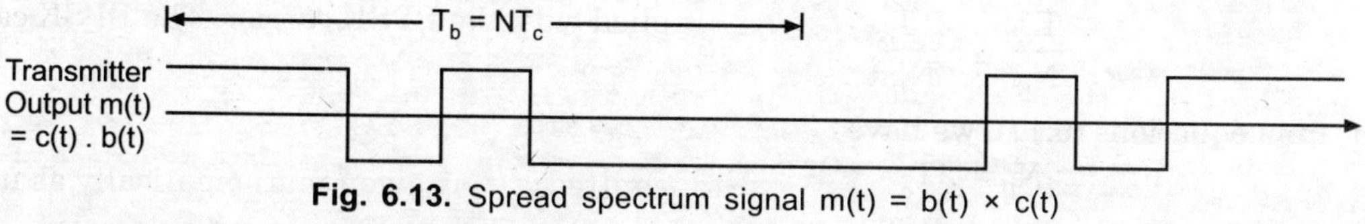

Fig. 6.13. Spread spectrum signal m(t) = b(t) × c(t)

The one bit period of the spread signal m(t) is given by T_c. The bandwidth of a NRZ signal is equal to the reciprocal of its one bit period.

Therefore, $$\text{BW of spread signal} = \frac{1}{T_c} \quad ...(6.12)$$

Bandwidth of Unspread Signal

The unspread signal b(t) is an NRZ signal and the bandwidth of the NRZ signal is reciprocal of the bit period.

Therefore, $$\text{BW (unspread signal)} = \frac{1}{T_b} \quad ...(6.13)$$

Substituting equations (6.12) and (6.13) into equation (6.11), we get

$$\text{Processing gain} = \frac{1/T_c}{1/T_b} = \frac{T_b}{T_c} \quad ...(6.14)$$

Thus, the longer we make the PN sequence (i.e., if the chip time T_C is reduced) then larger will be the processing gain.

6.8.3.2. Probability of Error

We have already proved that the error probability P_e of a coherent BPSK system is given by,

$$P_e = \frac{1}{2} \text{erfc} \sqrt{E_b / N_0} \quad \text{...(6.15)}$$

Where, E_b = Energy per bit

and $N_0/2$ = Power spectral density of white noise

In a direct-sequence spread binary PSK system the interference may be treated as a wideband noise signal with a power spectral density of $N_0/2$.

Hence,

$$\frac{N_0}{2} = \frac{J\,T_c}{2} \quad \text{...(6.16)}$$

or

$$N_0 = JT_c \quad \text{...(6.17)}$$

where, J = Average interference power

and T_c = Chip duration

Substituting this expression for N_0 into equation (6.15), we get error probability as under:

$$P_e = \frac{1}{2} \text{erfc} \sqrt{\frac{E_b}{J\,T_c}} \quad \text{...(6.18)}$$

This is the required expression for the error probability. The complementary error function is a monotonically decreasing function. Therefore, the error probability decreases as the value of $\sqrt{E_b / JT_c}$ increases. But, P_e increases if $\sqrt{E_b / JT_c}$ decreases due to increase in the average interference power J or the chip period T_c.

6.8.3.3. Antijam Characteristics (i.e., Jamming Margin)

Since the energy per bit i.e., $E_b = P_s T_b$. Where P_s is the average signal power and T_b is the bit duration, we can express the bit energy to noise density ratio as follows:

$$\frac{E_b}{N_0} = \frac{P_s T_b}{N_0} \quad \text{...(6.19)}$$

But, from equations (6.17), we have

$$N_0 = JT_c$$

Therefore,

$$\frac{E_b}{N_0} = \frac{P_s\,T_b}{JT_c}$$

$$\frac{E_b}{N_0} = \left(\frac{T_b}{T_c}\right)\left(\frac{P_s}{J}\right) \quad \text{...(6.20)}$$

In equation (6.14), we have defined the processing gain PG as under :

$$PG = \frac{T_b}{T_c}$$

Therefore,

$$\frac{E_b}{N_0} = PG\left(\frac{P_s}{J}\right)$$

or,

$$\frac{J}{P_s} = \frac{PG}{E_b / N_0} \quad \text{...(6.21)}$$

The ratio J/P_s is called as the **jamming margin**. Therefore, the jamming margin may be defined as the ratio of average interference power J and the signal power P_s. If the jamming margin and the process gain both are expressed in dB then,

$$(\text{Jamming margin})\ \text{dB} = (\text{Processing Gain})\ \text{dB} - 10\log_{10}[E_b/N_0]_{min} \quad ...(6.22)$$

Where, $(E_b/N_0)_{min}$ is minimum bit energy to noise density ratio required to support a prescribed average error probability.

EXAMPLE 6.2. A spread spectrum communication system is characterised by the following parameters:

Duration of each information bit, T_b = 4.095 ms

Chip duration of a PN sequence, T_c = 1 μs.

Determine the processing gain and jamming margin if (E_b/N_0) = 10 and the average probability of error $P_e = 0.5 \times 10^{-5}$.

Solution: It has been given that,

$$T_b = 4.095\text{ ms},\ T_c = 1\ \mu\text{s},\ (E_b/N_0) = 10,\quad P_e = 0.5 \times 10^{-5}$$

Processing gain PG

$$PG = \frac{T_b}{T_c} = \frac{4.095\times 10^{-3}}{1\times 10^{-6}} = 4095 \quad \textbf{Ans.}$$

We know that

$$T_b = N\,T_c$$

$$PG = \frac{NT_c}{T_c} = N$$

Hence, $N = 4095$

Jamming Margin

$$(\text{Jamming Margin})_{dB} = (PG)_{dB} - 10\log_{10}([E_b/N_0])$$

Therefore, $(\text{Jamming Margin})_{dB} = 10\log_{10}4095 - 10\log_{10}[10] = 36.1 - 10 = 26.1\text{ dB}$ **Ans.**

6.8.4. Advantages of DS-SS System

(i) This system has a very high degree of discrimination against the multipath signals. Therefore the interference caused by the multipath reception is minimized successfully.

(ii) The performance of DS-SS system in presence of noise is superior to other systems such as FH-SS system.

(iii) This system combats the intentional interference (jamming) most effectively.

6.8.5. Drawbacks of DS-SS System

(i) With the serial search system, the acquisition time is too large. This makes the DS-SS system slow.

(ii) The sequence genereted at the PN code generator output must have a high rate. The length of such a sequence needs to be long enough to make the sequence truly random.

(iii) The channel bandwidth required, is very large. But this bandwidth is less than that of a FH-SS system.

(iv) The synchronization is affected by the variable distance between the transmitter and receiver.

6.8.6. Few Important Salient Features of DS-SS System

Some of the important features of DSSS are as follows:

(i) It provides good security againt potential jamming or interception.

(ii) The DSSS is extremely effective against narrowband jamming signals.

(iii) The narrowband communication signals can coexist with the DSSS signals.

(iv) The DSSS signal is not very effective against broadband interference.

6.8.7. Few Important Applications of DS-SS System

Some of the important applications of the DS-SS system are as follows:

(i) To combat the intentional interference (jamming)

(ii) To reject the unintentional interference

(iii) To minimize the self interference due to multipath propagation

(iv) In the low probability of intercept (LPI) signal

(v) In obtaining the message privacy

(vi) Code division multiple access with DS-SS.

6.9. PSD OF DS SPREAD SPECTRUM

We assume that the type of modulation used along with DS is QPSK. This QPSK signal corresponding to the nth data symbol is shaped with the amplitude shaping pulse to obtain the DS/QPSK signal. If the nth data symbol has a zero mean value and it is uncorrelated symbol, then the power spectral density of the complex envelope is given by,

$$S(f) = \frac{A^2}{T}\sigma_x^2 |H_a(f)|^2 \quad \text{...(6.23)}$$

where A = Amplitude,

T = Symbol duration,

σ_x = Variance.

$H_a(f)$ = F.T. of $h_a(t)$ i.e., the amplitude shaping pulse.

The amplitude shaping pulse is given by,

$$h_a(t) = \sum_{k=0}^{N-1} a_k h_c(t - kT_c) \quad \text{...(6.24)}$$

Taking the Fourier transform of $h_a(t)$, we get

or
$$F[h_a(t)] = H_a(f) = H_c(f) \sum_{k=0}^{N-1} a_k e^{-j2\pi fk T_c} \quad \text{...(6.25)}$$

And
$$|H_a(f)|^2 = |H_c f|^2 \sum_{k=0}^{N-1} \sum_{l=0}^{N-1} a_k a_l e^{-j2\pi f(k-l)T_c} \quad \text{...(6.26)}$$

This expression may be converted in the form of DTFT (Discrete-Time Fourier Transform) as under:

$$|H_a(f)^2| = |H_c(f)|^2 \, 2N\phi_{k,k}(f) \quad \text{...(6.27)}$$

where, $\phi_{k,k}(f)$ = DTFT and is defined as under:

$$\phi_{k,k}(f) = \sum_{n=-N+1}^{N-1} \phi_{k,k}^{a}(n) e^{-j2\pi fn T_c} \quad \text{...(6.28)}$$

Substituting $T = nT_c$ and $\sigma_x^2 = \frac{1}{2}E[|x_i|^2] = 1/2$, we get

$$S(f) = \frac{A^2}{T}|H_c(f)|^2 \phi_{k,k}(f) \quad \text{...(6.29)}$$

This is the required expression for the PSD.

6.10. FREQUENCY HOPPING SPREAD SPECTRUM (FH-SS) SIGNALS

1. Basic Aspects

(i) In the DS-SS system discussed in the previous section, the NRZ data sequence b(t) modulates the PN sequence. The product signal [m (t) = b(t) × c(t)] is spread instantaneously in the frequency domain due to this process.

(ii) The capacity of DS-SS system to reject the intentional interference (jamming) is dependent on the processing gain (PG). The processing gain PG = (T_b/T_c).

(iii) Therefore, PG increases if the chip period T_c is decreased which in turn permits a greater transmission bandwidth and more chips per bit.

(iv) But the capabilities of the physical devices used for the generation of PN spread spectrum signal put a practical limitation to the maximum value of *processing gain* and hence on the capability to combat jamming.

(v) Under some operating conditions, the maximum attainable processing gain $PG_{(max)}$ is not sufficiently high for combating the jamming. Under such conditions, an alternative system called Frequency Hop (FH) spread spectrum.

2. Principle of Operation of (FH-SS) System

In this system the data is used to modulate a carrier. The data modulated carrier is then randomly hopped from one frequency to the other. Because of this, spectrum of transmitted signal is spread sequentially rather than instantaneously.

3. Types of Modulation

A common modulation technique used is the M-ary frequency shift keying (MFSK). The combination of frequency hopping (FH) and MFSK is shown as FH/MFSK.

Important Point: It is important to understand that the frequency hopping does not cover the entire spread spectrum instantaneously. Rather it covers the entire spectrum sequentially. Hence, we have to consider the rate at which the frequency hops occur. Based on the rate of hopping, the FH/MFSK system has been classified into two categories.

6.10.1. Types of Frequency Hopping

Depending upon the rate of frequency hopping, the FH/MFSK systems are classified into following two categories:

(i) Slow frequency hopping

(ii) Fast frequency hopping

(i) Slow Frequency Hopping

In slow frequency hopping, the symbol rate R_s of the MFSK signal is an integer multiple of the hop rate R_h. This means that several symbols are transmitted corresponding to each frequency hop i.e., each frequency hop ⇒ several symbols

Thus, frequency hopping takes place slowly.

(ii) Fast Frequency Hopping

In the fast frequency hopping, the hop rate R_h is an interger multiple of the MFSK symbol rate R_s. This means that during the transmission of one symbol, the carrier frequency will hop several times.

Therefore, each symbol transmission ⇒ several frequency hops

Thus, the frequency hopping takes place at a fast rate.

6.11. SLOW FREQUENCY HOPPING

Figure 6.14. shows the block diagram of a slow-frequency hopping FH-MFSK transmitter.

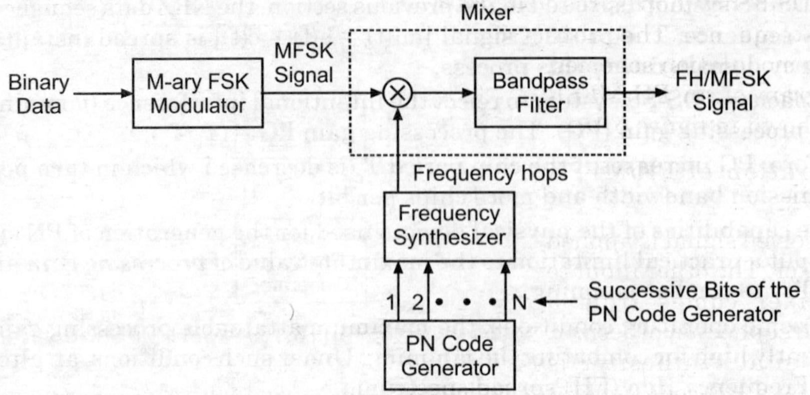

Fig. 6.14. Frequency hop spread M-ary FSK transmitter

6.11.1. Working Operation of the FH/MFSK Transmitter

(i) The binary data sequence b(t) is applied to the M-ary FSK modulator the output of which goes to the input of the mixer. The other input to the mixer block is obtained from a digital frequency synthesizer.

(ii) The mixer consists of a multiplier followed by a band pass filter.

(iii) At the multiplier output, we get the two input frequencies, their sum and their difference frequency components. The bandpass filter is designed to select only the sum frequency component rejecting all other components. This sum components of frequency is then transmitted.

(iv) Successive K-bits of the input binary data sequence will form one symbol. M such symbols can be transmitted using the M-ary FSK system with $M = 2^k$.

(v) The M-ary FSK modulator will assign a distinct frequency for each of these M symbols. Thus, the frequency of mixer input obtained from MFSK modulator is changing continuously. The other input to the mixer is obtained from the digital frequency synthesizer. The synthesizer output at a given instant of time is the *frequency hop*.

(vi) Each frequency hop is mixed with the MFSK signal to produce the transmitted signal as explained earlier. The frequency hops at the output of the synthesizer are controlled by the successive bits at the output of the PN code generator.

(vii) The output bits of the PN generator change randomly. Hence, the synthesizer output frequency will also change randomly.

(viii) Therefore, the frequency hops produced will vary in a random manner. If the number of successive bits at the output of PN generator is n, then the total number of frequency hops will be 2^n.

(ix) The total bandwidth of the transmitted FH/MFSK signal is equal to the sum of all the frequency hops. Therefore, the bandwidth of the transmitted FH/MFSK signal is very large of the order of few GHz.

6.11.2. FH/MFSK Receiver

Due to the large bandwidth occupied by the FH/MFSK signal, the coherent detection of this signal is possible within each hop. This is because for coherent detection, the phase synchronization of the locally generated carrier with the transmitted carrier is essential. But the frequency synthesizers used in FH/MFSK receiver are unable to maintain this phase coherence over successive hops. Hence, most frequency hop spread spectrum systems use the non-coherent M-ary modulation schemes. The block diagram of an FH-MFSK receiver is as shown in figure 6.15.

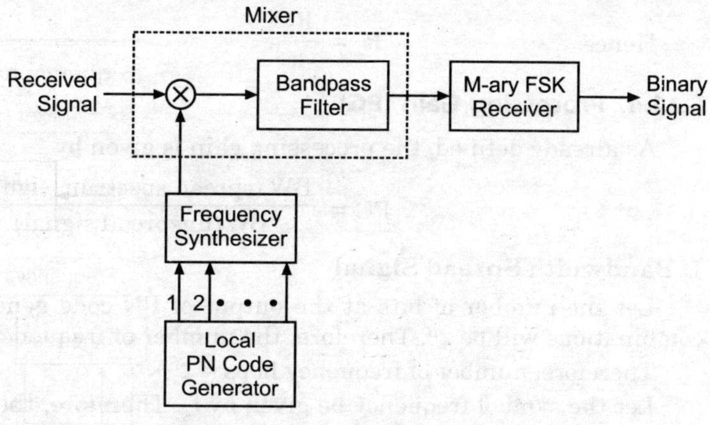

Fig. 6.15. An FH-MFSK receiver

Working Operation of FH/MFSK Receiver

(i) The received signal is applied to a mixer, The other input to the mixer comes from a digital frequency synthesizer. This digital synthesizer is driven by a PN code generator which is synchronized with the PN code generator at the transmitter and generates the same code sequence.

(ii) Therefore, the frequency hops produced at the synthesizer output will be identical to those at the synthesizer output at the transmitter.

(iii) At the output of the multiplier, we get the input signals, their sum and difference (as far as frequency is concerned).

(iv) Out of these frequency components, the difference frequency component is selected by the bandpass filter that follows the multiplier. This difference signal is the MFSK signal.

(v) Thus, the mixer removes the frequency hopping. The MFSK signal at the mixer output is then applied to a non-coherent MFSK demodulator.

(vi) At the output of the MFSK detector we obtain the digital modulating signal b(t).

(vii) The non-coherent M-ary FSK detector can be implemented by using a bank of M, non coherent matched filters. Each matched filter is matched to one of the tones of th e MFSK signal.

(viii) The largest output out of the M available outputs fo filters is selected to obtain the digital modulating signal.

6.11.3. Chip Rate (R_c) of FH/MFSK System

The individual tone (frequency) of shortest duration is defined as a **chip** in the FH/MFSK system. This is completely different from the definition of chip is DS/BPSK system. The **chip rate** R_c for an FH/MFSK system is defined as under:

$$\text{Chip rate, } R_c = \max(R_h, R_s) \quad ...(6.30)$$

where R_h = hop rate and R_s = symbol rate

Thus, chip rate R_c is higher than R_h or R_s for the slow hopping FH/MFSK signal. Let the R_b represent the bit rate of the incoming signal, R_s be the symbol rate of the MFSK signal and R_h be the hop rate. Then, the relation between them is given by

$$\text{Chip rate, } R_c = R_s = \frac{R_b}{K} \geq R_b \quad ...(6.31)$$

where $K = \log_2 M$

or $M = 2^k$ (M-ary FSK, k bit message).

Equation (6.31) implies that the chip rate is equal to the symbol rate R_s, because the symbol rate is higher than the hop rate for a slow hopping system, therefore, $R_s \geq R_h$ and each symbol is made of k bits

Hence, $R_s = \frac{R_b}{k}$.

6.11.4. Processing Gain (PG)

As already defined, the processing gain is given by

$$PG = \frac{\text{BW (spread spectrum signal)}}{\text{BW (unspread signal)}} \quad ...(6.32)$$

1. Bandwidth Spread Signal

Let the number of bits at the output of PN code generator be n. Therefore, the number of combinations will be 2^n. Therefore, the number of frequency hops will be 2.

Therefore, number of frequency hops = 2^n ...(6.33)

Let the symbol frequency be given by f_s. Therefore, the bandwidth of the FH/MFSK signal is given by,

BW(Spread spectrum signal) = $2^n \times f_s$ ...(6.34)

2. BW of Unspread Signal

The bandwidth of the unspread signal is equal to the symbol frequency f_s. Substituting the bandwidths into equation (6.32), we get,

$$\text{Processing gain} = \frac{2^n f_s}{f_s} = 2^n \quad ...(6.35)$$

EXAMPLE 6.3. The FH/MFSK transmitter has the following parameters:

1. **Number of bits per MFSK symbol: K = 2**
2. **Number of MFSK tones = M = $2^K = 2^2 = 4$**
3. **Length of PN sequence per hop: n = 3**
4. **Total number of frequency hops: $2^n = 2^3 = 8$**

Draw the FH/MFSK signal showing the variation in frequency with respect to time for one complete period of the PN sequence. Assume the PN sequence to be:

PN sequence	0	0	1	1	1	0	0	1	1	0	0	1	0	0	1

Therefore the period of one PN sequence is $2^4 - 1 = 15$ digits.

Solution:

(i) In this example, as the number of bits per symbol is K = 2, the carrier is hopped to a new frequency after transmitting two symbols or four information bits.

(ii) As the length of PN sequence, n = 3 there are 2^n i.e., 8 different possible combinations. Therefore there should be 8 different hops at the output of the frequency synthesizer. But as shown in figure 6.16(a), only, three FH carriers which are being used here because the combinations at the outputs of PN generator are 001, 110 and 011 only.

(iii) The number of symbols transmitted in one frequency hop is two. The total number of symbols is M = 4 i.e., 00, 01, 10, 11, corresponding to each symbol the MFSK modualtor generates four frequencies i.e., f_1, f_2, f_3, f_4 respectively. Thus, corresponding to each hop two of these frequencies will be transmitted as shown in figure 6.16(b). For example, during the first hop period, the messages transmitted are:

Symbol	Transmitted frequency
01	FH + f_2
11	FH + F4

where FH = carrier frequency for the hop which is controlled by the output of the PN code generator.

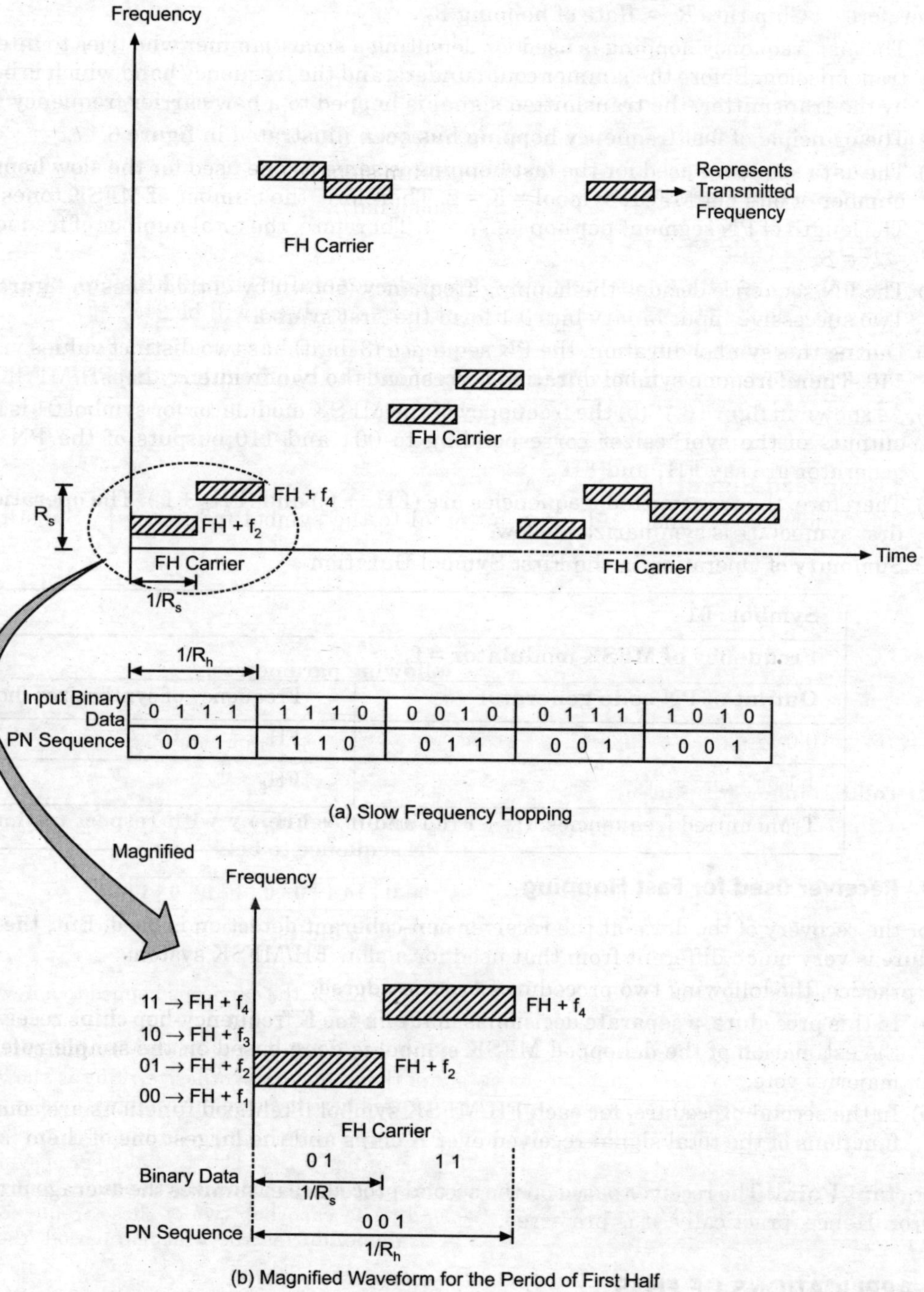

Fig. 6.16.

6.12. FAST FREQUENCY HOPPING

(i) As explained earlier, the fast FH/MFSK system is different than the slow FH/MFSK system. Because in the fast FH/MFSK system, there are multiple hops for each M-ary symbol. Hence, each hop is a **chip**.

Therefore, Chip rate R_c = Rate of hopping R_h

(ii) The fast frequency hopping is used for defeating a smart jammer who tries to interface the transmission. Before the jammer could understand the frequency band which is being used by the transmitter, the transmitted signal is hopped to a new carrier frequency.

(iii) The principle of fast frequency hopping has been illustrated in figure 6.17.

(iv) The data sequence used for the fast hopping is same as one used for the slow hopping. The number of bits per MFSK symbol = K = 2. Therefore, the number of MFSK tones = $2^K = 4$. The length of PN segment per hop i.e., n = 3. Therefore, the total number of frequency hops = $2^3 = 8$.

(v) The PN sequence decides the hopping frequency (shown by dotted lines in figure 6.17(a)). Two successive input binary bits 0 1 form the first symbol.

(vi) During this symbol duration, the PN sequence (3 digit) has two distinct values viz 001 and 110. Therefore, one symbol duration corresponds to two frequency hops.

(vii) As shown in figure 6.17(b) the frequency of the MFSK modulator for symbol 01 is f_2 and the outputs of the synthesizer corresponding to 001 and 110 outputs of the PN sequence generator are say FH_1 and FH_6.

(viii) Therefore, the transmitted frequencies are $(FH_1 + f_2)$ and $(FH_6 + f_2)$. The operation for the first symbol 01 is symmarized below.

(ix) Summary of Operation in the First Symbol Duration

Symbol : 01	
Frequency of MFSK modulator = f_2	
Output of PN code generator	Frequency of synthesizer (hop)
0 0 1	FH_1
1 1 0	FH_6
Transmitted frequencies: $(f_2 + FH_1)$ and $(f_2 + FH_6)$	

6.12.1. Receiver used for Fast Hopping

For the recovery of the data, at the receiver non-coherent detection is used. But, the detection procedure is very much different from that used for a slow FH/MFSK system.

In practice, the following two procedures are considered:

(i) In this procedure, a separate decision is made on the K frequency-hop chips received. Then, the estimation of the dehopped MFSK symbol is done based on the simple rule based on majority vote.

(ii) In the second procedure, for each FH/MFSK symbol likelihood functions are computed, as functions of the total signal received over K chips and the largest one of them is selected.

Important Point: The receiver based on the second procedure minimizes the average probability of error. Hence, practically, it is preferred.

6.13. APPLICATIONS OF FHSS

As a matter of fact, FHSS is being used in a number of applications. The two most important of them are as under:

1. Wireless local area networks (WLAN) standard for Wi-Fi.
2. Wireless personal area network (WPAN) standard of Bluetooth.

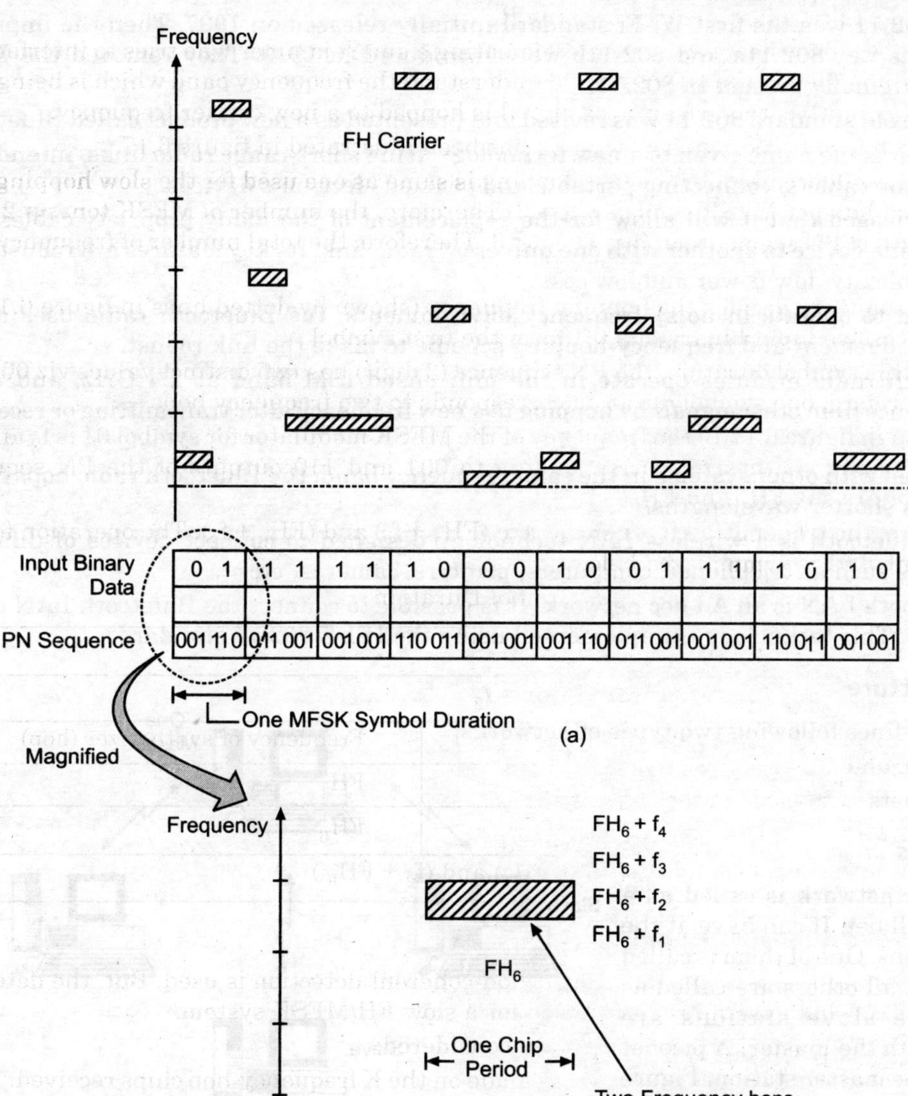

Fig. 6.17. Waveforms of fast hopping system

(i) IEEE 802.11 was the first Wi-Fi standard initially released on 1997. Then the improved standards i.e., 802.11a and 802.11b were introduced in 1999. They removed the FHSS option originally present in 802.11.

(ii) Then the old standard 802.11 was revised and presented as a new product called Bluetooth.

(iii) Bluetooth is the name given to a new technology using short-range radio links, intended to replace the cable(s) connecting portable and/or fixed electronic devices.

(iv) It is envisaged that it will allow for the replacement of the many propriety cables that connect one device to another with one universal radio link. Its key features are robustness, low complexity, low power and low cost.

(v) Designed to operate in noisy frequency environments, the Bluetooth radio uses a fast acknowledgement and frequency-hopping scheme to make the link robust.

(vi) Bluetooth radio modules operate in the unlicensed ISM band at 2.4 GHz. and avoid interference from other signals by hopping to a new frequency after transmitting or receiving a packet.

(vii) Compared with other systems in the same frequency band, the Bluetooth radio hops faster and uses shorter wavelengths.

(viii) Thus, Bluetooth is a wireless LAN technology designed to connect devices of different functions such as telephones, computers, printers, cameras, etc.

(ix) A Bluetooth LAN is an Ad hoc network. It is possible to connect the Bluetooth LAN to the Internet. This technology is implemented using the IEEE 802.15 standard.

6.13.1. Architecture

Bluetooth defines following two types of networks:

(i) Piconets, and

(ii) Scatternets

6.13.2. Piconets

A Bluetooth network is called as a piconet or a small net. It can have at the most eight stations. One of them is called as a master and all others are called as slaves. All the slave stations are synchronised with the master. A piconet can have only one master station. Figure 6.18. shows a piconet. The communication between a master and slaves can be one-to-one or one-to-many.

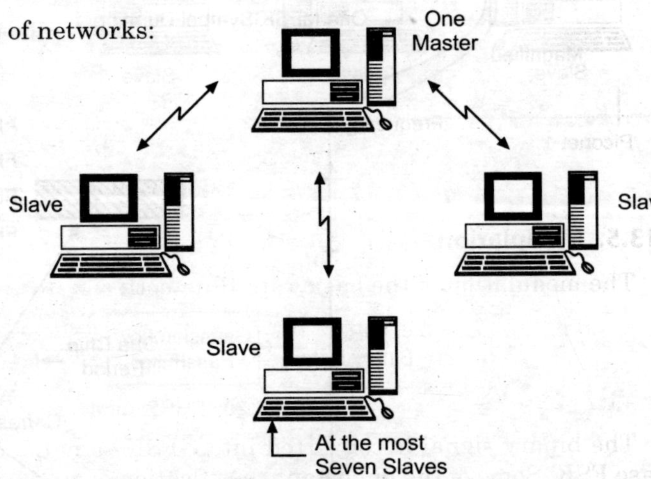

Fig. 6.18. A piconet

6.13.3. Scatternet

Piconets are combined to form a scatternet. It has been shown in figure 6.19.

Figure 6.19 illustrates a scatternet consisting of two piconets. A slave in the first piconet can act as a master in the second piconet. It will receive the message from the master in the first piconet by acting as a slave and then delivers the message to the slaves in the second piconet as shown in figure 6.19.

6.13.4. Bluetooth Devices

Every Bluetooth device consists of a built in short range radio transmitter. The current data rate is 1 Mbps and the bandwidth is 2.4 GHz. So an interface between the IEEE 802.11 wireless LAN and Bluetooth LAN is possible. The Bluetooth specification standard defines a short-range (10-meter) radio link. The devices carrying Bluetooth enabled chips can easily transfer data at a rate of about 1 Mbps. (Megabits per second) within 10 meters (33 feet) or range through walls, clothing and luggage bags. The interaction between devices occurs by itself without direct human intervention whenever they are within each others range. In this process, the software technology embedded in the Bluetooth transceiver chip triggers an automatic connection to deliver and accept the data flow. Since Bluetooth is of short range with limited speed and low-power technology. It is less attractive to corporate wireless local area networks that are generally powered with the 802.11 wireless LAN technologies. Each Bluetooth-enabled device contains a 1.5 inch square transceiver chip operating in the ISM (industrial, scientific, and medical) radio frequency band of 2.40 GHz to 2.48 GHz. This frequency is generally available worldwide for free without any licensing restrictions. The ISM band is divided into 79 channels with each carrying a bandwidth of 1 MHz.

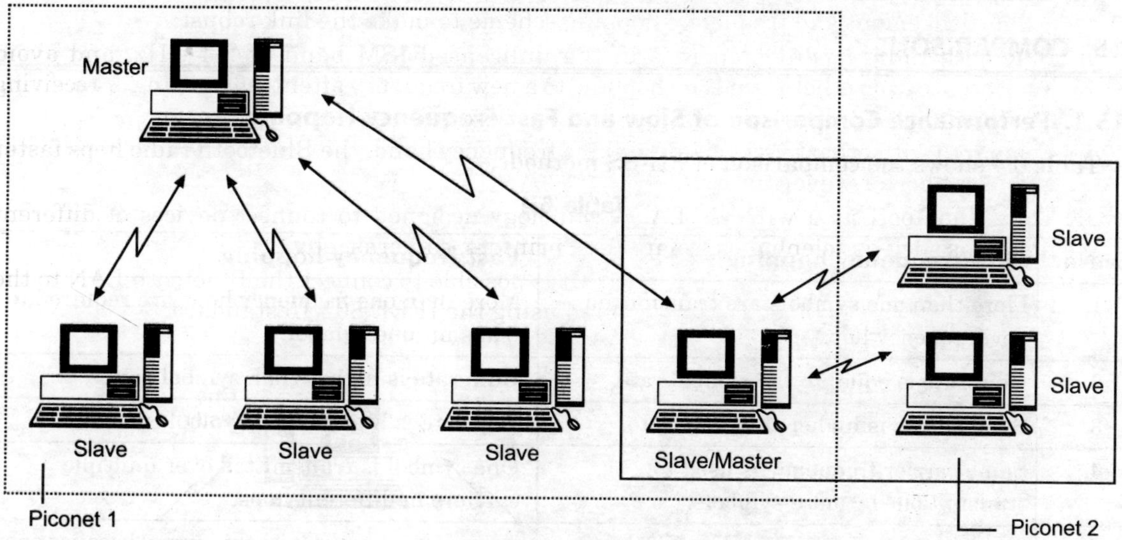

Fig. 6.19. Scatternet

6.13.5. Modulation

The modulation of the basic rate Bluetooth signal has been shown in figure 6.20.

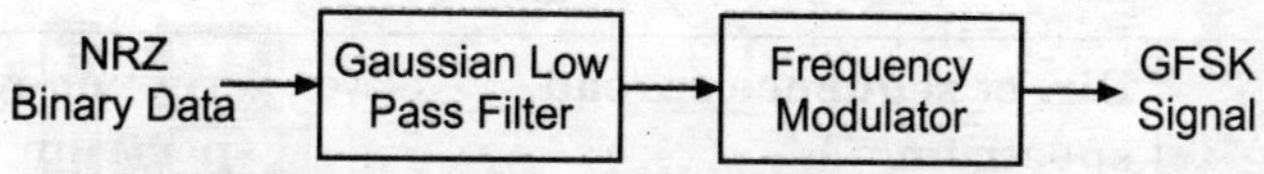

Fig. 6.20. FHSS modulation in Bluetooth

The binary signal is converted into GFSK signal before transmission GFSK is a continuous phase FSK. Some of the important specification of 802.11 FHSS and Bluetooth are as given below.

Table 6.3. Specification of 802.11 FHSS and Bluetooth

S. No.	Specification	802.11 FHSS	Bluetooth
1.	Frequency band	ISM 2.4-2.4835 GHz	ISM 2.4-2.4835 GHz
2.	Bandwidth per channel	1 MHz	1 MHz
3.	Number of overlapping channels	79	79
4.	Minimum hopping distance	6	6
5.	Modulation	GFSK-2, GFSK-4	GFSK-2
6.	Hopping rate	2.5-160 Hz	1600 Hz
7.	Data rate	1 M bit/s and 2 M bit/s	723.1 kbits/s

6.14. ADVANTAGES AND DRAWBACKS OF FH-SS SYSTEM

Advantages

(i) The synchronization is not greatly dependent on the distance.
(ii) The serial search system with FH-SS needs shorter time for acquisition.
(iii) The processing gain PG is higher than that of DS-SS system.

Drawbacks

(i) The bandwidth of FH-SS system is too large (in GHz).
(ii) Complex and expensive digital frequency synthesizers are required to be used.

6.15. COMPARISONS

6.15.1. Performance Comparison of Slow and Fast Frequency Hopping

Table 6.4 shows the comparison of FH-SS methods.

Table 6.4.

Sr.No.	Slow frequency hopping	Fast frequency hopping
1.	More than one symbols are transmitted per frequency hop.	More than one frequency hops are required to transmit one symbol
2.	Chip rate is equal to the symbol rate.	Chip rate is higher than symbol rate.
3.	Symbol rate is higher than hop rate.	Hop rate is higher than symbol rate.
4.	Same carrier frequency is used to transmit one or more symbols.	One symbol is transmitted over multiple carriers in different hops.
5.	A jammer can detect this signal if the carrier frequency in one hop is known.	A jammer cannot detect this signal because one symbol is transmitted using more than one carrier frequencies.

6.15.2. Performance Comparison of DSSS and FHSS

Sr.No.	Parameter	Direct sequence spread spectrum	Frequency hopping spread spectrum
1.	Definition	PN sequence of large bandwidth is multiplied with narrow band data signal.	Data bits are transmitted in different frequency slots which are changed by PN sequence.
2.	Chip rate	It is fixed $R_c = \frac{1}{T_c}$	$R_c = \max(R_h, R_s)$
3.	Modulation technique	BPSK	M-ary FSK
4.	Processing gain	$PG = \frac{T_b}{T_c} = N$	$PG = 2^{\prime}$
5.	Error probability	$P_e = \frac{1}{2} \text{erfc} \sqrt{\frac{E}{J\,T_c}}$	$P_e = \frac{1}{2} e^{-rb} R_c/2$

6.	Acquisition time	Long	Short
7.	Effect of distance	This system is distance relative.	Effect of distance is less.

6.16. APPLICATIONS OF SPREAD SPECTRUM TECHNIQUES

Some of the important applications of the spread spectrum technique may be described as under:

(i) **For combating the intentional interference (jamming)**: This can be used in the military applications as well as other commercial applications to avoid the intentional interference.

(ii) For reducing the unintentional interference.

(iii) For suppressing the interference due to multipath reception.

(iv) In the low probability of intercept (LPI) application.

(v) Due to large bandwidth of a spread spectrum signal, the ***fading*** (specially, the frequency selective fading) does not affect the entire spectrum. Infact, only a small portion of the complete spectrum is affected in the fading process. Therefore, the spread-spectrum signals are used in the mobile communication.

(vi) Due to the use of pseudo-noise code sequence, the spread spctrum signal can be recognised only by the authorised receiver. All other receivers consider this signal as noise. Thus the SS communication is a **secured** communication.

(vii) The spread spectrum signals are used in the RADAR and other navigation systems for ranging or distance measurement. We know that wideband signals are *time-limited*. Hence, they can be used for the measurement of time delays very precisely.

(viii) The most important application of the spread spectrum technique is code division multiple access (CDMA). This is a multiuser communication system in which many users can access the available channel bandwidth simultaneously. To avoid their interference each user is allotted a particular code sequence, using the PN code generators. These code sequences are used by the receivers to detect their signals in presence of other unwanted signals. CDMA is used for the cellular mobile telephone systems.

6.17. FEW SOLVED EXAMPLES

EXAMPLE 6.4. A PN sequence is generated using a feedback shfit register of length 4. Find the generated output sequence if the initial contents of the shift register are 1000. If the chip rate is 10^7 chips/sec., calculate the chip and PN sequence duration and period of output sequence. Draw its scheme arrangement.

Solution: One of the possible schematic diagram of the PN generator has been shown in figure. 6.21.

The PN sequence generated by the above generator has been shown in table 6.5.

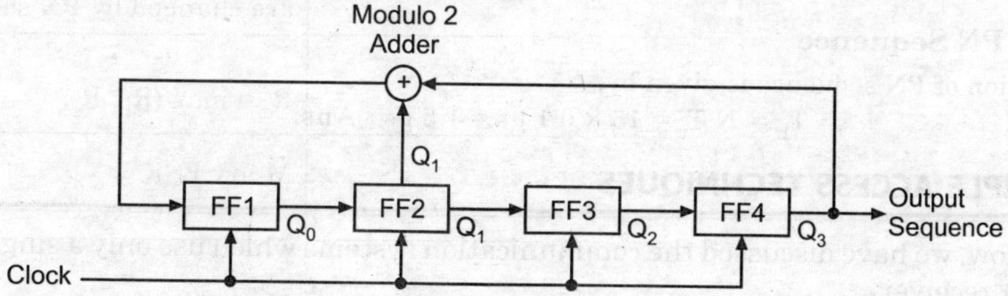

Fig. 6.21. A four stage shift register to generate PN sequnce

Table 6.5. PN sequence generator by the generator shown in figure 6.21.

Clock Pulse Number	Shift register outputs				EX-OR gate output	PN sequence
	Q_3	Q_2	Q_1	Q_0	$Q_3 \oplus Q_2$	Q_3
0	0	0	0	1	$0 \oplus 0 = 0$	0
1	0	0	1	0	$0 \oplus 1 = 1$	0
2	0	1	0	1	$0 \oplus 0 = 0$	0
3	1	0	1	0	$1 \oplus 1 = 0$	1
4	0	1	0	0	$0 \oplus 0 = 0$	0
5	1	0	0	0	$1 \oplus 0 = 1$	1
6	0	0	0	1	$0 \oplus 0 = 0$	0
7	0	0	1	0	$0 \oplus 1 = 1$	0
8	0	1	0	1	$0 \oplus 0 = 0$	0
9	1	0	1	0	$1 \oplus 1 = 0$	1
10	0	1	0	0	$0 \oplus 0 = 0$	0
11	1	0	0	0	$1 \oplus 0 = 0$	1
12	0	0	0	1	$0 \oplus 0 = 0$	0
13	0	0	1	0	$0 \oplus 1 = 1$	0
14	0	1	0	1	$0 \oplus 0 = 0$	0
15	1	0	1	0	$1 \oplus 1 = 0$	1
The sequence repeats after this						

Chip Duration

The chip rate $R_c = 1 \times 10^7$ chips/sec. Hence, the chip duration T_c is given by,

$$T_c = \frac{1}{R_c} = \frac{1}{1 \times 10^7} = 0.1\ \mu\text{sec.} \quad \textbf{Ans.}$$

Length of PN Sequence

The length of the PN sequence is given by,

$$N = 2^m - 1$$

But, m = 4 therefore, $N = 2^4 - 1 = 15$ digits **Ans.**

Duration of PN Sequence

The duration of PN sequence is given by,

$$T_b = N\,T_c = 15 \times 0.1\ \mu s = 1.5\ \mu s \quad \textbf{Ans.}$$

6.18. MULTIPLE ACCESS TECHNIQUES

(i) Till now, we have discussed the communication systems which use only a single transmitter and a receiver.

(ii) But, in this section, we shall focus our attention on multiple users and multiple receiver communication links.

(iii) Different ways in which multiple users can access a common channel for transmitting information shall be discussed in this article.

(iv) The multiple access, methods are used in the satellite networks, cellular and mobile communication networks and underwater acoustic networks.

(v) One type of multiple access system in which a large number of users sharing the same communication channel has been shown in figure 6.22.

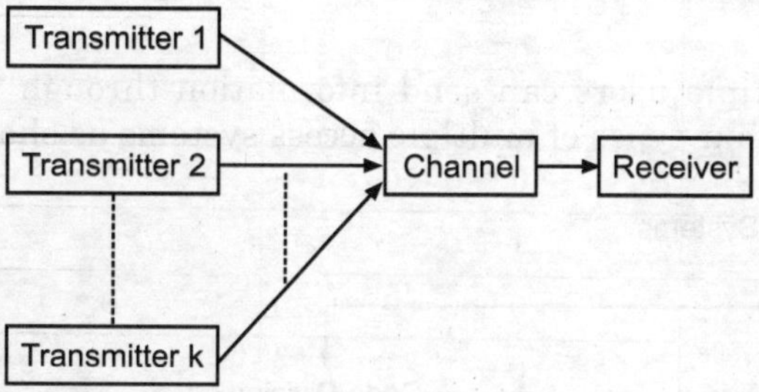

Fig. 6.22. A multiple access system

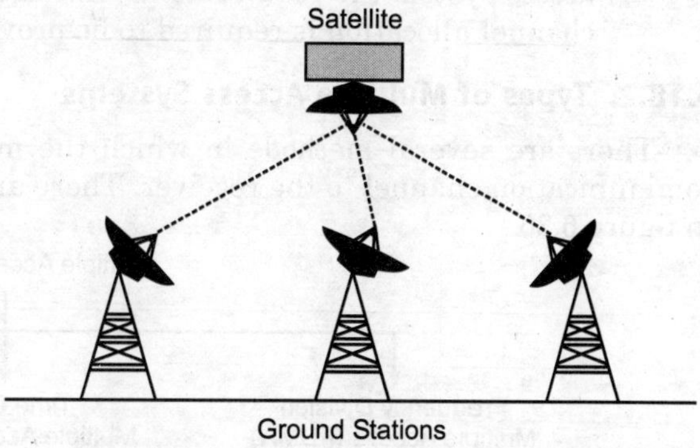

Fig. 6.23. Second type of multiuser system

(vi) The common channel can be uplink in a satellite communication system or it can be a cable or it can be a frequency band.

(vii) For example in a mobile communication system, the users of the network will be the transmitters and the receiver is in the base station.

(viii) The second type of multiuser communication system is as shown in figure 6.23.

(ix) In the second type multiuser system of figrue 6.23, a single transmitter (satellite) sends information to multiple receivers.

(x) The other examples of such system are the radio and TV broadcast systems.

(xi) The multiple access and broadcast networks are the most common multiuser communication system. The third type of multiuser system is a store and forward network shown in figure 6.24.

(xii) In this system the communication takes place in all the possible directions. The fourth type of system is shown in figure 6.25.

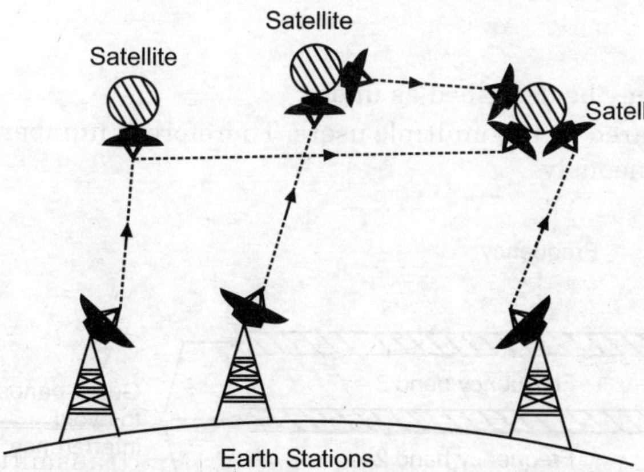

Fig. 6.24. A store and forward communication network with satellite relays

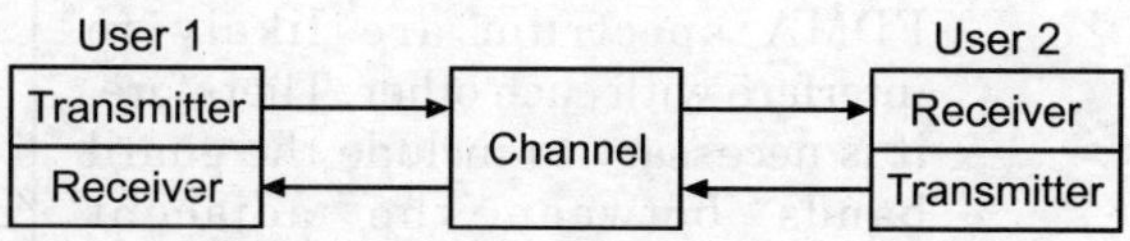

Fig. 6.25. A two way communication channel

6.18.1. Difference between Multiple Access and Multiplexing

There are very small differences between the multiple access and multiplexing. The differences are as follows:

(i) In multiple access the users will share the remote communication channels such as a satellite or radio channel. But in multiplexing the channel such as a telephone channel is shared by users confined to a local site.

(ii) In a multiplexed system, the requirements of users are generally fixed. But in a multiple access system the requirements of a user change dynamically with time. So the dynamic channel allocation is required to be provided.

6.18.2. Types of Multiple Access Systems

There are several methods in which the multiple users can send information through the communication channel to the receiver. There are four types of multiple access systems as shown in figure 6.26.

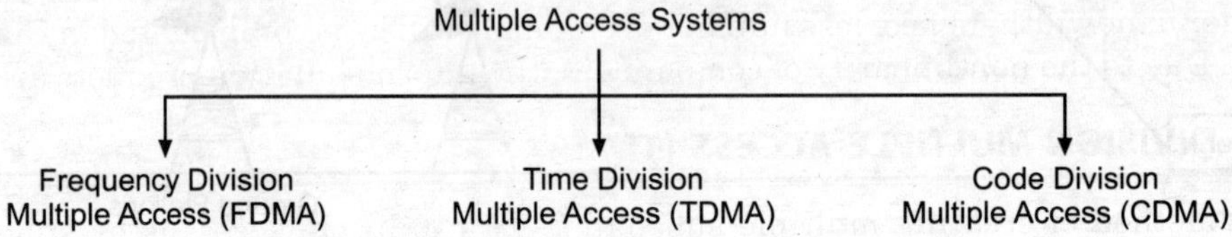

Fig. 6.26. Types of the multiple access systems

6.19. FREQUENCY DIVISION MULTIPLE ACCESS (FDMA)

One of the simplest multiple access methods is the frequency division multiple access (FDMA). In this method, the channel bandwidth is subdivided into a number division subchannels as shown in figure 6.27. There are k non-overlapping subchannels, as shown in figure 6.27. This method is therefore, called as frequency division multiple access and it is commonly used for the voice and data transmission.

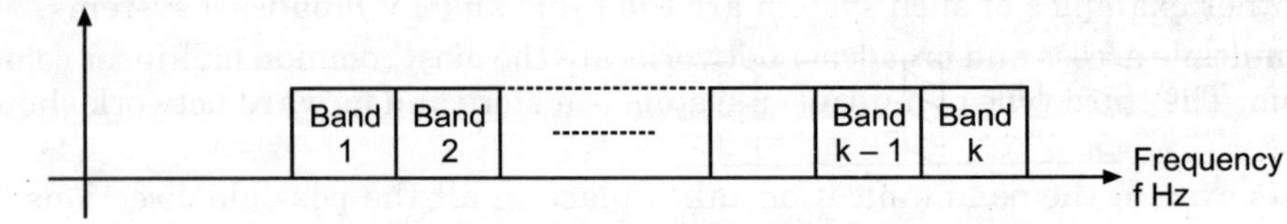

Fig. 6.27. Subdivision of channel into a number of subchannels

6.19.1. Salient Features of FDMA

Some of the important features of the FDMA may be described as under:

(i) The overall channel bandwidth is being shared by the multiple users. Therefore, a number of users can transmit information simultaneously.

(ii) The adjacent frequency bands in the FDMA spectrum are likely to interfere with each other. Therefore, it is necessary to include the guard bands between the adjacent frequency bands as shown in figure 6.28. The guard bands are also required because practically it is impossible to achieve the ideal filtering to separate the different users.

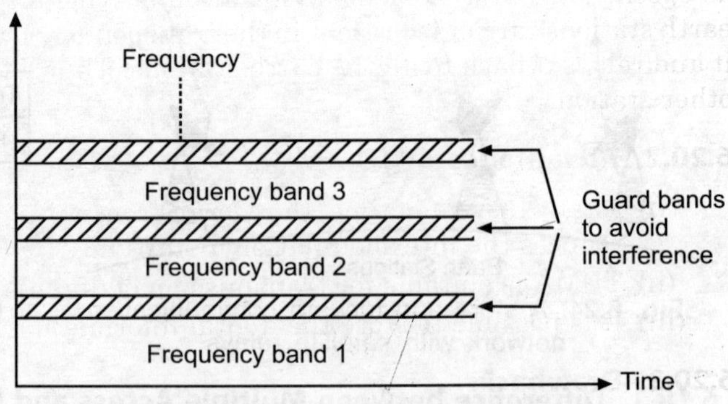

Fig. 6.28. The guard bands included to avoid the interference

(iii) No code words and synchronization is not required.

(iv) Power efficiency is reduced.

(v) FDMA is an old and proven system and is used for the analog signals.

6.19.2. Advantages of FDMA

All the stations can operate continuously all 24 hours without having it wait for their turn to come. The power required for transmission depends on the number of channels being transmitted. The signal to noise ratio is improved due to the use of FM. No synchronization is necessary.

6.19.3. Drawbacks of FDMA

Each channel or earth station can use only a part of the total satellite bandwidth. In spite of guard bands being provided, there is some adjacent channel interference present. As FM is used, it requires larger bandwidth, hence, less number of channels will be accommodated in the bandwidth of a satellite. Due to the non-linearity of companders, the intermodulation products are generated.

6.20. TIME DIVISION MULTIPLE ACCESS (TDMA)

Another method of creating multiple subchannels for multiple access is by subdividing the time duration T_f called as the frame duration, into say k non-overlapping subintervals each of duration T_f/k. After that each user who want to transmit information is assigned a particular slot, within each frame. This method is known as time division multiple access TDMA.

6.20.1. Salient Features of TDMA

TDMA is used for the transmission of data and digital voice signals. It is necessary to include ***guard times*** between the adjacent channels as shown in figure 6.29.

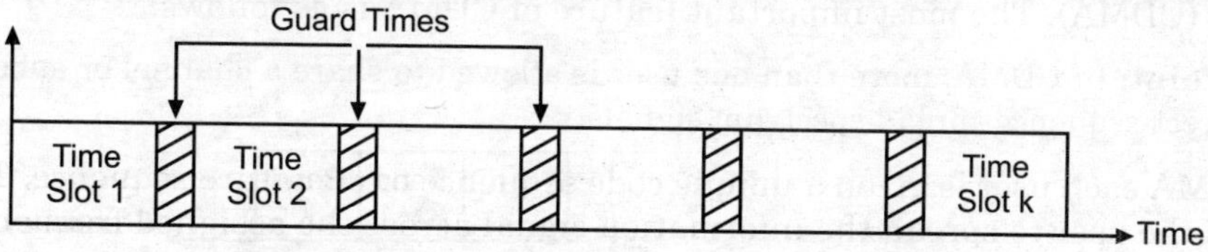

Fig. 6.29. Guard times included in TDMA

Synchronization is necessary in TDMA

Power efficiency of TDMA is better than that of the FDMA. TDMA is a method of time division multiplexing, the digitally modulated carriers between various earth stations in a satellite network through a common satellite transponder. Each earth station transmits a short burst of digitally modulated carrier during the time slot assigned to it in the TDMA frame. Such a time slot is called as epoch. The burst of each earth station is synchronised so that at any instant of time, only one earth stations carrier is present in the transponder. The transponder receives this carrier, amplifies it and relays it back to all the earth stations. Thus, each earth station receives the bursts from all other stations.

6.20.2. Advantages of TDMA

(i) At any instant of time, the carrier from only one station is present at the transponder. This reduces the intermodulation distortion.

(ii) TDMA is suitable for transmission of digital information.

(iii) It is possible to store the digital information, change the rate etc. in TDMA.

6.20.3. Drawbacks

(i) Precise synchronization is required

(ii) Bit and frame timing must be achieved and maintained.

6.20.4. Problems with FDMA and TDMA

(i) The problem with the FDMA and TDMA system is that, the channel is basically partitioned into independent single user subchannels.

(ii) This means that each subchannel in the FDMA is alloted to a single user and each time slot in TDMA has been allotted to a separate single user.

(iii) The FDMA and TDMA systems however prove to be inefficient when the data from the users is bursty in nature as shown in figure 6.30.

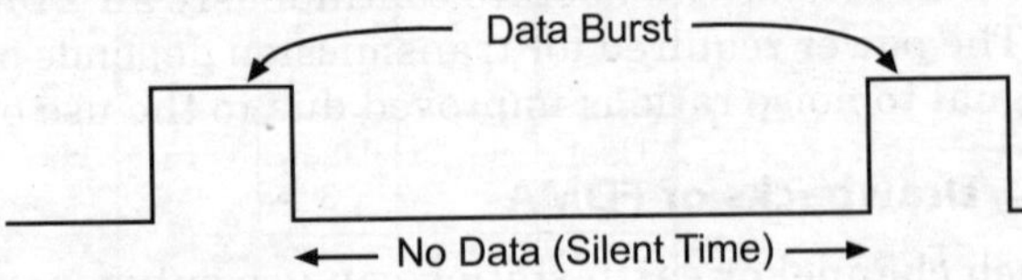

Fig. 6.30. Bursty data signal with low duty cycle

(iv) This type of data has low value of duty cycle i.e., the time for which data is being transmitted is much shorter than the silent time.

(v) Under such circumstances where the transmission from different users is bursty and low duty cycle, the FDMA and TDMA system will be inefficient.

(vi) This is because a large percentage of the available time or frequency slots do not convey any information.

(vii) Such a type of data is observed in computer communication networks and to some extent, in the mobile cellular communication systems carrying digitized voice.

6.21. CODE DIVISIONS MULTIPLE ACCESS (CDMA)

(i) An alternative to FDMA and TDMA is an another system called code division multiple access (CDMA). The most important feature of CDMA is as follows:

Important Point: In CDMA, more than one user is allowed to share a channel or subchannel with the help of direct-sequence spread spectrum signals.

(ii) In CDMA each user is given a unique code sequence or signature sequence. This sequence allows the user to spread the information signal across the assigned frequency band.

(iii) At the receiver the signal is recovered by using the same code sequence. At the receiver, the signals received from various users are separated by checking the cross-correlation of the received signal with each possible user signature sequence.

(iv) In CDMA the users access the channel in a random manner. Hence, the signals transmitted by multiple users will completely overlap both in time and in frequency.

(v) The CDMA signals are spread in frequency. Therefore, the demodulation and separation of these signals at the receiver can be achieved by using the pseudorandom code sequene.

(vi) CDMA is sometimes also called as spread spectrum multiple access (SSMA).

(vii) In CDMA as the bandwidth as well as well as time of the channel is being shared by the users, it is necessary to introduce the guard times and guard bands as shown in figure 6.31.

(viii) CDMA does not need any synchronization, but the code sequences or signature waveforms are required. In CDMA one channel carries all the transmission simultaneously.

6.21.1. Analogy

In CDMA transmission takes place with different codes. It is analogous to a room full of people who talk different languages but only one pair knows English, another pair knows French, third pair knows Hindi, etc. Thus, communication will take place only between the persons who know the same language.

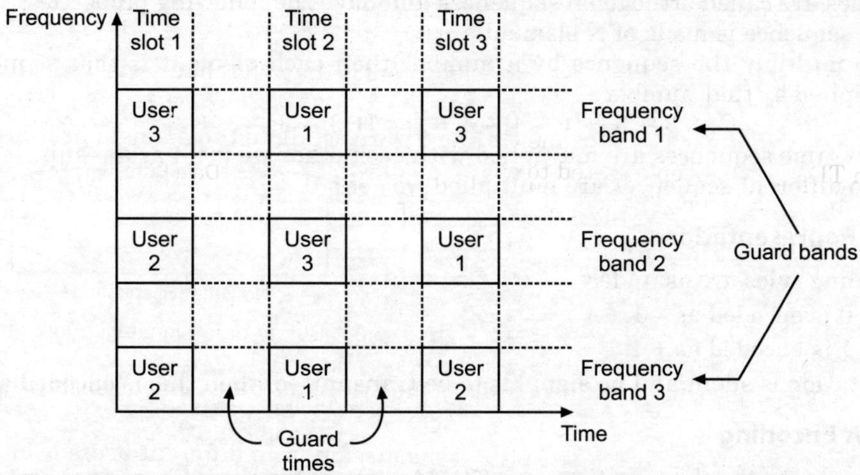

Fig. 6.31. Structure of CDMA showing the guard bands and the guard times

6.21.2. Idea

Now, let us assume that there are four stations 1, 2, 3 and 4. Let the data from these stations be denoted by D_1, D_2, D_3 and D_4 respectively. Let the code assigned to this data be C_1, C_2, C_3 and C_4 respectively. The important properties of codes are as under:

1. These codes are orthogonal. This means that if we multiply two different codes, we get a 0.
2. If any code is multiplied by itself, we get 4(i.e., the number of stations).

Station 1 performs multiplication $D_1.C_1$, station-2 performs multiplication $D_2.C_2$ etc. as shown in figure 6.32. All the product terms are then added and the total signal is sent over the common channel.

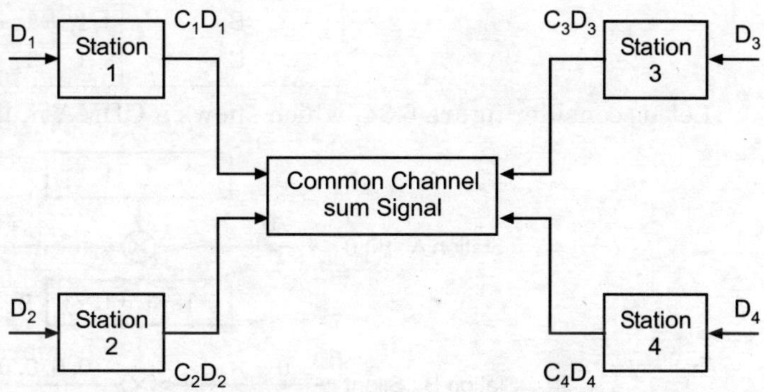

Fig. 6.32. Idea of CDMA

Each station receives the sum signal. If stations 2 and 3 are talking then station 3 will multiply the sum signal by the code of station-2 i.e., C_2. At station 3, the multiplication takes place as under:

$$\text{Sum signal} \times C_2 = [C_1\, D_1 + C_2\, D_2 + C_3\, D_3 + C_4\, D_4] \times C_2$$
$$= C_1\, C_2\, D_1 + C_2\, C_2\, D_2 + C_2\, C_3\, D_3 + C_2\, C_4\, D_4 = 0 + 4\, D_2 + 0 + 0$$

Station-3 divides the result by 4 and gets D_2 i.e., data from station –2.

6.21.3. Chips

In CDMA, each station is assigned a code. It is a sequence of numbers called as a chip. Let there be four stations A, B, C and D and let the codes assigned to them be as shown in figure 6.33.

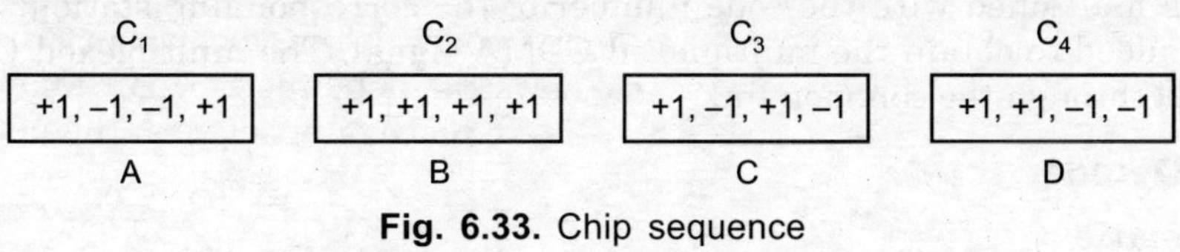

Fig. 6.33. Chip sequence

These codes are called orthogonal sequences and have the following properties:

(i) Each sequence is made of N elements.

(ii) If we multiply the sequence by a number then each element is that sequence will be multiplied by that number.

e.g., $C_1 \times 3 = 3\,[+1, -1, -1 +1] = [+3, -3, -3, +3]$

(iii) If two same sequences are multiplied with each other we get 4 as a result.

(iv) If two different sequences are multiplied we get a 0.

6.21.4. Data Representation

The encoding rules are as under:

(i) A bit 0 is encoded as – 1.

(ii) A bit 1 is encoded as + 1.

(iii) If a station is silent and no signal is to be transmitted, then this is encoded as 0.

6.22.5. CDMA Encoding

Let A, B and C be the three stations in a CDMA system. Let the unique codes assigned to them be as follows:

Table 6.6. Codes assigned to different stations

Stations	Code
A	+ 1, – 1, – 1, + 1
B	+ 1, + 1, + 1, + 1
C	+ 1, – 1, + 1, – 1

Let us consider figure 6.34, which shows a CDMA multiplexer.

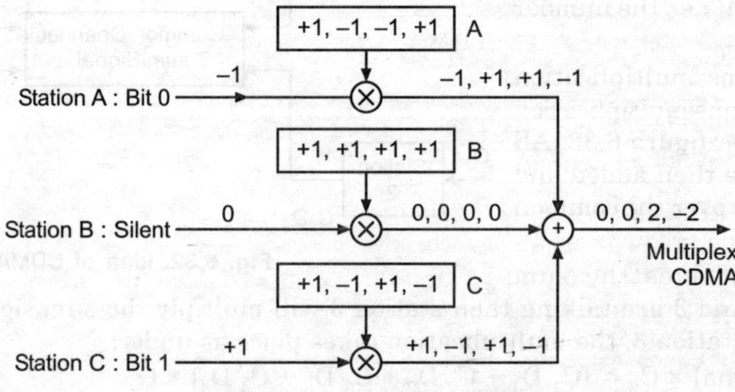

Fig. 6.34. CDMA multiplexer

We assume that station A wants to transmits a 0 bit, station B wants to remain silent and station C wants to transmit a 1.

Working Operation

The multiplexer receives one encoded number from each station (– 1, 0, + 1) as shown in figure 6.34. Each bit is multiplied with the code number of the corresponding station. The outputs of multipliers are added to obtain the multiplexed CDMA signal. The multiplexed CDMA signal is then transmitted through the common link.

6.21.6. CDMA Decoder

1. Block Schematic

Figure 6.35 illustrates the block schematic of a CDMA demultiplexer.

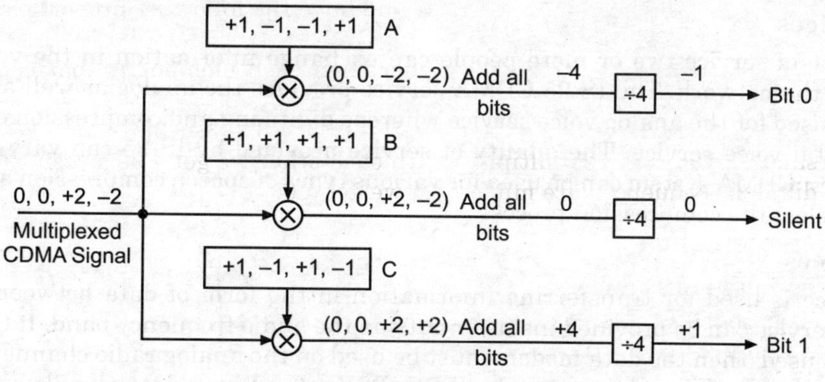

Fig. 6.35. CDMA Demultiplexer

2. Working Operation

The Multiplexed CDMA signal is applied to all the multipliers. It is multiplied with the code numbers assigned to the stations A, B and C. The bits available at the output of each multiplier are added together. This addition will always be either – 4, 0 or + 4. The result of addition at the output of each multiplier is divided by 4 to obtain the originally transmitted bits.

3. Orthogonal Sequences

The chip sequences i.e., the codes assigned to various stations are not selected randomly. They are chosen carefully. The codes used in the multiplexer and demultiplexer are called as orthogonal sequences.

6.21.7. Performance Comparison of FDMA, TDMA and CDMA

S. No.	FDMA	TDMA	CDMA
1.	Overall bandwidth is shared among many stations.	Time sharing takes palce.	Sharing of bandwidth and time both takes place.
2.	Due to non-linearity of devices inter modulation products are generated due to interference between adjacent channels.	Due to incorrect synchronization, there can be an interference between the adjacent time slots	Both type of interferences will be present.
3.	Synchronization is not necessary	Synchronization is essential	Synchronization is not necessary
4.	Codeword is not required	Codeword is not required	Codewords are required
5.	Guard bands between adjacent channels are necessary	Guard times between adjacent time slots are necessary	Guard bands and Guard time both are necessary

6.21.8. CDMA Services

Following are some of the important CDMA services:

1. Voice services
2. Data services
3. Circuit switched data
4. Packet switched data
5. Message services
6. CDMA radio
7. Location based services
8. CDMA radio channel

1. Voice Services

In this type of service two or more people can exchange information in the voice frequency band, through the network. The IS-95 CDMA service provides the analog as well as digital voice service. FM is used for the analog voice service whereas digitizing and compression of audio signal is used in digital voice service. The quality of service provided by IS-95 can vary depending on different factors. CDMA system can be used for various types of speech compression and the service provider can select the compression process.

2. Data Services

This service is used for transferring information in the form of data between two or more devices. This service can be provided inside or outside the audio frequency band. If the audio band (upto 4 kHz) is used, then the data modem must be used on the analog radio channel. This modem converts the digital data to analog signals. (FSK, PSK, etc.) When data signals are transmitted over CDMA, a data transfer adapter (DTA) is used. DTA converts the data bits into a format which is suitable for mobile device operating on a CDMA digital radio channel. Each CDMA channel can provide upto 14.4 kbps. If higher data speeds are required, then multiple data channels (upto 8) are combined together.

3. Circuit Switched Data

This method maintains a dedicated path between two devices. A connection needs to be established before initiating communication. For this the address is sent first and a path is established. After setting up this path, data is transferred continuously. The connection is disconnected after the data transfer is complete.

4. Packet Switched Data

The data is transferred from sender to receiver in the form of small packets. Packet switching principle is used. Each packet contains the data and the destination address. Each packet can take different path to reach the destination. CDMA packet switched data service is an always on type service.

5. Messaging Services

This service allows the transfer of short message between two persons. Each message contains a few hundred characters.

6. Location Based Services (LBS)

These are the information or advertisement services which are based on the location of the user. The CDMA allows LBSs through GSM (Global positioning system).

7. CDMA Radio

In IS-95 system, there are three types of radio channel used:

1. Wideband digital radio channel
2. Narrowband voice (audio) channel.
3. Narrowband audio control channel.

Some of the important characteristics of IS-95 CDMA radio system are, it uses a wide radio channel, multiple frequency bands, frequency reuse, channel multiplexing, full duplex radio channel pairs, RF power control, variable rate speech coding, discontinuous transmission for increased battery life.

8. CDMA Radio Channel

The radio channel used by the IS-95 which uses a 1.23 MHz slot, is divided into 64 channels. Some of these are used for traffic (voice or data), and some are used for control purposes. The spread spectrum technique is used in order to ensure better quality of reception.

6.22. CODE DIVISION MULTIPLE ACCESS (CDMA) WITH DS-SS

An alternative to FDMA and TDMA is an another system called code division multiple access (CDMA). The most important feature of CDMA is as under:

In CDMA, more than one user is allowed to share a channel or subchannel with the help of direct-sequence spread spectrum signals.

In CDMA each user is given a unique code sequence or signature sequences. This sequence allows the user to spread the information signal across the assigned frequency band. At the receiver the signal is recovered by using the same code sequence. At the receiver, the signals received from various users are separated by checking the cross-correlation of the received signal with each possible user signature sequence. In CDMA the users access the channel in a random manner. Hence, the signals transmitted by multiple users will completely overlap both in time and in frequency. The CDMa signals are spread in frequency. Therefore the demodulation and separation of these signals at the receiver can be achieved by using the pseudorandom code sequence. CDMA is sometimes also called as spread spectrum multiple access (SSMA).

6.22.1. A CDMA System Based on FH Spread Spectrum Signals

In addition to a CDMA system based on the DS-SS signals, it is also possible to have a CDMA system based on the FH-SS signals. Some of the important features of such a system are as follows: CDMA systems using FH-SS signal are used for the mobile users on the land, air and sea. This is because of the relaxed timing requirements as compared to those in the DS-SS systems. Each transmitter-receiver pair in FH-SS system is assigned itw own pseudorandom FH (frequency hopping) pattern. The frequency hopping pattern for every transmitter receiver pair is unique and different from that of other pairs. Frequency hop over significantly larger bandwidths as compared to those possible with DS-SS are possible. Therefore, the processing gain PG of FH-SS system is higher than that of a DS-SS system. The capacity of CDMA with FH-SS is higher than that of a DS-SS system. It may be noted that all the advantages stated above are applications only to the CDMA system based on FH-SS only.

6.22.2. Error Probability for DS CDMA on AWGN Channels

In the DS/CDMA system, the multiple access capability is obtained by assigning a unique PN spreading sequence to each user. But, the transmissions from different users are not synchronized. So they arrive at the receiver with different amplitudes, delays and phases. The DS CDMA signaling on AWGN channel is shown in figure 6.36. The signals from various transmitters will arrive at the receiver with different power levels. However, in DS-CDMA systems, the power control is exercised such that all the signals arrive at the receiver with same power level. This is essential to reduce the far-near effect. This means that to avoid the capture effect imposed by the strong signals. The receiver shown in figure 6.36, is a correlation receiver where the composite received signal is multiplied by a synchronized replica of the spreading sequence. The two sequences (original and its replica) will cancel each other and the desired data sequence is obtained at the output of the correlator. The exact error probability with a conventional correlator depends on the particular spreading sequence used and the random amplitudes, delays and phases of the signals arriving at the receiver. Unfortunately the exact error probability is difficult to derive and evaluate.

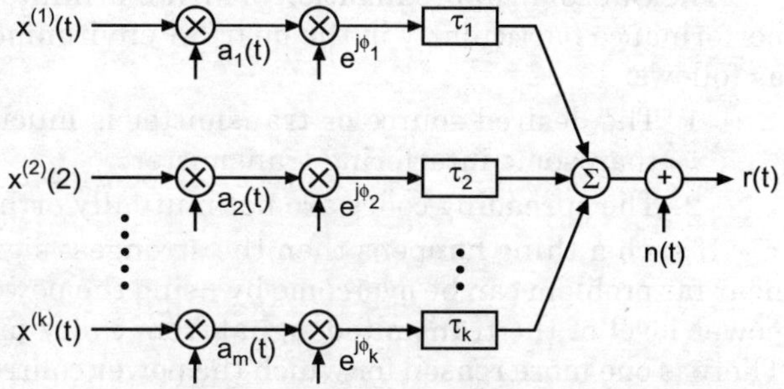

Fig. 6.36. DS CDMA signaling on an AWGN channel

6.23. CDMA OF DSSS

(i) The DSSS system has advantages against the problems of narrowband jamming and attempts at interception.

(ii) But use of DSSS for the transmission of only one signal would be the waste of a large bandwidth.

(iii) Hence, CDMA of DSSS can be achieved by allowing multiple users each one given a distinct PN spreading signal $C_i(t)$ as shown in figure 6.37.

(iv) All the users can access the broad bandwidth of LBs simultaneously.

(v) Each user makes use of a single-user optimum receiver which is an optimum matched filter or a correlation receiver.

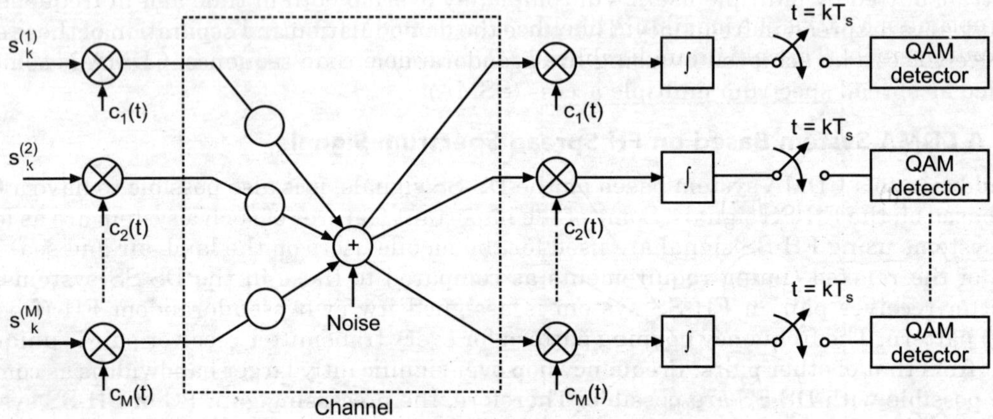

Fig. 6.37. A CDMA system based on DSSS

(vi) As these CDMA users transmit without any multiplexing scheme such as FDM or TDM, the multiple access interference (MAI) will be present at each receiver. The MAI can be approximated as a Gaussian function.

6.23.1. The Near-Far Problem

The Gaussian approximation of MAI has limitations when it is to be used for predicting system performance particularly in the near-far environment. The near-far environment can be described as follows:

1. The desired source or transmitter is much farther away from its destination or receiver than some interfering transmitter.
2. The spreading codes are not mutually orthogonal.

If such a thing happens then the strongest signal will capture the demodulation process. The near far problem can be overcome by using the power control at the base station so as to control the power level of the transmited signal from every mobile which is coming under that base station. There is one more reason for which the power control in CDMA is important. Every multiple access system is supposed to maximize the system capacity. The system capacity can be defiend as the maximum number of users that can be served reliably by a system. The system capacity can be maximized if the transmitter of each mobile is under control of the serving base station. The transmitter power of each subscriber is controlled tightly to reduce the problem of interference. A combination of open loop and closed loop power control is used for adjusting the transmitter power of each subscriber.

6.24. MULTIUSER DETECTION (MUD)

Multiuser detection (MUD) is another technique to counter the near-far effect. MUD can equalize the received signal power without feedback from the receiver to transmitter. This is different from the closed loop power control technique. MUD is a centralized receiver. Its purpose is to jointly receive (detect) all the user signals of different received signal strengths. The MUD is assumed to have an access to all M signal samples. It is also assumed that the MUD has the knowledge of the user signal strengths, spreading sequence cross correlation and statistics of noise samples. The goal of MUD receivers is to determine the unknown user data vector based on the received signal vector value. It is also possible to derive different joint MUD receivers according to different criteria.

6.24.1. Optimum MUD

The optimum MUD is the maximum likelihood detector (MLD) under the assumption of equally likely input symbols. The maximum likelihood MUD receiver is as shown in figure 6.38.

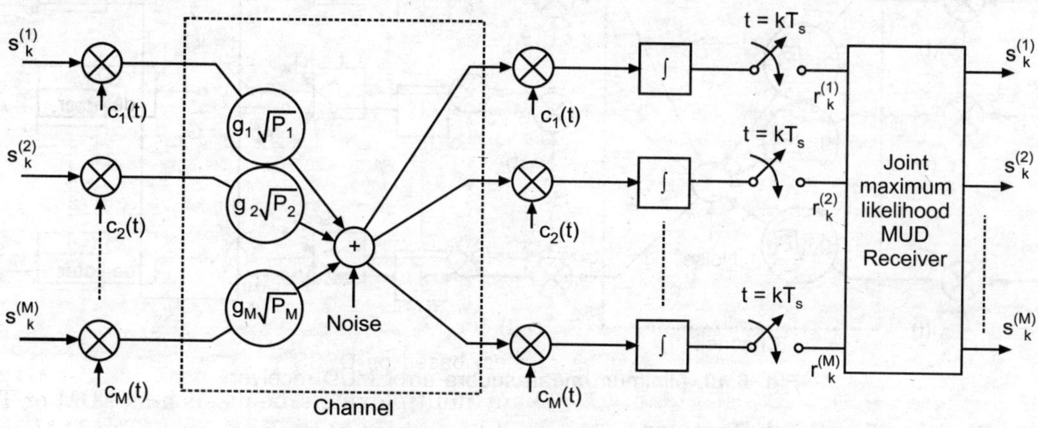

Fig. 6.38. Maximum likelihood MUD receiver

If each user uses 16 QAM to modulate its data then the optimum MUD receiver requires 16^M evaluations before making the final decision. This indicates that the maximum likelihood MUD has a very high complexity. The complexity increases exponentially with the number of CDMA users. But then this receiver has the advantages of optimum reception and near-far resistance.

6.24.2. Decorrelator Receiver

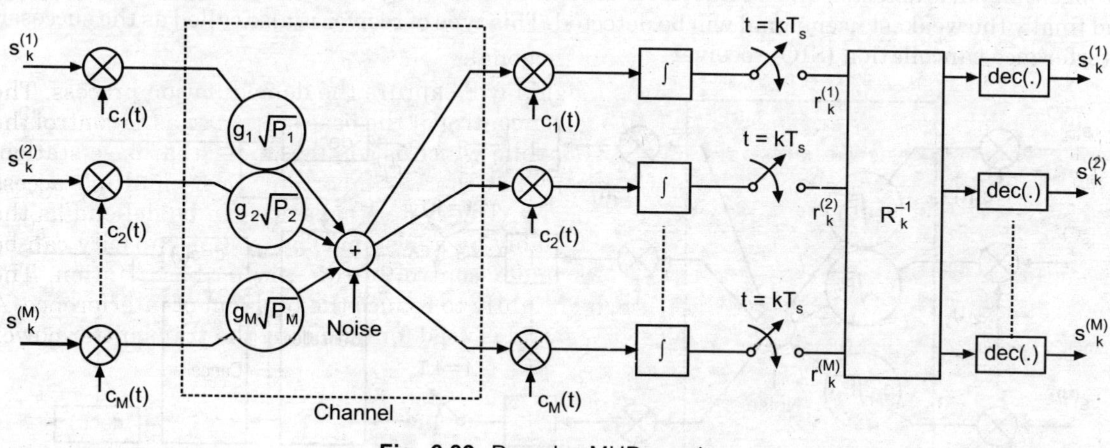

Fig. 6.39. Decoder MUD receiver

Due to high complexity the maximum likelihood MUD receiver is not used in many applications. Instead several low complexity and suboptimum MUD receivers are developed. The decorrelator type MUD receiver is one of such low complexity receivers. Figure 6.39. shows the block diagram of such a receiver. It simply uses the matrix multiplication to remove the MAI among different users. The decorrelator receiver is near-far resistant. The main block in the decoder MUD receiver is the matrix multiplier block R_k^{-1}.

6.24.3. Minimum Mean Square Error (MSE) Receiver

The noise performance of decorrelator receiver is inferior. This drawback can be overcome by using a more robust MUD which reduces the mean square error. Such a receiver is called as the minimum mean square error (MSE) receiver and its block diagram is as shown in figure 6.40. This receiver is also near-far resistant.

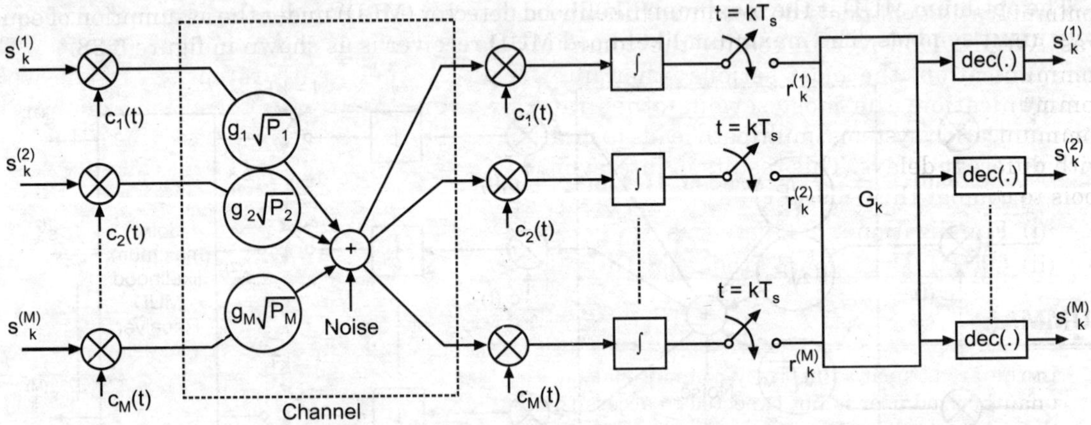

Fig. 6.40. Minimum mean square error MUD receiver

6.24.4. Decision Feedback Receiver

Both the decorrelator and the MMSE MUD receivers use linear matrix processing. So they are known as the linear MUD receivers and have less complexity. There is another suboptimum non-linear receiver called as the decision feedback MUD receiver and its operation is based on successive interference cancellation. Figure 6.41 shows the block schematic of the decision feedback MUD receiver. Once the strongest user symbol is known, its interference effects on the remaining symbols can be cancelled. Thus out of M symbols, now only M-I symbols remain. Among them the next strongest signal is detected and its interference with the others is cancelled. This process is continued and finally the weakest user signal will be detected. This type of rceiver is also called as the successive interference cancellation (SIC) receiver.

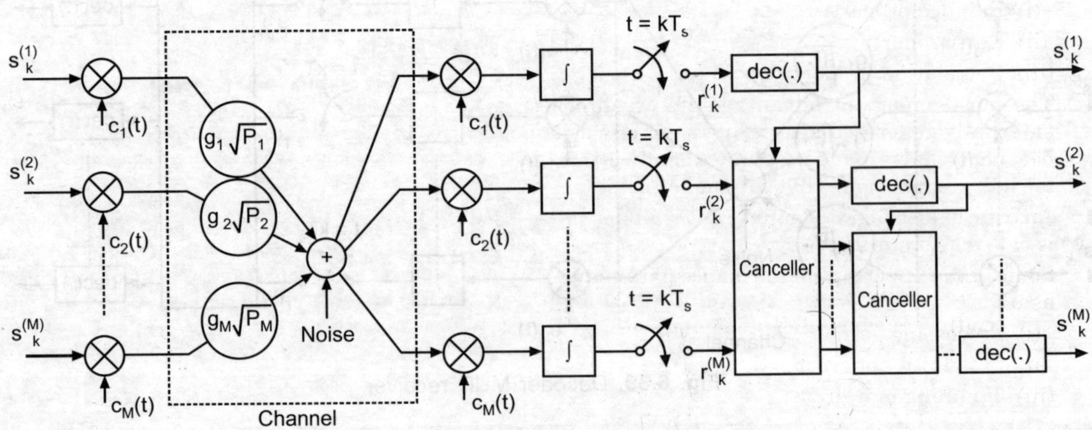

Fig. 6.41. Decision feedback MUD receiver

Advantages

(i) This receiver needs a little computation. So less complexity.
(ii) It is very sensible receiver.
(iii) Near-far problem is completely eliminated.

Drawbacks

(i) Error propagation takes place.
(ii) Decision errors due to error propagation.

6.25. OFDM

In our earlier discussion and analysis of digital communication systems it was assumed that the communication channels do not introduce any distortion. The only distortion that we considered was AWGN. However in reality the communication channels are not ideal. In the wireless communication, the most serious other interferance is multi-path reception. In the wireless communication, the most serious other interference is multipath reception. In the digital communication systems, multipath leads to multiple copies of the same signal arriving at the receiver with different delays. This results in intersymbol interference (ISI). There are two highly effective tools to combat this ISI:

(i) Equalization
(ii) OFDM (Orthogonal Frequency Division Multiplexing)

SUMMARY

- In the areas such as military communication, the information has to be *secured*. This means that an unauthorized user is not expected to access the information. Also he should not be allowed to interfere the communication by any means.
- Sometimes a hostile transmitter (say used by terrorists) can **jam** the transmission. To avoid this the channel should be **immune** to any external interference.
- Even in the non-military communications an unintentional interference is caused by a user who is transmitting its information through a channel which is already being used.
- A pseudo-noise (PN) sequence may be defined as a coded sequence of 1s and 0s with certain auto-correlation properties.
- The maximum length sequence is a type of cyclic code which represents a commonly used periodic PN sequence. Such a sequence has long periods and requires a linear feedback shift register. A PN sequence generated by an m-register linear feedback shift register is of length $2^m - 1$. A shift register of length m consists of m flip flops and all of them operate on the same clock.
- Maximum length sequences have many properties possessed by a truly rendom sequence. Some of the important properties are as under:
  (i) Balance property
  (ii) Run property
  (iii) Correlation property
- The most important advantage of spread spectrum modulation is that it provides protection against externally generated interfering signals. Such signals are called as a jamming signals. The information bearing signal is made to occupy a bandwidth which is much larger than the minimum bandwidth required for its transmission. This will make the signal to appear like a noise and blends into the background.
- An externally generated interfering signal is called as a jamming signal and such an intentional interference is called as jamming.
- Some of the important performance parameters of a direct sequence spread spectrum system may be listed as under:
  (i) Processing gain
  (ii) Probability of error
  (iii) Jamming margin

- This system has a very high degree of discrimination against the multipath signals. Therefore the interference caused by the multipath reception is minimized successfully.
- The performance of DS-SS system in presence of noise is superior to other systems such as FH-SS system.
- This system combats the intentional interference (jamming) most effectively.
- With the serial search system, the acquisition time is too large. This makes the DS-SS system slow.
- The sequence genereted at the PN code generator output must have a high rate. The length of such a sequence needs to be long enough to make the sequence truly random.
- The channel bandwidth required, is very large. But this bandwidth is less than that of a FH-SS system.
- The synchronization is affected by the variable distance between the transmitter and receiver.
- Depending upon the rate of frequency hopping, the FH/MFSK systems are classified into following two categories:
  - (i) Slow frequency hopping
  - (ii) Fast frequency hopping
- FHSS is being used in a number of applications. The two most important of them are as under:
  1. Wireless local area networks (WLAN) standard for Wi-Fi.
  2. Wireless personal area network (WPAN) standard of Bluetooth.
- A Bluetooth network is called as a piconet or a small net. It can have at the most eight stations. One of them is called as a master and all others are called as slaves.
- One of the simplest multiple access methods is the frequency division multiple access (FDMA). In this method, the channel bandwidth is subdivided into a number division subchannels
- All the stations can operate continuously all 24 hours without having it wait for their turn to come. The power required for transmission depends on the number of channels being transmitted. The signal to noise ratio is improved due to the use of FM. No synchronization is necessary.
- In CDMA, more than one user is allowed to share a channel or subchannel with the help of direct-sequence spread spectrum signals.
- Multiuser detection (MUD) is another technique to counter the near-far effect. MUD can equalize the received signal power without feedback from the receiver to transmitter. This is different from the closed loop power control technique. MUD is a centralized receiver. Its purpose is to jointly receive (detect) all the user signals of different received signal strengths.
- The optimum MUD is the maximum likelihood detector (MLD) under the assumption of equally likely input symbols.
- Both the decorrelator and the MMSE MUD receivers use linear matrix processing. So they are known as the linear MUD receivers and have less complexity.

SHORT QUESTIONS WITH ANSWERS

Q.1. Explain the features of DSSS?

Ans. Some of the important features of DSSS are as follows:

(i) It provides good security againt potential jamming or interception.

(ii) The DSSS is extremely effective against narrowband jamming signals.

(iii) The narrowband communication signals can coexist with the DSSS signals.

(iv) The DSSS signal is not very effective against broadband interference.

Q.2. Give The Applications of DS-SS System.

Ans. Some of the important applications of the DS-SS system are as follows:

(i) To combat the intentional interference (jamming)

(ii) To reject the unintentional interference

(iii) To minimize the self interference due to multipath propagation

(iv) In the low probability of intercept (LPI) signal

(v) In obtaining the message privacy

(vi) Code division multiple access with DS-SS.

Q.3. Define CDMA Services.

Ans. Following are some of the important CDMA services:

1. Voice services
2. Data services
3. Circuit switched data
4. Packet switched data
5. Message services
6. CDMA radio
7. Location based services
8. CDMA radio channel

Q.4. Define Multiuser Detection.

Ans. Multiuser detection (MUD) is another technique to counter the near-far effect. MUD can equalize the received signal power without feedback from the receiver to transmitter. This is different from the closed loop power control technique. MUD is a centralized receiver. Its purpose is to jointly receive (detect) all the user signals of different received signal strengths.

REVIEW QUESTIONS

1. State the problems encountered by a communication system.
2. How is SS signal different from the normal one?
3. State true or false. The SS signal is pseudorandom in nature.
4. State some applications of SS modulation.
5. Name various types of SS systems.
6. Which modulation techniques are used in SS?
7. Define a PN sequence.
8. What is meant by a maximum length sequence?
9. State the balance property.
10. State the run property.
11. Define chip duration.
12. What is chip rate?
13. State the performance paramters of a DS-SS system.
14. Define the processing gain.
15. What is the BW of spread signal?
16. Which factors decide the error probability?
17. Describe a modulation technique that offer certain advantages over band limited, non-linear channels.
18. Explain with block diagram, direct sequence spread spectrum technique.
19. Describe a modulation technique that offer certain advantages over bandlimited non-linear channels.
20. Explain with block diagram, direct sequences spread spectrum technique.
21. Describe FH-FSK system in brief.
22. Draw the block diagram for FH-SS system and explain the working. Differentiate between slow frequency hopping and fast frequency hopping.
23. Explain the principle and working of FH-SS in detail. Differentiate between slow frequency hopping and fast frequency hopping.
24. Explain the principle of FDMA, TDMA.
25. Explain the principle of CDMA.
26. Compare FDMA, TDMA and CDMA.
27. Explain the following: (i) GPS (ii) IS-95.

❑❑❑

UNIT– V

Measure of Information, Source Encoding, Error Free Communication over a Noisy Channel capacity of a discrete and Continuous Memory less channel Error Correcting codes: Hamming distance and error detecting and correcting capability, Linear block codes, encoding & syndrome decoding; Cyclic codes, encoder and decoders for systematic cycle codes; convolution codes, code tree & Trellis diagram, Viterbi and sequential decoding, burst error correction.

Above Syllabus has been covered in following chapters:

CHAPTER—7 : Information Theory

CHAPTER—8: Error Control Coding

CHAPTER 7

INFORMATION THEORY

SYLLABUS

Measure of Information, Source Encoding, Information Sources, Entropy, Information rate, Information sources, BSC, Error Free Communication over a Noisy Channel capacity of a discrete and Continuous Memory less channel.

7.1 INTRODUCTION

As discussed in chapter 1, the purpose of a communication system is to carry information-bearing baseband signals from one place to another place over a communication channel. In the last few chapters, we have discussed a number of modulation schemes to accomplish this purpose. But what is the meaning of the word *Information*'. To answer this question we need to discuss information theory.

Infact, information theory is a branch of probability theory which may be applied to the study of the communication systems. This broad mathematical discipline has made fundamental contributions, not only to communications, but also computer science, statistical physics and probability and statistics.

Further, in the context of communications, information theory deals with mathematical modelling and analysis of a communication system rather than with physical sources and physical channels. As a matter of fact, Information theory was invented by communication scientists while they were studying the statistical structure of electronic communication equipments. When the communique is readily measurable, such as an electric current, the study of the communication system is relatively easy. But, when the communique is information, the study becomes rather difficult. How to define the measure for an amount of

Inside this Chapter

information? And also having described a suitable measure, how can it be applied to improve the communication of information? Information theory provides answers to all these questions. Thus, this chapter is devoted to a detailed discussion of information theory.

7.2 WHAT IS INFORMATION?

Before discussing the quantitative measure of information, let us review a basic concept about the amount of information in a message. Few messages produced by an information source contain more information than others. This may be best understood with the help of following example:

Consider you are planning a tour a city located in such an area where rain fall is very rare. To know about the weather forecast you will call the weather bureau and may receive one of the following information:

(i) It would be hot and sunny,

(ii) There would be scattered rain,

(iii) There would be a cyclone with thunderstorm.

It may be observed that the amount of information received is clearly different for the three messages. The first message, just for instance, contains very little information because the weather in a desert city in summer is expected to be hot and sunny for maximum time. The second message forecasting a scattered rain contains some more information because it is not an event that occurs often. The forecast of a cyclonic storm contains even more information compared to the second message. This is because the third forecast is a rearest event in the city. Hence, on an conceptual basis the amount of information received from the knowledge of occurrence of a event may be related to the likelihood or probability of occurrence of that event. The message associated with the least likelihood event thus consists of maximum information. The above amount of information in a message depends only upon the uncertainty of the underlying event rather than its acutal content. Now, let us discuss few important concepts related to Information theory in the sections to follow.

7.3 INFORMATION SOURCES

(i) Definition

An information source may be viewed as an object which produces an event, the outcome of which is selected at random according to a probability distribution. A practical source in a communication system is a device which produces messages, and it can be either analog or discrete. In this chapter, we deal mainly with the discrete sources since analog sources can be transformed to discrete sources through the use of sampling and quantization techniques. As a matter of fact, a discrete information source is a source which has only a finite set of symbols as possible outputs. The set of source symbols is called the source alphabet, and the elements of the set are called symbols or letters.

(ii) Classification of Information Sources

Information sources can be classified as having memory or being memoryless. A source with memory is one for which a current symbol depends on the previous symbols. A memoryless source is one for which each symbol produced is independent of the previous symbols.

A discrete memoryless source (DMS) can be characterized by the list of the symbols, the probability assignment to these symbols, and the specification of the rate of generating these symbols by the source.

DO YOU KNOW?

A discrete information source consists of a discrete set of letters or alphabet of symbols. In general, any message emitted by the source consists of a string or sequence of symbols.

To make your children capable of honesty is the beginning of education. – **John Reuskin**

Pioneer in Communications: Claude Elwood Shannon (1916–1972)

Born in Gaylord, Michigan, Claude Shannon graduated from Michigan University with degrees in mathematics and engineering and, in 1940, from the Massachusetts Institute of Technology (MIT) with master's and doctoral degrees in mathematics. Shannon's theories effectively provided the mathematical foundation for designing digital electronic circuits, which form the basis of modern-day information processing.

After graduating from MIT, Shannon served as a National Research Fellow at the Institute for Advanced Study at Princeton University and, in 1941, joined Bell Telephone Laboratories as a research mathematician. While at Bell Labs, Shannon applied his information processing theories as part of his research for improving the transmission and reliability of information across long-distance telephone and telegraph lines. He described his research in the article, "Mathematical Theory of Communication," which was published in the Bell System Technical Journal in 1948. This paper, which quickly became known as "Information Theory" (or IT for short), contained the first published account of the term bit to denote the abbreviation of "binary digit" as a representation of the individual information units a computer processes. The paper also extended Nyquist's theory and the effect noise has on a communications channel. Information theory served as the basis for which the theoretical limit of any channel's capacity could be determined and eventually led to the design of efficient, error-free transmission over noisy channels. Shannon continued working for Bell Labs until 1972. The IEEE Information Theory Society's Claude E. Shannon Award has since been instituted to honor exemplary contributions to the field of information theory.

7.4 INFORMATION CONTENT OF A DISCRETE MEMORYLESS SOURCE (DMS)

The amount of information contained in an event is closely related to its uncertainty. Messages containing knowledge of high probability of occurrence convey relatively little information. We note that if an event is certain (that is, the event occurs with probability 1), it conveys zero information. Thus, a mathematical measure of information should be a function of the probability of the outcome and should satisfy the following axioms:

(i) Information should be proportional to the uncertainty of an outcome.
(ii) Information contained in independent outcomes should add.

7.5 INFORMATION CONTENT OF A SYMBOL (i.e., LOGARITHMIC MEASURE OF INFORMATION)

(i) Definition

Let us consider a discrete memoryless source (DMS) denoted by X and having alphabet $\{x_1, x_2, \dots x_m\}$. The information content of a symbol x_i, denoted by $I(x_i)$ is defined by

$$I(x_i) = \log_b \frac{1}{P(x_i)} = -\log_b P(x_i) \quad \text{...(7.1)}$$

where $P(x_i)$ is the probability of occurrence of symbol x_i.

(ii) Properties of $I(x_i)$

The information content of a symbol x_i, denoted by $I(x_i)$ satisfies the following properties:

$$I(x_i) = 0 \text{ for } P(x_i) = 1 \quad \text{...(7.2)}$$

$$I(x_i) \geq 0 \quad \text{...(7.3)}$$

$$I(x_i) > I(x_j) \text{ if } P(x_i) < P(x_j) \quad \text{...(7.4)}$$

$$I(x_i, x_j) = I(x_i) + I(x_j) \text{ if } x_i \text{ and } x_j \text{ are indepedent} \quad \text{...(7.5)}$$

(ii) Unit of $I(x_i)$

The unit of $I(x_i)$ is the bit (binary unit) if b = 2, Hartley or decit if b = 10, and nat (natural unit) if b = e. It is standard to use b = 2. Here the unit bit (abbreviated "b") is a measure of information content and is not to be confused with the term 'bit' meaning binary digit. The conversion of these units to other units can be achieved by the following relationships.

$$\log_2 a = \frac{\ln a}{\ln 2} = \frac{\log a}{\log 2} \quad ...(7.6)$$

(iii) Few Important points about $I(x_i)$

According to our concept, the information content or amount of information of a symbol x_i, denoted by $I(x_i)$, must be inversely related to $P(x_i)$. Also from our intuitive concept $I(x_i)$ must satisfy the following requirements:

(i) $I(x_i)$ must approach 0 as $P(x_i)$ approaches infinity. For example, consider the message 'Sun will rise in the east'. This message does not contain any information since sun will rise in the east with probability 1.

(ii) The information content $I(x_i)$ must be a non-negative quantity since each message contains some information. In the worst case $I(x_i)$ can be equal to zero.

(iii) The information content of a message having higher probability of occurrence is less than the information content of a message having lower probability.

Now let us discuss few numerical examples to illustrate all the above concepts.

EXAMPLE 7.1. A source produces one of four possible symbols during each interval having probabilities $P(x_1) = \frac{1}{2}, P(x_2) = \frac{1}{4}, P(x_3) = P(x_4) = \frac{1}{8}$. Obtain the information content of each of these symbols.

Solution: We know that the information content $I(x_i)$ of a symbol x_i is given by

$$I(x_i) = \log_2 \frac{1}{P(x_i)}$$

Thus, we can write

$$I(x_1) = \log_2 \frac{1}{\frac{1}{2}} = \log_2 (2) = 1 \text{ bit} \; ; \qquad I(x_2) = \log_2 \frac{1}{\frac{1}{4}} = \log_2 2^2 = 2 \text{ bits}$$

$$I(x_3) = \log_2 \frac{1}{\frac{1}{8}} = \log_2 2^3 = 3 \text{ bits} \; ; \qquad I(x_4) = \log_2 \frac{1}{\frac{1}{8}} = 3 \text{ bits} \quad \textbf{Ans.}$$

EXAMPLE 7.2. Calculate the amount of information if it is given that $P(x_i) = \frac{1}{4}$.

Solution: We know that amount of information $I(x_i)$ of a discrete symbol x_i is given by,

$$I(x_i) = \log_2 \frac{1}{P(x_i)}$$

The above expression may be written as under:

$$I(x_i) = \frac{\log_{10} \frac{1}{P(x_i)}}{\log_{10} 2}$$

The hour of the greatest triumph is the hour of the greatest humility. – **Mahatma Gandhi**

Substituting given value of $P(x_i)$ in above expression, we obtain

$$I(x_i) = \frac{\log_{10} 4}{\log_{10} 2} = 2 \text{ bits} \quad \textbf{Ans.}$$

EXAMPLE 7.3. Calculate the amount of information if binary digits (binits) occur with equal likelihood in a binary PCM system.

Solution: We know that in binary PCM, there are only two binary levels i.e., 1 or 0. Since, these two binary levels occur with equal likelihood, their probabilities of occurrence will be,

$$P(x_1) \text{ ('0' level)} = P(x_2) \text{ ('1' level)} = \frac{1}{2}$$

Therefore, the amount of information content will be given as

$$I(x_1) = \log_2 \frac{1}{P(x_1)}$$

and $$I(x_2) = \log_2 \frac{1}{P(x_2)} \quad \text{or} \quad I(x_1) = \log_2 2$$

and $$I(x_2) = \log_2 2 \quad \text{or} \quad I(x_1) = I(x_2) = \frac{\log_{10} 2}{\log_{10} 2} = 1 \text{ bit} \quad \textbf{Ans.}$$

Hence, the correct identification of binary digit (binit) in binary PCM carries 1 bit of information.

EXAMPLE 7.4. In a binary PCM if '0' occur with probability $\frac{1}{4}$ and '1' occur with probability equal to $\frac{3}{4}$, then calculate the amount of information carried by each binit.

Solution: Here, given that the binit '0' has $P(x_1) = \frac{1}{4}$ and binit '1' has $P(x_2) = \frac{3}{4}$

Then, amount of information is given as,

$$I(x_i) = \log_2 \frac{1}{P(x_i)}$$

with $$P(x_1) = \frac{1}{4},$$

we have $$I(x_1) = \log_2 4 = \frac{\log_{10} 4}{\log_{10} 2} = 2 \text{ bits}$$

and with $$P(x_2) = \frac{3}{4},$$

we have $$I(x_2) = \log_2 \left(\frac{4}{3}\right) = \frac{\log_{10}(4/3)}{\log_{10} 2} = 0.415 \text{ bits} \quad \textbf{Ans.}$$

✍ **Important Point:** *Here, it may observed that binit '0' has probability $\frac{1}{4}$ and it carries 2 bits of information whereas binit '1' has probability $\frac{3}{4}$ and it carries 0.415 bits of information.*

Thus, this reveals the fact that if probability of occurrence is less, then the information carried is more, and vice-versa.

EXAMPLE 7.5. If there are M equally likely and independent symbols, then prove that amount of information carried by each symbol will be,

$I(x_i) = N$ bits.

where $M = 2^N$ and N is an integer.

Solution: Since, it is given that all the M symbols are equally likely and independent, therefore, the probability of occurrence of each symbol must be $\frac{1}{M}$.

We know that amount of information $I(x_i)$ of a discrete symbol x_i is given by,

$$I(x_i) = \log_2 \frac{1}{P(x_i)} \quad ...(i)$$

Here, probability of each message is, $P(x_i) = \frac{1}{M}$.

Hence, equation (i) takes the following form:

$$I(x_i) = \log_2 M \quad ...(ii)$$

Further, it is given that $M = 2^N$

Substituting above value of M in equation (ii), we obtain

$$I(x_i) = \log_2 2^N = N \log_2 2 = N \frac{\log_{10} 2}{\log_{10} 2} = N \text{ bits} \quad \textbf{Hence Proved.}$$

DO YOU KNOW?

If the units of information are taken to be binary digits or bits, then the average information rate represents the minimum average number of bits per second needed to represent the output of the source as a binary sequence.

Important Point: *Hence, amount of information carried by each symbol will be 'N' bits. We know that $M = 2^N$. This means that there are 'N' binary digits (binits) in each symbol. This indicates that when the symbols are equally likely and coded with equal number of binary digits (binits), then the information carried by each symbol (measured in bits) is numerically same as the number of binits used for each symbol.*

EXAMPLE 7.6. Prove the statement stated as under:

"If a receiver knows the message being transmitted, the amount of information carried will be zero."

Solution: Here it is stated that receiver knows the message. This means that only one message is transmitted. Thus, probability of occurrence of this message will be $P(x_i) = 1$. This is because only one message and its occurrence is certain (probability of certain event is '1'). The amount of information carried by this type of message will be,

$$I(x_i) = \log_2 \frac{1}{P(x_i)} = \frac{\log_{10} 1}{\log_{10} 2} \qquad \text{substituting } P(x_i) = 1$$

or

$$I(x_i) = 0 \text{ bits}$$

This proves the statement that if the receiver knows message, the amount of information carried will be zero. Also, as $P(x_i)$ is decreased from 1 to 0, $I(x_i)$ increases monotonically from 0 to infinity. This shows that amount of information conveyed is greater when receiver correctly identifies less likely messages.

EXAMPLE 7.7. If $I(x_1)$ is the information carried by symbols x_1 and $I(x_2)$ is the information carried by message x_2, then prove that the amount of information carried compositely due to x_1 and x_2 is $I(x_1, x_2) = I(x_1) + I(x_2)$.

Solution: We know that the amount of information is expressed as

$$I(x_i) = \log_2 \frac{1}{P(x_i)}$$

The individual amounts carried by symbols x_1 and x_2 are,

$$I(x_1) = \log_2 \frac{1}{P(x_1)} \quad ...(i)$$

and
$$I(x_2) = \log_2 \frac{1}{P(x_2)} \quad ...(ii)$$

Here $P(x_1)$ is probability of symbol x_1 and $P(x_2)$ is probability of symbol x_2. Since messages x_1 and x_2 are independent, the probability of composite message is $P(x_1)\,P(x_2)$. Therefore, information carried compositely due to symbols x_1 and x_2 will be,

$$I(x_1, x_2) = \log_2 \frac{1}{P(x_1)\,P(x_2)} = \log_2 \left[\left(\frac{1}{P(x_1)}\right)\left(\frac{1}{P(x_2)}\right)\right]$$

or
$$I(x_1, x_2) = \log_2 \left(\frac{1}{P(x_1)}\right) + \log_2 \left(\frac{1}{P(x_2)}\right)^{*}$$

Using equations (i) and (ii), we can write RHS of above equation as,

$$I(x_1, x_2) = I(x_1) + I(x_2)$$ **Hence Proved.**

7.6 ENTROPY (i.e., AVERAGE INFORMATION)

(i) Definition

In a practical communication system, we usually transmit long sequences of symbols from an information source. Thus, we are more interested in the average information that a source produces than the information content of a single symbol. In order to get the information content of the symbol, we take notice of the fact that the flow of information in a system can fluctuate widely because of randomness involved into the selection of the symbols. Thus, we require to talk about the average information content of the symbols in a long message.

(ii) Assumptions

Thus, for quantitative representation of average information per symbol we make the following assumptions:

(i) The source is stationary so that the probabilities may remain constant with time.

(ii) The successive symbols are statistically independent and come from the source at an average rate of r symbols per second.

(iii) Mathematical Expression

The mean value of $I(x_i)$ over the alphabet of source X with m different symbols is represented by H(X) and given by

$$H(X) = E\,[I(x_i)] = \sum_{i=1}^{m} P(x_i) I(x_i) = -\sum_{i=1}^{m} P(x_i) \log_2 P(x_i) \text{ bits/symbol} \quad ...(7.7)$$

The quantity H(X) is known as the **entropy** of source X. It is a measure of the average information content per source symbol. The source entropy H(X) can be considered as the average amount of uncertainty within source X that is resolved by use of the alphabet.

(iv) Entropy for Binary Source

It may be noted that for a binary source X which generates independent symbols 0 and 1 with equal probability, the source entropy H(X) is

$$H(X) = -\frac{1}{2}\log_2\frac{1}{2} - \frac{1}{2}\log_2\frac{1}{2} = 1 \text{ b/symbol} \quad ...(7.8)$$

* Since log [AB] = log A + log B

(v) Lower and Upper Bounds on Entropy H(X)

The source entropy H(X) satisfies the following relation:

$$0 \le H(X) \le \log_2 m \quad ...(7.9)$$

where m is the size (number of symbols) of the alphabet of source X. The lower bound corresponds to no uncertainty, which occurs when one symbol has probability $P(x_i) = 1$ while $P(x_j) = 0$ for $j \neq i$, so X emits the same symbol x_i all the time. The upper bound corresponds to the maximum uncertainty which occurs when $P(x_i) = 1/m$ for all i, that is, when all symbols are equal likely to be emitted by X.

DO YOU KNOW?

A simple interpretation of the source entropy is the following : on the average, we can expect to get H bits of information per symbol in long messages from the information source even through we cannot say in advance what symbol sequences will occur.

7.7 INFORMATION RATE

Definition and Mathematical Expression

If the time rate at which source X emits symbols is r (symbols s), the information rate R of the source is given by

$$R = rH(X) \text{ b/s} \quad ...(7.10)$$

Here R is information rate.

H(X) is Entropy or average information

and r is rate at which symbols are generated.

Information rate R is represented in average number of bits of information per second. It is calculated as under:

$$R = \left[r \text{ in } \frac{\text{symbols}}{\text{second}} \right] \times \left[H(X) \text{ in } \frac{\text{Information bits}}{\text{symbol}} \right] \quad ...(7.11)$$

or $$R = \text{Information bits/second} \quad ...(7.12)$$

EXAMPLE 7.8. Verify the following expression:

$$I(x_i x_j) = I(x_i) + I(x_j) \text{ if } x_i \text{ and } x_j \text{ are independent}$$

Solution: If x_i and x_j are independent, then we know that

$$P(x_i x_j) = P(x_i)P(x_j)$$

Also $$I(x_i x_j) = \log \frac{1}{P(x_i x_j)} = \log \frac{1}{P(x_i)P(x_j)} = \log \frac{1}{P(x_i)} + \log \frac{1}{P(x_j)}$$

or $$I(x_i x_j) = I(x_i) + I(x_j)$$ **Hence Proved.**

EXAMPLE 7.9. A Discrete Memoryless Source (DMS) X has four symbols x_1, x_2, x_3, x_4 with probabilities $P(x_1) = 0.4$, $P(x_2) = 0.3$, $P(x_3) = 0.2$, $P(x_4) = 0.1$.

(i) Calculate H(X).

(ii) Find the amount of information contained in the messages $x_1\ x_2\ x_1\ x_3$ and $x_4\ x_3\ x_3\ x_2$, and compare with the H(X) obtained in part (i).

Solution: (i) We know that entropy is given by

$$H(X) = -\sum_{i=1}^{4} P(x_i)\log_2[P(x_i)]$$

Substituting values and simplifying, we get

$$H(X) = -0.4 \log_2 0.4 - 0.3 \log_2 0.3 - 0.2 \log_2 0.2 - 0.1 \log_2 0.1$$

Solving, we get

$$H(X) = 1.85 \text{ b/symbol}$$ **Ans.**

(ii) Now, we have

$$P(x_1 x_2 x_1 x_3) = (0.4)(0.3)(0.4)(0.2) = 0.0096$$

Therefore, $I(x_1x_2x_1x_3) = -\log_2 0.0096 = 6.70$ b/symbol* **Ans.**

Thus, $I(x_1x_2x_1x_3) < 7.4\ [= 4H(X)]$ b/symbol

Also, $P(x_4x_3x_3x_2) = (0.1)(0.2)^2(0.3) = 0.0012$

Therefore $I(x_4x_3x_3x_2) = -\log_2 0.0012 = 9.70$ b/symbol

We conclude that

$$I(x_4x_3x_3x_2) > 7.4[= 4H(X)] \text{ b/symbol}$$ **Ans.**

EXAMPLE 7.10. Consider a binary memoryless source X with two symbols x_1 and x_2. Prove that H(X) is maximum when both x_1 and x_2 equiprobable.

Solution: Here, let us assume that

$$P(x_1) = \alpha \text{ so that } P(x_2) = 1 - \alpha.$$

We know that entropy is given by

$$H(X) = -\sum_{i=1}^{m} P(x_i)\log_2 P(x_i) \text{ bits/symbol}$$

Thus, $$H(X) = -\alpha\log_2\alpha - (1-\alpha)\log_2(1-\alpha) \quad \text{...(i)}$$

Differentiating above equation with respect to α, we get

or $$\frac{dH(X)}{d\alpha} = \frac{d}{d\alpha}[-\alpha\log_2\alpha - (1-\alpha)\log_2(1-\alpha)]$$

Using the relation

$$\frac{d}{dx}\log_b y = \frac{1}{y}\log_b e\frac{dy}{dx},$$

We get $$\frac{dH(X)}{d\alpha} = -\log_2\alpha + \log_2(1-\alpha) = \log_2\frac{1-\alpha}{\alpha}$$

Note that the maximum value of H(X) requires that

$$\frac{dH(X)}{d\alpha} = 0$$

This means that

$$\frac{1-\alpha}{\alpha} = 1 \text{ or } \alpha = \frac{1}{2}$$

Note that H(X) = 0 when α = 0 or 1.

When $P(x_1) = P(x_2) = \frac{1}{2}$, H(X) is maximum and is given by

$$H(X) = \frac{1}{2}\log_2 2 + \frac{1}{2}\log_2 2 = 1 \text{ b/symbol} \quad \text{...(ii)}$$

Hence Proved

EXAMPLE 7.11. Verify the following expression:

$$0 \le H(X) \le \log_2 m$$

where m is the size of the alphabet of X.

Solution: Proof of the lower bound

Since $$0 \le P(x_i) \le 1,$$

$$\frac{1}{P(x_i)} \ge 1 \quad \text{and} \quad \log_2\frac{1}{P(x_i)} \ge 0$$

*Using the relation $I(x_i) = \log 2\ \frac{1}{P(x_i)}$.

Then, it follows that

$$P(x_i)\log_2 \frac{1}{P(x_i)} \geq 0$$

Thus, $$H(X) = \sum_{i=1}^{m} P(x_i)\log_2 \frac{1}{P(x_i)} \geq 0 \quad ...(i)$$

Next, we note that

$$P(x_i)\log_2 \frac{1}{P(x_i)} = 0$$

If and only if $P(x_i) = 0$ or 1. Since

$$\sum_{i=1}^{m} P(x_i) = 1$$

when $P(x_i) = 1$ then $P(x_j) = 0$ for $j \neq i$. Thus, only in this case, $H(X) = 0$.

Proof of the upper bound

Let us consider two probability distributions $[P(x_i) = P_i]$ and $[Q(x_i) = Q_i]$ on the alphabet $\{x_i\}$, i = 1, 2, ... m, such that

$$\sum_{i=1}^{m} P_i = 1 \text{ and } \sum_{i=1}^{m} Q_i = 1 \quad ...(ii)$$

We know that

$$\sum_{i=1}^{m} P_i \log_2 \frac{Q_i}{P_i} = \frac{1}{\ln 2}\sum_{i=1}^{m} P_i \ln \frac{Q_i}{P_i}$$

Next, using the inequality

$$\ln \alpha \leq \alpha - 1 \quad \alpha \geq 0 \quad ...(iii)$$

and noting that the equality holds only if $\alpha = 1$, we obtain

$$\sum_{i=1}^{m} P_i \ln \frac{Q_i}{P_i} \leq \sum_{i=1}^{m} P_i \left(\frac{Q_i}{P_i} - 1\right) = \sum_{i=1}^{m} (Q_i - P_i)$$

$$= \sum_{i=1}^{m} Q_i - \sum_{i=1}^{m} P_i = 0 \text{ by using equations (ii).} \quad ...(iv)$$

Hence, we have $$\sum_{i=1}^{m} P_i \log_2 \frac{Q_i}{P_i} \leq 0 \quad ...(v)$$

where the equality holds only if $Q_i = P_i$ for all i.

Setting $$Q_i = \frac{1}{m}, \quad i = 1, 2, ..., m \quad ...(vi)$$

We get $$\sum_{i=1}^{m} P_i \log_2 \frac{1}{P_i m} = -\sum_{i=1}^{m} P_i \log_2 P_i - \sum_{i=1}^{m} P_i \log_2 m$$

$$= H(X) - \log_2 m \sum_{i=1}^{m} P_i = H(X) - \log_2 m \leq 0$$

Therefore, we get $H(X) \leq \log_2 m$

and the quality holds only if the symbols in X are equiprobable.

The human intellect delights in inventing specious arguments in order to support injustice itself.

– **Mahatma Gandhi**

EXAMPLE 7.12. A high-resolution black-and-white TV picture consists of about 2×10^6 picture elements and 16 different brightness levels. Pictures are repeated at the rate of 32 per second. All picture elements are assumed to be independent, and all levels have equal likelihood of occurrence. Calculate the average rate of information conveyed by this TV picture source.

Solution: We know that entropy is given by

$$H(X) = -\sum_{i=1}^{m} P(x_i)\log_2 P(x_i) \text{ bits/symbol} \quad \text{...(i)}$$

Here, given that $m = 16, \; P(x_i) = \frac{1}{16}$

Substituting all the values in equation (i), we get

$$H(X) = -\sum_{i=1}^{16} \frac{1}{16}\log_2 \frac{1}{16} = 4 \text{ bits/element}$$

The rate, r at which symbols are generated is given by

$$r = 2(10^6)(32) = 64(10^6) \text{ elements/second}$$

Hence, the average rate of information conveyed is given by

$$R = rH(X) = 64(10^6)(4) = 256(10^6) \text{ b/s} = 256 \text{ Mb/s} \quad \textbf{Ans.}$$

EXAMPLE 7.13. Given a telegraph source having two symbols, dot and dash. The dot duration is 0.2 s. The dash duration is 3 times the dot duration. The probability of the dot's occurring is twice that of the dash, and the time between symbols is 0.2 s. Calculate the information rate of the telegraph source.

Solution: Given that

$$P(\text{dot}) = 2P(\text{dash})$$
$$P(\text{dot}) + P(\text{dash}) = 3P(\text{dash}) = 1$$

From above two equations, we have

$$P(\text{dash}) = \frac{1}{3} \quad \text{and} \quad P(\text{dot}) = \frac{2}{3}$$

Further, we know that the entropy is given by

$$H(X) = -\sum_{i=1}^{m} P(x_i)\log_2 P(x_i) \text{ bits/symbol}$$

Expanding accordingly, we get

$$H(X) = -P(\text{dot})\log_2 P(\text{dot}) - P(\text{dash})\log_2 P(\text{dash})$$

Substituting all the values, we get

$$H(X) = 0.667(0.585) + 0.333(1.585) = 0.92 \text{ b/symbol}$$

Also, given that

$$t_{dot} = 0.2\text{s}, \; t_{dash} = 0.6\text{s}, \; t_{space} = 0.2 \text{ s}$$

Thus, the average time per symbol will be

$$T_s = P(\text{dot})t_{dot} + P(\text{dash})t_{dash} + t_{space} = 0.5333 \text{ s/symbol}$$

Further, the average symbol rate will be given by

$$r = \frac{1}{T_s} = 1.875 \text{ symbols/s}$$

Therefore, the average information rate of the telegraph source will be

$$R = rH(X) = 1.875(0.92) = 1.725 \text{ b/s} \quad \textbf{Ans.}$$

EXAMPLE 7.14. An analog signal is bandlimited to f_m Hz and sampled at Nyquist rate. The samples are quantized into 4 levels. Each level represent one symbols. Thus, there are 4 symbols. The

probabilities of occurrence of these 4 levels (symbols) are $P(x_1) = P(x_4) = \frac{1}{8}$ and $P(x_2) = P(x_3) = \frac{3}{8}$. Obtain information rate of the source.

Solution: We are given four symbols with probabilities $P(x_1) = P(x_4) = \frac{1}{8}$ and $P(x_2) = P(x_3) = \frac{3}{8}$.

Average information H(X) (or entropy) is expressed as,

$$H(X) = P(x_1)\log_2\frac{1}{P(x_1)} + P(x_2)\log_2\frac{1}{P(x_2)} + P(x_3)\log_2\frac{1}{P(x_3)} + P(x_4)\log_2\frac{1}{P(x_4)}$$

Substituting all the given values, we get

or $$H(X) = \frac{1}{8}\log_2 8 + \frac{3}{8}\log_2\left(\frac{8}{3}\right) + \frac{3}{8}\log_2\left(\frac{8}{3}\right) + \frac{1}{8}\log_2(8)$$

or $H(X) = 1.8$ bits/ symbol **Ans.**

It is given that the signal is sampled at Nyquist rate. Nyquist rate for f_m Hz bandlimited signal is,

Nyquist rate $= 2f_m$ samples/sec.

Since every sample generates one source symbol,

Therefore, symbols per second, $r = 2f_m$ symbols/sec.

Information rate is given by

$$R = rH(X)$$

Putting, values of r and H(X) in this equation, we get

$$R = 2f_m \text{ symbols/sec} \times 1.8 \text{ bits/symbol}$$
$$= 3.6\, f_m \text{ bits/sec}$$ **Ans.**

DO YOU KNOW?

Messages produced by information sources consists of sequences of symbols. While the receiver of a message may interpret the entire message as a single unit, communication system often have to deal with individual symbols.

In this example, there are four levels. Those four levels may be coded using binary PCM as shown in Table 7.1.

Table 7.1.

S.No.	Symbol or level	Probability	Binary digits
1.	Q_1	$\frac{1}{8}$	0 0
2.	Q_2	$\frac{3}{8}$	0 1
3.	Q_3	$\frac{3}{8}$	1 0
4.	Q_4	$\frac{1}{8}$	1 1

Hence, two binary digits (binits) are required to send each symbol. We know that symbols are sent at the rate of $2f_m$ symbols/sec. Therefore, transmission rate of binary digits will be,

Binary digits (binits) rate $= 2$ binits/symbol $\times 2f_m$ symbols/sec.

$= 4f_m$ binits/sec.

Because one binit is capable of conveying 1 bit of information therefore, the above coding scheme is capable of conveying $4f_m$ bits of information per second. But in this example, we have

An optimist sees an opportunity in every calamity; a pessimist sees calamity in every opportunity.

– **Anon**

obtained that we are transmitting 9.6 f_m bits of information per second. This means that the information carrying ability of binary PCM is not completely utilized by the transmission scheme.

EXAMPLE 7.15. In the transmission scheme of last example, obtain the information rate if all symbols are equally likely.

Also, comment on the result.

Solution: We are given that there are four symbols. Since they are equally likely, therefore, their probabilities will be equal to $\frac{1}{4}$ i.e.,

$$P(x_1) = P(x_2) = P(x_3) = P(x_4) = \frac{1}{4}$$

We know that, average information per symbol (Entropy) is given as,

$$H(X) = P(x_1)\log_2\frac{1}{P(x_1)} + P(x_2)\log_2\frac{1}{P(x_2)} + P(x_3).\log_2\frac{1}{P(x_3)} + P(x_4)\log_2\frac{1}{P(x_4)}$$

since $P(x_1) = P(x_2) = P(x_3) = P(x_4)$

Therefore, $H(X) = 4P(x).\log_2\frac{1}{P(x)}$

Again, since $P(x_1) = P(x_2) = P(x_3) = P(x_4) = P(x) = \frac{1}{4}$

Thus, $H(X) = \log_2(4) = 2$ bits/symbol

We know that the information rate is expressed as,

$$R = rH(X)$$

Here, $r = 2f_m$ symbols/sec. as obtained in last example. Substituting, these values in above equation, we get,

$$R = 2f_m \text{ symbols/sec.} \times 2 \text{ bits/symbol} = 4\ f_m \text{ bits/sec.}$$ **Ans.**

Important Point: *Just before this example we have seen that a binary coded PCM with 2 binits per message is capable of conveying 4B bits of information per second. The transmission scheme discussed in above example transmits 4B bits of information per second. This has been made possible since all the messages are equally likely. Thus with binary PCM clding, the maximum information rate is achieved if all messages are equally likely.*

EXAMPLE 7.16. A discrete source emits one of five symbols once every millisecond with probabilities 1/2, 1/4, 1/8, 1/16 and 1/16 respectively. Determine the source entropy and information rate.

Solution: We know that the source entropy is given as

$$H(X) = \sum_{i=1}^{m} P(x_i)\log_2\frac{1}{P(x_i)} = \sum_{i=1}^{5} P(x_i)\log_2\frac{1}{P(x_i)} \text{ bits/symbol}$$

or $$H(X) = \frac{1}{2}\log_2(2) + \frac{1}{4}\log_2(4) + \frac{1}{8}\log_2(8) + \frac{1}{16}\log_2(16) + \frac{1}{16}\log_2(16)$$

or $$H(X) = \frac{1}{2} + \frac{1}{2} + \frac{3}{8} + \frac{1}{4} + \frac{1}{4} = \frac{15}{8} = 1.875 \text{ bits/symbol}$$ **Ans.**

The symbol rate $r = \frac{1}{T_b} = \frac{1}{10^{-3}} = 1000$ symbols/sec.

Therefore, the information rate is expressed as

$$R = rH(X) = 1000 \times 1.875 = 1875 \text{ bits/sec.}$$ **Ans.**

EXAMPLE 7.17. The probabilities of the five possible outcomes of an experiment are given as

$$P(x_1) = \frac{1}{2},\ P(x_2) = \frac{1}{4},\ P(x_3) = \frac{1}{8},\ P(x_4) = P(x_5) = \frac{1}{16}$$

Determine the entropy and information rate if there are 16 outcomes per second.

(UPTU, Sem. Exam., 2005-06)

Solution: The entropy of the system is given as

$$H(X) = \sum_{i=1}^{5} P(x_i)\log_2\frac{1}{P(x_i)} \text{ bits/symbol}$$

or $$H(X) = \frac{1}{2}\log_2 2 + \frac{1}{4}\log_2 4 + \frac{1}{8}\log_2 8 + \frac{1}{16}\log_2 16 + \frac{1}{16}\log_2 16$$

or $$H(X) = \frac{1}{2} + \frac{2}{4} + \frac{3}{8} + \frac{4}{16} + \frac{4}{16} = \frac{1}{2} + \frac{1}{2} + \frac{3}{8} + \frac{1}{4} + \frac{1}{4}$$

or $$H(X) = \frac{4+4+3+2+2}{8} = \frac{15}{8} = 1.875 \text{ bits/outcome} \quad \textbf{Ans.}$$

Now, rate of outcomes r = 16 outcomes/sec.

Therefore, the rate of information R will be

$$R = rH(X) = 16 \times \frac{15}{8} = 30 \text{ bits/sec} \quad \textbf{Ans.}$$

EXAMPLE 7.18. An analog signal bandlimited to 10 kHz is quantized in 8 levels of a PCM system with probabilities of 1/4, 1/5 1/5, 1/10, 1/10, 1/20, 1/20 and 1/20 respectively. Find the entropy and the rate of information.

Solution: We know that according to the sampling theorem, the signal must be sampled at a frequency given as

$$f_s = 10 \times 2 \text{ kHz} = 20 \text{ kHz}$$

Considering each of the eight quantized levels as a message, the entropy of the source may written as

$$H(X) = \frac{1}{4}\log_2 4 + \frac{1}{5}\log_2 5 + \frac{1}{5}\log_2 5 + \frac{1}{10}\log_2 10 + \frac{1}{10}\log_2 10 + \frac{1}{20}\log_2 20 + \frac{1}{20}\log_2 20 + \frac{1}{20}\log_2 20$$

or $$H(X) = \frac{1}{4}\log_2 4 + \frac{2}{5}\log_2 5 + \frac{7}{20}\log_2 20 = 2.84 \text{ bits/message}$$

As the sampling frequency is 20 kHz then the rate at which message is produced, will be given as

or r = 20000 messages/sec.

Hence, the information rate is

$$R = rH(X) = 20000 \times 2.84 = 56800 \text{ bits/sec.} \quad \textbf{Ans.}$$

7.8 THE DISCRETE MEMORYLESS CHANNELS (DMC)

In this section, let us discuss different aspects related to discrete memoryless channels (DMC).

7.8.1. Channel Representation

A communication channel may be defined as the path or medium through which the symbols flow to the receiver end. A discrete memoryless channel (DMC) is a statistical model with an input X and an output Y as shown in figure 7.1. During each unit of the time (signaling interval), the channel accepts an input symbol from X, and in response it generates an output symbol from Y. The channel is said to be "discrete" when the alphabets of X and Y are both finite. Also, it is said to be "memoryless" when the current output depends on only the current input and not on any of the previous inputs.

DO YOU KNOW?

The reader should be aware of the fact that the data rate and information rate are two distinctly different quantities.

A diagram of a DMC with m inputs and n outputs has been illustrated in figure 7.1. The input X consists of input symbols $x_1, x_2, \ldots x_m$. The probabilities of these source symbols $P(x_i)$ are assumed to be known. The outputs Y consists of output symbols $y_1, y_2, \ldots y_n$. Each possible input-to-output path is indicated along with a conditional probability $P(y_j \mid x_i)$, where $P(y_j \mid x_i)$ is the conditional probability of obtaining output y_j given that the input is x_i, and is called a **channel transition probability.**

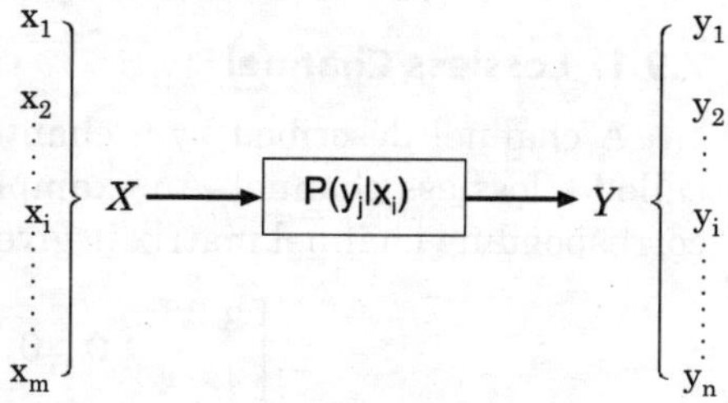

Fig. 7.1. Representation of a discrete memoryless channel (DMC)

7.8.2. The Channel Matrix

A channel is completely specified by the complete set of transition probabilities. Accordingly, the channel in figure 7.1 is often specified by the matrix of transition probabilities $[P(Y|X)]$. This matrix is given by

$$[P(Y|X)] = \begin{bmatrix} P(y_1 \mid x_1) & P(y_2 \mid x_1) & \cdots & P(y_n \mid x_1) \\ P(y_1 \mid x_2) & P(y_2 \mid x_2) & \cdots & P(y_n \mid x_2) \\ \cdots & \cdots & \cdots & \cdots \\ P(y_1 \mid x_m) & P(y_2 \mid x_m) & \cdots & P(y_n \mid x_m) \end{bmatrix} \quad ...(7.13)$$

This matrix $[P(Y|X)]$ is called the **channel matrix.**

Since each input to the channel results in some output, each row of the channel matrix must sum to unity. This means that

$$\sum_{j=1}^{n} P(y_j \mid x_i) = 1 \text{ for all } i \quad ...(7.14)$$

Now, if the input probabilities P(X) are represented by the row matrix, then we have

$$[P(X)] = [P(x_1)\ P(x_2) \ldots P(x_m)] \quad ...(7.15)$$

Also, the output probabilities P(Y) are represented by the row matrix as under:

$$[P(Y)] = [P(y_1)\ P(y_2) \ldots P(y_n)] \quad ...(7.16)$$

then

$$[P(Y)] = [P(X)][P(Y|X)] \quad ...(7.17)$$

Now, if P(X) is represented as a diagonal matrix, then we have

$$[P(X)]_d = \begin{bmatrix} P(x_1) & 0 & \cdots & 0 \\ 0 & P(x_2) & \cdots & 0 \\ \cdots & \cdots & \cdots & \cdots \\ 0 & 0 & \cdots & P(x_m) \end{bmatrix} \quad ...(7.18)$$

then

$$[P(X,Y)] = [P(X)]_d\ [P(Y|X)] \quad ...(7.19)$$

where the (i, j) element of matrix $[P(X,Y)]$ has the form $P(x_i, y_j)$.

The matrix [P(X, Y)] is known as the joint probability matrix, and the element $P(x_i, y_j)$ is the joint probability of transmitting x_i and receiving y_j.

7.9 TYPES OF CHANNELS

Other than continuous and discrete channels, there are some special type of channels with their own channel matrices. In this subsection, let us discuss various special channels.

7.9.1. Lossless Channel

A channel described by a channel matrix with only one non-zero element in each column is called a lossless channel. An example of a lossless channel has been shown in figure 7.2, and the corresponding channel matrix is given in equation (7.20) as under:

$$[P(Y|X)] = \begin{bmatrix} \frac{3}{4} & \frac{1}{4} & 0 & 0 & 0 \\ 0 & 0 & \frac{1}{3} & \frac{2}{3} & 0 \\ 0 & 0 & 0 & 0 & 1 \end{bmatrix} \quad ...(7.20)$$

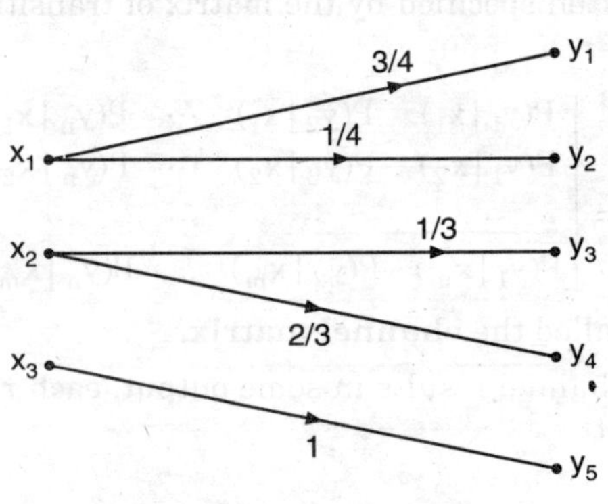

Fig. 7.2. Lossless Channel

It can be shown that in the lossless channel, no source information is lost in transmission.

7.9.2. Deterministic Channel

A channel described by a channel matrix with only one non-zero element in each row is called a deterministic channel. An example of a deterministic channel has been shown in figure 7.3, and the corresponding channel matrix is given by equation (7.21) as under:

$$[P(Y|X)] = \begin{bmatrix} 1 & 0 & 0 \\ 1 & 0 & 0 \\ 0 & 1 & 0 \\ 0 & 1 & 0 \\ 0 & 0 & 1 \end{bmatrix} \quad ...(7.21)$$

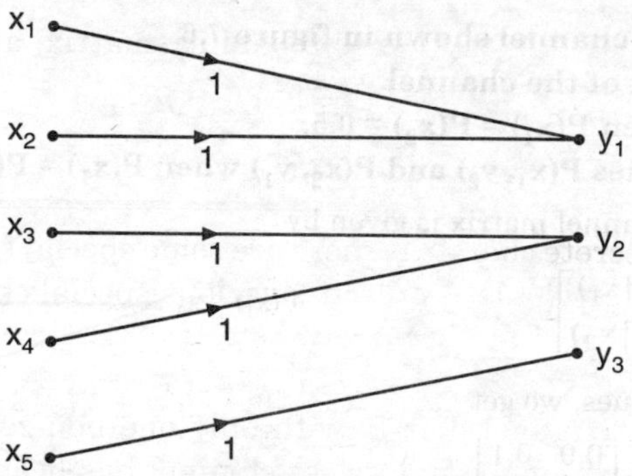

Fig. 7.3. Deterministic Channel

Important Point: It may be noted that since each row has only one non-zero element, therefore, this element must be unity by equation (7.21). Thus , when a given source symbol is sent in the deterministic channel, it is clear which output symbol will be received.

7.9.3. Noiseless Channel

A channel is called noiseless if it is both lossless and deterministic. A noiseless channel has been shown in figure 7.4. The channel matrix has only one element in each row and in each column, and this element is unity. Note that the input and output alphabets are of the same size, that is, m = n for the noiseless channel.

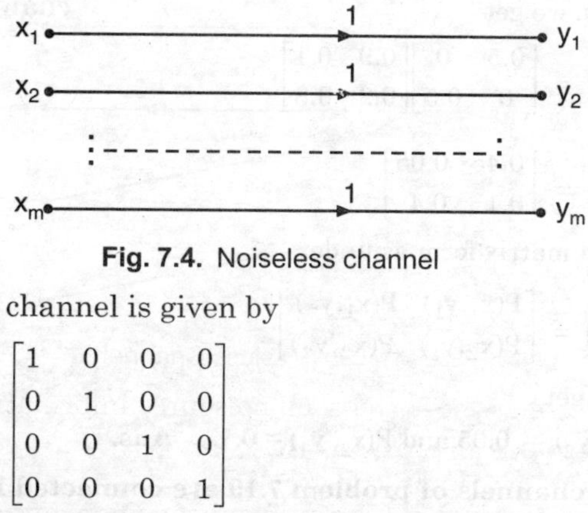

Fig. 7.4. Noiseless channel

The matrix for a noiseless channel is given by

$$\begin{bmatrix} 1 & 0 & 0 & 0 \\ 0 & 1 & 0 & 0 \\ 0 & 0 & 1 & 0 \\ 0 & 0 & 0 & 1 \end{bmatrix}$$

7.9.4. Binary Symmetric Channel (BSC)

The binary symmetric channel (BSC) is defined by the channel diagram shown in figure 7.5, and its channel matrix is given by

$$[P(Y|X)] = \begin{bmatrix} 1-p & p \\ p & 1-p \end{bmatrix} \quad ...(7.22)$$

A BSC channel has two inputs ($x_1 = 0, x_2 = 1$) and two outputs ($y_1 = 0, y_2 = 1$). This channel is symmetric because the probability of receiving a 1 if a 0 is sent is the same as the probability of receiving a 0 if a 1 is sent. This common transition probability is denoted by p as shown in figure 7.5.

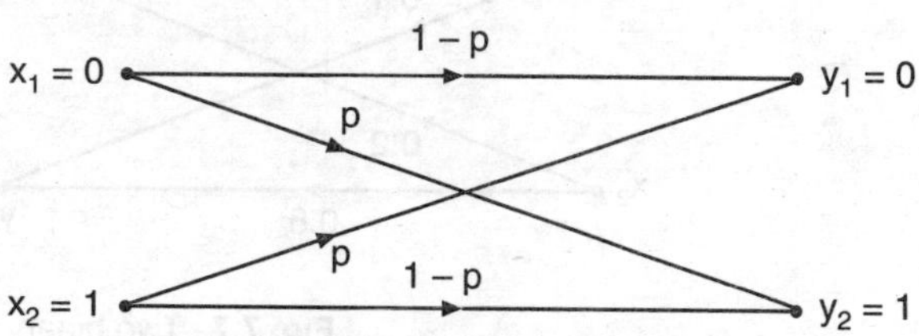

Fig. 7.5. Binary symmetric channel

EXAMPLE 7.19. Given a binary channel shown in figure 7.6.

(i) Find the channel matrix of the channel.

(ii) Find $P(y_1)$ and $P(y_2)$ when $P(x_1) = P(x_2) = 0.5$.

(iii) Find the joint probabilities $P(x_1, y_2)$ and $P(x_2, y_1)$ when $P(x_1) = P(x_2) = 0.5$.

Solution: (i) We know that the channel matrix is given by

$$[P(Y|X)] = \begin{bmatrix} P(y_1|x_1) & P(y_2|x_1) \\ P(y_1|x_2) & P(y_2|x_2) \end{bmatrix}$$

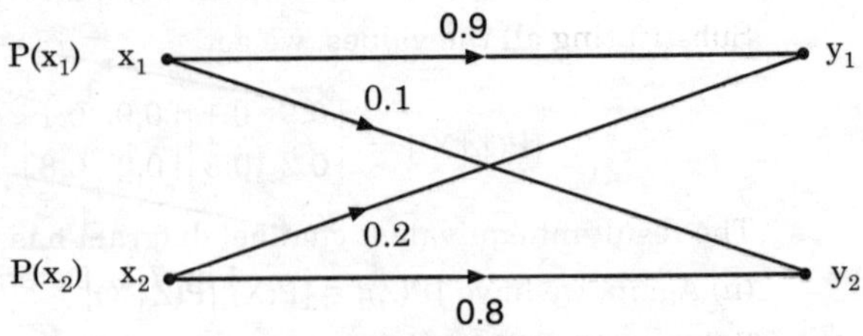

Fig. 7.6

Substituting all the given values, we get

$$[P(Y|X)] = \begin{bmatrix} 0.9 & 0.1 \\ 0.2 & 0.8 \end{bmatrix}$$

(ii) We know that

$$[P(Y)] = [P(X)]\,[P(Y|X)]$$

Substituting all the values, we get

or $$[P(Y)] = [0.5 \ \ 0.5]\begin{bmatrix} 0.9 & 0.1 \\ 0.2 & 0.8 \end{bmatrix}$$

or $$[P(Y)] = [0.55 \ \ 0.45] = [P(y_1)\ P(y_2)]$$

Hence, $P(y_1) = 0.55$ and $P(y_2) = 0.45$ **Ans.**

> **DO YOU KNOW?**
>
> The capacity of a noisy (discrete, memoryless) channel is defined as the maximum possible rate of information transmission over the channel.

(iii) Again, we have

$$[P(X, Y)] = [P(X)]_d\,[P(Y|X)]$$

Substituting all the values, we get

$$[P(X, Y)] = \begin{bmatrix} 0.5 & 0 \\ 0 & 0.5 \end{bmatrix}\begin{bmatrix} 0.9 & 0.1 \\ 0.2 & 0.8 \end{bmatrix}$$

or $$[P(X, Y)] = \begin{bmatrix} 0.45 & 0.05 \\ 0.1 & 0.4 \end{bmatrix} \quad \text{...(i)}$$

Now, let us write standard matrix form as under:

or $$[P(X, Y)] = \begin{bmatrix} P(x_1,y_1) & P(x_1,y_2) \\ P(x_2,y_1) & P(x_2,y_2) \end{bmatrix} \quad \text{...(ii)}$$

Comparing (i) and (ii), we get

$$P(x_1, y_2) = 0.05 \text{ and } P(x_2, y_1) = 0.1.$$ **Ans.**

EXAMPLE 7.20. Two binary channels of problem 7.19 are connected in cascade as shown in figure 7.7.

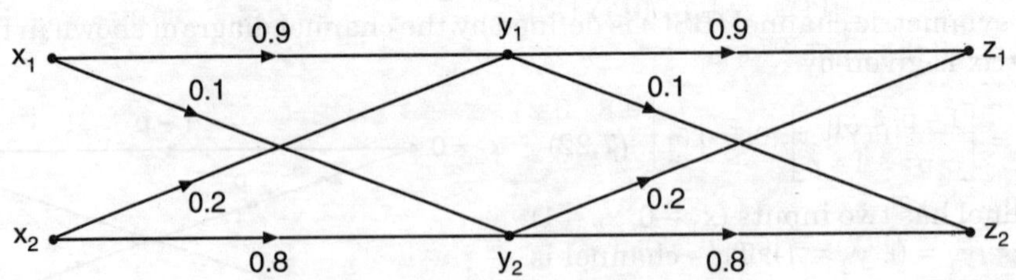

Fig. 7.7. Two binary channels in cascade

(i) Find the overall channel matrix of the resultant channel, and draw the resultant equivalent channel diagram.

(ii) Find $P(z_1)$ and $P(z_2)$ when $P(x_1) = P(x_2) = 0.5$

Solution: (i) We know that

$$[P(Y)] = [P(X)][P(Y|X)] = [P(Y)][P(Z|Y)]$$
$$[P(Z)] = [P(X)][P(Y|X)][P(Z|Y)] = [P(X)][P(Z|X)]$$

Thus, from figure 7.7, we have

$$[P(Z|X)] = [P(Y|X)][P(Z|Y)]$$

Substituting all the values, we get

$$[P(Z|X)] = \begin{bmatrix} 0.9 & 0.1 \\ 0.2 & 0.8 \end{bmatrix}\begin{bmatrix} 0.9 & 0.1 \\ 0.2 & 0.8 \end{bmatrix} = \begin{bmatrix} 0.83 & 0.17 \\ 0.34 & 0.66 \end{bmatrix}$$

The resultant equivalent channel diagram has been shown in figure 7.8.

(ii) Again, we have $[P(Z)] = [P(X)][P(Z|X)]$

Substituting all the values, we get

$$[P(Z)] = [0.5\ \ 0.5]\begin{bmatrix} 0.83 & 0.17 \\ 0.34 & 0.66 \end{bmatrix}$$
$$= [0.585\ \ 0.415] \quad \text{...(i)}$$

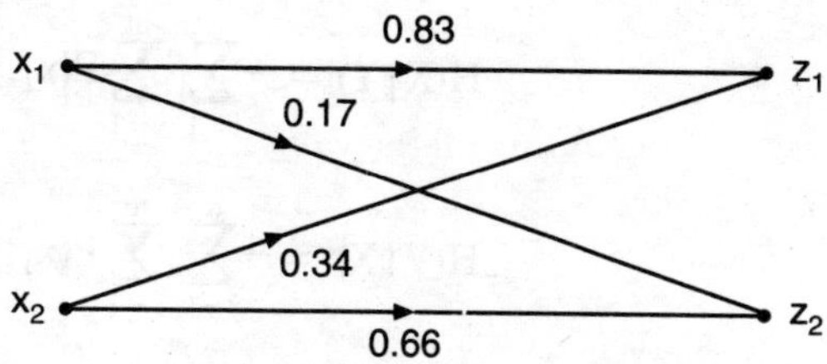

Fig. 7.8. The resultant equivalent channel diagram

Let us write standard expression as under:

$$[P(Z)] = [P(z_1)\ P(z_2)] \quad \text{...(ii)}$$

Comparing equation (i) and equation (ii), we obtain

$$P(z_1) = 0.585 \text{ and } P(z_2) = 0.415. \quad \textbf{Ans.}$$

EXAMPLE 7.21. A channel has the following channel matrix:

$$[P(Y|X)] = \begin{bmatrix} 1-p & p & 0 \\ 0 & p & 1-p \end{bmatrix} \quad \text{...(i)}$$

(i) Draw the channel diagram.

(ii) If the source has equally likely outputs, compute the probabilities associated with the channel outputs for p = 0.2.

Solution: (i) The channel diagram has been shown in figure 7.9. It may be noted that the channel represented by equation (i) is known as the binary erasure channel.

The binary erasure channel has two inputs $x_1 = 0$ and $x_2 = 1$ and three outputs $y_1 = 0$, $y_2 = e$, and $y_3 = 1$, where e indicates an erasure. This means that the output is in doubt, and hence it should be erased.

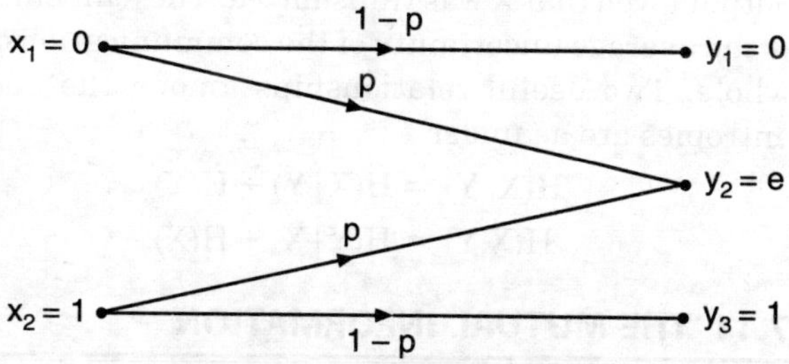

Fig. 7.9. Binary erasure channel

(ii) We know that [P(Y)] is expressed as

$$[P(Y)] = [P(X)][P(Y|X)]$$

Substituting all the values, we get

$$[P(Y)] = [0.5\ \ 0.5]\begin{bmatrix} 0.8 & 0.2 & 0 \\ 0 & 0.2 & 0.8 \end{bmatrix}$$

Simplifying, we get

$$[P(Y)] = [0.4\ \ 0.2\ \ 0.4]$$

Thus, we get $P(y_1) = 0.4$,

$P(y_2) = 0.2$,

and $P(y_3) = 0.4$ **Ans.**

Empty mind is a devil's workshop. – **Anonymous**

7.10 THE CONDITIONAL AND JOINT ENTROPIES

Using the input probabilites $P(x_i)$, output probabilities $P(y_j)$, transition probabilities $P(y_j | x_i)$, and joint probabilities $P(x_i, y_j)$, let us define the following various entropy functions for a channel with m inputs and n outputs:

$$H(X) = -\sum_{i=1}^{m} P(x_i)\log_2 P(x_i) \quad ...(7.23)$$

$$H(Y) = -\sum_{j=1}^{n} P(y_j)\log_2 P(y_j) \quad ...(7.24)$$

$$H(X|Y) = -\sum_{j=1}^{n}\sum_{i=1}^{m} P(x_i, y_j)\log_2 P(x_i | y_j) \quad ...(7.25)$$

$$H(Y|X) = -\sum_{j=1}^{n}\sum_{i=1}^{m} P(x_i, y_j)\log_2 P(y_j | x_i) \quad ...(7.26)$$

$$H(X, Y) = -\sum_{j=1}^{n}\sum_{i=1}^{m} P(x_i, y_j)\log_2 P(x_i, y_j) \quad ...(7.27)$$

These entropies can be interpreted as under:

H(X) is the average uncertainty of the channel input, and H(Y) is the average uncertainty of the channel output. The conditional entropy H(X | Y) is a measure of the average uncertainty remaining about the channel input after the channel output has been observed. Also, H(X | Y) is sometimes called the equivocation of X with respect to Y. The conditional entropy H(Y | X) is the average uncertainty of the channel output given that X was transmitted. The joint entropy H(Y | Y) is the average uncertainty of the communication channel as a whole. Two useful relationships among the above various entropies are as under:

DO YOU KNOW?

The maximum rate of transmission occurs when the source is matched to the channel.

$$H(X, Y) = H(X|Y) + H(Y) \quad ...(7.28)$$

$$H(X,Y) = H(Y|X) + H(X) \quad ...(7.29)$$

7.11 THE MUTUAL INFORMATION

The mutual information denoted by I(X ; Y) of a channel is defined by

$$I(X ; Y) = H(X) - H(X|Y) \text{ b/symbol} \quad ...(7.30)$$

Since H(X) represents the uncertainty about the channel input before the channel output is observed and H(X | Y) represents the uncertainty about the channel input after the channel output is observed, the mutual information I(X ; Y) represents the uncertainty about the channel input that is resolved by observing the channel output.

Properties of Mutual Information I(X; Y)

(i) $I(X;Y) = I(Y;X)$ ...(7.31)

(ii) $I(X;Y) \geq 0$ ...(7.32)

(iii) $I(X;Y) = H(Y) - H(Y|X)$ ...(7.33)

(iv) $I(X;Y) = H(X) + H(Y) - H(X,Y)$ ...(7.34)

EXAMPLE 7.22. For a lossless channel, prove that

$$H(X|Y) = 0 \quad ...(i)$$

Solution: When we observe the output y_j in a lossless channel shown in figure 7.10, it is clear which x_i was transmitted. This means that

$$P(x_i|y_j) = 0 \text{ or } 1 \quad ...(ii)$$

Further, we know that

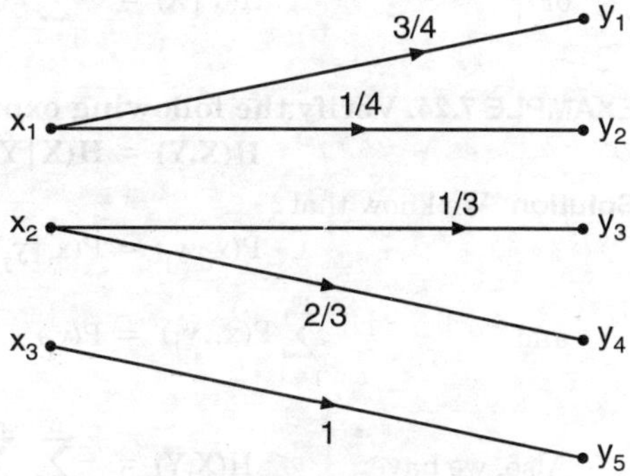

Fig. 7.10. Lossless channel

$$H(X|Y) = -\sum_{j=1}^{n}\sum_{i=1}^{m} P(x_i,y_j)\log_2 P(x_i|y_j)$$

or
$$H(X|Y) = -\sum_{j=1}^{n} P(y_j)\sum_{i=1}^{m} P(x_i|y_j)\log_2 P(x_i|y_j)$$

It may be noted that all the terms in the inner summation are zero because they are in the form of $1 \times \log_2 1$ or $0 \times \log_2 0$. Hence, we conclude that for a lossless channel

$$H(X|Y) = 0 \qquad \textbf{Hence Proved.}$$

EXAMPLE 7.23. Given a noiseless channel with m input symbols and m output symbols as shown in figure 7.11. Prove that

$$H(X) = H(Y) \quad ...(i)$$

and
$$H(Y|X) = 0 \quad ...(ii)$$

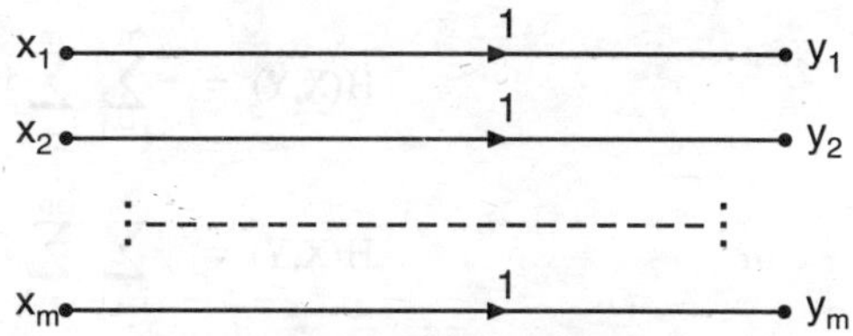

Fig. 7.11. Noiseless channel

Solution: For a noiseless channel, the transition probabilities are given by

$$P(y_j|x_i) = \begin{cases} 1 & \text{for } i = j \\ 0 & \text{for } i \neq j \end{cases} \quad ...(iii)$$

Hence,
$$P(x_i, y_j) = P(y_j|x_i)P(x_i) = \begin{cases} P(x_i) & \text{for } i = j \\ 0 & \text{for } i \neq j \end{cases} \quad ...(iv)$$

and
$$P(y_j) = \sum_{i=1}^{m} P(x_i,y_j) = P(x_j) \quad ...(v)$$

Also, we have
$$H(X) = -\sum_{i=1}^{m} P(x_i)\log_2 P(x_i) \text{ bits/symbol} \quad ...(vi)$$

Thus, using equations (v) and (vi), we have

$$H(Y) = -\sum_{j=1}^{m} P(y_j)\log_2 P(y_j)$$

or
$$H(Y) = -\sum_{i=1}^{m} P(x_i)\log_2 P(x_i) = H(X)$$

Further, we know that

$$H(Y|X) = -\sum_{j=1}^{n}\sum_{i=1}^{m} P(x_i, y_j)\log_2 P(y_j|x_i) \quad ...(vii)$$

Now, using equations (iii), (iv) and (vii), we have

$$H(Y|X) = -\sum_{j=1}^{m}\sum_{i=1}^{m} P(x_i, y_j)\log_2 P(y_j | x_i) = -\sum_{i=1}^{m} P(x_i)\sum_{j=1}^{m} \log_2 P(y_j | x_i)$$

or $$H(Y|X) = -\sum_{i=1}^{m} P(x_i)\log_2 1 = 0 \quad \textbf{Hence Proved.}$$

EXAMPLE 7.24. Verify the following expression:

$$\mathbf{H(X,Y) = H(X|Y) + H(Y)}$$

Solution: We know that

$$P(x_i, y_j) = P(x_i | y_j)\, P(y_j)$$

and $$\sum_{i=1}^{m} P(x_i, y_j) = P(y_j)$$

Also, we have $$H(X,Y) = -\sum_{j=1}^{n}\sum_{i=1}^{m} P(x_i, y_j)\log P(x_i, y_j)$$

$$H(X,Y) = -\sum_{j=1}^{n}\sum_{i=1}^{m} P(x_i, y_j)\log[P(x_i | y_j)P(y_j)]$$

or $$H(X,Y) = -\sum_{j=1}^{n}\sum_{i=1}^{m} P(x_i, y_j)\log P(x_i | y_j) - \sum_{j=1}^{n}\left[\sum_{i=1}^{m} P(x_i, y_j)\right]\log P(y_j)$$

or $$H(X,Y) = H(X|Y) - \sum_{j=1}^{n} P(y_j)\log P(y_j)$$

or $$H(X,Y) = H(X|Y) + H(Y) \quad \textbf{Hence Proved}$$

EXAMPLE 7.25. Show that the mutual information I(X;Y) of the channel described, by equation (7.11) with the input probabilities $P(x_i)$, i = 1, 2,... m and the output probabilities $P(y_j)$, j = 1, 2, ...,n, can be expressed as

$$\mathbf{I(X;Y) = \sum_{i=1}^{m}\sum_{j=1}^{n} P(x_i, y_j)\log_2 \frac{P(x_i | y_j)}{P(x_i)}} \qquad \text{...(i)}$$

Solution: We know that

$$I(X;Y) = H(X) - H(X|Y)$$

Also, we have

$$I(X;Y) = \sum_{i=1}^{m} P(x_i)\log_2 \frac{1}{P(x_i)} + \sum_{j=1}^{n}\sum_{i=1}^{m} P(x_i, y_j)\log_2 P(x_i | y_j)$$

$$= \sum_{i=1}^{m}\left[\sum_{j=1}^{n} P(x_i, y_j)\right]\log_2 \frac{1}{P(x_i)} + \sum_{j=1}^{n}\sum_{i=1}^{m} P(x_i, y_j)\log_2 P(x_i | y_j)$$

$$= \sum_{i=1}^{m}\sum_{j=1}^{n} P(x_i, y_j)\left[\log_2 \frac{1}{P(x_i)} + \log_2 P(x_i | y_j)\right]$$

$$= \sum_{i=1}^{m}\sum_{j=1}^{n} P(x_i, y_j)\log_2 \frac{P(x_i | y_j)}{P(x_i)} \quad \textbf{Hence Proved.}$$

EXAMPLE 7.26. Verify the following expression:

$$I(X;Y) = I(Y;X)$$

Solution: We can express I(Y; X) as

$$I(Y;X) = \sum_{i=1}^{m}\sum_{j=1}^{n} P(y_j, x_i)\log_2 \frac{P(y_j \mid x_i)}{P(y_j)} \quad \text{...(i)}$$

We know that $P(y_j, x_i) = P(x_i, y_j)$

and $$\frac{P(y_j \mid x_i)}{P(y_j)} = \frac{P(x_i \mid y_j)}{P(x_i)}$$

Thus, comparing equation (i) with equation (i), we conclude that

$$I(X;Y) = I(Y;X)$$ **Hence Proved**

EXAMPLE 7.27. Verify the following expression:

$$I(X;Y) \geq 0$$

Solution: We have

$$-I(X;Y) = \sum_{i=1}^{m}\sum_{j=1}^{n} P(x_i, y_j)\log_2 \frac{P(x_i)}{P(x_i \mid y_j)} \quad \text{...(i)}$$

Using Bayes' rule, we have

$$\frac{P(x_i)}{P(x_i \mid y_j)} = \frac{P(x_i)P(y_j)}{P(x_i, y_j)}$$

We can write equation (i) as under:

$$-I(X;Y) = \frac{1}{\ln 2}\sum_{i=1}^{m}\sum_{j=1}^{n} P(x_i, y_j)\ln \frac{P(x_i)P(y_j)}{P(x_i, y_j)} \quad \text{..(ii)}$$

Also, we know that $\ln \alpha \leq \alpha - 1$

Therefore, we have

$$-I(X;Y) \leq \frac{1}{\ln 2}\sum_{i=1}^{m}\sum_{j=1}^{n} P(x_i, y_j)\left[\frac{P(x_i)P(y_j)}{P(x_i, y_j)} - 1\right]$$

or $$-I(X;Y) \leq \frac{1}{\ln 2}\left[\sum_{i=1}^{m}\sum_{j=1}^{n} P(x_i)\,P(y_j) - \sum_{i=1}^{m}\sum_{j=1}^{n} P(x_i, y_j)\right] \quad \text{...(iii)}$$

Since $$\sum_{i=1}^{m}\sum_{j=1}^{n} P(x_i)P(y_j) = \sum_{i=1}^{m} P(x_i)\sum_{j=1}^{n} P(y_j) = (1)(1) = 1$$

$$\sum_{i=1}^{m}\sum_{j=1}^{n} P(x_i, y_j) = \sum_{i=1}^{m}\sum_{j=1}^{n} P(x_i, y_j) = \sum_{i=1}^{m} P(x_i) = 1$$

Equation (iii) reduces to

$$-I(X;Y) \leq 0$$

$$I(X;Y) \geq 0.$$ **Hence Proved.**

EXAMPLE 7.28. Given a binary symmetric channel (BSC) (figure 7.12) with $P(x_1) = \alpha$.

(i) Show that the mutual information I(X;Y) is given by

$$I(X;Y) = H(Y) + p\log_2 p + (1-p)\log_2(1-p) \quad \text{...(i)}$$

(ii) Calculate I(X;Y) for α = 0.5 and p = 0.1.

(iii) Repeat part (ii) for a = 0.5 and p = 0.5, and comment on the result.

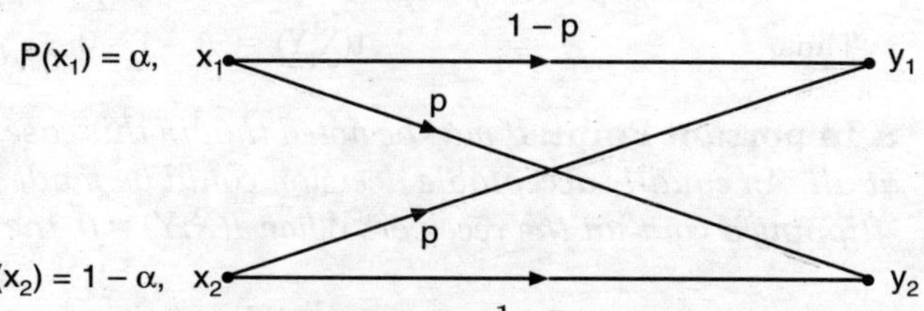

Fig. 7.12. Binary symmetric channel (BSC)

Figure 7.12 shows the diagram of the binary symmetric channel (BSC) with associated input probabilities.

Solution: (i) Using equations (7.18), (7.19) and (7.22), we have

$$[P(X,Y)] = \begin{bmatrix} \alpha & 0 \\ 0 & 1-\alpha \end{bmatrix}\begin{bmatrix} 1-p & p \\ p & 1-p \end{bmatrix}$$

or
$$[P(X,Y)] = \begin{bmatrix} \alpha(1-p) & \alpha p \\ (1-\alpha)p & (1-\alpha)(1-p) \end{bmatrix} = \begin{bmatrix} P(x_1,y_1) & P(x_1,y_2) \\ P(x_2,y_1) & P(x_2,y_2) \end{bmatrix}$$

We know that
$$H(Y|X) = -\sum_{j=1}^{n}\sum_{i=1}^{m} P(x_i, y_j)\log_2 P(y_j | x_i)$$

By above expression, we have

$$H(Y|X) = -P(x_1, y_1)\log_2 P(y_1|x_1) - P(x_1, y_2)\log_2 P(y_2|x_1) - P(x_2, y_1)\log_2 P(y_1|x_2) - P(x_2, y_2)\log_2 P(y_2|x_2)$$

or
$$H(Y|X) = -\alpha(1-p)\log_2(1-p) - \alpha p\log_2 p - (1-\alpha)p\log_2 p - (1-\alpha)(1-p)\log_2(1-p)$$

or
$$H(Y|X) = -p\log_2 p - (1-p)\log_2(1-p) \quad \text{...(ii)}$$

We know that
$$I(X;Y) = H(Y) - H(Y|X) = H(Y) + p\log_2 p + (1-p)\log_2(1-p)$$

(ii) We know that

$$P(Y)] = [P(X)]\,[P(Y|X)]$$

when $\alpha = 0.5$ and $p = 0.1$, then, we have

$$[P(Y)] = [0.5 \quad 0.5]\begin{bmatrix} 0.9 & 0.1 \\ 0.1 & 0.9 \end{bmatrix} = [0.5 \quad 0.5]$$

Thus,
$$P(y_1) = P(y_2) = 0.5$$

Now, using the expression

$$H(Y) = -\sum_{j=1}^{n} P(y_j)\log_2 P(y_j), \text{ we have}$$

$$H(Y) = -P(y_1)\log_2 P(y_1) - P(y_2)\log_2 P(y_2)$$

or
$$H(Y) = -0.5\log_2 0.5 - 0.5\log_2 0.5 = 1$$

$$p\log_2 p + (1-p)\log_2(1-p) = 0.1\log_2 0.1 + 0.9\log_2 0.9 = -0.469$$

Thus,
$$I(X;Y) = 1 - 0.469 = 0.531$$ **Ans.**

(iii) When $\alpha = 0.5$ and $p = 0.5$, we have

$$[P(Y)] = [0.5 \quad 0.5]\begin{bmatrix} 0.5 & 0.5 \\ 0.5 & 0.5 \end{bmatrix} = [0.5 \quad 0.5]$$

$$H(Y) = 1$$

$$p\log_2 p + (1-p)\log_2(1-p) = 0.5\log_2 0.5 + 0.5\log_2 0.5 = -1$$

Thus,
$$I(X;Y) = 1 - 1 = 0$$ **Hence Proved.**

DO YOU KNOW?

Telephone channels that are affected by switching transients and dropouts, and microwave radio links that are subjected to fading are examples of channels with memory.

Important Point: *It may be noted that in this case, (p = 0.5) no information is being transmitted at all. An equally acceptable decision could be made by dispensing with the channel entirely and "flipping a coin" at the receiver. When I(X;Y) = 0, the channel is said to be useless.*

Money is a good servant but a bad master. – **Bacon**

7.12 THE CHANNEL CAPACITY

In this section, let us discuss various aspects regarding channel capacity.

7.12.1. Channel Capacity Per Symbol C_s

The channel capacity per symbol of a discrete memoryless channel (DMC) is defined as

$$C_s = \max_{\{P(x_i)\}} I(X;Y) \text{ b/symbol} \quad ...(7.35)$$

where the maximization is over all possible input probability distributions $\{P(x_i)\}$ on X. Note that the channel capacity C_s is a function of only the channel transition probabilities which define the channel.

7.12.2. Channel Capacity Per Second C

If r symbols are being transmitted per second, then the maximum rate of transmission of information per second is rC_s. This is the channel capacity per second and is denoted by C(b/s), i.e.,

$$C = rC_s \text{ b/s} \quad ...(7.36)$$

7.12.3. Capacities of Special Channel

In this subsection, let us discuss capacities of various special channel.

7.12.3.1. Lossless Channel

For a lossless channel, $H(X|Y) = 0$, and

$$I(X;Y) = H(X) \quad ...(7.37)$$

Thus, the mutual information (information transfer) is equal to the input (source) entropy, and no source information is lost in transmission.

Consequently, the channel capacity per symbol will be

$$C_s = \max_{\{P(x_i)\}} H(X) = \log_2 m \quad ...(7.38)$$

where m is the number of symbols in X.

7.12.3.2. Deterministic Channel

For a deterministic channel, $H(Y|X) = 0$ for all input distributions $P(x_i)$, and

$$I(X;Y) = H(Y) \quad ...(7.39)$$

Thus, the information transfer is equal to the output entropy. The channel capacity per symbol will be

$$C_s = \max_{\{P(x_i)\}} H(Y) = \log_2 n \quad ...(7.40)$$

where n is the number of symbols in Y.

7.12.3.3. Noiseless Channel

Since a noiseless channel is both lossless and deterministic, we have

$$I(X;Y) = H(X) = H(Y) \quad ...(7.41)$$

and the channel capacity per symbol is

$$C_s = \log_2 m = \log_2 n \quad ....(7.42)$$

7.12.3.4. Binary Symmetric Channel (BSC)

For the binary symmetric channel (BSC), the mutual information is

$$I(X;Y) = H(Y) + p \log_2 p + (1 - p) \log_2 (1 - p) \quad ...(7.43)$$

and the channel capacity per symbol will be

$$C_s = 1 + p\log_2 p + (1-p)\log_2(1-p) \quad \text{...(7.44)}$$

EXAMPLE 7.29. Verify the following expression:

$$C_s = \log_2 m$$

where C_s is the channel capacity of a lossless channel and m is the number of symbols in X.

Solution: For a lossless channel, we have

$$H(X|Y) = 0$$

Then, by equation (7.30), we have

$$I(X;Y) = H(X) - H(X|Y) = H(X) \quad \text{...(i)}$$

Hence, by equations (7.35) and (7.39), we have

$$C_s = \max_{\{P(x_i)\}} I(X;Y)$$

or $$C_s = \max_{\{P(x_i)\}} H(X) = \log_2 m \qquad \textbf{Hence proved.}$$

EXAMPLE 7.30. Verify the following expression:

$$C_s = 1 + p\log_2 p + (1-p)\log_2(1-p)$$

where C_s is the channel capacity of a BSC (figure 7.12)

Solution: We know that the mutual information I(X: Y) of a BSC is given by

$$I(X;Y) = H(Y) + p\log_2 p + (1-p)\log_2(1-p)$$

which is maximum when H(Y) is maximum. Since, the channel output is binary, H(Y) is maximum when each output has a probability of 0.5 and is achieved for equally likely inputs.

For this case H(Y) = 1, and the channel capacity is

$$C_s = \max_{\{p(X)\}} I(X;Y) = 1 + p\log_2 p + (1-p)\log_2(1-p) \qquad \textbf{Hence Proved.}$$

EXAMPLE 7.31. Find the channel capacity of the binary erasure channel of figure 7.13.

Solution: Let $P(x_1) = \alpha$. Then $P(x_2) = 1 - \alpha$.

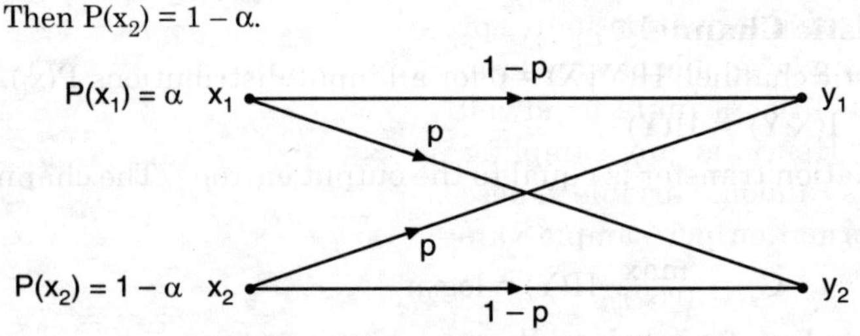

Fig. 7.13.

We have $$[P(Y|X)] = \begin{bmatrix} 1-p & p & 0 \\ 0 & p & 1-p \end{bmatrix} = \begin{bmatrix} P(y_1|x_1) & P(y_2|x_1) & P(y_3|x_1) \\ P(y_1|x_2) & P(y_2|x_2) & P(y_3|x_2) \end{bmatrix}$$

Using equation (7.17), we have

$$[P(Y)] = [\alpha \;\; 1-\alpha]\begin{bmatrix} 1-p & p & 0 \\ 0 & p & 1-p \end{bmatrix}$$

$$= [\alpha(1-p) \;\; p \;\; (1-\alpha)(1-p)] = [P(y_1) \; P(y_2) \; P(y_3)]$$

Money cannot buy friends, but you can get a better class of enemy. – Spike Millingan

By using equation (7.19), we have

$$[P(X, Y)] = \begin{bmatrix} \alpha & 0 \\ 0 & 1-\alpha \end{bmatrix} \begin{bmatrix} 1-p & p & 0 \\ 0 & p & 1-p \end{bmatrix}$$

or $$[P(X, Y)] = \begin{bmatrix} \alpha(1-p) & \alpha p & 0 \\ 0 & (1-\alpha)p & (1-\alpha)(1-p) \end{bmatrix}$$

or $$[P(X, Y)] = \begin{bmatrix} P(x_1,y_1) & P(x_1,y_2) & P(x_1,y_3) \\ P(x_2,y_1) & P(x_2,y_2) & P(x_2,y_3) \end{bmatrix}$$

In addition, from equations (7.24) and (7.26), we can calculate

$$H(Y) = -\sum_{j=1}^{3} P(y_j)\log_2 P(y_j)$$

$$= -\alpha(1-p)\log_2 \alpha(1-p) - p\log_2 p - (1-\alpha)(1-p)\log_2[(1-\alpha)(1-p)]$$

$$= (1-p)[-\alpha\log_2\alpha - (1-\alpha)\log_2(1-\alpha)] - p\log_2 p - (1-p)\log_2(1-p)$$

Also, we have $$H(Y|X) = -\sum_{j=1}^{3}\sum_{i=1}^{2} P(x_i,y_j)\log_2 P(y_j|x_i)$$

$$= -\alpha(1-p)\log_2(1-p) - \alpha p\log_2 p - (1-\alpha)p\log_2 p$$

$$-(1-\alpha)(1-p)\log_2(1-p)$$

$$= -p\log_2 p - (1-p)\log_2(1-p)$$

Thus, by equations (7.33) and (7.37), we have

$$I(X;Y) = H(Y) - H(Y|X) = (1-p)[-\alpha\log_2\alpha - (1-\alpha)\log_2(1-\alpha)] = (1-p)H(X)$$

And by equations (7.35) and (7.38), we have

$$C_s = \max_{\{P(x_i)\}} I(X;Y) = \max_{\{P(x_i)\}}(1-p)H(X) = (1-p)\max_{\{P(x_i)\}} H(X) = 1-p \quad \textbf{Ans.}$$

7.13 ENTROPY RELATIONS FOR A CONTINUOUS CHANNEL

In a continuous channel, an information source produces a continuous signal x(t). The set of possible signals is considered as an ensemble of waveforms generated by some ergodic random process. It is further assumed that x(t) has a finite bandwidth so that x(t) is completely characterized by its periodic sample values. Hence, at any sampling instant, the collection of possible sample values constitutes a continuous random variable X described by it probability density function $f_X(x)$. The average amount of information per sample value of x(t) (i.e., entropy of a continuous source) is measured by

$$H(X) = -\int_{-\infty}^{\infty} f_X(x)\log_2 f_X(x)\,dx \text{ bits/sample} \quad ....(7.45)$$

The entropy H(X) defined by equation (7.45) is known as the differential entropy of X.

Also, the average mutual information in a continuous channel is defined (by analogy with the discrete case) as

$$I(X;Y) = H(X) - H(X|Y) = H(Y) - H(Y|X)$$

where $$H(Y) = -\int_{-\infty}^{\infty} f_Y(y)\log_2 f_Y(y)\,dy \quad ...(7.46)$$

$$H(X|Y) = -\int_{-\infty}^{\infty}\int_{-\infty}^{\infty} f_{XY}(x,y)\log_2 f_X(x|y)\,dxdy \quad ...(7.47)$$

$$H(Y|X) = -\int_{-\infty}^{\infty}\int_{-\infty}^{\infty} f_{XY}(x,y)\log_2 f_Y(y|x)\,dxdy \quad ...(7.48)$$

7.14 CAPACITY OF AN ADDITIVE WHITE GAUSSIAN NOISE (AWGN) CHANNEL: SHANNON-HARTLEY LAW

In an additive white Gaussian noise (AWGN) channel, the channel output Y is given by

$$Y = X + n$$

where X is the channel input and n is an additive bandlimited white Gaussian noise with zero mean and variance σ^2.

The capacity C_s of an AWGN channel is given by

$$C_s = \max_{\{f_X(x)\}} I(X;Y)$$

$$= \frac{1}{2}\log_2\left(1+\frac{S}{N}\right) \text{ b/sample} \quad ...(7.49)$$

where S/N is the signal-to-noise ratio at the channel output. If the channel bandwidth B Hz is fixed, then the output y(t) is also a bandlimited signal completely characterized by its periodic sample values taken at the Nyquist rate 2B samples/s.

Then the capacity C(b/s) of the AWGN channel is given by

$$C = 2B \times C_s = B\log_2\left(1+\frac{S}{N}\right) \text{b/s} \quad ...(7.50)$$

Equation (7.50) is known as the **Shannon-Hartley law.**

Important Point: *The Shannon-Hartley law underscores the fundamental role of bandwidth and signal-to-noise ratio in communication. It also shows that we can exchange increased bandwidth for decreased signal power for a system with given capacity C.*

7.15 CHANNEL CAPACITY: A DETAILED STUDY

We know that the bandwidth and the noise power place a restriction upon the rate of information that can be transmitted by a channel. It may be shown that in a channel which is disturbed by a white Gaussian noise, one can transmit information at a rate of C bits per second, where C is the the channel capacity and is expressed as

$$C = B\log_2\left(1+\frac{S}{N}\right) \quad ...(7.51)$$

In this expression, B = channel bandwidth in Hz
S = Signal power
N = Noise power

It may be noted that the expression (equation 7.50) for channel capacity is valid for white Gaussian noise. However, for other types of noise, the expression is modified.

Proof: Let us present a proof of channel capacity formula based upon the assumption that if a signal is mixed with noise, the signal amplitude can be recognized only within the root mean square noise voltage. In other words, we can say that the uncertainty in recognizing the exact signal amplitude is equal to the root mean square noise voltage. Again, let us assume that the average signal power and the noise power are S watts and N watts respectively. This means that the root mean square value of the received signal is $\sqrt{S+N}$ volts and the root mean square value of the noise voltage is $\sqrt{N}$ volts. Now, we have to distinguish the received signal of the amplitude $\sqrt{S+N}$ volts in the presence of the noise amplitude $\sqrt{N}$ volts. As a matter of fact, the input signal variation of less than $\sqrt{N}$ volts will not be distinguished at the receiver end. Therefore, the number of the distinct levels that can be distinguished without error can be expressed as

$$M = \frac{\sqrt{S+N}}{\sqrt{N}}$$

$$M = \sqrt{1+\frac{S}{N}} \quad \text{...(7.51)}$$

Thus, equation (7.51) expresses the maximum value of M.

Now, the maximum amount of information carried by each pulse having $\sqrt{1+\frac{S}{N}}$ distinct levels is given by

$$I = \log_2 \sqrt{1+\frac{S}{N}}$$

$$I = \frac{1}{2}\log_2\left(1+\frac{S}{N}\right) \text{ bits} \quad \text{...(7.52)}$$

Now, after establishing expression in equation (7.15), we can determine the channel capacity. In fact, the channel capacity is the maximum amount of information that can be transmitted per second by a channel. If a channel can transmit a maximum of K pulses per second, then, the channel capacity C is given by

$$C = \frac{K}{2}\log_2\left(1+\frac{S}{N}\right) \text{ bits per second} \quad \text{...(7.53)}$$

DO YOU KNOW?

The communication system is designed to reproduce at the receiver either exactly or approximately the message emitted by the source.

Recall that for bandwidth requirements of PAM signals, it has been shown that a system of bandwidth nf_m Hz can transmit $2n f_m$ independent pulses per second. Further, under these conditions, the received signal will yield the correct values of the amplitudes of the pulses but will not reproduce the details of the pulse shapes.

Now, since, we are interested only in the pulse amplitudes and not their shapes, it is concluded that a system with bandwidth B Hz can transmit a maximum of 2B pulses per second. Further, since, each pulse can carry a maximum information of $\frac{1}{2}\log_2\left(1+\frac{S}{N}\right)$ bits, it follows that a system of bandwidth B can transmit the information at a following maximum rate:

$$C = B\log_2\left(1+\frac{S}{N}\right) \text{ bits per second} \quad \text{...(7.54)}$$

Therefore, the channel capacity C is limited by the bandwidth of the channel (or system) and the noise signal.

For a noiseless channel, N = 0 and the channel capacity will be infinite. However, practically, N is always finite and therefore, the channel capacity is finite.* The expression in equation (7.54) is also known as the **Hartley-Shannon law** and is treated as the central theorem of information theory. From Hartley-Shannon law, it is obvious that the bandwidth and the signal power can be exchanged for one another. To transmit the information at a given rate, we may reduce the signal power transmitted provided that the bandwidth is increased correspondingly. In a similar manner, the bandwidth may be reduced if we have to increase the signal power. As a matter of fact, the process of modulation is actually a means of effecting this exchange between the bandwidth and the signal-to-noise ratio.**

Important Point: *It may be noted that the channel capacity represents the maximum amount of information that can be transmitted by a channel per second. To achieve this rate of transmission, the information has to be processed properly or coded in the most efficient manner.*

7.16 TRANSMISSION OF CONTINUOUS SIGNALS

Now, let us further illustrate the Hartley-Shannon law for the exchange of bandwidth and signal-to-noise ratio by a continuous signal which is bandlimited to f_m Hz. According to the sampling theorem, the information of a continuous-time signal, which is bandlimited to f_m Hz, is completely specified by $2f_m$ samples per second.

Hence, to transmit the information of such a signal, it is necessary to transmit only these discrete samples.

Theoretical Aspect

Now, one important question arises. How much information does each sample contain. It depends upon how many discrete levels or values the samples may assume. In fact, these samples can assume any value and hence to transmit such samples, we require pulses capable of assuming infinite levels. Clearly, the information carried by each sample is infinite bits. Therefore, the information contained in a continuous bandlimited signal is infinite.

* This is true even if the bandwidth B is infinite. The noise signal is a white noise with a uniform power density spectrum over the entire frequency range. Therefore, as the bandwidth B is increased, N also increases and hence the channel capacity remains finite even if B = ∞.

If $\eta/2$ is the power density, then we have

$$N = \eta B, \text{ and}$$

$$C = B \log_2\left(1+\frac{S}{\eta B}\right)$$

and $$\lim_{B\to\infty} C = \frac{S}{\eta}\frac{\eta B}{S}\log_2\left(1+\frac{S}{\eta B}\right)$$

The above limit may be found with the help of following standard expression:

$$\lim_{x\to 0}\frac{1}{x}\log_2(1+x) = \log_2 e = 1.44$$

Therefore, we have

$$\lim_{B\to\infty} C = \frac{S}{\eta}\log_2 e = 1.44\frac{S}{\eta}$$

** The improvement in the signal-to-noise ratio in wideband FM and PCM can be properly understood.

In the presence of noise*, the channel capacity is finite. Therefore, it is impossible to transmit complete information in a bandlimited signal by a physical channel in the presence of noise**.

In the absence of noise, N = 0, the channel capacity is infinite and hence, any desired signal can be transmitted. It is quite obvious that it is impossible to transmit the complete information contained in a continuous signal unless the transmitted signal power is made infinite. Because of presence of noise, there is always a certain amount of uncertainty in the received signal. In fact, the transmission of complete information in a signal will mean a zero amount of uncertainty. Actually, the amount of uncertainty can be made arbitrarily small by increasing the channel capacity***. However, it can never be made zero.

7.17 UNCERTAINTY IN THE TRANSMISSION PROCESS

As a matter of fact, the uncertainty is introduced in the process of transmission. Therefore, although, it is possible to transmit the complete information in a continuous signal at the transmitter end, it is impossible to recover this infinite amount of information at the receiver end. The amount of information that can be recovered per second at the receiver is C bits per second where C is the channel capacity.

In place of transmitting all of the information at the transmitter, we can approximate the signal so that its information contents are reduced to C bits per second and transmit this approximated signal which has a finite information content. Now, it will be possible to recover all of the information that has been transmitted. In fact, this happens in pulse code modulation (PCM) system.

Now, the question arises: How can we approximate a signal so that the approximated signal has a finite information content per second?

In fact, this can be done by a process known as **quantization**. For illustration, let us consider the continuous signal bandlimited to f_m Hz as shown in figure 7.14.

Now, to transmit the information in this signal, we require to transmit only $2f_m$ samples per second. Figure 7.14 also shows samples.

The samples can take any value, and to transmit them directly, we require pulses which can assume an infinite number of levels.

Therefore, instead of transmitting the exact values of these pulses, we round off the amplitudes to the nearest one of the finite number of permitted values.

In this example, all of the pulses are approximated to the nearest tenth of a volt.

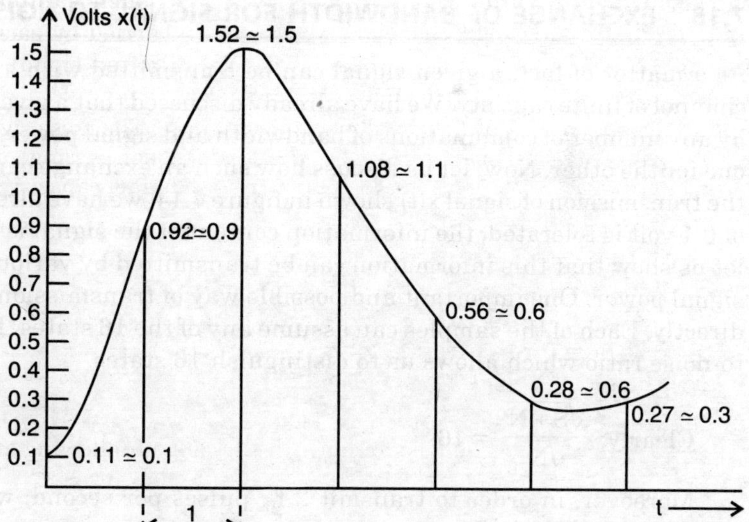

Fig. 7.14.

*That finite value of N.

** Existing over the same band.

*** By increasing the bandwidth and/or increasing the signal power.

It may be noted from the figure that each of the pulses transmitted assumes any one of the 16 levels, and thus, carries an amount of information of $\log_2 16 = 4$ bits. Also, since, there are $2f_m$ samples per second, the total information content of the approximated signal is $8 f_m$ bits per second. It the channel capacity is greater than or equal to $8 f_m$ bits per second, all of the information that has been transmitted will be recovered completely without any uncertainty. This means that the received signal will be an exact replica of the approximated signal that was transmitted. It can be shown that if the channel capacity is $8 f_m$ bits per second, the process of transmission does not introduce an additional degree of uncertainty. Now, let us consider that we are using a channel of bandwidth f_m Hz to transmit these samples, then, since the channel capacity required will be $8 f_m$ bits per second, the required signal-to-noise power ratio will be given by

$$8 f_m = f_m \log_2 \left(\frac{S+N}{N}\right)$$

Therefore, $\frac{S+N}{N} = 256$

We have already discussed that the number of levels that can be distinguished at the receiver is $\sqrt{\frac{S+N}{N}}$.

> **Important Point:** *It is obvious that in this case, the receiver can distinguish the 16 states without error. Thus, although the process of transmission introduces some noise in the desired signal, the levels are far enough apart to be distinguishable at the receiver. In other words, we can say that a channel of the capacity of $8f_m$ bits per second can transmit information of $8 f_m$ bits per second virtually without error.*

7.18 EXCHANGE OF BANDWIDTH FOR SIGNAL-TO-NOISE RATIO

As a matter of fact, a given signal can be transmitted with a given amount of uncertainty by a channel of finite capacity. We have already discussed that a given channel capacity may be obtained by any number of combinations of bandwidth and signal power. Actually, it is possible to exchange one for the other. Now, let us discuss how such an exchange can be affected. Again, let us consider the transmission of signal x(t) shown in figure 7.14, we have already observed that if an uncertainty of 0.1 volt is tolerated, the information content of the signal is given by $8 f_m$ bits per second. Now, let us show that this information can be transmitted by various combinations of bandwidths and signal power. One important and possible way of transmission is to send $2 f_m$ samples per second directly. Each of the samples can assume any of the 16 states. In this case, we must have a signal-to-noise ratio which allows us to distinguish 16 states.

Clearly, $\frac{\sqrt{S+N}}{\sqrt{N}} = 16.$

Moreover, in order to transmit $2 f_m$ pulses per second, we require a channel of bandwidth f_m Hz.

Therefore, the required channel capacity C is expressed as

$$C = f_m \log_2 \frac{S+N}{N}$$

or $$C = f_m \log_2 (16)^2 = 8 f_m \text{ bits per second.}$$

Hence, the channel capacity C is exactly equal to the amount of information per second in the signal x(t). In other method of transmission, we may transmit the samples in figure 7.14 by quaternary pulse (pulses that can assume four states). Thus, it is obvious that we require a group of two quaternary pulses to transmit each sample that can assume 16 states. In this case, the signal-to-noise ratio required at the receiver to distinguish pulses that assume four distinct states is

$$\frac{\sqrt{S+N}}{\sqrt{N}} = 4.$$

Hence, in this mode of transmission, the required signal power is reduced. But, now, we have to transmit twice as many pulses per second i.e., 4 f_m pulses per second. Hence, the required bandwidth is 2 f_m Hz. In this case, the channel capacity C will be

$$C = 2f_m \log_2 \frac{S+N}{N} = 2f_m \log_2 (4)^2 = 8f_m \text{ bits per second.}$$

EXAMPLE 7.32. Find the differential entropy H(X) of the uniformly distributed random variable X with the following probability density function (pdf):

$$f_X(x) = \begin{cases} \dfrac{1}{a} & \text{for } 0 \le x \le a \\ 0 & \text{elsewhere} \end{cases}$$

for (i) a = 1, (ii) a = 2, and (iii) a = $\frac{1}{2}$.

Solution: We know that the differential entropy of X is given by

$$H(X) = -\int_{-\infty}^{\infty} f_X(x) \log_2 f_X(x)\, dx \quad \text{bits/sample}$$

Making use of given pdf, we have

$$H(X) = -\int_0^a \frac{1}{a} \log_2 \frac{1}{a}\, dx = \log_2 a \qquad \text{...(i)}$$

Now, we have

(i) $a = 1,\ H(X) = \log_2 1 = 0$

(ii) $a = 2,\ H(X) = \log_2 2 = 1$

(iii) $a = \frac{1}{2},\ H(X) = \log_2 \frac{1}{2} = -\log_2 2 = -1$

It may be noted that the differential entropy H(X) is not an absolute measure of information.

EXAMPLE 7.33. The differential entropy of a random variable X is defined by following equation:

$$H(X) = -\int_{-\infty}^{\infty} f_X(x) \log_2 f_X(x)\, dx$$

Find the probability density function $f_X(x)$ for which H(X) is maximum.

Solution: We know that $f_X(x)$ must satisfy the following two conditions:

$$\int_{-\infty}^{\infty} f_X(x)\, dx = 1 \qquad \text{...(i)}$$

$$\int_{-\infty}^{\infty} (x-\mu)^2 f_X(x)\, dx = \sigma^2 \qquad \text{...(ii)}$$

where μ is the mean of X and σ^2 it its variance. Since the problem is the maximization of H(X) under constraints of equations (i) and (ii), therefore, we use the method of Lagrange multipliers as under:

First, we form the function:

$$G[f_X(x), \lambda_1, \lambda_2] = H(X) + \lambda_1 \left[\int_{-\infty}^{\infty} f_X(x)dx - 1\right] + \lambda_2 \left[\int_{-\infty}^{\infty} (x-\mu)^2 f_X(x)dx - \sigma^2\right]$$

$$= \int_{-\infty}^{\infty} [-f_X(x)\log_2 f_X(x) + \lambda_1 f_X(x) + \lambda_2 (x-\mu)^2 f_X(x)]\, dx - \lambda_1 - \lambda_2\sigma^2 \quad \text{...(iii)}$$

where the parameters λ_1 and λ_2 are the Lagrange multipliers. Then the maximization of H(X) requires that

$$\frac{\partial G}{\partial f_X(x)} = -\log_2 f_X(x) - \log_2 e + \lambda_1 + \lambda_2(x-\mu)^2 = 0 \quad \text{...(iv)}$$

Thus, $\log_2 f_X(x) - \log_2 e + \lambda_1 + \lambda_2(x-\mu)^2$

or $$\ln f_X(x) = -1 + \frac{\lambda_1}{\log_2 e} + \frac{\lambda_2}{\log_2 e}(x-\mu)^2$$

Hence, we obtain

$$f_X(x) = \exp\left[-1 + \frac{\lambda_1}{\log_2 e} + \frac{\lambda_2}{\log_2 e}(x-\mu)^2\right] \quad \text{...(v)}$$

In view of the constraints of equations (i) and (ii), it is required that $\lambda_2 < 0$. Let

$$\exp\left(-1 + \frac{\lambda_1}{\log_2 e}\right) = a \quad \text{and} \quad \left(\frac{\lambda_2}{\log_2 e}\right) = -b^2$$

Then, equation (v) can be rewritten as

$$f_X(x) = a e^{-b^2(x-\mu)^2} \quad \text{...(vi)}$$

Substituting equation (vi) into equations (i) and (ii), we get

$$a\int_{-\infty}^{\infty} e^{-b^2(x-\mu)^2} dx = a\frac{\sqrt{\pi}}{b} = 1 \quad \text{...(vii)}$$

$$a\int_{-\infty}^{\infty} (x-\mu)^2 e^{-b^2(x-\mu)^2} dx = a\frac{\sqrt{\pi}}{2b^3} = \sigma^2 \quad \text{...(vii)}$$

Solving equations (vii) and (viii) for a and b^2, we get

$$a = \frac{1}{\sqrt{2\pi}\sigma} \quad \text{and} \quad b^2 = \frac{1}{2\sigma^2}$$

Substituting these values in equation (vi), we observe that the desired $f_X(x)$ is given by

$$f_X(x) = \frac{1}{\sqrt{2\pi}\sigma} e^{-(x-\mu)^2/(2\sigma^2)}$$ **Hence proved.**

which is the probability density function of Gaussian random variable X of mean μ and variance σ^2.

EXAMPLE 7.34. Show that the channel capacity of an ideal AWGN channel with infinite bandwidth is given by

$$C_\infty = \frac{1}{\ln 2}\cdot\frac{S}{\eta} \cong 1.44\frac{S}{\eta} \quad \text{...(i)}$$

where S is the average signal power and $\eta/2$ is the power spectral density (psd) of white gaussian noise. (U.P.S.C. I.E.S. Examination-1999)

Solution: We know that the noise power N is given by $N = \eta B$.

Also, according to Shannon-Hartley law, the channel capacity is given by

$$C = B\log_2\left(1+\frac{S}{N}\right) \text{ b/s}$$

In this expression, substituting $N = \eta B$, we get

$$C = B\log_2\left(1+\frac{S}{\eta B}\right)$$

Let $S/(\eta B) = \lambda$. Then, we write

$$C = \frac{S}{\eta\lambda}\log_2(1+\lambda) = \frac{1}{\ln 2}\frac{S}{\eta}\frac{\ln(1+\lambda)}{\lambda} \quad \text{...(ii)}$$

Now,

$$C_\infty = \lim_{B\to\infty} B\log_2\left(1+\frac{S}{\eta B}\right) = \frac{1}{\ln 2}\cdot\frac{S}{\eta}\lim_{\lambda\to 0}\frac{\ln(1+\lambda)}{\lambda}$$

Since $\lim_{\lambda\to 0}\,[\ln(1+\lambda)]/\lambda = 1$, we get

$$C_\infty = \frac{1}{\ln 2}\cdot\frac{S}{\eta} \cong 1.44\frac{S}{\eta} \text{ b/s}$$

Hence proved.

> **DO YOU KNOW?**
>
> A second type of noise, impulse noise, is also encountered in the channel. Impulse noise is characterized by long quiet intervals followed by bursts of high amplitude noise pulses.

Important Point: *It may be noted that equation (i) can be used to estimate upper limits on the performance of any practical communication system whose transmission channel can be approximated by the AWGN channel.*

EXAMPLE 7.35. Given an AWGN channel with 4 kHz bandwidth and the noise power spectral density $\eta/2 = 10^{-12}$ W/Hz. The signal power required at the receiver is 0.1 mW. Calculate the capacity of this channel.

Solution: Given that

$$B = 4000 \text{ Hz}$$
$$S = 0.1(10^{-3})\text{W}$$
$$N = \eta B = 2(10^{-12})(4000) = 8(10^{-9}) \text{ W}$$

Thus,

$$\frac{S}{N} = \frac{0.1(10^{-3})}{8(10^{-9})} = 1.25(10^4)$$

And, by equation (14.50), we have

$$C = B\log_2\left(1+\frac{S}{N}\right) = 4000\log_2(1+1.25(10^4))$$

$$C = 54.44(10^3) \text{ b/s} \quad \textbf{Ans.}$$

EXAMPLE 7.36. An Analog signal having 4 kHz bandwidth is sampled at 1.25 times the Nyquist rate, and each sample is quantized into one of equally likely levels. Assume that the successive samples are statistically independent.

(i) What is the information rate of this source?

(ii) Can the output of this source be transmitted without error over an AWGN channel with a bandwidth of 10 kHz and an S/N ratio of 20 dB?

(iii) Find the S/N ratio required for error-free transmission for part (i).

(iv) Find the bandwidth required for an AWGN channel for error-free transmission of the output of this source if the S/N ratio is 20 dB?

Solution: (i) Here, $f_m = 4(10^3)$Hz

We know that Nyquist rate f_s is given by

$$f_s = 2f_m$$

$$\text{Nyquist rate} = 2f_m = 8(10^3) \text{ samples/s}$$

Also, we have

$$r = 8(10^3)(1.25) = 10^4 \text{ samples/s}$$

Further, we know that entropy is expressed as

$$H(X) = -\sum_{i=1}^{m} P(x_i) \log_2 P(x_i) \text{ bits / symbol}$$

Here $\quad P(x_i) = \dfrac{1}{256}$

Hence, $\quad H(X) = \log_2 256 = 8$ bits/sample

The information rate R of the source is given by

$$R = rH(X) = 10^4(8) \text{ b/s} = 80 \text{ kb/s} \quad \textbf{Ans.}$$

(ii) Again, we know that channel capacity is given by

$$C = B\log_2\left(1+\frac{S}{N}\right) \text{bits/s}$$

Hence, $\quad C = B\log_2\left(1+\dfrac{S}{N}\right) = 10^4 \log_2 (1 + 10^2) = 66.6(10^3)$ b/s

Here, since R > C, error-free transmission is not possible.

> **DO YOU KNOW?**
>
> The output of a discrete information sources is a message that consists of a sequence of symbols. The actual message that is emitted by the source during a message interval is selected at random form a set of possible messages.

(iii) The required S/N ratio can be found by

$$C = 10^4 \log_2\left(1+\frac{S}{N}\right) \geq 8(10^4)$$

or $\quad \log_2\left(1+\dfrac{S}{N}\right) \geq 8$

or $\quad 1+\dfrac{S}{N} \geq 2^8 = 256$

or $\quad \dfrac{S}{N} \geq 255$ (= 24.1 dB) **Ans.**

Thus, the required S/N ratio must be greater than or equal to 24.1 dB for error-free transmission.

(iv) The required bandwidth B can be found by

$$C = B \log_2 (1 + 100) \geq 8(10^4)$$

or $\quad B \geq \dfrac{8(10^4)}{\log_2(1+100)} \geq 1.2(10^4)$ Hz ≥ 12 kHz

and the required bandwidth of the channel must be greater than or equal to 12 kHz. **Ans.**

7.19 THE SOURCE CODING

(i) Definition

A conversion of the output of a discrete memoryless source (DMS) into a sequence of binary symbols (i.e., binary code word) is called **source coding.** The device that performs this conversion is called the **source encoder.** Figure 7.15 shows a source encoder.

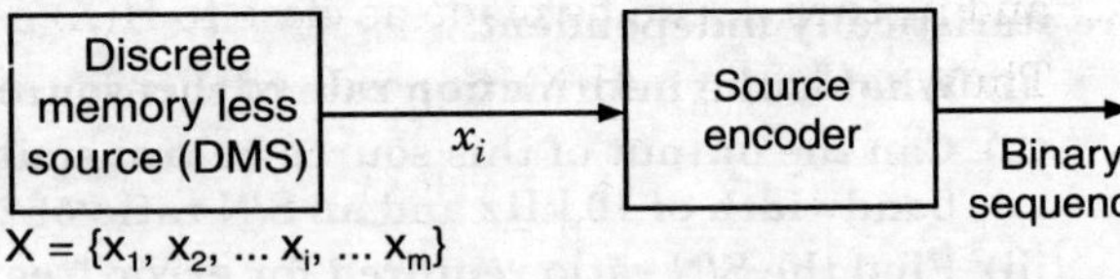

Fig. 7.15. Block diagram for source Coding

(ii) Objective of Source Coding

An objective of source coding is to minimize the average bit rate required for representation of the source by reducing the redundancy of the information source.

No country can be over populated, if there is work for everyone. – **J.L.Nehru**

7.19.1. Few Terms related to Source Coding Process

In this sub section, let us study the following terms which are related to source coding process:

(i) Codeword length

(ii) Average codeword length

(iii) Code efficiency

(iv) Code Redundancy

(i) Codeword Length

Let X be a DMS with finite entropy H(X) and an alphabet $\{x_1, ... x_m\}$ with corresponding probabilities of occurrence $P(x_i)(i = 1,...,m)$. Let the binary codeword assigned to symbol x_i by the encoder have length n_i, measured in bits. The length of a codeword is the number of binary digits in the codeword.

(ii) Average Codeword Length

The average codeword length L, per source symbol is given by

$$L = \sum_{i=1}^{m} P(x_i)n_i \qquad ...(7.51)$$

The parameter L represents the average number of bits per source symbol used in the source coding process.

(iii) Code Efficiency

The code efficiency η is defined as under:

$$\eta = \frac{L_{min}}{L} \qquad ...(7.52)$$

where L_{min} is the minimum possible value of L. When η approaches unity, the code is said to be efficient.

(iv) Code Redundancy

The code redundancy γ is defined as

$$\gamma = 1 - \eta \qquad ...(7.53)$$

7.19.2. The Source Coding Theorem

The source coding theorem states that for a DMS X, with entropy H(X), the average codeword length L per symbol is bounded as

$$L \geq H(X) \qquad ...(7.54)$$

and further, L can be made as close to H(X) as desired for some suitably chosen code.

Thus, with $L_{min} = H(X)$, the code efficiency can be rewritten as

$$\eta = \frac{H(X)}{L} \qquad ...(7.55)$$

7.19.3. Classification of Codes

Classification of codes is best illustrated by an example. Let us consider Table 7.2 where a source of size 4 has encoded in binary codes with symbol 0 and 1.

DO YOU KNOW?

By increasing the redundancy of the encoding, we can make the probability of error approach zero.

Table 7.2. Binary Codes

x_i	Code 1	Code 2	Code 3	Code 4	Code 5	Code 6
x_1	00	00	0	0	0	1
x_2	01	01	1	10	01	01
x_3	00	10	00	110	011	001
x_4	11	11	11	111	0111	0001

7.19.3.1. Fixed-Length Codes

A fixed-length code is one whose codeword length is fixed. Code 1 and code 2 of Table 7.2 are fixed-length with length 2.

7.19.3.2. Variable-Length Codes

A variable-length code is one whose codeword length is not fixed. All codes of Table 7.2 except codes 1 and 2 are variable-length codes.

7.19.3.3. Distinct codes

A code is distinct if each codeword is distinguishable from other codewords. All codes of Table (7.2) except code 1 are distinct codes – notice the codes for x_1 and x_3.

7.19.3.4. Prefix-Free Codes

A code in which no codeword can be formed by adding code symbols to another codeword is called a prefix-free code. Thus, in a prefix-free code, no codeword is a prefix of another. Codes 2, 4, and 6 of Table 7.2 are prefix-free codes.

7.19.3.5. Uniquely Decodable Codes

A distinct code uniquely decodable if the original source sequence can be reconstructed perfectly from the encoded binary sequence. It may be noted that code 3 of Table 7.2 is not a uniquely decodable code. For, example, the binary sequence 1001 may correspond to the source sequences $x_2x_3x_2$ or $x_2x_1x_1x_2$. A sufficient condition to ensure that a code is uniquely decodable is that no code word is a prefix of another. Thus, the prefix-free codes 2, 4, and 6 are uniquely decodable codes. Note that the prefix-free condition is not a necessary condition for unique decodability. For example, code 5 of Table 7.2 does not satisfy the prefix-free condition, and yet it is uniquely decodable since the bit 0 indicates the beginning of each codeword of the code.

7.19.3.6. Instantaneous Codes

A uniquely decodable code is called an instantaneous code if the end of any codeword is recognizable without examining subsequent code symbols. The instantaneous codes have the property previously mentioned that no codeword is a prefix of another codeword. For this reason, prefix-free codes are sometimes known as instantaneous codes.

7.19.3.7. Optimal Codes

A code is said to be optimal if it is instantaneous and has minimum average L for a given source with a given probability assignment for the source symbols.

7.19.4. The Kraft Inequality

Let X be a DMS with alphabet $\{x_i\}$ $(i = 1, 2, \ldots m)$. Assume that the length of the assigned binary codeword corresponding to x_i is n_i.

A necessary and sufficient condition for the existence of an instantaneous binary code is

$$K = \sum_{i=1}^{m} 2^{-n_i} \leq 1 \qquad \text{...(7.56)}$$

which is known as the **Kraft inequality.**

✍ **Important point:** *It may be noted that Kraft inequality assures us of the existence of an instantaneously decodable code with codeword lengths that satisfy the inequality. But it does not show us how to obtain these codewords, nor does it say any code satisfies the inequality is automatically uniquely decodable.*

EXAMPLE 7.37. Given a DMS X with two symbols x_1 and x_2 and $P(x_1) = 0.9$, $P(x_2) = 0.1$. Symbols x_1 and x_2 are encoded as follows (Table 7.3)

Table 7.3.

x_i	$P(x_i)$	Code
x_1	0.9	0
x_2	0.1	1

Find the efficiency η and the redundancy γ of this code.

Solution: We know that the average code length L per symbol is

$$L = \sum_{i=1}^{2} P(x_i)n_i = (0.9)(1) + (0.1)(1) = 1 \text{ b}$$

We know that

$$H(X) = -\sum_{i=1}^{2} P(x_i)\log_2 P(x_i)$$

$$= -0.9 \log_2 0.9 - 0.1 \log_2 0.1$$

$$= 0.469 \text{ b/symbol}$$

Also, the code efficiency η is

$$\eta = \frac{H(X)}{L} = 0.469 = 46.9\%$$

And, the code redundancy γ is given by

$$\gamma = 1 - \eta = 0.531 = 53.1\% \text{ Ans.}$$

EXAMPLE 7.38. The second-order extension of a DMS X, denoted by X^2, is formed by taking the source symbols two at a time. The coding of this extension has been shown in Table 7.4. Find the efficiency η and the redundancy γ of this extension code.

Solution:

Table 7.4.

a_i	$P(a_i)$	Code
$a_1 = x_1x_1$	0.81	0
$a_2 = x_1x_2$	0.09	10
$a_3 = x_2x_1$	0.09	110
$a_4 = x_2x_2$	0.01	111

We have $L = \sum_{i=1}^{4} P(a_i)n_i = 0.81(1) + 0.09(2) + 0.09(3) + 0.01(3) = 1.29$ b/symbol

The entropy of the second-order extension of X, $H(X^2)$, is given by

$$H(X^2) = -\sum_{i=1}^{4} P(a_i)\log_2 P(a_i)$$

$$= -0.81 \log_2 0.81 - 0.09 \log_2 0.09 - 0.09 \log_2 0.09 - 0.01 \log_2 0.01$$

or $H(X^2) = 0.938$ b/symbol

Therefore, the code efficiency η is

$$\eta = \frac{H(X^2)}{L} = \frac{0.938}{1.29} = 0.727 = 72.7\%$$

Also, the code redundancy γ will be

$$\gamma = 1 - \eta = 0.273 = 27.3\% \quad \textbf{Ans.}$$

Note that $H(X^2) = 2H(X)$.

EXAMPLE 7.39. Consider a DMS X with symbols x_i, i = 1, 2, 3, 4. Table 7.5 lists four possible binary codes.

Table 7.5.

x_i	Code A	Code B	Code C	Code D
x_1	00	0	0	0
x_2	01	10	11	100
x_3	10	11	100	110
x_4	11	110	110	111

(i) Show that all the codes except code B satisfy the Kraft inequality.

(ii) Show that codes A and D are uniquely decodable but codes B and C are not uniquely decodable.

Solution: We have

(i) For code A, $n_1 = n_2 = n_3 = n_4 = 2$

therefore, $$K = \sum_{i=1}^{4} 2^{-n_i} = \frac{1}{4} + \frac{1}{4} + \frac{1}{4} + \frac{1}{4} = 1$$

For code B, $n_1 = 1, n_2 = n_3 = 2, n_4 = 3$

therefore, $$K = \sum_{i=1}^{4} 2^{-n_i} = \frac{1}{2} + \frac{1}{4} + \frac{1}{4} + \frac{1}{8} = 1\frac{1}{8} > 1$$

For code C, $n_1 = 1, n_2 = 2, n_3 = n_4 = 3$

therefore, $$K = \sum_{i=1}^{4} 2^{-n_i} = \frac{1}{2} + \frac{1}{4} + \frac{1}{8} + \frac{1}{8} = 1$$

For code D, $n_1 = 1, n_2 = n_3 = n_4 = 3$

therefore, $$K = \sum_{i=1}^{4} 2^{-n_i} = \frac{1}{2} + \frac{1}{8} + \frac{1}{8} + \frac{1}{8} = \frac{7}{8} < 1$$

Thus, all codes except code B satisfy the Kraft inequality.

(ii) Codes A and D are prefix-free codes. They are, therefore, uniquely decodable. Code B does not satisfy the Kraft inequality, and it is not uniquely decodable. Although code C does satisfy the Kraft inequality, but it is not uniquely decodable. This can be seen by the following example:

Given the binary sequence 0110110. This sequence may correspond to the source sequences $x_1x_2x_1x_4$ or $x_1x_4x_4$.

EXAMPLE 7.40. Verify the following expression:

$$L \geq H(X)$$

where L is the average codeword length per symbol and H(X) is the source entropy.

Solution: We know that

$$\sum_{i=1}^{m} P_i \log_2 \frac{Q_i}{P_i} \leq 0$$

where the equality holds only if $Q_i = P_i$.

Let $$Q_i = \frac{2^{-n_i}}{K} \quad \text{...(i)}$$

where $$K = \sum_{i=1}^{m} 2^{-n_i} \quad \text{...(ii)}$$

Now, we have

$$\sum_{i=1}^{m} Q_i = \frac{1}{K}\sum_{i=1}^{m} 2^{-n_i} = 1 \quad \text{...(iii)}$$

and $$\sum_{i=1}^{m} P_i \log_2 \frac{2^{-n_i}}{KP_i} = \sum_{i=1}^{m} P_i \left(\log_2 \frac{1}{P_i} - n_i - \log_2 K \right)$$

$$= -\sum_{i=1}^{m} P_i \log_2 P_i - \sum_{i=1}^{m} P_i n_i - (\log_2 K) \sum_{i=1}^{m} P_i \quad \text{...(iv)}$$

$$= H(X) - L - \log_2 K \leq 0$$

From the Kraft inequality (7.56), we have

$$\log_2 K \leq 0 \quad \text{...(v)}$$

Thus, $$H(X) - L \leq \log_2 K \leq 0 \quad \text{...(vi)}$$

or $$L \geq H(X)$$

The equality holds when $K = 1$ and $P_i = Q_i$. **Hence proved.**

EXAMPLE 7.41. Let X be a DMS with symbols x_i and corresponding probabilities $P(x_i) = P_i$, $i = 1,2, \ldots m$. Show that for the optimum source encoding, we require that

$$\mathbf{K = \sum_{i=1}^{m} 2^{-n_i} = 1} \quad \text{...(i)}$$

and $$\mathbf{n_i = \log_2 \frac{1}{P_i} = I_i} \quad \text{...(ii)}$$

where n_i is the length of the codeword corresponding to x_i and I_i is the information content of x_i.

Solution: From the result of last problem, the optimum source encoding with $L = H(X)$ requires $K = 1$ and $P_i = Q_i$. Thus, equations using (ii) and (i) of last problem, we have

$$K = \sum_{i=1}^{m} 2^{-n_i} = 1 \quad \text{...(iii)}$$

and $$P_i = Q_i = 2^{-n_i} \quad \text{...(iv)}$$

Hence $$n_i = -\log_2 P_i = \log_2 \frac{1}{P_i} = I_i$$ **Hence proved.**

✍ **Important Point:** *Note that equation (ii) implies the following common sense principle. Symbols which occur with high probability should be assigned shorter codewords than symbols that occur with low probability.*

EXAMPLE 7.42. Consider a DMS X with symbols x_i and corresponding probabilities $P(x_i) = P_i$, i = 1, 2, ... m. Let n_i be the length of the codeword of x_i such that

$$\log_2 \frac{1}{P_i} \le n_i \le \log_2 \frac{1}{P_i} + 1 \quad ...(i)$$

Show that this relationship satisfies the Kraft inequality (7.56), and find the bound on K in equation (7.56).

Solution: Equation (i) can be rewritten as

$$-\log_2 P_i \le n_i \le -\log_2 P_i + 1 \quad ...(ii)$$

or $$\log_2 P_i \ge -n_i \ge \log_2 P_i - 1$$

Then, $$2\log_2 P_i \ge 2^{-n_i} \ge 2^{\log_2 P_i} 2^{-1}$$

or $$P_i \ge 2^{-n_i} \ge \frac{1}{2} P_i \quad ...(iii)$$

Thus, $$\sum_{i=1}^{m} P_i \ge \sum_{i=1}^{m} 2^{-n_i} \ge \frac{1}{2} \sum_{i=1}^{m} P_i \quad ...(iv)$$

or $$1 \ge \sum_{i=1}^{m} 2^{-n_i} \ge \frac{1}{2} \quad ...(v)$$

which indicates that the Kraft inequality (7.56) is satisfied, and the bound on K will be

$$\frac{1}{2} \le K \le 1 \qquad \textbf{Hence proved.} \quad ...(vi)$$

EXAMPLE 7.43. Consider a DMS X with symbols x_i and corresponding probabilities $P(x_i) = P_i$, i = 1, 2, ..., m. Show that a code constructed in agreement with equation (i) in last problem will satisfy the following relation:

$$H(X) \le L \le H(X) + 1 \quad ...(i)$$

where H(X) is the source entropy and L is the average codeword length.

Solution: Multiplying equation (ii) in last problem by P_i and summing over equation (i) yields

$$-\sum_{i=1}^{m} P_i \log_2 P_i \le \sum_{i=1}^{m} n_i P_i \le \sum_{i=1}^{m} P_i(-\log_2 P_i + 1) \quad ...(ii)$$

Now, $$\sum_{i=1}^{m} P_i(-\log_2 P_i + 1) = -\sum_{i=1}^{m} P_i \log_2 P_i + \sum_{i=1}^{m} P_i = H(X) + 1$$

Thus, equation (ii) reduces to

$$H(X) \le L \le H(X) + 1 \qquad \textbf{Hence proved.}$$

7.20 ENTROPY CODING

The design of a variable-length code such that its average codeword length approaches the entropy of DMS is often referred to as **entropy coding.** In this section, we present two examples of entropy coding. These are as under:

(i) Shannon-Fano coding

(ii) Huffman coding

7.20.1. Shannon-Fano Coding

An efficient code can be obtained by the following simple procedure, known as **Shannon-Fano algorithm:**

1. List the source symbols in order of decreasing probability.
2. Partition the set into two sets that are as close to equiprobables as possible, and assign 0 to the upper set 1 to the lower set.
3. Continue this process, each time partitioning the sets with as nearly equal probabilities as possible until further partitioning is not possible.

An example of Shannon-Fano encoding is shown in Table 7.6. Note in Shannon-Fano encoding the ambiguity may arise in the choice of approximately equiprobable sets.

Table 7.6. Shannon-Fano Encoding

x_i	$P(x_i)$	Step 1	Step 2	Step 3	Step 4	Code
x_1	0.30	0	0			00
x_2	0.25	0	1			01
x_3	0.20	1	0			10
x_4	0.12	1	1	0		110
x_4	0.08	1	1	1	0	1110
x_6	0.05	1	1	1	1	1111

$$H(X) = 2.36 \text{ b/symbol}$$
$$L = 2.38 \text{ b/symbol}$$
$$\eta = H(X)/L = 0.99$$

7.20.2. The Huffman Coding*

In general, Huffman encoding results in an optimum code. Thus, it is the code that has the highest efficiency. The Huffman encoding procedure is as follows:

1. List the source symbols in order of decreasing probability.
2. Combine the probabilities of the two symbols having the lowest probabilities, and reorder the resultant probabilities, this step is called reduction 1. The same procedure is repeated until there are two ordered probabilities remaining.
3. Start encoding with the last reduction, which consist of exactly two ordered probabilities. Assign 0 as the first digit in the codewords for all the source symbols associated with the first probability; assign 1 to the second probability.
4. Now go back and assign 0 and 1 to the second digit for the two probabilities that were combined in the previous reduction step, retaining all assignments made in Step 3.
5. Keep regressing this way until the first column is reached.

An example of Huffman encoding is shown in Table 7.7

$$H(X) = 2.36 \text{ b/symbol}$$
$$L = 2.38 \text{ b/symbol}$$
$$\eta = 0.99$$

* Explain Huffman coding with the help of an example.

Table 7.7. Huffman Encoding

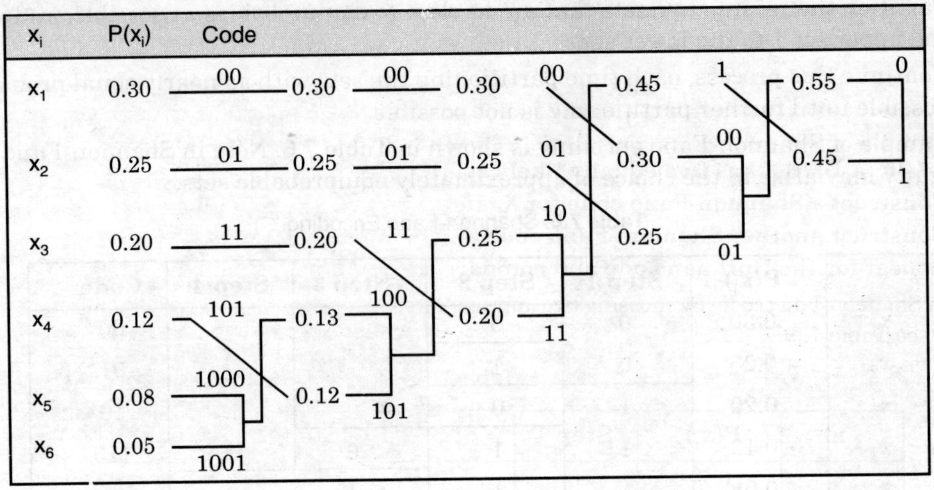

EXAMPLE 7.44 A DMS X has four symbols x_1, x_2, x_3, and x_4 with $P(x_1) = \frac{1}{2}$, $P(x_2) = \frac{1}{4}$, and $P(x_3) = P(x_4) = \frac{1}{8}$. Construct a Shannon-Fano code for X ; show that this code has the optimum property that $n_i = I(x_i)$ and that the code efficiency is 100 percent.

Solution: The Shannon-Fano code is constructed as follows (see Table 7.8).

Table 7.8.

x_i	$P(x_i)$	Step 1	Step 2	Step 3	Code
x_1	1/2	0			0
x_2	1/4	1	0		10
x_3	1/8	1	1	0	110
x_4	1/8	1	1	1	111

$$I(x_1) = -\log_2 \frac{1}{2} = 1 = n_1$$

$$I(x_2) = -\log_2 \frac{1}{4} = 2\ n_2$$

$$I(x_3) = -\log_2 \frac{1}{8} = 3 = n_3$$

$$I(x_4) = -\log_2 \frac{1}{8} = 3 = n_4$$

We know that, $$H(X) = \sum_{i=1}^{4} P(x_i) I(x_i)$$

or $$H(X) = \frac{1}{2}(1) + \frac{1}{4}(2) + \frac{1}{8}(3) + \frac{1}{8}(3) = 1.75$$

$$L = \sum_{i=1}^{4} P(x_i)\, n_i = \frac{1}{2}(1) + \frac{1}{4}(2) + \frac{1}{8}(3) + \frac{1}{8}(3) = 1.75$$

Also
$$\eta = \frac{H(X)}{L} = 1 = 100\% \quad \textbf{Ans.}$$

EXAMPLE 7.45. A DMS X has five equally likely symbols.

(i) Construct a Shannon-Fano code for X, and calculate the efficiency of the code.

(ii) Construct another Shannon-Fano code and compare the results.

(iii) Repeat for the Huffman code and compare the results.

Solution: A Shannon-Fano code [by choosing two approximately equiprobable (0.4 versus 0.6) sets] is constructed as follows (see Table 7.9).

Table 7.9.

x_i	$P(x_i)$	Step 1	Step 2	Step 3	Code
x_1	0.2	0	0		00
x_2	0.2	0	1		01
x_3	0.2	1	0		10
x_4	0.2	1	1	0	110
x_5	0.2	1	1	1	111

$$H(X) = \sum_{i=1}^{5} P(x_i)\log_2 P(x_i) = 5(-0.2\log_2 0.2) = 2.32$$

$$L = \sum_{i=1}^{5} P(x_i) n_i = 0.2(2+2+2+3+3)\ 2.4$$

The efficiency η is
$$\eta = \frac{H(X)}{L} = \frac{2.32}{2.4} = 0.967 = 96.7\% \quad \textbf{Ans.}$$

(ii) Another Shannon-Fano code [by choosing another two approximately equiprobable (0.6 versus 0.4) sets] is constructed as follows (see Table 7.10)

Table 7.10.

x_i	$P(x_i)$	Step 1	Step 2	Step 3	Code
x_1	0.2	0	0		00
x_2	0.2	0	1	0	010
x_3	0.2	0	1	1	011
x_4	0.2	1	0		10
x_5	0.2	1	1		11

$$L = \sum_{i=1}^{5} P(x_i) n_i = 0.2(2+3+3+2+2) = 2.4$$

Since the average codeword length is the same as that for the code of part (a), the efficiency is the same. **Ans.**

(iii) The Huffman code is constructed as follows (see Table 7.11)

$$L = \sum_{i=1}^{5} P(x_i)n_i = 0.2(2 + 3 + 3 + 2 + 2) = 2.4$$

Since the average code word length is the same as that for the Shannon-Fano code, the efficiency is also the same.

Table 7.11.

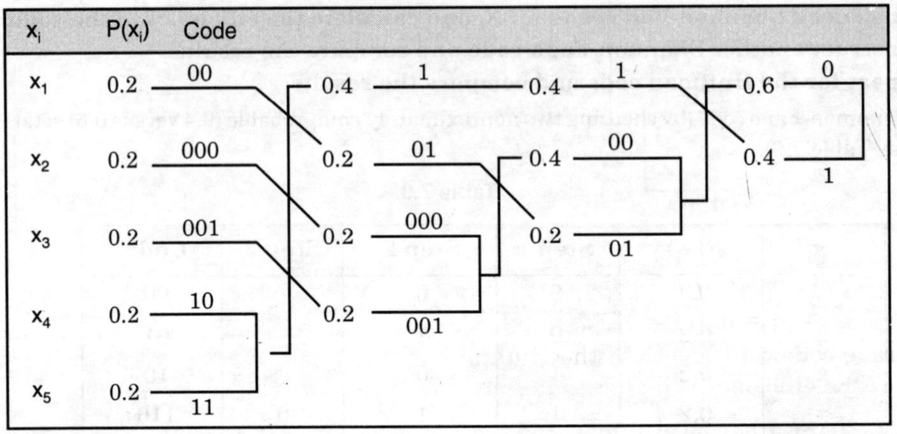

EXAMPLE 7.46 A DMS X has five symbols x_1, x_2, x_3, x_4, and x_5 with $P(x_1) = 0.4$, $P(x_2) = 0.19$, $P(x_3) = 0.16$, $P(x_4) = 0.15$, and $P(x_5) = 0.1$.

(i) Construct a Shannon-Fano code for X, and calculate the efficiency of the code.

(ii) Repeat for the Huffman code and compare the results.

Solution: The Shannon-Fano code is constructed as follows (see Table 7.12)

$$H(X) = \sum_{i=1}^{5} P(x_i)n_i = 0.4(1) + 0.19(2) + 0.16(2) + 0.15(3) + 0.1(3) = 2.25$$

Also, $$\eta = \frac{H(X)}{L} = \frac{2.15}{2.25} = 0.956 = 95.6\% \quad \textbf{Ans.}$$

Table 7.12.

x_i	P(x_i)	Step 1	Step 2	Step 3	Code
x_1	0.4	0	0		00
x_2	0.19	0	1		01
x_3	0.16	1	0		10
x_4	0.15	1	1	0	110
x_5	0.1	1	1	1	111

(ii) The Huffman code is constructed as follows (see Table 7.13):

$$L = \sum_{i=1}^{5} P(x_i)n_i$$

or $$L = 0.4(1) + (0.19 + 0.16 + 0.15 + 0.1)(3) = 2.2 \quad \textbf{Ans.}$$

Table 7.13.

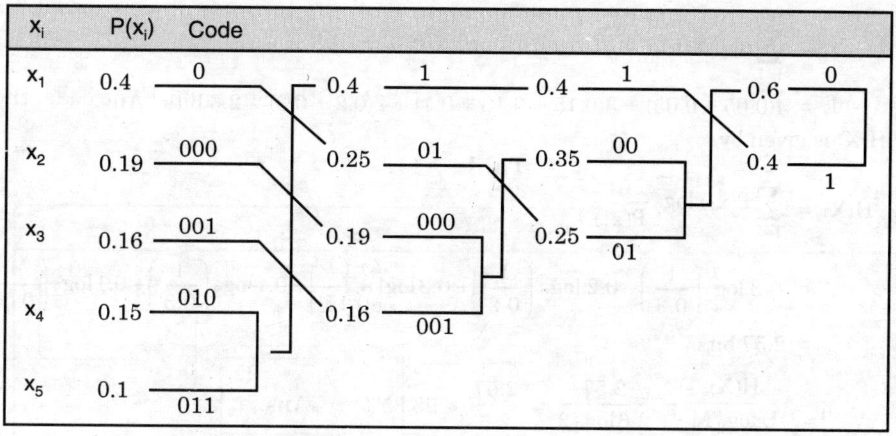

or $$\eta = \frac{H(X)}{L} = \frac{2.15}{2.2}$$

or $$\eta = 0.977 = 97.7\%$$

The average codeword length of the Huffman code is shorter than that of the Shannon-Fano code, and thus the efficiency is higher than that of the Shannon-Fano code. **Ans.**

EXAMPLE 7.47. Determine the Huffman code for the following messages with their probabilities given

x_1	x_2	x_3	x_4	x_5	x_6	x_7
0.05	**0.15**	**0.2**	**0.05**	**0.15**	**0.3**	**0.1**

Solution: Arranging and grouping of messages is done as shown below:

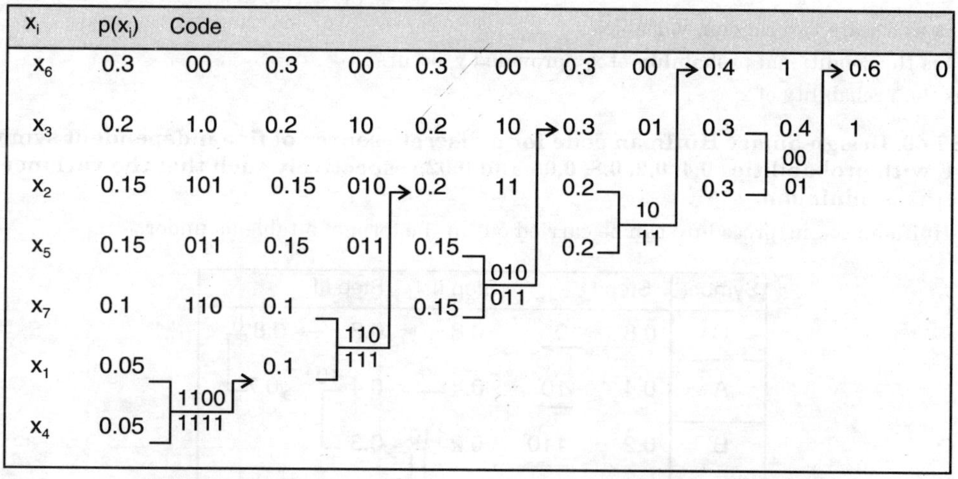

Message	Probability	Code	No. of bits in code
x_1	0.05	1110	4
x_2	0.15	010	3
x_3	0.2	10	2
x_4	0.05	1111	4
x_5	0.15	011	3
x_6	0.3	00	2
x_7	0.1	110	3

Therefore, average length L is given by

$$L = \sum_{i=1}^{7} n_i \, P(x_i)$$

or
$$L = 4(0.05 + 0.05) + 3(0.15 + 0.15 + 0.1) + 2(0.2 + 0.3) = 2.6 \text{ bits} \quad \textbf{Ans.}$$

Entropy H(X) is given by

$$H(X) = \sum_{i=1}^{7} P(x_i) \log_2 \frac{1}{P(x_i)}$$

$$= 0.3 \log_2\left(\frac{1}{0.3}\right) + 0.2 \log_2\left(\frac{1}{0.2}\right) + 0.3\log\left(\frac{1}{0.15}\right) + 0.1\log_2\left(\frac{1}{0.3}\right) + 0.1\log_2\left(\frac{1}{0.05}\right)$$

$$= 2.57 \text{ bits}$$

$$\eta = \frac{H(X)}{L \log_2 M} = \frac{2.57}{2.6 \log_2 2} = \frac{2.57}{2.6} = 98.85\% \qquad \textbf{Ans.}$$

EXAMPLE 7.48. Define Average Mutual Information between two discrete random variables and show that it is non-negative.

Solution: Average Mutual Information

The extreme values of mutual information are 0 and log $\left[\frac{1}{p(x_i)}\right]$. Most transmission channels fall somewhere between these extreme values. To analyze the general case, we define the average mutual information as under:

$$I(X ; Y) = \sum_{x,\,y} p(x_i, y_j) \,.\, I(x_i ; y_j) = \sum_{i=1}^{n} \sum_{j=1}^{m} p(x_i, y_j) \,.\, I(x_i ; y_j)$$

where x and y are two random variables

$p(x_i/y_j)$ is the conditional probability of x_i ; provided y_j occurs.

$p(x_i)$ is the probability of x_i.

EXAMPLE 7.49. Design binary Huffman code for a discrete source of five independent symbols A, B, C, D and E with probabilities 0.4, 0.2, 0.8, 0.08 and 0.02 respectively such that the variance of code-word lengths is minimum.

Solution: Huffman coding procedure can be carried out in the form of a table as under:

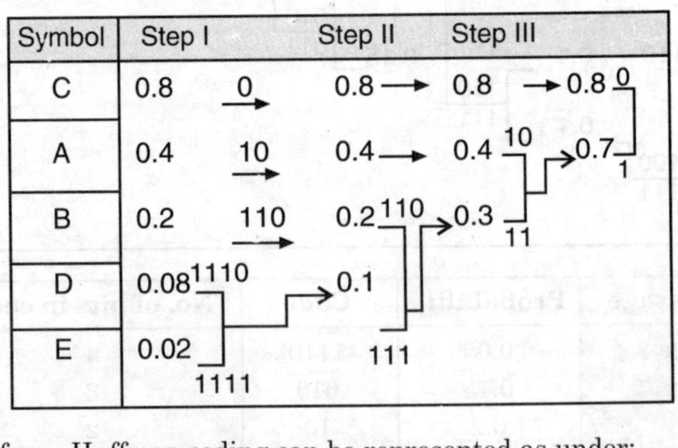

Therefore, in compact form, Huffman coding can be represented as under:

All progress has resulted from people who took unpopular position. – **Adlai Stevenson**

Symbol	Probability	Code word
C	0.8	0
A	0.4	10
B	0.2	110
D	0.08	1110
E	0.02	1111

We know that the average code word length is given by

$$L = \sum_{i=1}^{m} p_i\, n_i$$

or $$L = 0.8 \times 1 + 0.4 \times 2 + 0.2 \times 3 + 0.08 \times 4 + 0.02 \times 4$$

or $$L = 2.6. \quad \textbf{Ans.}$$

Now, variance is given by

$$\sigma^2 = \sum_{i=1}^{5} p_i (n_i - L)^2$$

or $$\sigma^2 = 0.8\,(1 - 2.6)^2 + 0.4\,(2 - 2.6)^2 + 0.2\,(3 - 2.6)^2 + 0.08\,(4 - 2.6)^2 + 0.02\,(4 - 2.6)^2$$

or $$\sigma^2 = 2.04 + 0.144 + 0.032 + 0.156 + 0.0392$$

or $$\sigma^2 = 2.4112. \quad \textbf{Ans.}$$

EXAMPLE 7.50. An event has six possible outcomes with probabilities

$$P_1 = \frac{1}{2},\ P_2 = \frac{1}{4},\ P_3 = \frac{1}{8},\ P_4 = \frac{1}{16},\ P_5 = \frac{1}{32}\ \text{and}\ P_6 = \frac{1}{32}$$

Find the entropy of the system. Also find the rate of information if there are 16 outcomes per second.

Solution: Given that

$$P_1 = \frac{1}{2},\ P_2 = \frac{1}{4},\ P_3 = \frac{1}{8},\ P_4 = \frac{1}{16},\ P_5 = \frac{1}{32}\ \text{and}\ P_6 = \frac{1}{32}$$

$$r = 16 \text{ outcomes/sec.}$$

We know that the entropy is given by

$$H = \sum_{i=1}^{m} P_i \log_2\left(\frac{1}{P_i}\right)$$

Substituting all the given values, we get

$$H = \frac{1}{2}\log_2 2 + \frac{1}{2}\log_2 4 + \frac{4}{8}\log_2 8 + \frac{1}{16}\log_2 16 + \frac{1}{32}\log_2 32$$

or $$H = \frac{1}{2} + \frac{2}{4} + \frac{3}{8} + \frac{4}{16} + \frac{10}{32}$$

or $$H = 1.9375 \text{ bits/message.}$$

We also know that information rate is given by

$$R = \text{Entropy} \times \text{message rate} = H \times r = 1.9375 \times 16$$

or $$R = 31 \text{ bits/sec.} \quad \textbf{Ans.}$$

EXAMPLE 7.51. A discrete source transmits messages x_1, x_2 and x_3 with probabilities 0.3, 0.4 and 0.3. The source is connected to the channel given in figure 7.16. Calculate H(X) and H(Y).

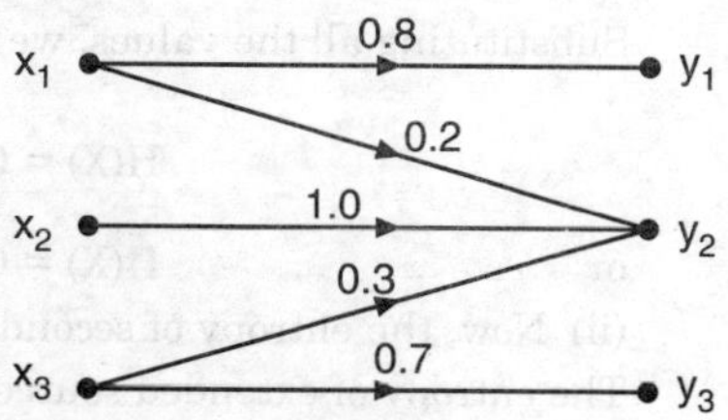

Fig. 7.16.

Solution: Given that

$$P(x_1) = 0.3,\ P(x_2) = 0.4 \text{ and } P(x_3) = 0.3$$

We know that the value of H(X) can be found as under:

$$H(X) = P(x_1) \log_2 \frac{1}{P(x_1)} + P(x_2) \log_2 \frac{1}{P(x_2)} + P(x_3) \log_2 \frac{1}{P(x_3)}$$

Substituting values, we get

$$H(X) = 0.3 \log_2 \left(\frac{1}{0.3}\right) + 0.4 \log_2 \left(\frac{1}{0.4}\right) + 0.3 \log_2 \left(\frac{1}{0.3}\right)$$

or $$H(X) = 0.52 + 0.52 + 0.52 = 1.568 \text{ bits/symbol.}$$

The given figure shows conditional probability matrix in the form of $P\left(\frac{Y}{X}\right)$ as,

$$P(Y|X) = \begin{matrix} & y_1 & y_2 & y_3 \\ & \downarrow & \downarrow & \downarrow \\ x_1 \rightarrow & 0.8 & 0.2 & 0 \\ x_2 \rightarrow & 0 & 1 & 0 \\ x_3 \rightarrow & 0 & 0.3 & 0.7 \end{matrix}$$

or $$P(X) = [0.3 \quad 0.4 \quad 0.3]$$

The P(X, Y) is obtained by multiplying P(Y | X) and P(X) as under:

$$P(X, Y) = \begin{bmatrix} 0.8\times0.3 & 0.2\times0.3 & 0 \\ 0 & 1\times0.4 & 0 \\ 0 & 0.3\times0.3 & 0.7\times0.3 \end{bmatrix} = \begin{bmatrix} 0.24 & 0.06 & 0 \\ 0 & 0.4 & 0 \\ 0 & 0.09 & 0.21 \end{bmatrix}$$

$P(y_1)$, $P(y_2)$ and $P(y_3)$ are obtained by adding column of P(X, Y).

Therefore, we have

$$P(y_1) = 0.24 + 0 + 0 = 0.24$$
$$P(y_2) = 0.06 + 0.4 + 0.09 = 0.55$$
$$P(y_3) = 0 + 0 + 0.21 = 0.21$$

Now, the value of H(Y) can be obtained as under:

$$H(Y) = -\sum_{j=1}^{n} P(y_j) \log_2 P(y_j) = -\sum_{j=1}^{3} P(y_j) \log_2 P(y_j)$$

or $$H(Y) = -[0.24 \log_2 0.24 + 0.55 \log_2 0.55 + 0.21 \log_2 0.21]$$

or $$H(Y) = -[-0.49 - 0.47 - 0.47]$$

or $$H(Y) = -1.43 \text{ bits/symbol}$$ **Ans.**

EXAMPLE 7.52. Consider a discrete memoryless source with source alphabet = $\{S_0, S_1, S_2\}$ and source statistics {0.7, 0.15, 0.15}

(i) Calculate the entropy of the source

(ii) Calculate the entropy of second order extension of the source.

Solution: (i) We know that the entropy of discrete memoryless source is given by

$$H(X) = \sum_{i=1}^{3} P_i \log_2 \frac{1}{P_i}$$

Substituting all the values, we get

$$H(X) = 0.7 \log_2 \left(\frac{1}{0.7}\right) + 0.15 \log_2 \left(\frac{1}{0.15}\right) + 0.15 \log_2 \left(\frac{1}{0.15}\right)$$

or $$H(X) = 0.36 + 0.41 + 0.41 = 1.1805 \text{ bits/symbol.}$$ **Ans.**

(ii) Now, the entropy of second order extension of the source can be evaluated as under:

The entropy of extended source is given by

$$H(X^n) = nH(X)$$

For second order, we have

$$H(X^2) = 2 \times H(X) = 2 \times 1.1805 = 2.361 \text{ bits/symbol.} \quad \textbf{Ans.}$$

EXAMPLE 7.53. State Hartley-Shannon Law.

Solution: The Hartley-Shannon law states that the channel capacity of a white, bandlimited Gaussian channel is given by

$$C = B \log_2 \left(1 + \frac{P}{N}\right) \text{ bits/sec} \quad \ldots \text{(i)}$$

where, B = channel bandwidth

P = signal power

N = noise within the channel bandwidth

> **DO YOU KNOW?**
>
> The shannon-Hartley theorem is of fundamental importance and has two important implications for communication system engineers.

Power, $P = \int_{-B}^{B}$ Power spectral density

Using above expression, the noise power N is as under:

$$\text{Noise power } N = \int_{-B}^{B} \frac{N_0}{2} df$$

where $\frac{N_0}{2}$ is the power spectral density (psd) of the white noise.

Therefore, we write

$$N = N_0 B \quad \ldots \text{(ii)}$$

Using equations (i) and (ii), we obtain

$$C = B \log_2 \left(1 + \frac{P}{N_0 B}\right)$$

EXAMPLE 7.54. A source emits one of four symbols S_0, S_1, S_2 and S_3 with probabilities $\frac{1}{3}, \frac{1}{6}, \frac{1}{4}$ and $\frac{1}{4}$ respectively. The successive symbols emitted by the source are independent. Calculate the entropy of source.

Solution: The given probabilities of the symbols are as under:

Symbol	S_0	S_1	S_2	S_3
Probability	$\frac{1}{3}$	$\frac{1}{6}$	$\frac{1}{4}$	$\frac{1}{4}$

We know that the expression for source entropy H(X) is given by

$$H(S) = \sum_{i=1}^{m} P(S_i) \log_2 \frac{1}{P(S_i)} \text{ bits/symbol}$$

Substituting all the given values, we get

$$H(S) = \frac{1}{3} \log_2 3 + \frac{1}{6} \log_2 6 + \frac{1}{4} \log_2 4 + \frac{1}{4} \log_2 4$$

or
$$H(S) = \frac{1}{3} \frac{\log_{10} 3}{\log_{10} 2} + \frac{1}{6} \frac{\log_{10} 6}{\log_{10} 2} + \frac{2}{4} \log_2 2 + \frac{2}{4} \log_2 2$$

or
$$H(S) = \frac{1}{3} \times 1.5849 + \frac{1}{6} \times 2.585 + \frac{2}{4} \times 1 + \frac{2}{4} \times 1$$

Prudent man walks warily under all circumstances. – Anonymouss

EXAMPLE 7.55. Find Shannon-Fano code for five messages given by probabilities $\frac{1}{2}, \frac{1}{4}, \frac{1}{8}, \frac{1}{16}, \frac{1}{16}$. Calculate the average number of bits/message.

Solution: Let the given messages be represented by x_1, x_2, x_3, x_4 and x_5.

With the help of given probabilities, we can develop Shannon-Fano code in the form of a table as follows:

Shannon-Fano Coding

Message	Corresponding Probability	Step I	Step II	Step III	Step IV	Code Word	Length n_i
x_1	1/2	0	Stop			0	1
x_2	1/4	1	0	Stop		10	2
x_3	1/8	1	1	0	Stop	110	3
x_4	1/16	1	1	1	0	1110	4
x_5	1/16	1	1	1	1	1111	4

We know that the average number of bits per message is given by

$$L = \sum_{i=1}^{m} P(x_i)\, n_i$$

Substituting all the values, we get

$$L = \left(\frac{1}{2}\times 1\right)+\left(\frac{1}{4}\times 2\right)+\left(\frac{1}{8}\times 3\right)+\left(\frac{1}{16}\times 4\right)+\left(\frac{1}{16}\times 4\right)$$

$$= \frac{1}{2}+\frac{1}{2}+\frac{3}{8}+\frac{1}{2} = 1.875 \text{ bits.} \qquad \textbf{Ans.}$$

EXAMPLE 7.56. A voice grade channel of the telephone network has a bandwidth of 3.4 kHz. Calculate the information capacity of the telephone channel for a signal-to-noise ratio of 30 dB.

Solution: Given that

$$B = 3.4 \text{ kHz, and}$$

$$\frac{S}{N} = 30 \text{ dB}$$

We know that $\frac{S}{N}$ ratio in dB is expressed as

$$\left(\frac{S}{N}\right)_{dB} = 10 \log_{10}\left(\frac{S}{N}\right) \quad \text{or} \quad 30 = 10 \log_{10}\left(\frac{S}{N}\right)$$

or

$$\frac{S}{N} = \text{Antilog } 30 = 1000$$

We know that the information capacity of the telephone channel is given by

$$C = B \log_2\left(1+\frac{S}{N}\right)$$

Substituting all the values, we get

$$C = 3.4 \times 10^3 \log_2 (1 + 1000)$$

or

$$C = 3.4 \times 10^3 \frac{\log_{10} 1001}{\log_{10} 2} = 3400 \times 9.9672$$

or

$$C = 33888.57 \text{ bits/sec} = 33.89 \text{ kbps.} \quad \textbf{Ans.}$$

GLOSSARY

1. **Information source:** May be viewed as an object which produces an event, the outcome of which is selected at random according to a probability distribution. A practical source in a communication system is a device which produces messages, and it can be either analog or discrete.
2. **DMS:** It can be characterized by the list of the symbols, the probability assignment to these symbols, and the specification of the rate of generating these symbols by the source.
3. **Information Rate:** If the time rate at which source X emits symbols is r (symbols).
4. **Communication channel:** it may be defined as the path or medium through which the symbols flow to the receiver end. A discrete memoryless channel (DMC) is a statistical model with an input X and an output Y. During each unit of the time (signaling interval), the channel accepts an input symbol from X, and in response it generates an output symbol from Y. The channel is said to be "discrete" when the alphabets of X and Y are both finite. Also, it is said to be "memoryless" when the current output depends on only the current input and not on any of the previous inputs.
5. **Noiseless Channel:** A channel is called noiseless if it is both lossless and deterministic. The channel matrix has only one element in each row and in each column, and this element is unity. Note that the input and output alphabets are of the same size, that is, m = n for the noiseless channel.
6. **Shannon-Hartley Law:** Underscores the fundamental role of bandwidth and signal-to-noise ratio in communication. It also shows that we can exchange increased bandwidth for decreased signal power for a system with given capacity C.
7. **Source Coding:** A conversion of the output of a discrete memoryless source (DMS) into a sequence of binary symbols (i.e., binary code word) is called **source coding.** The device that performs this conversion is called the **source encoder**.

SUMMARY

- The purpose of a communication system is to carry information-bearing baseband signals from one place to another place over a communication channel.
- Information theory is a branch of probability theory which may be applied to the study of the communication systems.
- In the context of communications, information theory deals with mathematical modeling and analysis of a communication system rather than with physical sources and physical channels.
- Information theory was invented by communication scientists while they were studying the statistical structure of electronic communication equipments. When the communique is readily measurable, such as an electric current, the study of the communication system is relatively easy.
- An information source may be an object produces an event, the outcome of which is selected at random according to a probability distribution. A practical source in a communication system is a device which produces messages, and it can be either analog or discrete.
- A discrete information source is a source which has only a finite set of symbols as possible outputs. The set of source symbols is called the source alphabet, and the elements of the set are called symbols or letters.
- Information sources can be classified as having memory or being memoryless. A source with memory is one for which a current symbol depends on the previous symbols. A memoryless source is one for which each symbol produced is independent of the previous symbols.
- A discrete memoryless source (DMS) can be characterized by the list of the symbols, the probability assignment to these symbols, and the specification of the rate of generating these symbols by the source.
- The amount of information contained in an event is closely related to its uncertainty. Messages containing knowledge of high probability of occurrence convey relatively little information.
- In a practical communication system, we usually transmit long sequences of symbols from an information source. Thus, we are more interested in the average information that a source produces than the information content of a single symbol.

Responsibility walks hand in hand with capacity and power. – **J.G. Holland**

- For quantitative representation of average information per symbol we make the following assumptions:
  (i) The source is stationary so that the probabilities may remain constant with time.
  (ii) The successive symbols are statistically independent and come from the source at a average rate of r symbols per second.
- The mean value of $I(x_i)$ over the alphabet of source X with m different symbols is given by

$$H(X) = E[I(x_i)] = \sum_{i=1}^{m} P(x_i)I(x_i) = -\sum_{i=1}^{m} P(x_i)\log_2 P(x_i) \text{ b/symbol}$$

- If the time rate at which source X emits symbols is r (symbols s), the information rate R of the source is given by

$$R = rH(X) \text{ b/s}$$

- A communication channel may be defined as the path or medium through which the symbols flow to the receiver. A discrete memoryless channel (DMC) is a statistical model with an input X and an output Y
- A channel is completely specified by the complete set of transition probabilities.
- A channel described by a channel matrix with only one non-zero element in each column is called a lossless channel.
- A channel described by a channel matrix with only one non-zero element in each row is called a deterministic channel.
- A channel is called noiseless if it is both lossless and deterministic.
- The mutual information: I(X ; Y) of a channel is defined by

$$I(X;Y) = H(X) - H(X|Y) \text{ b/symbol}$$

  Since H(X) represents the uncertainty about the channel input before the channel output is observed and H(X|Y) represents the uncertainty about the channel input after the channel output is observed, the mutual information I(X ; Y) represents the uncertainty about the channel input that is resolved by observing the channel output.
- The channel capacity per symbol of a DMC is defined as

$$C_s = \max_{\{P(x_i)\}} I(X;Y) \text{ b/symbol}$$

  where the maximization is over all possible input probability distributions $\{P(x_i)\}$ on X. Note that the channel capacity C_s is a function of only the channel transition probabilities which define the channel.
- In a continuous channel, an information source produces a continuous signal x(t). The set of possible signals is considered as an ensemble of waveforms generated by some ergodic random process. It is further assumed that x(t) has a finite bandwidth s_0 that x(t) is completely characterized by its periodic sample values. Thus, at any sampling instant, the collection of possible sample values constitutes a continuous random variable X described by it probability density function $f_X(x)$.
- The average amount of information per sample value of x(t) is measured by

$$H(X) = -\int_{-\infty}^{\infty} f_X(x)\log_2 f_X(x)\,dx \text{ b/sample}$$

  The entropy H(X) defined by above equation is known as the differential entropy of X.
- In an additive white Gaussian noise (AWGN) channel, the channel output Y is given by

$$Y = X + n$$

  where X is the channel input and n is an additive bandlimited white Gaussian noise with zero mean and variance σ^2.
- The Shannon-Hartley law underscores the fundamental role of bandwidth and signal-to-noise ratio in communication. It also shows that we can exchange increased bandwidth for decreased signal power for a system with given capacity C.
- A conversion of the output of a DMS into a sequence of binary symbols (i.e., binary codeword) is called source coding. The device that performs this conversion is called the source encoder

- An objective of source coding is to minimize the average bit rate required for representation of the source by reducing the redundancy of the information source.
- The average codeword length L, per source symbol is given by

$$L = \sum_{i=1}^{m} P(x_i)n_i$$

The parameter L represents the average number of bits per source symbol used in the source coding process.

- Also, the code efficiency η is defined as

$$\eta = \frac{L_{min}}{L}$$

where L_{min} is the minimum possible value of L. When η approaches unity, the code is said to be efficient.

- The code redundancy γ is defined as

$$\gamma = 1 - \eta$$

- The source coding theorem states that for a DMS X, with entropy H(X), the average codeword length L per symbol is bounded as

$$L \geq H(X)$$

and further, L can be made as close to H(X) as desired for some suitably chosen code.

Thus, with $L_{min} = H(X)$, the code efficiency can be rewritten as

$$\eta = \frac{H(X)}{L}$$

- A necessary and sufficient condition for the existence of an instantaneous binary code is

$$K = \sum_{i=1}^{m} 2^{-n_i} \leq 1$$

which is known as the Kraft inequality.

- The design of a variable-length code such that its average codeword length approaches the entropy of DMS is often referred to as entropy coding. In this section, we present two examples of entropy coding.
- An efficient code can be obtained by the following simple procedure, known as Shannon-Fano algorithm.

SHORT QUESTIONS WITH ANSWERS

Q.1. What is the meaning of word 'Information"?

Ans. The amount of information received from the knowledge of occurrence of a event may be related to the likelihood or probability of occurrence of that event. The message associated with the least likelihood event thus consists of maximum information. The above amount of information in a message depends only upon the uncertainty of the underlying event rather than its acutal content.

Q.2. What are Information Source? Explain.

Ans. An information source may be viewed as an object which produces an event, the outcome of which is selected at random according to a probability distribution. A practical source in a communication system is a device which produces messages, and it can be either analog or discrete. In this chapter, we deal mainly with the discrete sources since analog sources can be transformed to discrete sources through the use of sampling and quantization techniques. As a matter of fact, a discrete information source is a source which has only a finite set of symbols as possible outputs. The set of source symbols is called the **source alphabet**, and the elements of the set are called **symbols** or **letters.**

Information sources can be classified as having memory or being memoryless. A source with memory is one for which a current symbol depends on the previous symbols. A memoryless source is one for which each symbol produced is independent of the previous symbols.

A discrete memoryless source (DMS) can be characterized by the list of the symbols, the probability assignment to these symbols, and the specification of the rate of generating these symbols by the source.

Q.3. Define the Information content of a symbol.

Ans. The information content of a symbol x_i, denoted by $I(x_i)$ is defined by

$$I(x_i) = \log_b \frac{1}{P(x_i)} = -\log_b P(x_i)$$

where $P(x_i)$ is the probability of occurrence of symbol x_i.

Q.4. What do you mean by Information Rate? Explain.

Ans. If the time rate at which source X emits symbols is r (symbols s), the information rate R of the source is given by

$$R = rH(X) \text{ b/s}$$

Here R is information rate.

H(X) is Entropy or average information

and r is rate at which symbols are generated.

Information rate R is represented in average number of bits of information per second. It is calculated as under:

$$R = \left[r \text{ in } \frac{\text{symbols}}{\text{second}} \right] \times \left[H(X) \text{ in } \frac{\text{Information bits}}{\text{symbol}} \right]$$

R = Information bits/second

Q.5. What is a Discrete Memoryless Channels (DMC)? Explain.

Ans. A communication channel may be defined as the path or medium through which the symbols flow to the receiver end. A discrete memoryless channel (DMC) is a statistical model with an input X and an output Y as shown in figure. During each unit of the time (signaling interval), the channel accepts an input symbol from X, and in response it generates an output symbol from Y. The channel is said to be discrete when the alphabets of X and Y are both finite. Also, it is said to be memoryless when the current output depends on only the current input and not on any of the previous inputs.

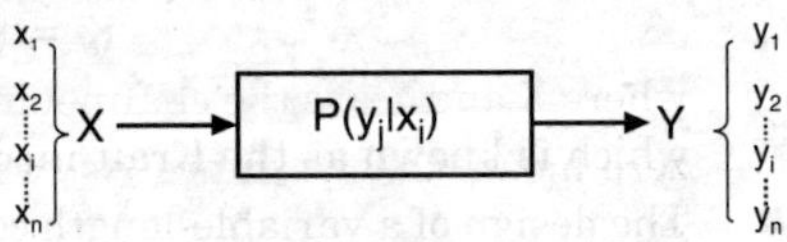

Fig. 7.17.

REVIEW QUESTIONS

1. Explain the concept of information.
2. What is Entropy? Explain.
3. What is information rate?
4. What are discrete memoryless channels (DMC)?
5. What is source coding?
6. Write short notes on the following:
   (i) Source-coding theorem, (ii) Kraft-inequality
7. Explain Shannon-Fano Algorithm.

NUMERICAL PROBLEMS

1. Consider a source X which produces five symbols with probabilities $\frac{1}{2}, \frac{1}{4}, \frac{1}{8}, \frac{1}{16}$ and $\frac{1}{16}$. Find the source entropy H(X). **[Ans.** 1.875 b/symbol]
2. Calculate the average information content in the English language, assuming that each of the 26 characters in the alphabet occurs with equal probability. **[Ans.** 4.7 b/character]

3. Two BSCs are connected in cascade, as shown in figure 7.18.

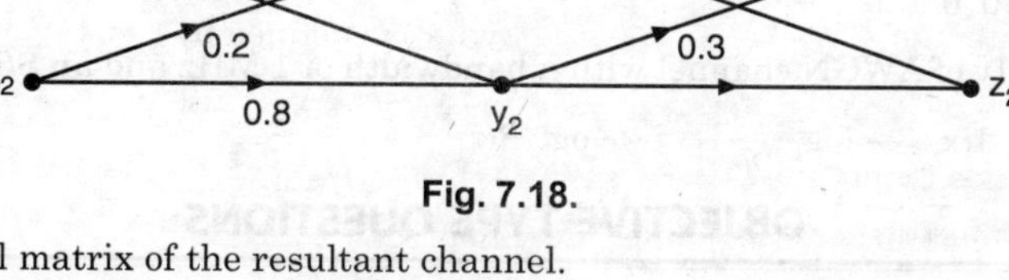

Fig. 7.18.

(i) Find the channel matrix of the resultant channel.

(ii) Find $P(z_1)$ and $P(z_2)$ if $P(x_1) = 0.6$ and $P(x_2) = 0.4$

[Ans. (i) $\begin{bmatrix} 0.62 & 0.38 \\ 0.38 & 0.62 \end{bmatrix}$ (ii) $P(z_1) = 0.524$, $P(z_2) = 0.476$]

4. Consider the DMC shown in figure 7.20.

(i) Find the output probabilities if $P(x_1) = \frac{1}{2}$ and $P(x_2) = P(x_3) = \frac{1}{4}$.

(ii) Find the output entropy H(Y). **[Ans.** (i) $P(y_1) = 7/24$, $P(y_2) = 17/48$, and $P(y_3) = 17/48$ (ii) 1.58 b/symbls]

5. Let X be a random variable with probability density function $f_X(x)$, and let $Y = aX + b$, where a and b are constants. Find H(Y) in terms of H(X). **[Ans.** $H(Y) = H(X) + \log_2 a$]

6. Find the differential entropy H(X) of a Gaussian variable X with zero mean and variance σ_X^2.

[Ans. $H(X) = \frac{1}{2}\log_2(2\pi e\sigma_X^2)$]

7. Consider an AWGN channel described by equations, that is,

$$Y = X + n$$

where X and Y are the channel input and output, respectively, and n is additive white Gaussian noise with zero mean and variance σ_n^2. Find the average mutual information I(X;Y) when the channel input X is Gaussian with zero mean and variance σ_X^2.

[Ans. $I(X;Y) = \frac{1}{2}\log_2\left(1 + \frac{\sigma_X^2}{\sigma_n^2}\right)$]

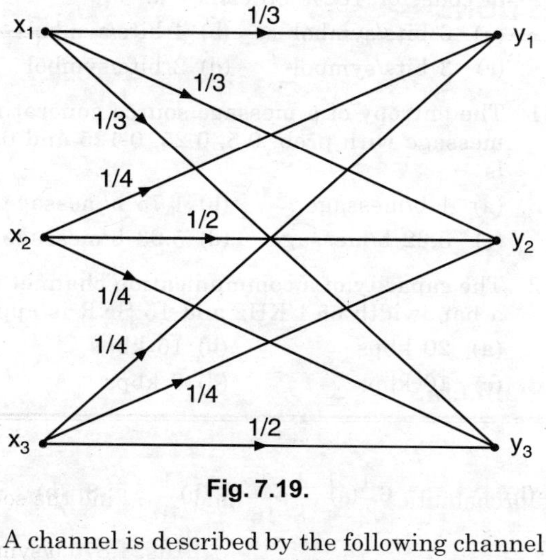

Fig. 7.19.

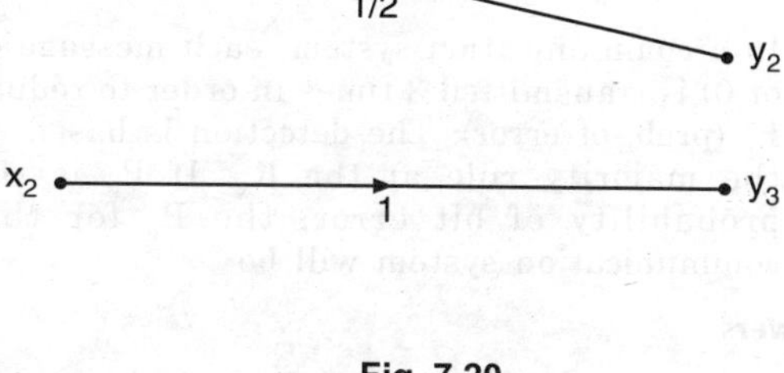

Fig. 7.20.

8. A channel is described by the following channel matrix:

(i) Draw the channel diagram.

(ii) Find the channel capacity.

$$\begin{bmatrix} \frac{1}{2} & \frac{1}{2} & 0 \\ 0 & 0 & 1 \end{bmatrix}$$

[**Ans.** (i) See figure 7.15, (ii) 1 b/symbol]

9. Calculate the capacity of AWGN channel with a bandwidth of 1 MHz and an S/N ratio of 40-dB.

[**Ans.** 13.29 Mb/s]

OBJECTIVE TYPE QUESTIONS

1. A binary symmetric channel (BSC) transmitting 1's and 0's with equal probability has an error rate of 10^{-2}. The channel transmission rate (in bit/symbol) will be
   (a) 0.99 (b) 0.919
   (c) 0.95 (d) 1
2. A communication system is used to transmit one of 16 possible signals. Assume that transmission is accomplished by encoding the signals into binary digits. If each binary digit requires 1 μ sec for transmission then how much information (in bits) is transmitted by the system in 8 μ sec?
   (a) 8 (b) 16
   (c) 4 (d) 2
3. The baud rate when using binary transmission is
   (a) always equal to the bit rate
   (b) equal to twice the BW of an ideal channel
   (c) not equal to the signaling rate
   (d) equal to one half of the BW of ideal channel
4. A zero source generates two messages with prob. 0.8 and 0.2. These are coded as 1 and 0.2. The code efficiency is
   (a) 0.2 (b) 0.5
   (c) 0.7 (d) 1.0
5. A communication channel with AWGN has a BW of 4 kHz and an SNR of 15. Its channel capacity is
   (a) 1.6 kbps (b) 16 kbps
   (c) 32 kbps (d) 456 kbps
6. In a communication system, each message (1 or 0) is transmitted 3 times in order to reduce P_e (prob. of error). The detection is based on the majority rule at the R_x. If P_e is the probability of bit error, the P_e for this communication system will be
   (a) $3P_e^2 - 2P_e^3$ (b) $-P_e - P_e^2$
   (c) P_e^2 (d) $P_e^2(1 - P_e)$
7. A source delivers symbols m_1, m_2, m_3 and m_4 with prob. 1/2, 1/4, 1/8 and 1/8 resp. The entropy of the system is
   (a) 1.7 bits/sec
   (b) 1.75 bits/symbol
   (c) 1.75 symbols
   (d) 1.75 symbol/bit
8. The channel capacity of a 5 kHz bandwidth binary system is
   (a) 10,000 bits/sec (b) 5000 bits/sec
   (c) 8000 bits/sec (d) 4000 bits/sec
9. A source generates 4 messages. The entropy of the source will be maximum when
   (a) all probabilities equal
   (b) one of the probabilities equal 1 and 2 others are zero
   (c) the probabilities are 1/2, 1/4 and 1/2
   (d) the two of the probabilities are 1/2 each and other zero
10. A source generates 4 equiprobable symbols. If source coding is employed, the average length of code for 100% efficiency is
    (a) 6 bits/symbol (b) 4 bits/symbol
    (c) 3 bits/symbol (d) 2 bits/symbol
11. The entropy of a message source generating 4 message with prob. 0.5, 0.25, 0.125 and 0.125 is
    (a) 1 b/message (b) 1.75 b/message
    (c) 3.32 b/message (d) 5.93 b/message
12. The capacity of a communication channel with a bandwidth of 4 KHz and 15 SNR is approx.
    (a) 20 kbps (b) 16 kbps
    (c) 10 kbps (d) 8 kbps

Answers

1. (a)	2. (a)	3. (a)	4. (a)	5. (b)	6. (a)	7. (b)	8. (b)
9. (a)	10. (d)	11. (b)	12. (b)				

□□□

CHAPTER 8

ERROR CONTROL CODING

SYLLABUS

Error Free Communication over a Noisy Channel Capacity of a Discrete and Continuous Memoryless Channel, Error Correcting Codes: Hamming distance and error detecting and correcting capability, Linear block codes, Encoding & syndrome decoding; Cyclic codes, Encoder and decoders for systematic cycle codes; Convolution codes, Code tree & Trellis diagram, Viterbi and sequential decoding, Burst error correction.

8.1 INTRODUCTION

When a digital signal is transmitted between two systems (such as computers), as shown in figure 8.1, the signal gets distorted because of the addition of noise to the transmitted signal.

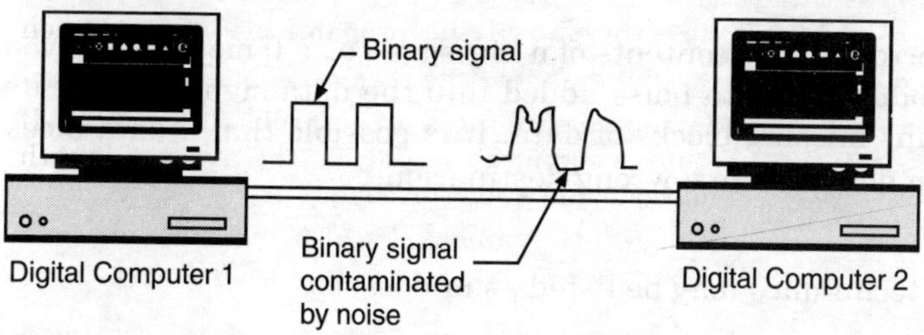

Fig. 8.1 Distortion of the binary signal by noise

The noise can introduce an error in the binary train of bit travelling from one system to the other. This means that a 0 may change to 1 or a 1 may change to 0. These errors can become a serious problem to the accuracy and performance of the digital system. Therefore, it is essential to detect and correct the errors. In digital communication, errors are introduced during the transmission of data from the transmitter to receiver due to noise or some other reasons. The reliability of data transmission gets severely affected because of these errors. To improve the reliability of data transmission, the designer has to increase the signal power or reduce the noise spectral density N_0 so as to

Chapter Outline

- Introduction
- Generation and Detection of Coded Signals
- Overview of Error Control Systems
- ARQ (Automatic Repeat Request) System
- Classification of Codes
- Random and Burst Errors
- Power and Bandlimited Channels
- Error Detection Methods
- Important Definitions Related to Codes
- Linear Block Codes
- Hamming Codes
- Syndrome Decoding (For Block Codes)
- Cyclic Codes
- Other Codes
- Convolutional Codes
- Decoding Methods of Convolutional Codes
- Sequential Decoding

maximize the ratio E_b/N_0. But, in practice, there is a limitation on the maximum value of the ratio E_b/N_0. We cannot increase the ratio beyond this limit. Hence, for a fixed value of E_b/N_0, we have to use some kind of coding in order to improve the quality of the transmitted signal.

Another advantage of using coding is that we can reduce the required value of E_b/N_0 for a fixed bit error rate (BER). This, in turn, reduces the transmitted power and the size of antenna. Error control coding is a practical way to achieve a very low bit error rate (BER) after transmission over a noisy bandlimited channel. This is theoretically possible to achieve as stated in the Shannon's channel theorem.

> **DO YOU KNOW?**
> Digital signal results in discrete, rather than continuous, changes in the modulated signal. The receiver examines the signal at specified times, and the state of the signal at each such time is called a symbol.

8.1.1. Detection and Correction of Errors

For the detection and correction of transmission errors, one or more than one extra bits are added to the data bits at the time of transmitting the signal. These extra bits are known as **parity bits**. These parity bits detect or sometimes correct the errors. In fact, the data bits alongwith the parity bits form a code word.

8.1.2. Classification of Codes

The codes are basically classified as under:

1. Error detecting codes.
2. Error correcting codes.
   (i) The error detecting codes are capable of only detecting the errors. They cannot correct the errors.
   (ii) The error correcting codes are capable of detecting as well as correcting the errors.
3. Error correction techniques.

8.1.3. Classification of Errors

The errors introduced in the transmitted data during their transmission may be categorized as under:

(i) Content errors
(ii) Flow integrity errors.

The content errors are nothing but errors in the contents of a message i.e., a 0 may be received as 1 or vice versa. Such errors are introduced due to noise added into the data signal during its transmission. Flow integrity errors means missing blocks of data. It is possible that a data block may be lost in the network as it has been delivered to a wrong destination.

8.1.4. Error Detection Techniques

Some of the popular error detection techniques may be listed as under:

(i) Parity checking
(ii) Check sum error detection
(iii) Cyclic redundancy check.

8.1.5. Error Correction Techniques

(i) The error correction techniques are based on generation of code words at the transmitter.
(ii) The coded words contain the data bits and check bits.

8.2 GENERATION AND DETECTION OF CODED SIGNALS

Error control for data integrity can be achieved by using a technique called Forward Error Correction (FEC). Figure 8.2 shows the model of a digital communication system using the FEC approach.

The discrete source generates information in the form of binary symbols. The channel encoder accepts these message bits and add redundancy (additional bits) according to a prescribed rule. We get the encoded data at the encoder output.

Important Point: The bit rate of this encoded data is higher than that of the source.

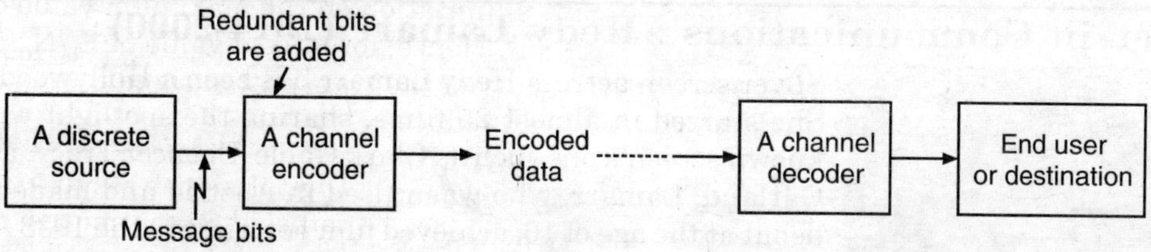

Fig. 8.2. Illustration of generation and detection of coded signals.

The channel decoder takes the help of the additional bits (i.e., redundancy) to decide whether the transmitted signal is a 1 or a 0. Thus, the combined objective of channel encoder and decoder is to minimize the effect of channel noise.

8.2.1. Drawbacks of Coding Technique

Some of the drawbacks of the coding techniques may be listed as under:

(i) An increased transmission bandwidth is required in order to transmit the encoded signal. This is because of the additional bits (i.e., redundancy) which is added by the encoder.

(ii) By the use of coding technique, the communication system becomes quite complex.

> **DO YOU KNOW?**
>
> The task facing the designer of a digital communication system is that of providing a cost-effective facility for transmitting information from one end of the system at a rate and a level of reliability and quality that are acceptable to a user at the other end.

8.2.2. Classification of Error-Correcting Codes

There are many different error correcting codes. These code words can be classified into two categories as under:

(i) Block codes and
(ii) Convolutional codes

The major difference between these codes is that the block codes do not need memory whereas the convolutional codes need memory.

Another way of classifying codes is:

(i) Linear codes,
(ii) Non-linear codes.

Important Point: *Linear codes have an important property that any two code words of a linear code can be added in modulo-2 adder to produce a third code word in the code. On the other hand, non-linear codes do not exhibit such a property. All the practically used codes are linear codes.*

8.3 OVERVIEW OF ERROR CONTROL CODING TECHNIQUES

Figure 8.3 shows the overview of error control coding techniques.

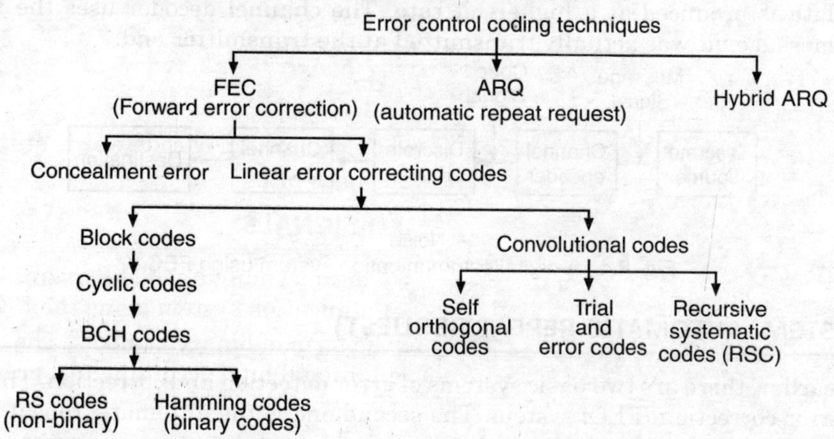

Fig. 8.3. Overview of error control coding techniques

Pioneer in Communications : Hedy Lamarr (1914-2000)

Silver screen actress Hedy Lamarr has been a Hollywood legend. She starred in almost 25 films, sharing the spotlight with well-known screen icons such as Clark Gable, Spencer Tracy, and Judy Garland. Lamarr, who was raised in Austria and made her film debut at the age of 16, achieved film notoriety in the 1933 Austrian art film Ecstacy, where she appeared briefly in the nude.

Known more for her looks than her acting skills, Lamarr featured in several sciences in which she simply stood around silently and looked beautiful.

Although she was best known for her Hollywood work, Lamarr was also credited with developing the concept of spread spectrum. In 1941, Lamarr and composer George Antheil received a patent for a secret communications system.

8.3.1. Types of Error Control

In this chapter, let us discuss two completely different approaches for the error control. They are as under:

(i) Automatic repeat request technique (ARQ).

(ii) Forward error correction (FEC).

(i) Automatic Repeat Request (ARQ) Technique

In an ARQ technique, the receiver can request for the retransmission of the complete or a part of message if it finds some error in the received message. This requires an additional channel called feedback channel to send the receiver's request for retransmission.

(ii) The FEC Technique

In an FEC technique, there is no such feedback path and these is no request is made for retransmission. Most of the coding techniques described in this chapter are related to FEC.

8.3.2. Forward Error Correction (FEC) Technique for Error Control

Error control to ensure the data integrity can be exercised by using the forward error correction technique (FEC). A digital modulation system which uses this technique is shown in figure 8.4. In FEC technique, the discrete source generates information in the form of binary symbols. The channel encoder accepts these message bits and will add redundancy according to the prescribed rule. Thus, the encoded data is produced at a higher bit rate. The channel decoder uses the redundancy to decide which message bit was actually transmitted at the transmitter end.

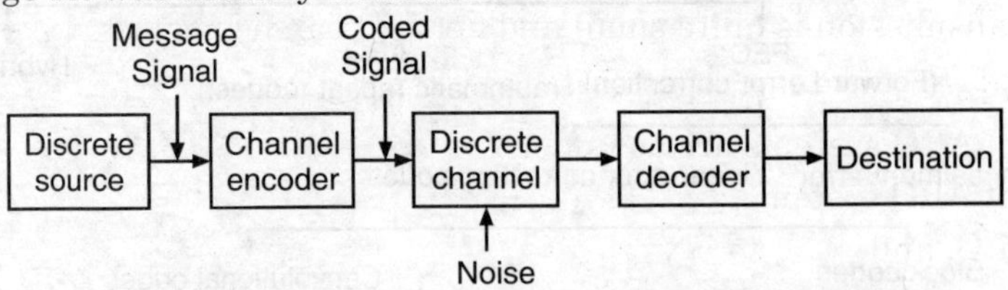

Fig. 8.4. A digital communication system using FEC

8.4 ARQ SYSTEM (AUTOMATIC REPEAT REQUEST)

As discussed earlier, there are two basic systems of error detection and correction. The first on being the forward error correction (FEC) system. The second one is the automatic repeat request (ARQ) system. In an ARQ system of error control, when an error is detected, a request is made for the retransmission of that signal. Therefore, a feedback channel is required for sending the request for retransmission. The ARQ systems differ from the FEC systems in three important respects. They are as under:

(i) In ARQ system less number of check bits (parity bits) are required to be sent. This increases the (k/n) ratio for an (n, k) block code if transmitted using the ARQ system.

(ii) A return transmission path and additional hardware in order to implement repeat transmission of codewords is required.

(iii) The bit rate of forward transmission must make allowance for the backward repeat transmission.

8.4.1. Basic ARQ System

The block diagram of a basic ARQ system is shown in figure 8.5.

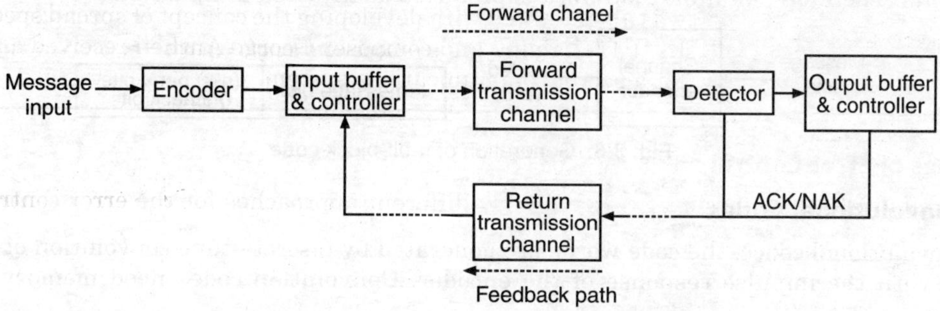

Fig. 8.5 Block diagram of a basic ARQ system.

Working Operation

The encoder produces codewords for each message signal at its input. Each codeword at the encoder output is stored temporarily and transmitted over the forward transmission channel. At the destination, a decoder decodes the code words and look for errors. The decoder will output a positive acknowledgment (ACK) if no errors are detected and it will output a negative acknowledgment (NAK) if errors are detected. On receiving a negative acknowledgment (NAK) signal via the return transmission path, the controller retransmits the appropriate word from the words stored by the input buffer. A particular word may be retransmitted only once or it may be retransmitted twice or more number of times. The output controller and buffer on the receiver side assemble the output bit stream from the code words accepted by the decoder.

Probability of Error on the Return Path

The bit rate of the return transmission which involves the return transmission of ACK/NAK signal is low as compared to the bit rate of the forward transmission. Therefore, the probability of error of the return transmission is quite small and hence it can be neglected.

Types of ARQ System

The three types of ARQ systems are as under:

(i) Stop-and-wait ARQ system

(ii) Go back N ARQ, and

(iii) Selective repeat ARQ.

8.5 CLASSIFICATION OF CODES

This shows that an error control code falls under one of the two basic types as under:

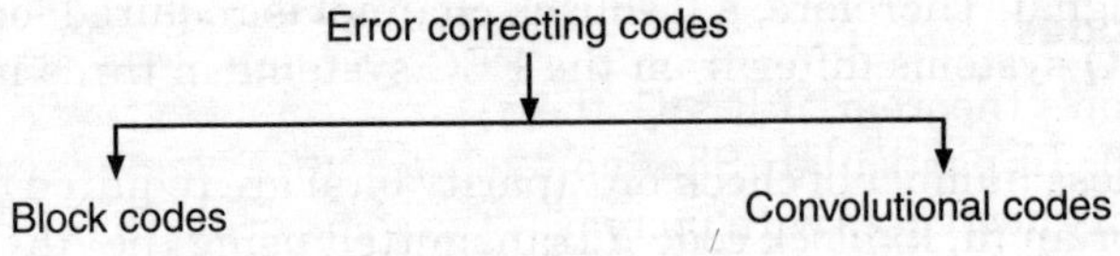

8.5.1. Block Codes

Figure 8.6 illustrates the generation of block codes. To generate an (n, k) block code, the channel encoder accepts the information in successive k bit blocks. At the end of each such block (of k message bits), it adds (n – k) parity bits as shown in figure 8.6. Since, these bits do not consist of any information, therefore, they are known as redundant bits.

It is important to note that the (n – k) parity bits are related algebraically depicted to the k message bits. Therefore, an n-bit code word is produced as depicted in figure 8.6.

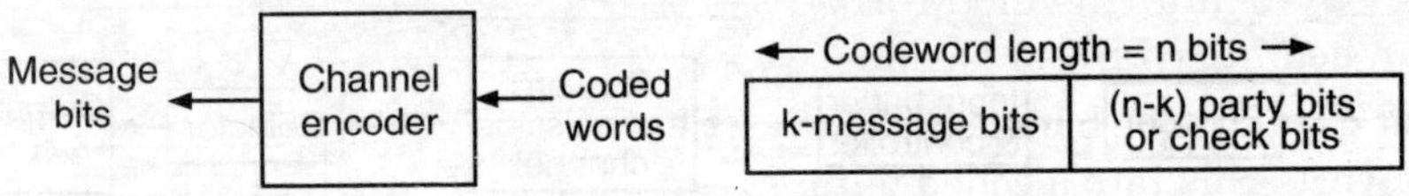

Fig. 8.6. Generation of n-bit-block code

8.5.2. Convolutional Codes

In convolutional codes, the code words are generated by discrete-time convolution of the input sequence with the impulse response of the encoder. Convolution codes need memory for their generation.

The encoder of a convolutional code operates on the incoming message sequence using a sliding window equal to the duration of its own memory. Thus in convolutional code unlike a block code, the channel encoder accepts message bits as a continuous sequence and generates a continuous sequence of encoded bits at its output.

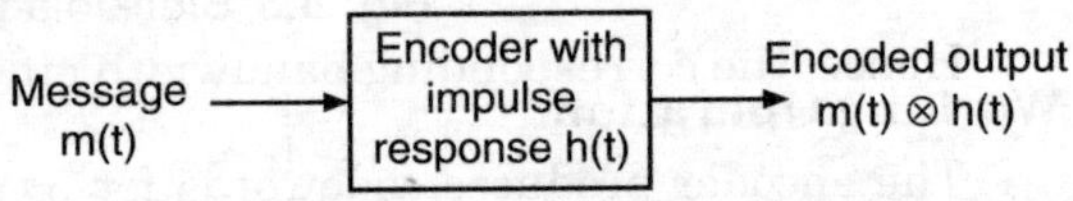

Fig. 8.7. Convolutional codes.

> **Important Point:** The major difference between the block codes and convolution codes is that the convolutional codes require memory whereas the block codes do not require any memory. Both types of codes are linear codes.

8.6 TRANSMISSION ERRORS: RANDOM AND BURST ERRORS

The transmission errors, basically, are of two types namely the **random errors** and **burst errors**. The errors that occurs in a purely random manner are known as **random errors**, whereas the errors which occur in the form of bunches and hence which the not independent are known as the **burst errors**. Codes like binary BCH codes (Bose, Chaudhari Hocquenghem code) are useful in dealing with the random errors. Because of impulse noise, more than one corresponding data bits change their state simultaneously, The error introduced in the received word is called as **burst error**. The convolutional error correcting codes are not effective in correcting the burst errors. Therefore, special codes are developed for detecting and correcting the burst errors. The best known codes of this category are the Fire Codes. These codes are a subclass of cyclic codes and they are named after P. Fire who discovered them.

> **DO YOU KNOW?**
>
> Sampling detection is optimal only if the received waveform is noise free.

8.6.1. Interleaving: Definition

Interleaving is a technique used as an alternative technique for correcting the burst errors.

8.6.2. Concatenated Codes

According to Shannon's theorem, if $R < C$, then it is possible to find a code such that the error probability is very small. In particular, Shannon showed that there exist certain block codes for which the error probability decreases with increase in the block length (n). These certain block codes are practical methods to obtain very long codes without increasing the decoder complexity.

8.7 POWER AND BANDLIMITED CHANNELS

8.7.1. Classical Error Control System

Figure 8.8 shows the classical error control system. It uses the conventional FEC system followed by a multilevel (M-ary) modulator. It can be M-ary PSK or QAM.

Let the input data rate be b bits/sec. The FEC encoder converts this into a coded data which has a rate b/R where R is the information rate.

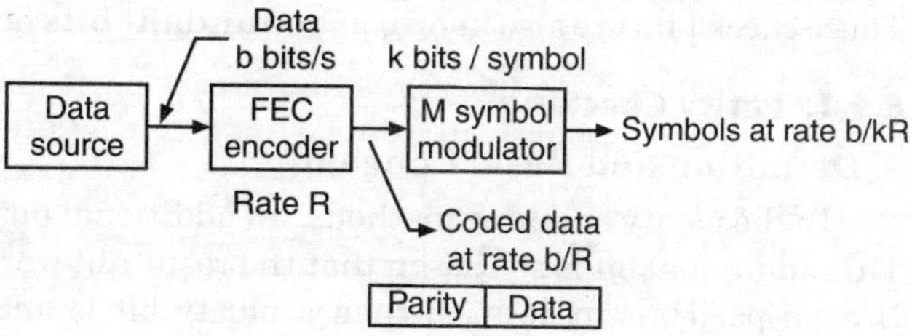

Fig. 8.8. Classical error control

The M symbol modulator converts the encoder output into M possible symbol constellation. The symbol rate at the output of the modulator is $r_s = b/kR$ baud.

Hence, the minimum system bandwidth required to transmit this signal successfully is,

$$B = r_s = (b/kR) \text{ Hz} \quad \text{... (8.1)}$$

Hence, the corresponding bandwidth efficiency is given by

$$\eta = \frac{b}{r_s} = kR = R \log_2 M \text{ bits/Hz.}$$

The use of FEC reduces the bandwidth efficiency of the system by a factor R.

8.7.2. Types of Channels

There are two types of channels namely:

(i) Bandlimited channel

(ii) Power limited channel

8.7.3. Bandlimited Channels

These channels have a fixed finite bandwidth. Therefore, signal which requires larger bandwidth cannot be transmitted over such channels without distorting them. The examples of bandlimited channels are telephone lines, parallel lines etc. When we add FEC to the bandlimited channels, then due to the additional parity bits, the data rate is reduced.

8.7.4. Power Limited Channels

The power limited channels have a limited power associated with them but have a large bandwidth. The examples of such channel is a satellite channel. Therefore, it is possible to accommodate the FEC inspite of increased data rate associated with it.

8.8 DISCRETE MEMORYLESS CHANNEL (DMC)

We have already discussed the meaning and significance of discrete memoryless channel (DMC) in previous chapter.

8.9 ERROR DETECTION METHODS

Before thinking of correcting the error introduced in the data bits, it is essential to first detect them. Some of the popular detection methods are as under:

(i) Parity checking
(ii) Check sum error detection
(iii) Cyclic Redundancy Check (CRC)

For error detection and correction, it is essential to add some check bits to a block of data bits. These check bits are also known as **redundant bits** because they do not carry any useful information.

8.9.1. Parity Checking

1. Definition and Basic Concept

In the parity checking methods, an additional bit called as parity bit is added to each data word. This additional bit is so chosen that the resultant word will have either an even parity or odd parity. If even parity is to be used then a parity bit is added to make the total number of 1 bits even. Similarly, for odd parity the total number of 1 bits is made odd by adding the parity bit. The examples of even and odd parity are as shown in figure 8.9 (a) and figure 8.9 (b).

P	Data Word
0	1001011

→ **Even parity,** as total number of 1 bits' in the data word is 4 i.e. even Therefore, parity bit P is 0.

P	Data Word
1	0010011

→ **Even parity.** The total number of 1 bits' in the data word is 3 originally. Add a parity bit P of 1 to make the total number of 1 bits to be 4 i.e. even.

(a) Inclusion of a parity bit to obtain the even parity

P	Data Word
1	1001011

→ **Odd parity.** The total number of 1 bits' in the data word is 4 i.e. even. To make it odd, the parity bit P = 1.

P	Data Word
0	1000110

→ **Odd parity.** The total number of 1 bits' in the data word is 3 i.e. odd. Therefore, parity bit P = 0.

(b) Inclusion of a parity bit to obtain the odd parity

Fig. 8.9.

2. Error Detection

The parity checking at the receiver can detect the presence of an error if the parity of the received signal is different from the expected parity. This means that if it is known that the parity of the transmitted signal is always going to be even and if the received signal has an odd parity then the receiver can conclude that the received signal is not correct. This has been shown in figure 8.10.

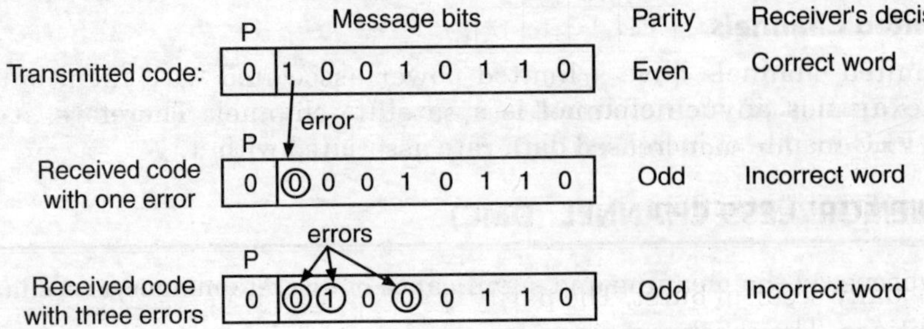

Fig. 8.10. The receiver detects the presence of error if the number of errors is odd i.e. 1, 3, 5.

When a single error or an odd number of errors occur during transmission the parity of the code word changes. Parity of the received code word is checked at the receiver and change in parity indicates that error is present in the received word. This is shown in figure 8.10.

3. Failure of Parity Checking Method

If the number of errors introduced in the transmitted code is two or any even number, then the parity of the received code word will not change. It will still remain even as shown in figure 8.11 and the receiver will fail to detect the presence of errors.

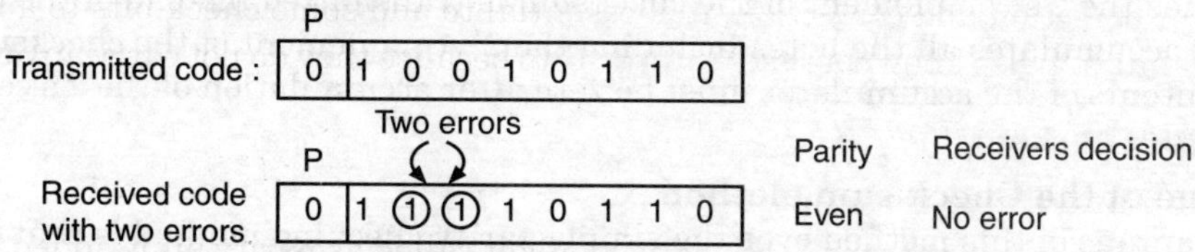

Fig. 8.11. The receiver cannot detect the presence of error if the number of errors is even i.e. 2, 4, 6.

4. Conclusions

(i) Double or any even number of errors in the received word does not change the parity. Therefore even number of errors will be unnoticed.

(ii) If one or odd number of errors occur then the parity of the received word will be different from the parity of transmitted signal. This error is noticed. However, this error cannot be corrected.

DO YOU KNOW?

One practical motivation for the use of coding is to reduce the required E_b/N_0 for a fixed bit error rate. This reduction in E_b/N_0 may, in turn, be exploited to reduce the required transmitted power or reduce the hardware costs by requiring a smaller antenna size in the case of radio communications.

5. Limitations of parity checking

(i) Thus, the simple parity checking method has its limitations. It is not suitable for detection of multiple errors (two, four, six etc.).

(ii) The other limitation of parity checking method is that it cannot reveal the location of errorneous bit. It cannot correct the error either.

EXAMPLE 8.1. Write the ASCII code of word HOLE using even parity.

Solution: The following table shows the ASCII code of word HOLE using even parity.

Table 8.1.

	P	7	6	5	4	3	2	1
H	0	1	0	0	1	0	0	0
O	1	1	0	0	1	1	1	1
L	1	1	0	0	1	1	0	0
E	1	1	0	0	0	1	0	1

Important Point: It may be noted that the parity bits are selected in order to obtain an even parity for each row (*i.e.* for each letter).

8.9.2. Check sum Error Detection

1. Definition

The errors usually occur in burst. The parity check method is not useful in detecting the errors under such conditions. The checksum error detection method can be used successfully in detecting such errors. In this method, a check sum is transmitted along with every block of data bytes. In this method an eight bit accumulator to used to add 8 bit bytes of a block of data to find the checksum byte. The carries of the MSB are ignored while finding out the checksum byte.

2. Detection of Error using the Check sum Byte

After transmitting a block of data bytes (say 8-data bytes) the check sum byte is also transmitted. The check sum byte is regenerated at the receiver separately by adding the received bytes. The regenerated check sum byte is then compared with the transmitted one. If both are identical then there is no error. If they are different then the errors are present in the block of received data bytes.

Sometimes the 2's complement of the checksum is transmitted instead of the check sum itself. The receiver accumulates all the bytes including the 2's complement of the checksum. If there is no error, the contents of the accumulator must be zero after accumulation of the 2's complement of the check sum byte.

3. Advantage of the Check sum method

The advantage of this method over the simple parity checking method is that the data bits are mixed up due to the 8 bit addition. Therefore, checksum represents the overall data block. In check sum, therefore, there is 255 to 1 chance of detecting random errors.

8.9.3. Error Detection using VRC and LRC

1. Basic Concept

The concept of parity checking has been discussed earlier. This concept can be extended for detection and correction of single errors. In order to do so, the data block is arranged in a rectangular matrix form as shown in figure 8.12, and we generate two sets of parity bits. They are as under:

(i) Longitudinal Redundancy Check (LRC)

(ii) Vertical Redundancy Check (VRC)

	Characters	C	O	M	P	U	T	E	R	
7 bit ASCII Codes (Message bits)	b_1	1	1	1	0	1	0	1	0	1
	b_2	1	1	0	0	0	0	0	1	1
	b_3	0	1	1	0	1	1	1	0	1
	b_4	0	1	1	0	0	0	0	0	0
	b_5	0	0	0	1	1	1	0	1	0
	b_6	0	0	0	0	0	0	0	0	0
	b_7	1	1	1	1	1	1	1	1	0
VRC bits (even parity)		1	1	0	0	0	1	1	1	1

These bits make the parity of each row even. (LRC bits (even parity))

These bits make the parity of each column even (VRC bits).

Fig. 8.12.

2. The Vertical Redundancy Check (VRC) Bits

As shown in figure 8.13, the VRC bits are parity bits associated with the ASCII code of each character. Each VRC bit makes the parity of its corresponding column an even parity. For example, we consider column 1 corresponding to character C. The ASCII code for the character C is,

	b_1						b_7	b_8
C =	1	1	0	0	0	0	1	1

← VRC bit to make parity even.

Fig. 8.13.

Therefore, the eighth bit b_8 which is a VRC bit is made 1 to make the parity even.

3. The Longitudinal Redundancy Check (LRC) Bits

The LRC bits are parity bits associated with the rows of the data block of figure 8.14. Each LRC bit makes the parity of the corresponding row, an even parity. For example, we consider row 1 of figure 8.14.

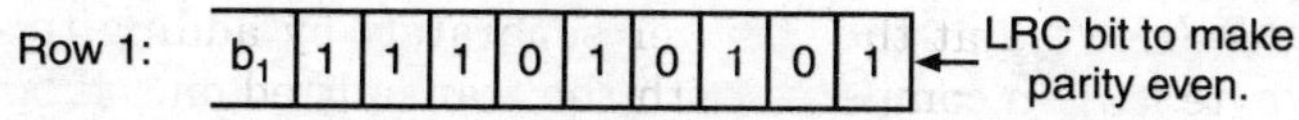

Fig. 8.14.

4. Signal Transmission

The ASCII characters of figure 8.15 are transmitted in the following sequence:

5. Identification of bits in Error

Even a single error in any bit will result in a non-correct LRC in one of the rows and an incorrect VRC in one of the columns. The bit which is common to the row and column is the bit in error. However, there is still a limitation on the Block parity code, which is that, multiple errors in rows and columns can be only detected but they cannot be corrected. This is because, it is not possible to locate the bits which are in error. This will be clear by the following example.

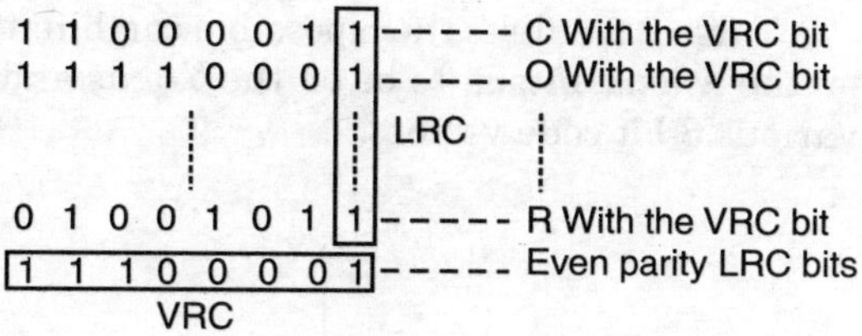

Fig. 8.15.

EXAMPLE 8.2. The following bit stream is encoded using VRC, LRC and even parity. Locate and correct the error if it is present.

11000011 11110011 10110010 00001010
00101010 00101011 10100011 01001011
11100001

Solution:

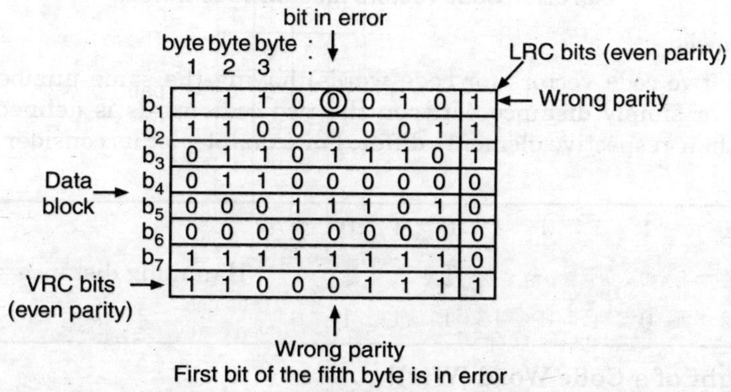

Fig. 8.16.

Figure 8.16 shows the received data block alongwith the LRC and VRC' bits. It may be noted that the parity bits corresponding to row 1 and column 5 indicate wrong parity. Therefore, the fifth bit in the first row (encircled bit) is incorrect. Thus, using VRC and LRC, it is possible to locate and correct the bits in error.

8.10 IMPORTANT DEFINITIONS RELATED TO CODES

1. Code Word

The code word is the n bit encoded block of bits. As already discussed, it contains message bits and parity or redundant bits.

2. Code Rate

The code rate is defined as the ratio of the number of message bits (k) to the total number of bits (n) in a code word.

$$\text{Therefore, Code rate (r)} = \frac{k}{n} \quad \text{... (8.2)}$$

3. Code Vectors

In previous article, we have used the code bits and code vectors as similar terms. Let us now define the code vectors. We can visualize an n-bit code word to be present in an n-dimensional space. The coordinates or elements of this code vector are the bits present in the code word. We consider figure 8.17 which shows the 3-bit code vectors.

Table 8.2 enlists the 8 possible combinations of the 3-bit code word. We can assume bits x_0 to be on the X-axis, bits x_1 to be on the Y-axis and bits x_2 to be on the Z-axis. Then figure 8.17 shows the various 3-bit code vectors.

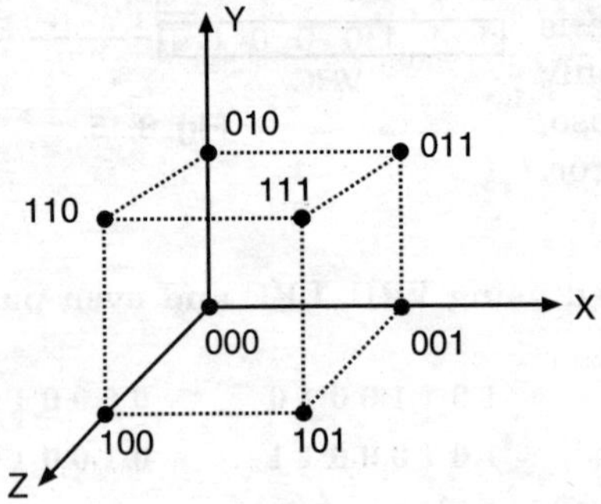

Table 8.2

S. No.	Code word		
	$x_2 = Z$	$x_1 = Y$	$x_0 = X$
0	0	0	0
1	0	0	1
2	0	1	0
⋮	⋮	⋮	⋮
7	1	1	1

Fig. 8.17. Code vectors for 3-bit code words.

4. Hamming Distance

Let us consider two code vectors (or code words) having the same number of elements. The Hamming distance or simply distance between the two code words is defined as the number of locations in which their respective elements differ. For example, let us consider the two code words given below:

code word No. 1	1 1 0 1 0 1 0 0	
	↕ ↕ ↕	Hamming distance = 3
code word No. 2	0 1 0 1 1 1 1 0	

5. Hamming Weight of a Code Word [W(x)]

The Hamming weight of a code word x is defined as the number of non-zero elements in the code word. Hamming weight of a code vector (code word) is the distance between that code word and an all-zero code vector (A code having all elements equal to zero).

6. Code Efficiency

The code efficiency is defined as the ratio of message bits to the number of transmitted bits per block.

$$\text{Therefore, code efficiency} = \text{code rate} = \frac{k}{n} \quad \text{... (8.3)}$$

EXAMPLE 8.3. Find the Hamming weight of the following code vector:

x = 1 1 0 1 0 1 0 0

Solution: As the number of non-zero elements in the above code word is 4, the Hamming weight W(x) = 4.

Minimum distance d_{min}

The minimum distance d_{min} of a linear block code is defined as the smallest Hamming distance between any pair of code vectors in the code.

Therefore, the minimum distance is same as the smallest Hamming weight of difference between any pair of code vectors. It can be proved that the minimum distance of a linear block code is the smallest Hamming weight of the non-zero code vectors in the code.

DO YOU KNOW?

The use of error-control coding adds complexity to the system, especially, for the implementation of decoding operations in the receiver.

Rate of d_{min} in Error detection and Correction

The error detection is always possible when the number of transmission errors in a code word is less than the minimum distance d_{min} because then the erroneous word is not a valid code word. But when the number of errors equals or exceeds d_{min}, the erroneous code word may correspond to another valid code word and errors cannot be detected. The error detection and correction capabilities of a coding technique depend on the minimum distance d_{min} as shown in Table 8.3.

Table 8.3. Error detection and correction capabilities

S.No.	Description	Expression
1.	Detect upto s errors per word	$d_{min} \geq (s + 1)$
2.	Correct upto t errors per word	$d_{min} \geq (2t + 1)$
3.	Correct upto t errors and detect s > t errors per word	$d_{min} \geq (t + s + 1)$

EXAMPLE 8.4. For a Hamming distance of 5, how many errors can be detected ? How many errors can be corrected ?

Solution: Let us consider table 8.3. Assuming minimum Hamming distance i.e. $d_{min} = 5$.

(i) Number of errors that can be detected, (s) can be obtained from

$$(s + 1) \leq d_{min}$$

or $$s \leq 5 - 1$$

or $$s \leq 4.$$ **Ans.**

Thus, at the most 4 errors can be detected.

(ii) Number of errors that can be corrected (t) can be obtained from

$$(2t + 1) \leq d_{min}$$

or $$t \leq 2.$$ **Ans.**

Thus, at the most, 2 errors can be corrected.

8.11 LINEAR BLOCK CODES

1. Now, let us discuss the linear block codes in detail and obtain various mathematical expressions related to these codes.
2. We consider an (n, k) linear block code in which k number of bits are identical to the message sequence which is to be transmitted.
3. The remaining (n – k) number of bits are called as the generalized parity check bits or simply parity bits.
4. These parity bits are computed from the message bits, according to the prescribed encoding rule which decides the mathematical structure of the code.

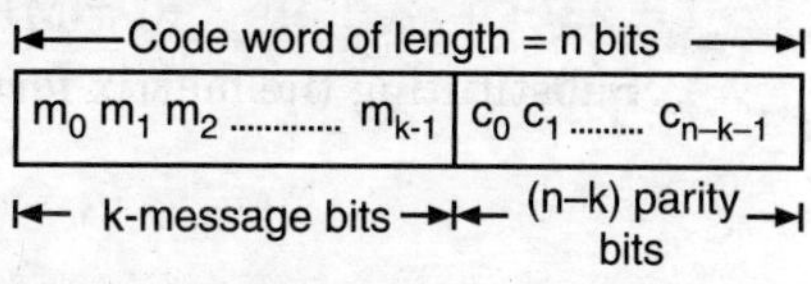

Fig. 8.18. Structure of the code word for a linear block code

5. Let us consider figure 8.18, which shows the structure of a code word.
6. A code word consists of k message bits which are denoted by $m_0, m_1, ..., m_{k-1}$ and (n – k) parity bits denoted by $c_0, c_1, ..., c_{n-k-1}$.
7. The sequence of message bits is applied to a linear block encoder to produce an n bit code word. The elements of this code word are $x_0, x_1, ..., x_{n-1}$.
8. As shown in figure 8.18, the first k bits of the code word are identical to the corresponding parity bits (c_0, c_1....). We can express this mathematically as under:

$$x_i = \begin{cases} m_i & \text{for } i = 0,1, \quad ....k-1 \\ c_{i-k} & \text{for } i = k, k+1, ...n-1 \end{cases} \quad ... (8.4)$$

9. The (n– k) parity bits are linear sums of the k message bits as will be discussed later on.
10. The code vector represented by equation (8.4) can be mathematically represented as:

$$X = [M : C] \quad ...(8.5)$$

where M = k - message vectors

and C = (n – k), parity vectors

11. A block code generator generates the parity vectors (or parity bits) required to be added to the message bits to generate the code words. The code vector X can also be represented as under :

$$X = MG \quad ... (8.6)$$

where X = Code vector of 1 × n size

M = Message vector of 1 × k size

and G = Generator matrix of k × n size.

Equation (8.6) can be represented in the matrix form as under:

$$[X]_{1\times n} = [M]_{1\times k}\,[G]_{k\times n} \quad ... (8.7)$$

12. The generator matrix is dependent on the type of linear block code used. The generator matrix is generally represented as under:

$$[G] = [I_k \mid P] \quad ... (8.8)$$

where I_k = k × k identity matrix

and P = k × (n – k) coefficient matrix.

Therefore,

$$I_k = \begin{bmatrix} 1 & 0 & ... & 0 \\ 0 & 1 & ... & 0 \\ \vdots & \vdots & & \vdots \\ 0 & 0 & & 1 \end{bmatrix}_{k\times k}$$

and

$$P = \begin{bmatrix} P_{00} & P_{10} & ... & P_{n-k,0} \\ P_{01} & P_{11} & ... & P_{n-k,1} \\ \vdots & \vdots & & \vdots \\ P_{0,k-1} & P_{1,k-1} & & P_{n-k,k-1} \end{bmatrix}_{k\times(n-k)} \quad ... (8.9)$$

13. The parity vector can be obtained as under:

$$C = MP \quad ... (8.10)$$

Substituting the matrix form, we obtain

$$[c_0, c_1 ... c_{n-k}]_{1\times n-k} = [m_0, m_1 ... m_k]_{1\times k} \begin{bmatrix} P_{00} & P_{10} & ... & P_{n-k,0} \\ P_{01} & P_{11} & ... & P_{n-k,1} \\ \vdots & \vdots & & \vdots \\ P_{0,k-1} & P_{1,k-1} & & P_{n-k,k-1} \end{bmatrix}_{k\times(n-k)} \quad ... (8.11)$$

By solving the above matrix equation, we get the parity vectors as under:

$$c_0 = m_0 P_{00} \oplus m_1 P_{01} \oplus m_2 P_{02} \oplus \ldots \oplus m_k P_{0,k-1}$$
$$c_1 = m_0 P_{10} \oplus m_1 P_{11} \oplus m_2 P_{12} \oplus \ldots \oplus m_k P_{1,k-1} \qquad \ldots (8.12)$$
$$c_2 = m_0 P_{20} \oplus m_1 P_{21} \oplus m_2 P_{22} \oplus \ldots \oplus m_k P_{2,k-1}$$

Similarly, we can obtain the expressions for the remaining parity bits.

8.12 HAMMING CODES

Hamming codes are linear block codes. The family of (n, k) Hamming codes for $q \geq 3$ is defined by the following expressions:

1. Block diagram: $n = 2^q - 1$
2. Number of message bits: $k = 2^q - q - 1$ ... (8.13)
3. Number of parity bits: $(n - k) = q$.
   where $q \geq 3$. i.e., minimum number of parity bits is 3.
4. The minimum distance $d_{min} = 3$.

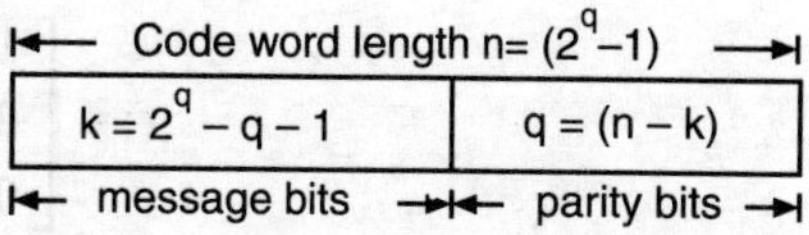

Fig. 8.19 Code word structure of Hamming code.

5. The code rate or code efficiency $= \dfrac{k}{n} = \dfrac{2^q - q - 1}{2^q - 1} = 1 - \dfrac{q}{2^q - 1}$ ... (8.14)

   If $q >> 1$, then code rate $r \cong 1$.

The general structure of Hamming code word is shown in figure 8.19.

8.12.1. Error Detection and Correction Capabilities of Hamming Code

Considering table 8.3 for the minimum distance, we have, $d_{min} = 3$,

1. The number of errors that can be detected per word = 2.
   since $d_{min} \geq (s + 1)$, therefore, $3 \geq s + 1$ or $s \leq 2$.
2. The number of errors that can be corrected per word = 1
   since $d_{min} \geq (2t + 1)$, therefore, $3 \geq (2t + 1)$ or $t \leq 1$
   Thus, with $d_{min} = 3$, it is possible to detect upto 2 errors and it is possible to correct upto only 1 error.

There is another way of expressing the relationship between the message bits and the parity check bits, of a linear block code. Let H denote an $(n - k) \times n$ matrix defined as under:

$$H = [P^T \mid I_{n-k}] \qquad \ldots (8.15)$$

where P^T = An $(n - k) \times k$ matrix representing the transpose of the coefficient matrix P and I_{n-k} is an $(n - k) \times (n - k)$ identity matrix.

The transpose of the coefficient matrix can be obtained by interchanging the rows and columns of the coefficient matrix P given by equation (8.9), i.e.,

$$P^T = \begin{bmatrix} P_{00} & P_{01} & \ldots & P_{0,k-1} \\ P_{10} & P_{11} & \ldots & P_{1,k-1} \\ \vdots & \vdots & & \vdots \\ P_{n-k,0} & P_{n-k,1} & & P_{n-k,k-1} \end{bmatrix}_{(n-k)\times k} \qquad \ldots (8.16)$$

Therefore, the matrix H is given by,

$$H = \left[\begin{array}{cccc|cccc} P_{00} & P_{01} & \cdots & P_{0,k-1} & 1 & 0 & 0 \cdots & 0 \\ P_{10} & P_{11} & \cdots & P_{1,k-1} & 0 & 1 & 0 \cdots & 0 \\ P_{20} & P_{21} & \cdots & P_{2,k-1} & 0 & 0 & 1 \cdots & 0 \\ \vdots & \vdots & & \vdots & \vdots & \vdots & \vdots & \vdots \\ P_{n-k,0} & P_{n-k,1} & \cdots & P_{n-k,k-1} & 0 & 0 & 0 \cdots & 1 \end{array}\right]_{(n-k)\times k} \quad \ldots (8.17)$$

The matrix H is called as the **parity check matrix**. If generator matrix G has been given, then we can obtain the parity check matrix and vice-versa.

EXAMPLE 8.5. The generator matrix for a (6, 3) block code is given below. Find all the code vectors of this code.

$$G = \left[\begin{array}{ccc|ccc} 1 & 0 & 0 & 0 & 1 & 1 \\ 0 & 1 & 0 & 1 & 0 & 1 \\ 0 & 0 & 1 & 1 & 1 & 0 \end{array}\right]$$

Solution: The general pattern of the block codes is (n, k), hence, in this case, n = 6 and k = 3. This means that the message block size k is 3 and the length of code vector i.e. n is 6. For obtaining the code vectors, we shall follow the steps given below:

(i) First, we separate out the identity matrix I and coefficient matrix

We know that the generator matrix is given by:

$$G = [I_k \mid P]$$

Comparing this with the given generator matrix, we obtain

$$I_k = I_{3\times 3} = \begin{bmatrix} 1 & 0 & 0 \\ 0 & 1 & 0 \\ 0 & 0 & 1 \end{bmatrix}$$

and

$$P = P_{3\times 3} = \begin{bmatrix} 0 & 1 & 1 \\ 1 & 0 & 1 \\ 1 & 1 & 0 \end{bmatrix}$$

As the size of message block is k = 3, there are eight possible message blocks: (0, 0, 0) (0, 0, 1), (0, 1, 0), (0, 1, 1) (1, 0, 0), (1, 0, 1), (1, 1, 0), (1, 1, 1).

Now, let us obtain the relation between parity vectors B, message vectors M and the coefficient matrix P as under:

$$[c_0, c_1, c_2] = [m_0, m_1, m_2] \begin{bmatrix} 0 & 1 & 1 \\ 1 & 0 & 1 \\ 1 & 1 & 0 \end{bmatrix} \quad \ldots \text{(i)}$$

Now, let us obtain parity words for all the message words.

The solution of equation (i) is given by,

$$c_0 = (m_0 \times 0) \oplus (m_1 \times 1) \oplus (m_2 \times 1) = m_1 \oplus m_2 \quad \ldots \text{(ii)}$$

$$c_1 = (m_0 \times 1) \oplus (m_1 \times 0) \oplus (m_2 \times 1) = m_0 \oplus m_2 \quad \ldots \text{(iii)}$$

$$c_2 = (m_0 \times 1) \oplus (m_1 \times 1) \oplus (m_2 \times 0) = m_0 \oplus m_1 \quad \ldots \text{(iv)}$$

By substituting the values of m_0, m_1 and m_2 into these equations, it is possible for us to obtain the parity bits c_0, c_1 and c_2.

1. For the message word $(m_0, m_1, m_2) = (0\ 0\ 0)$, we have

$$c_0 = 0 \oplus 0 = 0$$

$$c_1 = 0 \oplus 0 = 0$$
$$c_2 = 0 \oplus 0 = 0$$

Therefore, $(c_0, c_1, c_2) = (0, 0, 0)$

The complete code word for this message word is given by,

	m_0	m_1	m_2	c_0	c_1	c_2
Code word =	0	0	0	0	0	0
	message			parity		

DO YOU KNOW?

The two key system parameters available to the designer are transmitted signal power and channel bandwidth. These two parameters, together with the power spectral density of receiver noise, determine the signal energy per bit-to-noise power spectral density ratio E_b/N_0.

2. For the second message vector, i.e., $(m_0, m_1, m_2) = (0, 0, 1)$, we have

$$c_0 = 0 \oplus 1 = 1$$
$$c_1 = 0 \oplus 1 = 1$$
$$c_2 = 0 \oplus 0 = 0$$

Therefore, $(c_0, c_1, c_2) = (1, 1, 0)$

The complete code word for this message word is given by,

	m_0	m_1	m_2	c_0	c_1	c_2
Code word =	0	0	1	1	1	0

3. Similarly, we can obtain the code words for the remaining message words. All these code words have been given in Table 8.4

Table 8.4. Code vectors for (6, 3) block code

S.No.	Message vectors			Parity bits			Complete code vector					
	m_0	m_1	m_2	c_0	c_1	c_2	m_0	m_1	m_2	c_0	c_1	c_2
1.	0	0	0	0	0	0	0	0	0	0	0	0
2.	0	0	1	1	1	0	0	0	1	1	1	0
3.	0	1	0	1	0	1	0	1	0	1	0	1
4.	0	1	1	0	1	1	0	1	1	0	1	1
5.	1	0	0	0	1	1	1	0	0	0	1	1
6.	1	0	1	1	0	1	1	0	1	1	0	1
7.	1	1	0	1	1	0	1	1	0	1	1	0
8.	1	1	1	0	0	0	1	1	1	0	0	0

Now let us obtain the code words for a (7, 4) Hamming code.

EXAMPLE 8.6. The parity check matrix of a particular (7, 4) linear block code is given by,

$$[H] = \begin{bmatrix} 1 & 1 & 1 & 0 & 1 & 0 & 0 \\ 1 & 1 & 0 & 1 & 0 & 1 & 0 \\ 1 & 0 & 1 & 1 & 0 & 0 & 1 \end{bmatrix}$$

(i) Find the generator matrix (G).
(ii) List all the code vectors
(iii) What is the minimum distance between the code vectors ?
(iv) How many errors can be detected ? How many errors can be corrected ?

Solution: First, let us obtain the P^T matrix.

P^T is the transpose of the coefficient matrix P. The given parity check matrix H is (n – k) × n matrix. It is given that the code is (7, 4) Hamming code. Therefore, we have

$$n = 7$$

and $$k = 4$$

Hence, the parity check matrix [H] will be 3 × 7 matrix i.e.,

$$[H] = [P^T | I_{n-k}]$$

where P^T is (n – k) by k matrix and I_{n-k} is (n – k) × (n – k) matrix.

We have
$$[H]_{3\times3} = \underbrace{\begin{bmatrix} 1 & 1 & 1 & 0 \\ 1 & 1 & 0 & 1 \\ 1 & 0 & 1 & 1 \end{bmatrix}}_{P^T_{3\times4}} \underbrace{\begin{matrix} 1 & 0 & 0 \\ 0 & 1 & 0 \\ 0 & 0 & 1 \end{matrix}}_{I_{n-k}} {}_{3\times7}$$

The transpose matrix P^T is given by,

$$P^T = \begin{bmatrix} 1 & 1 & 1 & 0 \\ 1 & 1 & 0 & 1 \\ 1 & 0 & 1 & 1 \end{bmatrix}_{3\times4}$$

Let us obtain the coefficient matrix P from P^T.

The P matrix can be obtained by interchanging the rows and columns of the transposed matrix P^T.

We have
$$P = \begin{bmatrix} 1 & 1 & 1 \\ 1 & 1 & 0 \\ 1 & 0 & 1 \\ 0 & 1 & 1 \end{bmatrix}_{4\times3}$$

Let us add the identity matrix to P matrix to get generator matrix

The generator matrix G is a k × n matrix. So, here, it will be a 4 × 7 matrix.

Thus, we have $$G = [I_k | P]$$

where I_k is a k × k i.e. 4 × 4 matrix.

Substituting the 4 × 4 identity matrix and the coefficient matrix, we obtain,

$$G = \left[\begin{array}{cccc:ccc} 1 & 0 & 0 & 0 & 1 & 1 & 1 \\ 0 & 1 & 0 & 0 & 1 & 1 & 0 \\ 0 & 0 & 1 & 0 & 1 & 0 & 1 \\ 0 & 0 & 0 & 1 & 0 & 1 & 1 \end{array}\right]$$
$$\longleftarrow P_{4\times3} \longrightarrow \longleftarrow I_{4\times4} \longrightarrow$$

Ans.

This is the required generator matrix

Now, let us obtain the parity bits for each message vector.

The parity bits can be obtained by using the following expression:

$$C = MP$$

Therefore, $$[c_0, c_1, c_2]_{1\times3} = [m_0\ m_1\ m_2\ m_3]_{1\times4}\ [P]_{4\times3}$$

Therefore, $$[c_0, c_1, c_2]_{1\times3} = [m_0, m_1, m_2, m_3] \begin{bmatrix} 1 & 1 & 1 \\ 1 & 1 & 0 \\ 1 & 0 & 1 \\ 0 & 1 & 1 \end{bmatrix}_{4\times3}$$

Solving, we obtain

$$c_0 = (m_0 \times 1) \oplus (m_1 \times 1) \oplus (m_2 \times 1) \oplus (m_3 \times 0)$$

$$c_1 = (m_0 \times 1) \oplus (m_1 \times 1) \oplus (m_2 \times 0) \oplus (m_3 \times 1)$$
$$c_2 = (m_0 \times 1) \oplus (m_1 \times 0) \oplus (m_2 \times 1) \oplus (m_3 \times 1)$$

Simplifying the equations, we obtain

$$\begin{aligned} c_0 &= m_0 \oplus m_1 \oplus m_2 \\ c_1 &= m_0 \oplus m_1 \oplus m_3 \\ c_2 &= m_0 \oplus m_2 \oplus m_3 \end{aligned} \qquad \text{... (8.18)}$$

Using these equations, we can obtain the parity bits for each message vector. For example, let the message word be,

$$m_0 \quad m_1 \quad m_2 \quad m_3 = 0 \; 1 \; 0 \; 1$$

Then, we have

$$\begin{aligned} c_0 &= 0 \oplus 1 \oplus 0 = 1 \\ c_1 &= 0 \oplus 1 \oplus 1 = 0 \\ c_2 &= 0 \oplus 0 \oplus 1 = 1 \end{aligned}$$

Hence, the parity bits are $b_0\, b_1\, b_2 = 1\,0\,1$

Therefore, the complete code word for the message word 0 1 0 1 is

Complete code word = | 0 1 0 1 | 1 0 0 |

← Message →← Parity →

Similarly, we can obtain the code words for the other message vectors. All these message vectors and the corresponding parity bits and code words are given in table 8.5. The weight of the code word is also given.

Table 8.5. Code words for the (7, 4) Hamming code

S. No.	Message vector M				Parity bits B			Code words X							Weight of the code vector
	m_0	m_1	m_2	m_3	c_0	c_1	c_2	x_0	x_1	x_2	x_3	x_4	x_5	x_6	
1.	0	0	0	0	0	0	0	0	0	0	0	0	0	0	0
2.	0	0	0	1	0	1	1	0	0	0	1	0	1	1	3
3.	0	0	1	0	1	0	1	0	0	1	0	1	0	1	3
4.	0	0	1	1	1	1	0	0	0	1	1	1	1	0	4
5.	0	1	0	0	1	1	0	0	1	0	0	1	1	0	3
6.	0	1	0	1	1	0	1	0	1	0	1	1	0	1	4
7.	0	1	1	0	0	1	1	0	1	1	0	0	1	1	4
8.	0	1	1	1	0	0	0	0	1	1	1	0	0	0	3
9.	1	0	0	0	1	1	1	1	0	0	0	1	1	1	4
10.	1	0	0	1	1	0	0	1	0	0	1	1	0	0	3
11.	1	0	1	0	0	1	0	1	0	1	0	0	1	0	3
12.	1	0	1	1	0	0	1	1	0	1	1	0	0	1	4
13.	1	1	0	0	0	0	1	1	1	0	0	0	0	1	3
14.	1	1	0	1	0	1	0	1	1	0	1	0	1	0	4
15.	1	1	1	0	1	0	0	1	1	1	0	1	0	0	4
16.	1	1	1	1	1	1	1	1	1	1	1	1	1	1	7

We know that the minimum distance d_{min} is equal to the minimum weight of any non-zero code vector. Looking at table 8.5, we obtain

$$d_{min} = 3. \textbf{ Ans.}$$

Number of errors that can be detected is given by

$$d_{min} \geq s + 1$$

$$3 \geq s + 1$$

or $$s \leq 2$$

Hence, at the most two errors can be detected.

Number of errors that can be corrected is given by

$$d_{min} \geq 2t + 1$$

or $$3 \geq 2t + 1$$

or $$t \leq 1$$

This means that at the most one error can be corrected.

Thus, for the (7, 4) linear block code, at the most two errors can be detected and at the most only one error can be corrected.

8.12.2. Encoder of (7, 4) Hamming Code

The encoder for (7, 4) Hamming code is shown in figure 8.20. This encoder produces all the code words corresponding to various message words listed in table 8.5.

The parity or check bits (c_2, c_1, c_0) are being generated for each message word (m_3, m_2, m_1, m_0) with the help of equation 8.18. The parity bits are obtained from the message bits by means of modulo-2 adders*. The output switch S is first connected to the message register to transmit all the message bits in a code word. Then it is connected to the parity bit register to transmit the corresponding parity bits. Thus, we get a 7 bit code word at the output of the switch.

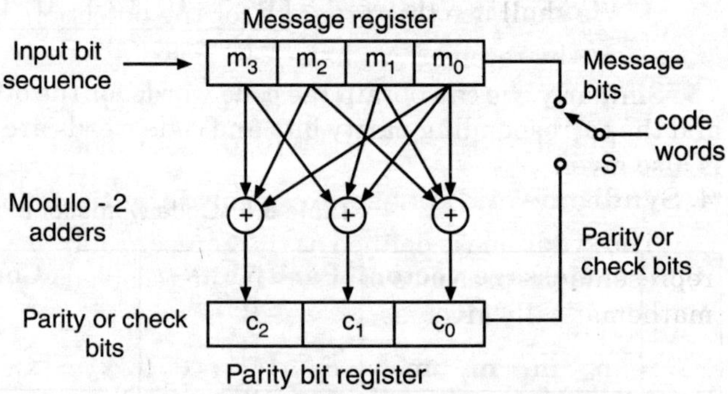

Fig. 8.20. Encoder for (7, 4) Hamming code

8.13 SYNDROME DECODING FOR BLOCK CODES

1. Basic Concept

In previous article, we have discussed about the encoder for the linear block code. Now, let us learn about the decoder. The two important functions of the decoder are as under:

(i) Error detection in the received code word.

(ii) Error correction.

The decoding of linear block codes is done by using a special technique called syndrome decoding, which reduces the memory requirement of the decoder to a great extent. The syndrome decoding technique is explained in the following headings.

2. Practical Assumptions

(i) Let X represent the transmitted code word and Y represent the received code word.

(ii) Then, if X = Y, no errors in the received signal and

if $X \neq Y$, then some errors are present.

*Modulo-2 adders are nothing but EX-OR gates.

3. Detection of Error

1. For an (n, k) linear block code, there exists a parity check matrix H of size (n – k) × n. We know that, parity check matrix,
$$H = [P^T : I_{n-k}]_{(n-k) \times n}$$
where P^T represents the transpose of P matrix and I_{n-k} is the identity matrix.
2. The transpose of the parity check matrix is given by,
$$H^T = \begin{bmatrix} P \\ \cdots \\ I_{n-k} \end{bmatrix}_{n \times (n-k)} \quad \text{... (8.19)}$$
where P is the coefficient matrix.
3. The transpose of parity check matrix (H^T) exhibits a very important property, i.e.
$$XH^T = (0, 0, ..., 0) \quad \text{... (8.20)}$$
This means that the product of any code vector X and the transpose of the parity check matrix will always be 0.

We shall use this property for the detection of errors in the received code words as under :

At the receiver, we have

If $\quad YH^T = (0, 0, ..., 0)$, then Y = X i.e. there is no error.

But, if $\quad YH^T \neq (0, 0, ..., 0)$, Then $Y \neq X$ i.e. error exists in the received code word

4. Syndrome and its use for error detection

The syndrome is defined as the non-zero output of the product YH^T. Thus, the non-zero syndrome represents the presence of errors in the received code word. The syndrome is represented by S and is mathematically given as,
$$S = YH^T \quad \text{... (8.21)}$$
or
$$[S]_{1 \times (n-k)} = [Y]_{1 \times n} \; [H^T]_{n \times (n-k)} \quad \text{... (8.22)}$$
Then, all zero elements of syndrome represent that there is no error, and a non-zero value of an element in syndrome represents the presence of error. But sometimes, even if all the syndrome elements have zero value, the error exists. This has been shown in figure 8.21.

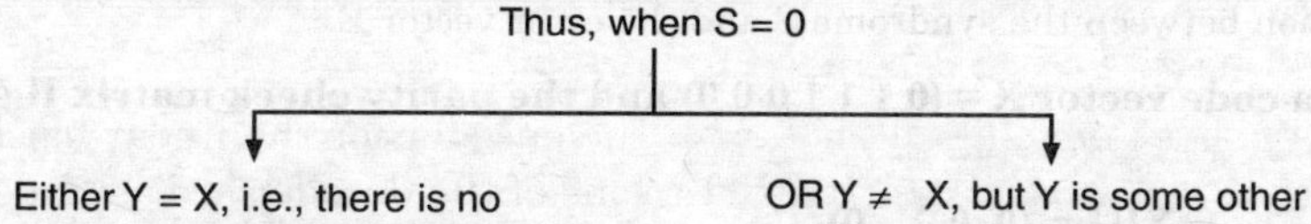

Fig. 8.21. Two different possibilities for S = 0.

5. Error Vector (E)

1. For the n-bit transmitted and received code words X and Y respectively, let us define an n-bit error vector E such that its non-zero elements represent the locations of errors in the received code vector Y as shown in figure 8.22.
2. The encircled entries in figure 8.22 indicate the presence of errors.

> **DO YOU KNOW?**
> The combined goal of the channel encoder and decoder is to minimize the effect of channel noise. That is, the number of errors between the channel encoder input and the channel decoder output is minimized.

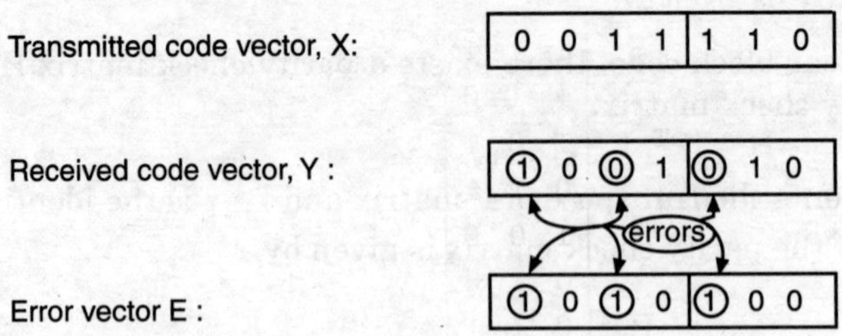

Fig. 8.22. The non-zero elements of error vector represent the locations of errors in the received code vector Y

The elements in the code word vector Y can be obtained by using the modulo-2 additions, as under:

$$Y = X \oplus E \qquad \text{... (8.23)}$$

From figure 8.22, we can write that,

$$Y = [0 \oplus 1, 0 \oplus 0, 1 \oplus 1, 1 \oplus 0, 1 \oplus 1, 1 \oplus 0, 0 \oplus 0]$$

or $$Y = [1, 0, 0, 1, 0, 1, 0]$$

3. **The principle of modulo-2 addition can be applied in a slightly different way as under:**

$$X = Y \oplus E \qquad \text{... (8.24)}$$

From figure 8.22, we can write that,

$$X = [1 \oplus 1, 0 \oplus 0, 0 \oplus 1, 1 \oplus 0, 0 \oplus 1, 1 \oplus 0, 0 \oplus 0] = [0, 0, 1, 1, 1, 1, 0]$$

6. Relation between syndrome and error vector

From equation (8.2), the syndrome is given by,

$$S = YH^T$$

Substituting the value of Y from equation (8.23), we obtain

$$S = [X \oplus E]\, H^T = XH^T \oplus EH^T$$

But, $$XH^T = 0$$

Therefore, $$S = 0 \oplus EH^T$$

or $$S = EH^T \qquad \text{.... (8.25)}$$

This is the relation between the syndrome S and the error vector E.

EXAMPLE 8.7. For a code vector X = (0 1 1 1 0 0 0) and the parity check matrix H given below, prove that

$$\mathbf{XH^T = (0, 0, ..., 0)}$$

$$\mathbf{H} = \begin{bmatrix} 1 & 1 & 1 & 0 & 1 & 0 & 0 \\ 1 & 1 & 0 & 1 & 0 & 1 & 1 \\ 1 & 0 & 1 & 1 & 0 & 0 & 1 \end{bmatrix}_{3\times7}$$

Solution: The transpose of the given parity matrix H, can be obtained by interchanging the rows and columns as under:

$$H^T = \begin{bmatrix} 1 & 1 & 1 \\ 1 & 1 & 0 \\ 1 & 0 & 1 \\ 0 & 1 & 1 \\ 1 & 0 & 0 \\ 0 & 1 & 0 \\ 0 & 1 & 1 \end{bmatrix}_{7\times3}$$

Also, the product XH^T is given by,

$$XH^T = [0\ 1\ 1\ 1\ 0\ 0\ 0]_{1\times 7} \begin{bmatrix} 1 & 1 & 1 \\ 1 & 1 & 0 \\ 1 & 0 & 1 \\ 0 & 1 & 1 \\ 1 & 0 & 0 \\ 0 & 1 & 0 \\ 0 & 1 & 1 \end{bmatrix}_{7\times 3}$$

or $$XH^T = [(0 \times 1) \oplus (1 \times 1) \oplus (1 \times 1) \oplus (1 \times 0) \oplus (0 \times 1) \oplus (0 \times 0) \oplus (0 \times 0),$$
$$(0 \times 1) \oplus (1 \times 1) \oplus (1 \times 0) \oplus (1 \times 1) \oplus (0 \times 0) \oplus (0 \times 1) \oplus (0 \times 1),$$
$$(0 \times 1) \oplus (1 \times 0) \oplus (1 \times 1) \oplus (1 \times 1) \oplus (0 \times 0) \oplus (0 \times 0) \oplus (0 \times 1)]$$

or $$XH^T = 0 \oplus 1 \oplus 1 \oplus 0 \oplus 0 \oplus 0 \oplus 0,\ 0 \oplus 1 \oplus 0 \oplus 1 \oplus 0 \oplus 0 \oplus 0,$$
$$0 \oplus 0 \oplus 1 \oplus 1 \oplus 0 \oplus 0 \oplus 0$$

or $$XH^T = [0, 0, 0]_{1\times 3}$$

Thus, we have proved that for a valid code word, the product $XH^T = (0, 0, 0)$ **Ans.**

EXAMPLE 8.8. The parity check matrix of a (7, 4) Hamming code is as under:

$$\mathbf{H} = \begin{bmatrix} 1 & 1 & 0 & 1 & 1 & 0 & 0 \\ 1 & 1 & 1 & 0 & 0 & 1 & 1 \\ 1 & 0 & 1 & 1 & 0 & 0 & 1 \end{bmatrix}_{3\times 7}$$

Calculate the syndrome vector for single bit errors.

Solution: We know that the syndrome vector is given by

$$S = EH^T = [E]_{1\times 7}\,[H^T]_{7\times 3}$$

Therefore, syndrome vector will represented by a 1 × 3 matrix.

Thus, $$[S]_{1\times 3} = [E]_{1\times 7}\,[H^T]_{7\times 3} \qquad \text{... (i)}$$

Now, let us write various error vectors.

Various error vectors with single bit errors are shown in table 8.6. The encircled bits represent the locations of errors.

Table 8.6.

S.No.	Error Vectors							Bit in Error
1.	(1)	0	0	0	0	0	0	First
2.	0	(1)	0	0	0	0	0	Second
3.	0	0	(1)	0	0	0	0	Third
4.	0	0	0	(1)	0	0	0	Fourth
5.	0	0	0	0	(1)	0	0	Fifth
6.	0	0	0	0	0	(1)	0	Sixth
7.	0	0	0	0	0	0	(1)	Seventh

Let us calculate the syndrome corresponding to each error vector.

(i) We have $[S]_{1\times 3} = [E]_{1\times 7}\,[H^T]_{7\times 3}$

Substituting [E] = (1 0 0 0 0 0 0) and H^T, we obtain

$$[S] = [1\,0\,0\,0\,0\,0\,0]_{1\times 7}\begin{bmatrix}1 & 1 & 1\\1 & 1 & 0\\1 & 0 & 1\\0 & 1 & 1\\1 & 0 & 0\\0 & 1 & 0\\0 & 1 & 1\end{bmatrix}_{7\times 3}$$

Therefore, $[S] = [1 \oplus 0 \oplus 0 \oplus 0 \oplus 0 \oplus 0 \oplus 0, 1 \oplus 0 \oplus 0 \oplus 0 \oplus 0 \oplus 0 \oplus 0,$
$1 \oplus 0 \oplus 0 \oplus 0 \oplus 0 \oplus 0 \oplus 0]$

Here, $[S] = [1, 1, 1]$

This is the syndrome for the first bit in error.

(ii) For the second bit in error, we have

$$[S] = [0\,1\,0\,0\,0\,0\,0]\begin{bmatrix}1 & 1 & 1\\1 & 1 & 0\\1 & 0 & 1\\0 & 1 & 1\\1 & 0 & 0\\0 & 1 & 0\\0 & 1 & 1\end{bmatrix}$$

Therefore, $[S] = [0 \oplus 1 \oplus 0 \oplus 0 \oplus 0 \oplus 0 \oplus 0 \oplus 0, 1 \oplus 0 \oplus 0 \oplus 0 \oplus 0 \oplus 0,$
$0 \oplus 0 \oplus 0 \oplus 0 \oplus 0 \oplus 0 \oplus 0]$

or $[S] = [1, 1, 0]$

(iii) Similarly, we can obtain the other syndromes as shown in table 8.7.

(iv) Here, it may be noted that the first row of table 8.7 represent an error vector with no errors. The corresponding syndrome is (0, 0, 0).

(v) Table 8.7 also shows that the syndrome vectors are same as the rows of the transpose matrix H^T.

Table 8.7. Syndromes for various error vectors

S.No.	Error vectors with single bit errors							Syndrome vector S			
1.	0	0	0	0	0	0	0	0	0	0	
2.	①	0	0	0	0	0	0	1	1	1	⟵ 1st Row of H^T
3.	0	①	0	0	0	0	0	1	1	0	⟵ 2nd Row of H^T
4.	0	0	①	0	0	0	0	1	0	1	⟵ 3rd Row of H^T
5.	0	0	0	①	0	0	0	0	1	1	⟵ 4th Row of H^T
6.	0	0	0	0	①	0	0	1	0	0	⟵ 5th Row of H^T
7.	0	0	0	0	0	①	0	0	1	0	⟵ 6th Row of H^T
8.	0	0	0	0	0	0	①	0	1	1	⟵ 7th Row of H^T

8.13.1. Error Correction using Syndrome Vector

In order to use the syndrome vector for the error correction, we follow a procedure as under:

Error correction using syndrome vector

(i) For a transmitted code vector X = (0 1 0 0 1 1 0), we obtain the received code vector Y = (0 1 ① 0 1 1 0), assuming the error to be in the 3rd position.

(ii) We calculate the corresponding syndrome vector as under:

$$S = YH^T$$

(iii) We obtain the error vectors as $S = EH^T$ i.e. $S = E$.

(iv) From the syndrome vector, we obtain the error vector.

(v) From the error vector, we obtain the transmitted signal as under

$$X = Y \oplus E$$

EXAMPLE 8.9. To clear above concept of error correction using syndrome vector, let us consider one particular example. For this, let us use the following parity check matrix:

$$[H] = \begin{bmatrix} 1 & 1 & 1 & 0 & 1 & 0 & 0 \\ 1 & 1 & 0 & 1 & 0 & 1 & 1 \\ 1 & 0 & 1 & 1 & 0 & 0 & 1 \end{bmatrix}_{3\times 7} \quad \text{... (i)}$$

Solution: (i) First, we obtain the received code vector Y.

Assuming X = (0 1 0 0 1 1 0) to be the transmitted code vector.

Let the received code vector be obtained by assuming that the third bit is in error.

Thus, Y = (0 1 ① 0 1 1 0), Here, encircled bit represents error.

(ii) Next, let us determine the corresponding syndrome vector.

We know that,

$$\text{Syndrome } S = YH^T$$

$$S = [0\ 1\ 1\ 0\ 1\ 1\ 0] \begin{bmatrix} 1 & 1 & 1 \\ 0 & 1 & 1 \\ 1 & 0 & 1 \\ 1 & 1 & 0 \\ 0 & 0 & 1 \\ 0 & 1 & 0 \\ 1 & 1 & 0 \end{bmatrix}$$

We have $S = (0 \oplus 1 \oplus 1 \oplus 0 \oplus 1 \oplus 0 \oplus 0,\ 0 \oplus 1 \oplus 0 \oplus 0 \oplus 0 \oplus 1 \oplus 0,$

$0 \oplus 0 \oplus 1 \oplus 0 \oplus 0 \oplus 0 \oplus 0)$

or $S = [1, 0, 1]$

This is the syndrome vector for the given received signal. This corresponds to the 3rd row of the transpose matrix H^T.

(iii) But $S = YH^T = EH^T$

Therefore, $EH^T = [1, 0, 1]$

(iv) Let us obtain the error vector for the syndrome vector S = [1, 0, 1].

From table 8.6, the error vector corresponding to the syndrome (1, 0, 1) is given by

$$E = (0\ 0\ 1\ 0\ 0\ 0\ 0)$$

The shows that the error is present in the third position of the received code vector Y.

(v) Now, let us correct the error.

The correct vector X can be obtained as under:

$$X = Y \oplus E$$

Similarly the values of Y and E, we obtain

$$X = [0\ 1\ 1\ 0\ 1\ 1\ 0] \oplus [0\ 0\ 1\ 0\ 0\ 0\ 0]$$

or $X = [0\ 1\ 0\ 0\ 1\ 1\ 0]$

This is same as the transmitted code vector.

Important Point: *Thus by modulo-2 addition of received signal with the error vector E, we can obtained the correct code word. Hence, one single error can be corrected.*

8.13.2. Syndrome Decoder for (n, k) Block Codes

1. Block Diagram

The block diagram of a syndrome decoder for (n, k) block codes for correcting errors is shown in figure 8.23.

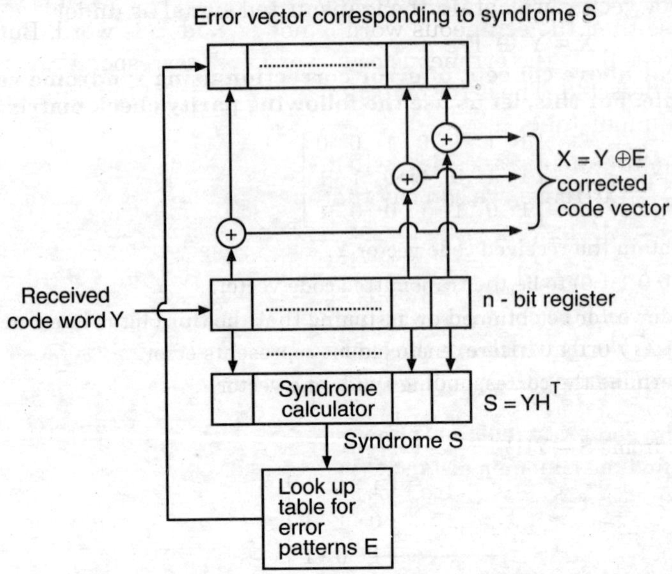

Fig. 8.23. Syndrome decoder for linear block code

2. Working Operation

The received n-bit code word Y is stored in an n-bit register. This code vector is then applied to a syndrome calculator to calculate syndrome $S = YH^T$. In order to obtain the syndrome, the transposed parity check matrix, H^T is stored in the syndrome calculator. The (n – k) bit syndrome vector S is applied to the look-up table containing the error patterns. An error pattern is selected corresponding to the particular syndrome S generated at the output of the syndrome calculator. The selected error pattern E is then added (modulo-2 addition) to the received signal Y to generate the corrected code vector X.

Therefore, $X = Y \oplus E$

> **DO YOU KNOW?**
>
> For a fixed modulation scheme, the addition of redundancy in the coded messages implies the need for increased transmission bandwidth.

8.13.3. Decoding of a Linear Block Code

1. Basic Objective

The basic objective of channel coding is to detect and correct errors when messages are being transmitted over a noisy channel. The noise can randomly alter some of the symbols in the transmitted code word, (i.e., from 1 to 0 or from 0 to 1). If only one symbol is changed as shown in figure 8.24, then the Hamming distance between the original code word and the erroneous code word is 1. If the noise transforms t symbols (i.e., if t symbols in the received code word are in error), then the Hamming distance of the received word will be t with respect to the correct code word.

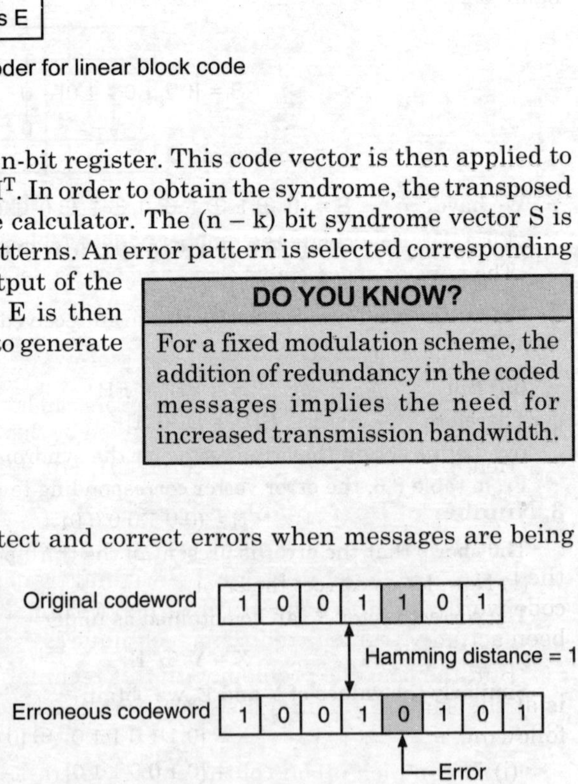

Fig. 8.24 Relation between number of errors and Hamming distance.

2. Number of Detectable Errors

An error will be detected as long as it does not transform one code word into another valid code word. If the minimum distance between the code word is d_{min} then the weight of the error pattern must be d_{min} or more to cause transformation of one code word to another. The error detection is always possible when the number of transmission errors in a code word is less than the minimum distance d_{min}, because then the erroneous word is not a valid code word. But when the number of errors equals or exceeds d_{min}, the erroneous code word may correspond to another valid code word and errors cannot be detected. The error detection and correction capabilities of a coding technique depend on the minimum distance d_{min}.

A code having a minimum distance d_{min} will detect all the non-zero error patterns of weight less than or equal to $(d_{min} - 1)$. Hence, we can say that upto s errors per word can be detected such that

$$d_{min} \geq (s + 1)$$

or $$s \leq (d_{min} - 1) \quad \text{... (8.26)}$$

Moreover, there is at least one error pattern of weight d_{min} which cannot be detected. This corresponds to two code words which are the closest. It is possible that some error patterns of weight d_{min} are detected but all error patterns of weight d_{min} will not be detected.

EXAMPLE 8.10. For the code X_1 = (000, 111) how many errors can be successfully detected ?

Solution: First, let us find the minimum distance.

It is given that the code consists of two code words 000 and 111. Hence, minimum distance is 3 as shown in figure 8.25.

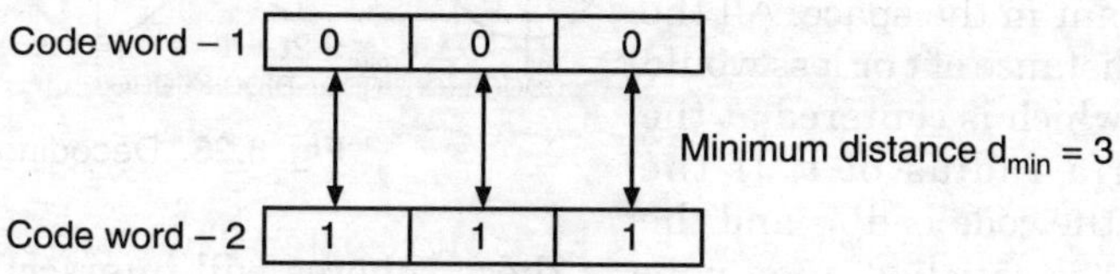

Fig. 8.25.

First, let us calculate the number of detectable errors.

The number of detectable errors is given by

$$s \leq (d_{min} - 1)$$

or $$s \leq 3 - 1$$

or $$s \leq 2$$

Thus, at the most two or upto two errors can be detected by this code. This means that any error pattern belonging to the following set can be detected by this code.

Hence, $$s = \{011, 101, 110, 001, 010, 100\}$$

3. Number of Correctable Errors

Let us now see, how many errors can be successfully corrected by a code. This will be based on the best possible guess about the original transmitted code word. It is logical to guess that a valid code word which is nearest (in terms of Hamming distance) to the received code word must have been actually transmitted. This technique is called as **Nearest Neighbour Decoding**

But, the possible problem with this technique is that it is possible that more than one code word is at the same Hamming distance from the received word. In this situation, the receiver does the following:

(i) It can pick up one of the equally distant code words on a random basis, or

(ii) It can request the transmitter to retransmit.

To ensure that the received word having at most t errors, is closest to the original code word, we put the following condition on the minimum distance.

$$d_{min} \geq 2t + 1 \quad \text{... (8.27)}$$

Hence, the number of errors that can be successfully corrected by a code is given by

$$t \leq \frac{d_{min} - 1}{2} \quad \text{... (8.28)}$$

4. Summary of error detecting and correcting capability

The error detecting and correcting capability of a code is summarized in Table 8.8.

Table 8.8. Role of d_{min} for detection and correction of errors

S.No.	Description	Expression
1.	Detect upto s errors per word	$d_{min} \geq (s + 1)$
2.	Correct upto t errors per word	$d_{min} \geq (2t + 1)$
3.	Correct upto t errors and detect s > t errors per word.	$d_{min} \geq (t + s + 1)$

5. Graphical representation for the condition $d_{min} \geq 2t + 1$

Figure 8.26 shows the graphical representation of the condition

$$d_{min} \geq 2t + 1$$

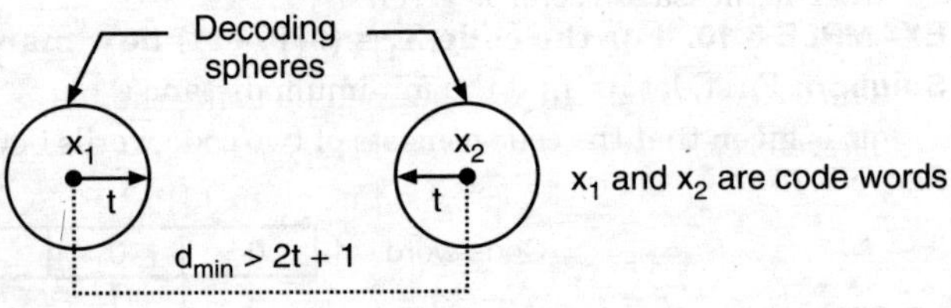

Fig. 8.26. Decoding spheres

Each n bit code word (i.e., code vector) can be represented as a point in the space. All the words at a Hamming distance of t or less would lie within the sphere which is centered at the code word and with a radius of t. If the minimum distance of the code is d_{min} and the condition $d_{min} \geq 2t + 1$ is satisfied, then none of these spheres will intersect. Any received vector which is basically a point within a specific sphere will be closest to its centre (which represents a code word) than any other code word. The spheres associated with each code word is called as its **Decoding sphere.** Hence, it is possible to decode the received vector using the nearest neighbour method. Figure 8.26 tells us that the code words within the sphere of radius t and centered at x_1 will be decoded as x_1. For decoding without any ambiguity, we have

$$d_{min} \geq 2t + 1$$

EXAMPLE 8.11. An error control code has the following parity check matrix.

$$H = \begin{bmatrix} 1 & 0 & 1 & 1 & 0 & 0 \\ 1 & 1 & 0 & 0 & 1 & 0 \\ 0 & 1 & 1 & 0 & 0 & 1 \end{bmatrix}$$

(i) Determine the generator matrix G

(ii) Find the code word that begin with 101 ...

(iii) Decode the received code word 1 1 0 1 1 0. Comment on error detection capability of this code.

Solution: From the parity check matrix $[H]_{3 \times 6}$, it obvious that this is a (6, 3) linear block code. Therefore, n = 6, k = 3 and (n – k) = 3.

(i) We know that the parity check matrix is given by,

$$[H] = [P^T : I_{n-k}]_{n-k \times n}$$

or

$$[H]_{3 \times 6} = [P^T : I_3]_{3 \times 6}$$

Using the given expression for H, we obtain

$$H = \left[\begin{array}{ccc|ccc} 1 & 0 & 1 & 1 & 0 & 0 \\ 1 & 1 & 0 & 0 & 1 & 0 \\ 0 & 1 & 1 & 0 & 0 & 1 \end{array}\right]$$

or $$P^T = \begin{bmatrix} 1 & 0 & 1 \\ 1 & 1 & 0 \\ 0 & 1 & 1 \end{bmatrix}_{3\times 3}$$

Therefore, the transpose of matrix P^T is given by,

$$P = \begin{bmatrix} 1 & 1 & 0 \\ 0 & 1 & 1 \\ 1 & 0 & 1 \end{bmatrix}$$

We know that the generator matrix is given by,

$$G = [I_k : P_{k \times (n-k)}]_{k \times n} = \left[\begin{array}{ccc|ccc} 1 & 0 & 0 & 1 & 1 & 0 \\ 0 & 1 & 0 & 0 & 1 & 1 \\ 0 & 0 & 1 & 1 & 0 & 1 \end{array}\right]_{3\times 6}$$

This is the required generator matrix.

(ii) The message vector is given by,

$$M = [1\ 0\ 1]$$

The three parity bits are obtained by using the following standard expression:

Therefore, $$C = MP$$

$$[c_0\ c_1\ c_2] = [m_0\ m_1\ m_2]\ [P]$$

or $$[c_0\ c_1\ c_2] = [m_0\ m_1\ m_2] \begin{bmatrix} 1 & 1 & 0 \\ 0 & 1 & 1 \\ 1 & 0 & 1 \end{bmatrix}$$

or $$c_0 = (m_0 \times 1) \oplus (m_1 \times 0) \oplus (m_2 \times 1) = m_0 \oplus m_2$$

Substituting $m_0 = 1$ and $m_2 = 1$, we obtain

$$c_0 = 1 \oplus 1 = 0$$

Similarly, $$c_1 = (m_0 \times 1) \oplus (m_1 \times 1) \oplus (m_2 \times 0) = m_0 \oplus m_1$$

Substituting, $m_0 = 1$ and $m_1 = 0$, we obtain

$$c_1 = 1 \oplus 0 = 1$$

and $$c_2 = (m_0 \times 0) \oplus (m_1 \times 1) \oplus (m_2 \times 1) = m_1 \oplus m_2$$

or $$c_2 = 0 \oplus 1 = 1$$

Therefore, the parity word is $C = [0\ 1\ 1]$.

Hence, the complete codeword is given by,

$$X = [\underbrace{1 0 1}_{M} \underbrace{0 1 1}_{C}] \quad \textbf{Ans.}$$

(iii) The received code word $Y = 1\ 1\ 0\ 1\ 1\ 0$

Therefore, the syndrome is given by

$$S = YH^T$$

Substituting for Y and H^T, we obtain

$$S = [110110] \begin{bmatrix} 1 & 1 & 0 \\ 0 & 1 & 1 \\ 1 & 0 & 1 \\ 1 & 0 & 0 \\ 0 & 1 & 0 \\ 0 & 0 & 1 \end{bmatrix}$$

or $S = [1 \oplus 0 \oplus 0 \oplus 1 \oplus 0 \oplus 0, 1 \oplus 1 \oplus 0 \oplus 0 \oplus 1 \oplus 0, 0 \oplus 1 \oplus 0 \oplus 0 \oplus 0 \oplus 0]$

or $S = [0, 1, 1]$

This is same as the second row of the transpose matrix H^T, which indicates that there is an error in the second bit of the received signal i.e.,

$$\begin{array}{c} \quad\downarrow \text{— Error} \\ Y = 1 \;\; 1 \;\; 0\,1\,1\,0 \end{array}$$

Therefore, the correct code word X = 1 0 0 1 1 0. **Ans.**

The correct code word is obtained by replacing the second bit by a 0.

(iv) It is possible to verify that this code has a minimum $d_{min} = 3$. The relation between d_{min} and the number of errors that can be detected is:

$$d_{min} \geq s + 1$$

For $d_{min} = 3$, we have

$$3 \geq s + 1$$

or $$s \leq 2$$

This means that upto two errors can be detected and $d_{min} \geq 2t + 1$

or $$3 \geq 2t + 1 \quad \text{or} \quad t \leq 1.$$

This means that upto one error can be corrected.

6. Hamming Bound

(i) The number of non-zero distinct syndromes for an (n, k) block codes is $(2^{n-k} - 1)$. If we substitute $(n - k) = q$, then the number of non-zero distinct syndromes will be $2^q - 1$.

(ii) The number of single error patterns will be ${}^nC_1 = n$.

Number of double error patterns will be ${}^nC_2$

Number of triple error patterns will be ${}^nC_3$ ...

Therefore, the corrections of t number of errors is possible if and only if the following expression is satisfied:

$$2^q - 1 \geq {}^nC_1 + {}^nC_2 + {}^nC_3 + \ldots\ {}^nC_t$$

or $$2^q \geq {}^nC_1 + 1 + {}^nC_2 + \ldots\ {}^nC_t$$

or $$2^q \geq \sum_{i=0}^{t} {}^nC_i$$

But, $q = n - k$. Hence, $2^{n-k} \geq \sum_{i=0}^{t} {}^nC_i$

Taking $\log_2$ of both the sides, we obtain

$$(n - k) \geq \log_2 \sum_{i=0}^{t} {}^nC_i$$

Dividing both the sides of above expression by n, we obtain

$$1 - \frac{k}{n} = \frac{1}{n} \log_2 \sum_{i=0}^{t} {}^nC_i$$

But, $\frac{k}{n} = r$, i.e., the code rate.

Therefore, $$(1 - r) = \frac{1}{n} \log_2 \sum_{i=0}^{t} {}^nC_i$$

Important Point: *Above expression gives the relation between the code rate r, number of correctable errors t, number of bits per word n. As the number of correctable errors is related to the* Hamming distance, the expression is called as the ***Hamming Bound.***

EXAMPLE 8.12. Given a (7, 4) linear block code whose generator matrix is given by:

$$G = \begin{bmatrix} 1 & 0 & 0 & 0 & 1 & 0 & 1 \\ 0 & 1 & 0 & 0 & 1 & 1 & 1 \\ 0 & 0 & 1 & 0 & 1 & 1 & 0 \\ 0 & 0 & 0 & 1 & 0 & 1 & 1 \end{bmatrix}$$

(i) Find all the code words. (ii) Find the parity check matrix.

Solution:

(i) First, we obtain the P matrix from the generator matrix.

(ii) Then, we obtain the parity bits for each message vector using the expression,

$$C = MP$$

(iii) Next, we obtain all the possible code words as

$$X = [M: C]$$

(iv) Lastly, we obtain the transpose of P matrix i.e. P^T and obtain the parity check matrix as :

$$[H] = [P^T \mid I_{n-k}]$$

(i) First, let us obtain the P matrix.

Given generator matrix
$$G = \left[\begin{array}{cccc:ccc} 1 & 0 & 0 & 0 & 1 & 0 & 1 \\ 0 & 1 & 0 & 0 & 1 & 1 & 1 \\ 0 & 0 & 1 & 0 & 1 & 1 & 0 \\ 0 & 0 & 0 & 1 & 0 & 1 & 1 \end{array}\right]$$

Therefore, the P matrix is given by

$$P = \begin{bmatrix} 1 & 0 & 1 \\ 1 & 1 & 1 \\ 1 & 1 & 0 \\ 0 & 1 & 1 \end{bmatrix}_{4\times3} \quad \text{... (i)}$$

(ii) Next, we obtain the parity (i.e., check) bits

The parity bits can be obtained by using the following expression:

$$C = MP$$

or
$$[c_0\ c_1\ c_2]_{1\times3} = [m_0\ m_1\ m_2\ m_3] \begin{bmatrix} 1 & 0 & 1 \\ 1 & 1 & 1 \\ 1 & 1 & 0 \\ 0 & 1 & 1 \end{bmatrix}_{4\times3}$$

Solving, we obtain

we
$$\left.\begin{aligned} c_0 &= m_0 \oplus m_1 \oplus m_2 \\ c_1 &= m_1 \oplus m_2 \oplus m_3 \\ c_2 &= m_0 \oplus m_1 \oplus m_3 \end{aligned}\right\} \quad \text{... (ii)}$$

Using these equations, we can obtain the parity bits for each message vector. For example, let the message word be

$$m_0\ m_1\ m_2\ m_3 = 0\ 1\ 0\ 1$$

Therefore, we write
$$c_0 = 0 \oplus 1 \oplus 0 = 1$$
$$c_1 = 1 \oplus 0 \oplus 1 = 0$$
$$c_2 = 0 \oplus 1 \oplus 1 = 0$$

Hence, the corresponding parity bits are $c_0\ c_1\ c_2 = 1\ 0\ 0$

Therefore, the complete codeword for the message word 0 1 0 1 is given by

0	1	0	1	1	0	0

Complete codeword = ← Message bits → ← Parity bits →

Similarly, we can obtain the codewords for the remaining message words. All the message vectors, the corresponding parity bits and codewords are given in table 8.9. The code weights are also given in the table 8.9.

Table 8.9. Code vectors for all the message vectors

S.No.	Message vector, M				Parity bits, C			Code words, X						
	m_3	m_2	m_1	m_0	c_2	c_1	c_0	X_6	X_5	X_4	X_3	X_2	X_1	X_0
1.	0	0	0	0	0	0	0	0	0	0	0	0	0	0
2.	0	0	0	1	1	0	1	0	0	0	1	1	0	1
3.	0	0	1	0	0	1	1	0	0	1	0	0	1	1
4.	0	0	1	1	0	1	0	0	0	1	1	0	1	0
5.	0	1	0	0	0	1	1	0	1	0	0	0	1	1
6.	0	1	0	1	1	1	0	0	1	0	1	1	1	0
7.	0	1	1	0	1	0	0	0	1	1	0	1	0	0
8.	0	1	1	1	0	0	1	0	1	1	1	0	0	1
9.	1	0	0	0	1	1	0	1	0	0	0	1	1	0
10.	1	0	0	1	0	1	1	1	0	0	1	0	1	1
11.	1	0	1	0	0	0	1	1	0	1	0	0	0	1
12.	1	0	1	1	1	0	0	1	0	1	1	1	0	0
13.	1	1	0	0	1	0	1	1	1	0	0	1	0	1
14.	1	1	0	1	0	0	0	1	1	0	1	0	0	0
15.	1	1	1	0	0	1	0	1	1	1	0	0	1	0
16.	1	1	1	1	1	1	1	1	1	1	1	1	1	1

(iv) Lastly, let us obtain the parity check matrix.

The parity check matrix [H] is a 3 × 7 matrix, i.e.,

$$H = [P^T : I_{n-k}]$$

The transpose matrix P^T is given by

$$P^T = \begin{bmatrix} 1 & 1 & 1 & 0 \\ 0 & 1 & 1 & 1 \\ 1 & 1 & 0 & 1 \end{bmatrix}_{3\times 4}$$

Therefore, we have

$$H = [P^T : I_{3\times 3}] = \left[\begin{array}{cccc|ccc} 1 & 1 & 1 & 0 & 1 & 0 & 0 \\ 0 & 1 & 1 & 1 & 0 & 1 & 0 \\ 1 & 1 & 0 & 1 & 0 & 0 & 1 \end{array}\right]_{3\times 7}$$ **Ans.**

This is the required parity check matrix.

EXAMPLE 8.13. A (7, 4) linear block code of which generator matrix is given by

$$G = \begin{bmatrix} 1 & 0 & 0 & 0 & 1 & 0 & 1 \\ 0 & 1 & 0 & 0 & 1 & 1 & 1 \\ 0 & 0 & 1 & 0 & 1 & 1 & 0 \\ 0 & 0 & 0 & 1 & 0 & 1 & 1 \end{bmatrix}_{4\times 7}$$

(i) Find code vector for any six messages

(ii) Write the parity check matrix of this code.

Solution: The generator matrix G is a k × n matrix. So, here it will be a 4 × 7 matrix in the following format

Therefore, we have $G = [I_k \mid P]$

where I_k is a k × k, i.e., 4 × 4 matrix.

Hence, we have

$$G = \begin{bmatrix} 1 & 0 & 0 & 0 & 1 & 0 & 1 \\ 0 & 1 & 0 & 0 & 1 & 1 & 1 \\ 0 & 0 & 1 & 0 & 1 & 1 & 0 \\ 0 & 0 & 0 & 1 & 0 & 1 & 1 \end{bmatrix}_{4\times 7} \qquad \text{... (i)}$$

$$\longleftarrow I_{4\times 4} \longrightarrow \; \longleftarrow P_{4\times 3} \longrightarrow$$

Now, let us obtain the parity bits for each message

The parity bits can be obtained by using the following expression:

$$C = MP \qquad \text{... (ii)}$$

or

$$[c_0, c_1, c_2] = [m_0, m_1, m_2, m_3]_{1\times 4}\, [P]_{4\times 3} \qquad \text{... (iii)}$$

Substituting the P matrix from equation (i), we can obtain the parity bits.

Hence,

$$[c_0, c_1, c_2] = [m_0, m_1, m_2, m_3] \begin{bmatrix} 1 & 0 & 1 \\ 1 & 1 & 1 \\ 1 & 1 & 0 \\ 0 & 1 & 1 \end{bmatrix}_{4\times 3}$$

Solving, we obtain

$$\left.\begin{aligned} c_0 &= m_0 \oplus m_1 \oplus m_2 \\ c_1 &= m_1 \oplus m_2 \oplus m_3 \\ c_2 &= m_0 \oplus m_1 \oplus m_3 \end{aligned}\right\} \qquad \text{... (iv)}$$

Using these equations, we can obtain the parity bits for each message vector. For example, let the message word be

$$m_0\, m_1\, m_2\, m_3 = 0\,1\,0\,1$$

Therefore, we have

$$c_0 = m_0 \oplus m_1 \oplus m_2 = 0 \oplus 1 \oplus 0 = 1$$

$$c_1 = m_1 \oplus m_2 \oplus m_3 = 1 \oplus 0 \oplus 1 = 0$$

$$c_2 = m_0 \oplus m_1 \oplus m_3 = 0 \oplus 1 \oplus 1 = 0$$

Hence, the parity bits are $c_0\, c_1\, c_2 = 1\,0\,0$

Therefore, the complete codeword for the message word 0 1 0 0 is

Complete codeword =

0 1 0 1	1 0 0
← Message →	← Parity →

Similarly, we can obtain the codewords for other message vectors shown in table 8.10.

Table 8.10.

S.No.	Message vector				Parity bits			Code Vectors						
	m_0	m_1	m_2	m_3	c_0	c_1	c_2	X_0	X_1	X_2	X_3	X_4	X_5	X_6
1.	0	0	0	0	0	0	0	0	0	0	0	0	0	0
2.	1	0	0	0	1	0	1	1	0	0	0	1	0	1
3.	0	1	0	0	1	1	1	0	1	0	0	1	1	1
4.	0	0	1	0	1	1	0	0	0	1	0	1	1	0
5.	0	0	0	1	0	1	1	0	0	0	1	0	1	1
6.	0	1	0	1	1	0	0	0	1	0	1	1	0	0

Now, the parity check matrix is given by

$$[H]_{3\times 7} = [P^T \mid I_{n-k}]$$

where P^T is the transpose of P and I_{n-k} is $I_{3\times 3}$ matrix.

We can obtain the transpose of P by interchanging the row and columns of P matrix in equation (i) i.e.,

$$P^T = \begin{bmatrix} 1 & 1 & 1 & 0 \\ 0 & 1 & 1 & 1 \\ 1 & 1 & 0 & 1 \end{bmatrix}_{3\times 4}$$

Therefore, the parity check matrix is given by

$$H = \begin{bmatrix} 1 & 1 & 1 & 0 & 1 & 0 & 0 \\ 0 & 1 & 1 & 1 & 0 & 1 & 0 \\ 1 & 1 & 0 & 1 & 0 & 0 & 1 \end{bmatrix}_{3\times 7} \quad \textbf{Ans}$$

This is the required parity check matrix.

EXAMPLE 8.14. For a systematic linear block code, the three parity check digits, c_4, c_5 and c_6 are given by

$$c_4 = m_1 \oplus m_2 \oplus m_3$$

$$c_5 = m_1 \oplus m_2$$

$$c_6 = m_1 \oplus m_3$$

(i) Construct generator matrix.

(ii) Construct code generated by this matrix.

(iii) Determine error correcting capability.

(iv) Prepare a suitable decoding table.

(v) Decode the received words 101100 and 000110.

Solution: (i) First, we parity matrix P and using it, obtain the generator matrix G.

(ii) Then, we obtain the values of c_4, c_5, c_6 for various combinations of m_1, m_2, m_3 and obtain all the possible code words.

(iii) Next, we obtain d_{min} and from the value of d_{min}, we calculate the error detecting and correcting capability.

(iv) After that, we obtain the decoding table by following the steps given below:

(a) We obtain the transpose of the parity check matrix H^T

(b) We calculate Syndrome $S = EH^T$

(c) We write the decoding table.

(v) Lastly, we decode the received words with the help of syndromes listed in the decoding table.

(i) First, let us obtain the parity matrix P and generator matrix G.

We know that the relation between the check (parity) bits, message bits and the parity matrix P is given by:

$$[c_4\, c_5\, c_6]_{1\times 3} = [m_1, m_2, m_3]_{1\times 3}\, [P]_{3\times 3} \quad \text{... (i)}$$

$$[c_4\, c_5\, c_6] = [m_1\, m_2\, m_3] \begin{bmatrix} P_{11} & P_{12} & P_{13} \\ P_{21} & P_{22} & P_{23} \\ P_{31} & P_{32} & P_{33} \end{bmatrix} \quad \text{... (ii)}$$

Therefore, we have

$$\begin{aligned} c_4 &= P_{11}\, m_1 \oplus P_{21}\, m_2 \oplus P_{31}\, m_3 \\ c_5 &= P_{12}\, m_1 \oplus P_{22}\, m_2 \oplus P_{32}\, m_3 \\ c_6 &= P_{13}\, m_1 \oplus P_{23}\, m_2 \oplus P_{33}\, m_3 \end{aligned} \quad \text{... (iii)}$$

Comparing equation (iii) with the given equations for c_4, c_5, c_6, we obtain

$$\begin{matrix} P_{11} = 1 & P_{12} = 1 & P_{13} = 1 \\ P_{21} = 1 & P_{22} = 1 & P_{23} = 0 \\ P_{31} = 1 & P_{32} = 0 & P_{33} = 1 \end{matrix}$$

Hence, the parity matrix is as shown below:

$$P = \begin{bmatrix} 1 & 1 & 1 \\ 1 & 1 & 0 \\ 1 & 0 & 1 \end{bmatrix}_{3\times 3}$$

This is the required parity matrix.

The generator matrix is given by:

$$G = [I_k : P] = [I_3 : P_{3\times 3}]$$

$$G = \left[\begin{array}{ccc|ccc} 1 & 0 & 0 & 1 & 1 & 1 \\ 0 & 1 & 0 & 1 & 1 & 0 \\ 0 & 0 & 1 & 1 & 0 & 1 \end{array}\right] \quad \textbf{Ans}$$

(ii) Now, let us obtain the codewords

It is given that,

$$c_4 = m_1 \oplus m_2 \oplus m_3$$

$$c_5 = m_1 \oplus m_2$$

$$c_6 = m_1 \oplus m_3$$

Using the equations, we can obtain the check bits for various combinations of the bits m_1, m_2 and m_3. After that the corresponding codewords are obtained as shown in table 8.11.

For, $m_1\, m_2\, m_3 = 0\;0\;1$, we have

$$c_4 = m_1 \oplus m_2 \oplus m_3 = 0 \oplus 0 \oplus 1 = 1$$

$$c_5 = m_1 \oplus m_2 = 0 \oplus 0 = 0$$

$$c_6 = m_1 \oplus m_3 = 0 \oplus 1 = 1$$

Therefore, $c_4\, c_5\, c_6 = 101$ and the codeword is given by:

Codeword for $m_1\, m_2\, m_3 = 0\;0\;1$ =

m_1	m_2	m_3	c_4	c_5	c_6
0	0	1	1	0	1

Similarly, the other codewords are obtained. They are listed in table 8.11.

Table 8.11. Codewords

S.No.	Message vector			Check bits			Code vectors or code words						Code weight W(X)
	m_1	m_2	m_3	c_4	c_5	c_6	m_1	m_2	m_3	c_4	c_5	c_6	
1.	0	0	0	0	0	0	0	0	0	0	0	0	0
2.	0	0	1	1	0	1	0	0	1	1	0	1	3
3.	0	1	0	1	1	0	0	1	0	1	1	0	3
4.	0	1	1	0	1	1	0	1	1	0	1	1	4
5.	1	0	0	1	1	1	1	0	0	1	1	1	4
6.	1	0	1	0	1	0	1	0	1	0	1	0	3
7.	1	1	0	0	0	1	1	1	0	0	0	1	3
8.	1	1	1	1	0	0	1	1	1	1	0	0	4

(iii) Now, let find the error correcting capacity.

The error correcting capacity depends on the minimum distance d_{min}. From table 8.11, $d_{min} = 3$.

Therefore, number of errors detectable is $d_{min} \geq s + 1$

or $3 \geq s + 1$ or $s \leq 2$.

So, at the most, two errors can be detected.

and $d_{min} \geq 2t + 1$

or $3 \geq 2t + 1$, or $t \leq 1$.

Thus, at the most one error can be corrected.

(iv) Now, let us find the decoding table.

To write the decoding table, we have to calculate the syndrome (S) i.e.,

$$S = EH^T \quad \text{... (iv)}$$

where E = Error vector and H^T = transpose of parity check matrix.

So, let us first obtain the transpose of matrix H.

$$H = [P^T : I_{n-k}]_{n-k,\, n} \quad \text{... (v)}$$

Therefore, the transpose of parity check matrix is given by

$$H^T = \begin{bmatrix} P \\ \cdots \\ I_{n-k} \end{bmatrix}_{n \times (n-k)} \quad \text{... (vi)}$$

Substituting the P matrix and the identity matrix, we obtain

$$H^T = \begin{bmatrix} 1 & 1 & 1 \\ 1 & 1 & 0 \\ 1 & 0 & 1 \\ 1 & 0 & 0 \\ 0 & 1 & 0 \\ 0 & 0 & 1 \end{bmatrix} \quad \text{... (vii)}$$

The error vector E is a 1 × 6 size vector. Assuming the second bit in error, we can get the error vector as under:

$$E = [0 \; \textcircled{1} \; 0\,0\,0\,0]$$

where the encircled bit shows the bit in error

Hence, the syndrome is given by,

$$S = EH^T = [0 \; \textcircled{1} \; 0\,0\,0\,0]\begin{bmatrix} 1 & 1 & 1 \\ 1 & 1 & 0 \\ 1 & 0 & 1 \\ 1 & 0 & 0 \\ 0 & 1 & 0 \\ 0 & 0 & 1 \end{bmatrix}$$

Therefore,

$$S = [0 \oplus 1 \oplus 0 \oplus 0 \oplus 0 \oplus 0, 0 \oplus 1 \oplus 0 \oplus 0 \oplus 0 \oplus 0, 0 \oplus 0 \oplus 0 \oplus 0 \oplus 0 \oplus 0]$$

or $$S = [1\,1\,0]$$

This is the syndrome corresponding to second bit in error. Observe carefully that it is same as the second row of the H^T matrix. The other syndromes can be obtained directly from the rows of H^T. The decoding table consisting of the error patterns and corresponding syndrome vectors is shown in Table 8.12.

Table 8.12. Decoding Table

S.No.	Error vector E (with single bit error pattern)						Syndrome vectors			Relation with H^T
1.	0	0	0	0	0	0	0	0	0	
2.	(1)	0	0	0	0	0	1	1	1	← 1st Row of H^T
3.	0	(1)	0	0	0	0	1	1	0	← 2nd Row of H^T
4.	0	0	(1)	0	0	0	1	0	1	← 3rd Row of H^T
5.	0	0	0	(1)	0	0	1	0	0	← 4th Row of H^T
6.	0	0	0	0	(1)	0	0	1	0	← 5th Row of H^T
7.	0	0	0	0	0	(1)	0	0	1	← 6th Row of H^T

(v) Let us obtain the decoding of the received words.

The first given codeword is 101100. But, this codeword does not exist in the code word table. This shows that the error must be present in the received code vector. Let us represent the received code word as under:

$$Y_1 = [1\,0\,1\,1\,0\,0]$$

The syndrome for this code word is given by,

$$S = Y_1 H^T$$

or $$S = [1\,0\,1\,1\,0\,0]\begin{bmatrix} 1 & 1 & 1 \\ 1 & 1 & 0 \\ 1 & 0 & 1 \\ 1 & 0 & 0 \\ 0 & 1 & 0 \\ 0 & 0 & 1 \end{bmatrix}$$

$$S = [1 \oplus 0 \oplus 1 \oplus 1 \oplus 0 \oplus 0, 1 \oplus 0 \oplus 0 \oplus 0 \oplus 0 \oplus 0, 1 \oplus 0 \oplus 1 \oplus 0 \oplus 0 \oplus 0]$$

or $$S = [1\,1\,0]$$

Thus, the syndrome of the received word is [1 1 0] which is same as the second syndrome in the decoding table. Hence, the corresponding error pattern is given by

$$E = [0 \; \textcircled{1} \; 0\,0\,0\,0]$$

And the correct word can be obtained as under

$$X_1 = Y_1 \oplus E = [1\ 0\ 1\ 1\ 0\ 0] \oplus [0\ 1\ 0\ 0\ 0\ 0]$$

or $$X_1 = [1\ 1\ 1\ 1\ 0\ 0]$$

This is the correct transmitted word.

Similarly, we can perform the decoding of 000110.

Let X_2 = 000110 -------- is the second received codeword. Even this is not the valid codeword listed in codeword table (Table 8.10). The syndrome for this can be obtained as under:

$$S = Y_2\, H^T$$

or $$S = [0\ 0\ 0\ 1\ 1\ 0] \begin{bmatrix} 1 & 1 & 1 \\ 1 & 1 & 0 \\ 1 & 0 & 1 \\ 1 & 0 & 0 \\ 0 & 1 & 0 \\ 0 & 0 & 1 \end{bmatrix}$$

or $$S = [1\ 1\ 0]$$

The error pattern corresponding to this syndrome is obtained from the decoding table as under :

$$E = [0\ \textcircled{1}\ 0\ 0\ 0\ 0]$$

Therefore, the correct codeword is given by,

$$X_2 = Y_2 \oplus E = [0\ 0\ 0\ 1\ 1\ 0] \oplus [0\ 1\ 0\ 0\ 0\ 0]$$

or $$X_2 = [0\ 1\ 0\ 1\ 1\ 0]$$

This is the correct transmitted word. **Ans.**

EXAMPLE 8.15. For a (6, 3) code, the generator matrix G is given by:

$$G = \begin{bmatrix} 1 & 0 & 0 & 1 & 0 & 1 \\ 0 & 1 & 0 & 0 & 1 & 1 \\ 0 & 0 & 1 & 1 & 1 & 0 \end{bmatrix}$$

(i) Realize an encoder for this code.

(ii) Verify that this code is a single error-correcting code.

(iii) If the received codeword is 100 011, find the syndrome.

Solution: (i) First, we obtain the expressions for the parity bits.

The parity bits can be obtained by using the expression:

$$C = MP$$

or $$[c_0, c_1, c_2]_{1\times 3} = [m_0\ m_1\ m_2]_{1\times 3}\ [P]_{3\times 3}$$

The parity matrix can be obtained from the generator matrix as under:

$$G = \left[\begin{array}{ccc|ccc} 1 & 0 & 0 & 1 & 0 & 1 \\ 0 & 1 & 0 & 0 & 1 & 1 \\ 0 & 0 & 1 & 1 & 1 & 0 \end{array}\right]$$

Therefore, $$[P] = \begin{bmatrix} 1 & 0 & 1 \\ 0 & 1 & 1 \\ 1 & 1 & 0 \end{bmatrix}_{3\times 3}$$

Also, $$[c_0, c_1, c_2] = [m_0, m_1, m_2] \begin{bmatrix} 1 & 0 & 1 \\ 0 & 1 & 1 \\ 1 & 1 & 0 \end{bmatrix}$$

From above, we get

$$c_0 = m_0 \oplus m_2$$

$$c_1 = m_1 \oplus m_2 \qquad \text{... (i)}$$

$c_2 = m_0 \oplus m_1$

Now, let us draw the encoder.

The encoder is obtained to implement the expressions given in equation (i) as shown in figure 8.27.

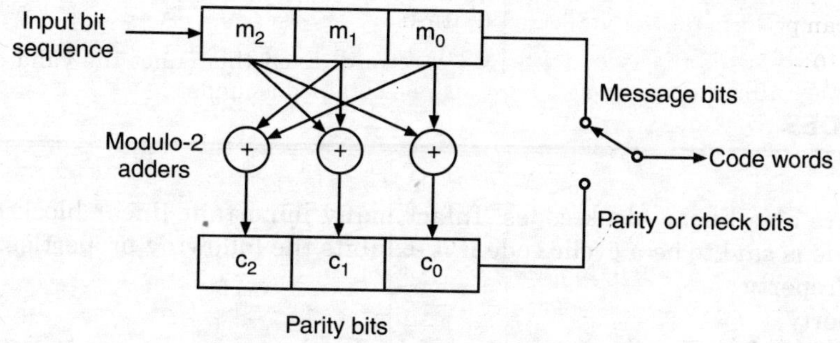

Fig. 8.27. Encoder block diagram

(ii) It can be proved that the minimum distance for the (6, 3) code is given by

$$d_{min} = 3$$

Therefore, number of errors can be corrected will be

$$d_{min} \geq 2t + 1$$

or $$3 \geq 2t + 1$$

or $$t \leq 1$$

Thus, at the most one error can be corrected.

(iii) To obtain the syndrome, let us obtain the transpose matrix H^T.

Again, first we have to obtain the parity check matrix H as under:

$$H = [P^T : I_{n-k}]$$

Here, $n - k = 3$ and let us obtain P^T from generator matrix.

$$G = \left[\begin{array}{ccc|ccc} 1 & 0 & 0 & 1 & 0 & 1 \\ 0 & 1 & 0 & 0 & 1 & 1 \\ 0 & 0 & 1 & 1 & 1 & 0 \end{array}\right]$$

Hence, we know the P matrix.

Thus, we have $$P^T = \begin{bmatrix} 1 & 0 & 1 \\ 0 & 1 & 1 \\ 1 & 1 & 0 \end{bmatrix}$$

Hence, the parity check matrix H will be

$$H = \left[\begin{array}{ccc|ccc} 1 & 0 & 1 & 1 & 0 & 0 \\ 0 & 1 & 1 & 0 & 1 & 0 \\ 1 & 1 & 0 & 0 & 0 & 1 \end{array}\right]$$

Now, let us obtain the transpose matrix H^T as under:

$$H^T = \begin{bmatrix} 1 & 0 & 1 \\ 0 & 1 & 1 \\ 1 & 1 & 0 \\ 1 & 0 & 0 \\ 0 & 1 & 0 \\ 0 & 0 & 1 \end{bmatrix}$$

We know that the syndrome is given by

$$\text{Syndrome } S = Y\,H^T$$

or
$$S = [1\ 0\ 0\ 0\ 1\ 1] \begin{bmatrix} 1 & 0 & 1 \\ 0 & 1 & 1 \\ 1 & 1 & 0 \\ 1 & 0 & 0 \\ 0 & 1 & 0 \\ 0 & 0 & 1 \end{bmatrix} = [1\ 1\ 1] \quad \textbf{Ans}$$

8.14 CYCLIC CODES

Definition

Cyclic codes are also linear block codes. Infact many important linear block codes are cyclic codes. A binary code is said to be a cyclic code if it exhibits the following properties:

(i) Linearity Property
(ii) Cyclic Property

(i) Linearity Property

A code is said to be linear if sum of any two code words also is a code word. This property states that the cyclic codes are linear block codes.

(ii) Cyclic Property

A code is said to be cyclic if any cyclic shift of a code word results in the formation of another code word. This is shown in figure 8.28. Let $(x_0, x_1, ..., x_{n-1})$ be an n-bit (n, k) linear block codes. This code is shifted right by 1 bit everytime in order to get the other codewords as shown in figure 8.28. All the n bit words obtained by the circular right shifting are new code words. This is called as the cyclic property of the cyclic codes.

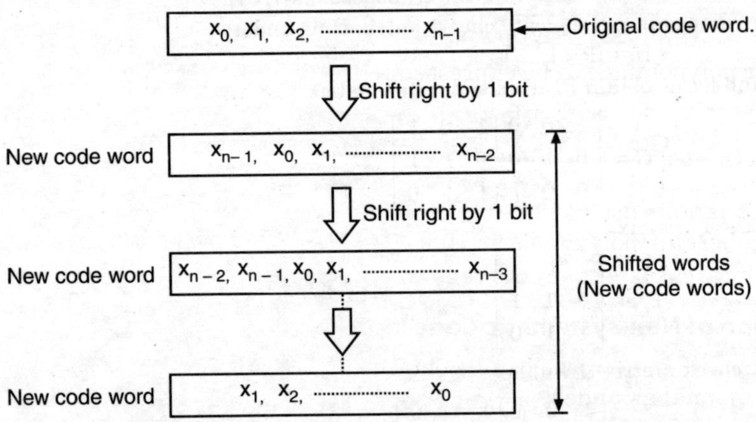

Fig. 8.28. Illustration of cyclic property of the cyclic codes

8.14.1. Code Word Polynomial

The cyclic property discussed earlier suggests that it is possible to treat the elements of code word of length n as the coefficients of a polynomial of a degree (n – 1). Thus, the code word:

$$[x_0, x_1, x_2, ..., x_{n-1}]$$

can be expressed in the form of a code word polynomial as under:

$$[X(p) = x_0 + x_1\, p + x_2\, p^2 + ... + x_{n-1}\, p^{n-1}] \quad ...\ (8.29)$$

where $\quad p$ = An arbitrary real variable.

Some important conclusions from the code word polynomial of equation (8.29) are as under :

(i) For binary codes, the coefficients p, p^2 ... are 1's and 0s.

(ii) Each power of p in the polynomial X(p) represents a one bit cyclic shift in time. Thus multiplication of the polynomial X(p) by p is equivalent to a cyclic shift or rotation to right provided that $p^n = 1$.

(iii) This special type of polynomial multiplication is known as Multiplication modulo $p^n - 1$. Thus if the polynomial X(p) is given as:

$$X(p) = x_0 + x_1 p + x_2 p^2 + ... + x_{n-1} p^{n-1}$$

then a single cyclic shift can be obtained by multiplying X(p) by p i.e.,

$$p\, X(p) \bmod (p^n - 1) = x_{n-1} + x_0 p + x_1 p^2 + ... x_{n-2} p^{n-1} \quad ...\ (8.30)$$

where mod is the abbreviation for modulo. The code word polynomial of equation (8.30) is nothing else but the polynomial representation of the code word $(x_{n-1}, x_0, x_1, ... x_{n-2})$.

(iv) In order to have two cyclic shifts, we must multiply X(p) by p^2 i.e.

$$p^2 X(p) \bmod (p^n - 1) = x_{n-2} + x_{n-1} p + ... + x_{n-3} p^{n-1} ...$$

This polynomial represents the code word $(x_{n-2}, x_{n-1}, x_0 ..., x_{n-3})$

Thus, each shifted word represents a different code word.

8.14.2. Generator Polynomial for the Cyclic Code

The generator polynomial of cyclic code is represented by G(p). It is used for the generation of cyclic code words, from the message bits. The generation of code word polynomial can be represented as under:

$$X(p) = M(p)\,.\,G(p) \quad ...\ (8.31)$$

where M(p) = Message signal polynomial of degree $\leq k$

G(p) = Generating polynomial of degree (n – k), which is given by,

$$G(p) = 1 + g_1 p + g_2 p^2 + ... + g_{n-k-1} p^{n-k-1} + p^{n-k} \quad ...\ (8.32)$$

The generator polynomial can be expressed in the summation form as under:

$$G(p) = 1 + \sum_{i=1}^{n-k-1} g_i p^i + p^{n-k} \quad ...\ (8.33)$$

This expression is quite useful while realizing the encoder for the cyclic codes. The other important point about the generator polynomial is that the degree of generator polynomial is equal to the number of parity bits (check bits) in the code word.

8.14.3. Generation of Non-systematic Code Words

The non-systematic code words can be obtained by multiplication of message polynomial with the generator polynomial as under:

$$X_1(p) = M_1(p)\,.\,G(p)$$

$$X_2(p) = M_2(p)\,.\,G(p)$$

$X_3(p) = M_3(p)\,.\,G(p)$ ... and so on.

8.14.4. Generation of Systematic Code Words

There are three steps involved in the encoding process for a systematic (n, k) cyclic code. They are as under:

(i) We multiply the message polynomial M(p) by p^{n-k} to get $p^{n-k} M(p)$. This multiplication is equivalent to shifting the message bits by (n – k) bits.

(ii) We divide the shifted message polynomial $p^{n-k} M(p)$ by the generator polynomial G(p) to obtain the remainder C(p).

$$\frac{p^{n-k}M(p)}{G(p)} = Q(p) \oplus \frac{C(p)}{G(p)} \quad ...(8.34)$$

where Q(p) = Quotient.

Note that all the additions are modulo – 2 additions.

(iii) We add the remainder polynomial C(p) to the shifted message polynomial p^{n-k} M(p) to obtain the code word polynomial X(p).

Therefore, code word polynomial $X(p) = [p^{n-k} M(p)] \oplus C(p)$ ... (8.35)

Therefore, the code word polynomial X(p) is exactly divisible by the generator polynomial G(p).

EXAMPLE 8.16. The generator polynomial for a (7, 4) cyclic Hamming code is given by,

$$G(p) = 1 + p + p^3$$

Determine all the systematic and non-systematic code vectors.

Solution: First, let us obtain the non-systematic code vectors.

(i) This is a (7, 4) cyclic Hamming code. Therefore, the message vectors are going to be 4 bit long. There will be total $2^4 = 16$ message words. Let us consider any message code word as under

$$M = (m_0\ m_1\ m_2\ m_3) = (1\ 0\ 1\ 0)$$

Therefore, the message polynomial is given by,

$$M(p) = m_0 + m_1\, p + m_2\, p^2 + m_3\, p^3 \quad ... (i)$$

Substituting the values of m_0 m_1, m_2 and m_3, we obtain

$$M(p) = 1 + p^2$$

(ii) The generator polynomial is given by,

$$G(p) = 1 + p + p^3$$

(iii) The non-systematic cyclic code is given by,

$$X(p) = M(p) \cdot G(p)$$
$$= (1 + p^2)(1 + p + p^3)$$
$$= 1 + p + p^3 + p^2 + p^3 + p^5$$

or $$X(p) = 1 + p + p^2 + p^3 (1 \oplus 1) + p^5$$

But, $1 \oplus 1 = 0$

Therefore, $$X(p) = 1 + p + p^2 + 0\, p^3 + 0\, p^4 + p^5 + 0\, p^6$$

Note that the degree of the code word polynomial is 6 i.e. (n – 1). The code word polynomial is given by,

$$X = (1\ 1\ 1\ 0\ 0\ 1\ 0)$$

This is the code word for the given message word. Similarly, we can obtain the other non-systematic code words.

Now, let us obtain the systematic code vectors.

To obtain the systematic code words, let us follow the following procedure:

(i) We multiply M(p) by p^{n-k}.

Therefore, $$p^{n-k} M(p) = p^3 (1 + p^2)$$

or $$p^{n-k} M(p) = p^3 + p^5$$

(ii) We divide by p^{n-k} M(p) by the generator polynomial.

Therefore, $$\frac{p^{n-k}M(p)}{G(p)} = \frac{p^3 + p^5}{1 + p + p^3}$$

The division is carried out as under:

$$\begin{array}{l} \qquad\qquad\qquad\quad p^2 + 0p + 0 \leftarrow \text{Quotient polynomial } Q(p) \\ p^3 + 0p^2 + p + 1 \,\big)\, p^5 + 0p^4 + p^3 + 0p^2 + 0p + 0 \end{array}$$

$$\begin{array}{ll} \text{Mod 2 addition} & p^5 + 0p^4 + p^3 + p^2 \\ & \oplus \quad \oplus \quad \oplus \quad \oplus \\ & 0 + 0 + 0 + 0 \\ & \quad p^2 + 0p + 0 \\ & \text{Remainder polynomial } C(p) \end{array}$$

Fig. 8.29. Division of $p^{n-k} M(p)$ by G(p)

Thus, the results of the division are as under:

Quotient polynomial $Q(p) = 0 + 0p + p^2$

Remainder polynomial $C(p) = 0 + 0p + p^2 \leftarrow$ Represents the parity bits.

(iii) We obtain the code word polynomial X(p).

The code word polynomial can be obtained by adding $p^{n-k} M(p)$ to the remainder polynomial C(p).

Therefore, $\quad X(p) = [p^{n-k} M(p)] \oplus B(p)$

$$= [0 + 0p + 0p^2 + p^3 + 0p^4 + p^5 + 0p^6] \oplus [0 + 0p + p^2]$$

or $\quad X(p) = [0 + 0p + p^2 + p^3 + 0p^4 + p^5 + 0p^6]$

The code word vector is given by: [0 0 1: 1 0 1 0]

Code word is given by

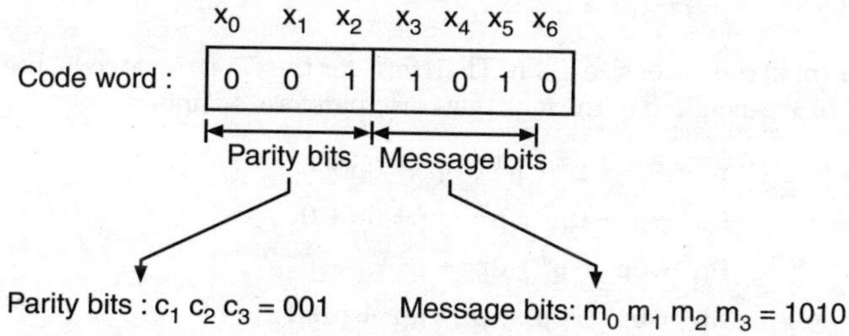

Fig. 8.30. Code word

Note that with this procedure, we get a code vector in the format as shown in figure 8.30 where the parity bits appear first followed by the message bits. This format of a code word is also a valid format. However, we can modify the code formats as shown in figure 8.31.

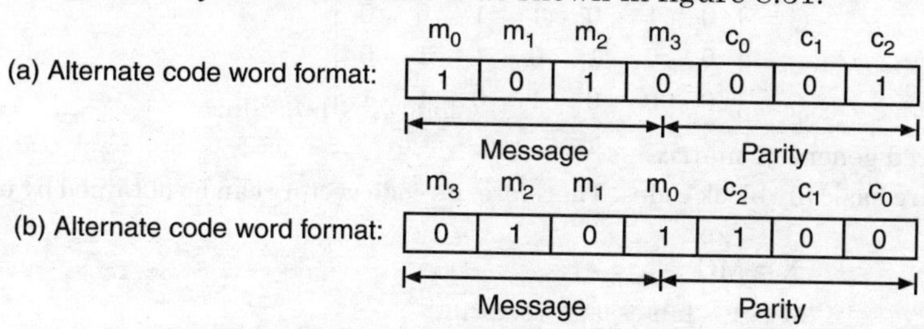

Fig. 8.31. Alternate formats of the code vectors

Figure 8.31 tells us that the values of m_0, m_1, m_2, m_3 or c_0, c_1 and c_2 remain unchanged. Only the way they are arranged is different in different codeword formats.

We can follow the same procedure to generate the remaining code words.

8.14.5. Generator and Parity Check Matrices of the Cyclic Codes

The cyclic codes are linear block codes. Therefore, we can define the generator and parity check matrices for the cyclic codes as well. The generator matrix G(p) has a size of k × n i.e., it has k rows and n columns. Let the generator matrix G(p) is given by,

$$G(p) = p^{n-k-1} + p^{n-k} g_{n-k-1} + ... + g_2 p^2 + g_1 p + 1 \quad ... (8.36)$$

We multiply both the sides by p^i, i.e.,

$$p^i G(p) = p^{n-k+i} + p^{n-k+i-1} + ... + g_2 p^{2+i} + g_1 p^{1+i} + p^i \quad ... (8.37)$$

where $i = (k-1), (k-2), ..., 2, 1, 0$

Equation (8.37) represents the polynomial for the rows of the generating polynomials. It is possible to obtain the generator matrix from this equation.

EXAMPLE 8.17. For a (7, 4) cyclic code, determine the generator matrix if $G(p) = 1 + p + p^3$.

Solution: Here, n = 7 and k = 4, hence, n – k = 3

$$G(p) = 1 + p + p^3$$

(i) We multiply both the sides of G(p) by p^i where i = (k – 1) ..., 1, 0.

Therefore, $p^i G(p) = p^i + p^{1+i} + p^{3+i}$, i = (k – 1) ..., 1, 0. ... (i)

But, i = 4, ∴ i = 3, 2, 1, 0.

(ii) By substituting these values of i into equation (i), we get four different polynomials as under:

These polynomials correspond to the four rows of the generator matrix as under:

Row No. 1: i = 3 → $p^3 G(p) = p^3 + p^4 + p^6$

Row No. 2: i = 2 → $p^2 G(p) = p^2 + p^3 + p^5$

Row No. 3: i = 1 → $p\ G(p) = p + p^2 + p^4$

Row No. 4: i = 0 → $G(p) = 1 + p + p^3$

The generator matrix for (n, k) code is of size k × n. Therefore, for the (7, 4) cyclic code, the generator matrix will be 4 × 7 matrix. The polynomials corresponding to the four rows are therefore, as under:

Row No. 1: i = 3 → $p^6 + 0p^5 + p^4 + p^3 + 0p^2 + 0p + 0$

Row No. 2: i = 2 → $0p^6 + p^5 + 0p^4 + p^3 + p^2 + 0p + 0$

Row No. 3: i = 1 → $0p^6 + 0p^5 + p^4 + 0p^3 + p^2 + p + 0$

Row No. 4: i = 0 → $0p^6 + 0p^5 + 0p^4 + p^3 + 0p^2 + p + 1$

These polynomials can be converted into generator matrix G as under:

$$G = \begin{bmatrix} p^6 & p^5 & p^4 & p^3 & p^2 & p^1 & p^0 \\ 1 & 0 & 1 & 1 & 0 & 0 & 0 \\ 0 & 1 & 0 & 1 & 1 & 0 & 0 \\ 0 & 0 & 1 & 0 & 1 & 1 & 0 \\ 0 & 0 & 0 & 1 & 0 & 1 & 1 \end{bmatrix}_{4\times 7}$$

This is the required generator matrix.

The cyclic codes are basically block codes. Therefore, its code vectors can be obtained by using the generator matrix as under:

$$X = MG \quad ... (8.38)$$

where, M = 1 × k message vector.

EXAMPLE 8.18. For the generator matrix of the previous example, determine all the possible code vectors.

Solution: All the code vectors can be obtained by using the following expression:

$$X = M\,G$$

Let $M = (m_3\, m_2\, m_1\, m_0) = (1\,0\,1\,0)$

Therefore, $X = [1\,0\,1\,0]\begin{bmatrix} 1 & 0 & 1 & 1 & 0 & 0 & 0 \\ 0 & 1 & 0 & 1 & 1 & 0 & 0 \\ 0 & 0 & 1 & 0 & 1 & 1 & 0 \\ 0 & 0 & 0 & 1 & 0 & 1 & 1 \end{bmatrix}$

Therefore, $X = [1 \oplus 0 \oplus 0 \oplus 0 \quad 0 \oplus 0 \oplus 0 \oplus 0 \quad 1 \oplus 0 \oplus 1 \oplus 0 \quad 1 \oplus 0 \oplus 0 \oplus 0$

$0 \oplus 0 \oplus 1 \oplus 0 \quad 0 \oplus 0 \oplus 1 \oplus 0 \quad 0 \oplus 0 \oplus 0 \oplus 0]$

Therefore, we have $X = [1\,0\,0\,1 : 1\,1\,0]$

Similarly, the other code vectors can be obtained.

> **Important Point:** *It may be noted that the generator matrix G is of non-systematic form and the code vector X also is of non-systematic form. Therefore, the first four bits of X are not the same as the message bits.*

8.14.6. Systematic Form of Generator Matrix

We know that the generator matrix in the systematic form is given by

$$G = [I_k : P_{k \times (n-k)}]_{k \times (n-k)} \quad \text{... (8.39)}$$

This means that there are k number of rows in the generator matrix. Let us represent the row number (in general) by i. The ith row of the generator matrix is represented by,

$$i^{th} \text{ row of } G = p^{(n-i)} + R_i(p) \quad \text{... where } i = 1, 2, ..., k. \quad \text{... (8.40)}$$

Now, we divide $p^{(n-i)}$ by the generator matrix G(p). The result of the division is expressed as,

$$\frac{p^{(n-i)}}{G(p)} = \text{Quotient} + \frac{\text{Remainder}}{G(p)} \quad \text{... (8.41)}$$

Let Remainder = $R_i(p)$

Quotient = $Q_i(p)$

Substituting this into equation (8.41), we obtain

$$\frac{p^{(n-i)}}{G(p)} = Q_i(p) + \frac{R_i(p)}{G(p)} \quad \text{... (8.42)}$$

Simplifying this equation, we obtain

$$p^{(n-i)} = Q_i(p)\, G(p) \oplus R_i(p); \quad \text{where } i = 1, 2, ..., k \quad \text{... (8.43)}$$

In mod-2 additions, the addition and subtraction will yield the same result.

Therefore, $p^{(n-i)} \oplus R_i(p) = Q_i(p)\, G(p)$ ... (8.44)

From equation (8.40), the above expression represents the ith row of the systematic generator matrix.

EXAMPLE 8.19. For systematic (7, 4) cyclic code, determine the generator matrix and parity check matrix. Given: $G(p) = p^3 + p + 1$.

Solution:

(i) The ith row of the generator matrix is given by equation (8.44) as under:

$$p^{(n-i)} \oplus R_i(p) = Q_i(p)\, G(p); \quad \text{where } i = 1, 2, ..., k \quad \text{... (i)}$$

(ii) It is given that the cyclic code is systematic (7, 4) code.

Therefore, $n = 7$, $k = 4$ and $(n - k) = 3$

Substituting these values into the above expression, we obtain

$$p^{(7-i)} \oplus R_i(p) = Q_i(p) \cdot (p^3 + p + 1) \text{ ... } i = 1, 2, 3, ...4.$$

(iii) With i = 1, the above equation is given by,

$$p^6 \oplus R_i(p) = Q_i(p)\,(p^3 + p + 1) \quad \text{... (ii)}$$

Let us obtain the value of $Q_i(p)$. The quotient $Q_i(p)$ can be obtained by dividing $p^{(n-i)}$ by G(p) as per equation (8.41). Therefore, to obtain $Q_i(p)$, let us divide p^6 by $(p^3 + p + 1)$.

The division takes place as under:

$$\begin{array}{rl} & p^2 + 0p + 0 \leftarrow \text{Quotient polynomial } Q(p) \\ p^3 + 0p^2 + p + 1 \big) & p^5 + 0p^4 + p^3 + 0p^2 + 0p + 0 \\ & p^6 + p^4 + p^3 \\ \text{Mod - 2} \rightarrow & \oplus \quad \oplus \quad \oplus \\ \text{additions} & \\ & p^4 + p^3 \\ & p^4 + 0p^3 + p^2 + p \\ \text{Mod - 2} \rightarrow & \oplus \quad \oplus \quad \oplus \quad \oplus \\ \text{additions} & p^3 + p^2 + p \\ & p^3 + 0p^2 + p + 1 \\ \text{Mod - 2 addition} \rightarrow & \oplus \quad \oplus \quad \oplus \quad \oplus \\ & p^2 + 0p + 1 \quad \leftarrow \quad \text{Remainder polynomial } R_i(D) \end{array}$$

Here the quotient polynomial $Q_i(p) = p^3 + p + 1$

and the remainder polynomial $R_i(p) = p^2 + 0p + 1$

Substituting these values into equation (ii), we obtain

$$p^6 \oplus R_i(p) = (p^3 + p + 1)(p^3 + p + 1)$$

$$= p^6 + p^4 + p^3 + p^4 + p^2 + p + p^3 + p + 1$$

or

$$= p^6 + 0p^5 + (1 \oplus 1)\,p^4 + (1 \oplus 1)\,p^3 + p^2 + (1 \oplus 1)\,p + 1$$

$$= p^6 + 0p^5 + 0p^4 + 0p^3 + p^2 + 0p + 1$$

Therefore, 1st Row polynomial $\Rightarrow p^6 + 0p^5 + 0p^4 + 0p^3 + p^2 + 0p + 1$

or, 1st Row Elements $\Rightarrow$ 1 0 0 0 1 0 1

Using the same procedure, we can obtain the polynomials for the other rows of the generator matrix as under:

2nd Row polynomial $\Rightarrow p^5 + p^2 + p + 1$

3rd Row polynomial $\Rightarrow p^4 + p^2 + p$

4th Row polynomial $\Rightarrow p^3 + p + 1$

These polynomials can be transformed into the generator matrix as under:

$$G = \begin{array}{r} \\ \text{Row 1} \rightarrow \\ \text{Row 2} \rightarrow \\ \text{Row 3} \rightarrow \\ \text{Row 4} \rightarrow \end{array} \begin{array}{c} p^6\ p^5\ p^4\ p^3\ p^2\ p^1\ p^0 \\ \left[\begin{array}{cccc|ccc} 1 & 0 & 0 & 0 & 1 & 0 & 1 \\ 0 & 1 & 0 & 0 & 1 & 1 & 1 \\ 0 & 0 & 1 & 0 & 1 & 1 & 0 \\ 0 & 0 & 0 & 1 & 0 & 1 & 1 \end{array}\right]_{4\times 7} \end{array}$$

This is the required generator matrix.

The parity check matrix is given by:

$$H = [P^T : I_{3\times 3}]$$

The transpose matrix P^T is given by interchanging the rows and columns of the P matrix.

$$P^T = \begin{bmatrix} 1 & 1 & 1 & 0 \\ 0 & 1 & 1 & 1 \\ 1 & 1 & 0 & 1 \end{bmatrix}_{3\times 4}$$

Hence, the parity check matrix is given by,

$$H = \left[\begin{array}{cccc|ccc} 1 & 1 & 1 & 0 & 1 & 0 & 0 \\ 0 & 1 & 1 & 1 & 0 & 1 & 0 \\ 1 & 1 & 0 & 1 & 0 & 0 & 1 \end{array}\right]_{3\times7}$$

This is the required parity check matrix.

EXAMPLE 8.20. Obtain the code vectors for the (7, 4) code having $G(p) = 1 + p + p^3$

Solution: (i) The code vectors in the systematic form can be obtained as under:

$$X = M\,G \quad \text{... (i)}$$

where M = Message matrix

and G = Generator matrix.

(ii) Let the message vector be 1 0 1 0. Let us use the generator matrix of the previous example to write,

$$M[m_3\ m_2\ m_1\ m_0] = [1\ 0\ 1\ 0]$$

Therefore,

$$X = MG = [1\ 0\ 1\ 0]\left[\begin{array}{cccc|ccc} 1 & 0 & 0 & 0 & 1 & 0 & 1 \\ 0 & 1 & 0 & 0 & 1 & 1 & 1 \\ 0 & 0 & 1 & 0 & 1 & 1 & 0 \\ 0 & 0 & 0 & 1 & 0 & 1 & 1 \end{array}\right]$$

Therefore, $X = [1\ 0\ 1\ 0 : 0\ 1\ 1]$ **Ans.**

This is the required code word.

8.14.7. Encoder for Cyclic Codes

The encoder for an (n, k) cyclic codes is as shown in figure 8.34. This encoder is useful for generating the systematic cyclic codes. The symbols used in the block diagram of the encoder are as shown in figure 8.32.

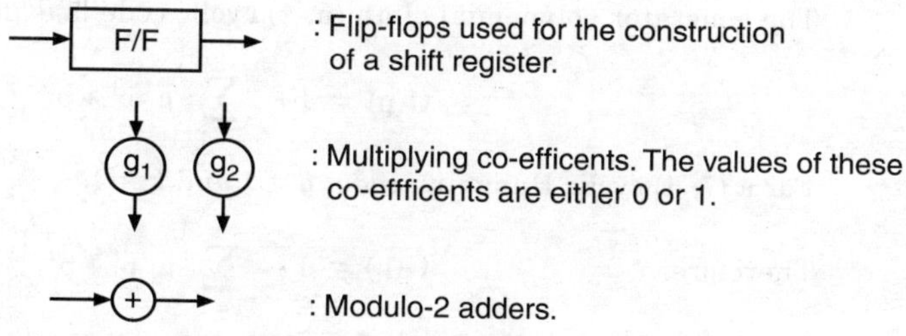

Fig. 8.32. Symbols used in the block diagram of encoder of cyclic codes.

Working operation of the encoder

The flip-flops (F/F) of figure 8.34 are used to construct a shift register. Operation of all these flip-flops is governed by an external clock, which is not shown in figure 8.34. The flip-flop contents will get shifted in the direction of the arrow corresponding to each clock pulse. The feedback switch is closed and the output switch is connected to the message input. All the flip-flops are initialized to zero state. First k message bits are shifted to the transmitter and also shifted into the shift register. After shifting the k message bits the shift register will contain the (n – k) parity (or check) bits. Therefore, after shifting the k message bits, the feedback switch is open circuited and the output switch is thrown to parity bit position. Now with every shift, the parity bits are transmitted over the channel. Thus, this encoder generates the code words in the format shown in figure 8.33.

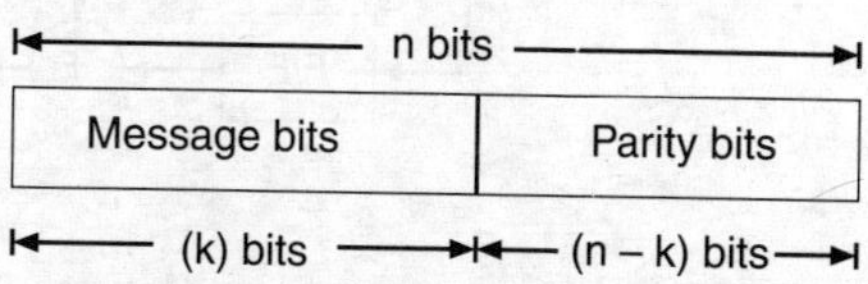

Fig. 8.33. Format of the code word

The encoder thus performs the division operations and generates the remainder. The remainder is nothing but the parity bits. When all the message bits are shifted out, but remains inside the shift register is remainder. The encoder also consists of modulo-2 adders. The output of the coefficient multipliers i.e., g_1, g_2 .... etc. are added to the flip-flop outputs to generate the parity bits.

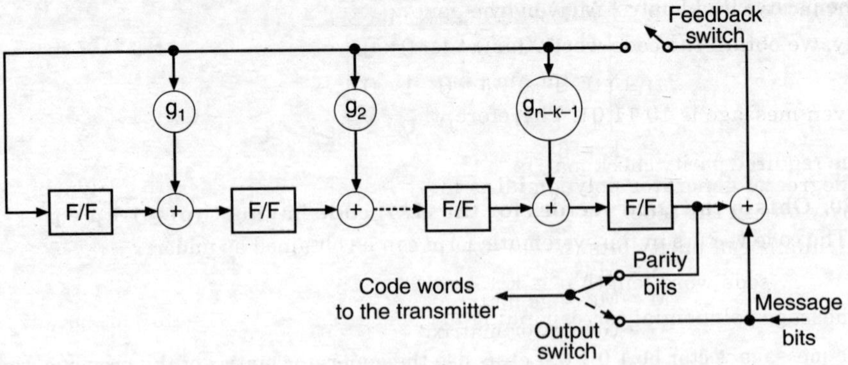

Fig. 8.34. Encoder for an (n, k) cyclic code

EXAMPLE 8.21. Draw the encoder for a (7, 4) cyclic Hamming code generated by the generator polynomial, $G(p) = 1 + p + p^3$.

Solution: The generator polynomial is given by

$$G(p) = p^3 + 0\,p^2 + p + 1 \qquad \text{...(i)}$$

The generator polynomial of an (n, k) cyclic code is expressed as under:

$$G(p) = 1 + \sum_{i=1}^{n-k-1} g_i\, p^i + p^{n-k} \qquad \text{...(ii)}$$

For a (7, 4) cyclic Hamming code, n = 7 and k = 4.

Therefore,

$$G(p) = 1 + \sum_{i=1}^{7-4-1} g_i\, p^i + p^{7-4} = 1 + g_1\, p + g_2\, p^2 + p^3$$

Hence,

$$G(p) = p^3 + g_2\, p^2 + g_1\, p + 1 \qquad \text{...(iii)}$$

Comparing equations (i) and (iii), we get obtain

$$g_1 = 1 \text{ and } g_2 = 0 \qquad \text{...(iv)}$$

Thus, the encoder for a (7, 4) Hamming code is shown in figure 8.35.

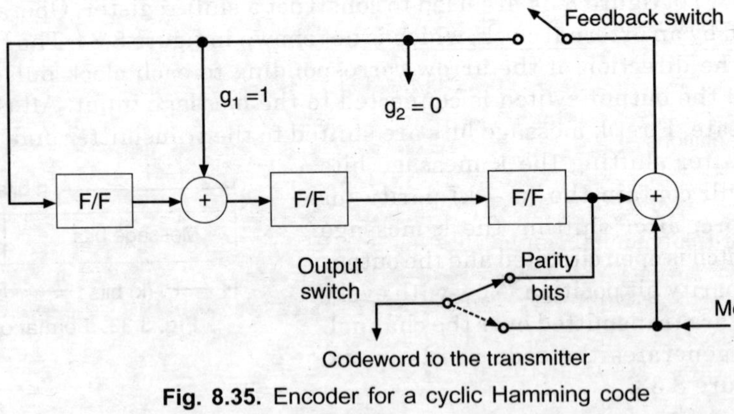

Fig. 8.35. Encoder for a cyclic Hamming code

EXAMPLE 8.22. A message 10 11 01 is to be transmitted in cyclic code with a generator polynomial $G(p) = p^4 + p^3 + 1$. Obtain the transmitted code word. How many check bits (i.e., parity bits) does the encoded message contain ? Draw the encoding arrangement for the same.

Solution: (i) First, we obtain the message polynomial M(p) i.e.,

$$M(p) = m_0 + m_1 p + m_2 p^2 + m_3 p^3 + m_4 p^4 + m_5 p^5$$

(ii) Then, we multiply M(p) by p^{n-k} to obtain $p^{n-k}\, M(p)$.

(iii) After that, we divide $p^{n-k} M(p)$ by the generator polynomial G(p).

(iv) Lastly, we obtain the code word X as under:

$$X = [m_5, m_4, m_3, m_2, m_1, m_0: c_3\ c_2\ c_1\ c_0]$$

(i) The given message is 10 11 01. Therefore, there are 6 message bits.

Thus, $k = 6$

(ii) The degree of generator polynomial is four. The degree of generator polynomial is equal to the number of parity bits in the code.

Therefore, number of parity bits $(n - k) = 4$

So, code word length $n = k + 4 = 10$

(iii) The message polynomial corresponding to the message 10 11 01 = $[m_5, m_4, m_3, m_2, m_1, m_0]$ is given as,

$$M(p) = m_0 + m_1 p + m_2 p^2 + m_3 p^3 + m_4 p^4 + m_5 p^5 \quad \text{...(i)}$$

Substituting the values of message bits, we obtain

$$M(p) = 1 + 0p + p^2 + p^3 + 0\,p + p^5 \quad \text{...(ii)}$$

(iv) The generator polynomial can be written in the form as under:

$$G(p) = 1 + 0\,p + 0\,p^2 + p^3 + p^4 \quad \text{...(iii)}$$

(v) Let us multiply the message polynomial M(p) by p^{n-k}

Since $(n - k) = 4$; so $p^{n-k} = p^4$

Therefore, $p^4 M(p) = p^4 [1 + 0\,p + p^2 + p^3 + p^4 + p^5]$

or $p^4 M(p) = p^4 + 0\,p^5 + p^6 + p^7 + 0\,p^8 + p^9 \quad \text{...(iv)}$

(vi) Now, we divide $p^4 M(p)$ by the generator polynomial

The division takes place as under:

$p^4 + p^3 + 0p^2 + 0p + 1$) $p^9 + 0p^8 + p^7 + p^6 + 0p^5 + p^4$ ($p^5 + p^4 + p^2$ ← Quotient polynomial

$p^9 + p^8 + 0p^7 + 0p^6 + p^5$

Modulo-2 additions → ⊕ ⊕ ⊕ ⊕ ⊕

$p^8 + p^7 + p^6 + p^5 + p^4$

$p^8 + p^7 + 0p^6 + 0p^5 + p^4$

Modulo-2 additions → ⊕ ⊕ ⊕ ⊕ ⊕

$p^6 + p^5$

$p^6 + p^5 + 0p^4 + 0p^3 + p^2$

Modulo-2 additions → ⊕ ⊕ ⊕ ⊕ ⊕

p^2 ← Remainder polynomial

(vii) The code word polynomial can be obtained by adding $p^{n-k} M(p)$ to the remainder polynomial C(p).

Thus, $X(p) = p^4 M(p) + C(p) \quad \text{...(v)}$

But, $C(p) = c_0 + c_1 p + c_2 p^2 + \ldots\ c_{n-k-1}\, p^{n-k-1} \quad \text{...(vi)}$

Here, $(n - k - 1) = (10 - 6 - 1) = 3$

Therefore, $C(p) = c_0 + c_1 p + c_2 p^2 + c_3 p^3 \quad \text{...(vii)}$

But we have obtained $c(p) = p^2$, so, comparing the two, we obtain $c_0 = 0$, $c_1 = 0$, $c_2 = 1$ and $c_3 = 0$.

Therefore, Parity bits $(c_3, c_2, c_1\ c_0) = (0\ 1\ 0\ 0) \quad \text{...(viii)}$

Therefore, the code word is given as under:

$$X = [m_5, m_4, m_3, m_2, m_1, m_0: c_3\ c_2\ c_1\ c_0]$$

Thus, Code word X = [1 0 1 1 0 1: 0 1 0 0]

And the corresponding code word polynomial is given as under:

$$X(p) = p^9 + 0\,p^8 + p^7 + p^6 + 0p^5 + p^4 + 0\,p^3 + p^2 + 0\,p + 0 \quad \textbf{Ans.}$$

The encoded message consists of 4 check bits or parity bits c_3, c_2, c_1 and c_0.

Now, let us draw the encoder for the cyclic code.

The generator polynomial is given as

$$G(p) = p^4 + p^3 + 1$$

or

$$G(p) = p^4 + p^3 + 0\,p^2 + 0\,p + 1 \quad ...(ix)$$

The generator polynomial of an (n, k) cyclic code is expressed as under:

$$G(p) = 1 + \sum_{i=1}^{n-k-1} g_i\, p^i + p^{n-k} \quad ...(x)$$

Substituting n = 10 and k = 6 and n – k = 4 in equation (x), we shall have

$$G(p) = 1 + \sum_{i=1}^{n-k-1} g_i\, p^i + p^{n-k}$$

or

$$G(p) = p^4 + g_3\, p^3 + g_2 p^2 + g_1 p + 1 \quad ...(xi)$$

Comparing equations (ix) and (xi) we get,

$$g_1 = 0$$

$$g_2 = 0$$

and

$$g_3 = 1 \quad ...(xii)$$

Therefore, the coding arrangement is shown in figure 8.36.

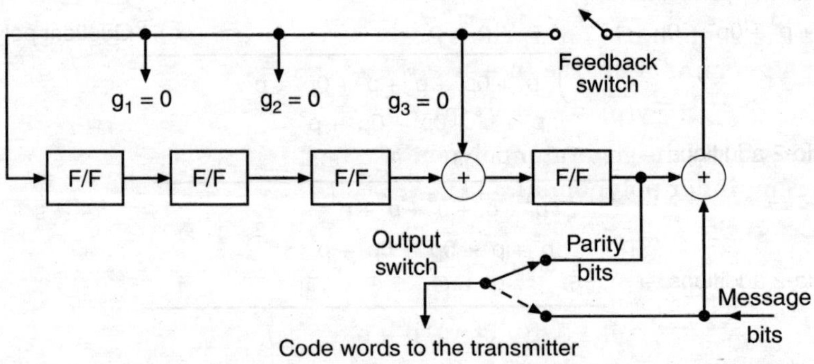

Fig. 8.36. Encoding arrangement

Advantages of cyclic codes

The advantage of cyclic codes over most of the other codes are as under:

(i) They are easy to encode.

(ii) They possess a well defined mathematical structure which has led to development of very efficient decoding schemes for them.

(iii) The methods that are to be used for error detection and correction are simpler and easy to implement.

(iv) These methods do not need look-up table decoding.

(v) It is possible to detect the error bursts using the cyclic codes.

Drawbacks of cyclic codes

Even though the error detection is simpler, the error correction is slightly more complicated. This is due to the complexity of the combinational logic circuit used for error correction.

8.14.8. Difference between Source Coding, Line Coding and Error Detection Codes

We have discussed various sources coding techniques such as PCM, Delta modulation, ADM etc. Source coding techniques are used just to convert an analog signal to its digitally coded equivalent signal. Infact, the output of a source coder is a train of binary digits i.e., 0s and 1s.

The line coding techniques convert the stream of binary digits into a format or code which is more suitable for transmission over a cable or any other medium. This is because the data transmission must satisfy certain requirements such as synchronization, elimination of dc component, low crosstalk etc. The line code format must be such that the bandwidth requirements must be as low as possible. Examples of line codes are RZ, NRZ AMI, Manchester and other formats. The error detection code words consist of two blocks. One block consists of the message bits and the other one consists of check bits or parity bits. Error detection and correction both will be possible at the receivers, which is not possible with the source coding or line coding. As we have already discussed, examples of error codes are Hamming codes, cyclic codes etc. The parity bits are related to message bits in a particular way.

8.14.9. Syndrome Decoding for the Cyclic Codes

When a code word X is transmitted over a noisy channel, errors are likely to get introduced into it. Thus, the received code word Y is different from X. For a linear block code, the first step in decoding is to calculate the syndrome for the received code word. If the syndrome is zero, then there are no transmission errors in the received code word. But, if the syndrome is non-zero, then received code word contains transmission errors which requires correction. In case of a cyclic code, in the systematic form, the syndrome can be calculated easily. Let the received code word be a polynomial of degree (n – 1) or less. Let it be given by,

$$y(p) = y_0 + y_1 (p) + ... + y_{n-1}\, p^{n-1} \qquad ..(8.45)$$

Now, we divide y(p) by the generator polynomial G(p). Let Q(p) represent the quotient polynomial and R(p) be the remainder polynomial.

Therefore,
$$\frac{y(p)}{G(p)} = Q(p) + \frac{R(p)}{G(p)} \qquad ...(8.46)$$

$$y(p) = Q(p)\,.\,G(p) + R(p) \qquad ...(8.47)$$

The remainder R(p) is a polynomial with degree (n – k – 1) or less. It is called as the syndrome polynomial. The coefficients of the syndrome polynomial will make up the (n – k) by 1 syndrome S.

Thus, syndrome polynomial S(p) = Remainder of $\dfrac{y(p)}{G(p)}$.

When the syndrome polynomial S(p) is non-zero, the presence of errors in the received code word is detected.

8.14.10. Syndrome Calculator for the Systematic Cyclic Codes

Figure 8.37 shows the syndrome calculator, that is identical to the encoder for the (n, k) cyclic code. The only difference between these two is that, here the received bits are fed into the (n – k) stages of the feedback shift register from the left. Initially, the output switch is connected to position 1 and all the flip-flops are in their reset mode. As soon, as all the received bits are shifted into the shift register, its contents will define the desired syndrome S. Once, we know the syndrome S, we can determine the corresponding error pattern E and then make the appropriate corrections. After shifting all the incoming bits of signal Y, the output switch is transferred to position 2 and clock pulses are applied to the shift register to out the syndrome. The following example will make the concept of syndrome calculation obvious.

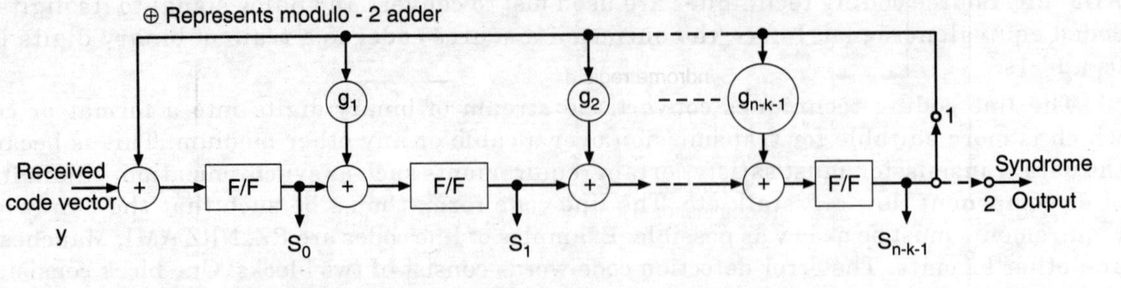

Fig. 8.37 Syndrome calculator for an (n, k) cyclic code.

EXAMPLE 8.23. Design a syndrome calculator for a (7, 4) Hamming code, generated by the generator polynomial G(p) = 1 + p + p³, if the transmitted and received code words are given by,

Transmitted code word, $X = (0\ 1\ 1\ 1\ 0\ 0\ 1)$

Received code word $Y = (0\ 1\ 1\ \textcircled{0}\ 0\ 0\ 1)$ — **Error**

Solution: The given generator polynomial is

$$G(p) = p^3 + 0\,p^2 + p + 1 \quad \text{...(i)}$$

We know that the general form of generator polynomial is as under:

$$G(p) = p^3 + g_2\,p^2 + g_1 p + 1 \quad \text{...(ii)}$$

Comparing equations (i) and (ii), we obtain

$$g_2 = 0$$
$$g_1 = 1 \quad \text{...(iii)}$$

Therefore, the required syndrome calculator is shown in figure 8.38.

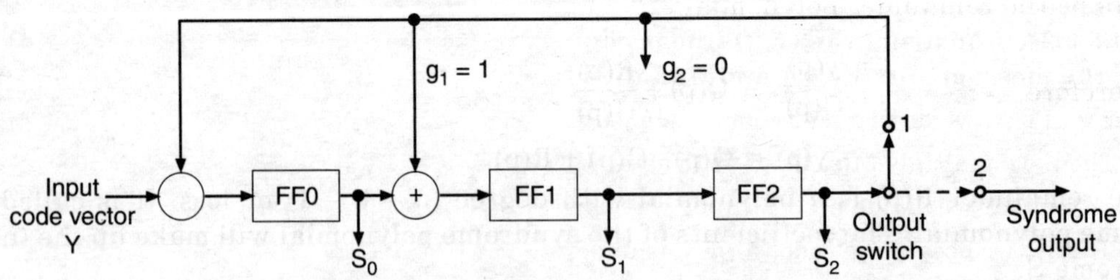

Fig. 8.38. Syndrome calculator for the (7, 4) cyclic code generated by the polynomial $G(p) = 1 + p + p^3$

8.14.11. Decoder for Cyclic Codes

As discussed earlier, once we calculate the syndrome, an error pattern E is detected corresponding to this syndrome.

This error vector is then added (modulo-2 addition) to the received code word Y, to get the corrected code vector X.

Therefore, Corrected code vector $X' = Y \oplus E$. ...(8.48)

The process of decoding has been illustrated in figure 8.39.

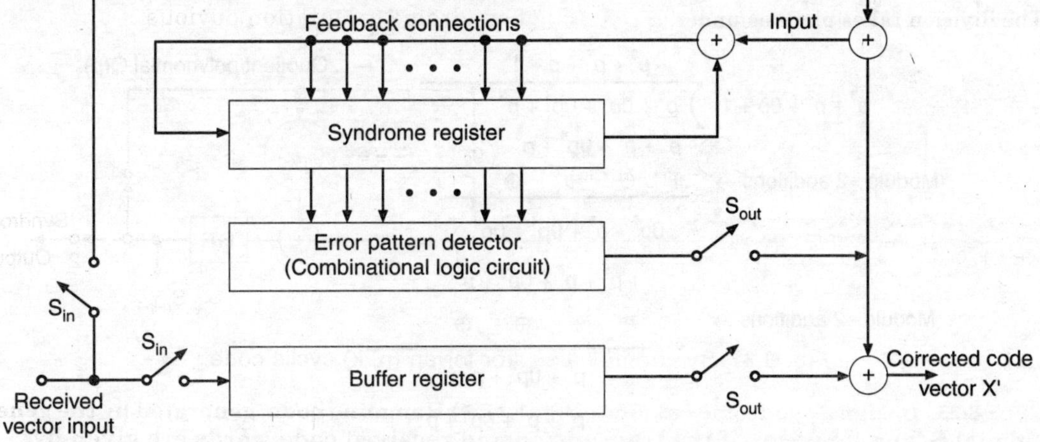

Fig. 8.39. General form of a decoder for cyclic code.

Working operation of the decoder

The switches S_{in} are closed and S_{out} are open circuited. The received code vector Y is then shifted into the buffer register and syndrome register. After shifting all the n bits of received code vector Y, the syndrome register holds the corresponding syndrome vector. Then, the contents of the syndrome register are read into the error pattern detector. This detector is a combinational logic circuit. A particular error pattern will be detected for each syndrome vector present in the syndrome register. Then the switches S_{in} are open circuited and S_{out} are closed. The contents of the buffer register, error register and syndrome register are then shifted. The received code vector Y (which is stored in the buffer register) , is then added with the error vector E(stored in the error register) bit by bit to obtain the corrected code word X at decoder output.

EXAMPLE 8.24. Construct the (7, 4) linear code word for the generator polynomial G(p) = 1 + p² + p³ for the message bits 1 0 0 1 and find the check sum for the same.

Solution: (i) First, we determine the message polynomial M(p).

(ii) Then, we multiply M(p) by p^{n-k} to obtain p^{n-k} M(p)

(iii) After that we divide p^{n-k} M(p) by the generator polynomial G(p).

(iv) Lastly, we obtain the code word X as under:

$$X = [m_3\ m_2\ m_1\ m_0 \colon c_2\ c_1\ c_0]$$

(i) The given message is 1 0 0 1, therefore, there are 4 message bits.

Thus, $[m_3\ m_2\ m_1\ m_0] = [1\ 0\ 0\ 1]$

Therefore, $M(p) = m_0 + m_1\, p + m_2\, p^2 + m_3\, p^3$

Substituting the values of message bits, we obtain

$$M(p) = 1 + 0\,p + 0\,p^2 + p^3$$

Therefore, $M(p) = 1 + p^3$ ...(i)

(ii) Let us multiply M(p) by p^{n-k}.

Given code is (7, 4) code. Therefore, n = 7 and k = 4. Hence, (n – k) = 3.

Therefore, we have

$$p^{n-k}\, M(p) = p^3\, [1 + 0\,p + 0\,p^2 + p^3]$$

$$p^3\, M(p) = p^3 + 0\,p^4 + 0\,p^5 + p^6 \quad \text{...(ii)}$$

(iii) Now, we divide $p^{n-k} M(p)$ by $G(p)$.

The division takes place as under:

$$p^3 + p^2 + p + 1 \quad \leftarrow \text{Quotient polynomial } Q(p)$$

$$p^3 + p^2 + 0p + 1 \;\overline{)\; p^6 + 0p^5 + 0p^4 + p^3}$$

$$p^6 + p^5 + 0p^4 + p^3$$

$$\text{Modulo - 2 additions} \rightarrow \quad \oplus \quad \oplus \quad \oplus \quad \oplus$$

$$0p^6 + p^5 + 0p^4 + 0p^3$$

$$+ p^5 + p^4 + 0p^3 + p^2$$

$$\text{Modulo - 2 additions} \rightarrow \quad \oplus \quad \oplus \quad \oplus \quad \oplus$$

$$p^4 + 0p^3 + p^2$$

$$p^4 + p^3 + 0p^2 + p$$

$$\text{Modulo - 2 additions} \rightarrow \quad \oplus \quad \oplus \quad \oplus \quad \oplus$$

$$p^3 + p^2 + p$$

$$p^3 + p^2 + 0p + 1$$

$$\text{Modulo -2 additions} \rightarrow \quad \oplus \quad \oplus \quad \oplus \quad \oplus$$

$$p + 1 \quad \leftarrow \text{Remainder polynomial } C(p)$$

Therefore, remainder polynomial $C(p) = p^2 + 0\,p + 0$ ...(iii)

(iv) Lastly, let us obtain the codeword polynomial:

Codeword polynomial $X(p) = p^3 M(p) + C(p)$

Substituting the values from equations (ii) and (iii) we get,

$$X(p) = [p^6 + 0p^5 + 0p^4 + p^3 + 0p^2 + 0p + 0] + [p^2 + 0p + 0]$$

or $$X(p) = p^6 + 0p^5 + 0p^4 + p^3 + p^2 + 0p + 0 \qquad \text{...(iv)}$$

Therefore, the codeword is given by

$$X = [m_3\, m_2\, m_1\, m_0 : c_2\, c_1\, c_0] = [\,1\,0\,0\,1 : 1\,0\,0\,] \quad \textbf{Ans.}$$

	m_3	m_2	m_1	m_0	c_2	c_1	c_0
X =	1	0	0	1	1	0	0
	Message bits				Parity bits		

8.14.12. Cyclic Redundancy Check (CRC)

This is a type of polynomial code is which a bit string is represented in the form of polynomials with coefficients of 0 and 1 only. Polynomial arithmetic uses a modulo-2 arithmetic i.e., addition and subtraction are identical to EX-OR. For CRC code, the sender and receiver must agree upon a generator polynomial G(x). A codeword can be generated for a given dataword (message) polynomial M(x) with the help of long division. This technique is more powerful than the parity check and checksum error detection. CRC is based on binary division. A sequence of redundant bits called CRC or CRC remainder is appended at the end of a data unit such as byte. The resulting data unit after adding CRC remainder becomes exactly divisible by another predetermined binary number. At the receiver, this data unit is divided by the same binary number There is no error if this division does not yield any remainder. But, a non-zero remainder indicates presence of errors in the received data unit. Such an erroneous data unit is then rejected.

Procedure to obtain CRC

The redundancy bits used by CRC may be derived by following the procedure given below:

(i) We divide the data unit by a predetermined divisor.

(ii) We obtain the remainder. It is the CRC.

Requirements of CRC

A CRC will be valid if and only if it satisfies the following requirements:

(i) It must have exactly one less bit than divisor.

(ii) Appending the CRC to the end of the data unit must result in the bit sequence which is exactly divisible by the divisor.

CRC generator

The CRC generator is shown in figure 8.40.

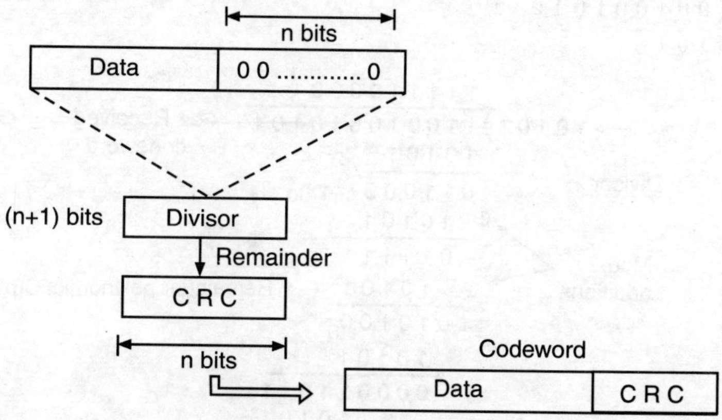

Fig. 8.40. CRC generator

The procedure for CRC generation is as under:

(i) We append a string of n 0s to the data unit where n is 1 less than the number of bits in the predecided divisor (n + 1 bit long).

(ii) We divide the newly generated data unit in step (i) by the divisor. This is a binary division.

(iii) The remainder obtained after the division in step (ii) is the n bit CRC.

(iv) This CRC will replace the n 0s appended to the data unit in step (i), to get the codeword to be transmitted as shown in figure 8.40.

CRC checker

Figure 8.41 shows the CRC checker.

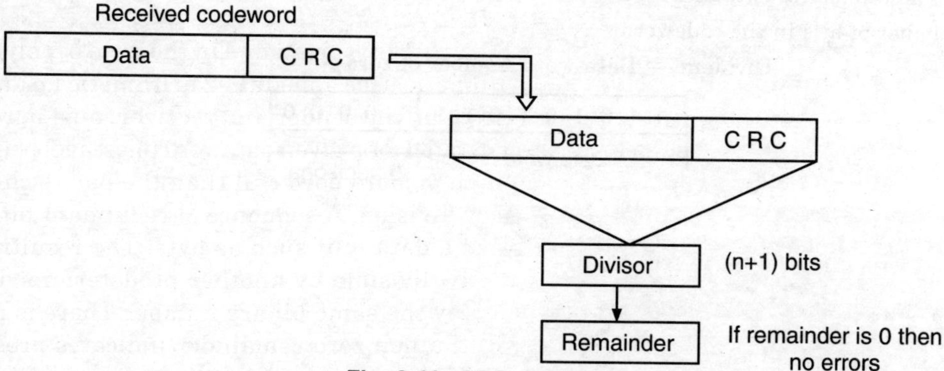

Fig. 8.41. CRC checker

> **Important Point:** *The codeword received at the receiver consists of data and CRC. The receiver treats it as one unit and divides it by the same (n + 1) bit divisor which was used at the transmitter. The remainder of this division is then checked. If the remainder is zero, then the received codeword is error free and hence must be accepted. But a non-zero remainder indicates presence of errors hence the corresponding codeword must be rejected.*

EXAMPLE 8.25. A codeword is received as 1100 1001 01011. Check whether there are errors in the received codeword, if the divisor is 10101. (The divisor corresponds to the generator polynomial).

Solution: We know that the codeword is formed by adding the dividend and the remainder. This codeword will have an important property that it will be completely divisible by the divisor. Thus, at the receiver we have to divide the received codeword by the same divisor and check for the remainder. If there is no remainder then there are no errors. But, if there is remainder after division, then there are errors in the received codeword. Let us use this technique and find if there are errors.

Data word: 1 1 0 0 1 0 0 1 0 1 0 1 1

Divisor: 1 0 1 0 1

```
                   1 1 1 1 1 0 0 0 0 1
          10101 ) 1 1 0 0 1 0 0 1 0 1 0 1 1  ← Received codeword
Divisor          ⊕1 0 1 0 1
                  0 1 1 0 0 0
                 ⊕  1 0 1 0 1
Mod-2               0 1 1 1 1 1
additions          ⊕  1 0 1 0 1
                      0 1 0 1 0 0
                     ⊕  1 0 1 0 1
                        0 0 0 0 1 1 0 1 1
                                1 0 1 0 1
                                0 1 1 1 0  ← Remainder
```

The non zero remainder shows that there are errors in the received codeword.

Generation of CRC code

The generation of CRC code is obvious after solving the following example.

EXAMPLE 8.26. Generate the CRC code for the data word of 1 1 0 0 1 0 1 0 1. The divisor is 1 0 1 0 1.

Solution: Given that

Data word: 1 1 0 0 1 0 1 0 1

Divisor: 1 0 1 0 1

The number of data bits = m = 9

The number of bits in the codeword = N

Dividend = Data word number of zeros.

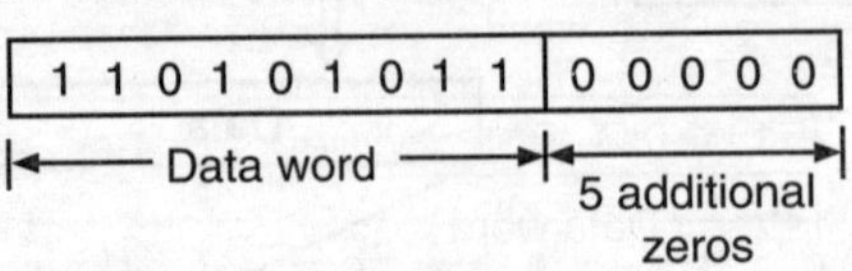

Now, let us carry out the division as under:

```
                1 1 1 1 1 0 0 1 1 1 1     ← Quotient
      ┌→10101 )1 1 0 0 1 0 1 0 1 0 0 0 0 0 ← Dividend
   Divisor      1 0 1 0 1
                ⊕⊕⊕⊕⊕
                0 1 1 0 0 0
                  1 0 1 0 1
                  0 1 1 0 1 1
                    1 0 1 0 1
                    0 1 1 1 0 0
                      1 0 1 0 1
                        1 0 0 1 1
                        1 0 1 0 1
                          1 1 0 0 0
                          1 0 1 0 1
                            1 1 0 1 0
                            1 0 1 0 1
                            1 1 1 1 0
                            1 0 1 0 1
                            1 0 1 1 0
                            1 0 1 0 1
                            0 0 1 1 0 ← Remainder
```

Codeword

In CRC the required codeword may be obtained by writing the data word followed by the remainder

Therefore, we have 1 1 0 0 1 0 1 0 1 0 0 0 0 0

1 1 0

Codeword = 1 1 0 0 1 0 1 0 1 0 0 1 1 0

Undetected errors in CRC

Important Point: *CRC cannot detect all types of errors. The probability of error detection and the types of detectable errors depends on the choice of divisor.*

EXAMPLE 8.27. Write the steps to compute the check sum in CRC code. Determine CRC for the frame 1 1 0 1 0 1 0 1 1 and the generator polynomial = $x^4 + x + 1$ and write the transmitted frame.

Solution: The generator polynomial actually acts as the divisor in the process of CRC generation.

Therefore, Data word : 1 1 0 1 0 1 0 1 1

Divisor : $x^4 + 0\,x^3 + 0\,x^2 + x + 1 = 1\,0\,0\,1\,1$

The number of data bits = m = 9

The number of bits in the codeword = N

Dividend = Data word + (N – m) number of zeros.

1 1 0 1 0 1 0 1 1	0 0 0 0 0
← Data word →	← 5 additional zeros →

Let us carry out the division as under:

```
                          1100001111
                    ___________________
             10011 ) 11010101100000
                     10011
Mod 2 addition→      ⊕⊕⊕⊕⊕
                     010011
                      10011
                      0000001100
                            10011
                            11111
                            10011
                            011000
                             10011
                             001010
                               10011
                               11001 ←— Remainder
```

Codeword

The codeword is given by

1 1 0 1 0 1 0 1 1	1 1 0 0 1
←— Data word —→	← Remainder →

EXAMPLE 8.28. A bit stream 10011101 is transmitted using the standard CRC method. The generator polynomial is $x^3 + 1$. Show the actual bit string transmitted. Suppose the third bit from left is inverted during transmission. Show that this error is detected at the receiver's end.

Solution: Given that, data word (Bit string): 1 0 0 1 1 1 0 1

Generator polynomial: $x^3 + 1 = x^3 + 0x^2 + 0x + 1 = 1\ 0\ 0\ 1$

We know that

$$\text{Dividend} = \text{Data word} + 4 \text{ zeros.}$$

The dividend will be as under:

Dividend →	1 0 0 1 1 1 0 1	0 0 0 0
	←— Data word —→	← 4 additional zeros →

Now, let us carry out the division as under:

```
                 100011001
          ________________
     1001 ) 100111010000
          ⊕ 1001
            00001101
                1001
                01000
                 1001
                 0001000
                    1001
                    0001 ← Remainder
```

The transmitted word may be obtained by writing the data word followed by the remainder as under:

Transmitted word = | 1 0 0 1 1 1 0 1 | 0 0 0 1 |

|←— Data word —→|←Remainder→|

Next let us write the erroneous received word.

The received word = 1 0 1 1 1 1 0 1 0 0 0 1

↑ Error

At the receiver, this word is divided by the same divider used at the transmitter i.e., 1001.

```
            101010000
      1001)101111010001  ← Received word
           1001
           001011
             1001
             001001
               1001
               00000001  ← Remainder
```

A non zero remainder shows that there is an error in the received codeword.

EXAMPLE 8.29. Why are cyclic codes effective in detecting error bursts ? The message 1001001010 is to be transmitted in a cyclic code with a generator polynomial g(x) = $x^2 + 1$.

(i) How many check bits does the encoded message contain ?

(ii) Obtain the transmitted code word.

(iii) Draw encoding arrangement to obtain remainder bits.

(iv) After the received word is clocked into the decoder input, what should be the content of the register stores ?

Solution: The message polynomial M(p) corresponds to the given message signal (1 0 0 1 0 0 1 0 1 0). This means that there are 10 message bits.

Hence, $k = 10$ ...(i)

Thus, the message polynomial is given by

$$M(p) = m_0 + m_1 p + m_2 p^2 + m_3 p^3 + m_4 p^4 + m_5 p^5 + m_6 p^6 + m_7 p^7 + m_8 p^8 + m_9 p^9.$$

$$= 0 + p + 0p^2 + 1p^3 + 0p^4 + 0p^5 + 1p^6 + 0p^7 + 0p^8 + 1p^9$$

or $M(p) = p^9 + p^6 + p^3 + p$ ...(ii)

The generator polynomial is given by

$G(p) = p^2 + 1$ ...(iii)

Hence, the degree of generator polynomial is 2. The degree of generator polynomial is equal to the number of parity (check) bits in the code.

so, number of check bits $= (n - k) = 2$ ...(iv)

Hence, length of the codeword $n = k + 2$

$n = 10 + 2 = 12$ ...(v)

Now, let us multiply M(p) by p^{n-k} i.e., by p^2 as under:

$$p^{n-k} . M(p) = p^2 [p^9 + p^6 + p^3 + p]$$

$$= p^{11} + p^8 + p^5 + p^3 \quad ...(vi)$$

Next, we divide p^{n-k} M([p) by the generator polynomial as under:

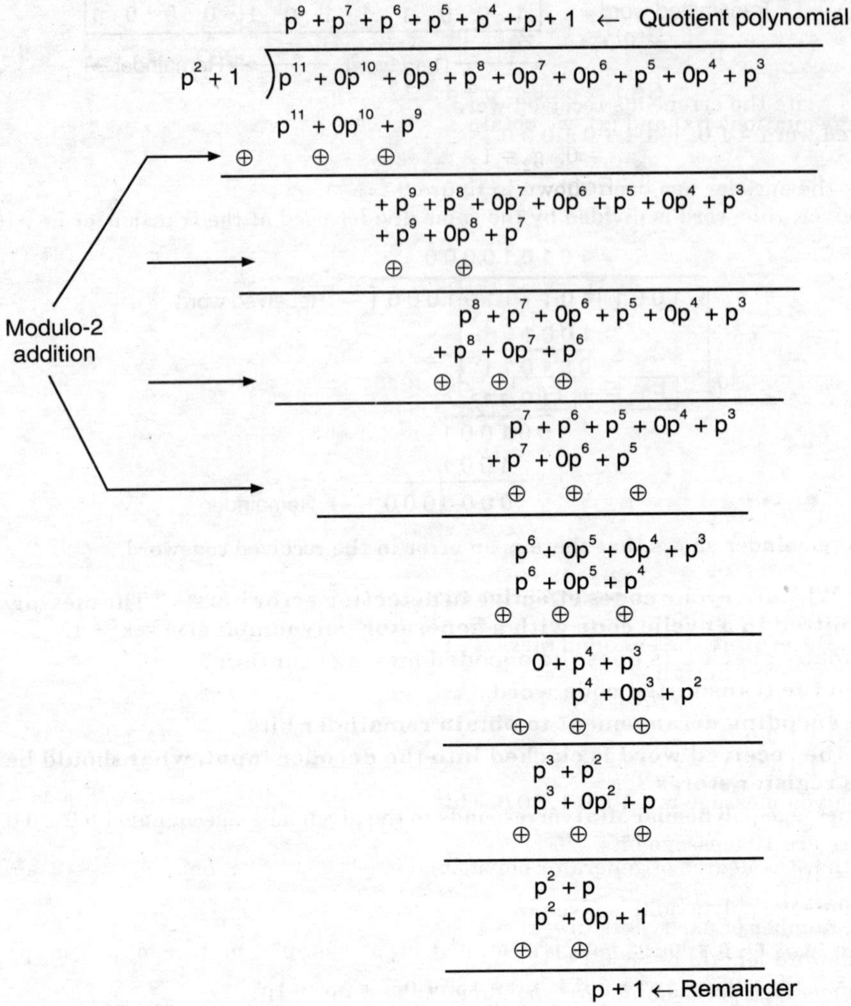

The codeword polynomial may be obtained by adding p^{n-k} M(p) to the remainder polynomial C(p).

Thus, $$X(p) = D^2 M(p) + C(p) \quad ...(vii)$$

The remainder polynomial C(p) = p + 1

Therefore, $$X(p) = [p^{11} + p^8 + p^5 + p^3] + [p + 1] \quad ...(viii)$$

Hence, the corresponding codeword is given by,

X =	1	0	0	1	0	0	1	0	1	0	1	1

← Message bits → ← Check bits →

Fig. 8.42.

Now let us draw the encoder for the cyclic code.

The given generator polynomial is

$$G(p) = p^2 + 1 = p^2 + 0p + 1 \quad ...(ix)$$

We know that the general form of the generator polynomial expressed as

$$G(p) = 1 + \sum_{i=1}^{n-k-1} g_i (p)^i + p^{n-k}$$

or $$G(p) = 1 + \sum_{i=1}^{1} g_i\,(p) + p^2 = 1 + g_1 p + p^2$$

or $$G(p) = p^2 + g_1\,p + 1 \qquad \text{...(x)}$$

Comparing equations (ix) and (x), we obtain

$$g_1 = 0,\ g_2 = 1$$

Therefore, the encoder has been shown in figure 8.43.

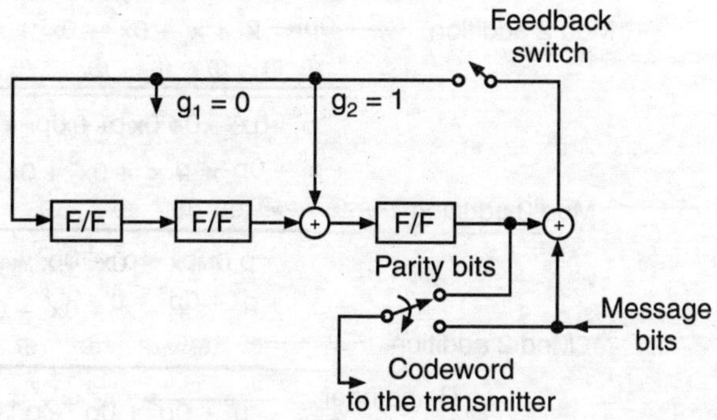

Fig. 8.43. Encoder for the cyclic code

EXAMPLE 8.30. Determine the encoded message for the following 8 bit data codes using the following CRC generating polynomial:

$$\mathbf{P(x) = x^4 + x^3 + x^0}$$

(i) 1 1 0 0 1 1 0 0

(ii) 0 1 0 1 1 1 1 1

Solution: The given message is 1 1 0 0 1 1 0 0. This means that there are 8 message bits.

Thus, $k = 8$

We know that the degree of generator polynomial is equal to the number of parity (check) bits in the code.

Therefore, number of parity bits $(n - k) = 4$

Hence, codeword length, $n = 8 + 4 = 12$

Hence the message polynomial corresponding to the message 1 1 0 0 1 1 0 0 is given as

$$M(x) = x^7 + x^6 + 0x + 0x^4 + x^3 + x^2 + 0x + 0$$

or $$M(x) = x^7 + x^6 + x^3 + x^2$$

Now, the generator polynomial is given by,

$$P(x) = x^4 + x^3 + 1$$

Next, let us multiply the message polynomial by x^{n-k} i.e. x^4

Thus, $$x^4\,[M(x)] = x^4\,[x^7 + x^6 + x^3 + x^2]$$

or $$x^4\,M(x) = x^{11} + x^{10} + x^7 + x^6$$

let us divide $x^4\,M(x)$ by the generator polynomial as under:

$$\begin{array}{r l}
 & x^7 + x^2 + x + 1 \\
x^4 + x^3 + 1 \;) & x^{11} + x^{10} + 0x^9 + 0x^8 + x^7 + x^6 + 0x^5 + 0x^4 + 0x^3 + 0x^2 + 0x^2 + 0 \\
 & x^{11} + x^{10} + 0x^9 + 0x^8 + x^7 \\
\text{Mod 2 addition} \dashrightarrow & \oplus \;\; \oplus \;\; \oplus \;\; \oplus \;\; \oplus \\
 & x^6 + 0x^5 + 0x^4 + 0x^3 + 0x^2 + 0x^1 + 0 \\
\text{Mod 2 addition} \dashrightarrow & x^6 + x^5 + 0x^4 + 0x^3 + x^2 \\
 & \oplus \;\; \oplus \;\; \oplus \;\; \oplus \;\; \oplus \\
 & 0 \oplus x^5 + 0x^4 + 0x^3 + x^2 + 0x + 0 \\
 & x^5 + x^4 + 0x^3 + 0x^2 + x \\
\text{Mod 2 addition} \dashrightarrow & \oplus \;\; \oplus \;\; \oplus \;\; \oplus \;\; \oplus \\
 & 0 + x^4 + 0x^3 + x^2 + x + 0 \\
 & x^4 + x^3 + 0x^2 + 0x + 1 \\
\text{Mod 2 addition} \dashrightarrow & \oplus \;\; \oplus \;\; \oplus \;\; \oplus \;\; \oplus \\
\text{Remainder polynomial } C(x) \longrightarrow & x^3 + x^2 + x + 1
\end{array}$$

Now, let us obtain the codeword polynomial X(x) as under:

$$X(x) = x^4 M(x) + C(x)$$

or $$X(x) = (x^{11} + x^{10} + x^7 + x^6) + (x^3 + x^2 + x + 1)$$

or $$X(x) = x^{11} + x^{10} + 0x^9 + 0x^8 + x^7 + x^6 + 0x^5 + 0x^4 + x^3 + x^2 + x + 1$$

Hence, the codeword will be

1 1 0 0 1 1 0 0	1 1 1 1
← Message bits →	← Parity or check bits →

Ans.

(ii) Similarly, we can obtain the codeword for the other message. The corresponding codeword for the message 0101111 is,

X(x) =

0 1 0 1 1 1 1 1	1 1 1 0
← Message bits →	← Parity or check bits →

EXAMPLE 8.31. Sketch the encoder and syndrome calculator for the generator polynomial $g(x) = 1 + x^2 + x^3$, and obtain the syndrome for the received codeword 1001011.

Solution:

1. Encoder

The received code word has 7 bits, hence, n = 7 and the degree of generator polynomial is 3, hence, n – k = 3, so, k = 4. Thus, the given code is a (7, 4) cyclic code. The generator polynomial an (n, k) cyclic code is expressed as,

$$g(x) = 1 + \sum_{i=1}^{n-k-1} g_i x^i + x^{n-k}$$

or $$g(x) = 1 + \sum_{i=1}^{3-1} g_i x^i + x^3 = 1 + g_1 x + g_2 x^2 + x^3 \qquad \text{...(i)}$$

The given generator polynomial is,

$$g(x) = 1 + 0x + x^2 + x^3 \quad \text{...(ii)}$$

Comparing equations (i) and (ii), we obtain

$$g_1 = 0, g_2 = 1$$

Hence, the encoder is shown in figure 8.44.

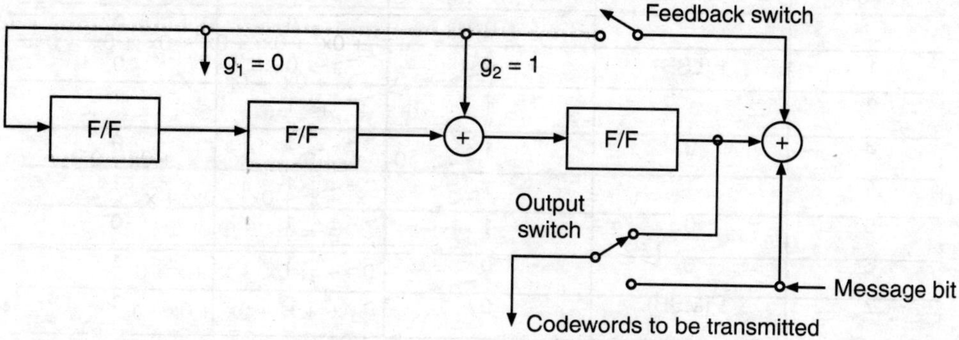

Fig. 8.44. Encoder for example 8.30

2. Syndrome calculator

The given generator polynomial is,

$$g(x) = 1 + x^2 + x^3$$

or

$$g(x) = x^3 + x^2 + 0x + 1 \quad \text{...(iii)}$$

We know that the general form of generator polynomial is given by

$$g(x) = x^3 + g_2x^2 + g_1x + 1 \quad \text{...(iv)}$$

Comparing equations (iii) and (iv), we get

$$g_1 = 0$$

$$g_2 = 1$$

The syndrome calculator is shown in figure 8.45.

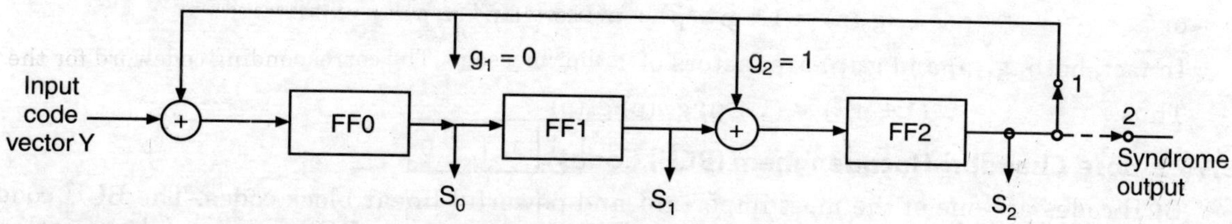

Fig. 8.45. Syndrome calculator

3. Calculation of syndrome

The output switch of figure 8.45 will be initially in position 1 until all the 7 bits of the received signal Y are shifted into the register. After that, the output switch is shifted to position 2. Clock pulses are then applied to the shift register to output the syndrome vector S. Table 8.13 explains the process of syndrome generation. Thus, the required syndrome will be

$$S = 100$$

Table 8.13.

Shift	Input bits (Y)	Contents of the shift register		
		$S_0 = Y \oplus S_2$	$S_1 = S_0$	$S_2 = S_1 \oplus S_0$
	—	0	0	0
1	1 (LSB)	1	0	0
2	1	1	1	1
3	0	1	1	0
4	1	1	1	1
5	0	1	1	0
6	0	0	1	1
7	1 (MSB)	0	0	1

Codeword LSB to MSB   Syndrome 100

8.15 OTHER CODES

8.15.1. Golay Codes

The (23, 12) Golay code is a very special type of binary code. This code is capable of correcting any combination of three or less than three random errors in the block of 23 bits. The number of message bits out of 23 is 12. This code has a minimum distance of 7. It satisfies the Hamming bound for t = 3. The (23, 12) Golay code is the only known code which is capable of correcting three errors. The (23, 12) Golay code is generated by one of the following two polynomials:

$$g_1(p) = 1 + p^2 + p^4 + p^5 + p^6 + p^{10} + p^{11}$$

or

$$g_2(p) = 1 + p + p^5 + p^6 + p^7 + p^9 + p^{11}$$

In fact, both, $g_1(p)$ and $g_2(p)$ are factors of $1 + p^{23}$.

Thus,
$$(1 + p^{23}) = (1 + p)\, g_1(p)\, g_2(p)$$

8.15.2. Bose-Chaudhri-Hocquenghem (BCH) Codes

BCH codes are one of the most important and powerful linear block codes. The BCH codes are cyclic codes with a wide variety of parameters. The most common BCH codes are characterized as under:

For any positive integer m (with $m \geq 3$) and $t < (2^m - 1)/2$, there exists a BCH code with the following parameters:

Block length : $n = 2^m - 1$

Number of message bits : $k \geq n - mt$

Minimum distance : $d_{min} \geq 2t + 1$

This type of BCH codes is called as primitive BCH codes. Each BCH code can correct t number of errors. This means that it can detect and correct upto t random errors per codeword. The Hamming codes with a single error correcting capability is a BCH code. The block lengths and code rates of BCH codes are variable. In that sense they provide a high flexibility. For block lengths of a few hundred, the BCH codes are among the best known codes of the same block length and code rate.

Generator polynomials of BCH codes

Table 8.14 presents the code parameters and generator polynomials for the binary BCH codes of length upto $2^5 - 1$.

Table 8.14. Binary BCH codes of length upto $2^5 - 1$.

n	k	t	Generator Polynomial
7	4	1	1 011
15	11	1	1 10 011
15	7	2	111 010 001
31	26	1	100 101
31	21	2	11 101 101 001
31	16	3	1 000 111 110 101 111
31	11	5	101 100 010 011 011 010 101
31	6	7	11 001 011 011 110 101 000 100 111

where n = block length

k = number of message bits

t = Maximum number of detectable errors

8.15.3. How to use table 8.14

Let us consider that we want to construct the generator polynomial for (15, 7) BCH code, then we refer to the third row of Table 8.17. The generator polynomial from the third row is given by

111 010 001

Hence, the generator polynomial is given by

$$g(p) = p^8 + p^7 + p^6 + 0p^5 + p^4 + 0p^3 + 0p^2 + 0p + 1$$

or $$g(p) = p^8 + p^7 + p^6 + p^4 + 1 \quad \textbf{Ans.}$$

8.15.4. Reed-Solomon codes (R-S Codes)

The Reed-Solomon codes are named after their inventors Reed and Solomon. The R-S codes are an important subclass of non binary BCH codes. The encoder for an RS code differs from a binary encoder. This encoder operates on multiple bits rather than the individual bits. For example, the encoder for an R–S (n, k) code on m symbols will group the incoming binary data stream into blocks. Each blocks is km bits long. Each block is considered as k number of symbols, with each symbol having m bits. The encoding algorithm then expands a block of k symbols to n symbols by adding (n – k) redundant symbols. When m is an integer power of two, the m-bit symbols are called **bytes.**

In practice, the most popular value of m is 8. The RS code with m = 8 is extremely powerful. The parameters of a t – error correcting R-S code may be listed as under:

Block length	:	$n = 2^m - 1$ symbols
Message size	:	k symbols
Parity check bit size	:	$n - k = 2t$ symbols

Minimum distance : $d_{min} = (2t + 1)$ symbols

Thus, the block length of the RS code is one less than the size of a code symbol and the minimum distance is always one greater than the number of parity symbols. It can be shown that no (n, k) linear block code can have a minimum distance greater than (n – k + 1). The (n, k) linear block code with a minimum distance equal to (n – k + 1) is called as **maximum distance separable code.** Hence, every RS code is maximum distance separable code.

Advantages of R-S codes

(i) The RS codes make the highly efficient use of redundancy.

(ii) It is possible to adjust the block lengths and the symbol sizes, to accommodate a wide range of message sizes.

(iii) The R-S codes provide a wide range of code rates. The code rate can be chosen to obtain the optimum performance.

(iv) The efficient decoding techniques are available for decoding the RS codes.

8.16 CONVOLUTIONAL CODES

The main difference between the block codes and the convolutional (or recurrent) codes may be listed as under:

(i) Block codes

In block codes, the block n bits generated by the encoder in a particular time slot depends only on the block of k message bits within that time slot.

(ii) Convolutional code

In the convolutional codes, the block of n bits generated by the encoder in a time slot depends not only on the k message bits within that time slot, but also on the preceding 'L' blocks of the message bits (L > 1). Generally, the values of k and n will be small.

Applications of convolutional code

Like block codes, the convolutional codes can be designed to either detect or correct errors. However, because data is usually retransmitted in blocks, the block codes are more suitable for error detection and the convolutional codes are more suitable for error correction.

Encoding and decoding

Encoding of the convolutional codes can be accomplished using simple shift register. Several practical methods have been developed for decoding. The convolutional codes perform as well or better than the block codes in many error control applications.

8.16.1. Encoder for Convolutional Coding

The fundamental hardware unit for the convolutional encoding is a taped shift register with (L + 1) stages, as shown in figure 8.46. Here, g_0, g_1 ..... etc. are the tap gains which are nothing but binary digits 0s or 1s. A tap gain of 0 represents an open circuit whereas a tap gain of 1 represents a short circuit.

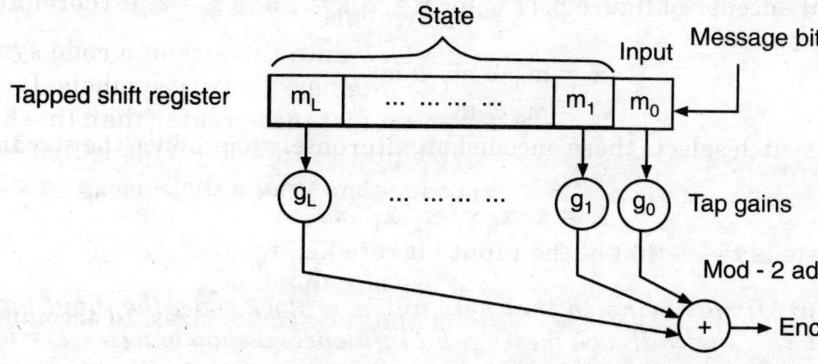

Fig. 8.46. Block diagram of a general convolutional encoder.

The message bits enter one by one into the tapped shift register, which are then combined by mod-2 addition to form the encoded bit x. Therefore, we have

$$x = m_L g_L \oplus \ldots\ldots\ldots \oplus m_1 g_1 \oplus m_0 g_0 \qquad \ldots(8.49)$$

or

$$x = \sum_{i=0}^{L} m_i \, g_i \text{ (mod-2 addition)}$$

The name convolutional encoding comes from the fact that equation (8.49) has a form of **binary convolution** which analogous to the convolutional integral. The message bit m_0 in figure 8.46 represents the **current input** whereas the bits m_1 to m_L represent the past input or **state** of the shift register. From equation (8.49), it is clear that a message bit x depends on the current message bit m_0 and the state of the shift register defined by the previous L message bits.

Important Point: It is important to note that a particular message bit influences a span of L + 1 successive encoded bits as it shift through the register.

8.16.2. Practical Convolutional Encoder

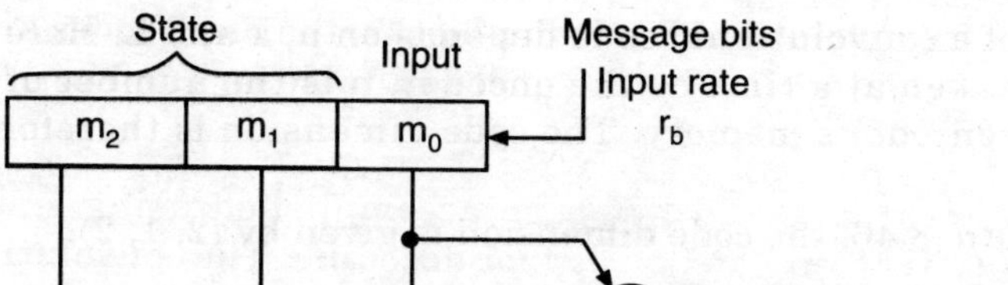

Fig. 8.47. Convolutional encoder with n = 2, k = 1 and L = 2

To provide an extra bit needed for error control, a complete convolutional encoder must generate output bits at a rate greater than the message bit rate r_b. This is achieved by using two or more mod-2 address to the registers as shown in figure 8.47, and interleaving the encoded bits via the commutator switch.

The convolutional encoder of figure 8.47 is for n = 2, k = 1 and L = 2. It therefore, generates n = 2 encoded bits as under:

$$\left.\begin{aligned} x &= m_0 \oplus m_1 \oplus m_2 \\ \text{and} \qquad x_2 &= m_0 \oplus m_2 \end{aligned}\right\} \qquad ...(8.50)$$

The commutator switch selects these encoded bit alternately to produce the stream of encoded bits as under:

$$X = x_1\ x_2\ x_1'\ x_2'\ x_1''\ x_2'' \qquad ...(8.51)$$

The output bit rate is twice that of the input bit rate i.e., r_b.

> **Important Point:** *It may be noted that here unlike a block code, the input bits have not been grouped into words. Instead, each message bit influences a span of n (L + 1) = 6 successive output bits.*

8.16.3. Few Definitions Related to convolutional Codes

1. Code rate (r)

The code rate of the encoder of figure 8.47 is expressed as

$$r = \frac{k}{n} \qquad ...(8.52)$$

Here, k = Number of message bits = 1

n = Number of encoded bits per message bits = 2

$$\text{Therefore,} \qquad r = \frac{1}{2} \qquad ..(8.53)$$

2. Constraint length (k)

Each message bit influences a span of n (L + 1) successive output bits. The quantity n (L + 1) is called as the constraint length. It is measured in terms of encoded output bits. For the encoder of figure 8.47, the constraint length is 6 bits as n = 2 and L = 2. Here, L is the encoder's memory measured in terms of input message bits.

3. Code Dimension

The code dimension of a convolutional code depends on n, k and L. Here k represents the number of message bits taken at a time by the encoder, n is the number of encoded bits per message bit and L is the encoder's memory. The code dimension is therefore represented by (n, k, L)

For the encoder of figure 8.46, the code dimension is given by (2, 1, 2).

8.16.4. Time Domain Approach

The time-domain behaviour of a binary convolutional encoder may be defined in terms of n-impulse responses. Let the impulse response of the adder generating x_1 in figure 8.47, be given by, the sequence $\{g_0^{(1)}, g_1^{(1)}, ... g_L^{(1)}\}$. Similarly, let the sequence $\{g_0^{(2)}, g_1^{(2)}, ... g_L^{(2)}\}$ denote the impulse response of the adder generating x_2 in figure 8.46. These impulse responses are also called as **generator sequences** of the code. Let (m_0, m_1, m_2 ...) denote the message sequence entering the encoder of figure 8.46 one bit at a time (starting from m_0). The encoder generates two output sequences by performing convolutions on the message sequences will the impulse responses. The bit sequence x_1 is given by,

$$x_1 = x_i^{(1)} = \sum_{l=0}^{L} g_l^{(1)}\ m_{i-1},\ i = 0, 1 \qquad ...(8.54)$$

Similarly, the other bit sequence x_2 is given by,

$$x_2 = x_i^{(2)} = \sum_{l=0}^{L} g_l^{(2)} m_{i-l}, \; i = 0, 1 \qquad ...(8.55)$$

Then, these bit sequences are multiplexed with the help of the commutator switch to produce the following output:

$$X = \{x_0^{(1)} x_0^{(2)} x_1^{(1)} x_1^{(2)} x_2^{(1)} x_2^{(2)} ...\} \qquad ...(8.56)$$

where $x_1 = x_0^{(1)} x_0^{(1)} x_1^{(1)}$ ......

and $x_2 = x_0^{(2)} x_1^{(2)} x_2^{(2)}$ ....

EXAMPLE 8.32. The convolutional encoder of figure 8.48 has the following two generator sequences each of length 3.

$$\left(g_0^{(1)}, g_1^{(1)}, g_2^{(1)} ....\right) = (1, 1, 1)$$

and $$\left\{g_0^{(2)}, g_1^{(2)}, g_2^{(2)}\right\} = (1, 0, 1)$$

Determine the encoded sequence for the following input message:

$$(m_0, m_1, m_2, m_3, m_4) = (1\ 0\ 0\ 1\ 1)$$

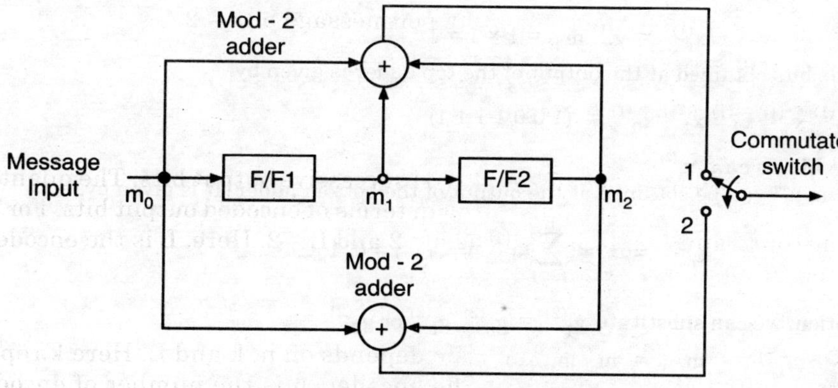

Fig. 8.48. Convolutional encoder

Solution: The given encoder can be drawn in the standard form as shown in figure 8.49.

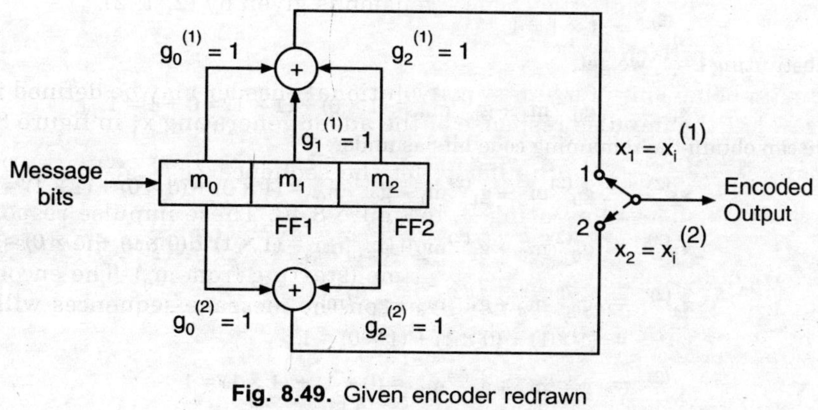

Fig. 8.49. Given encoder redrawn

(i) The output sequence may be obtained by using equations (8.54) and (8.55) stated to obtain the bit streams x_1 and x_2.

1. To obtain the bit stream $x_i^{(1)}$

We have $$x_1 = x_i^{(1)} = \sum_{l=0}^{L} g_l^{(1)} m_{i-l} \qquad i = 0, 1, 2, \ldots$$

Substituting i = 0, we get

$$x_0^{(1)} = g_{(0)}^{(1)} \quad m_0 = 1 \times 1 = 1$$

Similarly, substituting i = 1, we get,

$$x_1^{(1)} = g_0^{(1)} m_1 + g_1^{(1)} m_0 = (1 \times 0) + (1 \times 1) = 1$$

We can obtain the other code bits in a similar manner as under:

$$x_2^{(1)} = g_0^{(1)} m_2 + g_1^{(1)} m_1 + g_2^{(1)} m_0$$
$$= (1 \times 0) + (1 \times 0) + (1 \times 1) = 0 + 0 + 1 = 1$$

$$x_3^{(1)} = g_0^{(1)} m_3 + g_1^{(1)} m_2 + g_2^{(1)} m_1$$
$$= (1 \times 1) + (1 \times 0) + (1 \times 0) = 1$$

$$x_4^{(1)} = g_0^{(1)} m_4 + g_1^{(1)} m_3 + g_2^{(1)} m_2$$
$$= (1 \times 1) + (1 \times 1) + (1 \times 0) = 1 + 1 + 0 = 0 \quad \text{(Mod-2 addition)}$$

$$x_5^{(1)} = g_1^{(1)} m_4 + g_2^{(1)} m_3 = (1 \times 1) + (1 \times 1) = 1 + 1 = 0$$

$$x_6^{(1)} = g_2^{(1)} m_4 = 1 \times 1 = 1$$

Thus the code bits obtained at the output of the top adder is given by:

$$x_6^{(1)} x_5^{(1)} x_4^{(1)} x_3^{(1)} x_2^{(1)} x_1^{(1)} x_0^{(1)} = (1\,0\,0\,1\,1\,1\,1)$$

2. To obtain the bit stream $x_i^{(2)}$

The bit stream $x_i^{(2)}$ can be obtained at the output of the bottom adder. It is given by,

$$x_i^{(2)} = x_2 = \sum_{l=0}^{L} g_l^{(2)} m_{i-l}, \quad i = 0, 1, 2, 3 \ldots$$

In this equation, we can substitute $g_l^{(2)} = g_0^{(2)}, g_1^{(2)}$ or $g_2^{(2)}$

and $m_{i-l} = m_0, m_1, \ldots$

Substituting values of i into equation for $x_i^{(2)}$, we get,

$$x_0^{(2)} = g_0^{(2)} m_0$$

or $$x_0^{(2)} = 1 \times 1 = 1$$

Similarly substituting i = 1, we get,

$$x_1^{(2)} = g_0^{(2)} m_1 + g_1^{(1)} m_0 = (1 \times 0) + (1 \times 1) = 0 + 1 = 1$$

Similarly, we can obtain the remaining code bits as under:

$$x_2^{(2)} = g_0^{(2)} m_2 + g_1^{(2)} m_1 + g_2^{(2)} m_0 = (1 \times 0) + (0 \times 0) + (1 \times 1) = 1$$

$$x_3^{(2)} = g_0^{(2)} m_3 + g_1^{(2)} m_2 + g_2^{(2)} m_1 = (1 \times 1) + (0 \times 0) + (1 \times 0) = 1$$

$$x_4^{(2)} = g_0^{(2)} m_4 + g_1^{(2)} m_3 + g_2^{(1)} m_2$$
$$= (1 \times 1) + (0 \times 1) + (1 \times 0) = 1$$

$$x_5^{(2)} = g_1^{(2)} m_4 + g_2^{(2)} m_3 = (0 \times 1) + (1 \times 1) = 1$$

$$x_6^{(2)} = g_2^{(2)} m_4 = 1 \times 1 = 1$$

Hence, the code bits obtained at the output of the bottom adder is given by,

$x_6^{(2)}\ x_5^{(2)}\ x_4^{(2)}\ x_3^{(2)}\ x_2^{(2)}\ x_1^{(2)}\ x_0^{(2)} = (1\ 1\ 1\ 1\ 1\ 1\ 1)$

3. The output Encoded Sequence

By interleaving the code bits at the outputs of the two adders, we can get the encoded sequence at the encoder output as under:

Encoded sequence = $x_0^{(1)}\ x_0^{(2)}\ x_1^{(1)}\ x_1^{(2)}\ \ldots\ x_6^{(1)}\ x_6^{(2)}$

Substituting the values, we get,

Codeword = 11 11 11 11 01 01 11

It is important to note that the message sequence of length k = 5 bits produces an output coded sequence of length 14 bits.

8.16.5. Transform-Domain Approach

We know that the convolution in time domain is transformed into the multiplication of Fourier transforms in the frequency domain. We can use this principle in the transform domain approach.

In this process, the first step is to replace each path in the encoder by a polynomial whole coefficients are represented by the respective elements of the impulse response. For example, for the top-adder path in previous example (8.31), it is given that,

$$g_0^{(1)} = 1$$

$$g_1^{(1)} = 1$$

and

$$g_2^{(1)} = 1$$

Therefore, the input top adder output path of the encoder of figure 8.48 can be expressed in terms of the polynomial as under:

$$G^{(1)}(p) = g_0^{(1)} + g_1^{(1)}\,p + g_2^{(1)}\,p^2 \qquad \ldots(8.57)$$

Substituting the values we get,

$$G^{(1)}(p) = 1 + p + p^2 \qquad \ldots(8.58)$$

The general expression is given by,

$$G^{(1)}(p) = g_0^{(1)} + g_1^{(1)}\,p + g_2^{(1)}\,p^2 + \ldots + g_L^{(1)}p^L \qquad \ldots(8.59)$$

The variable p denotes a unit delay operation and the power of p defines the number of time units by which the associated bit in the impulse response has been delayed with respect to the first bit i.e., $g_0^{(1)}$.

Similarly the polynomial corresponding to the input-bottom adder-output path for the encoder shown in figure 8.48 is given by,

$$G^{(2)}(p) = g_0^{(2)} + g_1^{(2)}\,p + g_2^{(2)}\,p^2 \qquad \ldots(8.60)$$

Substituting, $g_0^{(2)} = 1$, $g_1^{(2)} = 0$ and $g_2^{(2)} = 1$ we obtain,

$$G^{(2)}(p) = 1 + p^2 \qquad \ldots(8.61)$$

The general form of polynomial is given by,

$$G^{(2)}(p) = g_0^{(2)} + g_1^{(2)} + g_2^{(2)}\,p^2 + \ldots + g_L^{(2)}\,p^L \qquad \ldots(8.62)$$

The polynomial $G^{(1)}(p)$ and $G^{(2)}(p)$ of equations (8.58) and (8.61) are called as the **generator polynomials** of the code.

From the generator polynomials, we can obtain the codeword polynomials as under:

Code word polynomial corresponding to top adder is given by,

$$x^{(1)}(p) = G^{(1)}(p)\cdot m(p) \qquad \ldots(8.63)$$

where $m(p)$ = Message polynomial

and the codeword polynomial corresponding to the bottom adder is given by,

$$x^{(2)}(p) = G^{(2)}(p) \cdot m(p) \qquad ...(8.64)$$

Once we get the two codeword polynomials, it is possible to obtain the corresponding output sequence by simply using the individual coefficients. This has been illustrated in the following example.

EXAMPLE 8.33. Determine the codeword for the cyclic encoder of figure 8.47 for the message signal (10011), using the transform domain approach. The impulse response of the input top adder output path is (1, 1, 1) and that of the input bottom adder output path is (1, 0, 1).

Solution: First, let us write the generator polynomial $G^{(1)}(p)$.

The impulse response of the input top adder output path of the convolutional encoder of figure 8.48. Therefore, we have

$$g_0^{(1)} = 1$$

$$g_1^{(1)} = 1$$

and

$$g_2^{(1)} = 1$$

Therefore, the generator polynomial $G^{(1)}(p)$ is given by,

$$G^{(1)}(p) = g_0^{(1)} + g_1^{(1)} p + g_2^{(1)} p^2$$

or

$$G^{(1)}(p) = 1 + p + p^2 \qquad ...(i)$$

The given message is ($m_0\ m_1\ m_2\ m_3\ m_4$ = 1 0 0 1 1). Therefore, the message polynomial is given by,

$$M(p) = m_0 + m_1 (p) + m_2 p^2 + m_3 p^3 + m_4 p^4$$

or

$$M(p) = 1 + p^3 + p^4 \qquad ...(ii)$$

Now, we find the code word polynomial for the top adder.

$$x^{(1)}(p) = G^{(1)}(p) \cdot M(p)$$

$$x^{(1)}(p) = (1 + p + p^2)(1 + p^3 + p^4)$$

$$= 1 + p^3 + p^4 + p + p^4 + p^5 + p^2 + p^5 + p^6$$

or

$$x^{(1)}(p) = 1 + p + p^2 + p^3 + (1 + 1) p^4 + (1 + 1) p^5 + p^6$$

or

$$x^{(1)}(p) = 1 + p + p^2 + p^3 + p^6 \text{ ... (since addition is Mod-2)} \qquad ...(iii)$$

Now, we obtain the generator polynomial $G^{(2)}(p)$

The impulse response of the input bottom adder output path of convolutional encoder of figure 8.48 is (1, 0, 1). Therefore,

$$g_0^{(2)} = 1,\ g_1^{(1)} = 0 \text{ and } g_2^{(2)} = 1$$

Therefore the generator polynomial $G^2(p)$ is given by,

$$G^{(2)}(p) = g_0^{(2)} + g_1^{(2)} p + g_2^{(2)} p^2$$

or

$$G^{(2)}(p) = 1 + p^2 \qquad ...(iv)$$

Codeword polynomial for the bottom adder is given by

$$x^{(2)}(p) = G^{(2)}(p) \cdot M(p)$$

Substituting equations (ii) and (iv) we get,

$$x^{(2)}(p) = (1 + p^2)(1 + p^3 + p^4)$$

or

$$x^{(2)}(p) = 1 + p^2 + p^3 + p^4 + p^5 + p^6 \qquad ...(v)$$

Next, let us obtain the code sequences.

The output sequence at the output of the top-adder can be obtained from the corresponding generator polynomial $x^{(1)}(p)$,

$$x^{(1)}(p) = 1 + p + p^2 + p^3 + p^6 = 1 + p + p^2 + p^3 + 0p^4 + 0p^5 + p^6$$

Therefore, the corresponding code sequence is (1 1 1 1 0 0 1)

Similarly, the code sequence for the bottom adder can be obtained from its generator polynomial.

$$x^{(2)}(p) = 1 + p^2 + p^3 + p^4 + p^5 + p^6$$

or $$x^{(2)}(p) = 1 + 0p + p^2 + p^3 + p^4 + p^5 + p^6$$

Thus, corresponding code sequence is (1 0 1 1 1 1 1)

Hence, the final code word at the output of the encoder is obtained by multiplexing (interleaving) the two code sequences.

$$\text{Code word} = 11\ 10\ 11\ 11\ 01\ 01\ 11$$

Advantage of Transform Domain Approach

Here, it may be noted that the codeword obtained using the time domain approach and the frequency domain approach. It may be observed that we get the same result. But, the computation using transform domain approach demands less efforts than the time-domain approach.

8.16.6. Graphical Representation for Convolutional Encoding

For the convolutional encoding, there are three different graphical representations that are widely used. They are related to each other. The graphical representations are as under:

(i) The code tree
(ii) The code trellis
(iii) The state diagram.

Let us discuss them one by one.

8.16.7. The code Tree

In this subsection, let us draw the code tree for the (2, 1, 2) cyclic encoder of figure 8.46. We assume that the register has been cleared so that it contains all zeros, when the first message bit m_0 arrives. Therefore, the initial state is $m_{-2}\, m_{-1} = 00$. Now, if $m_0 = 0$, then, the encoder output are as under:

$$x_1 = m_0 \oplus m_1 \oplus m_2 \text{ ...... According to equation (8.50)}$$

If $m_0 = 0$, we have

$$x_1 = 0 \oplus 0 \oplus 0 = 0 \text{ .....}$$

Also, $$x_2 = m_0 \oplus m_2 = 0 \oplus 0 = 0$$

Therefore, $$x_1\, x_2 = 0\ 0 \text{ ..... if } m_0 = 0$$

But, if $m_0 = 1$, then we have

$$x_1 = 1 \oplus 0 \oplus 0 = 1$$

and $$x_2 = 1 \oplus 0 = 1$$

Therefore $$x_1\, x_2 = 11 \text{ .. if } m_0 = 1$$

The code tree has been drawn in figure 8.50. It begins at a branch point on node a which represents the initial state. Hence, if $m_0 = 0$ we should take the upper branch from node a to obtain the output 00 and the next state $m_{-1}\, m_0 = 00$.

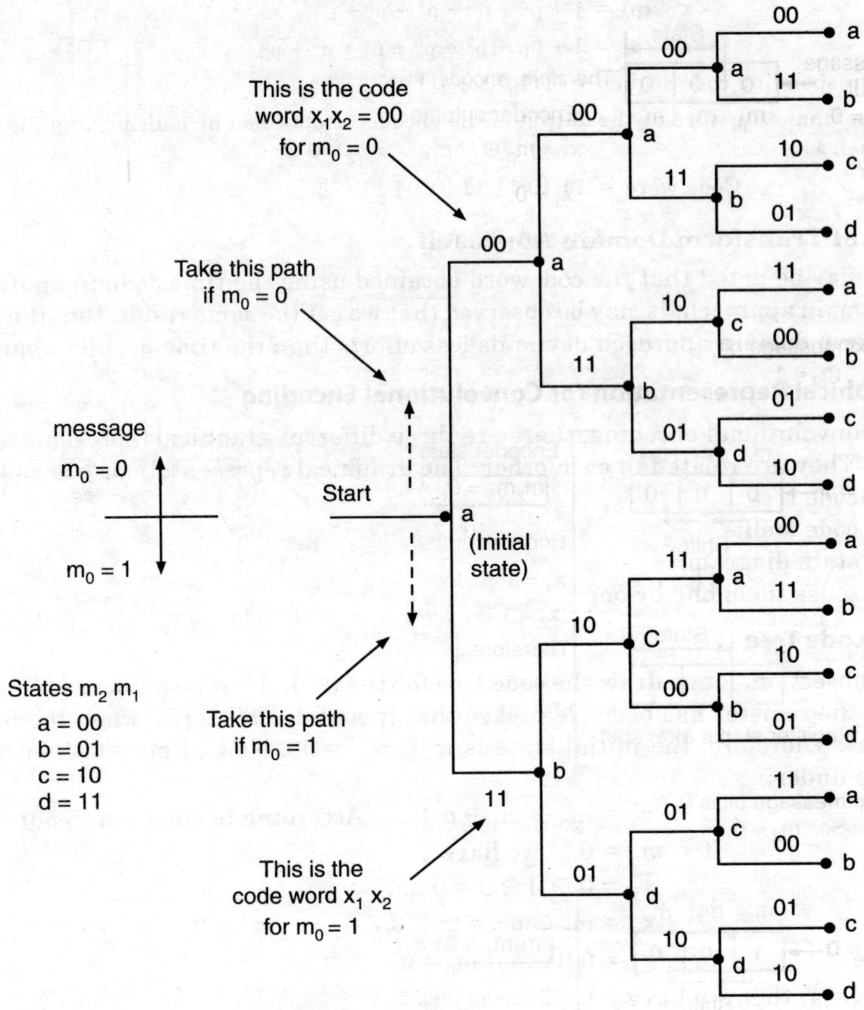

Fig. 8.50. Code tree for (2, 1, 2) encoder.

Therefore, even this state is labelled as a. But, if m_0 = 1, then we should take the lower branch from a to obtain the output 11 and the next state m_1 m_0 = 01. This state has been labelled as b. The code tree progresses in this manner for each new message bit. The nodes are labelled with letters a, b, c or d denoting the current state m_2 m_1 and we go up or down from a node depending on the value of m_0. Each branch shows the resulting encoded output x_1 x_2 calculated using equation (8.50). Each branch will terminate at another node which is labelled with the next state. There are 2^j possible branches for the jth message bit, but the branch pattern begins to repeat at j = 3 since the register length is L + 1 = 3. The procedure of drawing the code tree has been illustrated in figure 8.51.

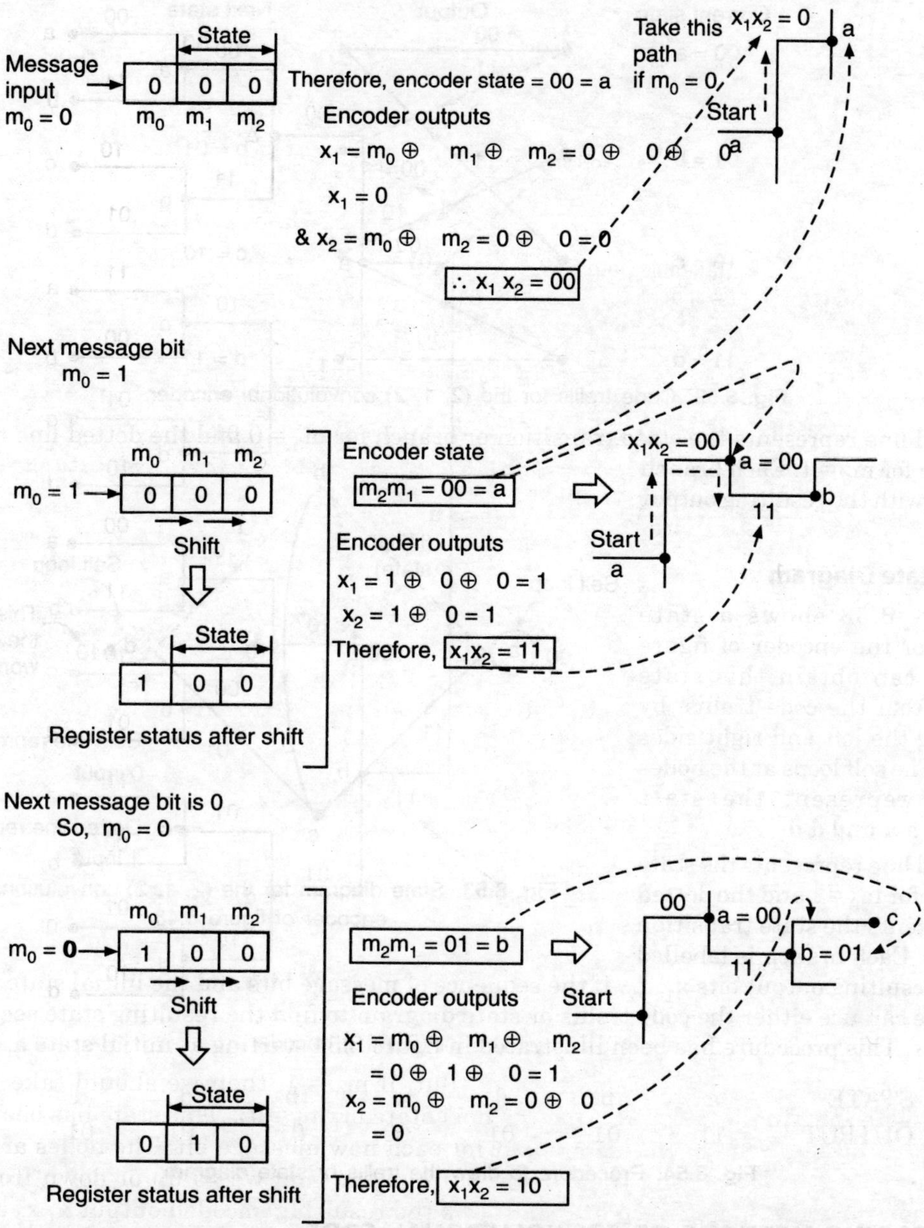

Fig. 8.51. Operation of the (2, 1, 2) encoder and development of code tree.

8.16.8. Code Trellis

Figure 8.52 shows a more compact graphical representation which is popularly known as code trellis. Here, the nodes on the left denote the four possible current states and the nodes on the right are the resulting next state.

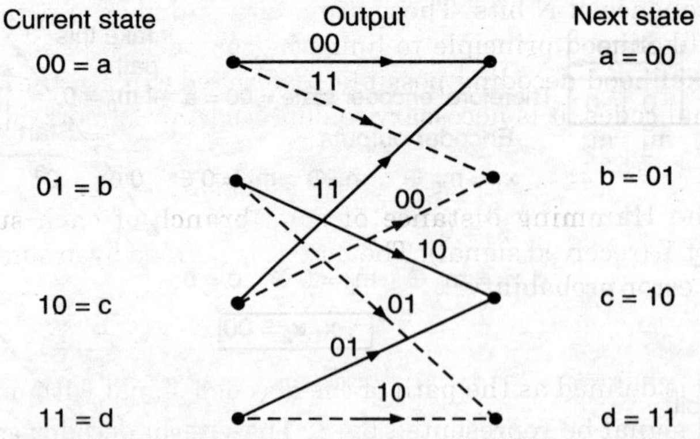

Fig. 8.52. Code trellis for the (2, 1, 2) convolutional encoder.

A solid line represents the state transition or branch for $m_0 = 0$ and the dotted line represents the branch for $m_0 = 1$. Each branch is labelled with the resulting output bits x_1 x_2.

8.16.9. State Diagram

Figure 8.53 shows a state diagram for the encoder of figure 8.48. We can obtain this state diagram from the code trellis, by connecting the left and right sides of trellis. The self loops at the nodes a and d represent the state transition a-a and d-d.

A solid line represents the state transition for $m_0 = 1$ and the dotted line represents the state transition for $m_0 = 1$. Each branch is labelled with the resulting output bits x_1, x_2. If the sequence of message bits and the initial state is given to us then we can use either the code trellis or state diagram to find the resulting state sequence and output bits. This procedure has been illustrated in figure 8.54 starting at initial state a.

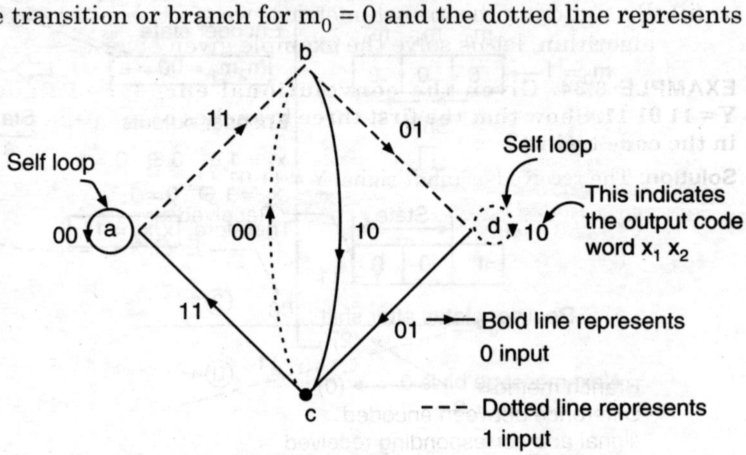

Fig. 8.53. State diagram for the (2, 1, 2) convolutional encoder of figure 8.48

STATE	a	b	d	c	b	d	d	c
OUTPUT	11	01	01	00	01	10	01	11

Fig. 8.54. Procedure to draw the trellis or state diagram

8.17 DECODING METHODS OF CONVOLUTIONAL CODES

There are three methods for decoding the convolutional codes. They are as under:

(i) Viterbi algorithm

(ii) Feedback decoding

(iii) Sequential decoding

8.17.1. Viterbi Algorithm

The viterbi algorithm operates on the principle of maximum likelihood decoding and achieves optimum performance. The maximum likelihood decoder has to examine the entire received sequence Y and find a valid path which has the smallest Hamming distance from Y. But there are 2^N possible paths for a message sequence of N bits. These are a large number of paths. The viterbi algorithm applies the maximum likelihood principle to limit the comparison of so many surviving paths, to make the maximum likelihood decoding possible. Before we explain the viterbi algorithm for the decoding of convolutional codes, it is necessary to define certain important terms.

Metric

It is defined as the Hamming distance of each branch of each surviving path from the corresponding branch of Y (received signal). The metric is defined by assuming that 0's and 1's have the same transmission error probability.

Surviving Path

The surviving path is defined as the path of the decoded signal with minimum metric.

(i) Let the received signal be represented by Y. The viterbi decoder assigns to each branch of each surviving path of a metric.

(ii) By summing the branch metrices we get the path metric. To understand more about viterbi algorithm, let us solve the example given below:

EXAMPLE 8.34. Given the convolutional encoder of figure 8.47 and for a received signal Y = 11 01 11. Show that the first three branches of the valid paths emerging from the initial node a_0 in the code trellis.

Solution: The received or input signal Y = 11 01 11.

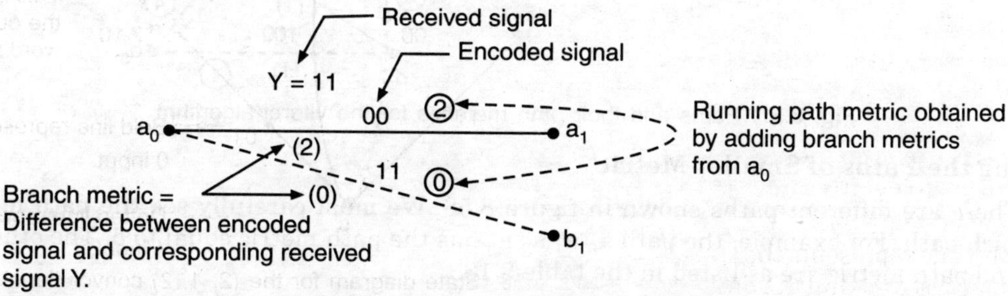

Fig. 8.55. First step in Viterbi's algorithm

Let us consider the code trellis diagram of figure 8.52 for the (2, 1, 2) encoder. It shows that for the current state a, the next state can be either a or b depending on the message bit m_0 = 0 or 1. We have redrawn these two branches in figure 8.55 where a_0 represents the initial state and a_1 and b_1 represent the next possible states. The solid line represents the branch for m_0 = 0 and the dotted line represents the branch for m_0 = 1.

In figure 8.55, the number in the brackets, written below the branches represent the **branch metric** which is obtained by taking the difference between the encoded signal and corresponding received signal Y. For example for the branch a_0 to a_1, encoded signal is 00 and received signal is 11, hence, the branch metric is (2), whereas for the path a_0 b_1, the encoded signal and received signal both are 11, hence, the branch metric is (0).

The encircled numbers at the right hand end of each branch represents the running path metric which is obtained by summing the branch metrices from a_0. For example, the running path metric for the branch a_0 a_1 = 2 and that for the branch a_0 b_1 is 0.

When the next part of input bits i.e., Y = 01 are received at the nodes a_1 and b_1, then four possible branches emerge from these two nodes are as shown in figure 8.56. The next four possible states are a_2, b_2, c_2 and d_2.

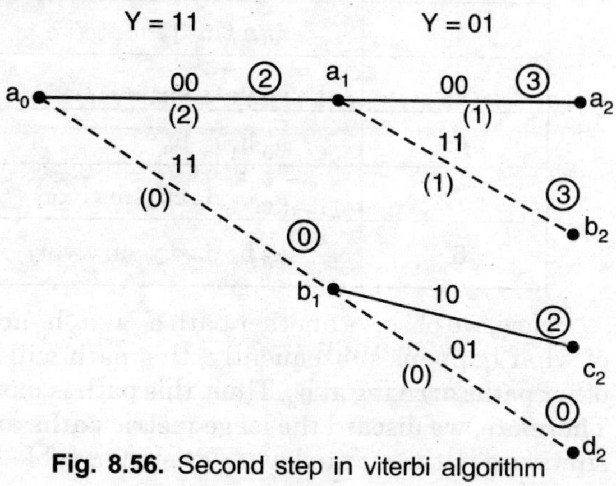

Fig. 8.56. Second step in viterbi algorithm

The numbers in the brackets written below each branch represent the branch metric. For example, for the branch a_1 a_2, the branch metric is (1) which is obtained by taking the difference between the encoded signal 00 and received signal 01. The running path metric for the same branch is 3 which is obtained by adding the branch metrics from a_0 [(2) + (1) = 3 from a_0 to a_1 and a_1 to a_2].

Similarly, the path metric for the path a_0 a_1 b_2 is 3, that of the path a_0 b_1 d_2 is 0 and so on. The viterbi's algorithm for all the input bits is as shown in figure 8.57.

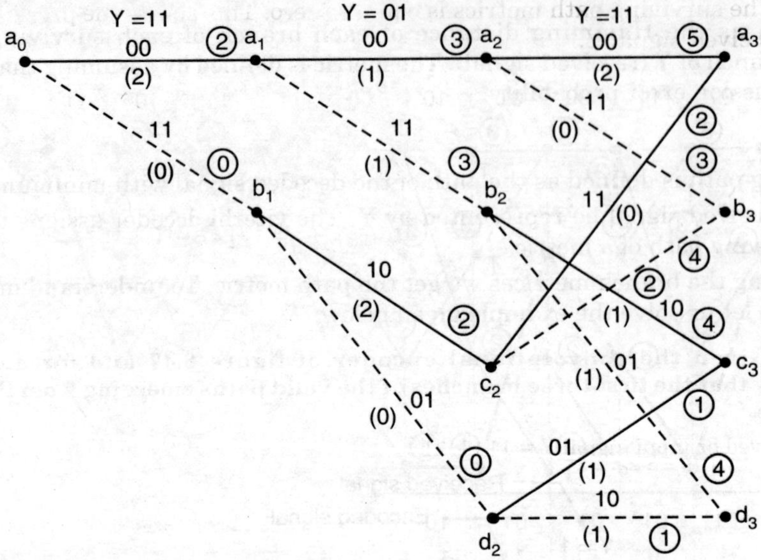

Fig. 8.57. Paths and their path metrices for the viterbi algorithm

Choosing the Paths of Smaller Metric

1. There are different paths shown in figure 8.58. We must carefully see the path metric for each path. For example, the path a_0 a_1 a_2 a_3 has the path metric equal to 5. The other paths and path metric are as listed in the table 8.18.

Table 8.15.

S. No.	Path	Path metric	Decision
1.	a_0 a_1 a_2 a_3	5	×
2.	a_0 a_1 a_2 b_3 (survivor)	3	✓
3.	a_0 a_1 b_2 c_3	4	×
4.	a_0 a_1 b_2 d_3	4	×
5.	a_0 b_1 c_2 a_3 (survivor)	2	✓
6.	a_0 b_1 c_2 b_3	4	×
7.	a_0 b_1 d_2 c_3 (survivor)	1	✓
8.	a_0 b_1 d_2 d_3 (survivor)	1	✓

Now, we observe another path a_0 a_1 a_2 b_3 arriving at node b_3 and has a metric 3. Hence regardless of what happens subsequently, this path will have a smaller Hamming Distance from Y than the other paths arriving at b_3. Thus, this path is more likely to represent the actual transmitted sequence. Therefore, we discard the large metric paths arriving at nodes a_3 c_3 and d_3, leaving the total $2^{kL} = 4$ surviving paths marked with (d) sign in table 8.15.

The paths marked by a (×) in table 8.18 and figure 8.57 are large metric paths and hence are discarded and the paths with smaller metric paths are declared as survivor at that node. Note that there is one survivor for each node (see table 8.15).

> **Important Point:** Thus, the surviving paths are: $a_0\ a_1\ a_2\ b_3$, $a_0\ b_1\ c_2\ a_3$, $a_0\ b_1\ d_2\ c_3$ and $a_0\ b_1\ d_2\ d_3$. None of the surviving path metrics is equal to zero. This shows the presence of detectable errors in the received signal Y.

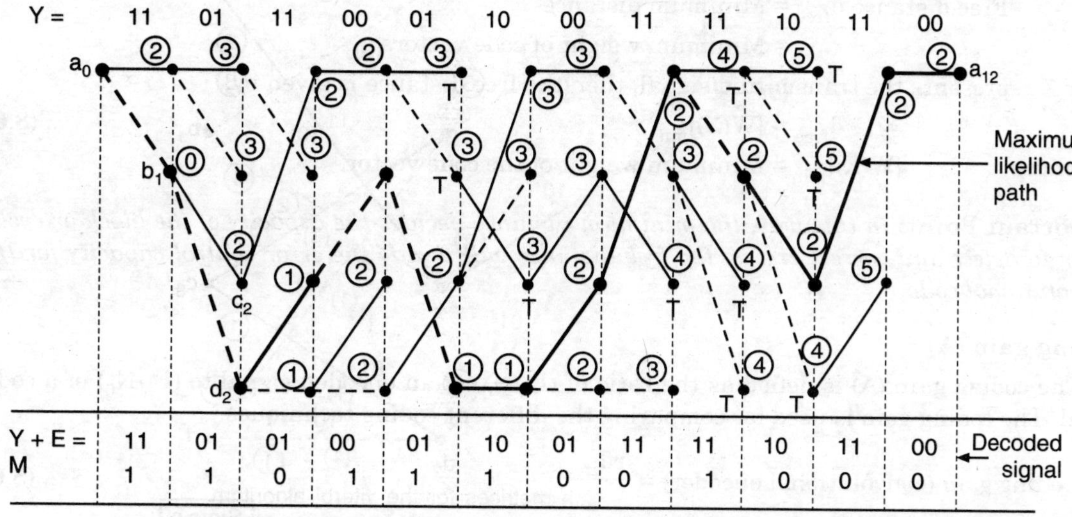

Fig. 8.58 Illustration of viterbi algorithm for maximum likelihood decoding.

Figure 8.58 depicts the continuation of figure 8.57 for a complete message of N = 12 bits. All the discarded branches and all labels except for running path metrics have been omitted for the sake of simplicity. The letter T under a node shows that the two arriving paths had equal running metrics. Under such circumstances the choice of survivor is arbitrary. The maximum likelihood path follows the solid line from a_0 to a_{12}, as shown in figure 8.59 . The final value of path metric is 2 which shows that there are atleast two transmission errors present in Y.

The decoder assumes the corresponding transmitted sequence Y + E and the message sequence M has been written below the trellis.

From figure 8.58, we observe that at node a_{12} only one path has arrived with metric 2. This path is shown by a dark line. Note that since this path has the lowest metric, it is the surviving path and signal Y is decoded from this path. Whenever this path is a solid line, the message is 0 and when it is a dotted line, the message is 1. This has been shown in figure 8.58.

> **Important Point:** *A viterbi decoder must calculate two metrics for each node (branch metric and path metric) and store 2^{kl} surviving paths, each consisting of N branches. Hence, the decoding complexity goes on increasing exponentially with L and linearly with N. Therefore, the viterbi algorithm is used only for codes with small values of L.*

8.17.2. Metric Diversion Effect

For a large number of message bits to be decoded, the storage requirement is going to be extremely large. This problem can be avoided by the metric diversion effect. The metric diversion effect is used for reducing the required memory storage.

8.17.3. Free Distance and Coding Gain

The error detection and correction capability of the block and cyclic codes is dependent on the minimum distance, d_{min} between the code vectors. But, in case of convolutional code the entire transmitted sequence is to be considered as a single code vector. Therefore, the free distance (d_{free}) is defined as the minimum distance between the code vectors. But, the minimum distance between the code vectors is same as the minimum weight of the code vector. Hence, the free distance is equal to minimum weight of the code vector.

Therefore,

$$\text{Free distance } d_{free} = \text{Minimum distance}$$
$$= \text{Minimum weight of code vectors}$$

If X represents the transmitted signal, then the free distance is given by,

$$d_{free} = [W(X)]_{min} \quad \text{... (8.65)}$$

where $[W(X)]_{min}$ = Minimum weight of the code vector.

Important Point: *In this way, the minimum distance decides the capacity of the block or cyclic codes to detect and correct errors, the free distance will decide the error control capacity for the convolutional code.*

Coding gain (A)

The coding gain (A) is defied as the ratio of (E_b/N_0) of an encoded signal to (E_b/N_0) of a coded signal. The coding gain is used for comparing the different coding techniques.

$$\text{Coding gain (convolutional encoder)} = \frac{rd_{free}}{2} \quad \text{... (8.66)}$$

where r = Code rate, and, d_{free} = The free distance.

8.17.4. Transfer Function of the Convolutional Codes

The distance properties and the error rate performance of a convolutional code can be obtained from its state diagram. The state diagram shown in figure 8.59 will be used to demonstrate the method of obtaining the distance properties of a convolutional code.

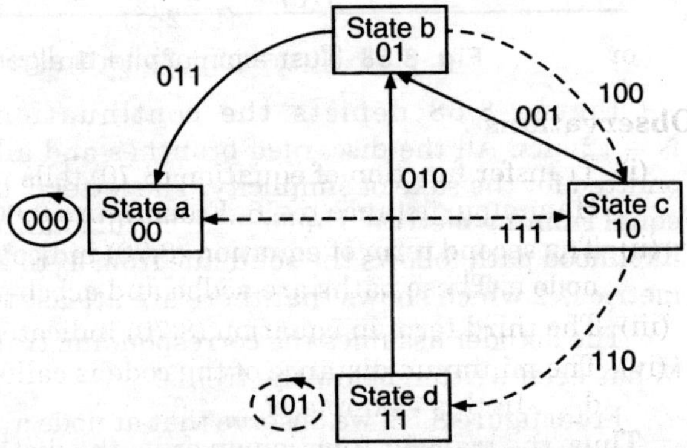

Fig. 8.59. State diagram for rate 1/3, k = 3 convolutional code

First, let us table the branches of the state diagram as either $p^0 = 1$ or p^1, p^2, p^3 etc. Here, p represents the Hamming distance between the sequence of output bits corresponding to each branch and the sequence of output bits corresponding to the all zero branch. This is because we assume that the input to the encoder is an all zero sequence.

The self loop at node (a) can be eliminated because it does not contribute to the distance properties of a code sequence relative to all zero code sequence. Node (a) is split into two nodes (a and e in figure 8.60). One of these nodes (a) represents the input of the state diagram and the other one node (e) represents the output of the state diagram of figure 8.60. After implementing all these modifications, we get the state diagram shown in figure 8.60.

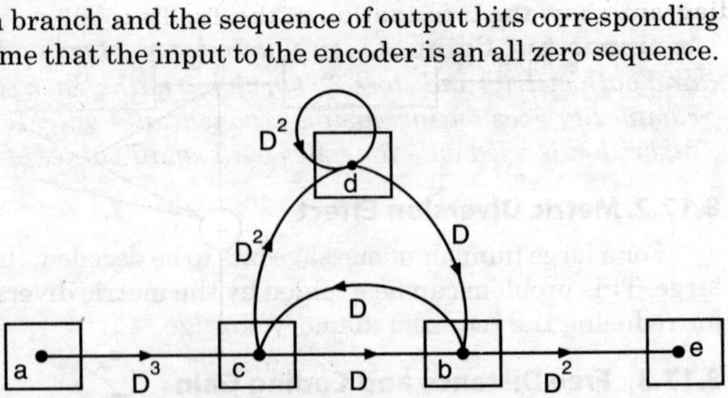

Fig. 8.60. State diagram for the rate 1/3, k = 3 convolutional code with distance labels on the branches

Carefully observe the distance labels marked on various branches in figure 8.60. For example, for the branch from a to c the output is 111 (figure 8.59). Hence, in figure 8.60, we have written a distance of p^3 on this branch (ac). On the similar lines, the output for the branch b to a in figure 8.59 is 011. Hence, it has been shown as branch (be) in figure 8.60 and the corresponding distance labels is p^2. Similarly, the other distance labels have been marked in figure 8.60.

The state diagram of figure 8.60 has five nodes (a to e) and the four state equations can be written as under:

$$X_c = p^3 X_a + p X_b$$
$$X_b = p X_c + p X_d$$
$$X_d = p^2 X_c + p^2 X_d$$
$$X_e = p^2 X_b \qquad \text{... (8.67)}$$

Here, it may be noted that these state equations have been written by considering the incident branches on each node. For example, for node c, the incident branches as ac and bc with distance labels of p^3 and p respectively. Therefore, the state equation is

$$X_c = p^3 X_a + p X_b \qquad \text{... (8.68)}$$

The transfer function of the code is defined as,

$$\text{Transfer function } T(p) = \frac{X_e}{X_a}$$

By solving the state equations given by equation (8.68), we obtain

$$T(p) = \frac{p^6}{(1-2p^2)} \qquad \text{... (8.69)}$$

or

$$T(p) = p^6 + 2p^8 + 4p^{10} + 8p^{12} + \ldots \qquad \text{... (8.70)}$$

Observations

(i) Transfer function of equation (8.70) tells us that the first term p^6 represents a single path of Hamming distance p = 6. From figure 8.60, it is clear that the p = 6 path is acbe.

(ii) The second term of equation (8.70) indicates that there are two paths of p = 8 from node **a** to node e. These paths are acdbe and acbcbe.

(iii) The third term in equation (8.70) indicates that there are four paths of distance p = 10.

(iv) The minimum distance of the code is called as the minimum free distance and is denoted by d_{free}. In this example, $d_{free} = 6$.

Thus, the transfer function gives us the distance properties of the convolutional code.

EXAMPLE 8.35. Determine the state diagram for the convolutional encoder shown in figure 8.61. Draw the Trellis diagram through the first set of steady state transitions. On the second Trellis diagram, show the termination of the Trellis to all-zero state.

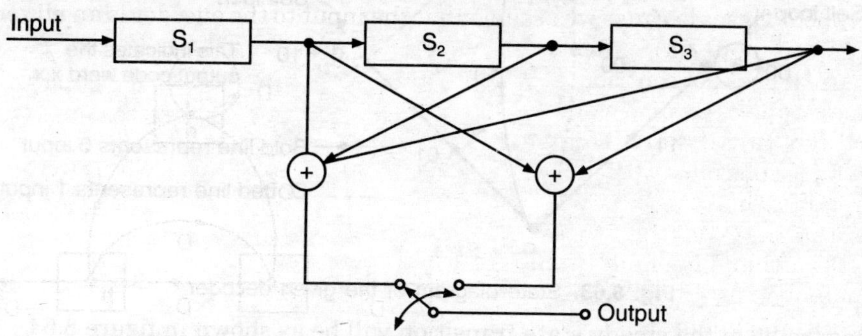

Fig. 8.61.

Solution:

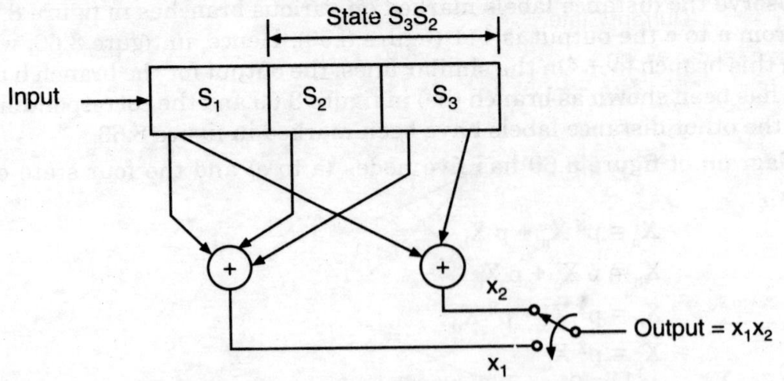

Fig. 8.62. Given encoder in redrawn form

Let us assume that all the registers have been cleared so S_1, S_2 and S_3 contain zeros in the beginning.

$$x_1 = x_i^{(1)} = S_1 \oplus S_2 \oplus S_3 \quad \text{... (i)}$$

$$x_2 = x_i^{(2)} = S_1 \oplus S_3 \quad \text{... (ii)}$$

and the encoder output = $x_1\, x_2$.

If the initial state is $S_3\, S_2$ = 00 and if $S_1 = 0$, then, we have

Output $x_1 = 0 \oplus 0 \oplus 0 = 0$

$x_2 = 0 \oplus 0 = 0$

Therefore, $x_1\, x_2 = 00$ ... if $S_1 = 0$

Next state $S_3\, S_2 = 00$

But, if $S_1 = 1$ then

$x_1 = 1 \oplus 0 \oplus 0 = 1$

and $x_2 = 1 \oplus 0 = 1$

Therefore, we have $x_1\, x_2 = 11$ ... if $S_1 = 1$

Next state $S_3\, S_2 = 01$

If we define the four different states as under:

00 = a, 01 = b, 10 = c and 11 = d

Then, the state diagram will be as shown in figure 8.63.

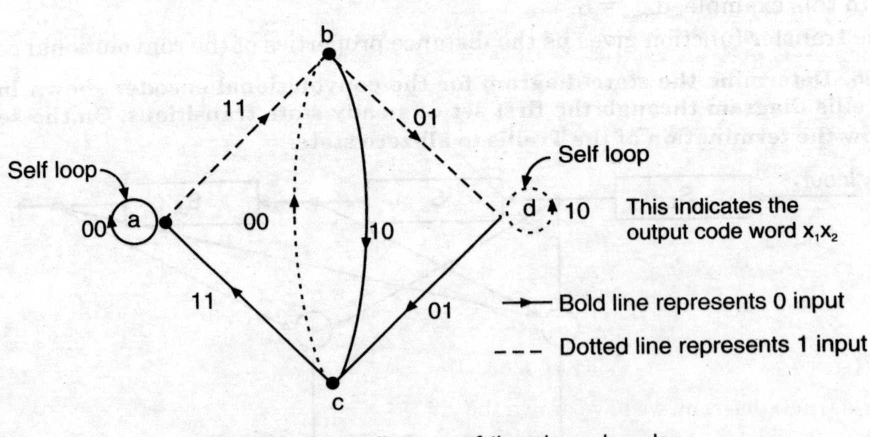

Fig. 8.63. State diagram of the given decoder

The code trellis in the steady state transition will be as shown in figure 8.64.

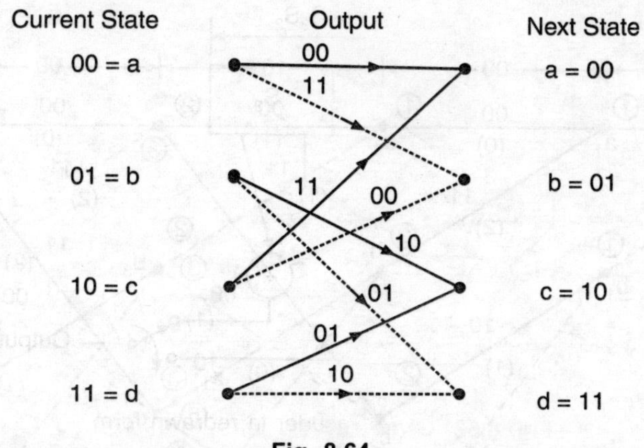

Fig. 8.64.

EXAMPLE 8.36. The encoder shown in figure 8.65 generates an all zero sequence which is sent over a binary symmetric channel. The received sequence 01001000.... There are two errors in this sequence (at second and fifth position). Show that this double error detection is possible with correction by application of viterbi algorithm.

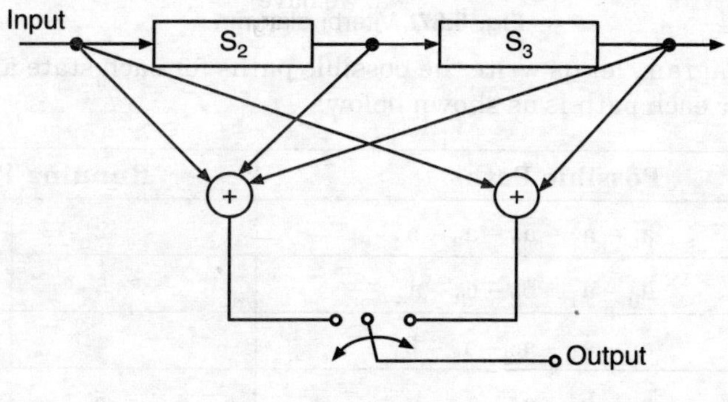

Fig. 8.65

Solution: The trellis diagram for the encoder shown in figure 8.65.

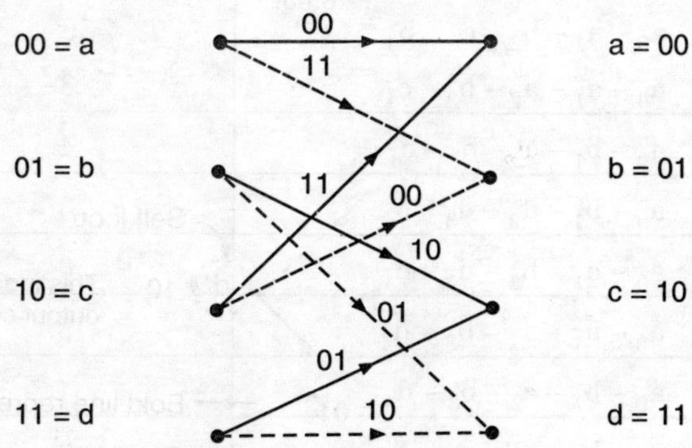

Fig. 8.66. The trellis diagram

From these trellis diagram, we have drawn the viterbi decoding diagram of figure 8.67. The input signal Y = 01 00 10 00... .

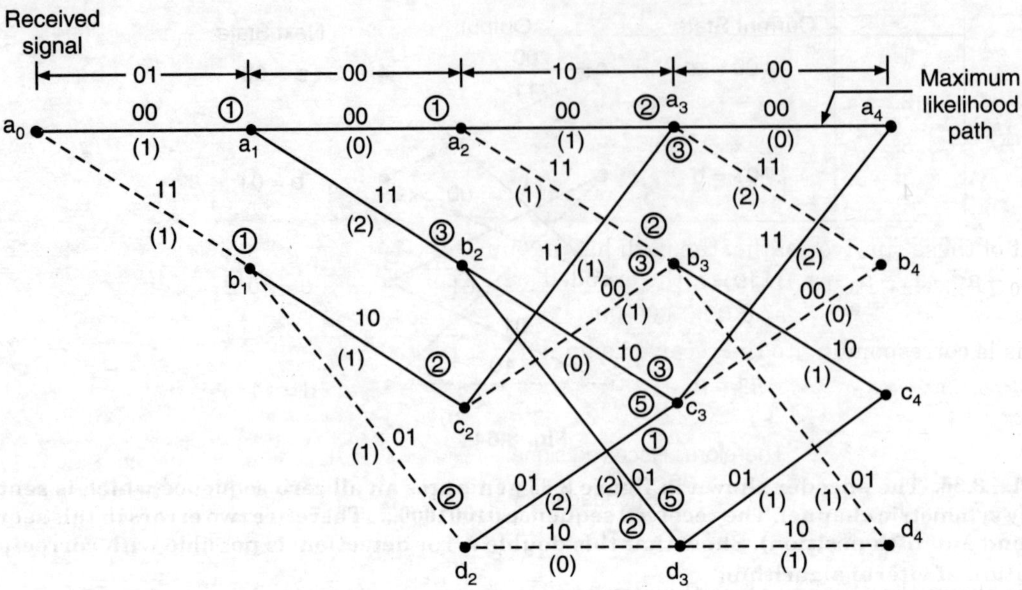

Fig. 8.67. Viterbi diagram

From the viterbi diagram, let us write the possible paths for each state a_4, b_4, c_4 and d_4 and the running path metric for each path is as shown below:

State	Possible Paths	Running Path Metric	
a_4	$a_0 - a_1 - a_2 - a_3 - a_4$	2	✓
	$a_0 - a_1 - b_2 - c_3 - a_4$	5	×
b_4	$a_0 - a_1 - a_2 - a_3 - b_4$	4	×
	$a_0 - b_1 - c_2 - a_3 - b_4$	5	×
	$a_0 - b_1 - d_2 - c_3 - b_4$	5	×
	$a_0 - a_1 - b_2 - c_3 - b_4$	3	✓
c_4	$a_0 - a_1 - a_2 - b_3 - c_4$	3	✓
	$a_0 - b_1 - c_2 - b_3 - c_4$	4	×
	$a_0 - b_1 - d_2 - d_3 - c_4$	3	×
	$a_0 - a_1 - b_2 - d_3 - c_4$	6	×
d_4	$a_0 - a_1 - a_2 - b_3 - d_4$	3	✓
	$a_0 - b_1 - c_2 - b_3 - d_4$	4	×
	$a_0 - b_1 - d_2 - d_3 - d_4$	3	×
	$a_0 - a_1 - b_2 - d_3 - d_4$	6	×

Out of the possible paths listed above, we select four survivor paths having the minimum value of running path metric. The survivor paths are marked by (✓) sign. They are as under:

S. No.	Paths	Path Metric
1.	$a_4 \rightarrow a_0 - a_1 - a_2 - a_3 - a_4$	2
2.	$b_4 \rightarrow a_0 - a_1 - b_2 - c_3 - b_4$	3
3.	$c_4 \rightarrow a_0 - a_1 - a_2 - b_3 - c_4$	3
4.	$d_4 \rightarrow a_0 - a_1 - a_2 - b_3 - d_4$	3

Out of these survivor paths, the path having minimum running path metric equal to 2, i.e., the path $(a_0 - a_1 - a_2 - a_3 - a_4)$. Hence, the encoded signal corresponding to this path is given by

a_4 d 00 00 00 00.

This is corresponding to the received signal 01 00 10 00.

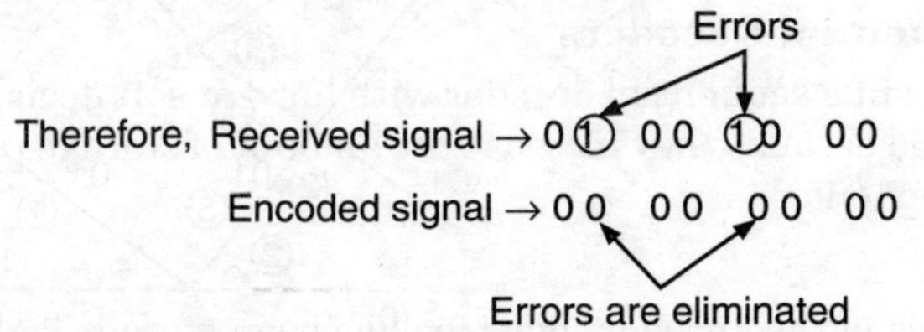

This shows that Viterbi algorithm can correct the errors present in the received signal.

EXAMPLE 8.37. Determine the code efficiency and code rate for the system given in example 8.35.

Solution: The code efficiency and code rate are one and the same.

Therefore, we write

$$\text{Code rate} = \text{Code efficiency} = \frac{k}{n}$$

where k = Number of messages bits in a codeword.

n = Number of transmitted bits.

For the encoder given in the example 8.35, for every input message bit, two encoded bits V_1 and V_2 are transmitted.

Hence, we have

$$\text{Code rate} = \text{Code efficiency} = \frac{1}{2} \quad \textbf{Ans.}$$

EXAMPLE 8.38. A convolutional encoder has a single shift register with two stages, three modulo-2 address and an output multiplexer. The generator sequences of the encoder are as under:

$$\mathbf{g^{(1)} = (1, 0, 1), g^{(2)} = (1, 1, 0) \text{ and } g^{(3)} = (1, 1, 1)}$$

Draw the block diagram of the encoder.

Solution: The generator sequences of the convolutional encoder are given.

The convolutional encoder has been shown in figure 8.68.

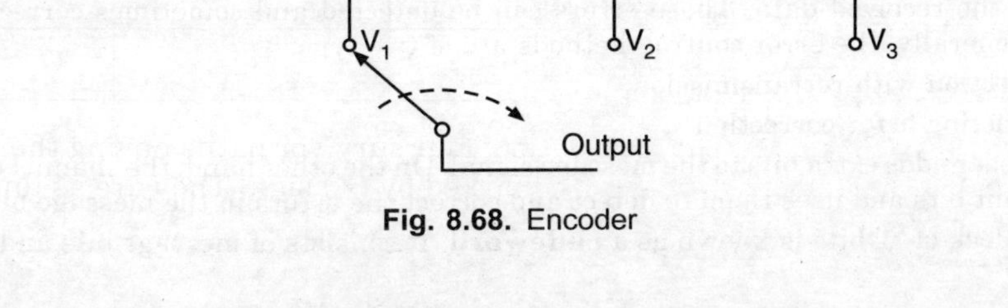

Fig. 8.68. Encoder

8.18 SEQUENTIAL DECODING

Before the discovery of Viterbi algorithm, other algorithms had been proposed for decoding of convolutional codes. Sequential decoding algorithm was the earliest and it was proposed by Wozencraft and modified by Fano. The principle of sequential decoding can be explained as under:

The sequential decoder generates hypotheses about the transmitted codeword sequence. It then computes a metric between these hypothesis and the received signal. The sequential decoder will go forward as long as the computed metric indicates that the choices made by decoder are likely. But, if the computed metric indicates that the choices made by decoder are unlikely, then the decoder will go backward and changes the hypothesis. The process of changing the hypothesis continues by trial and error until the decoder finds a likely hypothesis.

Implementation of Sequential Decoders

It is possible to implement a sequential decoder with hard or soft decisions. However, practically, the soft decisions are avoided because they increase the amount of storage required and the complexity of computation to a great extent.

Decoding Technique

At the decoder, a replica of the encoder code tree has been stored. The decoder follows that path on the tree which agrees with the received n code symbols. At every level in the code tree, the decoder generates both the possible paths and follows the path which agrees with the second group of n code symbols. Following this procedure, the decoder quickly penetrates the tree. But, if there is error in the received code word, then the decoder follows the most likely path. If two branches are equally likely, the receiver chooses one of them by using an arbitrary rule. At each new level in tree the decoder generates new branches and compares them with next set of n received symbols. This search continues to penetrate the tree along the most likely path and keeps the cumulative disagreement count. If the disagreement count exceeds a particular number, then the decoder understands that it is following a wrong path. So, it blocks that path and tries other.

Conclusions

(i) The sequential decoding is thus a trial and error technique used for searching out the correct path on the code tree.

(ii) The decoder searches the correct path in a sequential manner. It always operates on a single path at a time.

(iii) This is similar to an automatic traveller following a road map.

Limitations of Sequential Decoding

(i) The number of state metrics searched is a random variable.

(ii) The expected number of poor hypothesis and backward searches is a function of channel SNR. With reduction in SNR, the number of poor hypothesis increases.

(iii) If the average symbol arrival rate is higher than the average symbol decoder rate, the input buffer overflows and the errors are introduced due to loss of data.

SUMMARY

- The transmitted signal passes through some noisy channel. Due to noise interference, some errors are introduced in the received data. These errors can be detected and sometimes corrected using coding techniques. Generally, the error control methods are of two types:
  (i) Error detection with retransmission,
  (ii) Forward acting error correction.
- A channel encoder adds extra bits to the message signal. On the other hand, the channel decoder identifies these redundant bits and uses them to detect and correct the errors in the message bits if there is any.
- The encoded block of 'n' bits is known as a **codeword**. It consists of message bits and redundant bits.

- The number of bits 'n' after coding is known as the block length of the code.
- The ratio of message bits (k) and the encoder output bits (n) is known as **code rate**. Usually code rate is denoted by 'r' i.e.,

$$r = \frac{k}{n}$$

We find that $0 < r < 1$.

- Channel data rate is the bit rate at the output of encoder. If the bit rate at the input of encoder is R_S, then channel data rate would be,

$$\text{Channel data rate } (R_0) = \frac{k}{n} R_S$$

- An 'n' bit code word can be visualized in an n-dimensional space as a vector whose elements or coordinates are the bits in the code word.
- The Hamming distance between the two code vectors is equal to the number of elements in which they differ. As an example let X = 101 and Y = 110. Since the two code vectors differ in second and third bits therefore Hamming distance between X and Y is 'two'. Hamming distance is denoted by d(X,Y) or simply 'd'. i.e,

$$d(X,Y) = d = 2.$$

- Minimum distance (d_{min}) is the smallest Hamming distance between the valid code vectors.
- It may be noted that the error detection is possible if the received vector is not equal to some other code vector. This shows that the transmission errors in the received code vector must be less than minimum distance d_{min}.
- The code efficiency is expressed as the ratio of message bits in a block to the transmitted bits for that particular block by the encode i.e.,

$$\text{Code efficiency} = \frac{\text{Message bits in a block}}{\text{Transmitted bits for the block}}.$$

- The number of non-zero elements in the transmitted code vector is known as vector weight.
- Parity may be of two types:
  - **(i) Even Parity:** The parity of a binary word is known as 'even' if it contains even number of 1's. For example 01101001 has even parity since there are four number of 1's in this word.
  - **(ii) Even Parity:** The parity of a binary word is known as 'odd' if it contains odd number of 1's. For example 10010001 has odd parity since there are three number of 1s in this word.
- **Advantages of LRC over VRC:** Advantages of LRC method over VRC method :
  - (i) In LRC method, single errors can be detected and corrected.
  - (ii) In LRC method, double and triple errors can be detected.

  The drawback of LRC is that complexity in the method is increased.

SHORT QUESTIONS WITH ANSWER

Q.1. What are the types of error control methods?

Ans. Generally, the error control methods are of two types:

(i) Error detection with retransmission,

(ii) Forward acting error correction.

In case of first method, when an error is detected, the retransmission request (ARQ) is sent back to the transmitter. In the second method, the errors are detected and corrected by proper coding techniques at the receiver end. Whenever a single source transmits data to a number of receivers, the forward acting error correction is used. This is due to the fact that the retransmission is impossible in these cases.

Q.2. Define the following terms:

(i) Code word **(ii) Block length**

(iii) Code rate **iv) Channel data rate**

(v) Code vectors **(vi) Hamming distance**

(vii) Minimum distance

Ans. (i) Codeword: The encoded block of 'n' bits is known as a **code word**. It consists of message bits and redundant bits.

(ii) Block Length: The number of bits 'n' after coding is known as the block length of the code.

(iii) Code Rate: The ratio of message bits (k) and the encoder output bits (n) is known as **code rate**. Usually, code rate is denoted by 'r' i.e.,

$$r = \frac{k}{n}$$

we find that $0 < r < 1$.

(iv) Channel Data Rate: Channel data rate is the bit rate at the output of encoder. If the bit rate at the input of encoder is R_s, then channel data rate will be,

Channel data rate $(R_0) = \frac{k}{n} R_s$

(v) Code Vectors: An 'n' bit code word can be visualized in an n-dimensional space as a vector whose elements or coordinates are the bits in the code word.

(vi) Hamming Distance: The Hamming distance between the two code vectors is equal to the number of elements in which they differ.

As an example, let X = 101 and Y = 110. Since the two code vectors differ in second and third bits, therefore, Hamming distance between X and Y is 'two'. Hamming distance is denoted by d(X,Y) or simply 'd'. i.e,

$$d(X,Y) = d = 2$$

Hence, it may be observed that the Hamming distance between (100) and (011) is maximum, i.e. 3.

(vii) Minimum Distance (d_{min}): Minimum distance (d_{min}) is the smallest Hamming distance between the valid code vectors.

Q.3. What are block codes? Explain.

Ans. For the block of k message bits, (n – k) parity bits or check bits are added. This means that the total bits at the output of channel encoder are 'n'. Such types of codes are known as (n,k) block codes.

Q.4. Explain about Error Detection and Correction Capabilities of Hamming Codes.

Ans. Because the minimum distance (d_{min}) of Hamming code is 3, it can be used to detect double errors or correct single errors.

For detecting double (2) errors $\Rightarrow d_{min} \geq 2 + 1$ i.e. $d_{min} \geq 3$

and for correcting upto one (1) errors $\Rightarrow d_{min} \geq 2(1) + 1$ i.e. $d_{min} > 3$

Q.5. What are Repeated Codes? Explain.

Ans. In this code, only a single message bit is transmitted and q = 2t bit are parity bits. Hence k = 1.

Then the transmitted bits are,

$$n = 2t + 1$$

This code is known as **repeated code** because many redundant check bits are transmitted along with a single message bit. This code can correct 't' errors per block because this code uses many redundant check bits, therefore it requires a larger bandwidth.

Q.6. Write the Advantages and Disadvantages of Cyclic Codes?

Ans. Advantages:

1. The error correcting and detecting methods of cyclic codes are simpler and easy to implement. These methods eliminate the storage needed for lookup table decoding. Therefore, the codes becomes powerful and efficient.
2. The encoders and decoders for cyclic codes are simpler compared to noncyclic codes.
3. Cyclic codes also detect error burst that span many successive bits. Inspite of these advantages cyclic codes also have some advantages.

Disadvantage:

1. The error detection in cyclic codes is simpler but error correction is little complicated since the combinational logic circuits in error detector are complex.

REVIEW QUESTIONS

1. Which type of error is introduced in data bits during transmission ?
2. Define code efficiency.
3. Explain role of minimum distance in error correction and detection.
4. What are the error detection methods ?
5. Write the limitations of parity checking.
6. Write are the advantages of checksum method ?
7. Write the requirements of CRC ?
8. Explain the need of error detection and correction.
9. State the types of errors.
10. What do you mean by the error detection methods ?
11. Write the disadvantages off coding.
12. What is ARQ ? Explain its types.
13. What is FEC ? Explain.
14. What do you mean by power limited and band limited channels ?
15. When does the parity check fail to detect errors ?
16. Explain the concept of checksum error detection.
17. Define minimum distance d_{min} and explain its role in detecting and correcting errors.
18. Explain the principle and operation of encoder for Hamming code.
19. What do you mean by syndrome decoding of linear block codes ?
20. What is error vector (E) ? Explain.
21. State and explain the properties of cyclic codes.
22. Explain the format of code word polynomial.
23. How can you obtain the generator polynomial for the cyclic code ?
24. Explain the advantages of cyclic codes.
25. Explain the disadvantages of cyclic codes.
26. Write the syndrome calculator for systematic cyclic codes.
27. Write RS codes and state its advantages.
28. Write a note on BCH codes.

NUMERICAL PROBLEMS

1. Show that a block code with hamming distance of 2t + 1 can detect at most 2t bit errors and correct upto t bits of errors using minimum distance decoding rule.
2. Show that a cyclic code with (1 + x) as factor of its generator polynomial will detect all odd number of bit errors.
3. Let $g(x) = x^8 + x^6 + x^4 + x^2 + 1$ be a polynomial over the binary field. Find the lowest rate cyclic code whose generator polynomial is g(x).
4. Consider a (7, 4) linear block code with the parity check matrix H given by,

$$H = \begin{bmatrix} 1 & 0 & 1 & 1 & 1 & 0 & 0 \\ 1 & 1 & 0 & 1 & 0 & 1 & 0 \\ 0 & 1 & 1 & 1 & 0 & 0 & 1 \end{bmatrix}$$

   Construct code words for this (7, 4) code and show that this code is a Hamming code.
5. What is Hamming distance ? Give relation between minimum distance and error detecting and correcting capability. Describe a Hamming code. Also define Hamming spheres and Hamming bound.

❑❑❑

PREVIOUS YEARS QUESTION PAPER

B.Tech.

DIGITAL COMMUNICATIONS

Sixth Semester Theory Examination, 2010-11

Time : 3 hours ***Maximum Marks : 100***

Note : (i) Attempt all questions.
(ii) Assume the missing data if any.
(iii) Marks are indicated at the end of each Section.

1. **Attempt any four parts of the following:** **(5×4=20)**
   (a) The spectral range of a bandpass signal extends from 10.0 MHz to 10.04 MHz. Find the minimum sampling rate.
   (b) Calculate the capacity of a Gaussian channel with a bandwidth of 1 MHz and S/N ratio of 20 dB.
   (c) Discuss the term FSK, PSK and DPSK.
   (d) Discuss the role of Equalizer. Also explain the role of normalized equalization.
   (e) Explain the importance of Companding technique.

2. **Attempt any two parts of the following:** **(10×2=20)**
   (a) Differentiate between the Delta Modulation technique and Adaptive Delta Modulation technique. Explain how ADM improves the system's tolerance to slope overload.
   (b) Consider five messages given by the probability 1/2, 1/4, 1/8, 1/16, and 1/16. Calculate H. Use the Shannon-Fano algorithm to develop an efficient code, and for that code, calculate the average number of bits/message.
   (c) Write in brief on conditional Entropy and Redundancy. One of the four possible messages M_1, M_2, M_3 and M_4 having probabilities 1/8, 3/8, 3/8 and 1/8 respectively is transmitted. Calculate the average information per message.

3. **Attempt any two parts of the following:** **(10×2=20)**
   (a) What is the purpose of Hamming Code? A receiver receives the code 11001100111. When it uses the Hamming encoding algorithm, the result is 0101. Which bit is in error? What is the correct code?
   (b) Define the term Burst error and Error detection. How many types of redundancy checks are used in data communications? Discuss in brief VRC.
   (c) A sinusoidal signal with an amplitude of 5.25 V is applied to a uniform quantizer rod midtread type. The quantizer output takes on values 0, ± 1, ±2, ±3, ±4, and ±5 volts. Sketch the waveform of the resulting quantizer output for one complete cycle of the input.

4. **Attempt any two parts of the following:** **(10×2=20)**
   (a) Draw the block diagram of QPSK coherent receiver. Sketch the QPSK waveform for the sequence 1101010010, assuming the carrier frequency to be equal to the bit rate.
   (b) How does a binary PSK signal generated? Suggest a suitable synchronizing circuit arrangement for extracting the carrier at the receiver end for coherent detection of binary PSK.
   (c) Compare the noise performance of different digital modulation schemes.

5. **Write a brief on any four of the following :** **(5×4=20)**
   (a) Central limit theorem
   (b) Code Division Multiple Access of DS-SS
   (c) OFDM Communications
   (d) M-ary Digital Carrier Modulation
   (e) Encoder and decoders for systematic cycle codes.

❑❑❑

B.Tech.

DIGITAL COMMUNICATIONS (EEE 601)

Sixth Semester Theory Examination, 2011-12

Time : 3 hours **Maximum Marks : 100**

Note: 1. Answer all questions.
2. All questions carry equal marks.

1. Attempt any four parts of the following: **(5×4=20)**

(a) What are the main elements of the digital communication system? Sketch and explain the function of each element.

(b) Differentiate between Base-band and Band-pass data transmission system.

(c) If Z = X + Y – C, where X and Y are the independent random variables with variance σ_{x_2} and σ_{y_2} and *C* is constant. Find the variance of *Z*.

(d) Write short note on Kraft Inequality.

(e) Define the following terms:
(i) Mutual information,
(ii) Entropy,
(iii) Channel capacity,
(iv) Rate of Information.

(f) Verify equations that is I(X : Y) = H(Y) + H(X) – H(X, Y).

2. Attempt any two parts of the following: **(10×2=20)**

(a) Discuss the properties of matched filter. Explain the block diagram of optimum receiver for binary coded signal and derive the expression for Probability of Error (P_e) for optimum filter receiver.

(b) Explain the cause of ISI and discuss the Nyquist criterion for distortionless. Baseband Binary transmission. Design a binary baseband PAM system to transmit data at a bit rate of 3600 bits/sec with a bit error probability less than 10^{-4}. The channel response is given by:

$$H_c(f) = \{10^{-12} \quad \text{for } |f| < 2400$$
$$= \{0 \quad \text{elsewhere.}$$

The noise power spectral density is $G_n(f) = 10^{-14}$ watt/Hz.

(c) Give four desirable properties of baseband line codes. Draw the following data formats for the data bit stream 1 1 0 0 1 1 0:
(i) Unipolar RZ,
(ii) Polar NRZ,
(iii) Bipolar NRZ,
(iv) Manchester.

3. Attempt any two parts of the following: **(10×2=20)**

(a) Write short notes on the following:
(i) Byte Interleaving TI Carrier System,
(ii) T1 to T4 PCM TDM.

(b) Explain the different type of digital carrier modulation schemes giving their merits and demerits for transmission data on band-pass channel. Determine the performance of a QPSK receiver in the presence of AWGN channel.

(c) An FSK system transmits binary data at the rate of 2.5×10^6 bits per scond. During the course of transmission, white Gaussian noise of zero mean and power spectral density 10^{-20} W/Hz is added to the signal. In the absence of noise, the amplitude of the received sinusoidal wave for digit 1 or 0 is 1 mV. Determine the average probability of symbol error for the following system configurations:
  (i) Coherent binary FSK,
  (ii) Coherent MSK,
  (iii) Non-coherent binary FSK.

4. Attempt any two parts of the following: (10×2=20)

(a) What are the various characteristics of Spread Spectrum Signal? Explain the principle of direct sequence and frequency hopped spread spectrum communication system.

(b) Derive an expression for the probability of error of a DS spread spectrum using bianry PSK. In a fast FH spread spectrum system, the information is transmitted via FSK with non-coherent detection. Suppose there are N = 3 hops/bit. Determine the probability of, for this system in an AWGN channel with power density ½N0 and SNR = 13 dB (total SNR over the 3 hops).

(c) Explain the various types of multiuser systems. Determine and compare the capacity of FDMA, TDMA and CDMA in an ideal AWGN channel of bandwidth 'W'.

5. Attempt any four parts of the following: (5×4=20)

(a) A Gaussian channel has 1 MHz bandwidth. Calculate the channel capacity and maximum information rate if the signal power to noise spectral density ratio (S/N) is 10^5 Hz.

(b) The parity check matrix of a (7, 4) Hamming code is as given:

$$H = \begin{vmatrix} 1 & 1 & 0 & 1 & 1 & 0 & 0 \\ 1 & 1 & 1 & 0 & 0 & 1 & 1 \\ 1 & 0 & 1 & 1 & 0 & 1 & 1 \end{vmatrix}$$

Calculate the syndrome vector for single bit errors.

(c) Define the term Burst error and error detection. How many types of redundancy checks corrected?

(d) For a (7, 4) cyclic code determine the generator matrix if $G(P) = 1 + P + P^3$.

(e) Determine the Huffman code for the following message with their probabilities given:

$$\{X\}\begin{bmatrix} x1 & x2 & x3 & x4 & x5 & x6 & x7 \end{bmatrix}$$
$$\{P\}\begin{bmatrix} 0.05 & 0.15 & 0.2 & 0.05 & 0.15 & 0.3 & 0.1 \end{bmatrix}$$

(f) Write the advantage and disadvantage of cyclic codes.

□□□

B.Tech.

DIGITAL COMMUNICATIONS (EEE 601)

Sixth Semester Theory Examination, 2012-13

Time : 3 hours **Maximum Marks : 100**

SECTION-A

1. Attempt all questions. **(10×2=20)**

(a) Explain why digital communication is preferred over analog communications.

(b) A voice grade channel of the Telephone Network has a bandwidth of 3.4 kHz. Calculate the channel capacity of the telephone channel for a signal to noise ratio of 30 dB.

(c) Why power control is needed in CDMA?

(d) Define orthogonal signalling.

(e) Define Null-to-null Bandwidth of the transmitted signal.

(f) Explain geometric interpretation of signals.

(g) Calculate the channel capacity of the telephone channel of 3.4 kHz for a signal to noise ratio of 30 dB.

(h) Purpose of Eye diagram in digital communication.

(i) State Bays' Theorem and its purpose.

(j) Write different methods for Timing synchronization in digital receiver.

SECTION-B

2. Attempt any three questions: **(3×10=30)**

(a) With the help of block diagram, explain the signal processing operations involved in a digital communication system.

(b) Explain the working principle of direct sequence spread spectrum system. Discuss its applications in detail.

A slow FH/MSK system has following parameters:

The number of bits per MFSK symbol = 4

The number of MFSK symbols per hop = 5

calculate the processing gain of the system in decibels.

(c) We are required to trasmit 2.08×10^6 binary digits per second with $P_b \leq 10^{-6}$. Two possible schemes are considered:

(i) Binary,

(ii) 16-ary ASK.

The channel noise PSD is $S_n(\omega) = 10^{-8}$. Determine the transmission bandwidth and the signal power required at the receiver in each case.

(d) An audio signal of bandwidth 4 kHz is sampled at a rate 25% above the Nyquist rate and quantized. Quantisation error is not to exceed 0.1% of the signal peak amplitude. The resulting quantized samples are now coded and transmitted by 4-ary pulses.

(i) Determine minimum number of 4-ary pulses required to encode each pulse.

(ii) Determine minimum transmitted bandwidth required to transmit this data with zero ISI.

(e) (i) In an experiment, a trial consists of four successive tosses of a coin. If we define a random variable x as number of head appearing in a trial, determine Px(x) and Fx(x).

(ii) State Central Limit Theorem and explain purpose of this theorem.

(f) (i) Consider a noiseless channel with m input symbols and m output symbols as shown in Fig. 1. Show that

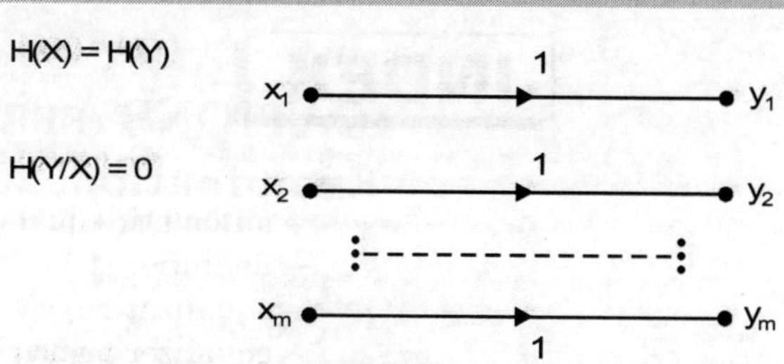

Fig. 1.

(ii) Define the term burst error and error detection. How many redundancy checks are used in data communication?

SECTION-C

Attempt all questions. **(5×10=50)**

3. Explain correlation receiver with a neat block diagram. Explain the function of each block. Also exlain why the correlator receiver is called as integrated and dump filter. 10

OR

Prove maximum output SNR of a method filter is $(SNR)_0 = 2E/N_0$.

4. Plot and compare the probabilities of error for non-coherent detection of Binary ASK, Binary FSK and Binary DPSK. 10

OR

Compare the probability of two Networks of Fig. 2. The probability of links s_1 and s_2 is p each.

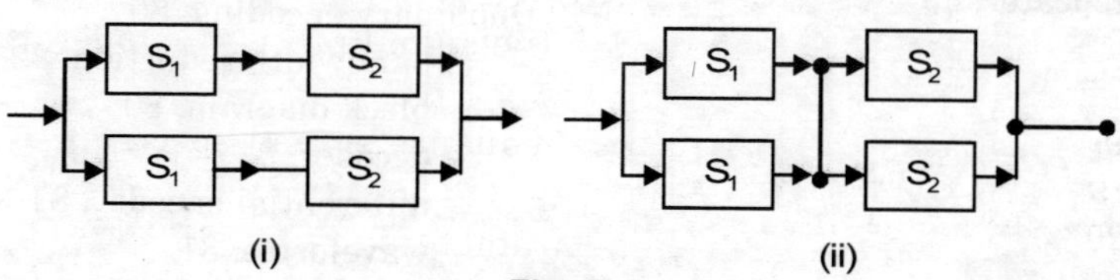

Fig. 2.

5. Explain principle of DSSS-CDMA. Derive an expression for the probability of error of a DSSS using PSK. 10

OR

Explain generation of FHSS with suitable block diagram. Show that advantages of FHSS lies in its ability of combat jamming.

6. With the help of suitable block diagram, explain the working of QPSK coherent receiver. Sketch the QPSK waveform for sequence 111001010010, assuming 10

   (i) Carrier frequency to be equal to bit rate.

   (ii) Carrier frequency is half the bit rate.

OR

Find the error probability for the multi-amplitude signalling and hence define the relationship between power and bandwidth of the same.

7. For a given generator polynomial $g(x) = 1 + x + x^3$. 10

   (i) Find the Generator matrix G for a symmetric (7, 4) cyclic code.

   (ii) Find systematic cyclic code for message bits 1010.

OR

A convolution encoder has two shift registers two modulo-2 adders and an output multiplexer. The generator sequences of the encoder are as follows: g(1) = (1, 0, 1); g(2) = (1, 1, 1). Assuming a 5-bit message sequence is transmitted. Using the state diagram and the message sequence when the received sequence is (11, 01, 00, 10, 01, 10, 11, 00, 00, ...).

□□□

INDEX

Win Prizes

ATTENTION READERS

Win Prizes

We request for your frank assessment regarding various aspects of the book as given below :

DIGITAL COMMUNICATIONS

By: Dr. SANJAY SHARMA

Please Fill up the Sheet in Neat and Clean Writing

(i) What topics of your syllabus, important for your examination point of view, are not covered in this book ?

(ii) What are the chapters wherein the treatment of the subject matter is not systematic or organised or updated ?

(iii) Have you come across misprints/mistakes/factual inaccuracies in the book? Please specify the chapters and the page numbers.

(iv) What other books on the same subject have you found/heard better than the present book? Please specify in terms of price and quality.

(v) Further suggestions and comments for the improvement of this book.

(PS: If need be please attach a separate sheet to write your views)

OTHER DETAILS

(i) Who recommended you this book ? (Please tick (✓) in the box near the option relevant to you.

Teacher ☐ Friends ☐ Bookseller ☐

....

....

....

....

(ii) Name of the recommending teacher, his designation and address :

....

....

....

....

(iii) Name and address of the bookseller you purchased the book from :

....

....

....

....

(iv) Name and address of your institution (Please mention the University or Board as the case may be)

....

....

....

....

(v) Your name and complete postal address :

....

....

....

....

(vi) Course you are enrolled in :

....

....

....

....

The best assessment will be awarded each month: The award will be in the form of our publications as decided by the Editorial Board.

Please mail the filed up coupon at your earliest to :

S.K.KATARIA & SONS®

4885/109, Prakash Mahal, Dr. Subhash Bhargav Lane,

Opposite Delhi Medical Association, Daryaganj, New Delhi–110002

☎: +91-11-23243489, 23269384

Email : katariabooks@yahoo.com

katariabook@gmail.com